Environmental Science and Engineering

Series Editors

Ulrich Förstner, *Buchholz, Germany*
Wim H. Rulkens, *Department of Environmental Technology, Wageningen University and Research Centre, Wageningen, Gelderland, The Netherlands*

The ultimate goal of this series is to contribute to the protection of our environment, which calls for both profound research and the ongoing development of solutions and measurements by experts in the field. Accordingly, the series promotes not only a deeper understanding of environmental processes and the evaluation of management strategies, but also design and technology aimed at improving environmental quality. Books focusing on the former are published in the subseries Environmental Science, those focusing on the latter in the subseries Environmental Engineering.

Dongfei Han · Mukesh Khare · Siyue Li
Editors

Environmental Crisis: Pollution and Governance

Selected Papers from 2024 4th International Conference on Environmental Pollution and Governance (ICEPG 2024)

Set 2

Springer

Editors
Dongfei Han
Department of Environmental Sciences
Suzhou University of Science
and Technology
Suzhou, China

Siyue Li
Wuhan Institute of Technology
Wuhan, Hubei, China

Mukesh Khare
Indian Institute of Technology (IIT) Delhi
New Delhi, Delhi, India

ISSN 1863-5520 ISSN 1863-5539 (electronic)
Environmental Science and Engineering
ISBN 978-3-031-86831-3 ISBN 978-3-031-86829-0 (eBook)
https://doi.org/10.1007/978-3-031-86829-0

This Springer imprint is published by the registered company Springer Nature Switzerland AG
The registered company address is: Gewerbestrasse 11, 6330 Cham, Switzerland

Preface

The 2024 4th International Conference on Environmental Pollution and Governance (ICEPG 2024) was held from October 25–27 in Zhengzhou, China. As environmental issues remain an important global problem, this conference has become an important platform for exchanging ideas and research results. More than 150 attendees from around the world gathered together to share their insights and engage in productive discussions.

Keynote Speakers brought invaluable perspectives to ICEPG 2024, enriching the dialogue with their expertise and innovative ideas. Among them were Professor Huijun Wang, from the Institute of Atmospheric Physics at the Chinese Academy of Sciences; Professor Shijie Han, from Henan University; and Professor Junguo Liu, from North China University of Water Resources and Electric Power. Apart from the keynote speeches, oral and poster presentation parts were also held and displayed by various scholars, leading to a warm atmosphere of academic discussion.

The conference proceedings is a valuable record of the conference which presents a selection of papers submitted to the conference from universities, research institutes, and industries. All the papers have been through a rigorous peer-reviewed process in order to meet international publication standards. The papers highlight the collaborative efforts among researchers, policymakers, and industry leaders, reflecting the interdisciplinary nature of tackling environmental challenges. As readers explore these works, they will gain valuable insights into cutting-edge solutions and methodologies that hold the potential to transform our approach to environmental stewardship in the years to come.

Gratitude extends to every participant, speaker, and organizer who contributed to making ICEPG 2024 a resounding success. Their enthusiasm and commitment to enhance environmental research and policy have created a lasting impact. It is hoped that the proceedings will not only commemorate the achievements of the conference but also inspire ongoing efforts to improve environmental governance for a sustainable future.

The Committee of ICEPG 2024

Committee Member

Conference General Chair

Huijun Wang	Institute of Atmospheric Physics, Chinese Academy of Sciences, China

Technical Program Committee Chairs

Junguo Liu	North China University of Water Resources and Electric Power, China
Baoyu Gao	Shandong University, China

Publication Chairs

Mukesh Khare	Indian Institute of Technology (IIT) Delhi, India
Siyue Li	Wuhan Institute of Technology, China
Dongfei Han	Suzhou University of Science and Technology, China

Technical Program Committee

Chunhui Li	Beijing Normal University, China
Shiliang Liu	Beijing Normal University, China
Rui Du	University of Chinese Academy of Sciences, China
Wenjing Lu	Tsinghua University, China
Hong Cui	Shanxi Xinhua Chemical Defense Equipment Research Institute, China
Snengr. Meichao Zhao	Shenyang Academy of Environmental Sciences, China
Xia Li	Inner Mongolia University of Science & Technology, China
Assoc. Qinzhong Feng	University of Chinese Academy of Sciences, China
Zhengguang He	Zhengzhou University, China

Contents

Soil, Waste, and Resource Management

Study on Sodium Persulfate Pretreatment Sludge Based on Different Activation Methods

Ting He, Tian Chai(✉), Jing Zhang, Zuocheng Yin, and Longjun Hua

Xiamen University of Technology, Xiamen 361024, China
chait@xmut.edu.cn

Abstract. With the acceleration of the urbanization process, many urban sewage treatment plants have been established in China, and the output of sludge has also increased rapidly. How to further improve the efficiency of sludge treatment, especially the efficiency of resource treatment, is a hot topic of current research. Sulfate radical has oxidation, and persulfate oxidation technology using sulfate radical is a new oxidation technology in the field of sludge treatment. It can be used for pretreatment of sludge, greatly reducing the moisture content of sludge, cracking the stable structure of sludge, so as to improve the utilization rate of sludge resources. In this study, sodium persulfate pretreatment sludge was activated by three activation methods: thermal activation, alkali activation and ultrasonic activation to achieve the effect of cracking the stability of sludge. In this paper, the effect of activated sodium persulfate pretreatment of sludge was studied under the conditions of 80 °C water bath heating, pH of 13 alkaline and ultrasonic frequency of 2000 Hz. The experimental results were indicated by the changes of chemical oxygen demand, sludge particle size, three-dimensional fluorescence and protein. The experimental results show that the three methods can promote the cracking of sludge, among which the ultrasonic activation at 2000 Hz frequency for 75 min and the water bath heating at 80 °C for 20 min can effectively promote the cracking of sludge, and are the most economical and efficient activation treatment methods in the pretreatment sludge.

Keywords: Sludge pretreatment · Different activation methods · Sodium persulfate

1 Introduction

The treatment and disposal of sludge pose significant challenges in environmental engineering. Various methods are available for sludge treatment, including conventional approaches like anaerobic digestion and composting [1], as well as innovative techniques such as microwave-enhanced advanced oxidation processes (MW-AOP) [2] and sludge treatment reed beds (STRB) [3]. However, there are limitations in terms of treatment efficiency and reduction, often resulting in treated sludge failing to meet environmental requirements for resource utilization standards. Therefore, investigating efficient and environmentally sustainable pretreatment technology for sludge holds practical implications for enhancing efficacy in its treatment and promoting resource utilization.

D. Han et al. (Eds.): ICEPG 2024, ESE, pp. 513–519, 2025.
https://doi.org/10.1007/978-3-031-86829-0_58

In recent years, Advanced Oxidation Processes (AOPs) have garnered significant attention owing to their robust oxidation capacity and exceptional treatment efficiency [4]. Sodium Persulfate (PDS), a widely utilized oxidizing agent, exhibits remarkable efficacy in the decomposition of recalcitrant organic compounds, thereby enhancing sludge biodegradability and reducing sludge volume. However, achieving efficient oxidation with PDS requires activation methods such as thermal activation, alkali activation, transition metal ion activation, and ultraviolet irradiation which significantly impact both PDS activation efficiency and sludge treatment efficacy [5]. This study aims to investigate the impact of different activation methods (thermal, alkali, and ultrasonic) on sodium persulfate pretreatment sludge in order to optimize operational conditions for effective treatment. Ultimately, a cost-effective and practical pretreatment scheme will be selected.

2 Materials and Methods

2.1 Experiment Design

Firstly, sodium persulfate solutions with concentrations of 2, 4, 6, 8, 10, 12, 14, 16, 18 and 20 g/L were individually added to the sludge. After centrifugation, the supernatant was collected for protein and COD analysis to generate a corresponding curve. The experiment was repeated three times to determine the optimal concentration of sodium persulfate. The selected optimal concentration solution of sodium persulfate was then added to the sludge and subjected to thermal activation using a heating stirrer as well as ultrasonic and alkali activations. For thermal activation at a temperature of 80 °C [7]: At specific time intervals (initially and at every subsequent interval of 5 min up until 50 min), samples of 50 mL solution were extracted. For ultrasonic activation at a temperature of 25 °C: At specific time intervals (initially and at every subsequent interval of 15 min up until 90 min), samples of 50 mL solution were extracted from an ultrasound machine operating at 2000 Hz. For alkali activation using sodium hydroxide with pH adjusted to13 [8]: At specific time intervals (initially and at every subsequent interval of 20 min up until 120 min), samples of 50 mL solution were extracted after adding the sodium persulfate solution. The obtained solutions from all three activation methods underwent centrifugation for protein and COD analysis in order to generate respective curves. The experiment was repeated three times to determine the optimal duration for thermal activation.

2.2 Analysis and Detection Method

The supernatant obtained from the three activation methods was filtered using a 0.45 filter membrane, and the protein and COD levels were determined in the filtered samples. The standard curves for COD were generated by measuring different concentrations of bovine serum protein solution at 280 nm using a UV spectrophotometer in test tubes with no protein (0 mg/mL). The absorbance of the sample was measured at 280 nm. COD was quantified using rapid resolution spectrophotometry (HJ/T 399-2007). After digestion for 10 min in COD digestion mode with a water quality detector, it was removed, cooled with air for 2 min, distilled water was added to the digestion tube and mixed well before cooling it in a water bath for another 2 min.

Additionally, particle size analysis of the sludge was performed using Malvern laser particle size analyzer (MS2000) with a dispersant solution containing sodium hexametaphosphate at a concentration of 0.1 mol/L. The excitation wavelength (Ex) ranged from 200 nm to 600 nm with an interval of every 3 nm. The emission wavelength (Em) remained fixed at range between 200–800 nm with an interval of every 1.17 nm. The slit width was set to 5 nm and the scanning speed was 1200 nm/min. Ultra-pure water was taken as a blank and Raman scattering as well as first and second order Rayleigh scattering were subtracted. Finally, internal rate correction was applied.

3 Results and Discussion

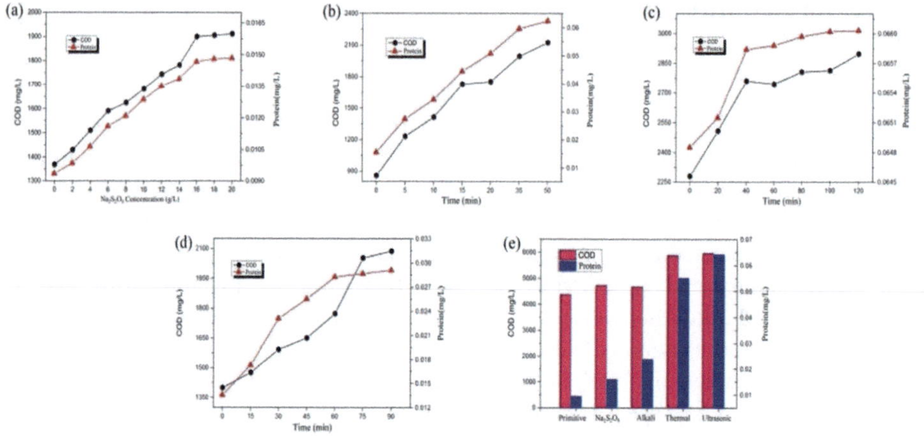

Fig. 1. (a) Effect of different sodium persulfate concentration on protein and COD (b) Effect of different heating time on protein and COD (c) Effect of different alkali activation time on protein and COD (d) Effect of different ultrasonic activation time on protein and COD (e) Comparison of protein and COD concentration of pretreatment sludge under optimal

3.1 Sodium Persulfate Concentration

As shown in Fig. 1(a), as sodium persulfate concentration increases from 0 to 20 g/L, both protein concentration and COD concentration increase. However, beyond a dosage exceeding 16 g/L, the growth rate of COD suddenly decelerates due to two potential factors: firstly, excessive amounts of sodium persulfate may cause free radicals $\cdot SO_4^-$ to combine with each other forming perdisulfate ions ($S_2O_8^-$), thereby impeding further degradation steps and reducing cracking efficiency; secondly, excessive quantities of sodium persulfate could lead to partial degradation of dissolved organic matter resulting in decreased organic content within the sludge [6]. Consequently, for subsequent experiments conducted hereinforth, the optimal concentration for sodium persulfate was determined as16 g/L.

3.2 Analysis of Different Activation Treatments

As depicted in Fig. 1(b) (c) (d), under thermal activation, an increase in protein and COD concentrations was observed with prolonged heating time at a temperature of 80 °C. However, considering the high energy consumption associated with heating and the similarity between the experimental increase rates at 20 min, it is concluded that the optimum condition for thermal activation is a heating temperature of 80 °C for 20 min. For alkali activation, maintaining a pH of 13 resulted in continuous increases in protein and COD concentrations over time; however, the rate of concentration increase was relatively small due to alkaline inhibition of persulfate oxidation. At a treatment time of 40 min, there was a turning point where changes in protein and COD concentrations transitioned from large to small amplitudes. Therefore, pH 13 and a treatment time of 40 min are considered as optimal conditions for alkali activation. Under ultrasonic activation at a frequency of 2000 Hz, both protein and COD concentrations steadily increased with increasing ultrasonic treatment time. A significant change occurred when reaching an ultrasonic treatment time of 75 min followed by gradual concentration changes thereafter. Henceforth, steady increases were observed in both protein and COD concentrations under ultrasonic activation conditions optimized at a frequency of 2000 Hz for an ultrasonic treatment duration lasting exactly 75 min.

As depicted in Fig. 1(e), the three optimal activation conditions (thermal activation at 80 °C for 20 min, alkali activation with a pH of 13 for 40 min, and ultrasonic activation at a frequency of 2000 Hz for 75 min) were compared to both the original sludge and the sludge group treated solely with sodium persulfate. From the figure, it is evident that whether sodium persulfate was added alone or under the three optimal activation conditions, there was an increase in protein and COD concentrations compared to the original sludge, indicating their significant impact on sludge degradation. Furthermore, Fig. 1(e) visually demonstrates that ultrasonic activated pretreatment under the best conditions exhibited higher release of protein and COD content than thermal activation; however, alkali activated pretreatment under these best conditions resulted in lower release compared to the original sludge. Thus, considering protein and COD as physicochemical indicators, it can be concluded that ultrasonic activation at a frequency of 2000 Hz for 75 min and heating at 80 °C for 20 min exhibit superior sludge degradation ability and more effective pretreatment.

3.3 Sludge Size Analysis

The Malvern Laser particle size analyzer (MS2000) laser particle size test system was employed to analyze the cracking degree of the sludge floc structure under three optimal activation conditions in this experiment, covering a range of 0.1–1000 μm. The results are presented in Fig. 2(a)–(e). The weighted average particle size volume of the sludge stock solution, sodium persulfate single addition, alkali activation, thermal activation, and ultrasonic activation were 52.169, 37.105, 24.926, 27.799, and 32.688 respectively as depicted in Fig. 2(f). Following the addition of sodium persulfate and the three activation methods mentioned above resulted in a decrease in the weighted average particle size volume compared to that of the initial sludge stock solution. Among these methods, alkali activation exhibited the smallest weighted average size followed by thermal activation and ultrasonic activation while single sodium persulfate addition showed relatively larger particles sizes indicating that alkali activation achieved superior reduction in sludge particle size compared to thermal and ultrasonic activations.

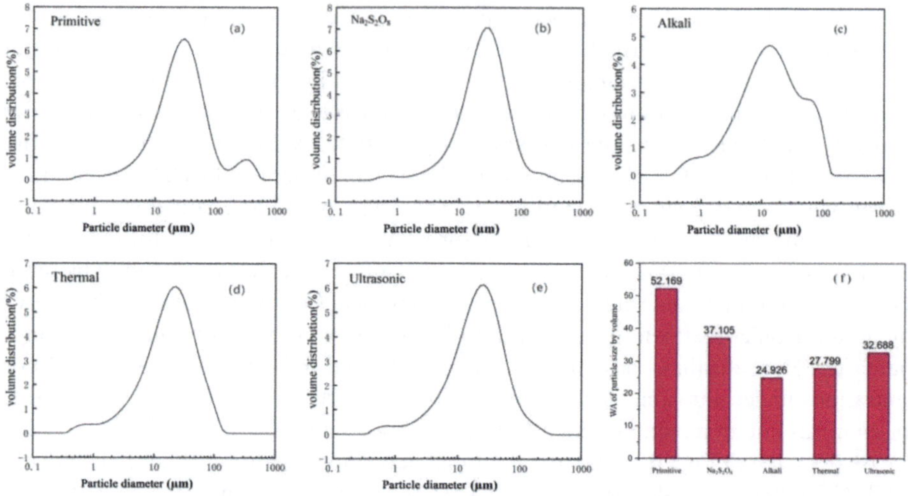

Fig. 2. Laser particle size distribution of different pretreatment methods (a) sludge stock solution (b) sodium persulfate alone (c) activation of alkali (d) heat activation (e) ultrasonic activation (f) Comparison of the weighted average size of different pretreatment methods

3.4 Three-Dimensional Fluorescence Analysis

The samples and sludge stock solution were subjected to three-dimensional fluorescence scanning under three optimal activation conditions, with the addition of only sodium persulfate. The results are presented in Fig. 3. According to the investigation on the three-dimensional fluorescence spectra of natural dissolved organic matter, the spectral response region can be divided into five distinct regions: region I (Φ1, n): tyrosine protein fluorescence region (Ex 220–250 nm/Em 280–330 nm), region II (Φ2, n): tryptophan protein fluorescence region (Ex 220–250 nm/Em 330–380 nm), region III (Φ3, n): ultraviolet fulvic acid fluorescence region (Ex 220–250 nm/Em 380–550 nm), region IV (Φ4, n): dissolved microbial metabolites fluorescence region (Ex 250–450 nm/Em 280–380 nm), and Region V (Φ5, n): visible light-type fulidic acid fluorescence region (Ex 250–450 nm/Em 380–550 nm) [9].

Revised sentence: As depicted in Fig. 3, the fluorescence response was observed not only in the sludge stock solution but also in the samples obtained through other treatment methods across all five regions, with a certain degree of increase. However, upon comparing Figs. 3(a), (b), and (c), it can be observed that the fluorescence response of samples obtained by alkali activation method exhibited minimal change compared to the stock solution. Furthermore, a comparison between Figs. 3(c), (d), and (e) revealed significant differences in fluorescence response among samples treated with alkali activation, ultrasonic activation, and thermal activation respectively. This indicates that alkali activation resulted in the least extent of sludge cracking while ultrasonic and thermal activations led to relatively higher degrees of sludge cracking. Additionally, when comparing Figs. 3(d) and (e), it is evident that ultrasonic activation induced more pronounced peak response enhancement across all five regions compared to thermal activation treatment, suggesting a greater degree of sludge cracking for ultrasonic activation than for thermal activation. Therefore, based on these comparisons within the four experimental groups mentioned above: ultrasonic activation > thermal activation > alkali activation > single addition of sodium persulfate.

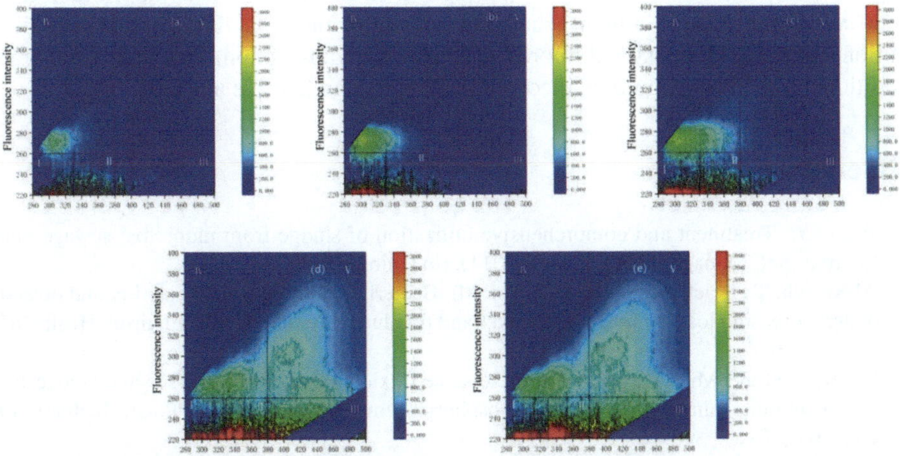

Fig. 3. Three-dimensional fluorescence spectra of different optimum activation conditions (a) sludge stock solution (b) sodium persulfate alone (c) activation of alkali (d) heat activation (e) ultrasonic activation

4 Conclusion

In this study, sodium persulfate was employed for sludge pretreatment using three different activation methods (alkali activation, thermal activation, and ultrasonic activation) to disrupt the stable structure of the sludge and extract extracellular polymers from the sludge suspension. The physicochemical parameters of protein and COD were measured as indicators. Additionally, laser particle size analysis and three-dimensional fluorescence spectrum analysis were conducted to investigate the impact of each method on sludge cracking degree. Finally, a comprehensive evaluation based on four indicators was performed to identify the optimal pretreatment method. Thermal activation at 80 °C for 20 min resulted in effective destruction of the stable structure of the sludge and achieved superior pretreatment performance. Alkali activation with a pH value of 13 and treatment time of 40 min demonstrated improved sludge cracking effect. Ultrasonic activation at a frequency of 2000 Hz for 75 min exhibited maximum ultrasonic efficiency leading to optimal sludge pretreatment. Through the comparison of protein and COD, particle size distribution, and three-dimensional fluorescence spectral absorption under the optimal conditions of the three activation methods, a comprehensive analysis revealed that the degree of sludge cracking induced by the pretreatment methods ranked as follows: ultrasonic activation (75 min at 2000 Hz) > thermal activation (20 min at 80 °C) > alkaline activation (40 min at pH 13).

Currently, the investigation of activated sodium persulfate pretreatment on sludge, both domestically and internationally, is in its nascent stage and requires further exploration and refinement. The composition of sludge is relatively intricate, with different activation methods yielding varying effects on distinct components. However, this experimental study solely focuses on comparing the effects of three activation methods on four indicators: protein content, chemical oxygen demand (COD), particle size distribution, and three-dimensional fluorescence spectrum absorption. To enhance accuracy, additional indicators should be incorporated for comprehensive analysis.

References

1. Wang, Y.: Treatment and comprehensive utilization of sludge from municipal sewage plant. Environ. Sci. Manag. **36**(01), 90–94 (2011). (in Chinese)
2. Al-Rashdi, T., Blackburn, D.M., Ahmed, M.: Bacterial community composition and diversity under increasing loading rates in reed wetland for sludge treatment. Soil Environ. Health **2**(2), 100083 (2024)
3. Lo, K.V., et al.: Microwave enhanced advanced oxidation treatment of sewage sludge from the membrane-enhanced biological phosphorus removal process. Sep. Purif. Technol. **197**, 202–209 (2018)
4. Diao, H., Cui, K., Liu, T., et al.: Biochar activation and degradation of acid orange G by persulfate. J. Hefei Univ. Technol. (Nat. Sci. Ed.) **46**(12), 1602–1609 (2023). (in Chinese)
5. Wang, R.G.: Preparation of hydrotalc-like catalysts and study on catalytic degradation of micropollutants in source water. University of Jinan, Jinan (2023). (in Chinese)
6. Zhou, S.X.: Effect of ultrasonic/persulfate pretreatment technology on sludge dewatering and anaerobic digestion. Nanchang University, Nanchang (2021). (in Chinese)
7. Lu, S.: Study on enhanced sludge dewatering performance by activated persulfate. Wuhan University of Science and Technology, Wuhan (2020). (in Chinese)
8. Lee, K.M., Kim, M.S., Lee, C.: Oxidative treatment of waste activated sludge by different activated persulfate systems for enhancing sludge dewaterability. Sustain. Environ. Res. **26**(4), 177–183 (2016)
9. Liu, W., Zhang, H., Liu, X., et al.: Application of mud film symbiosis/ozone catalytic oxidation process in upgrading wastewater treatment plants. Water Purif. Technol. **43**(04), 178–189 (2018). (in Chinese)

Fe_2O_3: A Kind of Starter for Autotrophic Nitrogen Removal Systems

Qi Wang[1,2], Linjie Wang[3], and Yongzhi Chen[1,2](✉)

[1] School of Environmental and Municipal Engineering, Lanzhou Jiaotong University, Lanzhou 730070, China
476411589@qq.com
[2] Technical Center of Sewage Treatment Industry in Gansu Province, Lanzhou 730070, China
[3] Gansu Mechanical and Electrical Vocational College, Tianshui 741000, China

Abstract. Fe^{3+} reduction-coupled anaerobic ammonia oxidation (Feammox) bacteria are a type of autotrophic nitrogen-removing bacteria discovered in recent years. In this study, Fe_2O_3 was added to the anammox system as a kind of fast starter for the Feammox, and the incubation process, nitrogen removal performance, and kinetic simulation process of Feammox were investigated. The data showed that the NH_4^+-N removal could reach up to 96% during the steady state period. The modified Boltzmann model fitted in the kinetic analysis of Feammox ($R^2 = 0.99632$), which could explain the NH_4^+-N removal and NO_3^--N generation. In summary, exploring the cultivation process and results of Feammox is crucial for NH_4^+-N removal performance and provides data support for constructing the Feammox process as the mainstream wastewater treatment process.

Keywords: Feammox · Ammonium removal · Kinetics

1 Introduction

In recent years, in the large-scale production of industries, a severe problem has been created: ammonia pollution in the water environment. Large quantities of nitrogen-containing substances can cause eutrophication of natural watercourses, with adverse effects on water quality and ecosystems [1]. The anammox process is considered to replace conventional nitrification and denitrification systems for autotrophic nitrogen removal using the direct conversion of NH_4^+-N and NO_2^--N to N_2 and NO_3^--N [2]. The main difficulty in anammox systems is achieving a stable source of NO_2^--N [3]. In contrast, conventional short-range nitrification or short-range denitrification is challenging to achieve [4]. Therefore, it is necessary to find electron acceptor substitutes other than NO_2^--N for autotrophic denitrogenation in anammox systems. Iron can exist in the valence states 0, +2, +3 and +6 [5]. Feammox bacteria can oxidize NH_4^+-N to N_2, NO_2^--N, and NO_3^--N while using Fe^{3+} as an electron acceptor and then converting it to Fe^{2+} (Eqs. 1, 2, 3). Compared to the nitrification-denitrification process, Feammox has excellent potential for application as it saves aeration energy, eliminates carbon addition,

D. Han et al. (Eds.): ICEPG 2024, ESE, pp. 520–525, 2025.
https://doi.org/10.1007/978-3-031-86829-0_59

and reduces sludge production. Therefore, a suitable electron acceptor is a prerequisite for the rapid initiation of Feammox to maximize the $NH4^+$-N removal efficiency.

$$3Fe(OH)_3 + 5H^+ + NH_4^+ \rightarrow 3Fe^{2+} + 9H_2O + 0.5N_2 \quad (1)$$

$$6Fe(OH)_3 + 10H^+ + NH_4^+ \rightarrow 6Fe^{2+} + 16H_2O + NO_2^- \quad (2)$$

$$8Fe(OH)_3 + 14H^+ + NH_4^+ \rightarrow 8Fe^{2+} + 21H_2O + NO_3^- \quad (3)$$

In this study, a new nitrogen removal process was developed in ASBR, and the main research contents are as follows: (1) exploring the nitrogen removal performance of the Feammox process; (2) analyzing the nitrogen transformations of Feammox; (3) and adopting the modified kinetic model to express the transformations of substrates. This study will provide a theoretical basis and scientific guidance for the initiation of practical wastewater treatment with the Feammox process.

2 Materials and Methods

2.1 Reactor Operation

The seed sludge used in this study was laboratory-grown mature granular anammox sludge. Fe_2O_3 (powder) (2g/L Fe) was added to the inoculated sludge, and no iron source was added to the control. These are referred to as the experimental group and the control group, respectively. The following compounds are used for configuring water: $NaHCO_3$, KH_2PO_4, $CaCl_2$, and $MgCl_2$. A 2L ASBR is protected from light and equipped with Stirrers, thermometers, pH meters, etc. The hydraulic retention time was set to 48 h, the temperature was maintained at 35 °C and pH at 6.5 ± 0.2, and the water exchange rate was 50%.

2.2 Chemical Analysis

Determination of NH_4^+-N, NO_2^--N and NO_3^--N by standard methods [6]. The concentration of Fe^{2+} in the effluent was determined using the o-phenanthroline method.

2.3 Kinetic Analysis

Modified of the Boltzmann, Logistic, and Gompertz kinetic models were used to describe the NH_4^+-N removal in typical cycles.

The modified Boltzmann model: $NRE = NRE_{max} + \frac{NRE_{min} - NRE_{max}}{1+\exp(\frac{t-t_c)}{t_d}}$.

The modified Logistic model: $NRE = \frac{NRE_{max}}{1+\exp\left[\frac{4R_m(\lambda-t)}{NRE_{max}}+2\right]}$.

The modified Gompertz model: $NRE = NRE_{max} \times \exp[-\exp\left(\frac{R_m \times e(\lambda-t)}{NRE_{max}}\right]$.

ARE is the NH_4^+-N removal efficiency (%); ARE_{max} is the maximum NH_4^+-N removal efficiency (%); ARE_{min} is the minimum NH_4^+-N removal efficiency; R_m is the maximum NH_4^+-N removal efficiency (%/h); t_d is the time constant (h); t_c is the time corresponding to 1/2 ARE_{max} (h); λ is the delay time of Iron-reducing bacteria ammonium removal (h).

3 Results and Discussion

3.1 Nitrogen Removal Performance of the Feammox Process

In order to explore the long-term treatment performance and operational characteristics of Feammox, the changes in key parameters, such as total inorganic nitrogen and effluent Fe^{2+} with operating time, were analyzed, as shown in Fig. 1. NO_2^--N was not discussed in this test because it was detected at concentrations below 1 mg/L throughout the test cycle.

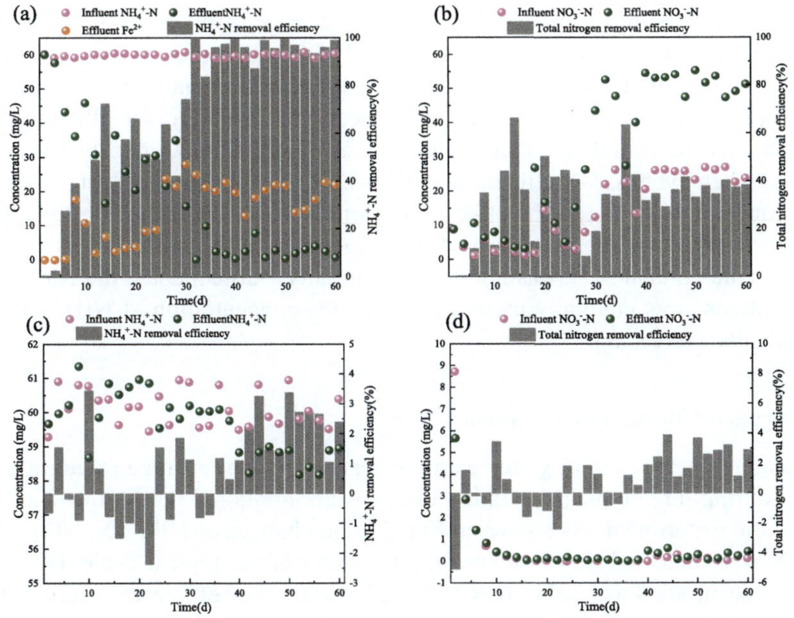

Fig. 1. The long-term performance of the Feammox system running for 60 days: (a) changes in NH_4^+-N, ARE, and Fe^{2+} effluent concentrations in the experimental group; (b) changes in NO_3^--N and NRE in the experimental group; (c) changes in NH_4^+-N and ARE in the blank group; (d) changes in NO_3^--N and NRE in the blank group.

As shown in Fig. 1c, during the first 36 d, the control group had higher NH_4^+-N effluent concentrations than the influent. One possible explanation for this is death due to a change in the environment in which the microorganisms live, and the NH_4^+-N concentration in the system increased after the autolysis of the organisms. Autolysis refers to the self-digestion of cells due to the breakdown of their own enzymes, which leads to the release of cellular contents, including nitrogen compounds. This process was observed in the control group but did not occur in the experimental group. Then, there was a gradual decrease in the NH_4^+-N effluent, and by the final cycle of the reaction, there was a removal of about 1 mg/L. This suggests that the system was gradually stabilizing and the microbial communities in the reactor were becoming more efficient at processing NH4+-N. In contrast, the effluent NH_4^+-N of the experimental group already had a significant decrease on the 4th day (Fig. 1a). Subsequently, it showed a wave-like decrease until 32 days, when the NH_4^+-N concentration dropped below 0.5 mg/L for the first time. 99.8% NH_4^+-N removal was achieved. This dramatic drop in NH4$^+$-N indicates the progressive efficiency of the Feammox process. Since the DO in the reactor was <0.1 mg/L at this time, the nitrification did not affect the NH_4^+-N concentration, indicating that the Feammox bacteria had been successfully cultured in the system.

The effect of Fe_2O_3 on NO_3^--N concentration was explored during the experiment. As shown in Fig. 1d, at the start of the control group, NO_3^--N gradually decreased with dilution by water exchange and denitrification by bacterial lysis. This suggests that bacterial populations in the control group may have undergone lysis, reducing the available nitrate for further reduction. By the late stage of the reaction, there was no significant production of NO_3^--N. The NO_3^--N of the experimental group, on the other hand, also gradually decreased at the early stage, indicating that although there was a similar situation with the control group, the concentration of NO_3^--N effluent gradually increased after the 16th d, and then maintained at about 55 mg/L after the reaction was stabilized when the NH_4^+-N effluent was below 3 mg/L. Fe^{2+} is a product of Feammox, and its concentration reflects the effect of Feammox and the extent of Fe_2O_3 reduction (Fig. 1a). In the reaction group 0–6 d, the experimental group Feammox reaction is weaker, so the effluent Fe^{2+} was not detected initially. Subsequently, the concentration of Fe^{2+} effluent was gradually increasing while the concentration of NH_4^+-N effluent was gradually decreasing.

3.2 Nitrogen Changes in a Typical Cycle

The changes in nitrogen during the Feammox reaction and the average values of nitrogen changes during three typical operation cycles were monitored during the stabilization phase of the experiment. As shown in Fig. 2d, the changes of NH_4^+-N, NO_2^--N, and NO_3^--N were detected at 4 h intervals. At the end of the typical cycle, the NH_4^+-N effluent concentration was reduced to 2.36 mg/L and removal efficiency reached 96.05%. No NO_2^--N was not detected in all cycles. The effluent concentration of NO_3^--N was 49.27 mg/L, and the total nitrogen removal rate reached 38.83%. This suggests that NO_3^--N production is negatively correlated with a decrease in NH_4^+-N, and the overall trend is the same as the long-term test results.

3.3 Kinetic Analysis

The removal efficiency of NH_4^+-N in a typical cycle was simulated using a kinetic model, fitted with a non-linear model. Boltzmann, Logistic, and Gompertz models were used to simulate microbial growth processes to explain the NH_4^+-N removal and NO_3^--N generation. The removal rate of NH_4^+-N in a typical cycle was fitted by a nonlinear model, and the ARE_{max} fitted by the modified Boltzmann model was 106.97%, with a t_c of 8.21 h, which was consistent with the trend of the typical cycle of NH_4^+-N. The results of the modified Gompertz and logistic model showed that λ was 1.49 h and 0.61 h, respectively, i.e., the modified logistic model was more suitable to describe the process of bacterial environmental adaptation. The correlation coefficients R^2 of the modified Boltzmann, Logistic, and Gompertz models for the experimental group were 0.99632,

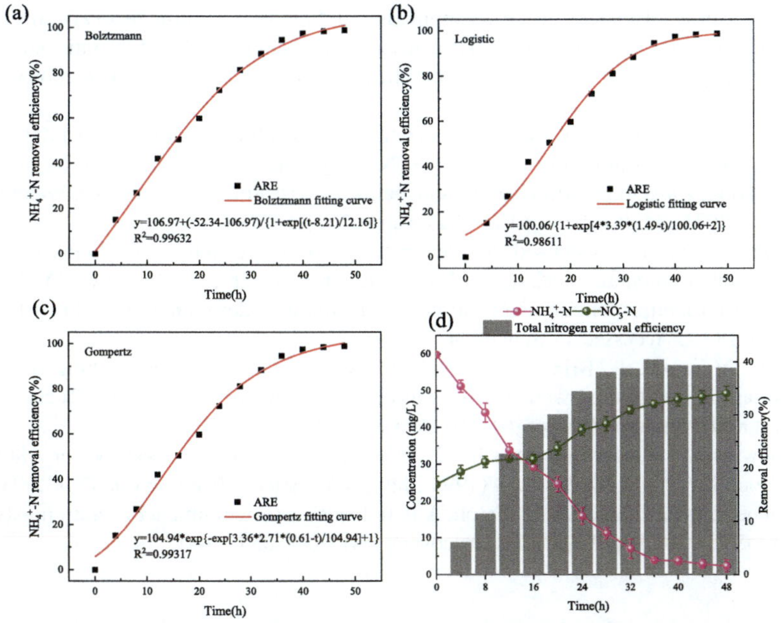

Fig. 2. Kinetic models reveal the ARE of Feammox: (a) modified Boltzmann model, (b) modified Logistic model (c) modified Gompertz model, (d) the changes of nitrogen in a typical cycle.

0.98611, and 0.99317, respectively. It was shown that the modified Boltzmann model better fitted the removal process of NH_4^+-N by Feammox.

4 Conclusions

In this study, Fe_2O_3 was used as a substrate for the Feammox in the ASBR. The conversion of NH_4^+-N, NO_3^--N, and Fe^{2+} yield and kinetics in the Feammox system were analyzed, and the main conclusions are as follows.

The long-term experiment concluded that:

(1) The addition of Fe_2O_3 to the anammox ASBR reactor produced Feammox, which significantly improved the NH_4^+-N removal efficiency to 96% (averaged over the last 5 cycles).
(2) It was shown that Fe_2O_3 can domesticate and enrich Feammox.
(3) It was shown that the modified models could accurately describe the transformation parameter of ammonia nitrogen in the Feammox system, among which the Boltzmann model was the best fitted.

References

1. Huang, J., Kankanamge, N.R., Chow, C., Welsh, D.T., Li, T., Teasdale, P.R.: Removing ammonium from water and wastewater using cost-effective adsorbents: a review. J. Environ. Sci. **63**, 174–197 (2018). https://doi.org/10.1016/j.jes.2017.09.009

2. Yuan, Z., et al.: Efficient nitrite accumulation and elemental sulfur recovery in partial sulfide autotrophic denitrification system: insights of seeding sludge, S/N ratio and flocculation strategy. Chemosphere **288**, 132388 (2022). https://doi.org/10.1016/j.chemosphere.2021.132388
3. Wu, J., et al.: A successful start-up of an anaerobic membrane bioreactor (AnMBR) coupled mainstream partial nitritation-anammox (PN/A) system: a pilot-scale study on in-situ NOB elimination, AnAOB growth kinetics, and mainstream treatment performance. Water Res. **207**, 117783 (2021). https://doi.org/10.1016/j.watres.2021.117783
4. Anammox bacteria disguised as denitrifiers: nitrate reduction to dinitrogen gas via nitrite and ammonium - Kartal - 2007 - Environmental Microbiology - Wiley Online Library (n.d.). https://enviromicro-journals.onlinelibrary.wiley.com/doi/10.1111/j.1462-2920.2006.01183.x. Accessed 17 Sept 2024
5. Wu, S., Vymazal, J., Brix, H.: Critical review: biogeochemical networking of iron, is it important in constructed wetlands for wastewater treatment? Environ. Sci. Technol. (2019). acs.est.8b07136. https://doi.org/10.1021/acs.est.8b07136
6. Pawlowski, L.: Standard methods for the examination of water and wastewater, 18th edition. Sci. Total Environ. **142**, 227–228 (1994). https://doi.org/10.1016/0048-9697(94)90332-8. In: Greenberg, A.E., Clesceri, L.S., Eaton, A.E. (eds.) Water Environment Federation, Alexandria, USA (1992). 1025 pp. US$120 (Hardcover); ISBN 0-87553-207-1

Facile Preparation of a PDMS, Nano-Fe_3O_4 and Graphene Decorated Melamine Sponge with Magnetism and Photothermal Conversion Properties for Effective Oil/Water Separation

Zichao Yin[1,2,3](✉), Zhenhua Sun[2,3], Jingjing Qi[2,3], Junyi Li[1,2,3], Xu Wei[2,3], and Weidong Wang[3]

[1] Post Doctorate Scientific Research Station, Shengli Petroleum Administrative Bureau, SINOPEC, Dongying 257000, China
yinzichao.slyt@sinopec.com
[2] SINOPEC Shengli Oilfield Technology Inspection Center, SINOPEC, Dongying 257000, China
[3] SINOPEC Shengli Oilfield Co., Dongying 257000, China

Abstract. With the further development of petroleum transportation, oil spills are becoming increasingly frequent and the leakage environment is becoming more and more complex. In order to cope with this trend, an oil adsorption material that combined magnetic and photothermal conversion functions was prepared in this study. Through experimental research, it can be found that the prepared material had a high porosity, and its skeleton surface was uniformly coated with Fe_3O_4 nanoparticle, graphene and PDMS. The adsorbent had excellent hydrophobicity and oleophilicity, and could efficiently recover oil substances in water. The adsorption capacity of the material for different oil substances was between 28.4–62.9 g/g. At the same time, the adsorbent also had magnetism and could be easily lifted by magnets. In addition, the adsorbent also possessed excellent photothermal conversion efficiency, and could adsorb crude oil with high viscosity under simulated sunlight irradiation. From the above conclusion, it can be concluded that this material has broad prospects in the field of oil spill treatment.

Keywords: Oil/water separation · melamine sponge · magnetism · photothermal conversion

1 Introduction

The intensification of offshore oil extraction and transportation has increased the occurrence frequency of oil spill. If an oil spill occurs, it will seriously damage the surrounding ecological environment [1] and cause a large amount of harmful substances [2] to leak. Currently, the leaked oil substances and the environments of the leakage are more and more complex. In order to cope with that trend, multifunctional oil adsorption materials have been researched and applied. Magnetic response function and photothermal conversion function are the two most common endowed functions. Oil adsorption materials

D. Han et al. (Eds.): ICEPG 2024, ESE, pp. 526–532, 2025.
https://doi.org/10.1007/978-3-031-86829-0_60

with magnetic response function can be recovered using magnets, effectively avoiding contact between workers and oil spill environments. Fe_3O_4 nanoparticle was the mostly used material to endow the adsorbent with magnetism. Adsorption materials with photothermal conversion ability can effectively adsorb the oil with high viscosity under sunlight. This type of material can reach a temperature of over 60 °C under sunlight exposure. At higher temperature, the viscosity of crude oil decreases and its flowability improves, which in turn increases the adsorption efficiency of the material. The materials with high photothermal conversion ability, such as graphene, MXene, carbon nanotubes and carbon black [3], have been coated on the substrates. Although the application scenarios of oil adsorbent have been largely expanded by these dual functional adsorbent materials, it is still difficult for them to cope with more complex oil spill environments and more diverse oil substances.

In this study, a melamine sponge-based adsorbent with both magnetic and photothermal conversion properties was successfully fabricated. Melamine sponge with excellent mechanical properties was used as the porous substrate, PDMS was used as the hydrophobic modifier, Fe_3O_4 nanoparticle was used to endow the material with magnetism, and graphene was used as the photothermal conversion material. PDMS, Fe_3O_4 nanoparticle and graphene were loaded onto the melamine sponge by dip-coating method. The prepared material was undergone a series of characterizations and performance tests, and the experimental results showed that the adsorbent had excellent hydrophobicity, magnetism, and photothermal conversion efficiency.

2 Experimental Procedure

Preparation of PDMS/Fe_3O_4/GR@MS. A facile dip-coating method was used to synthesize the adsorbent material. Firstly, 0.6 g of PDMS was added to 25 mL of n-hexane under stirring. Then, 0.02 g of graphene and 0.7 g of Fe_3O_4 nanoparticle were added to the solution, and the mixture was under an ultrasonic environment for 30 min. After the mixture was completely mixed, 5 pieces of melamine sponge with size of $2 \times 2 \times 1$ cm were added to the mixture and were also under ultrasonic treatment for 30 min. After that, the sponges were taken out and placed in an oven for heating treatment (60 °C) for 1 h. After the n-hexane was completely evaporated, the sponges were removed from the oven. Finally, the adsorbent was successfully fabricated and named as PDMS/Fe_3O_4/GR@MS.

Characterization. The surface morphology and pore structure of PDMS/Fe_3O_4/GR@MS were observed using a scanning electron microscope (SEM, Hitachi S-4800). The lattice structure of the adsorbent was analyzed using an X-ray diffractometer (Bruker D8 Advance). The elemental composition of the prepared adsorbent was measured using X-ray photoelectron spectroscopy (XPS, Thermo ESCALAB 250XI). The hydrophobicity of the material was tested using a contact angle analyzer (Dataphysics OCA15EC). The magnetic property of the fabricated adsorbent was analyzed by a vibrating sample magnetometer (VSM, Quantum Design). The temperature of the material under simulated sunlight was measured using an infrared camera.

Performance Testing. The adsorption capacity of PDMS/Fe_3O_4/GR@MS was determined by using it to adsorb different types of oil substances. The following formula was used to calculate its adsorption capacity for different oils. $Q = (m_1 - m_0)/m_0$. In the equation, Q is adsorption capacity of the adsorbent, m_0 is the weight of the adsorbent before adsorption, and m_1 is the mass of the adsorbent after adsorption. Different oil-water mixtures (1 g oil and 100 mL water) were prepared to test the oil-water separation efficiency (E) of PDMS/Fe_3O_4/GR@MS. The following formula was used for calculation. $E = (1000 - CV) \times 100\%/1000$. In the equation, C and V are the oil concentration (mg/L) and volume of the mixture after separation. The crude oil from Shengli Oilfield (Shandong, China) was used to test the material's adsorption capacity for viscous oil.

3 Results and Discussion

The SEM images of PDMS/Fe_3O_4/GR@MS is shown in Fig. 1. It can be seen from the figure that the adsorbent exhibits a high degree of porosity, and its skeleton structure is a three-dimensional network [4]. Further observation of its skeleton surface reveals that many granular substances are adsorbed on it, which are likely to be a composite of PDMS, Fe_3O_4 nanoparticle and graphene.

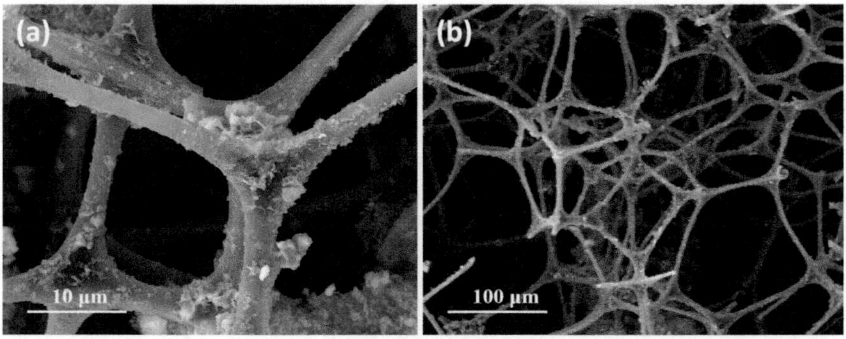

Fig. 1. SEM images of PDMS/Fe_3O_4/GR@MS at magnifications of (a) 1000 and (b) 100.

To demonstrate the successful coating of Fe_3O_4 nanoparticle, PDMS/Fe_3O_4/GR@MS was analyzed using an X-ray diffractometer and the result is shown in Fig. 2(a). In the spectrum of pure Fe_3O_4 nanoparticle, diffraction peaks are appeared at 30.2°, 35.5°, 43.1°, 57.1° and 62.6°, which belong to the typical (220), (311), (400), (511) and (440) planes of cubic magnetite (JCPDS card No. 19-0629). At the same position, diffraction peaks also appear in spectrum of PDMS/Fe_3O_4/GR@MS, proving the successful loading of Fe_3O_4 nanoparticle onto the sponge. To demonstrate the successful coating of PDMS, X-ray photoelectron spectroscopy was employed to analyze PDMS/Fe_3O_4/GR@MS. The analysis result (Fig. 2(b)) shows strong peaks at positions of 531.58 eV, 283.68 eV and 101.58 eV, indicating the presence of O, C, and Si elements. Among them, Si element belongs to PDMS, which proves the successful loading of PDMS.

According to the XRD and XPS results, it can be concluded that Fe_3O_4 nanoparticle had been successfully loaded onto the surface of PDMS/Fe_3O_4/GR@MS. Therefore,

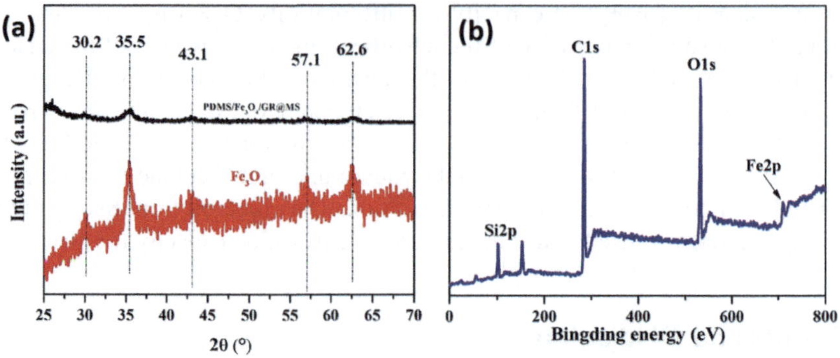

Fig. 2. (a) XRD spectra of pure Fe_3O_4 nanoparticle and PDMS/Fe_3O_4/GR@MS. (b) XPS spectrum of PDMS/Fe_3O_4/GR@MS.

it could be inferred that the material had magnetism. The magnetism of the material was tested by using a magnet to attract it. Figure 3(a) shows that the material can be easily lifted by a magnet, indicating that it had a certain degree of magnetism. To further explore its magnetism, the hysteresis curve of the material was measured. It can be seen from Fig. 3(b) that when the magnetic field varies between –20,000 and 20,000 Oe, the magnetic saturation value of the material reaches 16.8 emu/g, indicating that it had good magnetism. Compared with pure Fe_3O_4 nanoparticle (79.3 emu/g), the magnetic saturation value of the material was much more smaller. This is mainly due to the high content of non-magnetic substances in the material.

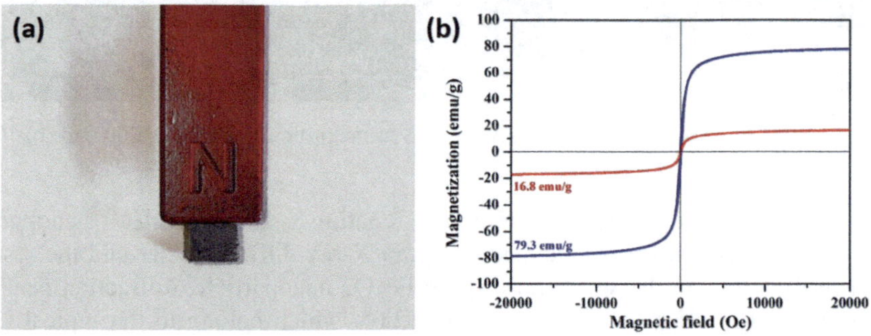

Fig. 3. (a) A piece of PDMS/Fe_3O_4/GR@MS was lifted by a magnet. (b) Magnetic hysteresis loops of PDMS/Fe_3O_4/GR@MS and pure Fe_3O_4 nanoparticle.

Due to the coating of graphene, the photothermal conversion ability of PDMS/Fe_3O_4/GR@MS was greatly improved [5]. By using simulated sunlight to irradiate the material and recording the temperature of the material with an infrared camera, the photothermal conversion ability of the material was explored. The temperature of the adsorbent at 7, 8, 9 and 10 cm away from the simulated light source was measured and the results are shown in Fig. 4. As shown in the figure, it can be concluded that the closer the material was to the simulated light source, the higher the temperature, which reached to 95.2, 96.7, 121.7 and 141.7 °C, respectively. The above test results prove the excellent photothermal conversion ability of PDMS/Fe_3O_4/GR@MS. The material can generate heat under sunlight to improve the viscosity and flowability of crude oil, thereby increasing the adsorption rate for viscous crude oil.

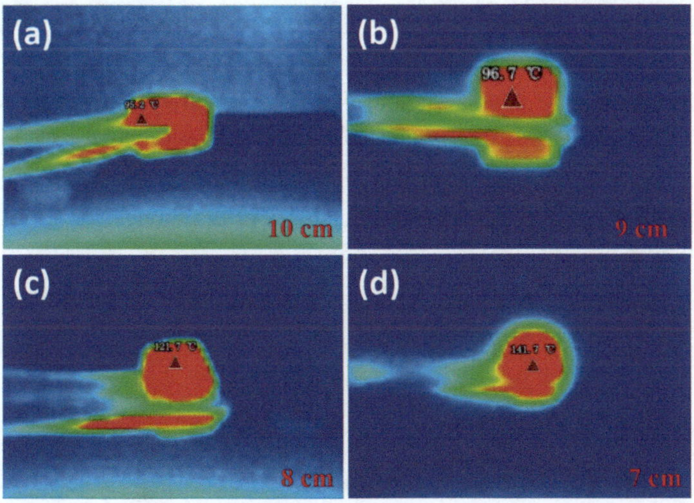

Fig. 4. Temperature of PDMS/Fe_3O_4/GR@MS at 7, 8, 9 and 10 cm away from light source.

According to the characterization results of PDMS/Fe_3O_4/GR@MS, it can be inferred that the material may have good hydrophobicity. Therefore, a series of tests were carried out on the material to explore its hydrophobicity and the results are shown in Fig. 5. As shown in the figure, the water contact angle of the material is 152°, which proves its good hydrophobicity. In addition, the water droplet on the surface of the material (Fig. 5(a)) shows a nearly spherical shape, more intuitively demonstrating its excellent hydrophobicity.

According to the hydrophobic tests of PDMS/Fe_3O_4/GR@MS, it was known that the material had excellent oil selective adsorption ability. In order to further test this ability of the material, it was used to adsorb the floating oil on the surface of the water and the sinking oil in the water. As shown in Fig. 5(b), the material can quickly absorb the n-hexane into its interior. As exhibited in Fig. 5(c), dichloromethane on the bottom of the beaker can be completely recovered by the material.

The adsorption capacity is an important indicator for judging an adsorbent. Therefore, the adsorption capacities of PDMS/Fe_3O_4/GR@MS for different organic reagents were tested, and the results are shown in Fig. 6. From the figure, it can be seen that the adsorption capacities of PDMS/Fe_3O_4/GR@MS for the testing organic reagents are from 28.4 to 62.9 g/g, proving excellent adsorption capacity. The difference in adsorption capacity is mainly caused by the difference in organic reagent density. In addition, the separation efficiencies of the sponge for different oil-water mixtures were also tested.

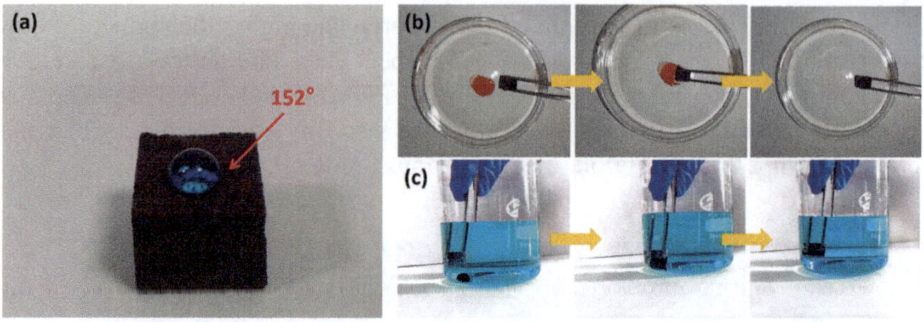

Fig. 5. (a) A water droplet on the surface of PDMS/Fe_3O_4/GR@MS. PDMS/Fe_3O_4/GR@MS was used to recover (b) floating oil (hexane) and (c) sinking oil (dichloromethane).

As shown in the figure, the separation efficiencies of the sponge for all mixtures were greater than 98%.

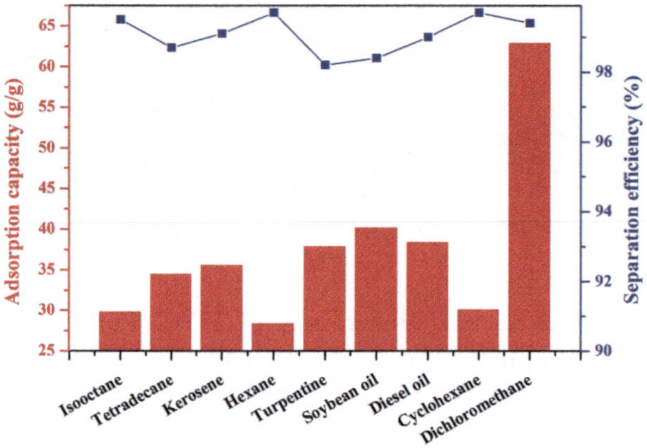

Fig. 6. Adsorption capacity and separation efficiency of PDMS/Fe_3O_4/GR@MS.

The above experimental results demonstrate that PDMS/Fe_3O_4/GR@MS had excellent hydrophobicity, adsorption capacity and photothermal conversion efficiency. Therefore, PDMS/Fe_3O_4/GR@MS was used for adsorption test for crude oil. The viscosity of the testing crude oil was between 57–78 mPa·s. The entire adsorption process is shown in Fig. 7. It can be seen from the figure that the crude oil placed on the sponge was difficult to be absorbed by the sponge without sunlight irradiation (Fig. 7(a)). It took about 304 s for the sponge to completely absorb the crude oil. In comparison, it was easier for the sponge under sunlight irradiation to complete the process. The whole adsorption process only took about 58 s (Fig. 7(b)). The reason is as follows. When the sponge was exposed to the simulated sunlight, its temperature slowly increased. Simultaneously, the crude oil was heated, causing its viscosity to decrease and its fluidity to become stronger. As a result, the crude oil was easily absorbed by the sponge.

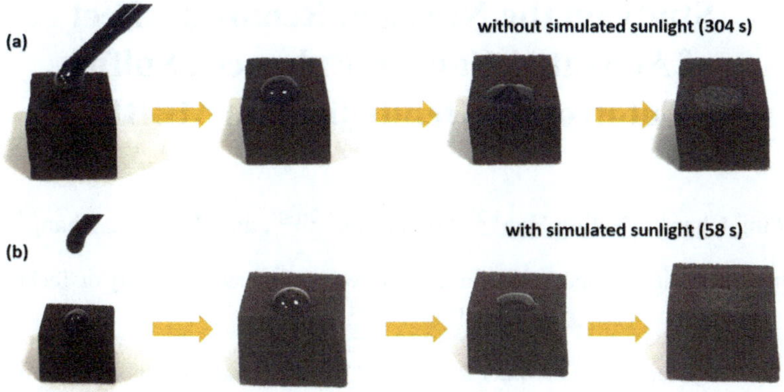

Fig. 7. Process of recovering crude oil by PDMS/Fe_3O_4/GR@MS (a) without simulated sunlight and (b) under simulated sunlight.

4 Conclusion

In summary, a adsorption material with oil-water separation ability, magnetism, and photothermal conversion ability was successfully synthesized through a simple dip-coating method. The prepared material had excellent hydrophobicity and oleophilicity, and could selectively recover oil substances from oil-water mixtures. The adsorption capacity of the material for different oil substances was as high as 28.4–62.9 g/g. At the same time, the material also had excellent magnetism (16.8 emu/g), which could be easily lifted by magnets. In addition, the material also had excellent photothermal conversion ability, which could heat the viscous crude oil under simulated sunlight irradiation and enhance its flowability to improve the adsorption efficiency. Due to its simple preparation process and multifunctionality, the material may have broad application prospects in the field of oil spill treatment.

References

1. Ewim, D.R.E., et al.: Survey of wastewater issues due to oil spills and pollution in the Niger Delta area of Nigeria: a secondary data analysis. Bull. Natl. Res. Centre **47**, 116 (2023)
2. Camilli, R., et al.: Tracking hydrocarbon plume transport and biodegradation at Deepwater Horizon. Science **330**, 201–204 (2010)
3. Wang, N., et al.: Eco-friendly MXene/enteromorpha prolifera biomass aerogel with excellent photothermal conversion and flame retardant properties for efficient separation of viscous oil/seawater mixture. Sep. Purif. Technol. **348**, 128259 (2024)
4. Shukla, V., Haq, N., Siddiqui, K.A.: Zinc MOF and melamine sponge composite: a cutting-edge solution for dose-dependent dye degradation and ultra-sensitive chemical sensing of nitro derivatives and sucrose. J. Inorg. Organomet. Polym. Mater. https://doi.org/10.1007/s10904-024-03350-7
5. Shang, M., et al.: Cellulose/poly(vinyl alcohol)/graphene composite photothermal aerogel membrane for solar-driven seawater desalination. Waste Biomass Valoriz. **15**, 5549–5558 (2024)

Study on the Nitrogen Removal Effect of Activated Carbon Enhanced Sulfur Autotrophic Denitrification MBBR

Ping Chen[1,2], Yaling Han[1,2], Mulan Zhu[1,2](✉), and Zhichong Zhang[1,2]

[1] School of Environmental Science and Engineering, Xiamen University of Technology, Xiamen 361024, China
292890310@qq.com

[2] The Key Laboratory of Water Resources Utilization and Protection of Xiamen, Xiamen 361024, China

Abstract. In order to solve the problem of comprehensive ecological management of the Daihai Sea in China, Inner Mongolia Autonomous Region has planned and implemented the project of diverting Yellow River to Daihai, however, due to the fact that the Yellow River diversion water is a micro-polluted surface water with nitrate nitrogen as the main pollutant, which has the characteristics of carbon and nitrogen ratio, DO high water quality, the sulphur autotrophic denitrification is a major technology that is expected to solve the problem at the present time. In this study, in order to strengthen the effect of nitrogen removal, coconut shell activated carbon, sulfur and limestone were used as the activated carbon enhanced sulfur autotrophic denitrification substrate package (ASL) and coupled with MBBR balls to form the ASL-MBBR process, the nitrogen nitrogen removal effects of four ASL-MBBR processes with different activated carbon percentages were investigated and compared with a blank control scheme. The results of the study showed that: 1 MBBR ball containing 18g of autotrophic denitrification substrate package was placed in 2L of experimental water under the following Under the conditions of, the four experimental protocols with the mass ratios of coconut shell activated carbon:sulfur:limestone in the substrate package were 1:1:1, 2:1:1, 3:1:1, 4:1:1, and the removals of TN were 66.40%, 60.19%, 71.36%, and 73.02%, respectively; and the removals of NO_3-N were 83.20%, 79.74%, and 87.01%, respectively, 88.08%. Among them, Scheme 4 with a mass ratio of 4:1:1 had the best nitrogen removal. This research result is expected to provide a new and more effective technology for nitrogen removal in solving the problem of micro-polluted surface water with high dissolved oxygen.

Keywords: activated carbon · nitrogen removal · MBBR

P. Chen and Y. Han Co-first author of the article.

D. Han et al. (Eds.): ICEPG 2024, ESE, pp. 533–539, 2025.
https://doi.org/10.1007/978-3-031-86829-0_61

1 Introduction

Located in Inner Mongolia Autonomous Region, Daihai Sea of China is one of the lakes listed in the national planning and protection plan, which has extremely important ecological functions. However, due to the decrease of precipitation in the watershed year by year, the dryness index increasing year by year, and the increase of water consumption in the neighbourhood year by year, the lake surface is increasingly shrinking [1]. However, due to the decreasing precipitation, increasing dryness index, and increasing water consumption, the lake surface is shrinking. Therefore, the Inner Mongolia government has taken a series of comprehensive measures to promote the comprehensive ecological management of Daihai Sea. Ecological water replenishment is an important part of the protection of the Daihai Sea, mainly through the diversion of Huang Jidai project by setting up the front reservoir, further settling of sediment and the purification of the Yellow River diversion, the purified Yellow River diversion into the Dai Sea. The longer the hydraulic retention time, the better the water purification effect, but the larger the storage capacity required by the pre-storage. Considering various factors, the proposed hydraulic retention time of the existing design scheme is 10d. The Yellow River water belongs to the slightly polluted surface water, there are mainly total nitrogen exceeds the standard, total nitrogen nitrate nitrogen as high as 4 mg/L, fail to meet the surface water quality standards of class V water. Secondly, the Yellow River water has low BOD5 (five-day biochemical oxygen demand), small carbon to nitrogen ratio, insufficient carbon source, high dissolved oxygen (DO), and heterotrophic denitrification, although it has been widely used for removing nitrate nitrogen from surface water, requires the addition of nitrate nitrogen in the reaction process [2]. Although heterotrophic denitrification has been widely used to remove nitrate-nitrogen from surface water, it requires the addition of an extra organic carbon source during the reaction process, which may lead to secondary pollution [3]. The use of autotrophic denitrification, which does not require an additional carbon source, is one of the keys to nitrogen removal.

In addition, denitrification process needs to be carried out under anaerobic conditions, but the DO of the Yellow River diversion water is high, so how to form an anaerobic environment is also the key to denitrification [4]. MBBR ball can form aerobic-anoxic-anaerobic environment by growing biofilm, which can overcome the problem of denitrification in surface water with high DO [5]. The MBBR ball can overcome the problem of high DO in surface water, which makes denitrification difficult. The sulfur autotrophic denitrification substrate is filled into the MBBR ball to form the sulfur autotrophic denitrification MBBR biofill ball, and then the biofill ball is thrown into the water body which needs nitrogen removal, it is expected to solve the double difficulties of insufficient carbon source and high DO difficult to denitrify and denitrify at the same time.

Since the sulphur autotrophic denitrification reaction produces acid, limestone cofactor is usually added to maintain the pH value. The research team previously used sulfur and limestone as the sulfur autotrophic denitrification substrate, and filled the substrate into MBBR packing ball (referred to as SL-MBBR), and carried out experimental research on the nitrogen removal effect, and the results showed that the SL-MBBR process can play a certain nitrogen nitrogen removal effect, and its removal rate of TN is about 18.35%. However, the nitrogen removal effect is not enough to make the Yellow

River diversion meet the water quality requirements of ecological water replenishment. On the other hand, some studies have shown that activated carbon can not only enrich microorganisms, but also has the role of catalysing sulphur autotrophic denitrification. Its good structural characteristics and non-toxic and harmless features make it one of the economic and efficient immobilisation carrier materials [6].

For this reason, in order to further strengthen the effect of nitrogen removal, this study proposes to use activated carbon to strengthen sulfur autotrophic denitrification, i.e., using sulfur, limestone, activated carbon as activated carbon to strengthen sulfur autotrophic (ASL) denitrification substrate coupled with the MBBR ball to form the ASL-MBBR process, to explore the influence of the different activated carbon ratio on the effect of denitrification in the ASL-MBBR process, and to select the best denitrification solution from it. The optimal denitrification programme was selected from which the TN concentration of the Yellow River diversion is expected to meet the surface water quality standard of Class III by applying the denitrification technology.

2 Materials and Methods

2.1 Experimental Materials

In this study, in order to improve the nitrogen nitrogen removal effect of the traditional MBBR process, a packet of activated carbon reinforced sulphur autotrophic denitrification substrate packet was further placed inside the MBBR biofiller ball, which is made of PP polypropylene, with built-in polyurethane sponge, and the diameter of 100 mm The MBBR biofiller ball was further injected with a packet of activated carbon enhanced sulphur autotrophic denitrification substrate packet (ASL) to form the ASL-MBBR biofiller sphere. The ASL denitrification substrate package consisted of coconut shell activated carbon with a particle size of 1–2 mm, sulphur with a particle size of 3–5 mm and limestone with a filtration diameter of 3–5 mm composition.

2.2 Experimental Programme

In order to investigate the effect of different activated carbon percentages on the nitrogen removal effect of ASL-MBBR process, in this study, four denitrification substrate schemes were formed by changing the percentage of activated carbon under the condition that the denitrification substrate of coconut shell activated carbon, sulphur, and limestone was put in a total of 18 g. At the same time, a blank control group, which did not put in any ASL-MBBR biofiller ball, was set up to carry out the comparative experimental study.

The specifics of the experimental programme are shown in Table 1.

The water for the experiment was taken from the lake water of the school, and potassium nitrate reagent was added to this experiment to increase the nitrate nitrogen concentration in the water body.

The experiment was conducted for a total of 10 days, and water samples were taken 2 cm below the surface of the water body between 16:00–17:00 PM daily for water quality index testing.

Table 1. Experimental programme.

Programme	proportion	Experimental material
1	1:1:1	ASL-MBBR Biofiller Balls (Placement Density: 1 Balls/2L) ASL denitrification substrate:activated carbon:sulphur:limestone (6g/ball: 6g/ball: 6g/ball)
2	2:1:1	ASL-MBBR biofiller balls (placement density:1 Balls/2L), ASL denitrification substrate:activated carbon:sulphur:limestone (9g/ball: 4.5g/ball: 4.5g/ball)
3	3:1:1	ASL-MBBR biofiller balls (placement density:1 Balls/2L), ASL denitrification substrate:activated carbon:sulphur:limestone (10.8g/ball:3.6g/ball:3.6g/ball)
4	4:1:1	ASL-MBBR biofiller balls (placement density:1 Balls/2L), ASL denitrification substrate:activated carbon:sulphur:limestone (12g/ball:3g/ball:3g/ball)
5	*Blank*	

The experimental water was filled into 5 experimental containers (the specification is 19.3 cm high, 13.3 cm diameter beaker), and each container was filled with 2 L of experimental water. Two water samples were taken for testing, and the average value of the two water samples was taken as the experimental results.

The experiment was carried out at room temperature (19–25°C).Initial water quality of experimental water: NO3-N 4.11 mg/L, TN 5.14 mg/L, COD 46.65 mg/L, DO 8.19 mg/L, and pH 8.02.

In this experiment, in addition to for 2 nitrogen indicators, nitrate nitrogen (NO_3-N) and total nitrogen (TN), chemical oxygen demand (COD), dissolved oxygen (DO), and acidity and alkalinity (pH) were also tested simultaneously.

3 Results and Discussion

3.1 Comparison of TN and NO_3-N Concentration Changes

Over the course of the 10-day experiment, the change in TN concentration of the blank control versus the four experimental protocols (protocols 1 to 4) was compared as in Fig. 1 shows; the comparison of the changes in NO_3-N concentration between the blank control and the four experimental protocols (Scheme 1–4) is shown in Fig. 2

By Fig. 1. It can be seen that during the 10-day experimental period, the TN concentration of the blank control not only did not decrease but also slightly increased from the 6th day onwards, whereas the TN concentrations of the schemes 1–4 all showed a significant downward trend. The initial TN concentration of the experiment was 5.14 mg/L, and the blank control TN concentration was 5.32 mg/L after 10 days, which increased by 0.18 mg/L, and the removal rate was −3.56%; whereas the TN concentrations of the schemes 1–4, namely, activated charcoal: sulphur: limestone were 1:1:1, 2:1:1, 3:1:1, 4:1:1 were reduced to 1.73 mg/L, 2.05 mg/L, 1.47 mg/L, 1.39 mg/L, and the TN removal

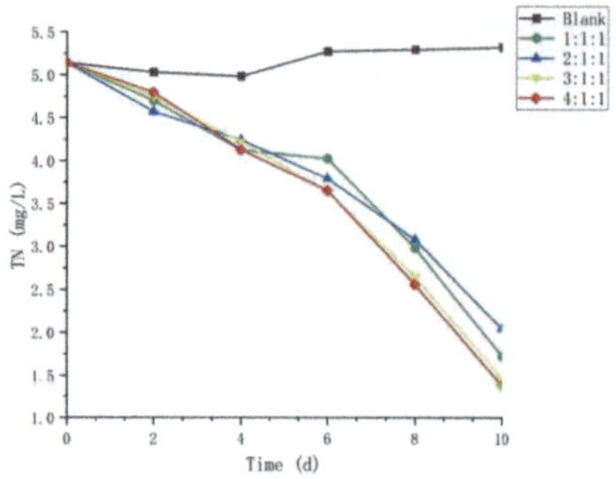

Fig. 1. TN concentration changes

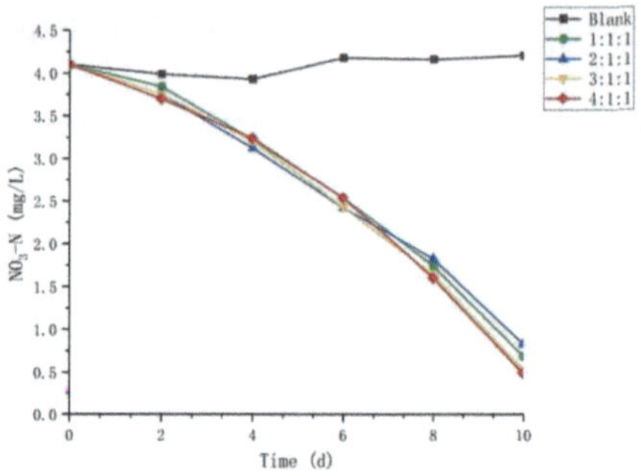

Fig. 2. NO_3-N concentration change

rate of was 66.40%, 60.19%, 71.36%, and 73.02%, respectively. The TN removal rate of was sorted as 4:1:1 > 3:1:1 > 1:1:1 > 2:1:1. It can be seen that the active carbon ratio has a certain effect on the removal of TN, and 4:1:1 of the Scheme 4 has the best effect on TN removal, This is due to the highest percentage of activated carbon, activated carbon catalyses autotrophic denitrification, and microorganisms in activated carbon have the largest aggregation, denitrification is the largest, and the best effect on TN absorption.

By Fig. 2. It can be seen that during the 10-day experimental period, the NO_3-N concentration of the blank control not only did not decrease but also slightly increased from the 6th day, and the changes were similar to those of TN. The NO_3-N concentrations in the protocols 1–4 showed a significant decreasing trend. The initial NO_3-N concentration of the experiment was 4.10 mg/L, and the blank control NO_3-N concentration was 4.21 mg/L after 10 days, which increased by 0.11 mg/L, and the removal rate was −2.61%, whereas in the schemes 1–4, namely, activated charcoal: sulphur: limestone with the mass ratios of 1:1:1, 2:1:1, 3:1:1, and 4:1:1 respectively, the NO_3-N concentrations decreased to 0.69 mg/L, 0.83 mg/L, 0.53 mg/L, and 0.49 mg/L, with removal rates of 83.20%, 79.74%, 87.01%, and 88.08%, respectively. Like TN, the order of removal rate of NO_3-N was also 4:1:1 > 3:1:1 > 1:1:1 > 2:1:1.The best removal rate of NO_3-N removal the best scheme is also scheme 4 with activated carbon: sulphur: limestone = 4:1:1.

3.2 Comparison of Other Relevant Water Quality Indicators

During the 10-day experiment, the DO concentrations of experiments protocols 1–4 showed a decreasing trend, from the initial 8.19 mg/L to 4.02 mg/L, 3.97 mg/L, 4.12 mg/L and 3.96 mg/L respectively, and the blank control raise to 8.29 mg/L.

Secondly, COD of scheme 1–4 and blank control scheme were reduced from the initial 46.65 mg/L to 30.76 mg/L, 33.58 mg/L, 35.19 mg/L, 32.26 mg/L 42.14 mg/L. The COD removal rates were 34.06%, 28.02%, 24.57%, 30.85%, and 9.67% respectively. The COD It can be seen that the experimental groups of Scheme 1–4 also have a certain removal rate of COD, and the percentage of activated carbon has little effect on the COD removal effect.

Finally, the pH of the experimental protocols 1–4 and the blank control protocol decreased from the initial 8.02 to 7.38, 7.27, 7.20, 7.14 and 7.55. The final pH values of protocols 1 to 4 were slightly lower than the blank control due to the acid production of the sulphur autotrophic denitrification process, which tends to lead to a decrease in pH [7], although alkaline limestone was added to the denitrification substrate, it could not completely neutralise the acid production of the sulphur autotrophic denitrification process. However, the pH of protocol 1 and 4 was basically kept between 7 and 8, and it was neutral, which is also the best range for sulfur autotrophic denitrifying bacteria. Therefore, the effect of pH on the denitrification effect is not considered.

4 Conclusions

In this study the dosage of biofiller balls was used 1 balls/2L balls/2L, and at the same substrate mass, comparing the mass ratio of coconut shell activated carbon:sulfur:limestone1:1:1(6 g/ball:6 g/ball:6 g/ball), 2:1:1(9 g/ball:4.5 g/ball:4.5 g/ball), 3:1:1(10.8 g/ball/3.6 g/ball:3.6 g/ball), 4:1:1(12 g/ball:3 g/ball:3 g/ball) activated carbon enhanced MBBR type sulphur autotrophic denitrification process for denitrification of micropolluted high dissolved oxygen surface water. The following main conclusions were obtained from the 10-day testing of a total of five water quality indexes, namely, NO_3-N, TN, COD, DO, and pH:

(1) Removal effect on nitrogen: The mass ratio of coconut shell activated carbon:sulphur:limestone was 1:1:1,2:1:1,3:1:1,4:1:1 The removal of TN was 66.40%, 60.19%, 71.36%, 73.02%, −3.56% for regimens 1–4 and blank control regimens, respectively; 83.20%, 79.74%, 87.01%, 88.08%, −2.61% for NO_3-N removal, respectively; 4:1:1 of Scheme 4 had the best nitrogen removal.
(2) Associated water quality indicators DO, COD, pH: the mass ratio of coconut shell activated carbon:sulfur:limestone were 1:1:1, 2:1:1, 3:1:1, 4:1:1 of the scheme 1–4 with the blank control scheme DO changed from the initial 8.19 changed to 4.02 mg/L, 3.97 mg/L, 4.12 mg/L, 3.96 mg/L, 8.29 mg/L respectively. The pH decreased from the initial 8.02 to 7.38, 7.27, 7.20, 7.14, 7.55, respectively. The COD was reduced from the initial 46.65 mg/L to 30.76 mg/L, 33.58 mg/L, 35.19 mg/L, 32.26 mg/L 42.14 mg/L.
(3) In conclusion, the best results in protocol 4. This study was conducted at room temperature (19–25 °C), but the temperature difference will gradually increase with the four seasons, so in order to further explore the effect of temperature on nitrogen removal, we will conduct experiments under different temperature conditions in the future.

Acknowledgements. This work was financially supported by "Fujian Provincial Science and Technology Project (2023I0042)".

References

1. Gao, Y., Yang, N.: Preliminary exploration of the formation mechanism of ecological water replenishment water price. Dev. Res. Center Ministry Water Resour. **49–51**, 59 (2020)

2. He, Q., et al.: Wood and sulfur-based cyclic denitrification filters for treatment of saline wastewaters. Biores. Technol. **328**, 124848 (2021)
3. Zhang, Q., Xu, X., Zhou, X., Chen, C.: Recent advances in autotrophic biological nitrogen removal for low carbon wastewater: a review. Water **14**, 1101 (2022)
4. Gong, L., et al.: Biomass characteristics and simultaneous nitrification–denitrification under long sludge retention time in an integrated reactor treating rural domestic sewage. Biores. Technol. **119**, 277–284 (2012)
5. Vinay Pratap, R.K., Kumar, S., Yadav, B.R.: Optimization of moving bed biofilm reactors for the treatment of municipal wastewater. CSIR-Na **241**, 117560 (2023)
6. Alsulaili, A.D., Refaie, A.A., Garcia, H.A.: Adsorption capacity of activated carbon derived from date seeds: characterization, optimization, kinetic and equilibrium studies. Chemosphere **313**, 137554 (2023)
7. Liang, J., Chen, N., Tong, S., Liu, Y., Feng, C.: Sulfur autotrophic denitrification (SAD) driven by homogeneous composite particles containing CaCO3-type kitchen waste for groundwater remediation. Chemosphere **212**, 954–963 (2018)

Bacillus Megaterium Enhances Phytoremediation of Cadmium-Contaminated Saline Soil Using Native Plant *Suaeda Salsa* L.

Congying Wang and Xianbin Liu(✉)

Tianjin University of Science and Technology, Tianjin 300457, China
lxb0688@tust.edu.cn

Abstract. For a long time, soil salinization and the heavy metal contamination in coastal areas have been global environmental issues. The phytoremediation efficiency of heavy metals is relatively low due to the high salt content of the soil. This study investigated the effectiveness and mechanisms of co-remediation of cadmium (Cd) contaminated saline soils by *Bacillus megaterium* (Bm) and *Suaeda salsa* L. (*S. salsa*) using a pot experiment system, providing a novel approach for addressing complex pollution through combined remediation techniques. The results indicated that Bm significantly promotes plant growth and accelerates the translocation of Cd from soil to plants and from the roots to the aboveground parts of *S. salsa.* Bm significantly promoted plant growth by enhancing the absorption of Ca from the soil, reducing MDA (malondialdehyde) content in plants, and increased the POD (peroxidase) activity under all salt stress treatments and the SOD (superoxide dismutase) activity under 4.00% NaCl concentration treatment. Under a 4.00% NaCl concentration treatment, the bioconcentration factor (BCF) of Cd inoculated with Bm increased by 27.19%. These findings demonstrated that Bm enhanced the efficiency of Cd removal from saline soils by *S. salsa*, laying the foundation for further research into plant-microbe combined remediation techniques for Cd-contaminated saline soils.

Keywords: Bohai Bay · saline soil · Bm bacteria · Cd · halophytes

1 Introduction

Over the past several decades, large-scale mining activities and industrial emissions have led to a significant increase in heavy metal pollution globally [1]. Among these pollutants, cadmium (Cd) is a critical industrial material and one of the most severe contaminants in the water and soil of the Bohai Sea [2]. Current research indicated that the Cd concentration in nearshore sediments of the Bohai Bay reaches 0.77 mg/kg, posing a moderate potential ecological risk [3, 4]. Additionally, factors such as climate change and anthropogenic activities have contributed to increased salinity in coastal soils [5]. This has resulted in a complex soil environment characterized by the coexistence of salinization and the Cd contamination, causing potential threats to biota through the food chain and emerging as a always global concern. *Suaeda salsa* L. (*S. salsa*),

D. Han et al. (Eds.): ICEPG 2024, ESE, pp. 540–547, 2025.
https://doi.org/10.1007/978-3-031-86829-0_62

exhibited significant dominance in the saline-alkali soils of Bohai Bay [6, 7]. Their robust adaptability allows it to thrive in both inland saline soils and the unique conditions of intertidal zones.

This study aims to investigate the effectiveness of the combined use of *S. salsa* and Bm in the remediation of Cd-contaminated saline soils under the Bohai Bay specific conditions. Researchers had demonstrated that salinity can significantly affect the absorption and translocation of Cd. However, the efficacy of Bm inoculation in soils with varying levels of Cd contamination and salinization remains uncertain. This study uniquely investigated the effects of Bm inoculation on the growth of *S. salsa* and the absorption and translocation of Cd under different salinity treatments. By examining various physiological indicators, this research provided insights into the mechanisms of remediation, thereby offering a scientific basis for the future restoration efforts.

2 Materials and Methods

2.1 Soil Sample

The soil samples were collected from the coastal area of the Bohai Bay (38°47′4.13″N, 117°37′6.25″E) at a depth of 0–15 cm. The fundamental physicochemical properties of the soil were as follows: pH, 9.88; NaCl, 1.00 g/kg; origin matter, 18.52 g/kg; alkaline hydrolysis nitrogen, 108.22 mg/kg; available phosphorus, 221.33 mg/kg; DTPA-Cd, 0.041 mg/kg; total Cd, 0.042 mg/kg. Initially, soil samples were air-dried naturally in a cool, ventilated area. The samples were sieved through a 2 mm mesh to ensure particle uniformity and consistency. To simulate varying degrees of the Cd-contaminated saline soil, solutions of cadmium chloride ($CdCl_2{\cdot}2.5H_2O$, with a Cd concentration of 2.12 mg/kg dry soil) and sodium chloride (NaCl) were added. After a 15-day period of natural aging and equilibrium, the soil samples were sterilized using an autoclave (BXM-30R, 120 °C, 2 h) to eliminate microorganisms and bacteria.

2.2 Plant and Strain

Seeds of *S. salsa*, a halophyte species, were procured from Qiao Huacheng Willow Cultivation Co., Ltd., China. The bacterial strain Bm utilized in this study was provided by the Shanghai Biological Resource Center (SHBCC), China. The process of liquid culture involved reviving the purchased strain, selecting single colonies, and inoculating them into nutrient broth medium. Cultivation was carried out in a shaking incubator at 30 °C and 220 revolutions per minute (r/min) for 12 h to achieve mid-logarithmic growth phase. The resulting bacterial suspension was centrifuged at 10,000 r/min for 10 min, resuspended in sterile water, and centrifuged three additional times to prepare a bacterial suspension (with a concentration of approximately 5×10^9 colony-forming units per milliliter, cfu/ml) for subsequent experiments.

2.3 Experimental Design

The present experiment employed a factorial experimental design involving two factors: the exogenous addition of Cd to the soil matrix and the inoculation of *Bacillus megaterium* (Bm). Four concentrations of salt treatment were established, resulting in a total of eight treatments: (1) Exogenous addition of 0 g/kg NaCl + no Bm inoculation (S0CdCK); (2) Exogenous addition of 0 g/kg NaCl + Bm inoculation (S0CdBm); (3) Exogenous addition of 10 g/kg NaCl + no Bm inoculation (S1CdCK); (4) Exogenous addition of 10 g/kg NaCl + Bm inoculation (S1CdBm); (5) Exogenous addition of 20 g/kg NaCl + no Bm inoculation (S2CdCK); (6) Exogenous addition of 20 g/kg NaCl + Bm inoculation (S2CdBm); (7) Exogenous addition of 40 g/kg NaCl + no Bm inoculation (S4CdCK); (8) Exogenous addition of 40 g/kg NaCl + Bm inoculation (S4CdBm). Each treatment was replicated three times. The pot experiment was conducted in a controlled environment chamber. The test period was 70 days. The seeds of *S. salsa* were surface sterilized using 0.5% $HgCl_2$ for 10 min, and then washed in sterilized double distilled water for three times. After 14 days of cultivation, they were transplanted into pots. Each treatment comprised three replicates. Seven days after transplantation, bacterial inoculation was performed by adding 30 ml of bacterial suspension obtained after centrifugation. For non-bacterial treatment group, 30 ml sterile water was used as a control. After an additional 14 days of growth, thinning was performed to ensure five uniformly sized seedlings per pot, followed by the application of NaCl solution corresponding to the respective concentration. Harvesting was conducted at 70 days after planting. Throughout the experiment, the average daytime temperature was maintained at 27 °C, nighttime temperature at 17 °C, and humidity ranged between 50% and 80%.

2.4 Sample Preparation and Determination Methods

Dry weight of the plants were determined by the method of weighing. The concentrations of Ca were measured using ICP-OES (Agilent 5110, USA). The concentration of Cd was determined by ICP-MS (Agilent 7800, USA). The activities of peroxidase (POD), superoxide dismutase (SOD), and catalase (CAT) were determined using reagent kits (Suzhou Keming Biotechnology Co, Ltd, China). The bioaccumulation factor (BCF) and translocation factor (TF) were calculated by follows.

$$\text{BCF} = \text{Cd concentration in the plant/Cd concentration in soil} \quad (1)$$

$$\text{TF} = \text{Cd concentration in the aboveground/Cd concentration in the root} \quad (2)$$

2.5 Data Processing and Analysis

Experimental data were initially processed and organized using Excel 2016 to ensure accuracy and completeness. Subsequently, the data were imported into SPSS 26.0 for one-way ANOVA to examine the effects of different treatments on various parameters. For cases with significant differences, Duncan's multiple range test was employed for further analysis, with the significance level set at $P < 0.05$ to ensure the reliability and statistical significance of the results. Finally, graphs were generated using Origin 2021 software.

3 Results and Discussions

3.1 Effects of Varying Salinity on the Growth Index of *S. Salsa*

Impact on Biomass

Salinity levels and Bm inoculation exhibited differential impacts on the biomass characteristics of *S. salsa.* (Table 1). Regardless of Bm inoculation, increased soil salinity reduced belowground biomass accumulation of *S. salsa* ($P < 0.05$). Compared to the CK treatment group, Bm inoculation alleviated the reduction in aboveground biomass under high salinity stress. Under S4 salinity conditions, the aboveground biomass of *S. salsa* increased by 59.72% compared to the CK group ($P < 0.05$). Therefore, Bm inoculation significantly enhanced the growth of *S. salsa* under the salinity and the Cd stress, increasing aboveground biomass and effectively diluting the concentrations of harmful ions such as Na and Cd, thereby mitigating the toxic effects of salinity and heavy metals on the plants.

Table 1. The effects of Bm on biomass characteristics of *S. salsa* under Cd stress under different salinity treatments

Treatments		Shoot dry Weight (g/plant)	Root dry Weight (g/plant)	Total dry Weight (g/plant)
CK	S0Cd	0.125 ± 0.014^{d}	0.024 ± 0.026^{c}	0.149 ± 0.062^{c}
	S1Cd	0.277 ± 0.161^{b}	0.007 ± 0.008^{ab}	0.284 ± 0.169^{c}
	S2Cd	0.310 ± 0.078^{c}	0.007 ± 0.009^{b}	0.317 ± 0.063^{d}
	S4Cd	0.216 ± 0.07^{a}	0.006 ± 0.006^{a}	0.222 ± 0.036^{a}
Bm	S0Cd	0.153 ± 0.012^{b}	0.055 ± 0.059^{c}	0.208 ± 0.054^{b}
	S1Cd	0.310 ± 0.105^{b}	0.008 ± 0.01^{c}	0.318 ± 0.114^{a}
	S2Cd	0.327 ± 0.156^{b}	0.004 ± 0.004^{b}	0.331 ± 0.157^{d}
	S4Cd	0.345 ± 0.019^{a}	0.004 ± 0.004^{a}	0.349 ± 0.015^{c}

Note: Different letters in the same column indicated that there were significant differences between the treatment groups ($P < 0.05$). The same below.

Influence of Cd Accumulation and Translocation

Under non-inoculated conditions, Cd accumulation in the aboveground parts of the plant initially increased and then decreased with rising salinity levels. Conversely, Cd accumulation in the roots showed a decreasing trend as salinity increased. Following Bm inoculation, Cd accumulation in the aboveground parts of the plant significantly increased with rising salinity ($P > 0.05$). Zhang demonstrated that Bm enhanced the growth of *Solanum nigrum* and mitigates Cd accumulation within the plant tissues [8]. Treatment with S4 followed by inoculation with Bm resulted in a 27.19% increase in the plant's BCF (see Fig. 1). The Bm inoculation significantly enhanced both the BCF and the TF of Cd across all salinity treatments (see Fig. 2). You also found that Bm significantly enhanced the BCF of *Solanum nigrum* for Cd compared to the control group without bacterial inoculation [9].

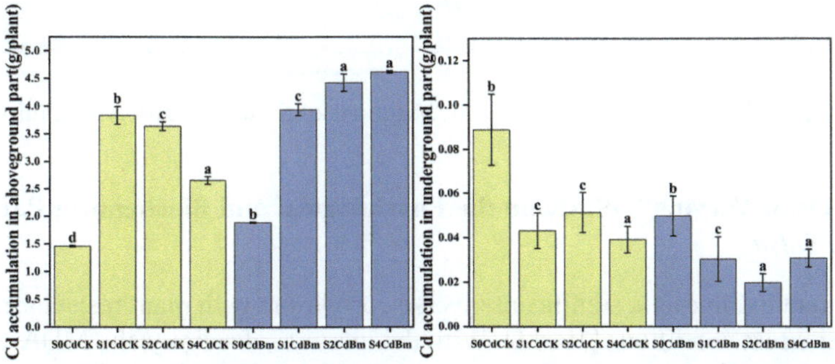

Fig. 1. The effect of Bm on Cd accumulation in *S. salsa* at different salinity. Note: The data in the figure are the mean ± standard error of three replicates. Plants are counted by dry weight. Different letters in the same figure indicate that there are significant differences between the treatment groups ($P < 0.05$). The same below.

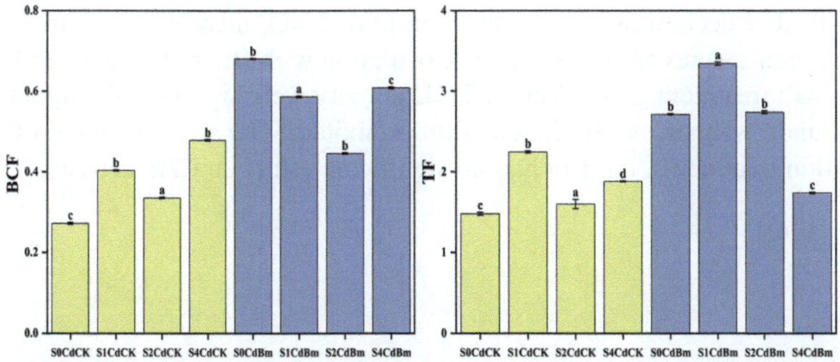

Fig. 2. The effect of Bm on Cd translocation of *S. salsa* at different salinity

The Influence on Mineral Element Concentration

After Bm inoculation, Ca concentration in the roots significantly decreases across all salinity treatments ($P > 0.05$). Following Bm inoculation, Ca concentration decreases by 68.08% (see Fig. 3). This reduction might be attributed to Bm enhancing the plant's ability to absorb mineral elements from the soil. Additionally, Ca may be selectively absorbed and transported into the plant by Bm, functioning as an osmotic agent and thereby preventing excessive Na uptake by the plant.

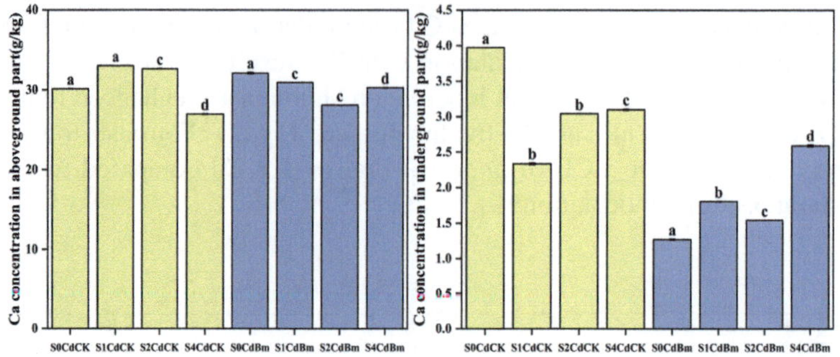

Fig. 3. The effect of Bm on Ca concentrations of *S. salsa* at different salinity

3.2 Effects of Varying Salinity on the Physiological and Biochemical Parameters of *S. Salsa*

MDA content exhibiting a significant inverse correlation with plant resistance, particularly under stress conditions [10, 11]. With the increase of salinity, the MDA content in plant cells showed a significant decreasing trend whether or not Bm was inoculated. This suggested that Bm may effectively mitigate the extent of lipid peroxidation in *S. salsa* under combined salt and Cd stress, thereby reducing oxidative damage (see Table 2). SOD primarily functions to convert superoxide anions, generated under stress conditions, into more stable H_2O_2 and $O_2^{\bullet -}$, thus alleviating cellular damage [12]. Post Bm inoculation, SOD activity significantly increased under high salinity conditions. CAT and POD further decompose H_2O_2 into non-toxic H_2O, thereby protecting cells from reactive oxygen species (ROS) damage. Inoculation with Bm reduced the POD activity of the non-salt treatment group, but the POD activity were significantly higher than that of the CK under salt stress. Bm inoculation promoted SOD activity under 4.00% NaCl concentration treatments, but Bm had no significant effect on CAT activity.

Table 2. The effects of Bm on antioxidant enzyme activity of *S. salsa* under Cd stress at different salinity treatments

Treatment		MDA (nmol/g FW)	POD (U/g FW)	SOD (U/g FW)	CAT (U/g·min FW)
CK	S0Cd	5.38 ± 1.595^{a}	2993.84 ± 1.596^{a}	1170.97 ± 1.597^{a}	0.36 ± 1.598^{a}
	S1Cd	4.39 ± 3.014^{a}	$1920 \pm 3.015a^{b}$	1459.74 ± 3.016^{a}	0.54 ± 3.017^{b}
	S2Cd	2.78 ± 2.595^{a}	1330.07 ± 2.596^{a}	303.35 ± 2.597^{a}	0.35 ± 2.598^{c}
	S4Cd	4.56 ± 2.391^{b}	3138.79 ± 2.392^{b}	127.32 ± 2.393^{b}	0.28 ± 2.394^{b}
Bm	S0Cd	4.35 ± 1.105^{c}	530.05 ± 1.106^{c}	1125.54 ± 1.107^{c}	0.43 ± 1.108^{c}
	S1Cd	3.99 ± 1.399^{b}	2652.47 ± 1.400^{b}	1220.3 ± 1.401^{d}	0.7 ± 1.402^{b}
	S2Cd	1.75 ± 2.963^{b}	4069.77 ± 2.964^{b}	87.97 ± 2.965^{b}	0.63 ± 2.966^{b}
	S4Cd	3.15 ± 0.097^{d}	3208.53 ± 0.098^{d}	154.91 ± 0.099^{d}	0.28 ± 0.061^{d}

Note: FW is the fresh weight of the plant.

4 Conclusion

Inoculation with Bm significantly promoted plant growth and accelerated the migration of Cd from soil to plant tissues as well as from the roots to the aboveground parts of *S. salsa* under saline-alkaline conditions. This promotion effect progressively intensified with increasing salinity levels. Furthermore, following Bm inoculation, the accumulation of Cd in the aboveground parts of plants notably increased with elevated salinity, possibly due to enhanced Cd absorption by plants under salt stress. In terms of physiological and biochemical indexes, inoculation of Bm under high salinity stress could significantly reduce the content of MDA in plants, and significantly increase the POD activity under all salt stress treatments and the SOD activity under 4.00% NaCl concentration treatment.

References

1. Li, Z., Ma, Z., Van Der Kuijp, T.J., et al.: A review of soil heavy metal pollution from mines in China: pollution and health risk assessment. Sci. Total Environ. **468–469**, 843–853 (2014)
2. Qin, Y.W., Meng, W., Zheng, B.H., et al.: Heavy metal pollution in tidal zones of Bohai Bay using the dated sediment cores. J. Environ. Sci. (China) **18**(3), 610–615 (2006)
3. Han, X., Wang, J., Cai, W., et al.: The pollution status of heavy metals in the surface seawater and sediments of the Tianjin Coastal Area, North China. Int. J. Environ. Res. Public Health **18**(21), 11243 (2021)
4. Håkanson, L.: The quantitative impact of pH, bioproduction and Hg-contamination on the Hg-content of fish (pike). Environ. Pollut. B. **1**, 285–304 (1980)
5. Röthig, T., Trevathan-Tackett, S.M., Voolstra, C.R., et al.: Human-induced salinity changes impact marine organisms and ecosystems. Glob. Change Biol. **29**(17), 4731–4749 (2023)
6. Bajji, M., Kinet, J.M., Lutts, S.: Osmotic and ionic effects of NaCl on germination, early seedling growth, and ion content of Atriplex halimus (Chenopodiaceae). Can. J. Bot. **80**(3), 297–304 (2002)

7. Labidi, N., Ammari, M., Mssedi, D., et al.: Salt excretion in Suaeda Fruticosa. Acta Biol. Hung. **61**(3), 299–312 (2010)
8. Zhang, Y., Zhao, S., Liu, S., et al.: Enhancing the phytoremediation of heavy metals by combining hyperaccumulator and heavy metal-resistant plant growth-promoting bacteria. Front. Plant Sci. **13**, 912350 (2022)
9. You, M., Wang, L., Zhou, G., et al.: Effects of microbial agents on cadmium uptake in Solanum nigrum L. and rhizosphere microbial communities in cadmium-contaminated soil. Front. Microbiol. **13**, 1106254 (2023)
10. Van Zelm, E., Zhang, Y., Testerink, C.: Salt tolerance mechanisms of plants. Annu. Rev. Plant Biol. **71**(1), 403–433 (2020)
11. Sun, S.M., Chang, W., Song, F.Q.: Research progress on antioxidant mechanism of arbuscular mycorrhizal fungi in plants under salt stress. J. Appl. Ecol. **31**(10), 8 (2020)
12. Khan, A.G., Kuek, C., Chaudhry, T.M., et al.: Role of plants, mycorrhizae and phytochelators in heavy metal contaminated land remediation. Chemosphere **41**(1–2), 197–207 (2000)

Study of ERW Contamination Based on SPRC Modeling

Xiao Wang(✉)

Trainingbase of Army Engineering, University of PLA, Xuzhou, Jiangsu Province, China
kelsi03@163.com

Abstract. There is a high probability that the environment will be deliberately targeted by explosive ordnance for certain purposes when conflicts happen. The contents of explosive ordnance, especially home-made explosives, are a potential source of chemical pollution of the environment, with serious adverse effects on ecosystems, the atmosphere, water sources, etc., and when left in the environment for a long period of time, they inevitably pose a threat to human beings, animals and the environment. Such threats include not only physical damage to ecosystems and the release of toxic substances during explosions, but also indirect collateral damage to the environment and subtle pollution of the environment, which can lead to both toxicological and ecological risks.

Keywords: ERW · SPRC · Environment

1 Introduction

In addition to the human casualties caused by the extensive use of explosive ordnance in the ongoing Russian-Ukrainian conflict, the impact on the environment should not be underestimated. Relevant media reports indicate that at least 10.5 million hectares of Ukrainian farmland have been contaminated with chemicals, with implications for global food security [1]. It is uncertain how the environment will deal with these toxic substances and what impact they will have on humans. Poor environmental management and untimely clearance of ERW during post-war mine action can have short- and even long-term adverse impacts on land, water, soil and air, as well as on communities in the vicinity of mined areas. Besides the ERW, The environmental impact of homemade explosives is more complex. Since the production of home-made explosives does not follow the same rigorous manufacturing processes as industrial or military explosives, there is no guarantee of their performance, reliability and safety. Common homemade explosives such as chlorates, perchlorates, nitrates and peroxides are capable of posing serious health hazards if exposed to the environment for long periods of time. The environmental impact is even more serious when certain precursor chemicals contain flammable, corrosive or toxic properties. Based on the SPRC model (source-pathway-receptor-consequence model), this paper describes how pollutants enter the environment from the source, such as explosive ordnance disposed of through open burning, and explosive ordnance made of metal buried in the ground or in water for a long period of

D. Han et al. (Eds.): ICEPG 2024, ESE, pp. 548–553, 2025.
https://doi.org/10.1007/978-3-031-86829-0_63

time, which ultimately move through the environment to the receptor via the air, soil, or water (the pathway),. Therefore, the construction of explosive ordnance pollution chain model study to determine its impact on the environment [2]. Among them, the source (S) represents the source material leading to environmental pollution; the pathway (P) represents the process of impacting the source of the impact on the receptor; the receptor (R) represents the possibility of contact with the pollutants and adversely affected by the "entity", including animals, Receptors (R) represent "entities" that may be adversely affected by exposure to pollutants, including animals, plants, humans, and regional ecosystems such as rivers and lakes, etc. Impact results (C) represent the environmental impacts of pollutants that reach and affect receptors. If the SPR link between the pollutant source and the receptor is broken at any point, then there will be no result (C) causing environmental impact. Therefore, the SPRC conceptual map can be used to analyze the pathways between sources and potential receptors in the area where the action is to be carried out in order to predict the possible impacts on the environment [3].

2 Origins of the International Landmine Problem

Explosive ordnance, such as landmines, is an efficient anti-mobility weapon and has been widely used for quite a long time. However, due to the long combat service of traditional landmines themselves, their inability to identify targets and their concealment, as well as the unfavorable clearance of landmines after the war and the difficulty of tracing the responsibility for mine clearance, they were left in place and formed explosive remnants of war, which harmed the local populace, constrained economic development, and caused turmoil, and thus gave rise to the international landmine problem. In Cambodia, for example, as one of the most mine-infested countries in the world, explosive remnants of war, mainly landmines, have caused a considerable number of casualties and severely constrained economic development. However, its impact on environmental pollution should not be underestimated. For one thing, explosive remnants of war had destroyed and contaminated vital resources and essential social and public places, such as agricultural land, water sources, roads, schools and hospitals. At the same time, when contaminated land, water and other natural resources cannot be used, the pressure on locally available resources increases dramatically, and overuse may also cause irreversible impacts on the environment [4]. Moreover, the lack of resources can easily lead to unrest, which restricts local development.

3 SPRC Modeling of ERWS

3.1 Source

The main sources of contamination in the ERW contamination chain include:

Munitions Internal Fillings
When a munition explodes, the explosives often burn insufficiently and produce residues, thus polluting the environment. Even if it does not explode, left in situ failure, but its internal charge of explosives either due to shell breakage direct contact with the soil

or still in the body of the bomb, will produce chemical decomposition, the decomposition rate and the surrounding conditions of the relevant. Many of the compounds and components of explosive ordnance charges can be categorized as energy-containing compounds, such as the majority of mines since the First World War, which have contained either TNT or HESO, or proportional mixtures of the two. However, their main charges, such as rissokin (1,3,5-trinitro-1,3,5-triazine) and TNT (2,4,6-trinitrotoluene), pose a potential threat to the current environment. TNT has been reported to be capable of severely impairing the ability of organisms to reproduce, grow and develop, while other explosives such as TNT are carcinogenic [4]. TNT can also be carried by the wind and dispersed over greater distances, and if mixed with rainwater for mobility, there is a high likelihood that the chemicals will seep into the groundwater, reaching even deeper into the subsurface, which means that there is a risk of drinking water contamination [5]. The PFM-1 antitriazines, for example, have been reported to have been used in a number of different ways. Another example is The PFM-1 anti-personnel mine, The mine has a small charge of 37 g of VS-6D liquid explosive and, as the contents are toxic, contact with skin and inhalation of fumes should be avoided. This liquid explosive is more likely to contaminate the environment than fused or pressed explosives. At the same time, these elements may enter surface water and contaminate streams, rivers and lakes, which in turn enter the food chain. People may be at risk if they consume contaminated fish.

Hazardous Gases and Dust

After an explosion is created by explosive ordnance such as landmines, toxic gases and flammable and explosive gases such as carbon monoxide, nitrogen oxides, and methane air mixtures are produced during the explosion. In addition to the production of large quantities of gases, large quantities of dust are also produced, such as asbestos from grenades, landmines and other munitions that are released into the environment when they explode. These dusts are highly dispersed and can remain suspended in the air for long periods of time, spreading with wind currents. Large quantities of pollutants are usually generated into the environment by means of chemical reactions. When used in urban areas, explosive ordnance inevitably leads to the destruction of infrastructure, releasing hazardous substances from the building itself into the air and on the ground, such as toxic fumes and heavy metals. [5] Crushed building material dusts are usually heterogeneous mixtures of a variety of materials, such as cement, metals, polychlorinated biphenyls (PCBs), silica, asbestos, and other synthetic fibers. Exposure to these dusts can have physical and chemical health effects.

Heavy Metals

The cartridge cases and internal components of many explosive ordnance are often made of metals or plastics, which may include heavy metals such as chromium, zinc, iron, and copper. The most common sources of heavy metal contamination are lead, antimony, tungsten, and bismuth, and soil samples from Kharkov, one of the main battlefields in the eastern part of the Russian-Ukrainian conflict, have been reported to show higher concentrations of lead and cadmium. Heavy metals have a negative effect on soil bacteria and can inhibit plant development and micronutrient availability. To make TNT explode faster, some detonators also contain heavy metals, such as mercury in the form of fulminate. As a heavy metal, mercury can damage nerve cells and cause birth defects in unborn babies. In addition, some explosives may contain specialized materials, such as depleted uranium and cobalt. These materials are corroded and weathered when buried in the soil and subsequently release various heavy metals into the soil. These hazardous substances are then transferred to the environment over time. Especially in agricultural areas, these heavy metals can easily penetrate the soil, reach the water table and enter the human food chain.

3.2 Pathway

Explosive ordnance remains in the environment for long periods of time and can enter the environment through weathering by rainfall, exposure to sunlight and other climatic conditions. There are three main mechanisms for the transfer of contaminants from the soil surface: gradual breakdown of solids into very small particles through erosion; chemical transformation of metals into other metal compounds and complexes through corrosion; and entry into water sources.

Explosive remnants of munitions left over time are susceptible to erosion or corrosion, which in turn contaminates the soil. Soil pollution refers to the soil to add some normal circumstances do not exist in the harmful substances, so that some of the inherent components of the content is too high, impairing the function of the soil, destroying the physical and chemical properties of the soil, affecting the growth of micro-organisms and crop production and yield; in addition, some toxic substances in the soil will be absorbed by crops, through the food chain harm to humans and animals, these harmful substances remain in the blasting area, in the reclamation of the land may be manifested in the soil pollution Characteristics.

But in general, the metal is eroded or decayed, and high-energy materials are more likely to be dissolved in water, converting the solid into smaller particles and chemical forms, which makes the pollutants much more mobile in the environment, not only are they able to be transported through the surface of the soil carried by water runoff and transported down through the soil through infiltration with rainfall, and the degree of contamination varies from pollutant to pollutant when dissolved. For example, TNT and 2,4-dinitrotoluene (DNT) tend to bind to organic matter in the soil and are therefore not as readily transported as RDx, which has a greater potential to act as a groundwater contaminant. Based on the mobility of water, hazardous substances will continue to contaminate soil and even cause chronic poisoning of plants, animals, and humans.

3.3 Receptor

Receptors in the mine contamination chain include mainly soil, water, air and even buildings, which in turn affect plants, animals and humans. The impact of a contaminant on the environment depends on how much of the contaminant reaches the receptor and how long the receptor is exposed to the contaminant. This relates to the pathway processes described above. For example, explosive ordnance may damage vegetation and the internal properties of the soil may be altered, reducing fertility and affecting the ability of the vegetation to take root and hold water, which in turn can lead to soil erosion and affect local agriculture. Ammunition installed in the water, such as mines, will produce solid pollutants that increase suspended solids over a long period of time, making the water turbid and decreasing its transparency, affecting the respiration of aquatic organisms and their vicarious emissions, and even causing the death of aquatic organisms. And if human beings ingest contaminated plants, animals and water sources, it is likely to increase the risk of disease and impact. Therefore, once the environment suffers from pollution, ecological damage, human beings are inevitably affected.

3.4 Consequence

In the case of the Battle of Mosul, for example, the United Nations Environment Programme estimates that the city has about 8 million tons of trash highly contaminated with unexploded ordnance and improvised explosive devices. In Nuremberg, Germany, after World War II, about 10 million tons of war waste piled up in a huge pit within the city and no safety measures were taken to contain these contaminants. Over time, hazardous materials seeped out and severely contaminated the nearby Silver Lake, causing severe exceedances of hydrogen sulfide concentrations. In Ukraine, agricultural production in the Donetsk region of the country declined by nearly 35% through 2016. Faced with increasing pressure on their livelihoods, some residents resort to deforestation and farming. Deforestation, in turn, affects soil and water levels, which can have an impact on fish and other wildlife, which in turn affects the population of entire species and even changes the food chain. Also the polluted environment can have an impact on human health. Secondly, damage to buildings is likely to release hazardous substances into the air and on the ground, which can also affect local residents in terms of health care, education and other public resources, and undermine local economic development. Therefore, the ultimate impact of environmental contamination by explosive ordnance will be on people, affecting health, education, the economy and other aspects.

3.5 SPRC Chain Diagram Using the Example of Anti-tank Mines

Long-term legacy anti-tank mines can degrade in the soil and leach their explosive contents. Over time, the contents enter the soil and groundwater. This in turn affects flora, fauna and human livelihoods as shown in Fig. 1 below.

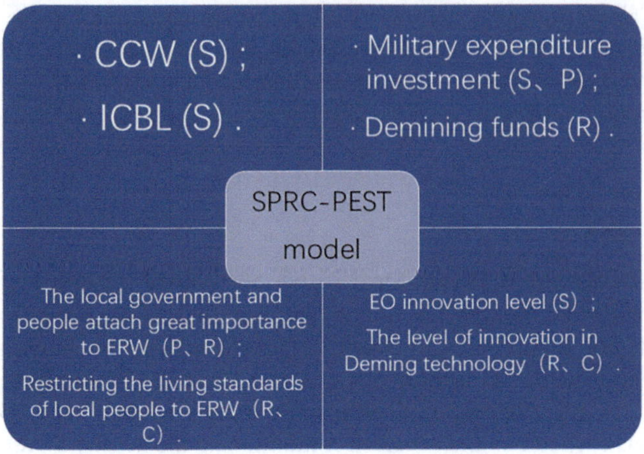

Fig. 1. SPRC contamination linkage

4 Conclusion

Explosive ordnance contamination can have an impact on the environment, therefore, examining the sources, pathways, receptors, and impacts through the SPRC contamination chain can provide a clearer understanding of how contamination occurs and how to break the links between SPRCs in order to be able to minimize the impacts that cause contamination.

In summary, the ultimate goal is to minimize the risk of environmental contamination by explosives as far as is reasonably practicable.

References

1. UNEP: From Conflict to Peacebuilding: The Role of Natural Resources and the Environment (UNEP, 2009), 15
2. UN Security Council Resolution 1925, UN Doc. S/RES/1925, 28 May 2010, para. 12s
3. Zemin, Z.: Ecological vulnerability assessment based on SPR modeling. Qinghai Environ. **30**(4), 186–190 (2020)
4. Nao, S.-Y.: Linking demining to post-conflict peacebuilding: a case study of Cambodia' in Jensen and Lonergan (4) 181.
5. GICHD, Guide to Explosive Ordnance Pollution of the Environment (2021)

Effects of Different Colonization Methods on the Adaptation of *Ruellia Simplex* to Three Mining Wasteland Substrates

Hongbin Huang[1,2](✉)

[1] School of Tourism and Landscape Architecture, Guilin University of Technology, Guilin 541006, Guangxi Zhuang Autonomous Region, China
2447008465@qq.com

[2] Guangxi Key Laboratory of Karst Plant Conservation and Restoration Ecology, Guangxi Institute of Botany, Chinese Academy of Sciences, Guilin 541006, Guangxi Zhuang Autonomous Region, China

Abstract. In this study, we investigated the adaptability of *Ruellia simplex* in three kinds of mining wasteland (quarry rock, manganese tailing slag, and improved red mud), and analyzed its physiological characteristics and growth by comparing the three methods of bare-root colonization, water-protecting mats colonization and bag planting. The results showed that: the *Ruellia simplex* did not die after 100 days of growth on the three substrates; the bare-root colonization on the improved red mud was the worst growth condition due to the strong alkali effect, while the quarry rock was the best growth environment due to its high organic matter content; the bag planting and water-protecting sponges colonization could significantly improve the organic matter of the substrate and reduce the pH of the improved red mud compared with the bare-root colonization; the *Ruellia simplex* were suitable for ecological restoration of the three kinds of wasteland, and the best growth environment was the quarry rock, and the best growth environment was the quarry rock. The best growing environment was quarry rock, the worst was improved red mud, and manganese tailing slag was in the middle; the best planting method was bag planting, followed by water-protecting sponge planting, and bare-root planting was the most ineffective.

Keywords: *Ruellia simplex* · quarry rock · manganese tailings · red mud · adaptation

1 Introduction

Mining wasteland refers to land damaged or occupied by mining and smelting activities, and some of the land has lost its utilization value as a result of various mining activities. At present, it is particularly important and urgent to select suitable plants and colonization methods for the effective restoration and management of these abandoned lands.

This paper aims to study the mining waste sites mainly abandoned quarries, manganese tailing ponds, and red mud dumps. Abandoned quarry refers to the abandoned

D. Han et al. (Eds.): ICEPG 2024, ESE, pp. 554–560, 2025.
https://doi.org/10.1007/978-3-031-86829-0_64

land left after the end of quarrying activities, in which the waste rock piles are mainly composed of a mixture of stripped topsoil and rock fragments, and therefore, the substrate of the quarry rock is generally high in calcium, barren, and alkaline; Manganese tailing ponds are the abandoned land used to store the tailings and slag from manganese ores after beneficiation and washing, and their characteristics include weak water-holding and retention capacity, acidity, barrenness, and high heavy metal content; Red mud dumps refer to the abandoned land used for open-air mining, and red mud dumps refer to the abandoned land used for open-air mining. Its characteristics include weak water-holding and water-retaining capacity, acidity, barrenness, and high heavy metal content; red mud dumps refer to sites used for open storage of red mud, the red residue produced in the process of refining alumina from bauxite ores, which is strongly alkaline, barren, and relatively poorly permeable and permeable to water and air, and so on. It can be seen that although there are significant differences in the physicochemical properties of these three types of mining wasteland substrates, they share the common characteristic of having a very limited number of plant species that can survive and grow on them.

Bluebonnet, scientifically known as *Ruellia simplex*, is a perennial evergreen clumping herb in the genus Ruellia in the family Jurassicaceae. It is also known as *Ruellia brittoniana*, a synonym for *Ruellia brittoniana*. Ruellia simplex, as an amphibious plant, not only thrives on land. But also demonstrates a certain degree of waterlogging tolerance, which makes it well-adapted to specific environments, such as sunken greenspace [1, 2]. *Ruellia simplex* are not only saline and heavy metal tolerant, but also show a wide range of resistance to mowing and insect pests [3–6]. The *Ruellia simplex* is tolerant and enriched to the environment. *Ruellia simplex* can tolerate and enrich heavy metal cadmium [4]; *Ruellia simplex* leaves are less injured than other plants under salt stress, and its strong salt-excluding ability may be one of the reasons for its high salt tolerance [5] Its underground rhizome can expand horizontally, new plants can grow at the nodes, and multiple plants often share the same root system, forming a complex rhizome network [7]. At the same time, *Ruellia simplex* can tolerate a strong alkaline environment up to pH 9.7, but its growth will be affected to a certain extent, manifested in reduced growth rate, stem bending, old leaf abscission, and secretion of white material at the edge of the leaf and other symptoms [8]. The plant has been used in various locations in the world, including the United States. At present, there are few comparative studies on different planting methods and adaptations of *Ruellia simplex*.

2 Materials and Methods

2.1 Makings

Quarry rock, manganese tailing slag, and red mud were extracted from a quarry rock pile in Lianhua Town, Gongcheng County, Guilin, a manganese tailing pond in Xialei Town, Daxin County, Chongzuo, and a red mud pile in Xin'an Town, Pingguo City, Baise, and the red soil used for red mud improvement was extracted from the Botanical Garden of Guilin, and passed through a 2 mm sieve for spare use.

Given the strongly alkaline character of red mud (which has a pH as high as 11.08), the survival of *Ruellia simplex* in it was low, and to improve the situation, the red mud was mixed with red soil (pH 5.62) in the ratio of 1:1 by volume and allowed to stand for at least one week for improvement, hereinafter referred to as improved red mud. The experiments were designed as three types of mining wasteland substrate treatment options, namely quarry rock, manganese tailing slag, and improved red mud, in which three methods of planting *Ruellia simplex* were designed, namely, direct bare-root planting, planting with water-retaining cotton pads at the bottom, and planting with seedling bags with cultivation soil.

Ruellia simplex is cuttings, specifically robust 5–10 cm long stem segments of *Ruellia simplex*, cuttings in the cuttings pool with the seedling substrate, watered thoroughly and placed in the semi-shade management, after 2 months to select a similar size (plant height 20–25 cm) of robust seedlings to use.

2.2 Methodologies

In the rain-shaded greenhouse, the three kinds of substrates were loaded into PP two-color pots (outer caliber of 16.8 cm, inner caliber of 15 cm, ground diameter of 11.8 cm, height of 14 cm), and the bottom of the pots was placed on a tray (outer caliber of 15 cm, inner caliber of 18 cm, height of 2.2 cm), which kept the heat preservation and humidity, and at the same time prevented pollutants from leaking out of the pots. Before potting, a layer of non-woven fabric was placed at the drainage hole at the bottom of the pot, and then 2.0 kg of substrate was added and planted into the *Ruellia simplex* seedlings. Repeat 5 pots for each different method of fixing the substrate, and plant 1 bluebell plant in each pot.

A total of nine treatments, as follows: (1) quarry rock bare root planting (Qr-C); (2) quarry rock pad water retention cotton planting (Qr-P); (3) quarry rock bag seedling bag cultivated soil planting (Qr-B); (4) manganese tailings slag bare root planting (Mn-C); (5) manganese tailings slag pad water retention cotton planting (Mn-P); (6) bag seedling bag cultivated soil planting (Mn-B); (7) modified red clay bare root planting (Rm-C); (8) modified red clay pad water retention cotton planting (Rm-P); (9) modified red clay bag seedling bag cultivated soil planting (Rm-B).

Before transplanting the *Ruellia simplex*, their initial aboveground and belowground part fresh weights and number of leaves were measured, and at the end of the experiment, the plants were removed from the pots, washed, and divided into roots and aboveground parts. After being put into envelopes, they were fixed and killed at 105 °C for 20 min, then baked at 80 °C for at least 24 h. Root biomass, aboveground biomass, relative growth rate (RGR), and root-shoot ratio (R/S) were weighed. The initial dry weight (m_0) was estimated based on the dry-to-fresh ratio of the alternate plants, and the relative growth rate (RGR, mg g^{-1} d^{-1}) of each plant was calculated according to the formula ($\ln m_1$-$\ln m_0$)/100 $\times$ 1000.

pH was measured using an ST 3100 laboratory pH meter (USA); organic matter was measured using the potassium dichromate-sulfuric acid heating method.

2.3 Data Processing

The data processing of this experiment was organized and analyzed by Excel 2018 software, SPSS 27.0.1 statistical software was used for statistical analysis, origin2021 was used for graphing, and the data in the graphs are mean $\pm$ standard error. To offset the influence of individual size effect, the differences in root length and leaf number of *Ruellia simplex* between different substrate treatments were analyzed by one-way ANCOVA with initial values as covariates, and subsequent multiple comparisons were tested by the LSD method. Other indices (relative growth rate, root-shoot ratio, pH, organic matter) were analyzed by one-way ANCOVA, LSD multiple comparisons for analysis, and Duncan's method for multiple comparisons.

3 Results and Analysis

3.1 Effect of Different Colonization Methods on Relative Growth Rate, Root-Shoot Ratio, Maximum Root Length, and Leaf Number of *Ruellia Simplex* on Three Mining Wasteland Substrates

Table 1 shows that there is no significant difference between the relative growth rates of the three colonization methods in the quarry rock group, bag planting is significantly greater than bare root colonization in the manganese tailings slag group, while there is no significant difference between padded water retaining cotton colonization and the other two colonization methods, bag planting is significantly greater than the other two colonization methods in the amended red mud group, and there is no significant difference between bare root colonization and padded water retaining cotton colonization, and the three different colonization methods of the three types of mining substrates are bag planting > padded water retaining cotton colonization > bare root colonization. > There was no significant difference between the three planting methods of *Ruellia simplex* on quarry rock, bare root planting on manganese tailing slag was significantly smaller than the other two substrates, bag planting on amended red mud was significantly larger than the other two planting methods, and there was no significant difference between the other two planting methods. The maximum root length under bag planting on quarry rock and manganese tailings slag was significantly greater than that under bare root colonization, and there was no significant difference between bag planting on padded water retaining cotton and the other two colonization methods, and the maximum root length under bag planting on amended red mud was significantly greater than that on the other two substrates, and there was no significant difference between the other two substrates; there was no significant difference in the number of leaves between different colonization methods on quarry rock, and there was no significant difference in the number of leaves between different colonization methods, and there was no significant difference in the number of leaves between the bag and padded water retaining cotton and the other two substrates. There were no significant differences between the different colonization methods; the number of leaves on quarry rock was not significantly different between the different colonization methods; the number of leaves on manganese tailing slag was significantly greater with water-retaining cotton colonization and bag planting than bare root colonization; the number of leaves on manganese tailing slag was not significantly different between the two other colonization methods.

Table 1. Effect of different colonization methods on relative growth rate, root-shoot ratio, maximum root length, and leaf number of *Ruellia simplex* on three mining wasteland substrates

deal with	relative growth rate ($mg \cdot g^{-1} \cdot d^{-1}$)	root-shoot ratio	Maximum root length (cm)	leaf number
Qr-C	2.71 ± 0.36a	0.30 ± 0.06a	7.03 ± 0.95b	20.07 ± 0.93a
Qr-P	2.89 ± 0.52a	0.36 ± 0.06a	8.41 ± 1.00ab	21.07 ± 0.93a
Qr-B	3.72 ± 0.40a	0.25 ± 0.02a	10.16 ± 1.02a	21.07 ± 0.93a
Mn-C	1.96 ± 0.31b	0.11 ± 0.01b	14.29 ± 1.39b	19.11 ± 1.02b
Mn-P	2.81 ± 0.28ab	0.21 ± 0.03a	17.06 ± 1.44ab	22.34 ± 1.03a
Mn-B	3.13 ± 0.26a	0.23 ± 0.02a	18.85 ± 1.42a	22.35 ± 1.04a
Rm-C	0.18 ± 0.06b	0.09 ± 0.02b	13.14 ± 2.20b	12.25 ± 1.18b
Rm-P	0.23 ± 0.12b	0.09 ± 0.01b	15.60 ± 2.27b	13.75 ± 1.18ab
Rm-B	1.76 ± 0.36a	0.22 ± 0.03a	22.66 ± 2.10a	16.41 ± 1.21a

Note: Different letters indicate significant differences ($P < 0.05$) between different colonization methods on the same substrate

3.2 Effect of Different Colonization Methods of *Ruellia Simplex* on pH and Organic Matter of Three Mining Wasteland Substrates

As can be seen from Fig. 1 A, the pH of quarry rock was not significantly different between the different colonization methods; the pH of manganese tailing slag, bag planting, and padded water retaining cotton colonization was significantly greater than bare root colonization; the pH of amended red mud was the highest and significantly different between the different colonization methods, and the magnitude of the pH was shown to be bag planting < padded water retaining cotton colonization < bare root colonization.

As can be seen from Fig. 1 B, the quarry rock has the highest organic matter content, and the bag planting method makes its organic matter content significantly larger than the other two colonization methods; manganese tailings slag organic matter content in the pad water retention cotton colonization and bag planting no significant difference between the organic matter content, but bag planting organic matter content is significantly larger than bare-root colonization; the organic matter content of the improved red mud, and the organic matter content of the different methods of colonization between the different methods of colonization are significantly different. The organic matter content of the amended red mud was significantly different between the different planting methods. The organic matter content of the three mining substrates was bag planting > water retaining cotton planting > bare root planting.

4 Discussion and Conclusions

Ruellia simplex grew better on quarry rock and had a higher growth rate, root-to-crown ratio, maximum root length, and leaf number in bag planting than in padded water retaining cotton colonization and bare root colonization, Cross et al. [9] pointed out that

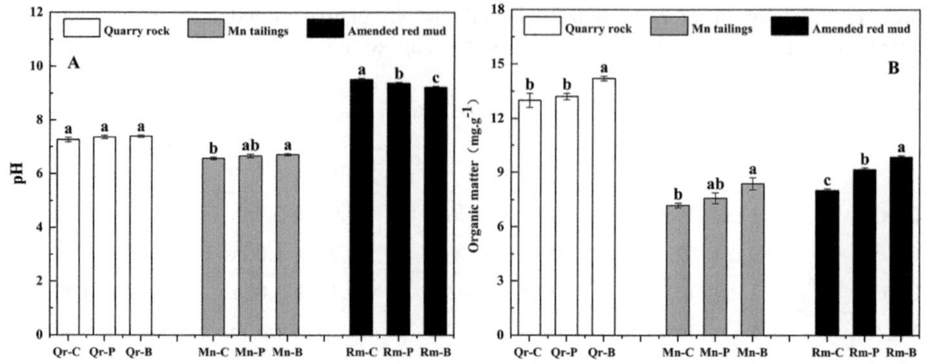

Fig. 1. Effect of different colonization methods of *Ruellia simplex* on pH and organic matter of three mining wasteland substrates. Note: Different letters indicate significant differences ($P < 0.05$) between different colonization methods on the same substrate.

the effect of manganese tailings slag on the root system of the plant was more significant than that on the aboveground part of the plant, in keeping with the results of the present study. This difference is most likely due to the large accumulation of heavy metals in the roots of the plants, which in turn leads to much more oxidative damage to the root system than to the above-ground part of the plant [10]. The difference is likely to be due to the large accumulation of heavy metals in the roots of the plants, which in turn resulted in the roots receiving much more oxidative damage than the aboveground part. In contrast, the pH of the amended red mud was also higher than the general plant tolerance range of 4.0–9.0 after planting *Ruellia simplex* in different colonization methods [11]. The pH of the amended red mud was higher than the tolerance range of 4.0–9.0. Because of the high pH, the relative growth rate, root-shoot ratio, and leaf number were lower than the other two substrates. However, there was no mortality after planting, suggesting that the plant is tolerant to strong alkalis, and bag planting and water-retaining sponge planting were more effective in lowering the pH of the amended red mud compared to bare-root planting.Liao et al. found that when *Ruellia simplex* was in red mud mixed with 4% horsetail pine leaves, the plants showed abnormalities such as purplish-red coloration of leaf abaxial veins, shedding of old leaves, and bending of stems, which was consistent with the results of this study [8]. High pH may directly affect root biomass and root-shoot ratio, leading to a decline, specifically, when pH exceeds 7.5, root growth tends to be inhibited due to the scarcity of protons in the exosome [12]. The organic matter content of quarry rock was significantly higher than that of the other two substrates, and bag planting also had the highest organic matter content among the same substrates, followed by padded water-retaining sponge bag planting and bare-root planting, respectively. Plant root metabolism and the death of a large number of fibrous roots will increase the organic matter content of the substrate, considering that the environment of bag planting and water-protected sheep planting is more conducive to the metabolism and growth of the root system, which leads to a higher content of organic matter compared to bare-root planting.

In summary, *Ruellia simplex* grew best on high calcium, alkaline quarry rock pile substrate, followed by manganese tailings slag with high heavy metal content, and was least adapted to strongly alkaline amended red mud. The best growth of *Ruellia simplex* under bag planting was followed by pad water retaining cotton colonization and finally bare root colonization, while bare root colonization performed poorly on substrates such as amended red mud. This suggests that *Ruellia simplex* has potential application in ecological restoration of mining wasteland, while bag planting and padded water retention cotton planting can ensure the best growth and ecological restoration of *Ruellia simplex* on various mining wastelands.

Acknowledgments. Funded by: Guangxi Key Research and Development Program Project (Gui Ke AB22035038); Guangxi Natural Science Foundation Project (2023GXNSFBA026332); Open Subject of the Technical Innovation Center for Mining Geo-environmental Remediation Engineering in the Southern Stone Mountain Region (NFSS2023012).

References

1. Liao, J.X., Zeng, D.J., Yao, Y.F., et al.: Plant diversity and composition of ecological floating bed affect domestic sewage purification. Guihaia **39**(1), 117–125 (2019)
2. Zhao, Q.M., Feng, C.T., Jian, X.Y.: Study on the purification effect of *Ruellia brittoniana* on wastewater and its practical application. Guangdong Landscape Archit. **41**(2), 64–68 (2019)
3. Fardous, M.A., Hegazi, M.A., El-Bably, S.Z., Hana, M.R.: Response of Mexican Petunia (Ruellia Brittoniana L.) to salinity, organic and bio materials. Appl. Ecol. Env. Res. **18**(4), 5789–5802 (2020). https://doi.org/10.15666/aeer/1804_57895802
4. Wang, S.M., Chen, N.H., Liu, X.G., et al.: Utilization of municipal sludge passivated by coal ash on yellow garden soil. Fujian J. Agricult. Sci. **33**(10), 1097–1103 (2018)
5. Fang, X.H., Yang, Z., Liu, X.W., Wei, D.Y.: Application of *Ruellia brittoniana* in Remediation of Cadmium Pollution in Soil and / or Wate: ZL201110373310.4[P]. 2013.12.25 (2013)
6. Sun, Y.P., Niu, G.H.: Relative salt tolerance of seven Texas superstar perennials. HortScience **50**(10), 1562–1566 (2015)
7. Sc, B.I.N.: Comparison of Morphological Structure of Root and Leaf of Three Garden Plants Under Various Soil Water Conditions. Guangxi University, Nanning (2021)
8. Liao, J.X., Huang, Y.Q., Xu, G.P., et al.: Effects of red mud mixed with pine needles on growth of Arundo donax and *Ruellia brittoniana.* Guihaia **38**(6), 704–709 (2018)
9. Cross, A.T., Ivanov, D., Steven, J.C., et al.: Nitrogen limitation and calcifuge plant strategies constrain the establishment of native vegetation on magnetite mine tailings. Plant Soil **461**, 181–201 (2021)
10. Tongtavee. N., LOISRUANGSIN, A., Mclaren, R.G.: Lead desorption and its potential bioavailability in soil used for disposing lead-contaminated pomelo peel: effects of contact time and soil pH. Effects of contact time and soil pH. Water, Air Soil Pollution **232**, 1–10 (2021)
11. Shu, W.S., Ye, Z.H., Zhang, Z.Q., et al.: Natural colonization of plants on five lead/zinc mine tailings in Southern China. Natural colonization of plants on five lead/zinc mine tailings in Southern China
12. Liu, K., Yu, F., Chen, M., et al.: A newly found manganese *hyperaccumulator-Polygonum lapathifolium* Linn. Int. J. Phytorem. **18**(4), 348–353 (2016)

Study on Chemical Leaching of Soils Contaminated with Chromium Medium Sand Using EDTA-2Na

Xiaolong Zhu[1,2], Qiang Shan[1,2], Xizhao Tian[1,2], Yanling Lai[1,2], Ying Wang[1,2], Bingtai Ma[1,2], and Zhihui Qu[3,4,5](✉)

[1] Hebei Geological Environment Monitoring Institute, Shijiazhuang 050021, China
[2] Hebei Key Laboratory of Geological Resources and Environment Monitoring and Protection, Shijiazhuang 050021, China
[3] Institute of Disaster Prevention, Sanhe 065201, China
qzh-19831001@163.com
[4] Hebei Key Laboratory of Resource and Environmental Disaster Mechanism and Risk Monitoring, Sanhe 065201, China
[5] Beijing Disaster Prevention Science and Technology Co., Ltd., Beijing, China

Abstract. In this study, for Cr(VI)-contaminated soil, chemical leaching remediation technology was used to investigate the removal effect of different concentrations of EDTA-2Na solution as a leaching agent on the removal of Cr(VI) from the contaminated medium sand and the removal effect of EDTA-2Na solution on Cr(VI) under different pH conditions. The results showed that the adsorption efficacy and removal efficiency of Cr(VI) were optimal when the concentration of the EDTA-2Na solution was 0.05 mol/L; Under the pH level of 4.17, the removal efficiency of Cr(VI) was superior to that of both pH 1.30 and 7.00.

Keywords: Chemical leaching · EDTA-2Na · Cr(VI)

1 Introduction

The predominant valence states of chromium (Cr) are Cr(III) and Cr(VI), with chromium primarily occurring as Cr(III) in its natural state. In the soil environment, these two distinct valence states of chromium can undergo interconversion [1, 2]. The toxicity of chromium is related to the valence state in which it exists. The Cr(III) can be readily adsorbed and immobilized by soil particles, exhibiting limited migration capability, while posing minimal risks to human health. Cr(VI) has the characteristics of not easy to be adsorbed, high solubility and strong fluidity, and is toxic, corrosive, and carcinogenic, which has potential harm to the surrounding environment and human health [3–6]. Therefore, it is of great practical significance to seek methods to reduce the Cr(VI) content in soil for the remediation of chromium-contaminated soil in our country.

The chemical leaching technique enables the separation of Cr(VI) from contaminated soil by introducing a leaching agent, which facilitates the transfer of Cr(VI) into a

D. Han et al. (Eds.): ICEPG 2024, ESE, pp. 561–566, 2025.
https://doi.org/10.1007/978-3-031-86829-0_65

centralized collection solution. Subsequently, other physical and chemical methods are employed to recover Cr(VI) from leachate, thereby achieving the objective of soil remediation [7, 8]. The utilization of EDTA-2Na as a leaching agent for remediating Cr(VI) contamination in soil represents an efficient and cost-effective approach, demonstrating promising prospects for practical applications [9, 10]. The purpose of this study was to explore the optimal leaching concentration of EDTA-2Na and the appropriate use environment, to provide a scientific basis for EDTA-2Na leaching of chrome-contaminated soil.

2 Materials and Methods

2.1 Test Materials

The test material was medium sand taken from a sand mining field in Langfang City, Hebei Province, Classification standards for the "Code for design of building foundation (GB 50007-2011)" [11]. The main reagents were K_2CrO_4, EDTA-2Na, HCl.

2.2 Test Methods

The test device was a plexiglass column of Φ10.00 cm, the column was filled with medium sand, the height was 30.00 cm, and the top and bottom were filled with Φ4.00–8.00 mm pure white quartz sand, the height was 2.50 cm. The cleaning process was carried out from the top down using a water distribution device to ensure uniformity, and a sampling port was provided underneath the column. The specific test device is shown in Fig. 1. The test was divided into 6 groups as shown in Table 1. Groups 1–4 carried out a comparative test on the leaching effect of EDTA-2Na at different concentrations, and groups 5–6 explored the optimal leaching effect of EDTA-2Na at various pH.

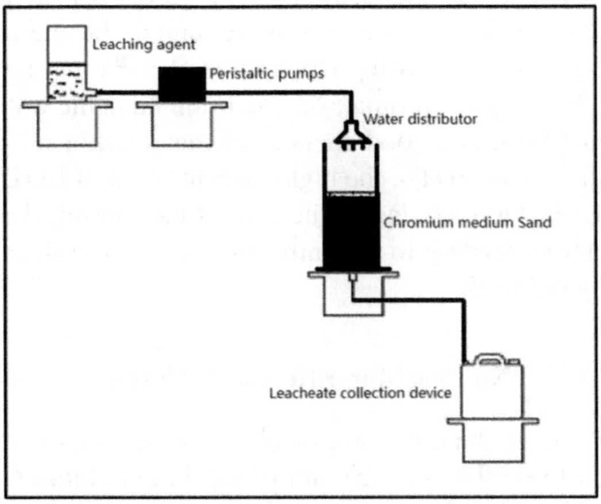

Fig. 1. Schematic diagram of the test device.

Table 1. Test conditions.

No.	Leaching column	EDTA-2Na concentration (mol/L)	pH
1	A	0.005	7.00
2	B	0.010	7.00
3	C	0.050	7.00
4	D	0.100	7.00
5	E	0.050	1.30
6	F	0.050	4.17

2.3 Detection Method

Chromium concentration was determined according to "Water quality-Determination of chromium(VI)-Flow injection analysis (FIA) and diphenylcarbazide spectrometric method (HJ 908-2017) [12]".

3 Results and Analysis

3.1 Analysis of Leaching Effect of EDTA-2Na with Different Concentrations

As can be seen from Table 2 and Fig. 2, the concentration of EDTA-2Na in column A was low, and the leaching effect on Cr(VI) in middle sand was poor, the removal rate was only 80.51%, and 116.909 mg Cr(VI) remains in the middle sand at the end of leaching. The leaching effect of column C was the best, the removal rate reached 98.65%, and only 8.073 mg Cr(VI) remained in the soil at the end of leaching. There was no significant difference in the effect of leaching columns D and C. The leaching effect did not increase but decreased, but it was still slightly better than that of the leaching column B, and 53.984 mg Cr(VI) remained in the sand at the end of the leaching. This is consistent with the findings of Wang Donghui et al. [13] that low concentrations of EDTA can promote the entry of chromate and dichromate in the water-soluble state into the liquid phase, but EDTA is a good chelator of metal cations, and by continuing to increase the concentration of EDTA, the high concentration of EDTA will form a stable complex with Cr(III) in the soil, and at the same time, inhibit the release of Cr(VI), which exists as an anion, leading to a significant reduction in the soil Cr(VI) removal rate was significantly reduced.

3.2 Analysis of EDTA-2Na Leaching Effect at Different pH Levels

As shown in Fig. 3, under the condition of pH 7.00, the removal effect of Cr(VI) in medium sand tended to be flat after 20 min of leaching column C, showing a tailing phenomenon. After 150 min of cumulative rinsing, the concentration of Cr(VI) in the rinsing solution was 0.042 mg/L, which was lower than the Class III standard limit in "Standard for groundwater quality (GBT-14848-2017)" [14]. After the leaching column

Table 2. Results of EDTA-2Na leaching at different concentrations.

No.	Leaching column	Accumulation of chromium in the leachate (mg)	Removal rate (%)
1	A	483.091	80.51
2	B	544.092	90.68
3	C	591.927	98.65
4	D	546.016	91.00
5	E	/	93.47
6	F	/	99.94

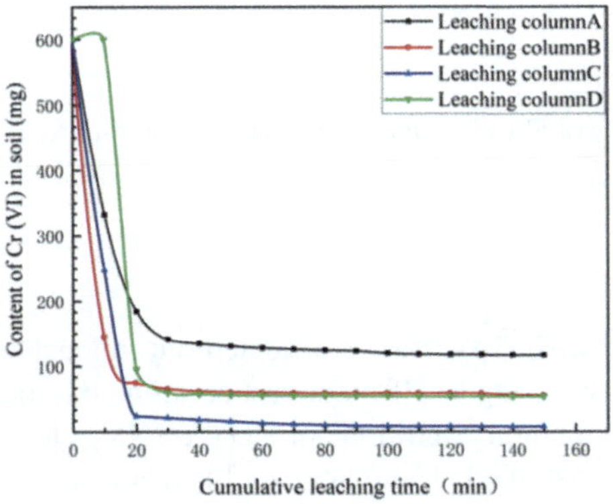

Fig. 2. Changes of Cr(VI) content with time in columns A to D leaching.

E was leached for a total of 80 min, the removal effect of Cr(VI) in medium sand tended to be flat, and the leaching curve showed a trailing phenomenon when the leaching column E was leached for a total of 150 min. The concentration of Cr(VI) in the leachate was 0.063 mg/L, which exceeded the Class III standard (0.05 mg/L) of "Standard for groundwater quality". This concentration was not suitable as a reagent for leaching and remediation. The same trailing phenomenon was observed when column F was cumulatively leached for 20 min, and the concentration of Cr(VI) in the leachate was 0.044 mg/L, which was lower than the Class III standard. Leaching column F achieved contaminant removal in a shorter time than leaching column C, with higher leaching efficiency. Di Palma [15] showed that when H^+ is present in the soil solution, it promotes both the dissolution of metal compounds and the chelation of metal ions, so the stronger the acidity, the better it is for the extraction of heavy metals from the soil. The presence of Cr varies according to the environmental pH, and Cr(VI) is easily reduced to Cr(III) under acidic conditions, and the Cr(III) generated is easily adsorbed and not easily transported by soil particles [16]. Therefore, the leaching rate of Cr(VI) under acidic conditions was greater than that under neutral conditions and the effect was better, and the leaching effect in weak acidic environment was better than that in strong acidic environment, which was consistent with the results of Du Pei [17] and Zhang Xiaowei [18].

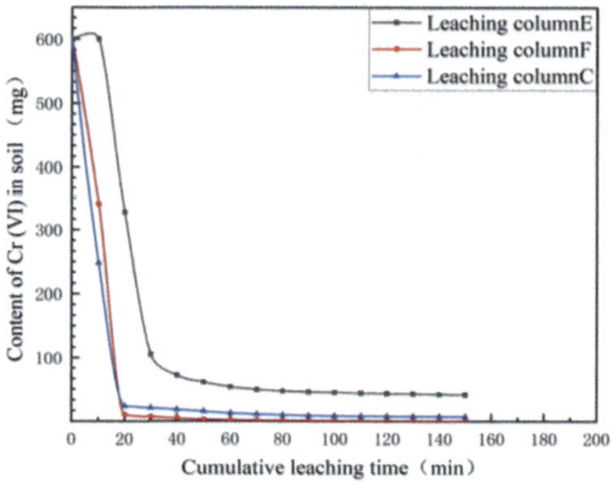

Fig. 3. Changes of Cr(VI) content with time in columns C, E and F leaching.

4 Conclusion

In this study, 6 groups of EDTA-2Na were carried out to leached Cr(VI) contaminated medium sand columns by using a self-made test column, and the optimal concentration and appropriate environmental conditions of EDTA-2Na as leachate to remove Cr(VI) from medium sand were studied. The main conclusions were as follows:

1. The test on the leaching effect of different concentrations of EDTA-2Na showed that the leaching effect of low concentration of EDTA-2Na on Cr(VI) contained in the sand was poor, and only a small part of Cr(VI) can be absorbed into the leachate. As the concentration of the leaching agent increased, the leaching effect was enhanced. 0.05 mol/L EDTA-2Na can adsorb Cr(VI) in the sand, and the removal efficiency can reach 98.65%, and the cumulative content of Cr(VI) in the leachate is 591.927 g, but the effect of too high concentration of leaching agent on the removal of Cr(VI) does not increase but rather decreased.
2. The tests on the effect of EDTA-2Na leaching at different pH showed that when the concentration of EDTA-2Na was the same, the complexation ability of EDTA-2Na on Cr(VI) was weakened under the condition of stronger acidity (pH = 1.30), and the removal effect was the weakest; The complexation ability of EDTA-2Na on Cr(VI) under the condition of weak acidity (pH = 4.17) was enhanced compared with that at pH = 7.0, and the removal effect was the best.

Acknowledgment. This work was supported by Open Fund for Hebei Key Laboratory of Geological Resources and Environment Monitoring and Protection "Research on electric and chemical oxidation combined remediation technology for coking polluted soil" (JCYKT202005) and "Research on the strength deterioration of refuse soil and slope stability under the coupling of temperature and leachate" of the Natural Science Foundation of Hebei Province (E2021210007).

References

1. Shuang, S., et al.: Study on the effectiveness of edta/humic acid for removal of typical heavy metals in humus soil. Environ. Sci. Technol. **46**(1), 47–50 (2023)
2. Zhuang, J.Y., Cheng, Y.Z., Hou, Y.Z., et al.: Simulation of dynamic leaching characteristics of chromium (VI) from contaminated soil and its influence on groundwater. South-to-North Water Transfers Water Sci. Technol. **22**(2), 1–10 (2024)
3. Zeng, Y., Deng, H., Li, W., et al.: Research progress of remediation technology of chromium-contaminated soil . Environ. Dev. **31**(03), 47–49 (2019)
4. Montes, M.O., Peralta-Videa, J.R., Parsons, J.G., et al.: Spectroscopic determination of the toxicity, absorption, reduction, and translocation of Cr(VI) in two magnoliopsida species. Int. J. Phytorem. **15**(2), 168–187 (2013)
5. Coetzee, J.J., Bansal, N., Chirwa, E.M.N.: Chromium in environment, its toxic effect from chromite-mining and ferrochrome industries, and its possible bioremediation. Exposure Health **12**(1), 51–62 (2020)
6. Wu, X., Ye, T., Xie, C., et al.: Experimental and model-ing study on Cr(VI) migration from slag into soil and groundwater. Processes **10**(11), 2235 (2022)
7. Yang, Y., Li, Z.: Influence of different leaching agents on the remediation effect and development trend of heavy metal contaminated soil. Shaanxi Geol. **41**(02), 10–15 (2023)
8. Guo, Y., Huo, Y., Wang, X.: Application examples of citric acid leaching for remediation of heavy metal contaminated soil. Henan Chem. Industry **41**(02), 25–28 (2024)
9. Tran, H.T., Lin, C., Hoang, H.G., et al.: Soil leacheding for the remediation of dioxin-contaminated soil:a review. J. Hazard. Mater. **421**, 126767 (2022)
10. Wasay, S.A., Barrington, S., Tokunaga, S.: Organic acids for the in situ re – mediation of soils polluted by heavy metals: soil flushing in columns. Water Air Soil Pollut. **127**, 301–314 (2001)
11. Water quality-Determination of chromium (VI) – Flow injection analysis (FIA) and diphenylcarbazide spectrometric method (HJ 908-2017)
12. Ministry of Urban and Rural Construction and Environmental Protection of the People's Republic of China. Code for Design of Building Foundation GB 50007-2011. China Construction Industry Press, 1991. Soil Henan Chem. Ind. **41**(02), 25–28 (2024)
13. Wang, D., Li, G., Qin, S., et al.: Analysis of remediation effect and economic cost of Cr(VI)-contaminated fine-grained soil by chemical leaching. J. Ecol. **39**(07), 2309–2315 (2020)
14. GBT-14848–2017. Standard for groundwater quality
15. Di Palma, L.: Influence of indigenous and added iron on copper extraction from soil. J. Hazard. Mater. **170**(1), 96–102 (2009)
16. Zhao, S., Li, J., Du, Y.: A review on the evaluation of morphology and valence of heavy metal chromium in soil. J. Agronomy **12**(04), 24–28 (2022)
17. Du, P.: Remediation of Soils Polluted by Heavy Metal Chrome Using Organic Acids (Master Thesis). Shenyang Aerospace University, Shenyang (2011)
18. Zhang, X.W.: The Study on Adsorption and Desorption of Cr(VI) in Soil (Master thesis). China University of Geosciences, Beijing (2017)

Study on Preparation and Properties of Foam Ceramics from Coal Gangue

Yipei Shen and Shaolei Song(✉)

School of Civil Engineering, Xuzhou University of Technology, Xuzhou 221018, China
songshaolei@xzit.edu.cn

Abstract. Coal gangue is a solid waste generated during the coal mining process, which has caused significant damage to the environment with the increasing accumulation over the years. Foam ceramics were prepared with coal gangue and waste glass as the main raw materials. The results show that the apparent density, porosity, compressive strength and thermal conductivity of the foam ceramics obtained under the conditions of sintering temperature of 775 °C, holding time of 30 min, sodium carbonate content of 5%, and boron oxide content of 5% are 0.35 ± 0.04 g/cm^3, 66.9 ± 0.7%, 6.2 ± 0.4 MPa, 0.042 ± 0.006 W/(m·°C), respectively. The excellent comprehensive performance makes it a potential inorganic insulation material for the construction industry.

Keywords: Coal gangue · foam ceramics · high strength · thermal conductivity

1 Introduction

Coal gangue is an associated product in the process of coal mining, accounting for about 10% of coal production, and as a major coal producer, China not only has billions of tons of coal gangue accumulation, but also grows at a rate of 100 million tons per year [1]. In order to effectively deal with this challenge, coal gangue has been used in the fields of backfill, power generation, refractory materials in mining areas, and has achieved certain results. With the proposal of the "dual carbon" goal, the high-value and efficient utilization of coal gangue has been put on the agenda. In view of the fact that the main chemical composition of coal gangue is similar to that of foam ceramics, it has become one of the potential raw materials for the production of foam ceramics.

Foam ceramics are widely used in construction, thermal insulation, biochemistry, environmental protection and other fields because of their excellent properties. However, when using pure raw materials to produce foam ceramics, the foaming process is more complex and difficult to control, and the actual prepared finished product has poor performance and is rarely adopted, and it is also faced with high production costs, which seriously affects its application. The use of coal gangue to prepare foam ceramics can not only reduce the environmental pressure, but also effectively reduce the production cost and preparation difficulty of foam ceramics. Fu et al. [2] showed the best comprehensive performance when using phosphate tailings and coal gangue as raw materials and adding 2 wt% ferric oxide as a foaming agent. The pore structure was uniform, with a porosity

D. Han et al. (Eds.): ICEPG 2024, ESE, pp. 567–577, 2025.
https://doi.org/10.1007/978-3-031-86829-0_66

of 71.4%, a density of 0.78 g/cm^3, and a compressive strength of 3.5 MPa. Xie et al. [3] used coal gangue and waste glass as raw materials, SiC as foaming agent, and albite as flux to prepare foam ceramic glass. Finally, they found that the volume density, compressive strength, and thermal conductivity of foam ceramic glass prepared under the conditions of 40 g coal gangue, 60 g waste glass, 1120 °C sintering temperature, and 5 °C/min heating speed were the best. In order to reduce the preparation cost. Ren et al. [4] successfully prepared foam ceramics with bulk density of 0.41 g/cm^3, compressive strength of 1.6 MPA, and a high porosity of 74.9% by adjusting the synthesis parameters at a sintering temperature of 1230 °C and a holding time of 60 min. In addition, the absorption and thermal conductivity are also reduced to 1.48% and 0.197 W/(M·°C), respectively. Li et al. [5] prepared alkali activated coal gangue foam by microwave foaming process with spontaneous combustion coal gangue and no added minerals. Produced high porosity (68.1–87.4 vol%), compressive strength (0.8–3.6 MPa), and low thermal conductivity (0.090–0.150 W/m·°C).In order to effectively utilize resources and further optimize the composition of raw materials, Zhang et al. [6] also used coal gangue and waste glass as the main raw materials to prepare foam ceramics, and further studied the effects of calcination temperature and Na_3PO_4 addition on the phase structure, apparent porosity, compressive strength and apparent density of foam ceramics. When the amount of Na_3PO_4 added is 3% (mass fraction) and the calcination temperature is 1140 °C, the sample pore size is uniform, with an average pore size of 0.92 mm, a compressive strength of 4.7 MPa, and an apparent density of 0.69 g/cm^3. Hu et al. [7] used untreated phosphorus tailings and coal gangue as raw materials to synthesize a foam glass ceramics that can be used in the construction field by using the self foaming technology. The results showed that under the condition of 1150 °C, the FGC with 1150 °C had the best comprehensive performance, with a porosity of 61.6%, a bulk density of 1.04 g/cm^3, and a compressive strength of 10.11 MPa.

In order to better reduce costs, further promote the development of coal gangue "reduction, resource utilization, recycling, and high-value", and achieve green and sustainable development. In this paper, foam ceramics were prepared from coal gangue and waste glass, and the effects of sintering temperature, holding time, Na_2CO_3 content and B_2O_3 content on apparent density and microstructure were explored. Under the optimal conditions, the apparent density of sintered foam ceramics is 0.35 ± 0.04 g/cm^3, the porosity is 66.9 ± 0.7%, the pore size is suitable, the uniformity is good, and the compressive strength is 6.2 ± 0.4 MPa. It can be used in construction and other fields, can realize the comprehensive utilization of coal gangue, and the lower price of waste glass can bring certain environmental and economic benefits.

2 Experimentation

2.1 Raw Material

The coal gangue comes from a mine in Xuzhou City, and the waste glass is formed during the demolition and construction process. The chemical composition of coal gangue and waste glass is shown in Table 1.

Table 1. Chemical composition of coal gangue and waste glass

sample	Mass fraction/%								
	SiO_2	Al_2O_3	$K_2O + Na_2O$	Fe_2O_3	CaO	MgO	TiO_2	SO_3	Others
Coal gangue	59.70	28.30	1.77	6.70	1.43	0.57	0.98	0.11	0.44
Waste glass	68.02	1.16	17.77	0.36	9.02	3.40	–	0.11	0.16

2.2 Preparation of Samples

In the experiment, the crushed coal gangue and waste glass were used as raw materials, and wet grinds and mixes them in a mass ratio of 3:7. The weight ratio of grinding beads to materials is 3:1, and the ball milling time is 2 h. The mixed powder after ball milling is placed in a 100 °C oven and dried for later use. The preparation process of foam ceramic green body is as follows: a certain amount of mixed powder is added with 3%–5% boron oxide and 2%–5% sodium carbonate solution for mixing. The mixed powder is placed in the mold and pressed into a green body with a diameter of 20 mm and a height of 10 mm. The firing process of foam ceramics is as shown: First, the room temperature is raised to 450 °C at the rate of 5 °C/min, then the temperature is raised to the specified temperature (750 °C, 775 °C, 800 °C) at the rate of 8 °C/min, and the heat preservation is conducted at the set temperature (10 min, 20 min, 30 min), then the temperature is lowered to room temperature at the rate of 10 °C/min, and the samples obtained are taken out for relevant performance tests.

2.3 Characterization

The crystalline phases of raw materials and samples were analyzed by XRD (Rigaku D/max 2200PC, scanning range: 10–80°, scanning rate: 8°/min). The chemical compositions of raw materials were analyzed by XRF (Shimadzu XRF-1800). The compressive strength of foam ceramics was tested by Meters CMT5305 microcomputer controlled electronic universal testing machine. The microstructure of the foam ceramics was analyzed by a Scanning Electron Microscope (FESEM, JSM-7500, JEOL). The apparent density (ρ) of the sample was obtained through the mass divided by the volume. The porosity of foam ceramics is measured using the Archimedean principle, as shown in (1):

$$\omega = (G2 - G1)/(G2 - G3) * 100\% \tag{1}$$

where, G1 is the dry weight of the sample in air (g), G2 is the floating weight of the sample in water (g), and G3 is the wet weight of the sample after absorbing water (g).

The thermal conductivity of foam ceramics was measured using a thermal conductivity analyzer (DRX-RL, China).and each thermal conductivity was measured three times to calculate the average value.

3 Results and Discussion

3.1 Effect of Sintering Temperature

The sintering temperature has a direct impact on the structure and performance of foam ceramics. Figure 1 (a-f) shows the pictures of foam ceramics obtained at 750 °C, 775 °C and 800 °C respectively. It can be seen from Fig. 1(a-c) that under the condition of fixed cosolvent, blowing agent and holding time, the number of pores overflow on the sample surface increases significantly with the increase of holding temperature. Moreover, in Fig. 1(c), obvious large pores can be observed, which is because the liquid phase viscosity decreases with the increase of temperature. The pressure inside the bubble is significantly higher than the surface tension of the liquid phase, resulting in porosity overflow and the formation of uneven pores on the surface [4]. In addition, as shown in Fig. 1 (d-f), the macro view of the polished surface of the sample shows that the internal holes are numerous and dense, indicating that most of the bubbles formed are retained in the inside of the sample, and this temperature range basically meets the synthesis requirements of foam ceramics. According to the shrinkage rate of the sample, with the increase of temperature, the shrinkage rate of the sample increases from 16.0% of 750 °C to 18.5% of 775 °C, and decreases to 15.8% when the temperature is further increased to 800 °C. The shrinkage rate shows a trend of first increasing and then decreasing with the increase of temperature, which further confirms that the synthesis temperature of foam ceramics should not be too high to ensure the formation of uniform bubbles and prevent a large number of bubbles from overflowing due to the increase of temperature.

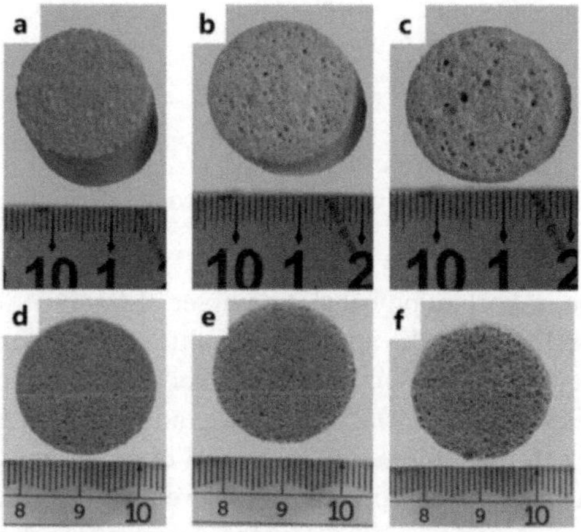

Fig. 1. Conditions of ceramic foam calcined at different temperatures. (a, d) 750 °C; (b, e) 775 °C; (c, f) 800 °C

Figure 2 shows the XRD patterns of foam ceramics obtained at different temperatures. It can be seen that the main crystal structures of the three samples are basically similar, with quartz and plagioclase being equal, and there are obvious amorphous phase characteristic peaks between 20° and 30°. As the temperature increases, the characteristic peak intensity of the amorphous phase decreases, indicating that some amorphous

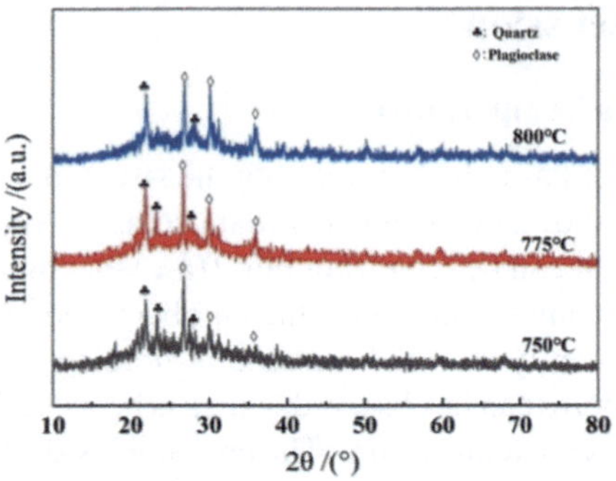

Fig. 2. XRD spectra of ceramic foam calcined at different temperatures

phases are transforming into crystalline, and the characteristic peak intensity of the plagioclase phase is significantly enhanced.

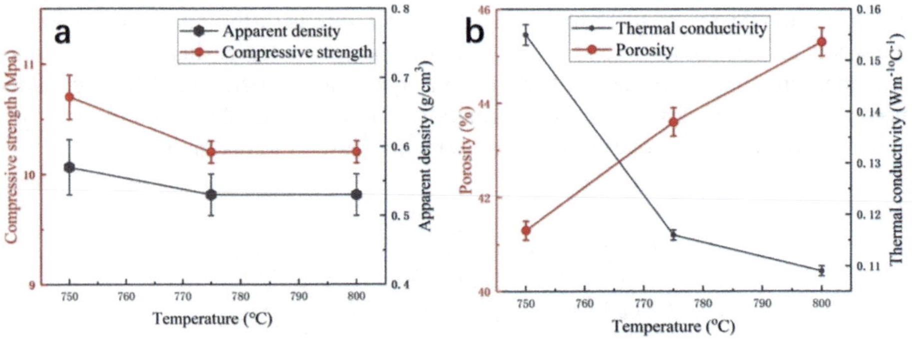

Fig. 3. Properties of foam ceramics obtained at different temperatures

Figure 3 shows the important physical parameter of foam ceramics at different temperatures. As shown in Fig. 3(a), the apparent densities of the foam ceramics are 0.57 ± 0.04, 0.53 ± 0.03, and 0.53 ± 0.03 g/cm^3, respectively. The compressive strengths of the corresponding samples were 10.7 ± 0.2, 10.2 ± 0.1 and 10.2 ± 0.1 MPa, respectively. The compressive strength can meet the requirements of commercial standards ($\geq$0.6 MPa) [1]. The apparent density shows a trend of first decreasing and then increasing with the increase of sintering temperature, while the change in pore size shows the opposite trend. Size and quantity of pores in foam ceramics directly affect the apparent density, and then affect the compressive strength. As temperature increases from 750 °C to 775 °C, the compressive strength decreased by 4.3%. When it further increased to 800 °C, the change in compressive strength was not significant. During the low-temperature sintering process (750–775°C), the pore size increases with the increase of temperature, while the apparent density and compressive strength decrease accordingly. When further heated, the trend of changes in compressive strength and apparent density is not obvious. The decrease in liquid viscosity facilitates the formation of a large number of interconnected pores, ultimately leading to uneven pore size in the sample. According to the above analysis, foam ceramics sintered at 775 °C with apparent density of 0.53 ± 0.03 g/cm^3, which is better than 0.98 g/cm^3 reported in the relevant literature [5]. From Fig. 3(b), it can be seen that the thermal conductivity of the sample is positively correlated with the apparent density, as samples with larger pore sizes have lower thermal conductivity. The thermal conductivity of sample samples under the best conditions is 0.116 ± 0.001 W/m·°C.

3.2 Effect of Holding Time

Figure 4 shows the physical and sectional morphology of foam ceramics prepared at different sintering times (10–30 min). As shown in Table 2, the average pore diameters of foam ceramics were 1.02 ± 0.12, 0.57 ± 0.04 and 0.46 ± 0.03 mm, respectively. The average pore size gradually increases with the increase of holding time. When further increased to 30 min, the aperture decreased. In addition, it can be seen from the section of the three foam ceramics that the upper part is more pores than the lower part. When the sintering temperature exceeds the decomposition temperature of the foaming agent, carbon dioxide bubbles are formed inside the matrix due to the decomposition of the foaming agent. As the sintering time increases, the number of carbon dioxide bubbles gradually increases and merges to form large bubbles. Then, the bubbles rise from the bottom to the top. However, prolonged insulation time causes the internal pressure of atmospheric bubbles to be higher than the surface tension provided by the surrounding liquid phase. The liquid phase will be relieved from the top surface of the foam ceramics, which will lead to the density process of sintering. In addition, from the cross section of foam ceramics, it can be found that the upper part has more pores than the lower part. As the temperature rises, Na_2CO_3 decomposes, a large number of carbon dioxide bubbles continue to grow over time, and the top bubbles are the first to break through the surface tension and release.

Figure 5 shows the variation curves of properties of foam ceramics at different holding times. The apparent densities of the three foam ceramics were 0.47 ± 0.05, 0.45 ± 0.05 and 0.36 ± 0.03 g/cm^3, respectively. The apparent density and the pore size change trend is opposite, which similar to the effect of temperature on the microstructure, With the change of sintering time, the pore size first decreases and then increases. The apparent density showed a decreasing trend. The compressive strength of samples was 8.3 ± 0.3, 6.4 ± 0.2 and 4.6 ± 0.1 MPa, respectively, showing a decreasing trend, which was positively correlated with the apparent density change. Considering that the increase of sintering time is likely to cause the combined growth of small holes, resulting in uneven pore structure [5], combined with the results in Fig. 5 (b), 30 min was selected as the optimal insulation time.

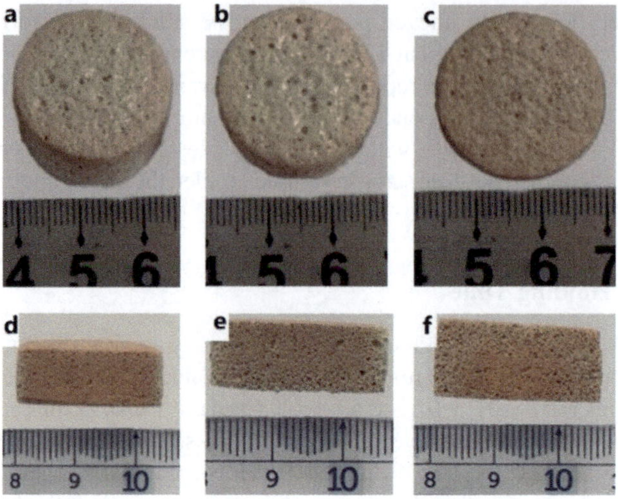

Fig. 4. Physical morphology and microstructure of ceramic foam with different holding time. (a, d) 10 min; (b, e) 20 min; (c, f) 30 min

Table 2. Structure and properties of foam ceramics obtained at different parameters

Sample	Average aperture size/mm	Apparent density/g/cm^3	Compressive strength/Mpa	Thermal conductivity/W/(m·°C)	Linear expansivity/%	porosity/%
F4Z4-10	1.02 ± 0.12	0.47 ± 0.05	8.3 ± 0.3	0.166 ± 0.008	18	47.5 ± 0.5
F4Z4-20	0.57 ± 0.04	0.45 ± 0.05	6.4 ± 0.2	0.134 ± 0.007	25	54.7 ± 0.6
F4Z4-30	0.46 ± 0.03	0.36 ± 0.03	4.6 ± 0.1	0.060 ± 0.005	28	65.5 ± 0.7
F2Z5-30	0.81 ± 0.06	0.38 ± 0.05	7.2 ± 0.6	0.153 ± 0.008	32	58.4 ± 0.5
F3Z5-30	0.76 ± 0.11	0.38 ± 0.05	6.8 ± 0.6	0.089 ± 0.007	34	62.4 ± 0.5
F4Z5-30	0.82 ± 0.07	0.36 ± 0.05	5.9 ± 0.4	0.060 ± 0.006	35	65.5 ± 0.5
F5Z3-30	0.73 ± 0.30	0.52 ± 0.07	7.5 ± 0.7	0.054 ± 0.007	20	57.4 ± 0.6
F5Z4-30	0.65 ± 0.07	0.39 ± 0.05	6.6 ± 0.6	0.047 ± 0.006	23	62.9 ± 0.6
F5Z5-30	0.78 ± 0.07	0.35 ± 0.04	6.2 ± 0.4	0.042 ± 0.006	37	66.9 ± 0.7

3.3 Effect of Different Contents of Na_2CO_3

The CO_2 formed by decomposition of Na_2CO_3 forms holes in foam ceramics, so the content of Na_2CO_3 plays an important role in the formation of foam ceramics. Figure 6 (a-d) shows the macroscopic physical diagram of 2%, 3%, 4% and 5% sodium carbonate dosage when the firing temperature is 775 °C and the holding time is 30 min. It can be seen from the diagram that the diameter of the sample generally increases with the increase of foaming agent dosage from 2% to 5%, from 24.6 to 26.2 mm, and pores left by bubble overflow can be observed on the surface. Figure 6 (e-h) shows the macroscopic surface grinding diagram. The samples all show a dense and uniform pore structure. According to Table 2, the average aperture size showed an increasing trend. This is due

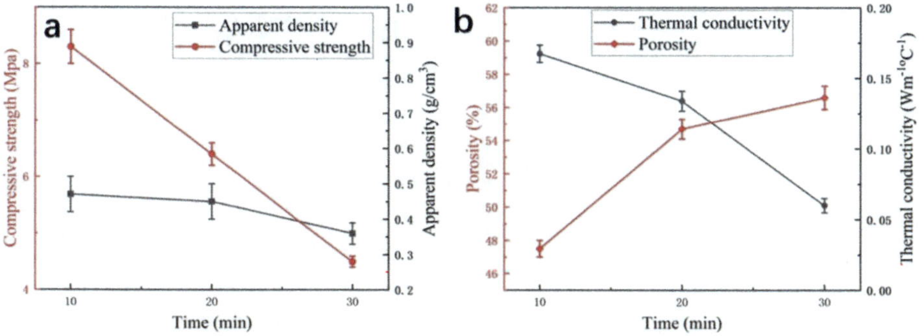

Fig. 5. Properties of foam ceramics obtained at different holding times

to the increase in the amount of blowing agent, more gas can be formed under the same conditions, thus forming a large number of holes inside the sample, and obtaining a larger volume of foam ceramics.

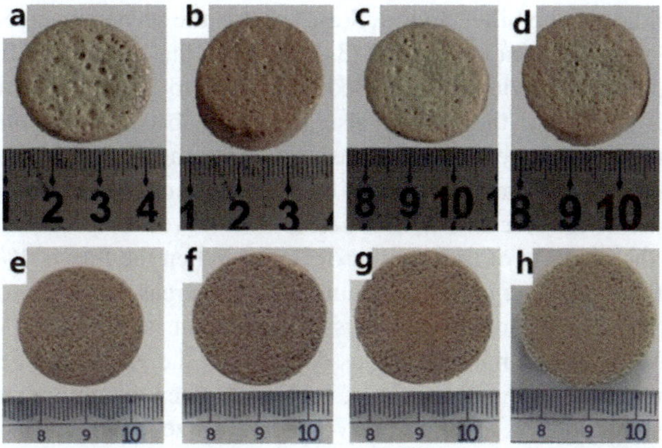

Fig. 6. Physical appearance and surface polishing of foam ceramic with different Na_2CO_3 content. (a, e) 2% (b, f) 3%; (c, g) 4%; (d, h) 5%

Figure 7 (a) shows the significant density of foam ceramics prepared with different content of Na_2CO_3 and the curve of the obvious density of foam ceramics. The apparent densities of the samples were 0.38 ± 0.05, 0.38 ± 0.05, 0.36 ± 0.05 and 0.35 ± 0.04 g/cm^3, respectively. The corresponding compressive strengths were 7.2 ± 0.6, 6.8 ± 0.6, 5.9 ± 0.4 and 6.2 ± 0.4 MPa, respectively. The pore size and quantity of the sample affect the presentation of its physical properties, such as apparent density and compressive strength. With the increase of Na_2CO_3 content, the growth rate of porosity shows a trend of first fast and then slow. Experiments have shown that when the Na_2CO_3 content is 5%, the significant density and available strength of the sample are the lowest. The Na_2CO_3 content is optimal.

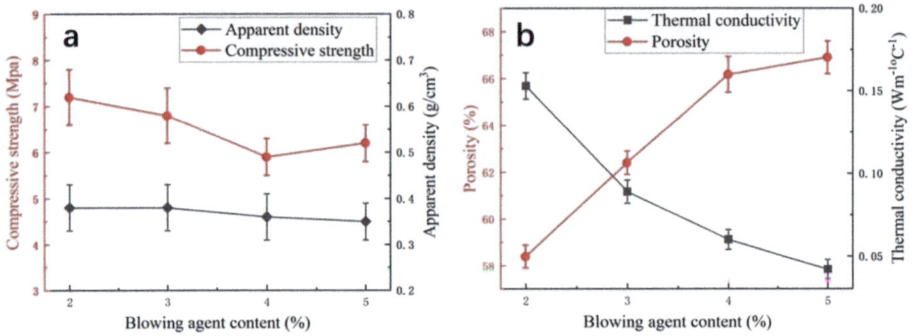

Fig. 7. Properties of ceramic foams obtained at different Na_2CO_3 contents

3.4 Effects of Different B_2O_3 Content

During the formation of foam ceramics, temperature has a great influence on the viscosity of green body. Reducing the viscosity and improving the fluidity of liquid phase are more conducive to the formation of bubbles. B_2O_3 can lower the softening temperature and increase the fluidity of the billet. Figure 8(a-c) shows the physical and SEM images of foam ceramics with different cosolvents (3–5%). With the increase of flux dosage, the diameter of the sample gradually increased from 24.0 to 26.9 mm, while the number of pores on the surface of the sample also increased, because the internal viscosity decreased, resulting in a large number of gases to overcome the action of surface tension rapidly expanded around, a large number of bubbles foaming rate increased, the diameter also increased. The pore size and distribution uniformity of the pores in the sample are improved with the increase of the amount of B_2O_3. When the flux is 3%, the pore size distribution of the sample is not uniform, but with the increase of the cosolvent, the pore size gradually changes to uniformity. This phenomenon is due to the increase in the amount of B_2O_3, resulting in a continuous decrease in the viscosity of the sample, the growth environment of large pores is more stable, so when the amount of B_2O_3 is 5%, the number of small pores is less but the average pore size is larger.

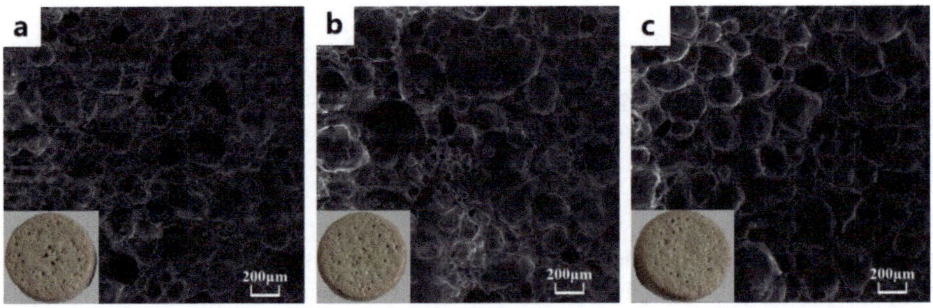

Fig. 8. Physical morphology and microstructure of ceramic foam with different B_2O_3 content. (a) 3%; (b) 4%; (c) 5%

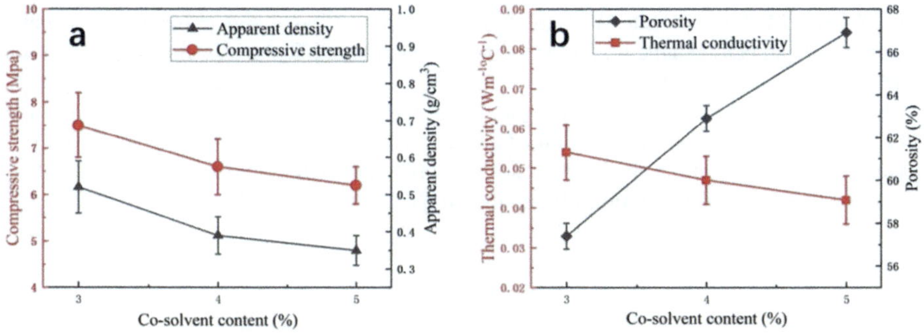

Fig. 9. Properties of ceramic foams obtained at different B_2O_3 contents

Properties of foam ceramics such as apparent density and compressive strength are shown in Fig. 9 (a). The apparent density of the samples were 0.52 ± 0.07, 0.39 ± 0.05 and 0.35 ± 0.04 g/cm^3, and the compressive strength were 7.5 ± 0.7, 6.6 ± 0.6 and 6.2 ± 0.4 MPa, respectively. With the increase of cosolvent, the compressive strength and apparent density decreased gradually, and the compressive strength and apparent density were roughly positively correlated. In addition, 5% of solvents have a relatively low heat conduction rate and high pore rate, which is better than other samples showing better overall performance (Fig. 9 (b)).

4 Conclusion

In this study, coal gangue and waste glass were used as raw materials, sodium carbonate was used as foaming agent, and foam ceramics with light weight, high strength and excellent insulation performance were prepared by high temperature decomposition foaming method. The effects of sintering temperature, holding time, dosage of foaming agent and co-solvent on pore structure and properties of foamed ceramics were systematically studied. The process of preparing foam ceramics from coal gangue and waste glass was optimized. The foam ceramics obtained under the conditions of sintering temperature of 775 °C, holding time of 30 min, Na_2CO_3 content of 5% and B_2O_3 content of 5% had relatively apparent density of 0.35 ± 0.04 g/cm^3, excellent compressive strength of 6.2 ± 0.4 MPa, thermal conductivity of 0.042 ± 0.006 W/(m·°C) and porosity 66.9 ± 0.7%.

References

1. Zhu, T., Wu, X., Xing, C., Ju, Q., Su, S.: Current situation and progress of coal gangue resource utilization. Coal Sci. Technol. **52**(1), 380–390 (2024)
2. Fu, F., Hu, N., Ye, Y., Chen, G., Jia, J.: Production of lightweight foam ceramics by adjusting sintering time and heating rate. Constr. Build. Mater. **394**, 132063 (2023)
3. Xie, C., Zhang, L.: Study on preparation and properties of foamed ceramics materials using coal gangue and waste glass. China Ceramics **58**(5), 51-5664 (2022)
4. Ren, P., Zhou, R., Liu, H., Huo, Y., Wang, Y.: Preparation of coal gangue based foamed ceramics with SiC as blowing agent and study on its thermal insulation performance. Ceramics International **50**, 31680–31690 (2024)

5. Li, X., Pan, M., Tao, M., Liu, W., Gao, Z., Ma, C.: Preparation of high closed porosity foamed ceramics from coal gangue waste for thermal insulation applications. Ceramics International **48**, 37055–37063 (2022)
6. Zhang, J., Meng, F., Wang, L., Dai, L., Yao, L., Li, Z.: Preparation and properties of coal gangue-based foamed ceramics. Bulletin of the Chinese Ceramic Society **42**(3), 960–969 (2023)
7. Hu, N., et al.: Preparation, characterization and self-foaming mechanism of total-tailings-based foamed glass-ceramics. Ceramics International **49**, 31881–31890 (2023)

Research on the Solid Waste Utilization and Hazard Control in Alumina Industry Driven by Sustainable Development

Jingwu Yao[1], Liwen Yao[2](✉), and Chang Huang[3](✉)

[1] Business School of Yulin Normal University, Yulin 537000, Guangxi, China
[2] Arts and Social Sciences School of Monash University Malaysia, 47500 Bandar Sunway, Selangor Darul Ehsan, Malaysia
lyao0021@student.monash.edu
[3] Business School of Imperial College London, South Kensington, London SW72AZ, UK
huangchang0619@outlook.com

Abstract. Alumina production is a complex chemical reaction with long process, frequent energy transfer and lots of equipments. It involves many kinds of hazards, and it generates a lot of solid waste(red mud) and ricks during producing alumina. Through the identification of alumina hazard sources and the investigation and analysis of hazards, this paper draws relevant conclusions; At the same time, through the investigation and analysis of alumina production process, equipment and raw materials, combined with the accident situation of alumina production enterprises, the environmental influencing factors and hazards of alumina production are put forward. Finally, driven by sustainable development, this paper puts forward a disposal scheme for recycling of red mud produced in the process of alumina production is proposed, And put forward new control measures for possible risks in the production process, It provides reference for alumina enterprises to achieve sustainable development, eliminate hidden dangers and achieve safe production.

Keywords: Alumina industry · solid waste utilization · hazard control · sustainable development

1 Introduction

Alumina has wide industrial usage, 90% of alumina are used for producing aluminum by electrolysis. It is applied widely in many industrial fields, such as metallurgy, chemical industry, medicine, national defense, aviation, construction materials and so on. But because of the complex alumina production, there are many solid waste and risks in the process, the accidents will take place easily. On June 22, 2016, a major accident occurred in Henan branch of Aluminum Corporation of China, killing 13 people and injuring 6 others. On October 20, 2023, an explosion occurred at an aluminum plant in Guangxi Province, Causing six people was dead and injured 4. Alumina industry accidents too numerous to enumerate, resulting in a large number of casualties and

D. Han et al. (Eds.): ICEPG 2024, ESE, pp. 578–592, 2025.
https://doi.org/10.1007/978-3-031-86829-0_67

property losses, Solid wastes and hazards need to be studied. This is the starting point of this paper. Driven by sustainable development, In this paper, hazard sources and hazards of alumina industry are analyzed in detail, effective control measures are discussed, and relevant conclusions are drawn.

2 Related Research Literature

2.1 Research on the Safety Management of Alumina Production Equipment

Guo Junxia (2009) pointed out that the production process of alumina has the characteristics of long process and numerous equipment, resulting in the complexity and uncertainty of unsafe factors [1]. Zhang Yan (2020) believes that operation errors in process will directly affect the safety of personnel and equipment, and modern science and technology must be used for process safety management [2]. Gao Shiyou et al. (2016) pointed out that in daily equipment management, in order to ensure safety and improve efficiency, equipment safety management must be standardized, scientific and systematic [3]. Wu Lin et al. (2021) pointed out that accidents in alumina production occur from time to time, and most of these problems are due to the lack of top-down safety awareness [4].

2.2 Study the Noise and Corrosion in the Process of Alumina Production

Li Pengchong (2018) pointed out that many equipment in the alumina production process are noise sources, which are easy to cause human anxiety and irritability, and seriously affect the physical and mental health of production personnel [5]. Tian Xinzhan (2016) believes that equipment will corrode due to the perforation effect, resulting in the loss of slurry medium and the termination of production equipment [6]. Jia Ye (2023) believes that it is necessary to analyze the risk factors in the whole process of alumina production, carefully identify the danger sources [7]. Gao Yuping (2019) pointed out that in terms of environmental protection, red mud is strongly alkaline and belongs to Class II general industrial solid waste due to the influence of alumina production process [8]. Bi Shiwen (2006) concluded that in the industrial production process, corrosion of equipment has a very obvious impact on production [9].

2.3 Research on Environmental Protection

Hart S L (1995) proposed that environmental strategy mainly includes three aspects: pollution prevention, product management and sustainable development [10]. Polas et al.(2021) pointed out that enterprise managers' awareness of the environment can not only identify potential opportunities in the external environment, but also improve the enterprise's ability to innovate in environmental protection, thus improving the sustainability of environmental governance[11]. Wu Lingbing et al. (2023) pointed out that wastewater has caused damage to groundwater resources and aquatic ecological environment, affecting people's life and health[12]. Liu Jing (2023) believes that in the process of environmental management, enterprises can seize any opportunity to carry out continuous innovation and improve their capabilities in the process of continuous innovation and transformation [13].

2.4 Research on the Identification and Management of Enterprise Risk Factors

Ma Yibo (2013) believes that enterprise risk is determined by both internal and external environment of an enterprise, and proposes corresponding management strategies for various risks [14]. Based on the characteristics of enterprise risk management, Arena M et al. (2017) divided enterprise risk into static risk and dynamic risk [15]. Acemoglu D et al. (2015) proposed that enterprise risk status should be evaluated and targeted risk prevention and control should be carried out according to the evaluation results, so as to effectively improve the level of enterprise risk management [16]. Zhang Jing (2018) analyzed the contents of enterprise risk management and the contents and standards of risk assessment and early warning, and took the results of risk assessment as the basis for risk management [17]. Tan Zhijia et al. (2022), through statistical analysis of the collected enterprise risk events, revealed the causes and rules affecting the occurrence of enterprise risk events, and proposed targeted measures and countermeasures for enterprise risk prevention and control [18].

3 Investigation and Analysis of the Solid Waste and Hazard Sources in an Alumina Plant

3.1 Investigation on Identification Degree of the Solid Waste and Hazard Sources

In order to realize how the danger sources and the degree of harm of alumina industry employees, a questionnaire survey was conducted in a large alumina enterprise. In this survey,1000 questionnaires were distributed, to understand employees' awareness of hazard sources and harmfulness. In this survey, a total of 1000 single-choice questionnaires were issued. In terms of the identifying hazardous materials. The results were as follows(See Fig. 1).

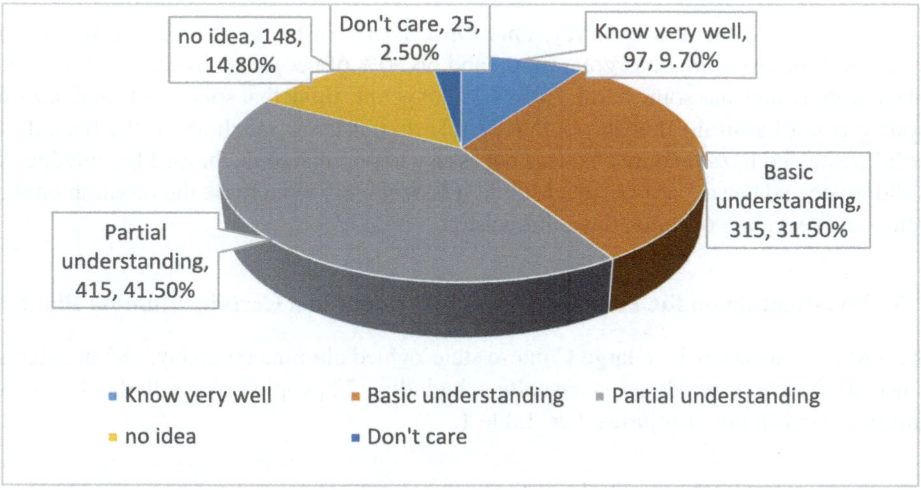

Fig. 1. Survey of the solid waste and hazard Identification.

According to the results of this survey, most employees in the alumina industry do not have a good understanding of hazard sources and hazards, Only 9.7% of people have a good understanding of alumina solid waste and dangerous substances, A staggering 14.8% of people knew nothing about alumina waste and hazardous materials, and it is urgent to change the scurrent situation.

3.2 Investigation on the Degree of Harm Caused by Solid Waste and Hazardous Sources

In terms of the harmful effects from hazardous materials. The results are as follows(See Fig. 2).

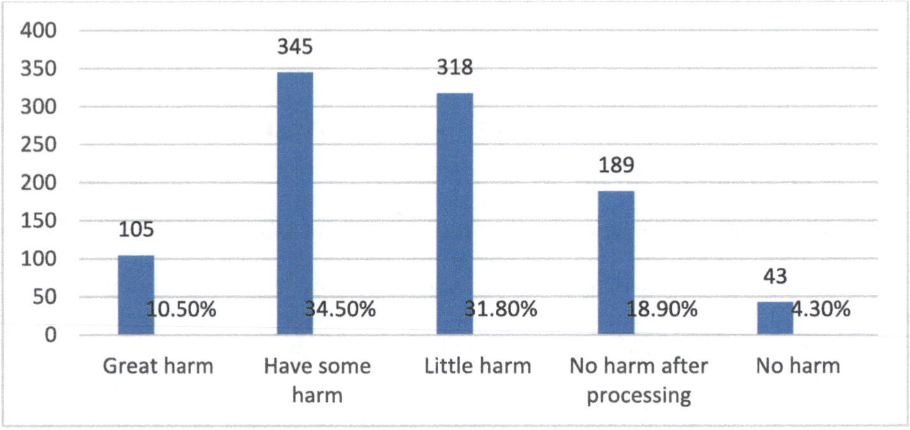

Fig. 2. Survey of the degree of degree of harm

According to the above survey, only Only 10.5% of the people realize that solid waste and hazard source has great harm, and 66.30% of people realize that solid waste and hazard source has some harm, Even 4.3% of people think that solid waste and hazard source is not harmful. This shows that people do not know much about the hazard of solid waste and hazard source, so it is necessary to popularize the hazard knowledge of solid waste and hazard source. In addition, it is necessary to increase the research on the reuse and recycling of industrial solid waste.

3.3 Investigation on the Personnel Injury Accident in a Certain Alumina Plant

According to a survey by a large Chinese state-owned alumina company, 187 accidents since 2002 have resulted in 194 casualties, including 22 people being killed, 14 serious injuries and 158 minor injuries. See Table 1.

Table 1. Summary of casualties of a large state-owned alumina enterprise.

Injury mode	casualties	Death	Serious injury	Minor injury
Object strike	30	1		29
Vehicle injury	14	3		11
Mechanical injury	46	3	4	39
Lifting injury	9	2		7
Get an electric shock	1			1
Scalding	44	3	3	38
Fall from high place	39	4	3	32
Collapse	10	5	4	1
Gunpowder	1	1		
Total	194	22	14	158

Source from: The author collated according to relevant data

4 Identification and Analysis of the Main Solid Waste in Alumina Industry

4.1 Analysis of the Solid Waste Influencing Factors in Alumina Industry

In the process of alumina production, a large number of pollutants such as smoke, waste water and red mud will be produced [19]. The analysis of solid waste influencing factors in alumina industry is shown in Table 2.

Table 2. Analysis of the solid waste influencing factors in alumina industry

Pollutant name	Hazard type	impact to environmental	effects to Health
Ore	dust	To Pollute air	Inhalation of high concentrations of gas is harmful to human health
Caustic alkali	corrosivity	To Pollute water	It burns the human skin and eyes, causes damage to the human respiratory system after inhalation, and can lead to death if accidentally taken

(*continued*)

Table 2. (*continued*)

Pollutant name	Hazard type	impact to environmental	effects to Health
coal gas	explosion hazard	To Pollute air	Inhaling high concentrations of the gas can cause dizziness, poisoning, and nausea
lime	irritability	To Contaminate water	It irritates the eyes, burns the skin, and is harmful to human health when inhaled
flocculant	Corrosiveness and instability	Spills can be harmful to plants and animals	to sear the eyes, irritates the skin, and affects the human respiratory system
Concentrated sulfuric acid	corrosivity	Slight hazard to aquatic life	to sear the skin, irritates the human respiratory system, and causes stomach damage after ingestion
Alkali liquid	corrosivity	To Contaminate water	It burns human skin and eyes and causes harm to human health after inhalation
Aluminium oxide	irritability	Contamination by residue	Cause irritation to the skin and eyes
Aluminium hydroxide	irritability	It is harmful to aquatic plants and animals, and then creates acidic soil	Irritation of the skin, eyes and respiratory system
Red mud with liquid	corrosivity	To Contaminate water	to sear human skin, eyes, if accidentally inhaled can cause human injury

Source from: The author collated according to relevant data

4.2 Reprofit of the Solid Waste of Red Mud in Alumina Industry

This paper takes red mud solid waste as an example to study the reuse of red mud solid waste. In the process of alumina production, a large amount of red mud must be produced. According to Li Na et al. (2020), the composition of red mud varies slightly according to the different sources of bauxite. In the red mud produced by the Bayer process in China, the main components are shown in Table 3 [20].

Table 3. Composition of the red mud from alumina production by Bayer process

Component name	Molecular formula	Content range (%)
hematite	Fe_2O_3	10–30
diaspore	AlOOH	0–5
Hydrated garnet	$3CaO \cdot AlO \cdot xSiO \cdot (6\text{-}2x)H_2O$	4–40
gibbsite	$Al(OH)_3$	2–20
calcite	$CaCO_3$	2–18
calciclinite	$Na_6CaAl_6Si_6(CO_3)O_{24}H_2O$	0–50
perovskite	$CaTi_4O_9$	0–12
magnetite	FeO	0–8

In the process of producing alumina by the Bayer process, the non-uniform composition of the red mud produced is very complex, which is very different from the original metal ore [21]. If the red mud is not transformed and utilized, the storage of a large amount of red mud will occupy precious farmland and land, and require some construction and maintenance costs of the storage yard. Due to the particularity of its composition, once it is leaked, it will cause irreversible pollution to the air, building surface, soil and groundwater. In addition, some radioactive substances such as radium are also found in the minerals contained in some red mud. Endangers the health of people, plants and animals. Therefore, fully exploiting the value of red mud and recycling some metal elements from it is the best way to treat red mud at present.

According to the Northeast Mineral Resources Supervision and Testing Center of the Ministry of Land and Resources of China, the red mud was analyzed by the all-element analysis, the chemical composition can be obtained, as shown in Table 4.

Table 4. The main chemical constituents of red mud

sample	Chemical composition (%)										
Analysis result	8.88	0.42	25.46	22.99	0.082	0.33	4.15	1.3	0.034	22.36	13.88

From the data in Table 5, it can be seen that the content of Al_2O_3 in the red mud used in the test is up to 25.46%, followed by SiO_2 and Fe_2O_3, which are 22.99% and 22.36% respectively, and the sum of the contents of the three is 70.81%, which is more than 70% of the weight of the red mud, and the value of extraction and utilization is very high. At the same time, the content of Na_2O and CaO are more than 4% respectively, and it is also possible to extract and recycle.

4.3 Reduction of the Aluminum and Iron in Red Mud by Concentrated Hydrochloric Acid

According to Chen Laibao (2009), aluminum in red mud mainly exists in sodalite-type aqueous sodium alumino-silicate and illite. They react with concentrated hydrochloric acid, and aluminum enters the filtrate in the form of trivalent aluminum ions [22]. The reaction equation is as follows:

$$\begin{aligned} &Na_{7.6}(Al_6Si_6O24)(CO_3)_{0.93}(H_2O)_{2.93} + 25.76HCl = \\ &7.6NaCl + 6AlCl_3 + 6SiO_2 + 0.93CO_2 + 15.86H_2O \end{aligned} \tag{1}$$

$$KAl_2Si_3AlO_{10}(OH)_2 + 10HCl = KCl + 3AlCl_3 + 3SiO_2 + 6H_2O \tag{2}$$

In addition, Chen Laibao (2009) believes that iron in red mud mainly exists in hematite (α-Fe_2O_3). They react with concentrated hydrochloric acid and the iron enters the filtrate in the form of trivalent iron ions. The reaction equation is as follows:

$$Fe_2O_3 + 6HCl = 2FeCl_3 + 3H_2O \tag{3}$$

Most metal ions can form hydroxide precipitation, and if the solubility of the precipitation is very different, it is possible to separate them by controlling the acidity, To achieve the purpose of reducing aluminum and iron elements in red mud.

4.4 Reduction of the Aluminum, Iron and Silicon in Red Mud by High Temperature Roasting and Sodium Salt

Zhuang Jinqiang (2012) concluded that during the reduction of aluminum, iron and silicon oxides by roasting at high temperature, not only the iron oxides undergo a reduction reaction from high to low iron valence, but also the generated FeO can undergo solid phase reaction with SiO_2 and $A1_2O_3$ [23].

$$FeO + Al_2O_3 = .FeO \cdot Al_2O_3 \tag{4}$$

$$2FeO + SiO_2 = 2FeO \cdot SiO_2 \tag{5}$$

$$1/2Al_2O_3 + SiO_2 = (1/2)Al_2O_3.2SiO_2 \tag{6}$$

In addition, Zhuang Jinqiang (2012) concluded that aluminum and silicon oxides in the red mud of alumina may react with sodium salts (sodium carbonate and sodium sulfate) in the reduction process as follows:

$$N_{a2}SO4 + Al_2O_3 = N_{a2}O \cdot Al_2O_3 + SO_2 + 1/2O_2 \tag{7}$$

$$N_{a2}SO_4 + 1/2SiO_2 = 1/2(2Na_2O \cdot SiO_2) + SO_2 + 1/2O_2 \tag{8}$$

Through the above high-temperature roasting chemical reaction, the insoluble aluminum, iron and silicon oxides in the red mud can be reduced to soluble aluminum, iron and silicon oxides, so as to achieve the purpose of recycling.

5 Analysis of the Hazard Source Identification and Control Innovation in Alumina Industry

5.1 Hazard Source Identification and Harm Analysis of Alumina Industry

According to the summary, the hazard sources and hazards in the alumina industry are shown in Table 5.

Table 5. Analysis of the hazard sources and harm in alumina industry

Major hazard	Possible harm
Fire hazard	The types of fuels used in calcining equipment such as lime shaft furnace, rotary kiln, roaster and molten salt furnace are: gas, natural gas, heavy oil, diesel and so on. These fuels are inflammable and explosive substances and are important fire hazards. In the process of storage, transportation and use, accidental leakage, or improper control, easy to cause fire or explosion. In addition, some transformer, distribution system if short circuit, aging facilities, overload or improper operation and other faults, can also cause fire
Blow up	Alumina production system mainly uses pressure vessels and pressure pipelines. The pressure in the container or pipe may be too high and the pressure cannot be released, resulting in an explosion. Some dust collectors may encounter ignition sources or naturally cause combustion explosions due to the accumulation of combustible dust and combustible gases
Scalding	There is high temperature lye in containers, equipment and pipelines in the main processes of alumina production. Because the lye is highly corrosive and flows at high speed, when the equipment and pipes are corroded, or the connection points and sealing points fail, the running slot, the trough, the leakage or the high temperature lye sprayed under pressure may cause burning to people
Mechanical injury	Various mechanical equipment in alumina production such as: Transport machinery, conveying machinery, loading machinery, pumps, motors, crushing machinery, mills, air compressors, blowers, other parts or devices of machinery that rotate or reciprocate at high speed in each equipment, due to inadequate protective measures, poor protection or no protection, or in special circumstances such as accidents and maintenance, It may cause accidental clamping, collision, shearing, entanglement, hinge grinding, cutting and other mechanical injuries to operation, maintenance and inspection personnel

(*continued*)

Table 5. (*continued*)

Major hazard	Possible harm
Vehicle injury	Vehicles transporting alumina production raw materials and finished alumina products and motor vehicles in the plant drive in the factory, due to driver violations, road or vehicle defects, resulting in vehicle injuries
Toxic asphyxia	As a fuel, carbon monoxide in gas, methane in natural gas and carbon monoxide, hydrogen sulfide and other toxic gases produced in lime firing process accidentally leak, which is easy to cause human poisoning and asphyxia
Noise and vibration	A large number of motors, mills, Roots fans, air compressors, diaphragm pumps, high pressure pumps, slurry flash tanks and other equipment with continuous noise are used in the device, including fluid power noise generated by fans, air pipes, liquid pumps, mechanical noise generated by movement, friction, impact, vibration of mechanical equipment and electromagnetic noise generated by electric equipment such as motors and transformers
Dust	Aluminum oxide plant in the production of alumina under normal circumstances will produce a large amount of dust, dust refers to the diameter of the air in the range of 10 microns to 300 microns of solid micro-particles, long-term contact with high concentrations of dust can easily lead to "silicosis" and other respiratory diseases. In addition to harming the health of the staff, it will also cause damage to the ecological environment [24]. In the process of conveying aluminum hydroxide to baking furnace, there is a hazard of aluminum hydroxide dust. The output of baking furnace is powdery alumina, and there are hazards of alumina dust in the process of conveying, warehousing and packaging

Source from: The author collated according to relevant data

5.2 Detection and Analysis of Noise and Dust in Production Position of an Alumina Enterprise

This paper takes noise and dust as an example to detect the noise and dust in the production system of an alumina enterprise. The test results are shown in Tables 6 and 7.

From Tables 6 and 7, it can be seen that the qualified rate of noise in the production position of this alumina enterprise is only 67.91%, and the qualified rate of boiler system is only 57.7%. The qualified rate of dust in the post is only 82.1%, and the qualified rate of dust in the alumina packaging post is only 70.6%. Noise and dust are not up to standard. Will have adverse consequences for the physical and mental health of employees.

In short, as long as there is a danger, it is necessary to strictly control the danger source to ensure the safe and stable production of alumina enterprises and achieve sustainable development.

Table 6. Qualified rate of noise detection in production system of an alumina enterprise

Production system	Number of detection points	testing result [dB(A)]	Contact time(h)	Qualified rate(%)
Mining	30	75.2–96.4	0.5–6	76.7
Raw fuel preparation	30	57.0–105.0	0.5–4	66.7
Pulp grinding	27	63.4–98.0	0.5–4	88.9
High pressure stripping	8	54.1–102.3	0.5–1	75
mother liquor evaporation	15	66.5–101.5	0.5–4	93.3
Red mud settlement	10	68.7–90.3	0.5–2	90
Filtration and roasting	18	58.1–97.3	0.5–8	100
Alumina packaging	3	80.5–91.3	0.25–8	100
Boiler system	26	68.2–106.7	0.5–6	57.7
Thermodynamic system	15	67.0–96.9	0.5–1	93.3
total	215			67.91

Table 7. Qualified rate of dust in production system of an alumina enterprise

Production system	TWA(Time Weighted Average)			STEL (Short-term exposure limit)		
	Number of detection positions	testing result (mg/M^3)	Qualified rate (%)	Number of detection points	testing result (mg/M^3)	Qualified rate (%)
mining	11	1.6–5.7	100	42	1.1–16.5	90.5
Raw fuel preparation	7	0.5–7.9	100	44	1.2–26.8	72.7
Pulp grinding	6	1.5–13.6	50	42	2.6–19.3	88.1
Filtration and roasting	1	4.1	100	12	1.5–4.8	100

(*continued*)

Table 7. *(continued)*

Production system	TWA(Time Weighted Average)			STEL (Short-term exposure limit)		
	Number of detection positions	testing result (mg/M^3)	Qualified rate (%)	Number of detection points	testing result (mg/M^3)	Qualified rate (%)
Alumina packaging	5	2.3–3.7	100	17	3.3–24.1	70.6
Boiler system	9	0.8–9.1	77.8	44	0.3–44	77.3
total	39		87.2	201		82.1

5.3 Innovative Measures to Control Hazard Sources and Their Hazards of Alumina Industry

According to the hazard resources existed in the process of alumina production, it should be solved by management innovation.

Adopting Advanced Technology and Equipment, Eliminating and Reducing the Risk Fundamentally
Modern alumina enterprises continue to introduce new equipment and new processes. Only the process safety management will be used for scientific and reasonable analysis, to develop feasibility: safety management measures. In order to meet the safety management requirements of alumina enterprises [25].

Adopting advanced process technology and equipment could fundamentally overcome inherent and hard-eliminated hazard resources of original process, such as fire, explosion, high-temperature, dust, smoke fog and so on, which could reduce labor intensity, increase operating stability and improve the safety and sanitary conditions fundamentally.

Adopting the Automatic Control, Reducing the Accidental Operation
The high safety control equipments with high automation, microcomputer control and DCS (distributed control system) are adopted. Because of the integrative unified management of the central control room, the main process parameters of temperature, pressure, density, flow rate and material level are automatically detected, record, adjusted and controlled, which are more exact and timely than the manual control parameters. It effectively reduced faults of manual operation and decreased the frequency, time and range of operator reaching high temperature, high pressure and high alkaline by the greatest degree.

Installing the Detecting, Alarming and Linking System and the Emergency Schedule, Effectively Controlling the Hazards in Time
According to the need of process and safety, the important parameters of main equipments are on-line data detection, over-limit sound and light alarming, automatic protection & interlock control, emergency shutdown system, safety valve, emission vessel,

spill prevention cofferdam and so on. In the early production process of alumina production, a set of optimization control system is built to effectively improve the working efficiency of the process, save energy consumption, reduce the number of operators and work intensity, and reduce abnormal parking on site [26].

Improving the Safety and Occupation Protection, avoiding Accident and Occupation Hazard
There should be necessary guard rod, platform, ladder stand and cover board for the hole, groove, well and top on site; the lightning arrester for the high buildings and distribution facilities; fine insulation and earthing device for electric system;fire extinguishing system and equipment for the critical parts for fire control safety; some necessary safe marks for sites.

For the dust resource and poison resources of coal gas, acid and alkaline, there should be sprinkler, sealing, dust collector and ventilation; for the noised equipments, such as aircompressor, there should be isolated foundation and isolated machine room to reduce the propagation of noise and vibration, and isolated operating room, sound insulation room, control room, double glassed windows and double doors; for the radioactive source, the permanent or temporary shielding measure and alarming measure on site should be used.

Carrying on Each Safety Countermeasure, strengthening Accident Protecting Ability
During the daily production, each safe management measure should be carried out, which could prevent all kinds of accidents better. The main safe managing measures include: improving the regulation of each safe management and executing it strictly; establishing safe managing organization and carrying out responsibility system for safety in production; developing focused safety education training and safety inspection; insuring the safety input; preparing protective articles for labor insurance and strengthening individual protection; establishing rescuing organization and drafting emergency plans;enforcing the implementation of safety monitor of major hazards and inspection safety management.

6 Conclusion

In summary, alumina production will produce a large amount of solid waste, there will be more dangerous sources, with higher risks, greater harm, should cause the alumina industry to attach great importance to.

However, through the above research, the recycling of solid waste such as red mud can be realized, and the hazard sources can also be effectively controlled, so as to achieve the safe, stable and sustainable development of the alumina industry.

Acknowledgments. This paper is the phased result of the following project:

Project name: Research on the coupling development of bauxite resource economy between China and Laos under the background of the Belt and Road Initiative. Project principal: Jingwu Yao. Project number: G2020SK015.

References

1. Xia, G.J.: Analysis on safety management of alumina production process. China Sci. Technol. Info. (20), 114–115 (2009)
2. Yan, Z.: Analysis of process safety management in modern alumina enterprises. Chemical Design Bulletin **46**(02), 78+81 (2019)
3. Shiyou, G., Zhilin, W., Xiaoyu, F.: Discussion on safety management of alumina production equipment. World Nonferrous Metals **10**, 31–33 (2016)
4. Lin, W., Chenyu, W.: Discussion on the development status of alumina enterprises and research on safety production management measures. World Non-ferrous Metals **16**, 153–154 (2021)
5. Pengchong, L.: Study on environmental protection management countermeasures in alumina production process. Low Carbon World **06**, 35 (2018)
6. Xinzhan, T.: Analysis of environmental protection management measures in alumina production. Shanxi Metallurgy **39**(06), 125–126 (2016)
7. Ye, J.: Study on comparison and selection of red mud storage site discharge and dam stacking mode for an alumina project. Nonferrous Metals and Metallurgy **39**(03), 58–61 (2019)
8. Yuping, G.: Analysis of technical application and key points of implementation in hydraulic environmental geological exploration. World Nonferrous Metals (05), 214–215 (2019)
9. Shiwen, B., Haiyan, Y., et al.: Alumina Production Process, 2nd edition, pp. 8–12. Chemical Industry Press (2006)
10. Hart, S.L.: A natural resource based view of the firm. The Academy of Management Review **20**(4), 986–1014 (1995)
11. Polas, H., Tabash, M.I., Bhattacharjee, A., et al.: Knowledge management practices and green innovation in SMES: the role of environmental awareness towards environmental sustainability. Int. J. Organizat. Analy. **14**, 196–218 (2021)
12. Lingbing, W., Jia, J.: Problems and solutions of ecological environment protection in the production process of chemical enterprises. Liaoning Chemical Industry. **52**(08), 1161–1164 (2023)
13. Jing, L.: Analysis on the importance of enterprise environmental management to enterprise management. Leather Making and Environ. Technol. **4**(14), 165–167 (2023)
14. Yibo, M.: Research on growth risk warning of knowledge-intensive new ventures. Harbin Engineering University, Harbin (2013)
15. Arena, M., Arnaboldi, M., Palermo, T.: The dynamics of (dis)integrated risk management: a comparative field study. Accounting Organizations And Society **62**, 65–81 (2017)
16. Acemoglu, D., Ozdaglar, A., Tahbaz-Salehi, A.: Systemic risk and stability in financial networks. American Economic Review **105**(2), 564–608 (2015)
17. Jing, Z.: Research on financial risk identification and Evaluation of technology-based smes under the background of innovation drive. Yunnan University, Yunnan (2018)
18. Zhijia, T., Qilu, Z., Wuxiang, Z., et al.: From finance to entity: Microcontagion mechanism of liquidity risk and prevention measures based on a multi-case study of SME financing guarantee industry. Manage. World **38**(03), 35–58 (2022)
19. Zongzhi, W., Jindong, G., Lijun, W.: Risk assessment method and its application, pp. 58–62. Beijing Meteorological Press (2001)
20. Na, L., Ji, L., et al.: Research on recycling technology based on Alumina red mud by Bayer Process. World Nonferrous Metals. **19**, 1–2 (2020)
21. Shihong, L., Xiaolian, Z., et al.: Experimental study on preparation of wear resistant white cast iron from red mud by bayer process. Foundry **01**, 29–34 (2020)
22. Laibao, C.: Study on recycling process of red mud. Liaoning Technical University, No.03 (2012)

23. Jinqiang, Z.: Research on iron recovery technology from high iron alumina red mud. Central South University, No. 02 (2013)
24. Bin, L., Yanjun, W.: Hazards and treatment of dust pollution in alumina production. World Non-ferrous Metals (20), 273+275 (2019)
25. Bin, W.: Study on process safety management measures of alumina enterprises. Metallurgy and Materials. **41**(02), 15–16 (2021)
26. Guoliang, C.: Optimization control of alumina grinding process based on an aluminum mill. Min. Metall. **30**(05), 134–140 (2021)

Simulation Analysis and Experimental Verification of the Fragmented State of Retired Crystalline Silicon Solar Panels

Jialin Zhan[1,2](✉), Hui Chang[1], Xinbai Xue[1], and Fujing Zhou[1]

[1] Department of Intelligent Manufacturing, Anhui Vocational and Technical College, Hefei, China
954168378@qq.com

[2] Department of Materials Science and Technology, Nanjing University of Aeronautics and Astronautics, Nanjing, China

Abstract. This study focuses on the theoretical exploration and empirical investigation of the physical fragmentation method for photovoltaic (PV) modules. It aims to delve into the mechanism of PV module physical damage, elucidate its fragmentation principles, and validate the accuracy of theoretical models. The research examines the mechanical properties of PV module materials under stress and investigates the transfer and consumption of energy during the fragmentation process. Additionally, a physical simulation model of PV module damage is constructed to provide theoretical support for selecting fragmentation equipment and optimizing process parameters. A specific PV module physical fragmentation experimental apparatus was chosen, and extensive experiments were conducted using discarded PV modules as research samples. Various parameters related to the fragmentation equipment and fragmentation methods were adjusted during the experiments. The accuracy of the theoretical models was confirmed through SEM and EDS analysis of the experimental data. It was found that different types of PV modules exhibit systematic fragmentation characteristics and compositional changes during the fragmentation process. Moreover, the parameters and conditions of the fragmentation equipment significantly enhance the fragmentation results and recovery rates. This study provides empirical evidence and scientific basis for the sustainable recycling of PV module waste, underscoring its theoretical and practical significance in advancing the utilization of discarded PV resources.

Keywords: physical fragmentation · theoretical models · SEM and EDS analysis

1 Introduction

With the widespread application and accelerated upgrading of photovoltaic modules Traditional methods such as landfilling and incineration not only occupy significant land resources but also risk causing secondary environmental pollution. Therefore, seeking an efficient and environmentally friendly method for recycling and processing waste PV modules has become a research hotspot [1]. The physical fragmentation method for PV

D. Han et al. (Eds.): ICEPG 2024, ESE, pp. 593–601, 2025.
https://doi.org/10.1007/978-3-031-86829-0_68

modules is considered a potential waste management strategy [2]. This method primarily utilizes mechanical force to crush discarded PV modules into various sizes of fragments, thereby facilitating the separation and recovery of their constituent parts [3].

Previous studies have primarily focused on recovery technologies involving thermal treatment of ethylene-vinyl acetate (EVA) (Chitra et al., 2020) and wet metallurgical processes for decommissioned PV modules, including physical pretreatment processes, silicon leaching depths, and chemical processes (Ran Tao, 2023) [4]. The pretreatment process generally includes dismantling (Antonio Rubino et al., 2021; Qing Huang et al., 2021) [5]. Nearly all previous works have started from mechanical pretreatment processes such as crushing, screening, and laser and have involved dismantling or separating different components from retired modules (Teng-Yu Wang et al., 2024) [6].

This paper explores the feasibility of PV module recycling from two perspectives: the stress on PV module cells during crushing and the energy consumption involved. Through Ansys Workbench simulations of various common states of PV module cell crushing, the relationship between crushing efficiency and pressure is determined. Finally, crushing experiments are conducted to validate these simulations. SEM and EDS analysis of the crushed materials reveal good material states and clear component proportions, highlighting their positive role in subsequent separation and reuse of PV module components.

2 Photovoltaic Module Physical Crushing Method: Theoretical Analysis and Recycling Route

2.1 Fundamental Principles of Physical Crushing Method

The fundamental principle of the physical crushing method involves using mechanical force to break down materials, transforming them from large pieces into small fragments or granular materials. This process can be deduced using basic formulas of mechanics and energy conversion [7]. The following is a simplified derivation process:

Firstly, during the physical crushing process, an external force FFF is applied to the material. This force exerts stress δ on the material [8], which can be expressed as:

$$\delta = \frac{F}{A} \tag{1}$$

Here, A represents the cross-sectional area subjected to force. This stress causes the material to undergo strain ϵ, which is given by:

$$\varepsilon = \frac{\Delta L}{L} \tag{2}$$

where ΔL is the deformation and L is the original length.

When the material is stressed, it stores elastic potential energy. According to Hooke's Law, the elastic potential energy U can be represented as:

$$U = \frac{1}{2} \times \delta \times \varepsilon \tag{3}$$

Here, U is the elastic potential energy, δ is the stress, and ϵ is the strain.

In the physical crushing process, the external force performs work W, defined as:

$$W = F \times d \tag{4}$$

where W is the work done, F is the external force, and ddd is the displacement.

According to the law of conservation of energy, the work done by the external force equals the change in internal deformation energy ΔU plus the change in kinetic energy ΔK

$$W = \Delta U + \Delta K \tag{5}$$

Here, ΔU represents the change in internal deformation energy, and ΔK is the change in kinetic energy.

2.2 Energy Consumption and Efficiency Analysis During Crushing Process

In the physical crushing process of photovoltaic modules, analyzing energy consumption and efficiency is crucial for evaluating process performance and economic benefits. Meanwhile, crushing efficiency directly relates to the rate of material fragmentation, particle size distribution, and final recovery rate [9].

Energy consumption E during the physical crushing of photovoltaic modules can be expressed as:

$$E = P \times t \times \eta \tag{6}$$

Here, P denotes the actual power of the crushing equipment in watts (W), t is the crushing time in seconds (s), and η is the energy utilization efficiency in percentage (%).

The crushing efficiency η is given by:

$$\eta = Q/(P \times t) \tag{7}$$

where Q is the amount of material processed per unit time during crushing in kilograms per second (kg/s).

Particle size distribution can be evaluated using the formula:

$$D = \sqrt{D_1^2 + D_2^2} \tag{8}$$

Here, D represents the particle size after crushing in millimeters (mm), and D_1 and D_2 are the particle sizes before crushing in millimeters (mm).

Finally, the recovery rate R is expressed as:

$$R = m_1/m_0 \tag{9}$$

where m_1 is the recovered quantity of valuable material in kilograms (kg), and m_0 is the total material quantity before crushing in kilograms (kg).

These formulas and analytical methods contribute to assessing the efficiency and economic viability of the physical crushing method for recycling photovoltaic modules, offering theoretical guidance and technical support for optimizing the crushing process.

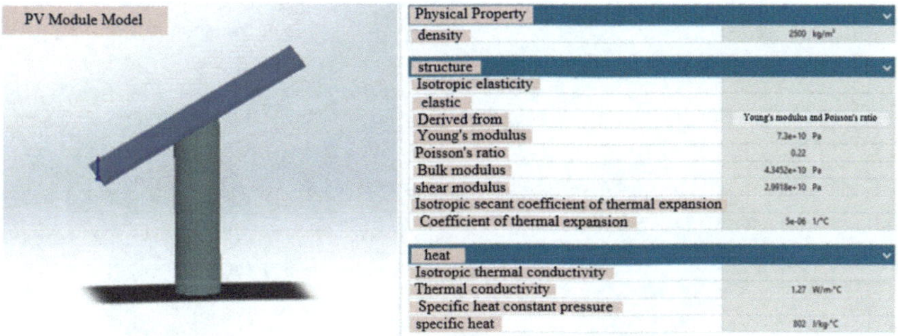

Fig. 1. PV Module Model and Material Properties

3 Finite Element Analysis of the Fragmentation State of Solar Panels

To study the equivalent stress, load, and constraints on the surface of photovoltaic modules during physical crushing, this study utilized Ansys Workbench finite element analysis software to analyze the solar panel components. The finite element analysis process consists primarily of preprocessing and postprocessing. Preprocessing involves defining part parameters, meshing, and applying constraints and loads,as shown in Fig. 1. Post-processing includes analyzing maximum deformation, maximum principal stress, and equivalent stress [10]. Based on literature review, it was determined that crushers exert pressures ranging from 3000 Pa to 5000 Pa during crushing operations. In this study, pressures of 3000 Pa, 4000 Pa, and 5000 Pa were applied to simulate the pressures experienced by photovoltaic modules during physical crushing. Initially, a three-dimensional model of the photovoltaic module was created, which was then imported into Ansys Workbench. Material properties of the photovoltaic module were defined as shown in Figs. 2, 3 and 4.

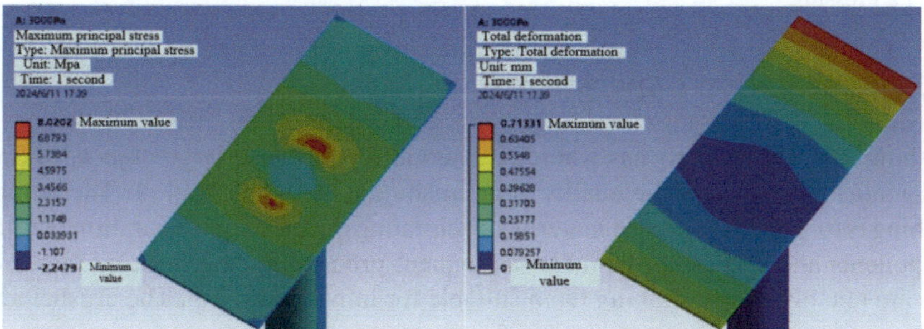

Fig. 2. Contour Plots of Maximum Principal Stress,and Maximum Total Deformation under 3000 Pa Pressure

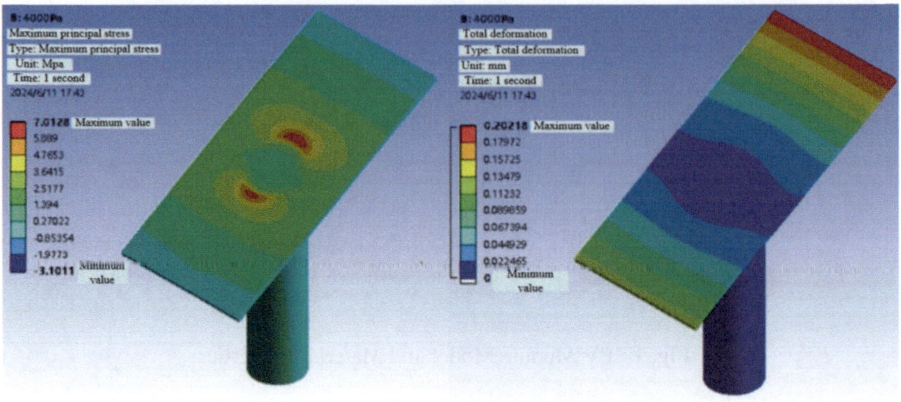

Fig. 3. Contour Plots of Maximum Principal Stress and Maximum Total Deformation under 4000 Pa Pressure

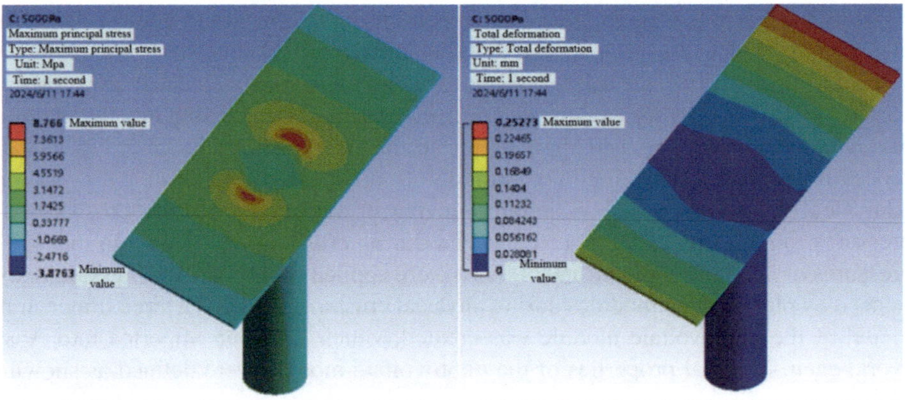

Fig. 4. Contour Plots of Maximum Principal Stress and Maximum Total Deformation under 5000 Pa Pressure

4 Fragmentation Experiments

Crushing refers to the process whereby external forces cause large objects to fracture into smaller pieces [11]. The quality after crushing is shown in Table 1. The purpose of crushing is to reduce the size of materials, facilitating the thorough separation of different components and preparing them for subsequent processing stages. Solar photovoltaic cells are notably brittle, making them suitable for impact crushing. The crusher and the crushing process are illustrated in Fig. 5.

Table 1. Sieve Analysis Results

Size (mm)	Weight (g)	Percentage (%)
<45	0.1	1.01
45–60	0.62	6.28
60–80	0.86	8.72
80–150	6.02	61.05
>150	2.26	22.94
Total	9.86	100

Fig. 5. Experimental materials and tools involve placing the solar panels into the fragmentation machine.

5 SEM and EDS Analysis

To determine the elemental composition of the battery powder, samples ranging from 80 to 150 mesh were subjected to quantitative chemical element analysis after physical crushing. The analysis revealed that silicon accounted for 51.3% of the composition, while carbon and oxygen constituted 33.3% and 9.6%, respectively. Trace amounts of other elements such as iron and aluminum were also detected, albeit at lower levels [12].

Further investigation into the phase composition and elemental distribution of the battery powder was conducted using SEM-EDS analysis [13]. As depicted in Fig. 6, the powder predominantly consisted of silicon, appearing in blocky structures. Oxygen was mostly present in granular forms, while carbon exhibited a flaky morphology. Other elemental constituents were negligible.

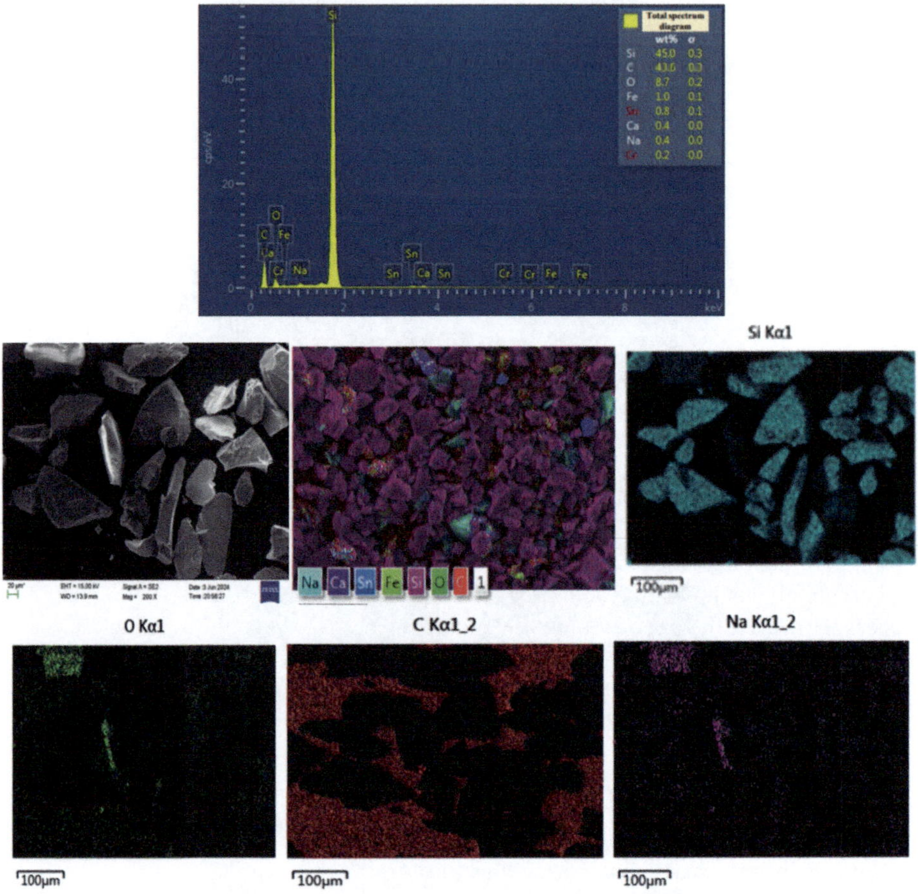

Fig. 6. SEM-EDS Elemental Analysis Result

6 Summary

Using fundamental principles of mechanics and energy conversion, explore the physical fragmentation mechanism of photovoltaic module crushing. During the crushing process, mechanical forces generate energy transfer and conversion in the material, with some energy used for material fragmentation and the remainder dissipated as heat and sound. Mechanistic studies contribute to optimizing the energy efficiency of crushing equipment during the implementation of physical fragmentation methods.

Through analysis of the equivalent stress, load, and constraints experienced by photovoltaic modules under different pressures during crushing, it is evident that higher pressures lead to more pronounced fragmentation effects on the modules. When external pressure exceeds the module's load-bearing limit, severe or complete fragmentation of the components is likely. Increasing pressure results in concentrated internal stresses within the modules, thereby intensifying the extent and range of fragmentation.

To ascertain the elemental composition of the battery powder obtained from experiments, samples ranging from 80 to 150 mesh were subjected to quantitative elemental analysis. SEM-EDS was employed to examine the battery powder, further confirming its phase composition and elemental distribution. The battery powder is predominantly composed of silicon elements in block form. Oxygen elements are primarily granular, while carbon exhibits a sheet-like appearance, indicating good condition and complete granularity, which facilitates subsequent deep separation and purification processes.

Funding. The research received the following financial support: the Natural Science Research Projects of Universities in Anhui Province (2022AH052063, 2023AH051438, 2023AH040199, 2023AH052102, 2024AH050871), the Quality Engineering Project of Anhui Province (2022kcsz156), and the Vocational and Adult Education Association Project of Anhui Province (AZCJ2023060).

References

1. Lee, S., Lee, K.D., Bae, H.S.: Prevention of potential-induced degradation using a moisture barrier in crystalline silicon photovoltaic modules. Progress in Photovoltaics **32**(6), 390–398 (2024)
2. Yao, L., Zhang, W., Huang, F., et al.: Experimental and simulation study on the thermoelectric performance of semi-transparent crystalline silicon photovoltaic curtain walls. J. Build. Eng. 90 (2024). https://doi.org/10.1016/j.jobe.2024.109421
3. Sah, D., Kumar, S.: Experimental, cost and waste analysis of recycling process for crystalline silicon solar module. Solar Energy, 273 (2024). https://doi.org/10.1016/j.solener.2024.112534
4. Li, X., Liu, H., You, J., et al.: Back EVA recycling from c-Si photovoltaic module without damaging solar cell via laser irradiation followed by mechanical peeling. Waste Management **137**, 312–318 (2022). https://doi.org/10.1016/j.wasman.2021.11.024
5. Husain, D., Tewari, K., Sharma, M., et al.: Ecological Footprint of Multi-silicon Photovoltaic Module Recycling. (2022). https://doi.org/10.1007/978-981-16-8426-5_3
6. Urbina, A.: Sustainability of photovoltaic technologies in future net-zero emissions scenarios. Progress in Photovoltaics (12), 31 (2023). https://doi.org/10.1002/pip.3642
7. Azeumo, M.F., Conte, G., Ippolito, N.M., Franco, M., Luigi, P., Settimio, S.: Photovoltaic module recycling, a physical and a chemical recovery process. Sol. Energy Mater. Sol. Cells **193**, 314–319 (2019). https://doi.org/10.1016/j.solmat.2019.01.035
8. Ohnishi, Y., Takahashi, Y., Fujiwara, K.M.H.: Modeling of I-V characteristics for crystalline silicon photovoltaic modules based on a simplified equivalent circuit and a temperature correction. Electrical Engineering in Japan **215**(4) (2022)
9. Cucchiella, F., D'Adamo, I., Rosa, P.: End-of-life of used photovoltaic modules: a financial analysis. Renew. Sustain. Energy Rev. **47**, 552–561 (2015). https://doi.org/10.1016/j.rser.2015.03.076
10. Yin, A., Zhang, Z., Jin, X., et al.: Carbon emission analysis of two crystalline silicon components throughout the life cycle. IOP Conference Series: Earth and Environmental Science, p. 983 (2022). https://doi.org/10.1088/1755-1315/983/1/012111

11. Fiandra, V., Sannino, L., Andreozzi, C., Corcelli, F., Graditi, G.: Silicon photovoltaic modules at end-of-life: Removal of polymeric layers and separation of materials. J. Waste Manag. **87**, 97–107 (2019). https://doi.org/10.1016/j.wasman.2019.02.004
12. Sharma, A., Mahajan, P., Garg, R.: End-of-life solar photovoltaic panel waste management in India: forecasting and environmental impact assessment. Int. J. Environ. Sci. Technol. (2), 21 (2024)
13. Dobra, T., Vollprecht, D., Pomberger, R.: Thermal delamination of end-of-life crystalline silicon photovoltaic modules. Waste Management & Research **40**(1), 96–103 (2022). https://doi.org/10.1177/0734242X211038184

Effect of Humic Acid on Saline Soil

Yanning Li, Wenwen Li(✉), Yanchao Yin, Fan Zhang, Hongbin Sun, and Ying Xu

Fifth Exploration Team of Shandong Coal Geological Bureau, Jinan, China
553649828@qq.com

Abstract. Adding humic acid of different concentrations to saline soil in Hekou District, Dongying, to collect and analyze soil samples. Through statistical analysis, it was found that the addition of humic acid not only reduced the SAR (sodium adsorption ratio) and total salt, but also promoted the supply of nitrogen, phosphorus and potassium and the absorption of organic matter. In addtion, the addition of humic acid to the saline soil reduced the Cr in the soil from 0 to 20 cm and the soil from 20 to 40 cm, but may cause Hg, Cu, Pb pollution in the soil.

Keywords: Humic acid · saline soil · soil

1 Introduction

In Hekou District, Dongying province, Saline soil is one of the main soil types in Dongying Hekou District, which has been affected by seawater intrusion during the land formation process. Saline soil has some bad physical and chemical properties, such as high salt content, high pH, poor soil structure, etc. [1]. Which seriously affects crop yield and even causes plant death. Soil salinization not only causes the destruction of land resources and the loss of agricultural production, but also threatens the biosphere and ecological environment. Humic acid is a kind of macromolecular organic weak acid mixture with many functional groups, which is formed and accumulated from animal and plant remains through a series of microbial decomposition and transformation, as well as a series of geochemistry and physical changes. The active functional groups make humic acid acidic, hydrophilic, interfacial activity, cation exchange ability, complexation and adsorption and dispersion ability, these characteristics make it have a good role in promoting crop growth [2–7]. Therefore, the field experiment of different concentrations of humic acid was carried out in the saline-alkali soil of Hekou District, Dongying province, and the effects of humic acid on soil were analyzed.

2 Materials and Methods

2.1 Time and Place of the Trial

Field trials were conducted in Hekou District, Dongying between July and November of the 2023. Dongying Hekou District is located in the middle latitude, the back land surface sea, by the Eurasian continent and the western Pacific common influence. Hekou

D. Han et al. (Eds.): ICEPG 2024, ESE, pp. 602–608, 2025.
https://doi.org/10.1007/978-3-031-86829-0_69

District, Dongying belongs to the warm temperate continental monsoon region, and there have significant differences between spring, summer, autumn and winter. Dongying Hekou District has an average annual temperature of 12.8 °C, and the annual average precipitation is 598.1 mm.

2.2 Test Materials

The soil tested was saline soil in Hekou District, Dongying. The pH of soil background value was 8.3. The Exchange Sodium Percentage of soil background value was was 11.8. And the total salt content of soil background value was 1.3 g·Kg^{-1}. The tested humic acid was renate alginate + peat source humic acid potassium, provided by REHNERT (Qingdao) Import & Export Co., Ltd..

2.3 Test Design

The experiment was controlled by field microzone, the length of microzone was 8 m, the width was 5 m, the area was 40 m^2. In order to prevent water and fertilizer from mixing in each microzone, an isolation zone of 3M was set in each zone.

The experiment design is 5 treatments, each treatment is 3 replicates, the experiment micro-area is set randomly. F0 is the soil without humic acid. F1 is the soil with humic acid concentration of 250 kg hm^{-2}. F2 is the soil with humic acid concentration of 350 kg hm^{-2}. F3 is the soil with humic acid concentration of 450 kg hm^{-2}. F4 is the soil with humic acid concentration of 550 kg hm^{-2}. A total of 15 test plots are required. In order to ensure the uniformity of humic acid and avoid the loss, the soil was mixed with humic acid and sprinkled into the corresponding test microzone.

In addition to applying humic acid, other agronomic measures and field management measures, such as sowing, fertilization, disease control and weeding, were applied at the same level.

2.4 Collection and Determination of Test Samples

When humic acid is added about 100 days, three to five sampling points were set up in each test area, and soil samples were collected from 0 to 20 cm, 20 to 40 cm, 40 to 60 cm below the surface by multi-point mixed sampling method. The soil pH, organic matter, available nitrogen, available phosphorus, available potassium, total salt content, cation exchange capacity, Na^+, Exchangeable Na, Cl, Ca, Mg, Cr, Hg, Cu and Pb were analyzedby using the Inductively coupled plasma emission spectrometer Agilent 5800, and UV visible photometer T9CS.

3 Results and Analysis

3.1 Effect of Humic Acid on Soil Salinity

Based on the pH value, exchange sodium percentage, total salt, SAR (sodium adsorption ratio) of soil has been analysed, origin was used to draw Fig. 1.

According to Fig. 1(a), it can be seen that the pH value of soil from 0 to 20 cm is higher and that of soil from 40 to 60 cm is lower. The pH value of soil F1, F2, F3, F4 was not lower than that of soil F0. The results showed that soil F1, F2, F3, F4 added humic acid could not decrease soil pH. According to Fig. 1(b), the exchange sodium percentage of soil F1 from 40 to 60 cm was lower than that of soil F0. But the other soils was not lower than that of soil F0. So soil F2, F3, F4 added humic acid could not reduce the Exchange Sodium Percentage of soil. According to Fig. 1(c), it can be seen that the total salt content of the soil from 0 to 20 cm is higher, the soil from 20 to 40 cm is lower, and the soil from 40 to 60 cm is the highest. The total salt content of soil F1 and F4 was lower than that of soil F0. In addition, the total salt content of soil F2 and F3 was lower than that of soil F0. Therefore, the soil added humic acid can effectively reduce the total salt content of soil. According to Fig. 1(d), the SAR(sodium adsorption ratio) in the soil from 0 to 20 cm and the soil from 40 to 60 cm were the highest, the soil from 20 to 40 cm was the lowest. The SAR of profile soil F1 was lower than that of profile soil F0. In addition, The SAR in the soil F4 and F2 from 40 to 60 cm was lower than that of soil F0. Therefore, the soil F1, F2, F4 added humic acid could decrease the SAR of soil.

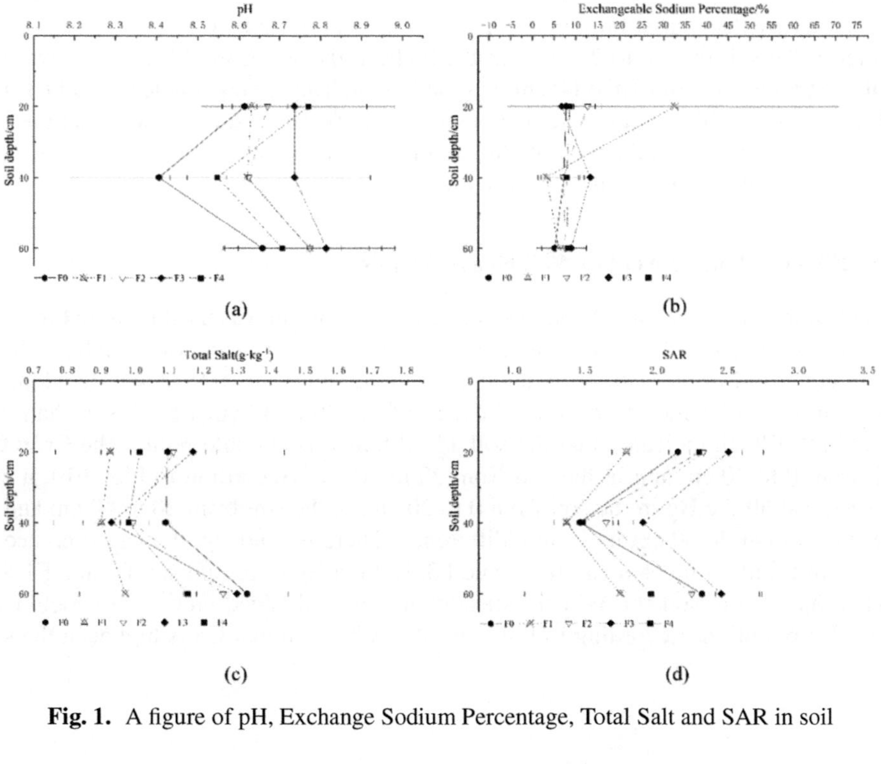

Fig. 1. A figure of pH, Exchange Sodium Percentage, Total Salt and SAR in soil

3.2 Effect of Humic Acid on Soil Nutrient Elements

Based on the alkaline hydrolytic nitrogen, available phosphorus, available potassium and organic matter of soil has been analysed, origin was used to draw Fig. 2.

According to Fig. 2(a), it can be seen that the alkaline hydrolytic nitrogen is the highest in the soil from 0 to 20 cm, the lowest in the soil from 20 to 40 cm and the lowest in the soil from 40 to 60 cm. The soil alkaline hydrolytic nitrogen of profile soil F1 and F4 was higher than that of profile soil F0. Furthermore, the alkaline hydrolytic nitrogen of middle and deep soil with soil F3 was higher than that of soil F0. Therefore, the soil F1, F3, F4 added of humic acid could improve the nitrogen supply capacity of soil. According to Fig. 2(b), the higher available phosphorus was found in the soil from 0 to 20 cm, the lower in the soil from 20 to 40 cm and the soil from 40 to 60 cm. The available phosphorus of surface soil F1, F2, F3, F4 was higher than that of soil F0. It indicated that the soil added humic acid could improve the supply of phosphorus in surface soil. According to Fig. 2(c), the soil available potassium is highest in the soil from 0 to 20 cm, lowest in the soil from 20 to 40 cm and higher in the soil from 40 to 60 cm. The content of available potassium in surface soil F2 and F3 was higher than that of soil F0. In addition, the content of available potassium in deep soil F1, F2, F3 was also higher than that of soil F0. Therefore, the soil F1, F2, F3 added humic acid could improve the potassium supply capacity of soil. According to Fig. 2(d), the organic matter in the soil from 0 to 20 cm was the highest, the soil from 20 to 40 cm was the higher and the soil from 40 to 60 cm was the lowest. The organic matter of soil F1, F2, F3, F4 from 0 to 20 cm was lower than that of soil F0. The results showed that the soil added of humic acid could promote the absorption of surface soil organic matter, and the organic matter of soil was relatively lower.

3.3 Effect of Humic Acid on Soil Heavy Metals

Based on the heavy metals of soil has been analysed, origin was used to draw Fig. 3.

According to Fig. 3(a), it can be seen that the Cr is higher in the soil from 0 to 20 cm, lower in the soil from 20 to 40 cm and higher in the soil from 40 to 60 cm. In addition, the Cr in soils F3 which from 0 to 20 cm and from 20 to 40 cm were lower than that of soil F0. Which indicated that the soil added humic acid could reduce the Cr in the soil from 0 to 20 cm and in the soil from 20 to 40 cm. According to Fig. 3(b), it can be seen that all the Hg in the soil from 0 to 20 cm, in the soil from 20 to 40 cm and in the soil from 40 to 60 cm have little difference. There was no significant difference in the Hg of profile soil between F0, F2 and F3. But the Hg of profile soil F1 and F4 was higher than that of soil F0. Which indicated that the soil added humic acid could may cause Hg pollution. According to Fig. 3(c), it can be seen that Cu is higher in the soil

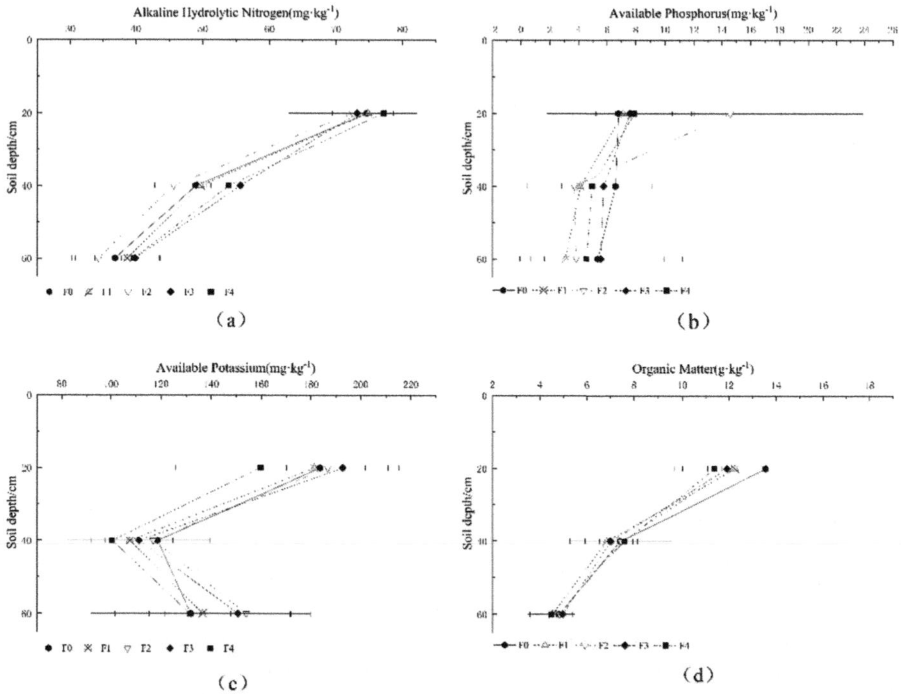

Fig. 2. A figure of alkaline hydrolytic nitrogen, available phosphorus, available potassium in soil

from 0 to 20 cm, lower in the soil from 20 to 40 cm and higher in the soil from 40 to 60 cm. There was no significant difference in the Cu of profile soil between F0, F1 and F2. But the Cu in soil of F3 and F4 was higher than that of soil in F0 except F4 in the soil from 20 to 40 cm. Which indicated that the soil added humic acid could may cause Cu pollution. According to Fig. 3(d), it can be seen that the Pb in the soil of F0 is not different. However, the Pb in the soil profile with humic acid was significantly different. In addition, the Pb of humic acid in the soil from 20 to 40 cm and in the soil from 40 to 60 cm were higher than that of soil F0. It indicated that the soil added humic acid may cause Pb pollution.

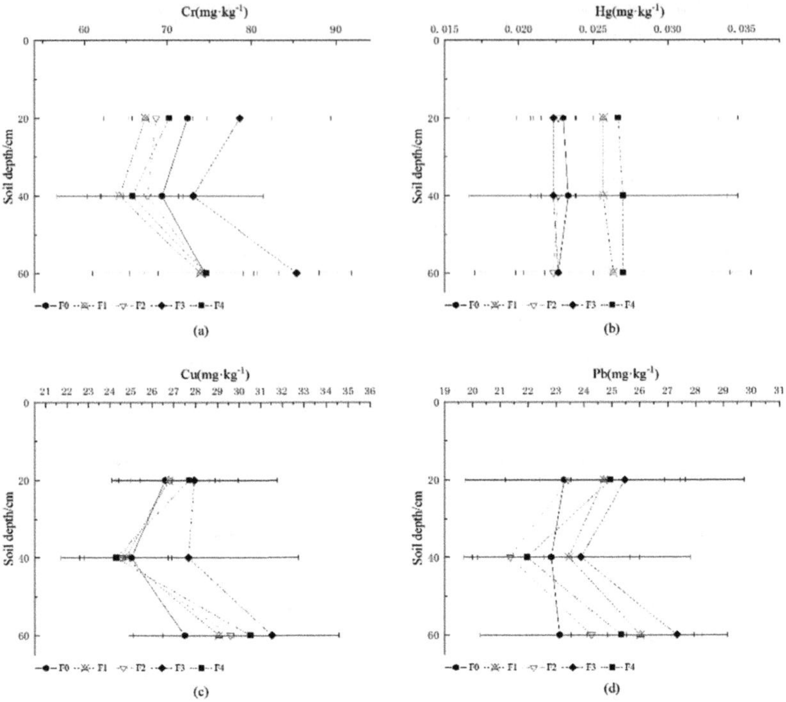

Fig. 3. A figure of heavy metalse in soil

4 Conclusion

(1) Addition of humic acid to the saline soil of Hekou District, Dongying, did not significantly reduce the soil pH and exchange sodium percentage, but it can decrease the SAR of soil and effectively reduce the total salt content of soil.

(2) Addition of humic acid to the saline soil of Hekou District, Dongying, enhanced the supply of nitrogen, phosphorus and potassium to the soil and could promote the absorption of surface soil organic matter, and the soil organic matter was relatively lower.

(3) Addition of humic acid to the saline soil of Hekou District, Dongying, reduced the Cr in the soils which from 0 to 20 cm and from 20 to 40 cm, but increased the Hg, Cu, Pb in the soil, and may cause Hg, Cu, Pb pollution in the soil.

Acknowledgments. Fund project: Shandong Province Coal Geological Bureau Scientific Research Special Award Fund Coal Land Branch Word (2023)9: Remediation of saline-alkali soil in Hekou District, Dongying, China by coal-based humic acid and its mechanism.

References

1. Wang, Z.R, Zhao, G.X., Gao, M.X., et al.: Spatial variation of soil water and salt and microscopic variation of soil salinity in summer in typical area of the Yellow River Delta in Kenli County. Acta Ecol. Sin. **36**(4), 1040–1049 (2016). (in Chinese)
2. Han, J., Sun, Y., Zhang, L., et al.: Effect of biochar and humic acid on physical and chemical properties of saline-alkali soil. Agricultural Research in the Arid Areas **38**(6), 121–127 (2020). (in Chinese)
3. Azcona, I., Pascual, I., Aguirreolea, J., et al.: Growth and development of pepper are affected by humic substances derived from composted sludge. J. Plant Nutr. Soil Sci. **174**(6), 916–924 (2011)
4. Zandonadi, D.B., Canellas, L.P., Facanha, A.R.: Indolacetic and humic acids induce lateral root development through a concerted plasmalemma and tonoplast H+ pumps activation. Planta **225**(6), 1583–1595 (2007)
5. Jing, J., Yuan, L., Zhang, S., et al.: Effects and mechanism of humic acid in humic acid enhanced phosphate fertilizer on fertilizer-phosphorus migration. Scientia Agricultura Sinica **54**(23), 5032–5042 (2021). (in Chinese)
6. Zhou, L., Yuan, L., Zhao, B., et al.: Response of maize roots to different additive amounts of weathered coal humic acids. Scientia Agricultura Sinica **52**(2), 285–292 (2019). (in Chinese)
7. Du, Q., Zhang, S.S., Antonietti, M., et al.: Sustainable leaching process of phosphates from animal bones to alleviate the world phosphate crisis. ACS Sustainable Chemistry & Engineering **8**(26), 9775–9782 (2020)

Restoration of Heavy Metals and Pollution Risk Analysis of Surface Soil in Coal-Based Solid Waste Dump Using Biochar, Viscous Sludge and Ryegrass

Mengmeng Jiang, Chongwei Zhong, Die Hu, and Jiu Huang(✉)

School of Environment and Spatial Informatics, China University of Mining and Technology, Xuzhou 221116, China
jhuang@cumt.edu.cn

Abstract. To explore the remediation of the surface soil of coal-based solid waste dumps by biochar, viscous sludge and ryegrass and the risk of HM pollution, five heavy metals (HMs), Pb, Zn, Cd, Ni and Cu, were represented by pot experiments. Different proportions of anaerobic and membrane-treated water jet loom sludge (0%, 5%, 15%) and biochar (0%, 3%) were added to the contaminated soil, and ryegrass was used as a soil HM extraction plant. The results showed that the addition of biochar and sludge increased the soil organic matter content by 7.7 times compared with the control group and reduced the activation rates of Pb and Ni by 62% and 71%, respectively. Under low sludge ratio, biochar had a significant passivation effect on Pb, Ni and Cu, which could reduce the pollution risk of sludge to soil HMs. The simultaneous application of biochar and sludge could improve the enrichment capacity of ryegrass for Zn, followed by Cd and Cu. The combined application of biochar, sludge and ryegrass could reduce the pollution risk index of Cd in the surface soil of coal-based solid waste dumps. In the actual sludge application process, its application ratio can be reduced, and biochar and plants can be used to control the HM risk at the same time.

Keywords: coal-based solid waste dump · biochar · heavy metal pollution · risk analysis · phytoremediation

1 Introduction

Long-term storage of coal-based solid waste can easily cause serious pollution to the ecological environment such as soil and water bodies. The coal mining areas are mostly distributed in the central and northwestern regions in China. Due to scarce precipitation and surface evaporation, HMs in the soil in the coal-based solid waste dump area tend to continue to migrate to the surface soil, resulting in an increase in the content of HMs in the surface soil, thus increasing the ecological risk. It was reported that viscous sludge contains high levels of nitrogen, phosphorus, organic matter and trace nutrients, which can increase crop yields and have a beneficial effect on soil microorganisms [1]. At the

D. Han et al. (Eds.): ICEPG 2024, ESE, pp. 609–615, 2025.
https://doi.org/10.1007/978-3-031-86829-0_70

same time, it can increase the clay content of sandy soil and improve the water holding capacity. However, the presence of potentially harmful substances such as HMs in sludge is an obstacle to its soil remediation. Environmentally non-toxic/neutral adsorbents to fix HMs is needed and the adsorbents need to have a positive impact on the soil. Biochar, as an environmentally friendly adsorption material, can significantly increase the organic matter and nutrient content of the soil [2], has a high affinity and passivation effect on HMs [3, 4], and is widely used in soil HM remediation research. Therefore, this study used biochar, water jet loom sludge and ryegrass for soil HM control. The changes in the content of five HMs (Pb, Zn, Cd, Ni, Cu) in the soil and the effect of plant extraction were explored, the risk of biological HM pollution in the remediated soil was also evaluated.

2 Materials and Methods

The soil was collected from Ningdong area, China. It was type of sandy loam with clay, silt and sand content of 14%, 28% and 58% respectively. The Pb, Zn and Cd content in the soil was higher than the local background value of uncontaminated soil (20.6, 58.8, 0.11, 36.6 and 22.1 mg·kg^{-1}) [5]. The water jet loom sludge was collected from Wujiang District, Jiangsu Province. It was naturally air-dried and crushed with a grinder as clay material. The biochar was made from rice husks pyrolyzed at 500 °C. The microbial inoculants used included *Bacillus subtilis* and *B. licheniformis*, with an effective viable count of ≥ 200 million per gram. The basic properties of the test materials of soil, water jet loom sludge and biochar are shown in Table 1.

Table 1. Basic characteristics of soil, water jet loom sludge and biochar

Parameters	Soil	Water jet loom sludge	Biochar	Parameters	Soil	Water jet loom sludge	Biochar
pH	9.1	8.2	9.7	Pb/(mg·kg^{-1})	29	26	ND
Conductivity /(mS·m^{-1})	127.0	92.5	22.5	Zn/(mg·kg^{-1})	70	540	19.1
Organic matter/(g·kg^{-1})	3.2	486	336	Cd/(mg·kg^{-1})	0.8	0.4	ND
Total N/(g·kg^{-1})	0.29	13.3	5.4	Ni/(mg·kg^{-1})	36	57	2.0
Total P/(g·kg^{-1})	0.80	4.26	0.41	Cu/(mg·kg^{-1})	20	226	3.2

A total of 6 treatments were set up, namely: S0 (control group), S0B (addition of 3% biochar), S1 (mass ratio of soil: sludge = 95:5, the same below), S1B (soil: sludge = 95:5, and 3% biochar), S2 (soil: sludge = 85:15), S2B (soil: sludge = 85:15, and 3% biochar). Each treatment was set up with 3 parallels, and 30 g·pot^{-1} of microbial agent was added as base fertilizer, then about 0.25 g of ryegrass seeds was sowed. The ryegrass growth for 72 days, after which the aboveground stems and leaves were harvested and dried to constant weight to test its HM content, and the soil was tested for HMs and basic physical and chemical properties. Two parameters, enrichment coefficient and accumulation amount, were introduced to evaluate the accumulation capacity of plants for HMs, and they were calculated according to Eqs. (1)–(3).

$$AR_i = \frac{C_{i,bioavailable}}{C_{i,total}} \times 100\% \tag{1}$$

$$BF_i = \frac{C_{i,plant}}{C_{i,total}} \tag{2}$$

$$\mathrm{Accumulation_i} = \mathrm{DW} \times \mathrm{C_{i,\,plant}} \tag{3}$$

where AR_i is the activation rate of HM i (%); $C_{i,bioavailable}$ and $C_{i,total}$ are the bioavailability and total content respectively of HM i in the soil (mg·kg^{-1});. BF_i and $Accumulation_i$ are the enrichment coefficient and accumulation respectively of HM i in the aboveground of the ryegrass; DW is the dry weight of the ryegrass (g).

The evaluation of soil HM pollution was by potential ecological risk index. The formulas for the single potential ecological risk index of a single HM and the comprehensive potential ecological risk index of multiple HMs are shown in Eqs. (4)–(6).

$$\mathrm{P_i} = \frac{\mathrm{C_i}}{\mathrm{S_i}} \tag{4}$$

$$E_r^i = T_r^i \times P_i \tag{5}$$

$$RI = \sum_{i=1}^{n} \mathrm{E_r^i} \tag{6}$$

where P_i is the single factor pollution index of HM i; C_i is the measured content of HM i (mg·kg^{-1}); S_i is the evaluation standard value of HM i (mg·kg^{-1}); E_r^i is the single potential ecological risk index of HM i; T_r^i is the toxicity corresponding coefficient of HM i (i.e. the evaluation reference value of non-polluted area); RI is the comprehensive potential ecological risk index. The risk levels corresponding to the single and comprehensive potential ecological risk indices are shown in Table 2.

Table 2. Potential ecological risk index and corresponded risk level

E_r^i	RI	Risk level
<40	<150	slight
40(included) ~ 80	150(included) ~ 300	normal
80(included) ~ 160	300(included) ~ 600	Relatively strong
160(included) ~ 320	≥600	strong
≥320	/	Very strong

3 Results and Discussion

3.1 Changes in Physical and Chemical Properties of Soil

The physical and chemical properties of soil in different treatment groups are shown in Table 3. The contents of SOM, TN and TP in the soil increased with the increase of sludge ratio, while the pH value and TK content decreased, especially in the groups where sludge and biochar coexisted. The pH value in S2B was 8.40, and SOM was about 7.7 times higher than that in the control group. This was closely related to the large amount of organic matter in the sludge. At the end of the experiment, the ryegrass in the control group grew sparsely, and the aboveground part was mostly withered and yellow. With the introduction of sludge and biochar, the ryegrass grew luxuriantly. The aboveground biomass of plants in each improved treatment was greater than the control, among which S1B, S2 and S2B reached significant levels, which were 3.3, 3.4 and 4.3 times that of S0, respectively. The remediation of soil by sludge and biochar promoted the growth of ryegrass and increased its biomass.

Table 3. Nutrients of soil and ryegrass biomass after remediated with water jet-loom sludge and biochar

	pH	SOM/(g·kg^{-1})	TN/(g·kg^{-1})	TP/(g·kg^{-1})	TK/(g·kg^{-1})	Ryegrass biomass (g)
SO	9.0 ± 0.2a	5.4 ± 0.2c	0.43 ± 0.01b	0.43 ± 0.03c	15.31 ± 0.63a	0.32 ± 0.08d
SOB	9.1 ± 0.0 a	8.8 ± 1.8c	0.46 ± 0.05b	0.46 ± 0.01bc	15.31 ± 0.21a	0.51 ± 0.26cd
S1	8.9 ± 0.2 ab	23.0 ± 2.6b	0.79 ± 0.07b	0.56 ± 0.03b	15.20 ± 0.16a	0.65 ± 0.27c
S1B	8.6 ± 0.1 bc	23.5 ± 5.5b	0.72 ± 0.15b	0.57 ± 0.05b	14.96 ± 0.29a	0.83 ± 0.31b
S2	8.6 ± 0.2 bc	54.9 ± 10.0a	1.48 ± 0.45a	0.78 ± 0.11a	14.02 ± 0.30b	1.08 ± 0.16b
S2B	8.4 ± 0.2 c	46.4 ± 7.3a	1.15 ± 0.18a	0.89 ± 0.09a	13.49 ± 0.07b	1.39 ± 0.32a

3.2 Total Amount and Activation Rate of HMs in Soil

The results of the full analysis of soil HMs Pb, Zn, Cd, Ni, and Cu after the experiment are shown in Fig. 1. The HM concentrations in all groups were lower than the pollution risk screening values in the Soil Environmental Quality Agricultural Land Pollution Risk Control Standard (GB 15618–2018). It can be found that the sludge increased the Zn and Cu content in the soil, which may be due to the high Zn and Cu content in the sludge itself. In addition, the treatment group with biochar addition had a lower Zn content than the group without addition, with a minimum of 40.8 mg·kg^{-1}. In the group with only biochar addition, the total amount of Pb and Cd increased slightly, but decreased with the introduction of sludge. A higher proportion of sludge addition combined with biochar reduced the total amount of Pb and Cd in the soil to a certain extent. Ni was less affected by the restoration conditions of this study.

The activation rate of HMs can reflect the impact of HM pollution on soil. As shown in Fig. 1b, although the content of Cd in the soil is low, its activation rate is the highest,

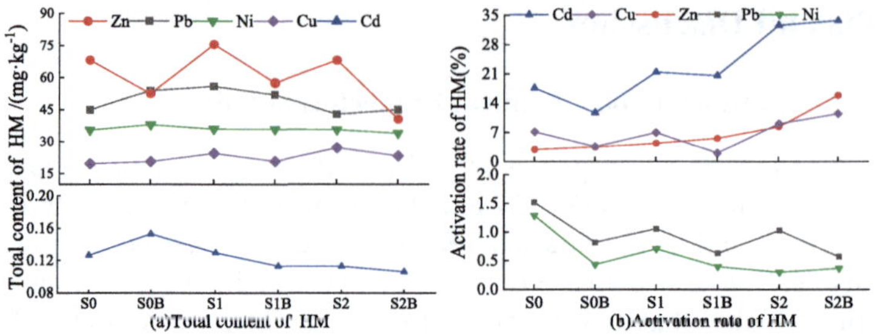

Fig. 1. Total amount, bioavailability and activation rate of soil HMs Pb, Zn, Cd, Ni and Cu

followed by Cu and Zn. The activation rates of the three HMs increased with the increase of sludge ratio. The Cd, Cu and Zn in the S2B group were 1.9, 1.6 and 5.3 times that of the control group, respectively. The activation rates of Pb and Ni were low and decreased with the increase of sludge ratio and the introduction of biochar. The activation rates of the treatment group S2B were reduced by 62% and 71% respectively compared with the control group S0, indicating that Pb and Ni have an impact on the presence of soil.

3.3 Effect of Ryegrass on the Extraction of HMs from Soil

The HM enrichment coefficients of ryegrass stems and leaves are shown in Fig. 2. The enrichment coefficients of ryegrass for Pb and Ni are less than 1, and decrease with the addition of sludge; the enrichment coefficient for Zn is greater than 1, and the enrichment coefficients for Cd and Cu are in the range of 0.8 ~ 1.6 and 0.6 ~ 1.2, respectively, indicating that ryegrass has a strong enrichment ability for Zn, followed by Cd and Cu. The Pb and Ni enrichment coefficients of each experimental group were significantly lower than those of the control group, and the smallest ones were 0.003 and 0.073, respectively, in the S2B group. After soil remediation, the trend of Pb and Ni transferring from soil to ryegrass stems and leaves was significantly weakened, and it was also affected by the activation rate of HMs in soil.

The enrichment coefficient of Zn in each experimental group was greater than that in the control group, and biochar could increase the enrichment of Zn in ryegrass, indicating that biochar may enhance the extraction efficiency of Zn by plants on the surface of coal-based solid waste storage soil; on the contrary, the addition of biochar promoted the passivation of soil Cu, Pb and Ni, and may reduce the extraction efficiency of three HMs by plants on the surface of coal-based solid waste storage soil.

3.4 Evaluation of HM Pollution in Soil After Phytoremediation

The evaluation and analysis of soil HM pollution is shown in Fig. 3. According to the single potential ecological risk index, Cd has a relatively strong pollution risk index among the five HMs and is at a medium risk level in the S0B treatment group. The activation rate of Cd in the soil is the highest, which may be the reason for the higher soil Cd pollution risk. The individual potential ecological risk indexes of the four HMs of Pb, Zn, Ni, Cu were all below 40, indicating a normal risk. The comprehensive potential

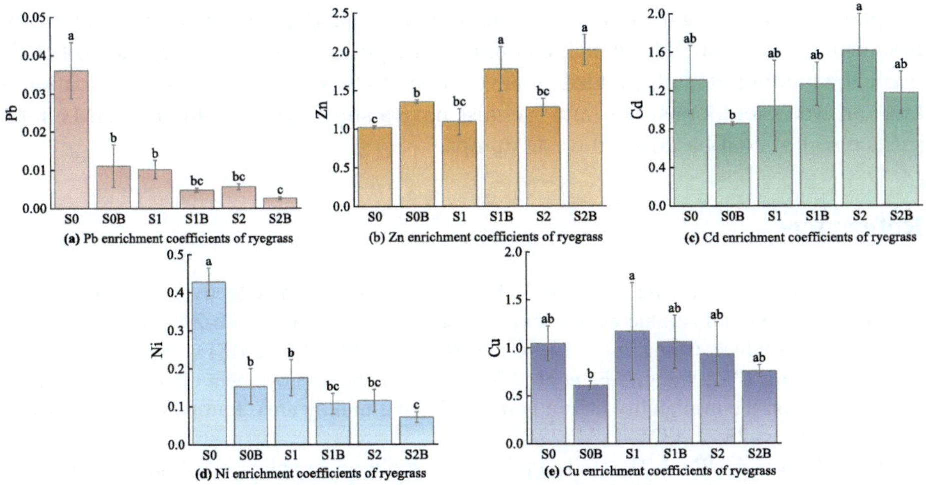

Fig. 2. Plant accumulation coefficient of heavy metals

ecological risk index showed that the comprehensive risk index of the five HMs was mild, and the comprehensive potential ecological risk index of soil HMs with only biochar (S0B) added was relatively high; after the addition of sludge, the significant changes in organic matter in the soil affected the plant growth characteristics and the extraction characteristics of HMs, thus affecting the risk of soil HM pollution. The combined application of biochar, water jet loom sludge and ryegrass can reduce the risk of HM pollution in the soil to a certain extent.

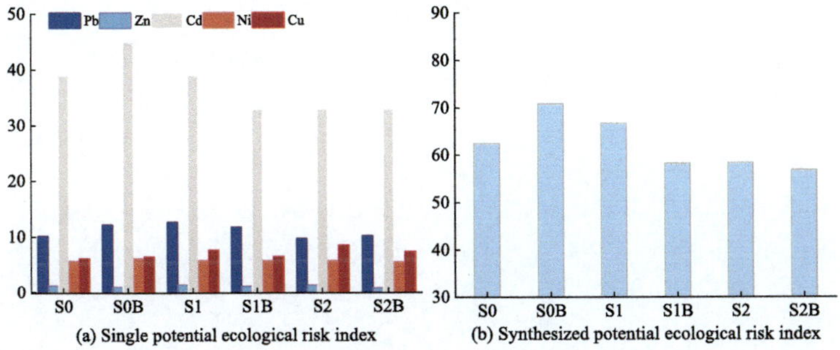

Fig. 3. Potential ecological risk index of HM pollution in soil remediation.

4 Conclusions

Biochar and viscous water jet loom sludge co-remediation of the surface soil of coal-based solid waste dumps, significantly increasing the soil organic matter content and facilitating the growth of ryegrass. When sludge and biochar are applied together, ryegrass has a strong enrichment capacity for Zn, followed by Cd and Cu. Sludge promotes the passivation of Pb and Ni, and the activation of Zn and Cd. Biochar has a significant passivation effect on HMs, which can reduce the pollution risk of sludge to soil HMs, while the amount of sludge added should not be too high. The combined application of biochar, water jet loom sludge and ryegrass have a slight ecological risk to HMs in the surface soil of coal-based solid waste dumps.

References

1. Li, Z.Y., Yang, S.X., Peng, X.Z., et al.: Field comparison of the effectiveness of agricultural and nonagricultural organic wastes for aided phytostabilization of a Pb-Zn mine tailings pond in Hunan Province, China. Int. J. Phytorem. **20**(12), 1264–1273 (2018)
2. Yue, F.X., Li, J.W., Wang, Y.F., et al.: Mechanism of the improvement effect by biochar and AM fungi on the availability of soil nutrients in coal mining area. Journal of Plant Nutrition and Fertilizers **25**(8), 1325–1334 (2019)
3. Hao, J.C., Li, Z., Wu, L.H., et al.: Preliminary study on cadmium and lead stabilization in soil highly polluted with heavy metals using different stabilizing agents. Soils **51**(4), 752–759 (2019)
4. Sun, T., Li, K., Fu, Y.T., et al.: Effect of modified biochar on immobilization remediation of weakly alkaline Cd-contaminated soil and environmental quality. Acta Sci. Circum. **40**(7), 2571–2580 (2020)
5. Xia, Z.S., Bai, Y.R., Wang, Y.Q., et al.: Spatial distribution and source analysis of soil heavy metals in a small watershed in the mountainous area of southern ningxia based on PMF model. Environmental Science **43**(1), 432–441 (2022)

Effect of Exogenous Activators Synergistic Chelating Agent on Remediation of Cd-Contaminated Soil with *Helianthus annuus* L.

Weishun Cheng, Xi Luo(✉), Li Jiang, Fengmei Jin, and Chen Gang

Institute of Environment and Safety, Wuhan Academy of Agricultural Science, Wuhan 430074, China
405650151@qq.com

Abstract. The impact of chelating agents and external activators, particularly tartaric acid and Methyl Jasmonate, on cadmium (Cd) accumulation in sunflowers was examined to improve the efficacy of sunflower-based remediation in Cd-polluted soil. Utilizing sunflowers as the experimental model, a pot experiment was conducted to evaluate the effects of the chelating agent EGTA alongside tartaric acid (TA) and Methyl Jasmonate (MeJA) on various parameters, including plant height, dry weight of both root and shoot systems, Cd uptake, and the rate of soil Cd extraction. The findings revealed that the application of TA or MeJA alone significantly enhanced sunflower growth, resulting in biomass increases of 18.85% and 5.28%, respectively, compared to the control group. Conversely, the use of EGTA alone suppressed growth, causing a biomass reduction of 23.13%. The individual and combined effects of EGTA with external activators on Cd concentration and accumulation across different parts of the sunflower plant exhibited variability. Notably, the individual applications of EGTA, TA, and the combination of EGTA with TA and MeJA considerably elevated Cd concentration and accumulation in sunflowers relative to the control. The most significant effects were observed with the combination of EGTA and 1.25 mmol/kg of TA, which produced increases in Cd concentration of 96.48%, 86.86%, and 502.19% in the roots, stems, and leaves, respectively, leading to an overall increase in Cd accumulation in the entire sunflower plant of 328.54% compared to the control. Moreover, the application of EGTA and external activators affected the Cd transport coefficients in different sunflower organs, with EGTA, TA alone, and the combination of EGTA with TA and MeJA enhancing the efficiency of Cd extraction from the soil. The highest extraction efficiency, reaching 1.15%, was achieved with the combination of EGTA and 1.25 mmol/kg of TA. These results suggest that the integration of external activators with chelating agents in the remediation of Cd-contaminated soil significantly boosts the ability of sunflowers to accumulate and remove Cd.

Keywords: Cadmium pollution · Sunflowers · EGTA · Exogenous activator · Phytoremediation

D. Han et al. (Eds.): ICEPG 2024, ESE, pp. 616–627, 2025.
https://doi.org/10.1007/978-3-031-86829-0_71

1 Introduction

Cadmium (Cd) is highly biotoxic and readily available to plants, posing significant threats to ecosystems [1, 2]. The rapid advancement of industrial and agricultural practices, particularly sewage irrigation, coupled with the extensive use of pesticides and fertilizers, and the weathering of waste liquids and residues from mining and smelting operations, has led to a Cd concentration exceeding acceptable limits in 7.0% of soils in China. Consequently, Cd has emerged as the most significant heavy metal pollutant in soil, critically jeopardizing food safety and representing a direct risk to human health [3, 4]. Addressing Cd contamination in soil is, therefore, of paramount research importance. Phytoremediation of Cd-contaminated agricultural land is a lengthy process, prompting keen interest in enhancing its efficiency. Current strategies to boost phytoremediation effectiveness primarily involve the application of soil amendments, fertilization, improved cultivation practices, and genetic enhancements [5]. The introduction of chelating agents has demonstrated potential in improving plant tolerance to heavy metal Cd and increasing its absorption by plants; however, these agents may pose environmental risks and stress plants, thereby inhibiting their growth and development [6]. Research has indicated that while chelation encourages heavy metal extraction by plants, employing exogenous activators can effectively mitigate the stress imposed by heavy metals and chelating agents, thereby promoting plant growth and increasing extraction efficiency. For instance, Tammam et al. found that a combination of gibberellin (GA_3), indole acetic acid (IAA), and EDTA counteracted the adverse effects of EDTA on plant growth, notably enhancing lead absorption by roots [7]. Similarly, Jan et al. reported that the combined application of GA_3, IAA, and EDTA significantly boosted growth and Cd extraction in ragweed, with the Cd content in each plant increasing by 6.72 mg compared to the control group [8]. Zhang et al. revealed that applying citric acid (CA), malic acid (MA), and EDTA did not adversely affect the growth and yield of honeysuckle, while conspicuously increasing the Cd enrichment coefficient in honeysuckle stems [9]. Additionally, Wang et al. demonstrated that the combination of diethylaminoethanol hexanoate (DA-6) and tetrasodium glutamate diacetate (GLDA) significantly enhanced both biomass and Cd content in switchgrass [10]. Furthermore, Chen et al. explored the regulatory effects of EGTA and tartaric acid on Cd stress and accumulation in castor plants, finding that their combined application mitigated the toxicity of the chelating agents to plants while substantially enhancing their capacity to accumulate cadmium [11].

In recent years, researchers have identified numerous hyperaccumulator plants capable of efficiently absorbing, transporting, and removing cadmium (Cd), including species such as *Noccaea caerulescens* [12], *Viola baoshanensis* [13], and *Solanum nigrum* [14]. However, these hyperaccumulator plants often face challenges such as limited height, slow growth rates, low biomass, and extended remediation periods, which ultimately lead to insufficient remediation efficacy. In contrast, sunflower (*Helianthus annuus* L.) is an energy crop noted for its strong Cd accumulation and tolerance as well as its high economic value. It possesses several advantageous traits, including broad distribution, strong environmental adaptability, rapid growth, and substantial biomass [15]. Nevertheless, there is a notable scarcity of research focused on the use of chelating agents in conjunction with exogenous activators to mitigate the phytotoxic effects of these agents and heavy metals, while simultaneously enhancing the ability of sunflowers to extract heavy metals such as Cd from contaminated soils. Therefore, this study was designed to investigate the synergistic effects of the chelating agent ethylenediaminetetraacetic acid (EDTA) and exogenous activators such as tannic acid (TA) and Methyl Jasmonate (MeJA) on Cd enrichment in sunflowers, thereby providing a scientific foundation for the remediation of Cd-contaminated agricultural land in China.

2 Materials and Methods

2.1 Tested Soil

The experimental soil was collected from the surface layer (0–20 cm) of cultivated land at the Wuhu base of the Wuhan Academy of Agricultural Sciences. This brown soil contained minimal gravel, along with small amounts of humus and plant residues. Following air-drying, the fundamental physical and chemical properties were assessed, revealing a pH of 5.59, organic matter content of 21.49 g/kg, total nitrogen at 0.176%, total phosphorus at 0.119%, total potassium at 0.318%, hydrolyzable nitrogen at 163.91 mg/kg, available phosphorus at 23.30 mg/kg, available potassium at 170.00 mg/kg, and total cadmium (Cd) concentration of 0.32 mg/kg.

2.2 Test Material

The exogenous activators evaluated in this study were TA and MeJA, while the chelating agent used was EGTA; both of these compounds were procured from Wuhan NewEast Biotechnology Co., Ltd. The plant under investigation was the sunflower variety 567DW, acquired from Hubei Fuyue Seed (Group) Co., Ltd. Wuhan Branch.

2.3 Research Methodology

The pot experiment was conducted in the greenhouse of the Crop Research Institute at the Wuhan Academy of Agricultural Sciences in 2021, using plastic flowerpots with a diameter of 80 cm and a height of 60 cm. Each pot was filled with 20 kg of sifted, air-dried soil. Analytical grade $CdCl_2 \cdot 2.5H_2O$ was utilized as the treatment agent, resulting in a heavy metal cadmium (Cd) concentration of 100 mg/kg in the test soil. The $CdCl_2 \cdot 2.5H_2O$ was thoroughly mixed with the air-dried soil, placed in a basin, and allowed to stand for complexation for 20 days. Mature sunflower seeds were disinfected with sodium hypochlorite, soaked in warm water for 24 h, sown in a Cd-free seedling substrate, and then transferred into the pots with a Cd concentration of 100 mg/kg after 15 days of emergence and growth. Following an additional 30 days, various exogenous activators and chelating agents were applied; specifically, TA and MeJA were sprayed on the leaves, while EGTA was applied to the roots. Three concentration gradients were employed for the exogenous activator TA (0.5, 1.25, and 2.5 mmol/kg) and MeJA (25, 50, and 75 μmol/L). Previous studies have indicated that high concentrations of EGTA can severely inhibit sunflower shoot growth and may even lead to plant death. However, a concentration of 2.5 mmol/kg EGTA has been shown to enhance the absorption rate of heavy metals in sunflowers and promote better remediation [16], thus this concentration was chosen for this study. Control treatments included pots without exogenous activators, EGTA, and pots sprayed with clean water (CK). The specific experimental design is summarized in Table 1. Totally, 10 treatments were established, each replicated three times, with each replicate consisting of one pot containing three plants. The plants were harvested after 20 days of treatment.

Table 1. Test design.

No	Treatment
1	CK
2	EGTA 2.5 mmol/kg
3	MeJA 50 μmol/L
4	TA 1.25 mmol/kg
5	EGTA 2.5 mmol/kg + TA 0.5 mmol/kg
6	EGTA 2.5 mmol/kg + TA 1.25 mmol/kg
7	EGTA 2.5 mmol/kg + TA 2.5 mmol/kg
8	EGTA 2.5 mmol/kg + MeJA 25 μmol/L
9	EGTA 2.5 mmol/kg + MeJA 50 μmol/L
10	EGTA 2.5 mmol/kg + MeJA 75 μmol/L

2.4 Determination Methodology

Biomass Determination

Measure the height of the sunflower plants and proceed to harvest the roots, stems, and leaves. Initially, rinse the samples with tap water, followed by a thorough rinse with deionized water. Absorb any excess water using filter paper, then dry the samples in an oven at 105 °C for 1 h. Subsequently, continue drying at 75 °C until reaching a constant weight, and record the dry weight of the sunflower samples.

Determination of Cd Content.

The 2 kg rhizosphere soil sample was collected for each replicate, and the samples were allowed to air dry naturally before being sieved for analysis. The plant samples were ground, crushed, and sieved prior to testing. Both the plant and soil samples were digested using a mixture of concentrated nitric acid and perchloric acid in a ratio of 4:1 (v:v). The concentration of cadmium (Cd) in the digested solution was then measured using atomic absorption spectrophotometry with a flame detector [17].

2.5 Data Processing and Analysis

The cadmium (Cd) accumulation and transport coefficient in plants was calculated with the following formula.

Shoot Cd accumulation (mg/plant) = Shoot biomass × Shoot Cd content.

Root Cd accumulation (mg/plant) = Root biomass × Root Cd content.
Transport coefficient = Cd content in shoot/Cd content in root.
Cd extraction rate in soil (%) = Cd accumulation of whole plant/(Cd content of contaminated soil × quality of contaminated soil) × 100.
WPS 2010 was used for data statistics and plotting, and SPSS 26.0 was used for analysis of variance of the test data.

3 Results and Analysis

3.1 Effects of Different Treatments on Growth of *Helianthus Annuus*

In this study, we measured the plant height and biomass of sunflowers subjected to various treatments, with the results presented in Table 2. Compared to CK, both the plant height and dry weight of sunflowers increased when treated with the exogenous activator alone. Additionally, the combination of EGTA and the activator resulted in greater plant height and dry weight compared to EGTA alone. The effects of the different treatments on sunflower plant height and biomass were inconsistent, revealing significant differences among the treatments. Notably, treatment 4 (TA at 1.25 mmol/kg) yielded the tallest plants and the heaviest dry weights of both root and shoot portions, showing increases of 14.76%, 16.33%, and 18.85%, respectively, when compared to CK. Treatment 3 (MeJA at 50 μmol/L) followed, demonstrating increases of 6.5%, 8.02%, and 5.28% relative to CK. Conversely, treatment 2 (EGTA at 2.5 mmol/kg) resulted in the lowest plant height and biomass, which were reduced by 12.31%, 18.91%, and 23.13%, respectively, compared to CK.

3.2 Effects of Sunflower on Cd Uptake Under Different Treatments

As shown in Fig. 1, the addition of the chelating agent alone, as well as the combined application of the chelating agent and activator, results in an increased cadmium content in sunflower plants compared to the control (CK). The range of cadmium accumulation in the roots, stems, and leaves was measured at 36.14 to 73.17 mg/kg, 18.98 to 35.54 mg/kg, and 5.96 to 48.34 mg/kg, respectively. When ethylene glycol tetraacetic acid (EGTA) was applied alone, there was no significant difference in cadmium content in the sunflower roots compared to CK; however, cadmium levels in the stems and leaves were significantly higher than those in CK, exhibiting increases of 21.77% and 417.94%, respectively, with the differences in stem and leaf cadmium contents being statistically significant ($P < 0.05$). The application of methyl jasmonate (MeJA) alone resulted in negligible differences in cadmium content across the roots, stems, and leaves in comparison to CK. In contrast, when thapsigargin (TA) was applied alone, cadmium content in the roots, stems, and leaves of sunflower was higher than that in CK, with increases of 12.43%, 5.64%, and 65.95%, respectively, and the differences in cadmium content for roots and leaves compared to CK were significant ($P < 0.05$). Moreover, simultaneous application of the chelating agent and activator led to higher cadmium content in the roots, stems, and leaves of sunflowers than CK, with significant differences observed ($P < 0.05$). Notably, the most effective treatment was EGTA at 2.5 mmol/kg combined with TA at 1.25 mmol/kg, resulting in cadmium contents of 75.17 mg/kg in the roots,

35.54 mg/kg in the stems, and 48.34 mg/kg in the leaves, which represented increases of 96.48%, 86.86%, and 502.19%, respectively, compared to CK. Additionally, cadmium absorption across all sunflower organs exhibited a pattern of root > leaf > stem under treatments 2, 5, 6, and 8, while the remaining treatments displayed a pattern of root > stem > leaf.

3.3 Effects of Different Treatments on Cd Enrichment Ability of *Helianthus Annuus*

The data presented in Table 3 indicate that cadmium (Cd) accumulation in the shoots was significantly higher than in the roots across various treatments. Notably, compared to the control (CK), a single application of EGTA at 2.5 mmol/kg resulted in a significant increase in Cd accumulation in the aerial parts and total biomass of sunflower plants, reaching 0.94 mg/plant and 1.05 mg/plant, respectively, while Cd accumulation in the roots decreased significantly. In contrast, no significant differences were observed in Cd accumulation in the roots, shoots, and overall biomass of sunflower plants following a single application of MeJA at 50 μmol/L when compared to CK. However, the single administration of TA at 1.25 mmol/kg yielded a significant increase compared to CK. Among all treatments, the combination of EGTA at 2.5 mmol/kg and TA at 1.25 mmol/kg yielded the highest Cd accumulation in the roots, shoots, and total biomass, measuring 0.27, 2.05, and 2.32 mg/plant, respectively, which represented increases of 107.69%, 400%, and 329.63% relative to CK. This was followed by the combination of EGTA at 2.5 mmol/kg and MeJA at 25 μmol/L, which resulted in Cd accumulations of 0.25, 1.84, and 2.09 mg/plant, respectively, corresponding to increases of 92.31%, 348.78%, and 287.04% compared to CK. In summary, a single application of EGTA, as well as the combined treatment of EGTA with specific activators, can enhance Cd accumulation in sunflowers, with the latter demonstrating a more pronounced effect.

The transport coefficient effectively reflects the impact of various treatment combinations on the ability of sunflowers to transport cadmium (Cd). As shown in Table 3, both EGTA treatment and the EGTA synergistic activator treatment significantly increased the transport coefficient for cadmium in sunflowers. With the exception of CK, MeJA at 50 μmol/L, and TA at 1.25 mmol/kg, the transfer coefficients for all other treatments exceeded 1, with the combination of EGTA at 2.5 mmol/kg and TA at 2.5 mmol/kg yielding the highest transfer coefficient at 2.73; in contrast, MeJA at 50 μmol/L exhibited the lowest transfer coefficient of 0.72.

3.4 Purification Effect of Different Treatments on Cd Contaminated Soil

Figure 2 illustrates that the single applications of EGTA and TA, as well as the combined application of EGTA and activator, significantly enhanced the extraction efficiency of cadmium (Cd) from soil by sunflowers to varying extents; however, the effect of the single application of methyl jasmonate (MeJA) was not pronounced. Specifically, the extraction rate increased from 0.26% to 0.35% with the application of TA alone, and from 0.26% to 0.52% with EGTA alone. When EGTA and TA were applied together (treatments 5, 6, 7), the Cd extraction initially increased and then decreased with rising TA concentration; the highest extraction efficiency was observed at a TA concentration of 1.25 mmol/kg, reaching 1.15%, while a concentration of 0.5 mmol/kg resulted in the lowest extraction efficiency of 0.66%, with all three treatments demonstrating statistically significant improvements compared to the control (CK) ($P < 0.05$). In contrast, when EGTA and MeJA were applied together (treatments 8, 9, 10), the extraction rate of soil Cd gradually declined as the concentration of MeJA increased, showing the highest extraction efficiency of 1.04% at a MeJA concentration of 25 μmol/L, and the lowest efficiency of 0.57% at 75 μmol/L; again, all three treatments exhibited significantly greater extraction efficiency than CK ($P < 0.05$).

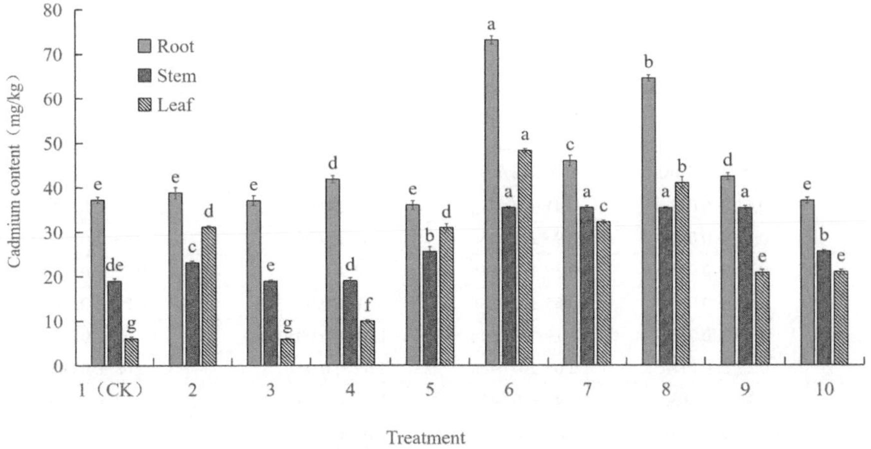

Note: Different lowercase letters indicate the significant difference in the same index under different treatments(P<0.05). The same as below.

Fig. 1. The content of Cd in *Helianthus annuus* under different treatments.

Table 2. Effects of different treatments on plant height and biomass of *Helianthus annuus*.

Treatment	Plant height		Subterranean part		Aerial part	
	Plant height(cm)	Compared with the CK(%)	Dry weight(g)	Compared with the CK(%)	Dry weight(g)	Compared with the CK(%)
1(CK)	46.54 ± 0.84de	-	3.49 ± 0.15d	-	14.96 ± 0.24cd	-
2	40.81 ± 0.52h	−12.31	2.83 ± 0.09f	−18.91	11.50 ± 0.59g	−23.13
3	49.57 ± 0.60b	6.5	3.77 ± 0.06c	8.02	15.75 ± 0.22bc	5.28
4	53.41 ± 0.34a	14.76	4.06 ± 0.11ab	16.33	17.78 ± 0.63a	18.85
5	44.45 ± 0.33f	−4.49	3.32 ± 0.04d	−4.87	13.74 ± 0.29e	−8.16
6	47.22 ± 0.67cd	1.46	3.69 ± 0.25c	5.73	14.18 ± 0.30de	−5.21
7	42.70 ± 0.43g	−8.25	3.04 ± 0.06e	−12.89	12.65 ± 0.36f	−15.44

(*continued*)

Table 2. (*continued*)

Treatment	Plant height		Subterranean part		Aerial part	
	Plant height(cm)	Compared with the CK(%)	Dry weight(g)	Compared with the CK(%)	Dry weight(g)	Compared with the CK(%)
8	48.13 ± 0.60c	3.42	3.87 ± 0.08bc	10.89	15.59 ± 0.56c	4.21
9	45.49 ± 0.33ef	−2.26	3.29 ± 0.01d	−5.73	15.29 ± 0.23c	2.21
10	41.93 ± 1.69gh	−9.91	2.96 ± 0.07ef	−15.19	12.77 ± 0.20f	−14.64

Note: Different lowercase letters in the same column indicate significant difference at 0.05 level. The same as below

Table 3. Effects of different treatments on Cd accumulation.

Treatment	Cd accumulation in root(mg/plant)	Cd accumulation in shoot(mg/plant)	Cd accumulation in whole plant(mg/plant)	Transfer coefficient
1(CK)	0.13 ± 0.007ef	0.41 ± 0.022g	0.54 ± 0.03g	0.82 ± 0.029f
2	0.11 ± 0.002h	0.94 ± 0.063e	1.05 ± 0.06e	2.43 ± 0.121bc
3	0.14 ± 0.005def	0.38 ± 0.005g	0.52 ± 0.02g	0.72 ± 0.025f
4	0.17 ± 0.006c	0.55 ± 0.028f	0.72 ± 0.01f	0.83 ± 0.029f
5	0.12 ± 0.003fg	1.19 ± 0.064d	1.31 ± 0.005d	2.62 ± 0.167ab
6	0.27 ± 0.017a	2.05 ± 0.046a	2.32 ± 0.04a	2.08 ± 0.041d
7	0.14 ± 0.003d	1.47 ± 0.045c	1.61 ± 0.04c	2.73 ± 0.148a
8	0.25 ± 0.007b	1.84 ± 0.074b	2.09 ± 0.06b	1.91 ± 0.067e
9	0.14 ± 0.003de	1.45 ± 0.009c	1.59 ± 0.02c	2.52 ± 0.015bc
10	0.11 ± 0.002gh	1.02 ± 0.026e	1.13 ± 0.02e	2.42 ± 0.051c

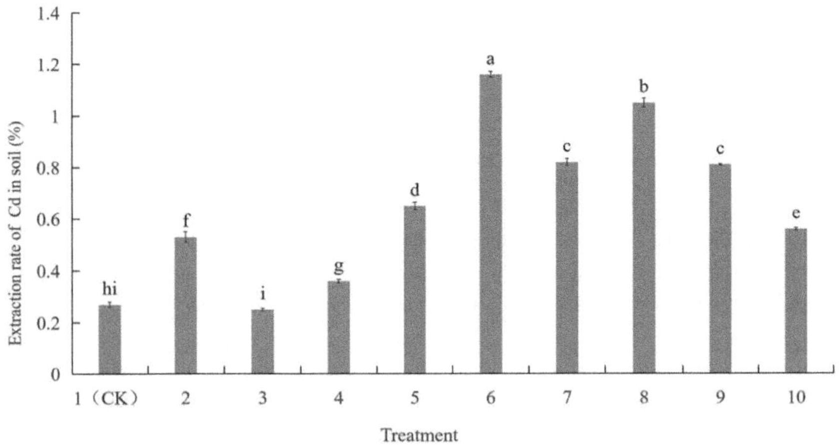

Fig. 2. Effect of different treatments on soil Cd phytoextraction rate.

4 Discussion

4.1 Effects of TA, MeJA and EGTA on Sunflower Growth

Biomass is a crucial factor in assessing the capacity of plants to remediate heavy metal pollution, often evaluated through indicators such as plant height, dry weight, and various other measurements. In this study, we observed and measured the growth, height, and dry weight of both the underground and aboveground parts of sunflowers subjected to different treatments. Our findings indicate that sunflowers exhibited normal growth under the experimental conditions, with variations observed across the different treatments; however, no significant signs of heavy metal toxicity were detected. Specifically, the application of the activators TA or MeJA alone resulted in increased plant height and dry weight, particularly under TA treatment, which aligns with previous research on its effects on crops such as elephant grass and amaranth [18, 19]. This improvement can be attributed to the ability of TA to alleviate the toxicity of heavy metals and chelating agents while also promoting cell differentiation and elongation. Conversely, while MeJA is known to enhance plant stress resistance and mitigate stress-related damage, it may inhibit growth, rendering it less effective in promoting sunflower growth compared to TA. The application of EGTA as a simple addition treatment resulted in decreased sunflower biomass compared to the control (CK) and the individual treatments with activators, suggesting that the combination of the activator and the chelating agent can mitigate the detrimental effects of both the chelating agent and heavy metals on plant biomass, consistent with findings by Xia et al. [20] In contrast, our results diverge from those of Zhang et al. on Xanthium, potentially reflecting different species' varying responses to chelating agent treatments [21].

4.2 Effects of TA, MeJA and EGTA on Cd Uptake by Sunflowe

The application of appropriate amounts of exogenous activating substances and chelating agents can enhance the availability of heavy metals in soil and increase the cadmium (Cd) content in various plant parts, a finding supported by numerous studies [22, 23], including the present research. When ethylene glycol tetraacetic acid (EGTA) is applied alone, the heavy metal content in sunflower roots is slightly lower than that of the control (CK), whereas the content in the aboveground parts is significantly higher, being approximately three times greater than that of CK. This can be attributed to EGTA's role as a widely utilized chelating agent, which forms complexes with metal ions, thereby increasing the solubility of heavy metals. Consequently, this enhances their diffusion range within the roots, improves bioavailability, and facilitates the absorption and accumulation of heavy metals in sunflowers, leading to elevated concentrations in the aboveground parts [24–26]. In comparison to CK, the Cd content in both the roots and shoots of sunflowers increased to varying degrees when tannic acid (TA) was applied alone or in combination with EGTA. This effect may be due to TA's ability to promote the growth of sunflower tissues, regulate root morphology, and modify the rhizosphere ecological environment, which collectively influence the uptake of available Cd from the soil [27].

4.3 Effects of TA, MeJA and EGTA on Cd Accumulation in Sunflower

The total accumulation of heavy metals in plants is closely linked to their absorption capacity and biomass. In this study, the sole application of TA significantly increased cadmium (Cd) accumulation in both the aerial and root parts of sunflower compared to the control (CK); however, the transfer coefficient decreased relative to CK. This effect suggests that TA enhances Cd accumulation in the aerial parts primarily by promoting tissue growth rather than improving the upward transport efficiency of Cd from the roots. When methyl jasmonate (MeJA) was applied alone, Cd accumulation in sunflower roots increased compared to CK, while both shoot and transfer coefficients declined, indicating that MeJA treatment may mitigate Cd accumulation in the shoots and somewhat alleviate the toxic effects of Cd on sunflower plants [28, 29]. Furthermore, the application of ethylenediaminetetraacetic acid (EGTA), either alone or in combination with exogenous activators, led to an increase in Cd accumulation in the aerial parts of sunflower and the entire plant, with the combined application demonstrating superior efficacy. This finding is consistent with the results reported by Chen et al. regarding castor plants, wherein the combination of EGTA and TA was more effective in promoting Cd accumulation in the aerial parts than EGTA alone [11]. This enhanced effectiveness is likely attributable to the combined application of exogenous activators facilitating cell division, thereby improving the movement of metal chelates to vacuoles and cell walls, which may alleviate the toxicity of heavy metals or chelating agents. This mechanism allows for increased heavy metal absorption without significantly harming plant growth and development, as noted in prior studies [30]. Additionally, both EGTA treatment and the synergistic application of EGTA with exogenous activators significantly improved the Cd transport coefficient in sunflower, corroborating the findings of Luo et al. [24].

5 Conclusion

In conclusion, while the exclusive use of chelating agents can significantly enhance the efficiency of phytoremediation, it presents certain drawbacks, including high toxicity, challenges associated with degradation, and elevated costs. Conversely, the application of plant hormones alone has not yielded substantial improvements in remediation efficiency. This study proposes a synergistic approach that combines exogenous activators and chelating agents to leverage their respective strengths and mitigate their weaknesses. This integrated strategy enhances the ability of plants to accumulate heavy metals while reducing the detrimental effects of both chelating agents and heavy metals on plant health, thereby ensuring that biomass is either maintained or only slightly diminished. Consequently, this approach aims to increase the overall accumulation of heavy metals while prioritizing environmental protection. However, it is important to note that this research is still confined to laboratory simulations and has not yet explored the underlying mechanisms in depth. Future work should focus on advancing research in soil field phytoremediation and elucidating the associated remediation mechanisms. This will provide a solid foundation for the practical application of phytoremediation technology in the treatment of large-scale heavy metal-contaminated soils, thus offering a theoretical basis for effective interventions.

References

1. Xu, Y.P., Yang, M., Guo, H.Y.: Phytoremediation potential of five industrial hemp varieties on five heavy metal polluted soils. Acta Agron. Sin. **46**(12), 1970–1978 (2020)
2. Long, C., Wang, D., Chen, L., Jiang, W.J., Xiang, M.W.: Effect of four kinds of phytohormones on U and Cd accumulation in *Helianthus annuus*. Chinese J. Environ. Eng. **11**(5), 3251–3257 (2017)
3. Chen, N.C., Zhen, Y.J., He, X.F., Li, X.F., Zhang, X.X.: Analysis of the report on the national general survey of soil contamination. J. Agro-Environ. Sci. **36**(9), 1689–1692 (2017)
4. Chen, W.P., Yang, Y., Xie, T., Wang, M.E., Peng, C., Wang, R.D.: Challenges and countermeasures for heavy metal pollution control in farmlands of China. Acta Pedol. Sin. **55**(2), 261–272 (2018)
5. Liao, X.Y., Chen, T.B., Yan, X.L., Nie, C.J.: Enhancement of heavy metal removal in phytoremediation of soils contaminated with heavy metals. Acta Sci. Circum. **6**, 881–893 (2007)
6. Liu, W.Y., Wu, G., Hu, H.Q.: Effect of EDDS application on Soil Cu/Cd availability and uptake/transport by castor. Environ. Sci. **45**(3), 1803–1811 (2024)
7. Tammam, A., Ei-Aggan, W., Abou-Shanab, R.: Improved of growth and phytostabilization potential of lead (Pb) in Glebionis coronaria L. under the effect of IAA and GA 3 alone and in combination with EDTA by altering biochemical attributes of stressed plants. Inte. J. Phytoremediation **23**(9), 958–968 (2021)
8. Jan, A.U., Hadi, F., Shah, A.: Plant growth regulators and EDTA improve phytoremediation potential and antioxidant response of *Dysphania ambrosioides* (L.) Mosyakin & Clemants in a Cd-spiked Soil. Environ. Sci. Pollut. Res. **28**(32), 43417–43430 (2021)
9. Zhang, Y.R., et al.: Effect of chelating agent and organic acid on strengthening remediation of cadmium and arsenic contamina-ted soil with *Lonicera japonica* Thunb. Chin. J. Ecol. **43**(5), 1408–1415 (2024)
10. Wang, Z., Sun, Z.J., Sameh, M., Wang, Z., He, J., Han, L.: DA-6 and GLDA Enhanced Pancium virgatum L. to Phytoextract Cd from Contaminated Soils. Environ. Scie. **41**(12), 5589–5599 (2020)
11. Chen, Y.H., Jun, L., Wang, M.X., Fang, S.P., Zhu, B.: Regulation of EGTA and tartaric acid on Cd stress and accumulation in *Ricinus communis* L. Acta Sci. Circum. **34**(5), 1025–1031 (2014)
12. Yung, L., Blaudez, D., Maurice, N., Azou-Barre, A., Sirguey, C.: Dark Septate Endophytes Isolated from Non-hyperaccumulator Plants Can Increase Phytoextraction of Cd and Zn by the Hyperaccumulator *Noccaea caerulescens*. Environ. Sci. Pollut. Res. **28**(13), 16544–16557 (2021)
13. Liu, W., Shu, W.S., Lan, C.Y.: *Viola baoshanensis*, a plant that hyperaccumulates cadmium. Chin. Sci. Bull. **49**(1), 29–32 (2004)
14. Wei, S.H., Clark, G., Doronila, A.I., Jin, J., Monsant, A.C.: Cd hyperaccumulative characteristics of Australia ecotype Solanum Nigrum L. and Its implication in screening hyperaccumulator. Int. J. Phytoremediation **15**(3), 199–205 (2013). https://doi.org/10.1080/15226514.2012.687019
15. January, M.C., Cutright, T.J., Keulen, H.V., Wei, R.: Hydroponic phytoremediation of Cd, Cr, Ni, As, and Fe: can helianthus annuus hyperaccumulate multiple heavy metals? Chemosphere **70**(3), 531–537 (2008)
16. Zou, X.J.: Study on Runoff Risk of Sunflower Remediation of Cd and Pb Contaminated Soil Induced by Chelating. Guangzhou University, Guangzhou (2021)
17. Da-Col, J.A., Domene, S.M.A., Pereira-Filho, E.R.: Fast determination of Cd, Fe, Pb, and Zn in food using AAS. Food Anal. Methods **2**(2), 110–115 (2009)

18. Ma, L.X., et al.: Tartaric acid and agronomic mowing to assist *Pennisetum purpureum* schum in cadmium-contaminated soil phytoremediation and its effect on soil microbial community. Res. Environ. Sci. **37**(2), 359–370 (2024)
19. He, M.Q., et al.: Effects of combined application of aspartic acid diethoxvsuccinate and tartaric acid on growth and cadmium enrichment characteristics of amaranthus hybridus L. J. Sichuan Agricult. Univ. **39**(6), 827–834 (2021)
20. Xia, J.W., Xu, X.X., Lu, X., Chen, Z.Y., Tang, Y., Zhang, S.R.: Effect of combined application of EGTA and organic acid on remediation of cd-contaminated soil by corchorus capsularis L. J. Agro-Environ. Sci. **38**(2), 333–341 (2019)
21. Zhang, Y.R., et al.: Effect of chelating agents and organic acids on remediation of cadmium and arsenic complex contaminated soil using Xanthium sibiricum. Environ. Sci. **43**(8), 4292–4300 (2022)
22. Saffari, V.R., Saffari, M.: Effects of EDTA, citric acid, and tartaric acid application on growth, phytoremediation potential, and antioxidant response of Calendula officinalis L. in a cadmium-spiked calcareous soil. Int. J. Phytoremediat. **22**(11), 1204–1214 (2020). https://doi.org/10.1080/15226514.2020.1754758
23. Guo, D., et al.: EDTA and Organic acids assisted phytoextraction of Cd and Zn from a smelter contaminated soil by potherb mustard (*Brassica juncea,* Coss) and evaluation of its bioindicators. Ecotoxicol. Environ. Saf. **167**(1), 396–403 (2019)
24. Luo, H.Z., Nan, Z.R., Hu, Y.H., Jin, C., Wang, N.: Chelate-induced uptake and accumulation of cd in soil by poplar (Populus bolleana Lauche). China Environ. Sci. **33**(3), 461–465 (2013)
25. Wang, S., Liu, J.N.: The effectiveness and risk comparison of EDTA with EGTA in enhancing Cd phytoextraction by *Mirabilis jalapa* L. Environ. Monit. Assess: An Int. J. **186**(2), 751–759 (2014)
26. Zhang, L.: Preliminary study on chelate-enhanced phytoremediation of cadmium-contaminated soil by cotton plants. J. Soil Water Conserv. **29**(4), 321–326 (2015)
27. Yuan, T.Y., Wan, D.J., Yang, W.J., Gu, J.F., Zhou, H.: Tartaric acid coupled with gibberellin improves remediation efficiency and ensures safe production of crops: a new strategy for phytoremediation. Sci. Total Environ. **908**(1), 726–735 (2024)
28. Yan, Z., Zhang, W., Chen, J., Li, X.: Methyl jasmonate alleviates cadmium toxicity in solanum nigrum by regulating metal uptake and antioxidative capacity. J. Soil Water Conserv. **59**(2), 373–381 (2015)
29. Wang, F.B., et al.: Exogenous methyl jasmonate (MeJA) enhances the tolerance to cadmium (Cd) stress of okra (*Abelmoschus esculentus* L.) plants. Plant Cell, Tissue and Organ Culture: An Int. J. in Vitro Cult. High. Plants **155**(3), 236–245 (2023)
30. Chen, H.C., et al.: Effects of exogenous organic acids on the characteristics of cd accumulation of *Salix variegata* under Cd stress. Acta Ecol. Sin. **39**(12), 4510–4518 (2019)

Recent Progress and Development of Titanium-Bearing Blast Furnace Slag

Guoyin Xu(✉) and Purong Wang

Panzhihua University, Panzhihua City, Sichuan, China
277785081@qq.com

Abstract. Titanium-containing blast furnace slag in China has nearly 100 million tons of accumulation, not only occupy a lot of land and pollute the environment, and because it contains a lot of titanium resources and other valuable components have not been effectively used, resulting in a serious waste of resources. In this paper, the structural properties of titanium-containing blast-furnace slag, the principles and advantages of wet, fire and selective precipitation of titanium components were reviewed. It is suggested that improving the viscosity and other physical properties of the slag, increasing the diffusion rate of titanium components during crystallization, expanding the crystallization temperature difference between titanium components and other components, and introducing crystal seeds or more effective auxiliaries to increase the crystalline particle size of titanium components are the key directions of future research.

Keywords: titanium-bearing blast furnace slag · crystallization · perovskite · rutile

1 Introduction

Vanadium titanium magnetite is a multicomponent symbiotic iron ore with vanadium, titanium and iron as the main elements, and accompanied by many valuable elements. However, titanium has a close symbiosis in the magnetite lattice, and vanadium exists in the titanium magnetite in the form of isotropy, so the suitable ore is called vanadium titanium magnetite. China is rich in titanium resources, and most titanium resources exist in Panxi area in the form of vanadium titanium magnetite. But Panxi vanadium titanium magnet mineral phase is complex, mineral structure is special, the main titanium-containing minerals are titanomagnetite and ilmenite; gangue minerals are titanium pyroxene, silicon oxide and so on. Vanadium-titanium-bearing magnetite concentrate, i.e. Vanadium-titanium-magnetite concentrate, is produced after mining and beneficiation, in which about 50% of titanium resources in the raw ore are inevitably recovered [1, 2]. The reason is that titanium magnetite is a composite mineral phase composed of magnetite as the main crystal and ilmenite as the guest crystal, and magnesia-aluminum spinel is occasionally found in titanium magnetite [3]. In titanium magnetite, minerals such as ilmenite and spinel produce fine particle size (less than 0.005 mm). From the point of view of grinding, it is difficult to dissociate the monomer between magnetite

D. Han et al. (Eds.): ICEPG 2024, ESE, pp. 628–637, 2025.
https://doi.org/10.1007/978-3-031-86829-0_72

and ilmenite crystals, which is the root cause of difficult separation between iron and titanium. At present, the comprehensive utilization process of vanadium titanium magnetite concentrate is mainly based on the blast furnace method, which better realizes the separation of iron, titanium and vanadium resources from each other and the recycling and utilization of iron and vanadium resources. However, the difficult separation of ferrotitanium leads to the inclusion of titanium in the lattice of vanadium titanium magnetite concentrate remaining in the blast furnace slag after smelting iron extraction, which is the so-called titanium-containing blast furnace slag. Titanium bearing blast furnace slag is generally composed of CaO-MgO-SiO_2-$A1_2O_3$-TiO_2 five components, according to the TiO_2 content from low to high can be divided into: Low-titanium blast-furnace slag ($TiO_2 < 10\%$), medium-titanium blast-furnace slag ($TiO_2$10%–15%) and high-titanium blast-furnace slag (TiO_2 in the slag reaches about 20–25%) [4].

Since the titanium component in the titania-containing blast furnace slag exists in the form of several microns, and the current main producing area of titania-containing blast furnace slag in Panxi, has accumulated nearly tons of titania-containing 25% of the slag body, and the annual increase of millions of tons, with a more and more rapid trend, resulting in a large amount of land waste and occupation and environmental pollution. Because the titanium content of titania-containing blast-furnace slag is high, there are nearly 30 million tons of TiO_2, and at least one million tons of TiO_2 are added every year, which has a high recovery value, for decades, it has attracted many scholars at home and abroad to study its titanium extraction and comprehensive utilization ways. But also because of its complex composition and mineral phase, titanium components are in the form of particles as small as a few microns, resulting in a long time, relevant research progress has made certain results, but due to limited processing capacity, secondary pollution, high process costs, can be industrialized results have not yet appeared.

To realize the resource utilization of titanium-containing blast-furnace slag and relieve the environmental pressure of iron and steel industry. The titanium resources in titanium-containing blast furnace slag were extracted to improve the utilization rate of resources. Explore the new technology of comprehensive utilization of titanium-containing blast-furnace slag to promote the sustainable development of the steel industry. It is the long-term expectation of metallurgical and environmental protection industries at home and abroad.

2 Study on Physicochemical Properties of Titanium-Containing Blast-Furnace Slag

The study of physical and chemical properties of titanium-containing blast-furnace slag is of great significance for the effective utilization of titanium resources in slag and environmental protection. The blast furnace slag containing titanium usually presents as black or gray blocky solid. Due to the adjustment of alkalinity in smelting, the basic density is generally between 2.5–3.5 g/cm^3, containing titanium, silicon, calcium, magnesium, aluminum and other elements. Titanium mainly exists in the form of titanium oxides, such as TiO_2.

Titanium-containing blast-furnace slag is a kind of "short slag" with high melting temperature and narrow fluidity. Viscosity is one of the most important physical and chemical properties affecting the stability and performance of slag. However, the viscosity of titanium-containing blast-furnace slag is not high at high temperature. The current research believes that the reason for the thickening of high titanium slag is mainly due to the existence of a certain amount (such as more than 1%) of titanium carbon and nitride in the slag, the melting point of these compounds is as high as 3000 °C, the density is similar to the slag, and does not melt in the slag, so these titanium carbon nitride in the form of a few microns of solid suspended matter dispersed in the liquid slag. The viscosity of slag is significantly increased [5]. Bai Chenguang [6] found that in a neutral or oxidizing atmosphere, the viscosity of titanic-containing blast-furnace slag would not become thick. The mechanism is that the thickening of slag is mainly related to the reduction degree of TiO_2. Titanium produced by deep reduction has more solid particles of carbonitride, and these titanium carbonitride will increase the apparent viscosity of slag. When the weight percentage of carbonitride produced exceeds 6%, the viscosity of slag increases sharply. The highly dispersed form of carbonitride in the slag also has a great positive effect on the viscosity. According to the calculation, the reaction kinetics and diffusion conditions also have important effects on the dispersion form of titanium carbonitride. The more important finding is that the molten state of titanium-containing blast-furnace slag is more consistent with the state of non-Newtonian fluid, and in the past, researchers have always used the state of Newtonian fluid to study titanium-containing blast-furnace slag, which is also the reason why the previous research results deviate greatly from the actual.

In addition, a large number of iron beads in high titanium slag is also considered to be one of the important reasons for the sharp increase in slag viscosity. The melting temperature of vanadium and titanium slag increases significantly with the increase of Al_2O_3 and TiO_2 content, decreases with the increase of FeO content, and first decreases and then increases with the increase of binary alkalinity [7, 8]. Feng C [9] studied that the content of CaO/SiO_2 and Al_2O_3 is negatively correlated with the viscosity and viscous flow activation energy of blast furnace slag, and positively correlated with the melting temperature.Hu Mengjun et al. [10] found that the crystalline form of titanium components in titanium-containing slag is closely related to their alkalinity. At low alkalinity, titanium components are crystallized out in the form of TiO_2, while at high alkalinity, they are crystallized out in the form of perovskite. However, the crystal size of titanium minerals obtained by the two crystallization paths is small and symbiosis with other substances, making it difficult to separate and enrich titanium resources. Liu J. H. [11] analyzed the effects of B_2O_3 and TiO_2 on slag viscosity and melting point temperature, and the results showed that the liquid phase regions contain magnesium titanate were changed when the percent of TiO_2 was enhanced. The melting rate of blast-furnace slag can be increased by reducing its viscosity, so the content of TiO_2 and B_2O_3 should be controlled.

3 Research on the Direct Use of Titanium-Containing Blast-Furnace Slag as Catalytic Material and Building Material

Due to the lack of viable recycling technologies, TBFS are mainly used as building materials or for outdoor stacking, leading to waste of resources and environmental concerns. The TiO_2 content of vanadium titanium-magnetite used in blast furnace smelting outside China generally does not exceed 3–4%, and the TiO_2 content of titanium-containing blast furnace slag is less than 10%, which has little impact on the strength of cement and can be directly used as building materials such as cement or ceramics. However, the TiO_2 content of titanium-containing blast furnace slag in China exceeds 20%, which limits the feasibility of direct use of titanium-containing blast furnace slag as building materials.

Li Shuqin et al. [12] studied the design of CaO-SiO_2-MgO(15%)-Al_2O_3 phase diagram to regulate the physical properties, mineral phase structure and other properties of the prepared samples, taking pyroxene, alkali basalt calc albitite, perovskite and ferroplate titanite as the research objects to regulate the growth of pyroxene while decomgenerating titanium minerals such as perovskite. The preparation of high strength ceramic material was realized. Zhao Xiaoteng [13] prepared calcium oxalate cement (COC) by mixing titanium-containing blast furnace slag with oxalic acid, and used it as matrix material to cure/stabilize heavy metals nickel and lead. However, the influence of titanium components on the performance of COC was not studied. Huo Hongying [14] studied the synthesis of photocatalyst by doping Ni with TiO_2 in titania-containing blast furnace slag to realize the comprehensive utilization of titania-containing blast furnace slag by non-titanium extraction method.

4 Research on Titanium Extraction from Titanium-Containing Blast-Furnace Slag

4.1 Leaching

In view of the complex composition and difficult treatment of titanium-containing blast-furnace slag produced by blast furnace smelting of vanadium titanium magnetite, Xie Siyuan et al. [15] used ammonium chloride roasting to activate the oxides of calcium, magnesium and other elements in the slag before acid leaching, and the chlorination roasting destroyed the complex and refractory mineral phase structure of calcium, magnesium and aluminum in the blast-furnace slag, making it transform into chloride or simple oxides easily soluble in hydrochloric acid. Most of the calcium, aluminum and magnesium were removed by acid leaching to achieve enrichment of titanium in the slag. However, the process did not achieve the removal of silicon element, which accounted for the largest proportion of gangue minerals in the titanium-containing blast-furnace slag, and the enrichment effect of titanium was not ideal. Shan Zhu [16] studied the influence of mechanical activation on the process of hydrochloric acid leaching of titanium-containing blast-furnace slag by using univariate method. The optimum leaching characteristics were determined by changing leaching temperature, leaching time, stirring intensity, mass fraction of hydrochloric acid solution and liquid-solid ratio. The optimum mechanical activation characteristics were determined by changing the mechanical activation time, activation speed and the ratio of grinding ball to material mass. The influence of different mechanical activation methods on the leaching rate of titanium from BLast-furnace slag was summarized. Liu Y et al. [17] studied the leaching rule of hydrochloric acid leaching of titanium-containing blast-furnace slag, and investigated the influence of technological conditions on the leaching rates of calcium, magnesium, aluminum and iron, as well as the phase of the leaching slag. The experimental results show that HCl solution can selectively leach elements from titanium-containing blast-furnace slag. The leaching slag phase is mainly perovskite, followed by magnesium silicate and tricalcium aluminate, and titanium-rich materials can be obtained by desiliconization. Zhou L et al. [18] studied the test of leaching after roasting titanium-containing blast-furnace slag by sulphation. After roasting by sulphation, the main metal components of the blast-furnace slag containing titanium can be converted into sulfate. It can be seen from the analysis that under the condition of strong acid, the main metal components can react to form sulfate, and the effective separation of Ti, Mg, Al from Ca and Si can be realized.

4.2 High Temperature Carbonization Process

Wu KH [19] studied the combined process of vacuum carbothermal reduction-acid leaching to treat titanium-containing blast furnace slag. It is found that the carbon thermal reduction temperature and vacuum degree are positively correlated with the reduction degree of each component in the slag. Under better parameters, silicon magnesium and titanium in slag can be completely separated, but the vacuum high-temperature acid-leaching process is complicated, especially the vacuum is difficult to maintain, and it is difficult to realize industrialization in production.

Many studies have been conducted on the titanium extraction process of titanium-containing blast-furnace slag by first carbonizing at high temperature and then chlorinating at low temperature [20–22]. In this process, the titanium oxide in the blast furnace slag is reduced to Ti (C, N) by high temperature carbonization, and then chlorinated to $TiCl_4$ at low temperature, so as to realize the separation of titanium slag. The process has the advantages of short process flow, high carbonization rate and large processing capacity, but the high dispersion of Ti (C, N) remelting slag formed in the high temperature carbonization process, and the difficulty of separating Ti (C, N) from the slag greatly restricts the subsequent chlorination process. After the study, Jian Tingfang [23, 24] proposed the idea of enriching Ti(C, N) by adding iron in the high-temperature carbonization process, which promoted the accumulation of Ti(C, N). However, there were still some problems in this process, such as grinding difficulties in the chlorination process, easy bonding of slag, and accumulation of a large amount of chlorinated tailings.

4.3 Selective Precipitation of Titanium Minerals

Because titanium elements are distributed in different mineral phases with fine particle size ($<10\mu m$) of titanium-containing blast-furnace slag, this is the main reason for the low recovery and high recovery cost of traditional methods such as leaching to prepare alloys. The process idea of selective precipitation of titanium minerals is to achieve a variety of fine-grained titanium components through mineral phase recombination, control crystallization to achieve the enrichment of titanium components to the same mineral phase, and then recover titanium components by physical or chemical methods. The essence of this process is "mineral processing and metallurgical combination".

There are many studies on selective enrichment of titanium from titanium-containing blast-furnace slag to perovskite phase. The main idea is to adjust the physical and chemical properties of titanium-containing blast furnace slag, so that the titanium component is enriched into the three ore phases of perovskite, rutile or black titanite, and induce the grain growth to about 70 microns, which is convenient for subsequent recovery. Sui Zhitong [25] proposed that since perovskite is the primary crystalline phase in the slag and has the highest titanium content, perovskite phase was selected as the designed titanium-rich phase. After modifying the physical and chemical properties of blast slag, the titanium components dispersed in various mineral phases were enriched into perovskite as much as possible by mineral phase recombination, so as to realize the transformation from dispersion to concentration and recycle perovskite by means of mineral processing. To achieve comprehensive recovery of titanium. Li Yuhai [26] showed that perovskite is the main titanium-rich phase through the process mineralogy analysis of titanium-bearing blast-furnace slag of Panzhihua Steel. The use of CaO and Fe_2O_3 to modify the blast furnace slag not only increases the precipitation temperature of perovskite, but also significantly reduces the precipitation temperature of the other two main titanium-containing minerals, Pantiodiopside and titanium-rich diopside. Under proper heat treatment conditions, nearly 80% of titanium components enter the perovskite phase, and the average crystal size of perovskite reaches 90 μm, which achieves selective enrichment of titanium components in the slag and creates conditions for mineral separation. However, perovskite is dendritic crystal and the density is similar to gangue, so the subsequent separation of perovskite is difficult. Compared with perovskite, rutile is a non-dendritic crystal with a higher density (4.2–4.3 g/cm^3) and high added value. Therefore, it is more valuable to study the enrichment of titanium elements in titanium-containing blaster slag to rutile than to perovskite [27]. Zhang Wu [28] made most of the titanium components in the slag form rutile phase through the modification treatment of titanium-containing blast-furnace slag, and realized the enrichment of titanium components in the slag. The titanium components in the modified slag were separated by a method combining shaking bed separation and hydrochloric acid leaching for impurity removal. Han Jiping [29, 30] obtained the raw material for converting most titanium components into rutile by modifying titaniferous blast-furnace slag with silicon oxide and compressed air at high temperature, and then realized selective separation of rutile crystals by gravity separation-leaching process to obtain high-quality artificial rutile with a TiO_2 content of 98.61%.

The third idea is to enrich the titanium components in the titanium-bearing blast furnace slag to black titanite. Ji Su [31] et al., by adjusting the binary alkalinity in the titanium-containing BLBF slag and conducting slow cooling treatment within the appropriate temperature range, realized that the titanium-rich phase in the slag was black ilmenite solid solution, and made the black ilmenite phase in the titanium-containing slag grow and coarser. According to the statistics, the average equivalent diameter of the grain section of black titanite is 100 ~ 130 μm, and the larger the mineral phase size, the better the monomer dissociation of the mineral phase. Li et al. [32] studied the effect of adding high-titanium slag as an additive to titanium-containing blast-furnace slag on the change and precipitation of titanium enriched phase. By adjusting the mixture of high-titanium slag and titanium-containing slag in different proportions, the titanium enriched phase was transformed from perovskite to black titanite. In addition, complete crystallization can be achieved when it is held at 1450 °C for 25 min. However, the current selective precipitation process of titanium minerals is limited by the fact that the titanium components in the titanium-containing blaster slag are too fine, requiring a certain holding time to realize the crystallization and growth of the titanium components, and the cooling rate also has a great impact on the crystallization particle size, resulting in a more complicated process. In addition, the crystallization particle size achieved by the current research process is not enough, which is uneconomical and unfriendly to the subsequent beneficiation process. Therefore, how to achieve faster and larger grain size crystallization of precipitated titanium minerals is the difficulty and focus of selective precipitation of titanium minerals. In general, the process of selective precipitation of titanium minerals is the most economical, the best to achieve industrialization and the most promising in the titanium extraction process containing many titanium blast-furnace slag. Liu Weizao et al. [33] proposed the idea of using waste to make waste, and a new process of co-roasting titanic-containing blast furnace slag and green alum to enrich rutile. At high temperature, sulfur dioxide from the thermal decomposition of green alum is sulfated with perovskite and titanium-bearing pyroxene in titanium-bearing blast furnace slag, and the calcium and magnesium components are transformed into sulfate, while the titanium components are enriched into rutile. It was found that adding Na_2SO_4 could significantly improve the enrichment efficiency of Ti.

4.4 Direct Preparation of Alloys

Zou Xingli [34] studied the direct extraction of titanium-silicon alloy from the titanium-containing blast furnace slag in Panzhihua by solid oxygen permeable film (SOM) method and direct electrolysis process. Ding Mantang [35] studied the preparation of titanium-silica-aluminum alloy by reducing titanium-containing blast furnace slag with aluminum under the protection of argon. Wang Chao [36] used the strong reductivity of low-purity silicon/silicon waste to extract titanium from titania-containing blast-furnace slag to obtain Si-Ti alloy, and used the oxide slag generated by the blast-furnace slag to remove the most important and most difficult oxygen impurities from the silicon waste. Li Zushu et al. [37] used direct current silicothermal method to produce titania-silicon alloy containing about 20% Ti, and the titanium recovery rate was less than 60%. The process of producing titanium silicon alloy by DC Si thermit was developed.

The titanium content in titanium-containing blast-furnace slag is up to 25%, and the silicon content is about 26%. If the silicon and titanium can be recovered at the same time, the processing cost per ton of titanium-containing blast-furnace slag can be reduced. Liu Y et al. [38] from Sichuan University prepared micron porous Si/Si-Ti alloy cathode materials for lithium batteries with good application prospects by thermit reduction method from titanic-containing blast furnace slag (TBFS). The process includes Al metal thermal reduction TBFS and acid etching. By adjusting the ratio of TBFS and Al, the specific surface of the anode material can be controlled to control the volume expansion of Si, and the electron transfer efficiency and cycle stability can be improved. The process realizes simultaneous recovery of silicon and titanium, which is economical and simple.

5 Conclusion and Prospect

(1) Inhibit the formation of titanium carbon nitride or further improve the viscosity and other physical properties of titanium containing blast furnace slag, and increase the diffusion rate of titanium components during crystallization.
(2) Develop new catalysts or melting aids, expand the crystallization temperature difference between titanium components and other components, achieve lower energy consumption and pollution emissions, and develop green and sustainable treatment technologies.
(3) The introduction of crystal seeds or more effective additives to increase the crystal size of titanium components, which is convenient for subsequent beneficiation or wet leaching recovery.

Acknowledgments. Supported by: Key Project of Science and Technology Plan of Sichuan Province, Number: 2022YFS0458, Name: Research and Application of Key technology of titanium extraction from titanium-containing Blast-furnace slag in Panzhihua.

References

1. Zhang, Q., Gao, L., Chen, X., et al.: Research progress of titanium extraction from titanium-containing blast furnace slag. Chem. Miner. Process. **52**(12), 28–36 (2023)
2. Yong, D., Zhengliang, Z., Junguo, L., et al.: Titanium enrichment technology and titanium resource utilization of titanium-containing blast furnace slag. China Metallurgy **32**(08), 25–31 (2022)
3. Chen, F., Yang, X., Cai, X., et al.:0 Experimental study on iron separation of Baima gabbro-type ultra-low grade vanadium-titanium magnetite in Panxi area. Compr. Utilization Miner. (06), 26–30 (2020)
4. Chen, F., Wen, Y., Wang, Z., et al.: Study on quality improvement technology of a vanadium-ilmenite concentrate in Panxi. Min. Metall. Eng. **43**(05), 77–79+84 (2023)
5. Ting, D., Hongjuan, S., Tongjiang, P., et al.: Study on mineral crystallization behavior during heating process of water-quenched titanium-containing blast-furnace slag. Chem. Miner. Process. **50**(02), 40–45 (2021)
6. Bai, C.G.: Research on some physicochemical Problems of Titanic-containing blast-furnace slag. Chongqing University (2003)
7. Qiu, G., Miao, D., Cai, M., et al.: Viscosity and melting property of Titanium-containing blast-furnace slag. Iron and Steel **57**(11), 42–52 (2022). (in Chinese)
8. He, S.Q.: Study on Extraction and Separation Principle of Valuable Components and Chemical Kinetics of High Titanium Blast Furnace Slag in Panzhihua. Southwest University of Science and Technology (2020)
9. Feng, C., Chu, M., Tang, J., et al.: Effect of CaO/SiO_2 and Al_2O_3 on viscous behaviors of the titanium-bearing blast furnace slag. Steel Res. Int. **87**(10), 1274–1283 (2016)

10. Hu Mengjun, Q., Zhengfeng, L.L., et al.: Effect of binary alkalinity on crystallization process of Titania-containing blast furnace slag. J. Chongqing Univ. Technol. (Nat. Sci. Edn.) **30**(6), 51–57 (2016)
11. Liu, J.H., Wang, Q., Guan, Z.G., et al.: Reaction mechanism analysis of the TiO_2 in vanadium-titanium blast furnace (BF) slag. Metalurgija **59**(1), 7–9 (2020)
12. Li, S., Hua, S., Li, X., et al.: Preparation of high strength ceramic materials from titanium-containing blast furnace slag. Compr. Utilization Mine. Resourc.1-12[2024-08-19]
13. Zhao, X.: Study on preparation and solidification of heavy metals Nickel and Lead by oxalate cement containing titanium blast furnace slag. Kunming University of Science and Technology (2023)
14. Hongying, H.: Preparation of photocatalyst by Ni-doped titanium-containing blast furnace slag and its properties. Rare Metals Cemented Carbide **50**(02), 64–69 (2022)
15. Siyuan, X., Wei, J., Shengdong, W., et al.: Activation test of ammonium chloride roasting of Titanium-containing blast-furnace slag. Nonferrous Metals (Smelting Part) **05**, 68–75 (2024)
16. Zhu, S., Hu, J., Zhang, C., et al.: Process optimization and kinetics of titanium leaching from mechanically activated titanium-bearing blast furnace slag. J. Sustain. Metall. **9**(1), 230–239 (2023)
17. Liu, Y, Chen, X, Mao, S., et al.: Extraction of valuable metals from titanium-bearing blast furnace slag by acid leaching. J. Wuhan Univ. Technol. Mater. Sci. Ed. **39**(2), 376–385 (2024)
18. Zhou, L., Peng, T., Sun, H., et al.: Thermodynamics analysis and experiments on Ti-bearing blast furnace slag leaching enhanced by sulfuric acid roasting. RSC Adv. **12**(54), 34990–35001 (2022)
19. Wu, K.H., Zhang, G.H., Chou, K.C.: Preparation technology of Ti-rich material from ilmenite via method of vacuum carbothermal reduction. Can. Metall. Q. **58**(2), 196–203 (2019)
20. Lv, X., Song, K., Xin, Y., Lv, X.: Novel process for deep removal of chlorine and recycling of chlorinated tailings from titanium-bearing blast-furnace slag. Process Safety Environ. Protect. **159**, 842–849 (2022)
21. Li, L., Jiang, T., Chen, B., et al.: Overall utilization of Ti-extraction blast furnace slag as a raw building material: removal of chlorine from slag by water washing and sintering. J. Sustain. Metall. **7**, 1116–1127 (2021)
22. You, H., Sun, H., Peng, T., et al.: Recovery of residual carbon from Ti-extraction blast furnace slag by flotation with simultaneous dechlorination. Energies **15**(18), 6777 (2022)
23. Tingfang, J., Lang, Z., Xiangyu, L., et al.: Study on the separation of titanium-slag in the carbonization process of Titanium-containing blast furnace slag. Vanadium and Titanium Steel **42**(06), 51–58 (2021)
24. Jian, T.: Study on Ti (C,N) Enrichment and Separation during Carbonization of Titanium-Containing Blaster Slag. Chongqing University (2022)
25. Yuhai, L., Taiping, L., Zhitong, S.: Selective enrichment of titanium in titanium-containing blast furnace slag and crystallization behavior of perovskite. Chinese J. Nonferrous Met. **10**(5), 4 (2000)
26. Sui, Z., Guo, Z., Li, L., et al.: Green separation technology of titanium from titanium-containing blast furnace slag. Chinese J. Mater. Metall. (02), 93–97 (2006)
27. Han, J., Li, L., Yin, H., et al.: Study on selective separation and leaching mechanism of rutile from modified Titanium-containing blast-furnace slag. Nonferrous Metals Eng. (2024)
28. Zhang, W.: Study on the precipitation and separation of rutile phase from modified titanium-containing blast-furnace slag. Northeastern University (2013)
29. Jiqing, H., Jing, Z., Jiahao, Z., et al.: Recovery of Iron, vanadium and titanium from modified blast furnace slag containing titanium. Trans. Nonferrous Metals Soc. China **32**(01), 333–344 (2022)
30. Jiqing, H., Jiahao, Z., Xiao, C., et al.: Effect of crystal shape on deposition of rutile. Trans. Nonferrous Metals Soc. China **30**(10), 2848–2860 (2020)

31. Ji, S., Qi, H., Qiuju, L., et al.: Experimental study on gravity separation of black-ilmenite phase from titanium-containing slag. Steel **39**(03), 63–68 (2018)
32. Li, J., Zhang, Z.T., Wang, X.D.: Precipitation behaviour of Ti enriched phase in Ti bearing slag. Ironmaking Steelmaking **39**(6), 414–418 (2012)
33. Weizao, L., Minyu, H., Qingcai, L., et al.: Study on the mechanism of enrichment of rutile by synergistic roasting of titanium-containing blast furnaceslag/green alum. Protect. Utilization Miner. Resour. **2**(03), 75–81 (2024)
34. Zou, X., Lu, X.: Direct preparation of titanium alloys from blast furnace slag containing titanium in Panzhihua. Chin. J. Nonferrous Metals **20**(09), 1829–1835 (2010)
35. Ding, M.: Synthesis of titanium silicon aluminum alloy by thermit reduction of titanium-containing blast furnace slag. Compreh. Utilization Miner. Resour. (2024)
36. Wang, C.: Separation and purification of Si-Ti alloy and clean utilization of titanium-containing blast furnace slag and silicon waste. Kunming University of Science and Technology (2021)
37. Li, Z., Lei, Y., Ma, W., et al.: Preparation of high-purity $TiSi_2$ and eutectic Si–Ti alloy by separation of Si–Ti alloy for clean utilization of Ti-bearing blast furnace slag. Sep. Purif. Technol. **265**, 118473 (2021)
38. Liu, Y., Zhong, Y., Zeng, Z., et al.: Scalable synthesis of a porous micro Si/Si-Ti alloy anode for lithium-ion battery from recovery of titanium-blast furnace slag. ACS Appl. Mater. Interfaces (2024)

Micromechanical Analysis on Fissures Propagation for Loess Under Unloading of Loess Cut Slopes

Zhongxiang Liu[1(✉)], Wanjun Ye[2], Qiujuan Zhou[1], Xihao Dong[2], and Minyi Wang[1]

[1] School of Municipal Engineering, Guangdong Institute of Water Conservancy and Electric Power Vocational and Technical College, Guangzhou 510635, China
1091191547@qq.com

[2] School of Architecture and Civil Engineering, Xi'an University of Science and Technology, Xi'an 710054, China

Abstract. Fissure extending in loess is one of the main causes of resulting in the collapse of loess, and unloading of loess cut slope impels fissure extend In order to simulate stress paths of loess under unloading of cut slopes, loess samples were carried on lateral unloading test by triaxial apparatus (GDS).). Combined with CT scanning image analysis, the change law of fissure extend of loess under unloading was disscured. The results show that with the increase of the unloading weight, the sensitivity of the loess was increased. The results show that with the increase of the unloading weight, samples change from elastic deformation to elastic-plastic and plastic deformation. With the increase of The larger the unloading velocity, the elastic-plastic deformation changes from elastic deformation to elastic-plastic and plastic deformation. With the increase of the larger the unloading velocity, the elastic deformation become smaller, the elastic-plastic and plastic deformation become larger, and shear stresses of samples become larger.

Keywords: loess · unloading · fissures · CT scanning

1 Introduction

Due to the special topographic conditions in the loess area, in the process of building factories, opening up Zhuangji, building kilns and other industrial and civil construction activities, or in the process of road construction, water conservancy project construction, often cut slopes and unloading in order to increase the area of land for construction or stabilization of slopes. Cutting slope unloading to form up to several meters, dozens of meters of steep cliffs and even hundreds of meters of steep slopes, increasing the construction land, but also lead to unloading cracks inside the slope, these cracks play a decisive role in the stability of the slope body, increasing the slope deformation and destruction of the hidden danger. In order to design these slopes in a safe, reliable and economically rational way, it is especially important to correctly understand the expansion law of the cracks inside the loess body under the unloading condition of excavation.

D. Han et al. (Eds.): ICEPG 2024, ESE, pp. 638–650, 2025.
https://doi.org/10.1007/978-3-031-86829-0_73

In recent years, domestic scholars have achieved certain results by using CT technology to study the structural changes of loess. Wang Jingming et al. [1–3] studied the types and characteristics of joint fractures in loess. He Shixiu et al. [4] found that by studying the lateral unloading test of the soil body of the foundation pit, the vertical pressure is unchanged, and the stress-strain relationship of the soil body during lateral unloading is transformed from strain-hardening type to strain-softening type with the increase of the consolidation pressure. Li Jiagui et al. [5–7] found that the stress-strain curve of unsaturated Q3 loess was of hardening type through the study of fine structure evolution of lateral unloading of Q3 loess. Pu Yibin et al. [8] scanned the in-situ loess in lateral limit compression, triaxial compression, water immersion and loading-immersion process by CT machine combined with a simple pressure chamber, and observed the characteristics of the fine structure changes of the soil samples in each test process. Lei Shengyou et al. [9] conducted CT scans on Q3 in-situ loess specimens during triaxial shear and triaxial wet subsidence tests, and preliminarily analyzed the change mechanism of the fine structure of loess during strain softening, strain hardening and wet subsidence. Chen Zhenghan et al. [10–12] analyzed the structural evolution characteristics of loess during loading, wet subsidence and post-wet subsidence loading with the help of CT-triaxial instrument. Ye Wanjun et al. [13] analyzed the structure of loess samples under different freezing rates in Luochuan Q2 loess, and obtained the influence of freezing rate on the density zones ofloess. Zhu Yuanqing et al. [14] carried out CT-triaxial experimental study on the dynamic evolution of the fine structure of the Q3 loess during loading and wet sinking to reveal the changes of the fine structure of the loess during loading and wet sinking. Lu Quanzhong et al. [15–18] studied the mechanical properties of fissured loess and its deformation and damage mechanism. Liang Yan et al. [19] studied the effect of water content change on the cohesion and internal friction angle of primary loess under unloading condition by taking the primary loess of Q2 in the south source of Shaanxi Province Trail as the research object.

The above conclusions are of great significance to the study of the development of loess fissures under unloading, but the mechanism of unloading of loess slopes on loess fissures is not clear yet. The influence of unloading rate, unloading volume and water content on the expansion of loess fissures has not been reported, and the characteristics of the stress-strain relationship curves of loess soils under different unloading rate and unloading volume conditions need to be further studied.

To this end, this paper adopts GDS unsaturated soil triaxial instrument to carry out lateral unloading test of primary loess with different unloading rate, unloading volume and water content, to analyze the characteristics of the secondary stress change of loess slope excavation, combined with CT scanning technology, to study the unloading fissure expansion law of primary loess, as well as the stress-strain relationship curves and the change rule of strength parameter of the soil under different unloading conditions, and to explore the effect of unloading rate and volume on the mechanical strength characteristics of loess. The influence of unloading rate and unloading amount on the mechanical strength of loess is investigated, and the evolution of loess fissures under unloading is explored.

2 Lateral Unloading Test on In-Situ Loess

2.1 Test Soil Samples

In the unloading fissure development Luochuan loess plateau side to choose a typical loess collapse point, take Q3 in-situ loess, the color is yellowish brown, no stratification, vertical joints, the soil quality is uniform, visible wormholes, locally rich in macropores, visible white calcium nodules, slightly wet, is hard plastic state. The physical property indexes of the soil samples taken are shown in Table 1.

Table 1. Indexof physical properties of loess samples

norm	Water content %	Density g/cm3	porosity ratio	Liquid limit %	Plastic limit %	relative density
numer	13.2	1.43	1.15	20.6	31.2	2.72

2.2 Test Equipment

The unloading test was carried out using the GDS Unsaturated Soil Triaxial Test System (Fig. 1(a)) in the Unsaturated Soil Triaxial Laboratory of Xi'an University of Science and Technology (XUSTech), which was developed by GDS (Geotechnical Digital Systems Instruments Ltd.), UK, and is capable of completing the static triaxial test of saturated and unsaturated soils under multi-stress paths.

The CT scanning tests were completed at the State Key Laboratory of Permafrost Engineering, Institute of Cold and Arid Regions Environment and Engineering, Chinese Academy of Sciences (CAS), using a SIEMENS SOMATOM plus type X-ray spiral CT machine (Fig. 1(b)), which has a resolution of 0.35 mm × 0.35 mm, a minimum recognizable volume of 0.12 mm^3 (layer thickness of 1 mm), and a density contrast resolution of 0.3%. The scanning voltage was 120 kV, the current was 165 mA, the scanning thickness was 5 mm, the scanning time was 2s, the post-positioning, the magnification factor was 10.0, and the temperature was room temperature (20 C). This CT scanning was divided into initial scanning and scanning after unloading test, each soil sample was scanned 1 section (middle) and its position was marked, and the marked position was scanned after unloading test.

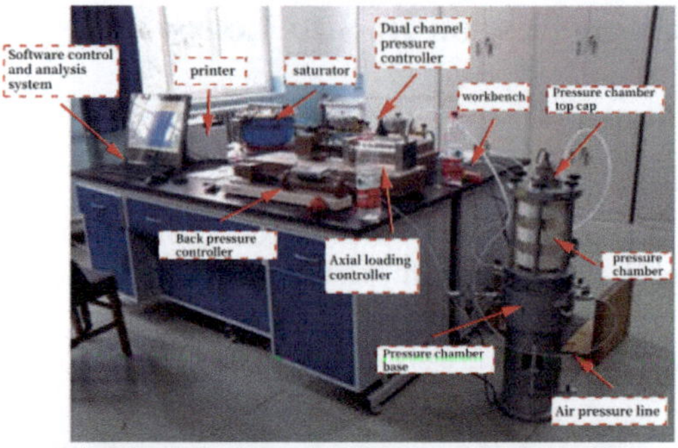

(a) GDS triaxial meter for unsaturated soils

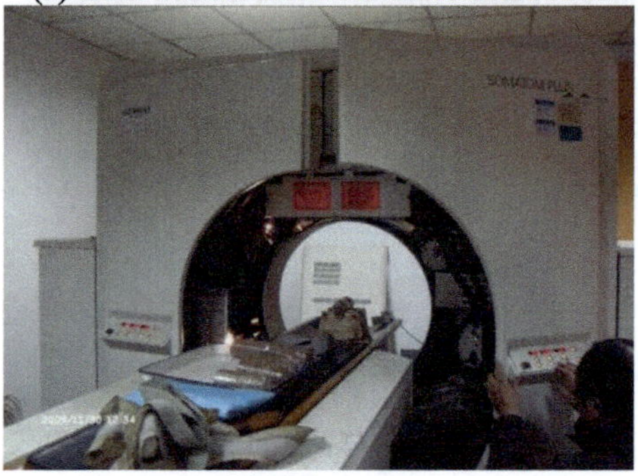

(b) CT scanner

Fig. 1. Instrument and equipment

2.3 Experimental Steps

The loess specimens were prepared in the laboratory to carry out the loess lateral unloading test and CT scan test. The experimental steps are as follows:

(1) In the laboratory to prepare φ39.1 mm × 80 mm in situ specimens, using natural air-drying method and artificial humidification method equipped with different moisture content w (7%, 13%, 18%, 22%, 27%) of the specimen, and grouped and numbered;
(2) An initial CT scan was performed on the middle (0.5 h) section of the specimen and the initial scan position was marked with a marker;
(3) so that the specimen in the 500 kPa pressure under the action of isobaric consolidation, such as consolidation is completed to maintain the axial pressure is unchanged, at a certain rate Δq (0.01 kPa/min, 0.02 kPa/min, 0.04 kPa/min, 0.08 kPa/min) for lateral unloading, when the amount of unloading (50 kPa, 100 kPa, 150 kPa, specimen completely destroyed) and then end the test. kPa, the specimen is completely destroyed) after the test is completed;
(4) The unloaded specimen was subjected to CT scanning again, with the scanning position remaining the same as in the initial scan.

The scanned section was partitioned (Fig. 2), from the center to the edge of the section, it was divided into three areas, namely, the center, the middle ring, and the outer ring, and two synthetic areas, i.e., the large area and the whole area, where the large area is the sum of both the center and the middle ring, and the whole area is the sum of the center, the middle ring, and the outer ring. For the initial scanning before the unloading test, since the size of the soil sample is φ39.1 mm × 80 mm, the corresponding area of the full zone is 12.00 cm^2, the diameter of the center is 1.60 cm, and its area is 1.96 cm^2, the diameter of the large zone is 2.90 cm, and its area is 6.50 cm^2, and the diameter of the middle ring is 4.54 cm^2, and its area is 5.50 cm^2.

3 Test Results and Analysis

3.1 Stress-Strain Characterization

Figure 3 shows the stress-strain relationship curves of the specimens ($w = 13\%$) under different unloading rate and unloading amount conditions. The test results show that the larger the unloading rate is, the smaller the elastic deformation of the specimen is, and the larger the elastic and plastic deformation is. When the unloading rate is 50 kPa, the stress-strain relationship is roughly a straight line segment, indicating that the soil body is still in the elastic deformation stage, but the elastic deformation of the soil body is different under different unloading rates, the larger the unloading rate is, the larger the slope of the straight line segment is, and the smaller the elastic deformation is. Continue to increase the unloading amount (100 kPa), the soil body appeared elastic-plastic deformation, the unloading rate is large elastic-plastic deformation is small, and viceversa, the deformation is larger. When the unloading amount reaches 150 kPa, the soil with large unloading rate is still in the plastic deformation stage, while the soil with small unloading rate has entered the plastic deformation stage. When the specimen is completely destroyed, the stress-strain relationship curve is a weak hardening type, which can be roughly divided into elastic, elastic-plastic and plastic three deformation stages, and the larger the unloading rate, the smaller the elastic deformation, elastic-plastic, plastic deformation is larger.

It can also be seen from Fig. 3 that the higher the unloading rate, the higher the shear stress of the specimen. The reason is that the effect of unloading rate on the strength of loess is mainly divided into two aspects: the weakening of soil material strength and the strengthening of friction strength. Different unloading rates have different degrees of deterioration of soil material strength and strengthening of friction strength. When the unloading rate is large, the soil body does not have sufficient time to carry out plastic deformation, the deformation of the soil body tends to localize, showing obvious elasticity, that is, the elastic deformation increases, the plastic deformation is delayed, and the strength of the soil body is also increased accordingly. In the unloading test soil samples before the deformation and destruction, the friction strength of the degree of excitation is enhanced, the degree of cohesion weakening with the unloading rate increased and weakened, i.e., the greater the unloading rate, the greater the cohesion and the angle of internal friction, which in the joint action of the two lead to the soil body's overall strength is greater.

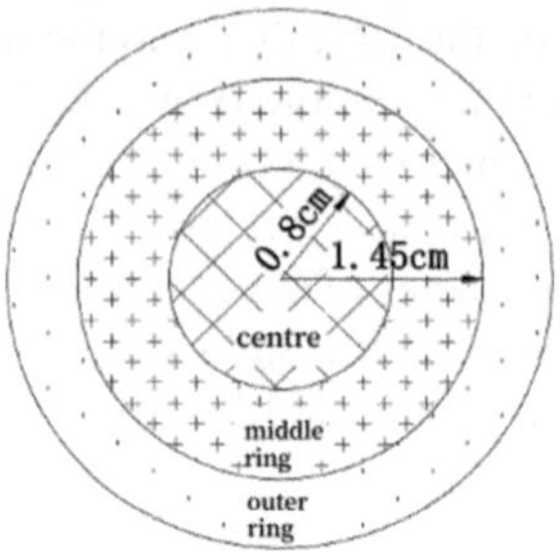

Fig. 2. Schematic diagram of scanning section zoning

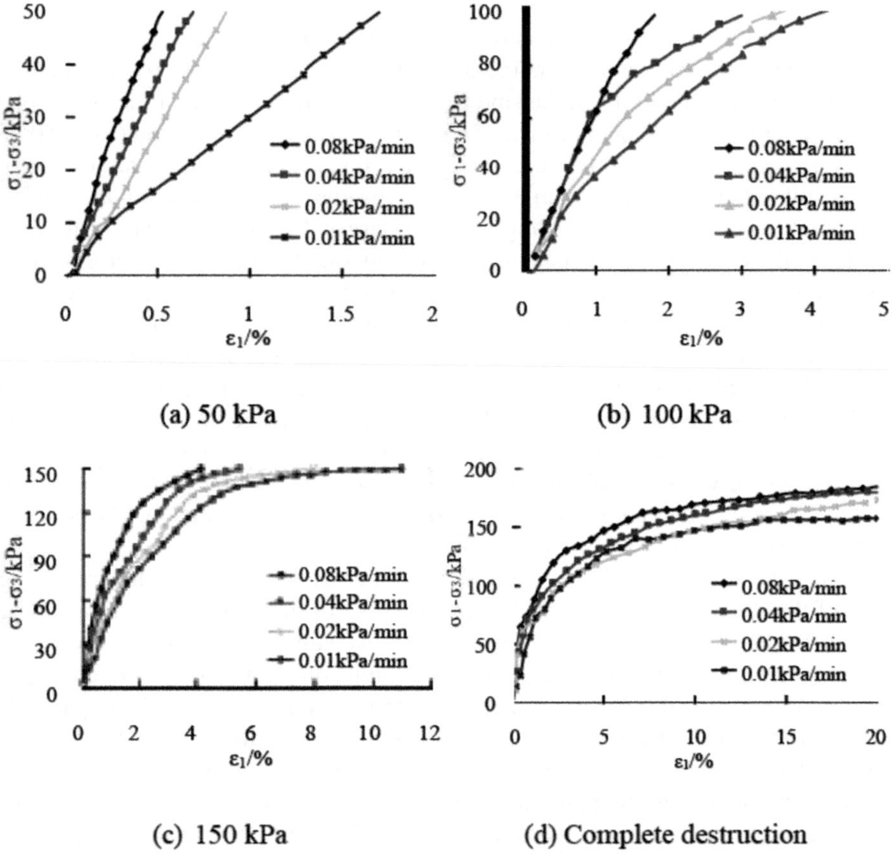

Fig. 3. The stress-stain relationship curve

3.2 Research on Fissure Extension Law

Expansion Law of Loess Cracks Under Different Unloading Amount

CT scanning image can qualitatively reflect the development of internal cracks, holes, etc. before and after unloading of soil samples from a fine view, while the CT number and its variance can quantitatively reflect the structural characteristics of the soil sample cross-section, complementing the principle that the larger the CT number is, the higher the density of the soil cross-section is, and the larger the variance of the CT number is, the more inhomogeneous is the internal part of soil cross-section, and the more strongly the development of cracks and holes, etc.

The CT scan results of loess specimens before and after unloading ($\Delta q = 0.08$ kPa/min, $w = 13\%$) under different unloading amounts were obtained by performing CT scanning tests with different unloading amounts (Fig. 4). From the initial CT scan images, it can be seen that there are many pores, fissures and groups of soil particles of large size inside the specimen, and these media are distributed in a disordered manner, with different sizes and locations; the connection between particles and particles, and between particles and cement is relatively weak, and microfissures tend to exist at the particle boundaries, which suggests that the particle boundaries in the soil specimen form an important initial fine structure and produce initial damage, and the initial damage exhibits a strong inhomogeneity. For example, the initial scanning CT numbers of the four specimens were 1084.98, 1083.53, 1095.84, and 1084.62, respectively, indicating that the overall density of the specimens had small differences, but the CT numbers of the different regions were different, with the outer ring being the largest, the middle ring being the second largest, and the center being the smallest, indicating that the density of the specimens was gradually increasing from the outside to the inside, which may be related to the disturbance of the soil body during the sample making; the squareness of the regions of the soil sample of the same section was also different. The variance of each region of the soil sample in the same section is also different, in general, the whole region is the largest, followed by the large region, and the center is the smallest, which indicates that there is a certain variability of the internal structure of the soil body under the initial state, especially the size and location of the pores. When the unloading amount is small (50 kPa, 100 kPa), the fine structure of the soil samples before and after the test changes very little, the CT number of each region increases slightly, but the CT number of each region increases with different magnitudes, the outer ring is the largest, the middle ring is the second, and the center is the smallest; when unloading amount reaches 150 kPa, the soil samples appear a small fissure, and the CT number decreases compared with that before the unloading; continue to increase unloading amount to the complete destruction of soil samples, and the internal soil samples appear to have a small fissure, and the CT number of the unloading is reduced. When the unloading amount reaches 150kPa, a small fissure appears in the soil sample, and the CT number decreases compared with that before unloading. For the CT number variance, the CT number variance decreases when the unloading amount is small, which indicates that the original holes and fissures are compressed and closed under the pressure; with the increase of the unloading amount, the CT number variance gradually appears to increase.

To summarize: only when a certain amount of unloading is reached, cracks will appear inside the soil body. When the unloading amount is small, the soil sample is still in the elastic deformation stage, and the internal structure of the soil body is not damaged, the change of the fine structure of the soil sample is mainly formed by the consolidation compression of the soil sample under the action of the consolidation perimeter pressure, because the perimeter pressure is directly applied from the edge of the soil sample, the stress in the center area is gradually increased, so that the degree of compression is the largest in the outer ring of the soil sample, and the degree of compression is the smallest in the center, moreover, due to the moisture content of the soil sample and the circumferential pressure is not large, the change of the internal structure of the soil sample is very weak, so the CT number and variance do not change much; as the unloading proceeds, combined with the stress-strain curve (Fig. 3), it can be seen that the soil sample enters into the elastic-plasticity from the elasticity, i.e., the soil body begins to appear plastic deformation, due to the unloading amount is not large, the circumferential pressure is relatively high, the plastic deformation inside the soil body is mainly due to the concentration of the local stresses in the parts of the soil particles connected to the weak particle alignment When a certain amount of unloading is reached, the adjustment of the order of soil particles can not meet the requirements of the stress state, and a large amount of energy gathered inside the soil body is released by tearing the soil body, resulting in the expansion of cracks and penetration. In the process of unloading, the lateral stress inside the soil sample gradually decreases from the edge to the center, and the soil particles or groups move axially from the two ends of the soil sample to the middle under the action of axial pressure, and radially from the center to the edge, and the soil in the outer ring is first subjected to tensile stress, while the soil in the center is still in the state of compression and density under the action of axial stress, so that the decrease of CT number of the outer ring is bigger than that of the middle ring, and the CT number of the center area is becomes larger.

Expansion Law of Loess Cracks Under Different Unloading Rates. Figure 5 shows the CT scan results of loess samples before and after unloading ($w = 13\%$, complete destruction) at different unloading rates. From the CT scan image in Fig. 5(a), it can be seen that the larger the unloading rate is, the more obvious the specimen fissure extension is. When the unloading rate is small, the microfissure development is rich and mainly distributed in the edge part of the soil sample, the fissure is short and narrow, the width of the fissure is large near the edge, and the width of the fissure is smaller the more toward the center, the fissure is interlaced with each other, which is in the form of dendrites, and the fissure expansion is evolving from the edge to the center. With the increase of unloading rate, the number of fissures in the section is less, but the length and width are larger, and the wide fissures are more developed. From Fig. 5(b), it can be seen that the CT number in the center of soil samples after unloading test becomes larger, and the CT number in the middle ring and outer ring become smaller, and the amount of change in the CT number of the specimens under different unloading rates is different, the larger the unloading rate, the larger the change, and the amount of change in the CT number of the different regions is also different, with the largest one in the outer ring, followed by that in the middle ring, and the smallest one in the center. From Fig. 5(c), it can be seen that the larger the unloading rate, the larger the amount of change in CT number variance, and vice versa, the smaller the amount of change.

To summarize: the larger the unloading rate is, the more significant the expansion of cracks is. The effect of unloading rate on the crack expansion of soil samples is related to the particle movement effect in the shear process, that is, under the action of unloading stress, the soil particles appear to move relatively, axially from the two ends of the soil samples to the middle of the movement, and radially from the center to

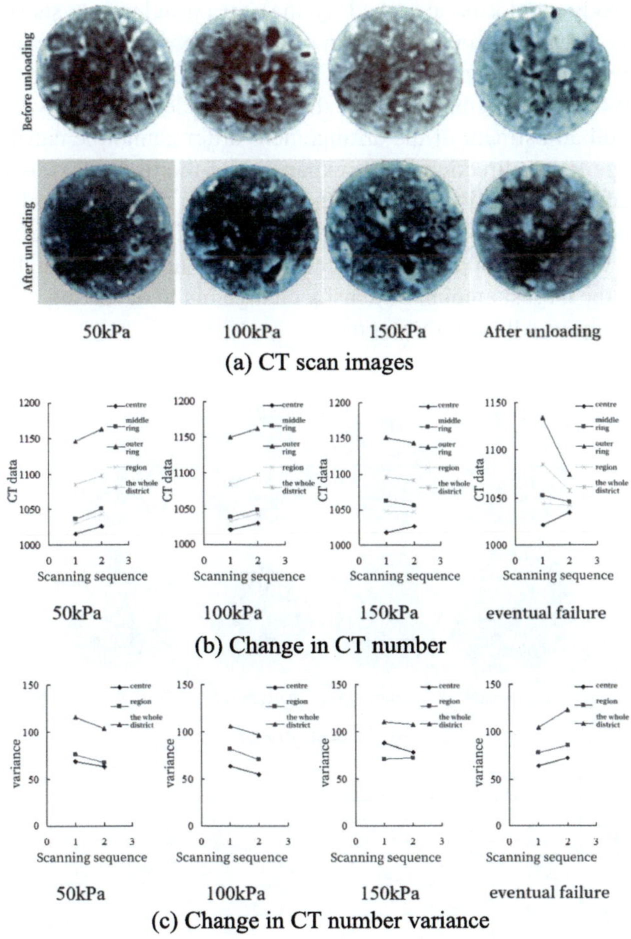

Fig. 4. CT scan results of loess samples under different unloading weight before and after unloading

the edge of the movement. The unloading rate is small, the soil particles have enough time to move, the density difference between the regions inside the soil body is small, so the change of CT number is relatively small, the concentrated stress generated at the tip of the fissure expansion can be rapidly weakened by the relative movement of the particles, when the fissure expansion encounters the defects such as pore, hole, and so on, the tip of the concentrated stress further weakened, so the width of the fissure is generally not large, with the unloading of the fissure, the fissure As the unloading proceeds, the fissure further begins to expand in the new direction, resulting in a more curved shape of the fissure, and the energy gathered at the tip of the fissure is transferred to the soil particles to overcome the resistance and produce movement, resulting in a larger deformation of the soil samples, which is macroscopically manifested as plastic flow damage. When the unloading rate is larger, the stress concentrated at the crack tip inside the soil body cannot be absorbed by the soil particles in a short time, and the energy is mainly released through the form of crack expansion, so the crack expansion is sharp and the width and length of the crack are larger. Due to the large unloading rate, the amount of movement is small because the soil particles do not have enough time to move, and the full adjustment of the arrangement order cannot be carried out, and the amount of change in the direction of the expansion of the fissure is relatively small, so that the shape of the fissure is smoother. In addition, in the process of unloading, the stress in the periphery is released first, and the soil particles move rapidly to the edge of the soil sample, while the soil particles in the center do not have enough time to move, which results in the largest amount of density change in the outer ring, the second in the middle ring, and the smallest in the center.

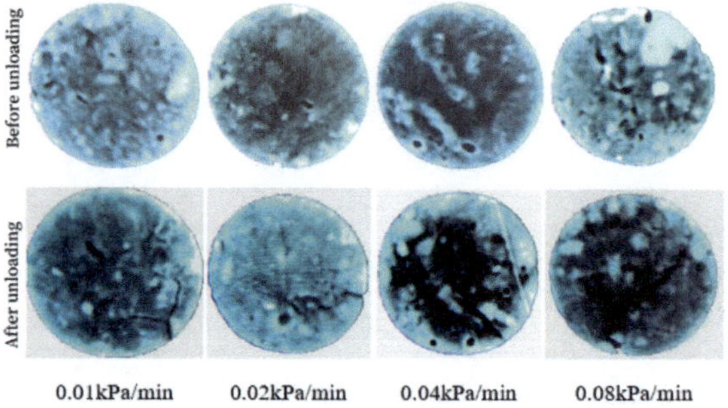

(a) CT scan images

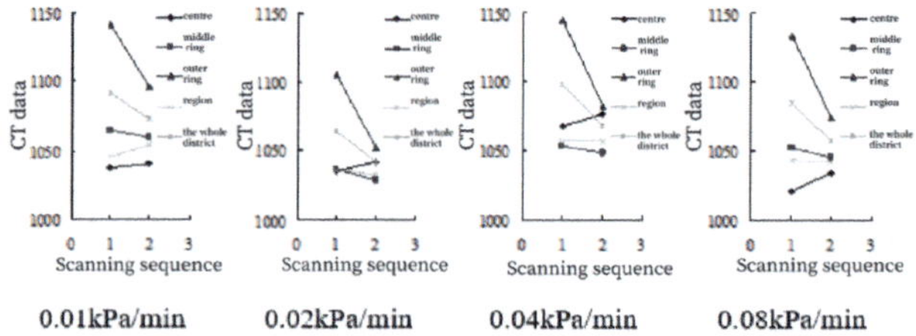

(b) Change in CT number

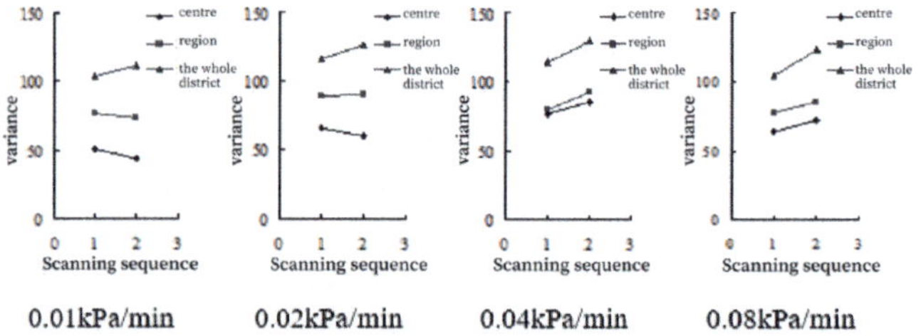

(c) Change in CT number variance

Fig. 5. CT scan results of loess samples under different unloading velocity before and after unloading

Expansion Law of Loess Fissures Under Different Water Contents

Figure 6 shows the CT scan results before and after unloading ($\sigma_i c = 200$ kPa, $\Delta q = 0.08$ kPa/min, complete destruction) of loess samples with different water content. From Fig. 6(a), it can be seen that when the water content is small ($w = 7\%$), there are more cracks inside the soil samples, and the cracks are interlocked with each other and have larger widths. With the increase of water content ($w = 18\%$), the internal shear damage zone appeared, the fracture and holes in the section were less, but the internal shear zone could be clearly seen, and the soil body located in the shear zone became looser. When the water content $w \geq 22\%$, the coarse fissures can hardly be seen inside the soil samples, while the fine microfissures are quite developed, and the fissures are mainly distributed in the outer ring and the middle ring, with small width and length, and the fissures cross each other. The CT number and its variance of the center increased after unloading of soil samples with different water content, the CT number of the middle ring increased and decreased, and the CT number of the outer ring decreased obviously, and the larger

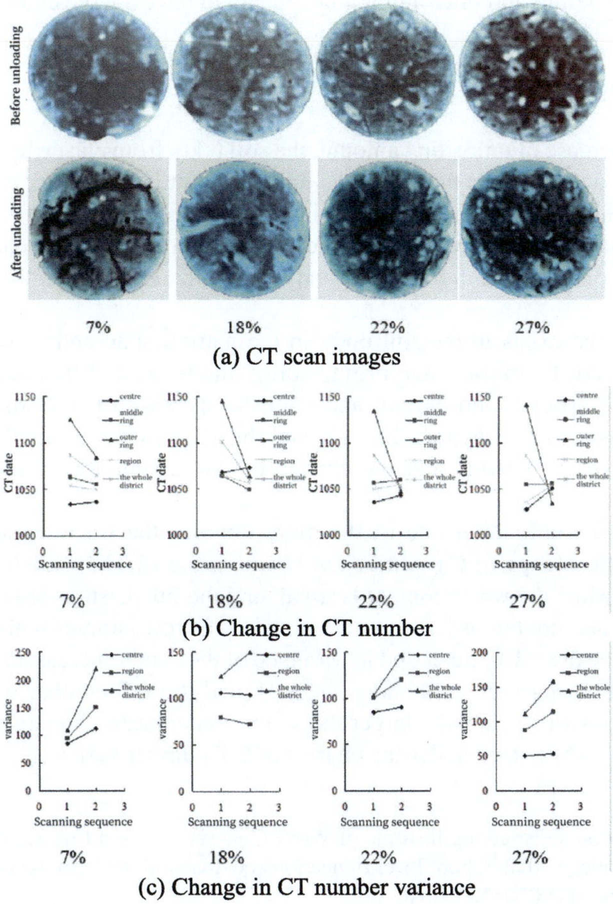

Fig. 6. CT scan results of loess samples with different moisture content before and after unloading

the water content, the larger the change of CT number, while the smaller the change of CT number variance.

To summarize: when the water content is small, the coarse fissures are more developed; when the water content is large, the fine fissures are more developed. When the water content is small, the strength of the soil body is high, the deformation in the unloading process is small, the ability to gather inside the soil body is released at a fast rate, the fissure expansion is obvious, the wide fissures are more developed, and it is mostly brittle damage, except for the fissures near the majority of the soil body skeleton structure is not destroyed or the destruction of the perturbation is small, and the density of each region is small, so that the change of the CT number is relatively small. When the water content is larger, due to the role of water, weaken the strength of the joints between the particles, so that the cementation between the soil particles is weakened, in the process of unloading, the original structure within the soil body is completely destroyed or most of them are destroyed, the soil particles are able to move more freely, and the soil body has a large deformation, often plastic flow damage, resulting in more fine cracks, and the CT number before and after unloading appears to have a large change.

4 Conclusions

(1) With the increase of unloading amount, the soil body from elastic into elastic-plastic and plastic deformation; the larger the unloading rate, the smaller the elastic deformation of the soil body, the larger the elastic-plastic and plastic deformation, and the larger the shear destructive stress, which is mainly due to the unloading rate of the weakening of the strength of the soil body material and the strengthening of friction strength.

(2) Initial damage exists in the soil body in the natural state and shows strong inhomogeneity. At the initial stage of unloading, the holes and fissures inside the soil body are compacted and closed, and with the increase of unloading, the fissures sprout, expand, and penetrate through, and the CT number gradually changes from increasing to decreasing, and the change of the CT number variance is just the opposite.

(3) The larger the unloading rate is, the more obvious the fracture extension is, and the larger the change of CT number and its variance is. The wide fissures are more developed when the water content is small, and the fine fissures are more developed when the water content is large. After unloading of soil samples with different water content rates, the CT number and its variance of the center increased, the CT number of the middle ring increased and decreased, and the CT number of the outer ring decreased obviously, and the larger the water content rate, the larger the change of CT number, while the smaller the change of CT number variance.

Acknowledgments. Guangdong Institute of Water Conservancy and Electric Power Vocational and Technical College 2022 School-level Education and Teaching Reform Research and Practice Project, Project No. GX0205JGYB040.

2023 Guangdong Province Curriculum Ideological and Political Team (Project Number: KCSZ03005, School of Municipal Engineering Curriculum Ideological and Political Team).

Supported by the National Natural Science Foundation of China (No. 41172262); New Century Talent Support Program of the Ministry of Education of China (No. NCET-12-1044); and Key Science and Technology Innovation Team Program of Shaanxi Province (No. 2014KCT-30).

References

1. Jingming, W., Jun, Z.: On loessial joints. J. Earth Sci. Environ. **02**, 30–41 (1985). (in Chinese)
2. Pan, D., Loess, Geological Publishing House (1958). (in Chinese)
3. Tang, D., Yao, X., Peng, J., et al.: Joint development characteristics, spatial division and stability of loess slope. Coal Geol. Explor. **03**, 86–90 (2015). (in Chinese)
4. He, S., Zhu, Z., Yang, X.: A study of true triaxial test on lateral unloading of soil mass of foundation pit. Rock Soil Mech. **06**, 869–872+892 (2005). (in Chinese)
5. Li, J., Chen, Z., Huang, X., et al.: CT-triaxial shear tests on the meso-structure and strength of unsaturated loess Q3 during unloading confining pressure. Rock Soil Mech. **04**, 1084–1091 (2010). (in Chinese)
6. Zhao, L., Chen, K., Guo, Z., et al.: Study on the macro and micro damage laws of intact loess cracks under slope cutting and unloading Roadbed Engineering (06), 46–51 (2023). (in Chinese)
7. Feng, X., Zhu, X., Sun, H., et al.: Microscopic Pore Structure Experiment of Shallow Loess in Jingyang. Hydrogeol. Eng. Geol. (2024). (in Chinese)
8. Pu, Y., Chen, W., Liao, Q.: Research on CT structure changing for damping process of loess in Longdong. Chin. J. Geotech. Eng. **22**(1), 52–57 (2000). (in Chinese)
9. Lei, S., Tang, W.: Analysis of microstructure change for loess in the process of loading and collapse with CT scanning. Chin. J. Rock Mech. Eng. **23**(24), 4166–4169 (2004). (in Chinese)
10. Chen, Z., Fang, X., Zhu, Y., et al.: Research on meso-structures and their evolution laws of expansive soil and loess. Rock Soil Mech **30**(1), 1–11 (2009). (in Chinese)
11. Zheng, J., Zhuang, J., Kong, J., et al.: Research on the pore structure characteristics of loess based on CT scanning Geological Science and Technology Bulletin **41**(06), 211-222 (2022). (in Chinese)
12. Yang, G., Li, C., Shang, Y., et al.: Experimental study on improving the pore characteristics of loess with slag micro powder based on CT scanning technology Water Resources and Hydropower Technology (Chinese and English) **54**(09), 201–209 (2023). (in Chinese)
13. Ye, W., Yang, G., Li, X., et al.: Test study of effect of freezing speed on behavior of Q_2 loess. Chin. J. Rock Mech. Eng. **30**(9), 1912–1917 (2011). (in Chinese)
14. Zhu, Y., Chen, Z.: Experimental study on dynamic evolution of meso-structure of intact Q_3 loess during loading and collapse using CT and triaxial apparatus. Chinese J. Geotech. Eng. **31**(8), 1219–1228 (2009). (in Chinese)
15. Lu, Q., Chen, S., Peng, J., et al.: Stress-strain features and deformation failure mechanisms of fissured loess under triaxial compression. China J. Highway Transport **01**, 8–16 (2015). (in Chinese)
16. Lu, Q., Peng, J., Wang, S., et al.: Double-parameter binary-medium model for fissured loess. Chinese J. Geotech. Eng. **05**, 93–898 (2012). (in Chinese)
17. Lu, Q., Ge, X., Peng, J., et al.: Failure characteristics of fissured loess under triaxial compression condition. Rock Soil Mech. **12**, 3689–3694 (2009). (in Chinese)
18. Sun, P., Peng, J., Wu, S., et al.: An experimental study on mechanical properties of fractured loess. J. Central South Univ. (Sci. Technol.) **06**, 2188–2195 (2015). (in Chinese)
19. Liang, Y., Du. X.: The influence of water content on the mechanical properties of undisturbed loess' cohesion and angle of internal friction under the lateral unloading condition. Sci. Technol. Eng. **18**, 34–237 (2016). (in Chinese)

Spatiotemporal Analysis of Soil Erosion and Environmental Impact Factors in the Fu River Basin

Yuanling Zhao(✉) and Wenjie Zhang

Department of Soil and Water Conservation, Changjiang River Scientific Research Institute, Wuhan 430010, China
zhaoyuanling@mail.crsri.cn

Abstract. Soil erosion poses a critical threat to the sustainability of river basins, with significant implications for ecological balance, agricultural productivity, and water quality. This study provides a comprehensive spatiotemporal analysis of soil erosion in the Fu River Basin, China, over the period 2000–2020, employing the Chinese Soil Loss Equation (CSLE) alongside multi-source data on precipitation, land use, and vegetation coverage. The results reveal marked fluctuations in soil erosion intensity, with a pronounced peak in 2010 linked to anomalous precipitation patterns. However, by 2020, effective land management practices, including reforestation and soil conservation techniques, had substantially mitigated erosion rates. These findings underscore the necessity for integrated land use and erosion control strategies to preserve the environmental integrity and sustainability of river basins under changing climatic conditions.

Keywords: Soil Erosion · Spatiotemporal Analysis · Fu River Basin · Land Use · Vegetation Coverage · Precipitation

1 Introduction

Soil erosion is a significant environmental concern that affects the sustainability of ecosystems and agricultural productivity, particularly in river basins where water and land resources are closely intertwined. The Fu River Basin, located in China, is one such region where soil erosion poses a substantial threat to the ecological balance and long-term agricultural viability.

The spatiotemporal analysis of soil erosion and environmental impact factors in river basins has been extensively studied, revealing intricate relationships between land use changes, topography, and hydrological processes. Gitima et al. [1] examined the Zoa watershed in Southwest Ethiopia, finding significant cropland expansion, especially on steeper slopes, which exacerbates soil degradation. Yohannes et al. [2] identified that changes in landscape patterns in the Beressa watershed of the Blue Nile Basin critically impact hydrological services, particularly water yield and sediment output, driven by agricultural expansion. Ludat and Kuebler [3] highlighted the role of tectonic activity in the Mara River Basin, East Africa, in shaping soil formation and vegetation, indirectly

D. Han et al. (Eds.): ICEPG 2024, ESE, pp. 651–657, 2025.
https://doi.org/10.1007/978-3-031-86829-0_74

influencing erosion processes. Ruiz Sinoga et al. [4] emphasized the importance of soil surface conditions in regulating runoff and erosion on a metamorphic hillside in Southern Spain, showing significant effects based on varied soil conditions under Mediterranean climatic influences. Collectively, these studies underscore the need for comprehensive land use planning, soil conservation practices, and ecological restoration to manage soil erosion and mitigate environmental impacts effectively.

This study aims to investigate the spatiotemporal variations in soil erosion within the Fu River Basin across the years 2000, 2010, and 2020. By analyzing the influence of three key environmental factors—rainfall, land use/cover, and vegetation cover—on soil erosion, this research provides valuable insights into the driving forces behind soil erosion in the region.

2 Research Area and Data Used

2.1 Research Area

The Fu River Basin, as shown in Fig. 1, is situated between 115°36′ and 117°10′E longitude and 26°30′ and 28°20′N latitude. The basin covers a total area of approximately 15,600 km^2, with the main stream extending about 350 km from south to north, traversing the entire basin. The topography of the basin is characterized by mountains surrounding it on three sides, with the northern part consisting of a lake plain. The region predominantly comprises hills, which account for 53.1% of the area, and mountains, which cover 29.5%. The basin experiences a subtropical monsoon climate with abundant rainfall.

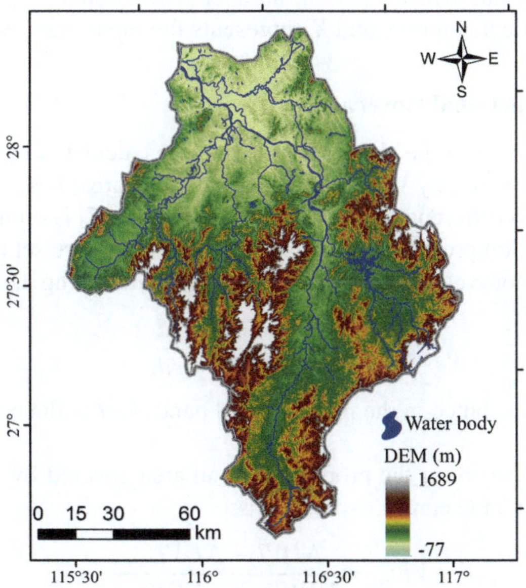

Fig. 1. Research area

2.2 Data Used

The land use and vegetation coverage data utilized in this study are derived from Landsat series satellite images. The Landsat satellites provide multispectral images with a 30 m resolution, and this research specifically selected images covering the Fu River Basin from the years 2000, 2010, and 2020. The preprocessing of these images involved steps such as radiometric correction, geometric correction, and atmospheric correction, ensuring that the image quality meets the requirements for subsequent analysis. Precipitation data were obtained from statistical records of rainfall observed at meteorological stations within the Fu River Basin.

3 Methodology

3.1 Land Use Classification

In this study, the Random Forest (RF) model [5] was utilized for remote sensing image classification. RF is a machine learning method based on an ensemble of decision trees, offering advantages in handling large-scale remote sensing data and high-dimensional data. RF classifies remote sensing images by constructing multiple decision trees and combining their predictions, which provides high accuracy and robustness against overfitting. The classification result of the RF can be expressed as:

$$Y = \frac{1}{N}\sum_{i=1}^{N} f_i(X) \tag{1}$$

where Y is the classification result, N is the number of decision trees, f_i is the classification function of the i-th decision tree, and X represents the input features.

3.2 Vegetation Fractional Coverage

Vegetation fractional coverage (VFC) [6] is typically calculated by analyzing the proportion of ground covered by vegetation within a given area, based on remote sensing data. The normalized difference vegetation index (NDVI) [7] is often used as a primary indicator for vegetation presence, where specific thresholds are set to classify pixels as either vegetation or non-vegetation. The NDVI is calculated using the following formula:

$$NDVI = \frac{NIR - Red}{NIR + Red} \tag{2}$$

where NIR is the reflectance in the near-infrared band, Red is the reflectance in the red band.

The VFC is a measure of the proportion of an area covered by vegetation, derived from the NDVI. The VFC can be expressed as:

$$VFC = \frac{NDVI - NDVI_{min}}{NDVI_{max} + NDVI_{min}} \tag{3}$$

where $NDVI_{max}$ and $NDVI_{min}$ correspond to the minimum and maximum NDVI values within the study area.

3.3 Soil Erosion Modulus

In this study, the Chinese Soil Loss Equation (CSLE) [8] was employed to calculate the soil erosion modulus in the Fu River Basin. The CSLE is a model specifically designed for assessing soil erosion in hydraulic erosion regions of China, taking into account factors such as rainfall erosivity, soil erodibility, slope length, slope gradient, vegetation cover, engineering measures, and tillage practices. The basic form of the CSLE equation is as follows:

$$A = R \times K \times L \times S \times B \times E \times T \quad (4)$$

where A represents the soil erosion modulus, R is the rainfall erosivity factor, K is the soil erodibility factor, L is the slope length factor, S is the slope gradient factor, B is the vegetation cover and biological measures factor, E is the engineering measures factor, and T is the tillage practices factor.

4 Results

4.1 Environmental Impact Factors

Land Use

The land use data for the Fuhe River Basin from 2000 to 2020 shows notable changes across different categories (Table 1). Arable land increased from 4207.49 km^2 in 2000 to 4377.68 km^2 in 2020. Forest land also saw a slight increase from 10251.66 km^2 in 2000 to 10264.77 km^2 in 2020. Grassland decreased from 437.06 km^2 in 2000 to 415.40 km^2 in 2010 but then increased to 489.56 km^2 by 2020. Urban land use dropped significantly from 388.10 km^2 in 2000 to 168.08 km^2 in 2020, indicating urban contraction. Unused land expanded slightly, from 0.96 km^2 in 2000 to 1.38 km^2 in 2020. These changes reflect dynamic shifts in land use patterns within the basin.

Table 1. Land use (km^2)

Land use	2000	2010	2020
Arable land	4207.49	4299.53	4377.68
Forest	10251.66	10346.91	10264.77
Grassland	437.06	415.40	489.56
Water body	314.74	316.34	298.53
Urban land	388.10	220.81	168.08
Unused land	0.96	1.01	1.38

VFC

The VFC data for the Fuhe River Basin shows significant changes from 2000 to 2020 (Table 2). The area with over 75% coverage increased from 8320.16 km^2 in 2000 to 11444.68 km^2 in 2020, indicating improved vegetation density. Conversely, areas with 60%–75% coverage decreased from 6088.84 km^2 in 2000 to 3646.28 km^2 in 2020. The area with 30%–45% coverage initially declined from 56.07 km^2 in 2000 to 27.91 km^2 in 2010, but rose sharply to 124.73 km^2 by 2020. These trends suggest a general improvement in high-density vegetation coverage, although mid-range coverage areas fluctuated over the period.

Table 2. VFC (km^2)

VFC	2000	2010	2020
<30%	0.99	0.99	0.49
30%–45%	56.07	27.91	124.73
45%–60%	1133.94	343.32	383.82
60%–75%	6088.84	4252.70	3646.28
>75%	8320.16	10975.08	11444.68

Precipitation

The annual precipitation data for the Fuhe River Basin reveals significant fluctuations between 2000 and 2020 (Table 3). The maximum precipitation increased from 2057.9 mm in 2000 to 2764.7 mm in 2010, before slightly decreasing to 2066.4 mm in 2020. Similarly, the minimum precipitation rose from 1544.5 mm in 2000 to 2109.9 mm in 2010, then fell to 1556.1 mm in 2020. The average precipitation also saw an increase from 1699.32 mm in 2000 to 2326.87 mm in 2010, followed by a decrease to 1712 mm in 2020. These trends suggest variability in rainfall patterns over the two decades, with a peak around 2010.

Table 3. Annual precipitation (mm)

Precipitation	2000	2010	2020
Maximum	2057.9	2764.7	2066.4
Minimum	1544.5	2109.9	1556.1
Mean	1699.32	2326.87	1712

4.2 Soil Erosion Intensity

Figure 2 shows the distribution of soil erosion modulus in the Fu River Basin for the years 2000, 2010, and 2020, as calculated using the CSLE. The soil erosion intensity from the Fu River Basin reveals significant changes between 2000 and 2020. The area of mild erosion increased from 5294.09 km^2 in 2000 to 6860.83 km^2 in 2010, then decreased to 3918.12 km^2 by 2020. Moderate erosion rose from 1125.56 km^2 in 2000 to 2116.01 km^2 in 2010, before dropping to 613.25 km^2 in 2020. Severe erosion increased from 416.84 km^2 in 2000 to 1058.87 km^2 in 2010, then reduced to 197.83 km^2 by 2020. The increase in soil erosion intensity in the year 2010 may be related to the higher rainfall during that year. These trends indicate initial worsening conditions followed by effective erosion control measures by 2020.

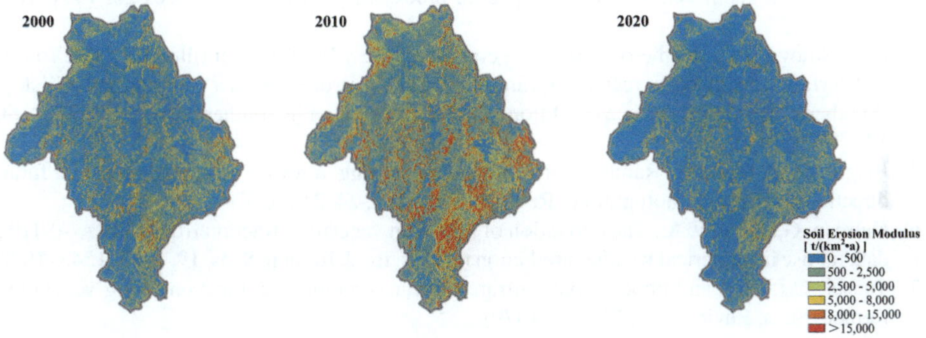

Fig. 2. Soil erosion intensity

5 Conclusions

The spatiotemporal analysis of soil erosion within the Fu River Basin over the past two decades reveals significant fluctuations in erosion intensity, driven primarily by variations in precipitation and land use practices. The peak erosion observed in 2010 highlights the vulnerability of the region to extreme weather events, underscoring the critical need for adaptive management strategies. However, the notable decline in erosion by 2020 demonstrates the efficacy of targeted land management practices, including reforestation and the implementation of soil conservation measures. These findings emphasize the importance of integrating environmental monitoring with proactive land use planning to mitigate erosion risks and preserve the ecological integrity of river basins. Future research should focus on refining predictive models to account for climate variability and human interventions, ensuring that erosion control measures are both effective and sustainable. The insights gained from this study provide a foundation for developing robust strategies to manage soil erosion, contributing to the long-term sustainability of river basins in similar regions globally.

Acknowledgements. This research was funded in part by SGSDJS00CWJS2310191, CKSF2024351/TB, CKSF2024341/TB, and CKSD2023929/TB.

References

1. Gitima, G., Teshome, M., Kassie, M., Jakubus, M.: Spatiotemporal land use and cover changes across agroecologies and slope gradients using geospatial technologies in Zoa watershed, Southwest Ethiopia. Heliyon 8 (2022)
2. Yohannes, H., Soromessa, T., Argaw, M., Dewan, A.: Impact of landscape pattern changes on hydrological ecosystem services in the Beressa watershed of the Blue Nile Basin in Ethiopia. Sci. Total Environ. **793**, 148559 (2021)
3. Ludat, A.L., Kuebler, S.: Tectonic controls on the ecosystem of the Mara River basin, East Africa, from geomorphological and spectral index analysis. Biogeosciences **20**, 1991–2012 (2023)
4. Ruiz Sinoga, J.D., Romero Diaz, A., Ferre Bueno, E., Martinez Murillo, J.F.: The role of soil surface conditions in regulating runoff and erosion processes on a metamorphic hillslope (Southern Spain) Soil surface conditions, runoff and erosion in Southern Spain. CATENA **80**, 131–139 (2010)
5. Belgiu, M., Drăguţ, L.: Random forest in remote sensing: a review of applications and future directions. ISPRS J. Photogramm. Remote. Sens. **114**, 24–31 (2016)
6. Gutman, G., Ignatov, A.: The derivation of the green vegetation fraction from NOAA/AVHRR data for use in numerical weather prediction models. Int. J. Remote Sens. **19**, 1533–1543 (1998)
7. Tucker, C.J.: Red and photographic infrared linear combinations for monitoring vegetation. Remote Sens. Environ. **8**, 127–150 (1979)
8. Zhang, H., Zhang, R., Qi, F., et al.: The CSLE model based soil erosion prediction: comparisons of sampling density and extrapolation method at the county level. CATENA **165**, 465–472 (2018)

Study on the Role of Plant Root Exudates in Soil Pollution Control

Jin Wang, Mingyan Cheng, and Jianwen Duo(✉)

Lanzhou Resources and Environment Voc-Tech University, LanzhouGansu 730030, China
38468507@qq.com

Abstract. Heavy metals in soil pollution pose a serious threat to human beings and ecosystems. Phytoremediation technology can effectively control soil pollution through plant root exudates, which shows good effect. The purpose of this study was to investigate the role of plant root exudates in soil pollution control. Through literature review and data analysis, it was revealed that plant root exudates have biogeochemical functions such as adsorption, chelation, reduction and stabilization, which contribute to the migration and transformation of soil pollutants and reduce environmental harm.

Keywords: phytoremediation · root exudates · soil pollution · heavy metal

1 Introduction

Heavy metals refer to metals with a specific gravity greater than 5 (metals with a density greater than 4.5 g per cubic centimeter), including copper, iron, lead, mercury, nickel, cadmium, chromium, etc. Heavy metal pollution is often more difficult to control than other organic compound pollution, many organic compounds can be degraded or transformed by nature's own chemical, physical or biological purification, so as to reduce or remove toxic and harmful effects, while heavy metals can not be biodegradable, and have enrichment, can directly threaten higher organisms including human beings [1].

The dissolved organic matter DOM (dissolved organic matter DOM) usually refers to the part of organic matter that can dissolve in water. Strictly speaking, this is only an operational definition, it refers to the material after water extraction, through the 0.45μm filter membrane, with different structures and molecular weight of organic matter (such as low molecular weight free amino acids, carbohydrates, organic acids, and macromolecular enzymes, amino sugars, polyphenols and humus, etc.) of the continuous or mixture. DOM is the most active organic carbon pool in the terrestrial ecosystem, and it is also an important form of material exchange between the soil sphere and related layers, which has a great impact on the activation, migration and ecotoxicity of heavy metals, nutrient elements and organic pollutants. A large number of studies have pointed out that because DOM acts as the main transport carrier for many trace organic or inorganic pollutants, many hydrophobic pollutants, such as heavy metals, pesticides and polycyclic aromatic hydrocarbons, can produce obvious migration or diffusion in soil and water environment.

D. Han et al. (Eds.): ICEPG 2024, ESE, pp. 658–665, 2025.
https://doi.org/10.1007/978-3-031-86829-0_75

Studies have shown that the composition and properties of DOM have a great impact on the environmental chemical behavior of pollutants, and root exude, as a kind of DOM, also has a corresponding effect on pollutants. Traditional biological remediation of organic pollutants is completed by microorganisms. However, phytoremediation is more suitable for on-site remediation than microbial remediation. There are many researches on phytoremediation at home and abroad, but there are few researches on the effect of root exudates on solubility of organic pollutants and chemical determination methods.

With the harm of soil pollution to the environment, it has brought serious threats [2] to the ecology and human health. The methods of soil pollution control mainly include chemical remediation, physical remediation, biological remediation and so on. 1) Chemical remediation is mainly the use of chemical decomposition or fixation to change the chemical structure of pollutants, so as to reduce their migration and toxicity. This technology involves precipitation, adsorption, oxidation-reduction, dechlorination, polymerization and hydrolysis and other methods, aiming to effectively reduce the migration, bioavailability and toxicity of pollutants in soil.

2) Physical remediation includes physical separation remediation technology, steam extraction remediation technology, stabilization/solidification remediation technology, heat treatment remediation technology, electrodynamic remediation technology [3], physical remediation is the use of pollutants in the soil adsorption, diffusion and other laws to remove pollutants. It has the characteristics of good effect, short repair time, easy operation, and insensitive to the nature and concentration of pollutants.

3) Bioremediation is the process of removing or degrading the pollutants in the soil through microorganisms to harmlessly. It includes phytoremediation technology, microbial remediation technology and microbial-plant combined remediation technology. Phytoremediation technology is an important means of bioremediation, which uses green plants to transport, store or transform pollutants, so that they are harmless to the environment.

2 Research Status at Home and Abroad

Wang Yuanyuan et[4] al. Mentioned in their studies that root exudates have significant biological activity and can transmit information about interactions between plants and microorganisms. Biological matter can improve the availability of soil nutrients and affect the composition of microbial communities. The secretion and composition of plant root exudates are affected by a variety of factors including plant light duration and intensity, soil temperature and humidity, mineral nutrients and soil microorganisms, which can all affect the secretion and composition of root exudates. Li Yueming et al. [5] found that both carbon secretion and root exudates composition would change under drought conditions. A study by the Institute of Soil Research in Nanjing, Chinese Academy of Sciences, found that plant roots secrete some substances as they grow, which have a great influence on soil microbes. Li Yuting et al. [6] showed that inoculation of degraded bacteria such as Brucella in PAHs contaminated soil for bioenhancement accelerated the removal of pollutants, and the residual amount of PAHs was 15.8% lower than that of the natural decay control group.

The types and amounts of root exudates released by different plant species and growth stages were different. Several types of plant root exudates have been identified in some studies. For example, rhizosphere mucous is an important type of root exudates, which is composed of hydrogel properties and compounds containing mixtures, of which polysaccharide is the main component. Afzal Y M et al. [7] mentioned that there is spatial variability in the soil microenvironment, and the organic substances secreted by plant roots are called root exudates, which is conducive to the absorption of these substances by bacteria. Tian Y et al. [8] showed that root exudates promoted the growth and chemotaxis of strain Za, which was related to the formation of biofilm. The importance of the degradation of diphenyl ether herbicides in shaping rhizosphere soil. In addition, the role of key root exudates in shaping rhizosphere microbial communities under stress conditions is also an important issue.

3 Research Significance

Plant root exudates play a certain role in phytoremediation, which can improve the soil environment, promote the absorption of nutrient elements by plants, and help the propagation of microorganisms. However, there are relatively few researches on the role and influencing factors of plant root exudates in soil pollution control, so it is necessary to pay more attention to and understand the role of plant root exudates in soil pollution control.

Among them, plant root exudates play an important role in phytoremediation, which is mainly reflected in:

1) Plant roots can absorb harmful substances in soil or water, and then send harmful substances to the above-ground parts of plants through plant root secretions, such as stems, leaves, etc.
2) Root exudates can be divided into low molecular weight exudates and high molecular weight exudates according to their molecular weight. The former mainly includes sugars, phenols, organic acids and various amino acids, while the latter mainly consists of polysaccharides, macromolecular mucous substances and extracellular enzymes [9].

4 Production of Plant Root Exudates and Its Main Influencing Factors

4.1 Production of Plant Root Exudates

Composition of Plant Root Exudates

The composition of root exudates is diverse and complex, including organic substances, inorganic substances and bioactive substances. The key compounds in root exudates play an important role in soil pollution control. Zhou Sanyang et al. have shown that organic substances in plant root exudes mainly include organic acids, amino acids, plant growth hormones and secondary metabolites [10].

Content of Plant Root Exudates

The amount of plant root exudates is not fixed, it is affected by many factors. For example, the type of plant, the stage of growth, the condition of the soil and some environmental conditions. Li Yanlin et al. [11] have shown that organic acids in root exudates are conducive to increasing the content of main nutrients in soil, and have a positive effect on soil enzyme activity, which is conducive to improving soil productivity. Zhang Invention et al. [12] mentioned that under atrazine stress, vetiver might adapt to the stressed environment by adjusting the quantity and relative content of compounds in root secretions. The results showed that the secretion content could affect the growth of plants and the efficiency of pollutant treatment.

4.2 Influence of Environmental Factors on Root Exudates

Illumination

Both the time and intensity of light have a big impact on the photosynthesis of plants, as well as the food (photosynthetic products) they make. Some studies have found that 28% to 59% of plant photosynthetic products are transferred to the ground through the secretion of roots, while most of the photosynthetic products enter the soil, affecting the secretion of plant root secretions. At the same time, the intensity of light also affects the physiological activity of plants, which in turn affects the secretion of roots.

Temperature

The level of soil temperature is an important factor to determine the physiological status of plant roots. Too high or too low soil temperature will cause stress, harm the physiological metabolism of roots, and then affect the type and quantity of root exudes. Wang Pengqiang et al. [13] found that under the conditions of low temperature and high temperature, the absorption of substances and trace elements of non-heat-resistant and heat-resistant tomato varieties would increase with the increase of rhizosphere temperature. It was further proved that the temperature had a great influence on plant roots.

Water

The influence of water on plant roots is critical. Usually the water content of the soil should be maintained at about 65%. If the water content is below this standard, the soil will appear to be short of water and dry, at which time the growth of plant roots will be inhibited and may become slow or even stop. This is because the lack of water will interfere with the metabolism and absorption process of the plant, preventing its normal growth. However, too much water can also have an adverse effect on plant roots. Peng noted that when soil water content exceeds 65% for a long time, too much water can clog pores in the soil, blocking root respiration and causing root cells to be starved of oxygen. This lack of oxygen can damage root cells, leading to root rot and eventually the death of the entire plant.

Role of Rhizosphere Microorganisms

Rhizosphere microorganisms can change root cell permeability and root metabolic activity. Rhizosphere microorganisms secrete organic acids, amino acids, hormones and other metabolites, which can regulate the permeability of root cell wall and make it more flexible and firm. High salt environment will lead to the decrease of water inside root cells, resulting in increased permeability of cell membrane, affecting plant growth and metabolism. Plant hormones such as abscisic acid (ABA), ethylene, etc. can affect the permeability of root cells. For example, ABA can induce stomatal closure, reduce water loss, and affect the permeability [14] of root cells by regulating the activity of ion channels and transporters.

Moreover, rhizosphere microorganisms play a crucial role in improving the efficiency of plant nutrient absorption, and play a positive role in stimulating the metabolic activity of roots. The metabolites produced by rhizosphere microorganisms can be used as signal molecules or nutrient sources to promote the growth and metabolic activities [15] of plant roots. For example, metabolites of Pseudomonas fluorescens and Fusarium fungi can significantly increase the amount of amino acids secreted by the roots of plants such as alfalfa and wheat, thus changing the metabolic activity of the roots.

There are also indigenous pathogenic microorganisms in the rhizosphere environment, which can cause plant diseases or produce harmful substances, which have a negative impact [16] on plant growth and root metabolic activity.

5 Repair Effect of Root Exudates on Contaminated Soil

5.1 Repair Effect of Root Exudates on Soil Contaminated by Heavy Metals

The repair effects of root exudates on heavy metal contaminated soil are mainly activation and solidification. The cases of phytoremediation of heavy metal pollution are shown in Table 1. Activation is to transform heavy metals in soil into effective states that can be absorbed by plants through complexation, acidification, dissolution and reductive activation, so as to improve the absorption and utilization capacity of heavy metals by plants. For example, the organic acids in the root secretions of the chrysanthemum japonicum combined with cadmium ions to improve its soil mobility and promote the absorptionof Cd by plants; Organic acids can acidify rhizosphere soil by reducing rhizosphere pH value, which can not only improve the extraction rate of heavy metal plants, but also reduce the content [17] of heavy metals in soil. Lian Huan's [18] research found that when ryegrass grows in the soil polluted by uranium, the soil pH value will decrease with the increase of the concentration of uranium pollution. When the addition of exogenous citric acid increases the content of active uranium in the soil, some inert uranium is activated, and it is easier to absorb uranium from the soil.

The mechanism of curing is that organic acids, amino acids and mucilagins in plant root secretions can react with metal ions in soil to form stable chelates, which is of great [19] significance for reducing the diffusion of heavy metal pollution sources, alleviating the heavy metal toxicity in local areas and improving the ecological environment of heavy metal pollution areas. Studies have shown that organic acids can inhibit the absorption of Pb by radish. After adding exogenous citric acid and tartaric acid, the conjugates formed are not easily absorbed, which reduces the toxicity [20] of Pb to plants.

Table 1. Heavy metal pollution cases of different phytoremediation

Heavy metal	Plant name	Repair action
Cd	beadweed	Cd-oa combination enhances soil mobility → through mass flow and diffusion → Cd contact increases in roots → Cd content increases in roots → Cd transfers to the above-ground part
U	ryegrass	The decrease of pH value and the addition of exogenous citric acid increased the content of active uranium in soil, promoted the conversion of inert uranium in plant roots, and increased the content of active uranium in rhizosphere soil
Pb	radish	Exogenous citric acid and tartaric acid were added to form a combination that was not easily absorbed
Cu	Sweet orange	The low molecular weight organic acids secreted by the roots fix Cu and reduce its bioavailability by complexing with Cu

5.2 Repair Effect of Root Exudates on Soil Polluted by Organic Matter

The plant-microbial combined remediation technology is to enrich, fix and degrade organic pollutants in soil, and to repair them through a complex system composed of plants and microorganisms. The found that when adding rhamnose and inoculating PAHs obligate degrading bacteria in the treatment of PAHs pollution, the two synergies could enhance the dehydrogenase activity and the number of PAHs degrading bacteria, thus improving the degradation efficiency of PAHS. The found that the removal efficiency of a single plant with rhizosphere growth-promoting bacteria added was higher than that without PAHS.

Enzymes in root secretions can also repair soil polluted by organic matter. For example, the study showed that enzymes released by plant roots in soil can directly degrade related compounds, and enzymes released to the environment after plant death can continue to perform decomposition function.

5.3 Repairing Effect of Root Exudates on Other Pollutants in Soil

Plant root exudates are an important means of bioremediation to repair persistent organic pollutants (POPs) in soil, as well as other pollutants such as pesticides, fertilizers, etc. Root exudates contain a variety of enzymes and organic acids, including prolinase, chitanase, cellulase, etc., which can promote the growth and metabolism of soil microorganisms and speed up the degradation and transformation of these pollutants. For example, prolinase can promote the growth and metabolism of soil microorganisms and accelerate the degradation of pesticides in the soil. Cellulase can decompose cellulose in the soil, release nutrients in the soil, promote the growth and metabolism of soil microorganisms, and accelerate the degradation of fertilizers in the soil.

In addition, organic acids in plant root secretions, such as citric acid, malic acid, etc., can promote the growth and metabolism of soil microorganisms, and accelerate the degradation and transformation of pollutants such as pesticides and fertilizers in soil.

6 Conclusions and Prospects

The purpose of this paper is to investigate the role of root exudates in phytoremediation and to provide scientific basis for practical application. The organic acids, proteins and enzymes released from root exudates can transform heavy metals in soil by means of analysis, tanning and chelation, thereby reducing their effectiveness and toxicity. Root exudates can not only promote soil microbial activity, enhance soil bioremediation ability, but also promote the biotransformation of heavy metals. These tiny molecules can not only effectively degrade heavy metals and organic pollutants, but also play an important role in shaping soil microbial community structure and promoting soil ecosystem restoration. Through the effects of heavy metal pollution and organic pollutants on soil and soil ecosystem, it was found that the accumulation and migration of heavy metal and organic pollutants would have an important impact on the physical and chemical properties and biological processes of soil. As an important medium between plants and soil, root exudates play an important role in repairing soil polluted by heavy metals and organic pollutants. Studies have shown that plant root exudates can have a good removal effect on heavy metals and organic pollutants, with the highest removal rate of about 50%. However, the adsorption and removal effect of root exudates on pollutants is a dynamic process, and the composition and quantity of root exudates will change with the change of plant growth conditions. Therefore, in the process of pollution control, It is necessary to determine the optimal growth conditions through the gradient experiment to achieve the expected adsorption treatment effect.

The research results of this paper confirm the significant role of root exudates in soil pollution control and the huge application potential. This discovery not only provides a valuable theoretical and practical basis for optimizing the pollution control technology of heavy metals and organic pollutants, but also takes a solid step to promote the healthy restoration of soil ecosystem.

Acknowledgments. Fund Projects: Mechanism of S. hizophila in response to Pb^{2+} stress (24JRRA741) Natural Science Foundation of Gansu Province; Study on the Key mechanism of Toxic Effects of Iron Oxide Activated persulfate System on Bacteria in water (X2024ZD-01), Lanzhou Resources and Environment Vocational Technical University key talent support plan research project.

References

1. Liu, X., Chen, X., Wu, L., et al.: GC-MS analysis of root exudates of Elymus plantata seedlings. J. Tropical Crops **37**(04), 835–843 (2016)
2. Chen, Z.Y.: Study on the impact of pollutant discharge permit system on environmental protection and its optimization strategy. Heilongjiang Environ. Bull. **36**(09), 10–13 (2023). (in Chinese)
3. Huang, S.: Risk Assessment and Chemical Leaching of Heavy Metal Contaminated Soil. Chongqing University (2013)
4. Wang, Y., Gao, J.: Effects of root exudates on heavy metal pollution and soil ecological restoration. Modern Horticult. **46**(05), 19–21 (2023)

5. Li, Y.M., Yang, F., Han, P.L., et al.: Advances in mechanism of plant root exudates responding to abiotic stress. Chin. J. App. Environ. Biol. **28**(5), 1384–1392 (2022)
6. Li, Y., Bao, W., Zhang, W., et al.: Effect of root exudates on bioenhanced remediation of PAHs contaminated soil. China Environ. Sci. **43**(08), 4183-4193 (2023)
7. Afzal, M.Y., Das, B.K., Valappil, V.T., Scaria, J., Brözel, V.S.: Root exudate compounds change the bacterial community in bulk soil. Rhizosphere **30**, 100885 (2024)
8. Tian, Y., Zhong, F., Shang, N., Yu, H., Mao, D., Huang, X.: Maize root exudates promote bacillus sp. Za detoxification of diphenyl ether herbicides by enhancing colonization and biofilm formation. Mol. Plant-Microbe Interactions® **37**(7), 552–560 (2024). https://doi.org/10.1094/MPMI-02-24-0020-R
9. Ya, W., Fye, F., Jing, G., et al.: Effect and mechanism of plant root exudates on soil pollution remediation. Acta Ecol. Sin. **42**(03), 829–842 (2022)
10. Zhou, S., Yang, K., Wu, X., et al.: Research progress on the removal of heavy metals from wastewater by constructed wetland plants. Wetland Sci. **14**(05), 717–724 (2016)
11. Li, Y., He, H., Liu, J., et al.: Analysis of organic acids secreted by different roots and their effects on soil nutrient and enzyme activities. Northern Horticult. (2024)
12. Zhang, I., Wang, Q., Yang, S., et al.: Response of root exudates of vetiver to atrazine stress and removal of atrazine residues in soil. J. Ecol. Rural Enviro. **39**(10), 1323–1331 (2019)
13. Wang, P.Q., Wu, Y.L., Li, Y.F., et al.: Research and prospect of plant root exudates and microdomain environment. Rural Sci. Technol. **12**(27), 96–98 (2021)
14. Ma, T., Luo, C., Li, S., et al.: Research status of woody plant response to drought stress. J. Sichuan Univ. (Nat. Sci. Edn.) **60**(05), 25-34 (2023)
15. Tang, Z.X.: Screening and Identification of Growth-Promoting Bacteria from Wormwood Root System and its Application effect. Northwest A&F University (2022)
16. Liu, J., Li, X., Yao, M.: Research progress in the construction of plant rhizosphere microbial community. Acta Microbiol. Sinica **61**(02), 231–248 (2021)
17. Zhao, K., Wan, X., Xing, D., et al.: Research progress on effects of low molecular weight organic acids on the release of soil available phosphorus and heavy metals. Chin. J. Soil Sci. **53**(05), 1228–1236 (2022)
18. Lian, H.: Rhizosphere effect of ryegrass on plant extraction and remediation of uranium contaminated soil. East China University of Technology (2018). (in Chinese)
19. Wang, Y., Chen, L., Zhang, P., et al.: Response and remediation of root exudates and rhizosphere microorganisms to soil heavy metal pollution. Jiangsu Agricult. Sci. **52**(05), 19–27 (2024)
20. Chen, H.-H., Chen, X.-F., Zheng, Z.-C., et al.: Characterization of copper-induced-release of exudates by Citrus sinensis roots and their possible roles in copper-tolerance. Chemosphere **308**, 136348 (2022)

Distribution Characteristics and Risk Assessment of PAHs in Soil of a Typical wet Coking Plant

Xuejiang Cao[1], Yanning Li[1], Wenwen Li[1], Qun Dong[2], Yanchao Yin[1], Fan Zhang[1], and Fengxia Sun[3(✉)]

[1] The Fifth Prospecting Team of Shandong Coal Geology Bureau, Jinan 250100, China
[2] Zhangqiu Branch of Jinan Ecological Environment Bureau, Jinan 250200, China
[3] College of Resources and Environment, Shandong Agricultural University, Tai'an 271000, China
sfx@sdau.edu.cn

Abstract. The content and composition characteristics of 16 kinds of polycyclic aromatic hydrocarbons (PAHs) in the soil of a coking plant were analyzed by gas chromatography-mass spectrometry (GC-MS). The source of PAHs was determined by ratio method, and the ecological risk of PAHs in different functional areas of the coking plant was analyzed by quality standard method. At the same time, according to the Technical Guidelines for Soil Pollution Risk Assessment of Construction Land (HJ 25.3-2019), the health risk assessment of 16 kinds of PAHs in the plant was carried out.The results showed that: (1) The total detection rate of Σ16 PAHs in the coking plant is 38.22%, and the PAHs in the plant are mainly 3-ring, accounting for 46.43%. 34.6% of the soil samples have Σ16 PAHs at the pollution level, and 25.7% of the soil samples have reached the severe pollution level. (2) Σ16 PAHs showed an obvious spatial distribution pattern, and the content of PAHs in different functional areas was in the order of chemical production area > coal field > Coke field > office area; In the vertical soil layer, 52.64% of PAHs concentrated in the soil of 0 ~ 3 m. (3) The main source of the chemical production area may be the tar leakage in the production process, and the PAHs in the coke field mainly comes from the settlement of soot substances produced by coke ovens; (4) According to the proportion of high ecological risk points, the ecological risk ranking of each functional area is chemical production area > coke field > coal field > office area. In summary, the soil in different regions of the coking plant has different degrees of PAHs ecological risk, and it is recommended to control and use the zoning of this plot.

Keywords: Polycyclic aromatic hydrocarbons · coking plant · soil · risk assessment · distribution characteristics

1 Introduction

Polycyclic aromatic hydrocarbons (PAHs) are widely distributed organic pollutants with physiological toxicity, mutagenicity, and carcinogenicity in the environment [1] that are primarily sourced from pollutants from human life and production activities [2]. The

D. Han et al. (Eds.): ICEPG 2024, ESE, pp. 666–681, 2025.
https://doi.org/10.1007/978-3-031-86829-0_76

US Environmental Protection Agency (EPA) has declared 16 PAHs monomers as priority environmental pollutants for control [3]; The coking industry is a crucial source of anthropogenic pollution of PAHs [4]. In 2014, the global emissions of 16 PAHs were 510000 tons, while that of China were approximately 125000 tons (From newspaper https://www.sohu.com/a/126579741_114731). PAHs pollution in soil environment comes from a wide range of sources, among which the leakage of petroleum compounds containing PAHs in coking plants is one of the important sources [5].

Although PAHs are volatile, their hydrophobicity and lipophilicity render it easier for them to adsorb soil particles and accumulate in the soil [6]. The amount of PAHs entering the human body through soil is substantially higher than through other pathways, rendering it one of the primary sources of PAHs entering organisms in the environment [7]. At present, domestic and foreign scholars have primarily studied the content, distribution characteristics, source analysis, and remediation techniques of PAHs in polluted soil. Yan [8] showed that the concentration of PAHs in the surface and deep soil of a certain abandoned coking plant for recycling, crude benzene refining, tar storage, and coking workshop was higher than 1.00 mg/kg, reaching a severe pollution level. The concentration of PAHs in the recycling workshop was as high as 140 mg/kg, and the soil PAHs pollution was primarily caused by three and four-ring PAHs. Lulu [9] found that the content of ΣPAHs in the soil of the Dagang Oilfield was as high as 11.76 mg/kg, which is severe pollution, with three-ring PAHs being the primary pollutant. The primary methods for identifying sources of PAHs are the ratio method and mathematical statistical analysis method. Huiling [10] used the ratio method to analyze the source of soil PAHs and found that the primary source of PAHs in the atmospheric environment of a coking plant was coal combustion. Mingyu [11] conducted a study using principal component analysis, and their results showed that the proportion of PAHs from coal combustion in the three coking plants in Beijing, Chongqing, and Taiyuan areas was approximately 88.46–90.38%, 83.56–98.17%, and 74.56–92.96%, respectively. A small number of pollution sources remain, such as oil leakage and combustion, in the soil of coking plants.

2 Research Area and Methods

2.1 General Situation

This study focuses on a typical early wet coking plant in the north. Soil samples were collected onsite to investigate and analyze the residual levels and pollution distribution characteristics of PAHs in the soil of the coking plant. The sources of PAHs in the soil of the coking plant were analyzed using the ratio method, and the ecological risk of PAHs in different functional areas of the plant was evaluated using the quotient method. These analyses were conducted to provide a theoretical basis for soil remediation and reuse at the coking plant site, as well as a reference and support for environmental investigation, risk assessment, and remediation of similar sites.

The large-scale coking plant uses a coal wet coking process to produce coke, covering an area of approximately 230000 m^2. It is primarily divided into a chemical production area, coal yard, coke yard, and office area (Fig. 1). The chemical production area primarily includes the tar tank area, desulfurization workshop, gas tank, ammonia tank area, and sewage station; the coal yard is primarily equipped with coal preparation workshops and purification tanks; the coke field is primarily equipped with coke ovens; and the office area is primarily equipped with office buildings and laboratories.

2.2 Soil Sample Collection

In this study, grid point distribution and professional judgment methods were used for the sampling and point distribution. More points were distributed in areas with a high pollution probability, and the density of points in areas with low pollution was sparse. In total, 247 soil points were established, including 54 coal yards, 75 coke fields, 97 chemical production areas, and 21 office areas (Fig. 1). A total of 1494 soil samples were collected.

The depth of the soil sample collection was 0–22 m, and 4–11 samples were collected at each point. One sample was collected at each point at depths of 0 –1 m and 1–3 m, and the depth of the surface soil collection was 0–0.5 m. After the sample was collected, it was placed in a 250 mL brown glass bottle and stored at a low temperature in an incubator containing dry ice. Samples were transported to the laboratory for testing on the same day.

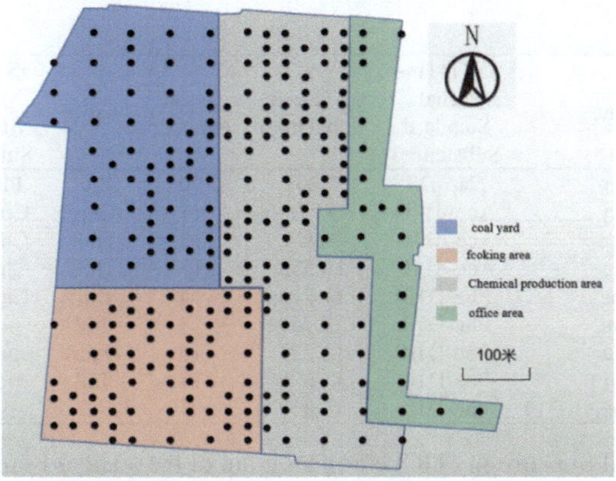

Fig. 1. Distribution map of soil testing points

2.3 Pretreatment of Soil Samples

The soil sample was placed in an air-dried tray, spread into a thin layer of 2–3 cm, crushed, and flipped, and the crushed stones, sand, and plant residues were removed. The dried sample was poured onto an organic glass plate, tapped, crushed, impurities were removed, mixed well, and the crushed sample was passed through a 0.25 mm (20 mesh) nylon sieve using a quartering method. After sieving, all samples were placed on a colorless polyethylene film, thoroughly stirred, mixed, and ready for use.

2.4 Pretreatment of Soil Samples

We accurately weighed 10.00 g of soil sample with a 1% balance, and an appropriate amount of quartz sand was added, mixed well, and placed in a rapid solvent extractor (ASE). The substitute 4,4-terphenyl-D14 was added, and a mixed solvent of acetone/dichloromethane (1:1, V:V) was used as the extractant. The ASE was used for extraction. After dehydration with anhydrous sodium sulfate, the internal standards (six internal standard mixed standards: 1,4-dichlorobenzene-D4, naphthalene-D8, acenaphthene-D10, phenanthrene-D10, tro-D12, and perylene D12) were added and concentrated to a constant volume (1.0 mL) and transferred to injection vials for testing.

2.5 Instrument Analysis Conditions for PAHs

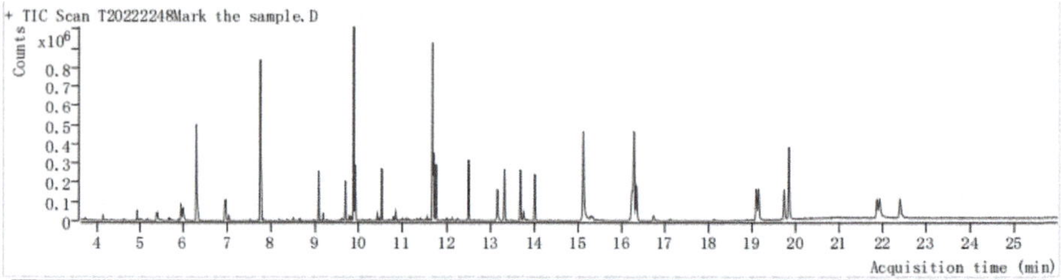

Serial Number	name	ISTD (Internal Standard Substance)	RT (Retention Time)	Serial Number	name	ISTD (Internal Standard Substance)	RT (Retention Time)
1	Nap	Nap-D8	7.801	10	BaA	Phe-D10	16.247
2	Acy	Ace-d110	9.722	11	Chry	Chry-D12	16.353
3	Ace	Ace-d110	9.944	12	BbF	Chry-D12	19.102
4	Flu	Ace-d110	10.555	13	BkF	Chry-D12	19.155
5	Phe	Phe-D10	11.716	14	BaP	Chry-D12	19.743
6	Ant	Phe-D10	11.769	15	InP	per-D12	21.866
7	Fla	Phe-D10	13.329	16	DahA	per-D12	21.929
8	Pyr	Phe-D10	13.690	17	InP	per-D12	22.401
9	4,4-Triphenyl-D14	Phe-D10	14.017	—	—	—	—

Fig. 2. Total ion current (TIC) chromatogram of the standard sample of PAHs

According to the national standard "Determination of polycyclic aromatic hydrocarbons in soil and sediment by gas chromatography-mass spectrometry" (HJ 805-2016), the Agilent 8860 + 5977B gas mass spectrometry detector was used for analysis. The gas chromatography conditions were as follows: injection port temperature of 280 °C; The initial temperature of the column was 40 °C, maintained for 2 min and then heated to 240 °C using a 20 °C/min program, followed by a 5 °C/min program to 310 °C, followed by a 10 °C/min program to 320 °C, and maintained for 4 min; The carrier gas was helium with a flow rate of 1.0 mL/min. The capillary chromatography column was DB-5MS (Agilent; 30.0 m × 0.25 mm × 0.25 μm). For the mass spectrometry conditions: we used the electron bombardment ion source, and the ion source and interface temperatures were 230 and 280 °C, respectively; The electron bombardment energy was 70 eV; The quality scanning range was from 45 to 450 amu. A total ion current (TIC) chromatogram of the PAHs was obtained after scanning (Fig. 2).

2.6 Accuracy and Precision of the Method

To ensure the accuracy of the analysis, each batch of collected samples included quality control samples, such as a transport blank and a full process blank. Blank sample PAHs were not detected, and the accuracy of the method was verified using the spiked recovery rate method. We added 4,4-terphenyl-D14 to the soil at concentrations of 0.2, 4, and 8 mg/kg, respectively. The spiked recovery rate of the sample was 51.2–89%, the standard deviation was approximately 2.08–4.45%, the coefficient of variation was approximately 7.65–13.76%, and the relative deviation range of parallel samples was approximately 0–13%. The precision satisfied the requirements for the relative deviation in the detection standard.

2.7 Ecological Risk Assessment Method for PAHs

Ecological risk assessment (ERA) is the process of predicting the harmful effects of environmental pollutants on ecosystems and is a crucial means of quantitatively studying the ecological hazards of pollutants [12]. Commonly used ERA methods include the risk quotient and probabilistic ecological risk assessment, which are used to characterize the severity and probability of the harm of pollutant exposure to the ecological environment.

The carcinogenic and non-carcinogenic risks of single pollutants in soil in the plant are based on six exposure paths: oral ingestion of soil, skin contact with soil, inhalation of soil particulate matter, inhalation of gaseous pollutants from surface soil in outdoor air, inhalation of gaseous pollutants from the underlying soil in outdoor air, and inhalation of gaseous pollutants from the underlying soil in indoor air. The health risks of 16 kinds of PAHs to human body were calculated according to the corresponding assessment models of different exposure pathways. The formulas and model parameter values were referred to HJ 25.3-2019 Technical Guidelines for Soil Pollution Risk Assessment of Construction Land. The total carcinogenic risk and total hazard quotient are calculated as the sum of the compound risk.

This study uses the commonly used quality benchmark method domestically and internationally to evaluate the ecological risks of PAHs in different functional areas Q_{ME} [13]. The formula for selecting the quality benchmark method is:

$$Q_{ME} = \sum \frac{C_i \div E_i}{n} \quad (1)$$

where Q_{ME} is the median quotient of the average effect interval; C_i is the measured concentration of PAH i in the soil (mg/kg); Ei is the median effect interval of polycyclic aromatic hydrocarbon I (mg/kg); *n* is the quantity of PAH i; This is 16. Table 1 lists the median effect ranges of the 16 PAHs [14]. In general, pollution levels are classified based on Q_{ME}:$Q_{ME} \leq 0.1$, no ecological risk; $0.1 < Q_{ME} \leq 0.5$, low ecological risk; $0.5 < Q_{ME} \leq 1.5$, moderate ecological risk; $Q_{ME} > 1.5$, high ecological risk [15].

Table 1. Reference values for median effect ranges of PAHs in soil (unit: mg/kg)

Pollutant	E_i	Pollutant	E_i
Nap	2.10	BaA	1.60
Acy	0.64	Chry	2.80
Ace	0.50	BbF	1.62
Flu	5.10	BkF	1.62
Phe	1.50	BaP	1.60
Ant	1.10	DahA	0.26
Fla	0.54	BghiP	1.60
Pyr	2.60	InP	1.60

2.8 Data Processing and Analysis

Microsoft Excel 2019 and SPSS22.0 statistical software were used to conduct data statistical analysis and draw bar charts. Origin v7.5 was used to map the distribution of PAHs concentration and ratio.

3 Results and Discussion

3.1 Residual Levels of PAHs in Soil

The detection and analysis of 16 PAHs in the soil of the coking plant showed that the detection rate of PAHs was 38.22%, and their residual levels are shown in Table 2. The proportion of PAHs with different ring numbers is indicated by Σ. The proportion of the 16 PAHs is shown in Fig. 3.

From Table 2 and Fig. 3, the average overall concentration of the 16 PAHs ranges from 0.27 to 6.49 mg/kg, with the highest average content of phenanthrene at 6.49 mg/kg and the lowest average content of dibenzo (a, h) anthracene at 0.27 mg/kg. The maximum concentration of the 16 PAHs was 2786 mg/kg for phenanthrene. The variation coefficient can reflect the average variation degree of pollutants between soil layers and the degree of pollutants affected by external factors [16], the coefficient of variation of the 16 PAHs was relatively high (756%–1376%), indicating significant heterogeneity in soil pollution at the site. In general, the proportion of three-ring PAHs was the highest, 46.43%; The proportion of PAHs in the sixth ring road was the lowest (3.88%). In the factory, low-ring (2–3 ring) PAHs dominated, accounting for 61.19%, and high-ring (4–6 ring) PAHs accounted for 38.61%, which was consistent with the research of Li Qun [17].

According to Malisewska-Kordybach [18], based on soil data from European countries, soil pollution assessment standards have been established (with a critical value of 0.2 mg/kg for soil pollution and 1 mg/kg for severe pollution). The total amount of PAHs in 517 soil samples at this site was greater than the pollution threshold of 0.2 mg/kg, while the total amount of PAHs in 384 soil samples was greater than the severe pollution

Table 2. Residual content of 16 PAHs in the soil of the coking plant

PAHs	average	standard deviation	minimum value	Maximum value	coefficient of variation	Proportion/% of Σ 16 PAHs
	w/(mg/kg)					
Nap	6.17	63.84	ND	1235	10.35	14.76
Acy	2.68	36.92	ND	1193	13.76	6.42
Ace	3.83	28.97	ND	486	7.56	9.17
Flu	4.19	39.17	ND	1163	9.35	10.02
Phe	6.49	84.45	ND	2786	13.01	15.54
Ant	2.29	29.48	ND	910	12.88	5.48
Fla	4.16	53.70	ND	1769	12.90	9.96
Pyr	2.98	37.10	ND	1182	12.46	7.13
BaA	1.81	20.26	ND	601	11.21	4.33
Chry	1.43	15.50	ND	447	10.83	3.42
BbF	1.40	14.57	ND	401	10.38	3.36
BkF	0.91	11.41	ND	382	12.49	2.19
BaP	1.54	16.32	ND	465	10.59	3.69
DahA	0.27	2.08	ND	50.7	7.65	0.65
BghiP	0.75	8.97	ND	288	11.93	1.80
InP	0.87	10.21	ND	317	11.76	2.08
two-ring PAHs	Nap					14.76
Σ three-ring PAHs	Acy, Ace, Flu, Phe, Ant					46.43
Σ four-ring PAHs	Pyr, Fla, BaA, Chry					24.84
Σ five-ring PAHs	BbF, BkF, BaP, DahA					9.89
Σ six-ring PAHs	BghiP, InP					3.88

threshold of 1 mg/kg, indicating that certain soil within the site has been severely polluted by PAHs, and the total residual amount is significantly higher than that of similar polluted sites [19].

The US EPA has listed seven types of PAHs with carcinogenic effects [benzo (a) anthracene, Chry, benzo (b) fluoranthene, benzo (k) fluoranthene, benzo (a) pyrene, indeno (1,2,3-cd) pyrene, and dibenzo (a,h) anthracene] [20], and from the perspective of carcinogenic PAHs, the highest content of seven PAHs in all soil layers of the plot was benzo (a) anthracene, with a maximum value of 601 mg/kg and an average value of 1.81 mg/kg, accounting for 4.33% of the total amount of the sixteen PAHs.

The "Soil Environmental Quality Construction Land Soil Pollution Risk Control Standard" (GB36600-2018) issued by the Ministry of Ecology and Environment of the People's Republic of China stipulates the screening values and control measures for eight of sixteen PAHs (benzo [a] anthracene, benzo [a] pyrene, benzo [b] fluoranthene, benzo [k] fluoranthene, Chry, dibenzo [a,h] anthracene, indeno [1,2,3-cd] pyrene, and

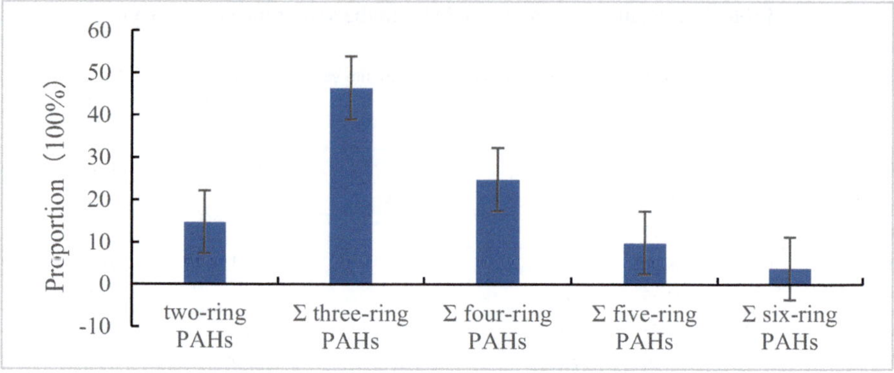

Fig. 3. Proportion of PAHs in each ring ratio of the Σ16 PAHs

naphthalene). The content of eight PAHs in the soil below this value is an ideal target for soil environmental quality, and its risk can be ignored; above this value, there may be an unacceptable risk, and further investigation and risk assessment are needed. The types and pollution levels of PAHs that exceeded national residual standards are listed in Table 3. As shown in Table 3, the detection rate of naphthalene among the eight types of PAHs in all soil layers of the study area was the highest at 27%. Among them, 1.1% of the soil content in the sample points was greater than the risk screening value for Class II land use, 0.3% of the soil content in the sample points was greater than the risk control value for Class II land use, and the detection rate of indeno [1,2,3-c, d] pyrene among the eight types of PAHs in all soil layers was the lowest at 9%. Among them, 0.9% of the sample sites had a soil content greater than the risk screening value for Class II land use, 0.1% of the sample sites had a soil content greater than the risk control value for Class II land use, none of the soil samples exceeded the risk screening and control values for Class II land use, and none of the soil samples of benzo [k] fluoranthene exceeded the control values for Class II land use risk.

Table 3. Screening and control values of eight national PAHs (unit: mg/kg)

Index	Nap	BaA	Chry	BbF	BkF	BaP	DahA	InP
Maximum value	1235	601	447	401	382	465	50.7	317
Average value	22.63	10.94	9.69	9.00	6.31	10.22	2.50	5.93
Filter Values	70	15	1293	15	151	1.5	1.5	15
Control Value	700	151	12900	151	1500	15	15	151
Number of detections	407	241	214	219	206	219	135	208
Detection rate%	27%	16%	14%	15%	14%	15%	14%	9%
Number of risk screening values for super second-class land use	16	23	0	23	2	87	36	13
Number of risk control values for super second-class land use	5	4	0	4	0	24	6	2

3.2 Vertical and Horizontal Distribution Characteristics of PAHs in Coking Plant Soil

Vertical Distribution Characteristics

The vertical distributions of PAHs across the entire factory area and various functional areas are shown in Fig. 4. From the perspective of the entire factory area, there were 12 samples with 16 PAHs greater than 1000 mg/kg, of which 33% were samples from a depth of less than 3 m. For the entire factory area, among the top ten samples containing sixteen PAHs, four samples had a soil depth of less than 3 m, accounting for 40%.

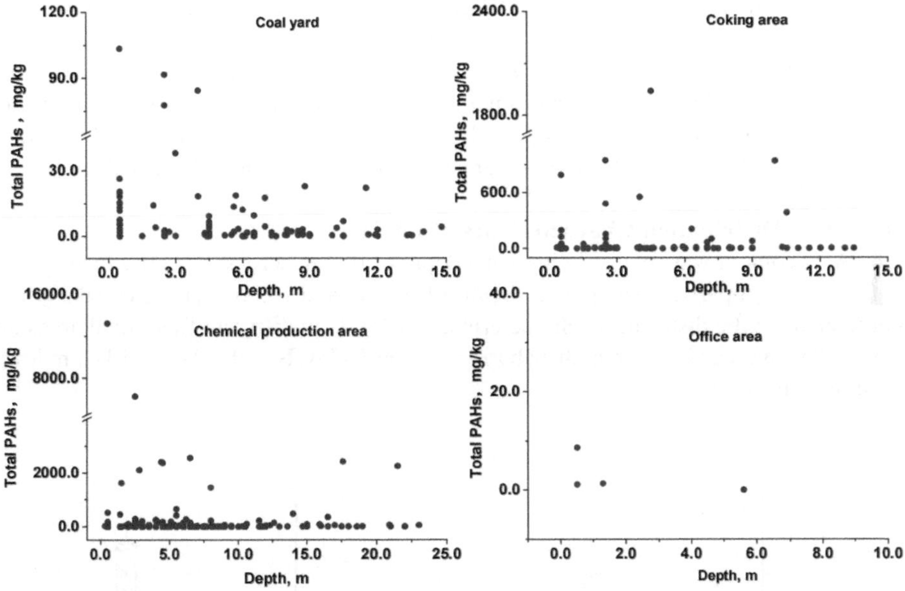

Fig. 4. Total amount of 16 PAHs in soil of different depth in different functional zones

From the perspective of each functional area, among the top 10 samples in the 16 PAHs rankings, the chemical production area had the highest proportion of samples in depths of less than 3 m, accounting for 90%, followed by the coking area, which accounted for 10%. From the perspective of the stratigraphic distribution, most areas of the plot with a depth of less than 3 m were covered by silty clay aquifers that were difficult to migrate. Therefore, the PAHs content at depths of less than 3 m was relatively high. In general, PAHs, as hydrophobic organic pollutants, primarily migrate by adsorbing organic matter from soil [21]. Under natural conditions, the texture of deep soil is hard, and the porosity is low, which is not conducive to the migration and diffusion of pollutants. They are typically enriched in shallow soils. Under undisturbed soil conditions, the PAHs content decreases with increasing depth [6]. Pei [4] showed that the soil PAHs in the coking plant area were enriched owing to the antifouling performance of the clay layer and its binding with organic matter. In addition, overall, most areas of the factory area (coal yards and chemical production areas) had surface soil (0–1 m) with a PAHs content higher than that at a depth of 2.5 m, such as the chemical production area. In a soil sample at a depth of 5 m, the highest content of 16 PAHs was 13218.7 mg/kg, which is substantially higher than that of soil samples at a depth of 2–2.5 m (6271.6 mg/kg). The reasons might be as follows: First, the sampling point is located in the tar tank area of the chemical production area, and during the long-term production process, there may be phenomena, such as "running, emitting, dripping, and leaking," which may lead to surface soil contamination. The content of PAHs is high, and pollutants enter the soil and vertically infiltrate and migrate downwards. In addition, the sample content was abnormally high at certain points greater than 3 m, such as the depth of the coking kiln in the coke field area (approximately 1–2 m) and the depth of tar pollution (primarily greater than 3 m). This also explains why the total amount of PAHs at depths greater than 3 m was higher than that in the surface soil. For the surface layers of the coal yard and office area (0–1.0 m), the content of PAHs was relatively high, primarily due to rainfall leaching in the coal yard and smoke generated in the coke oven area entering the surface soil through atmospheric sedimentation. From the perspective of production history, the surface soil of the coal yard and office areas in this plot were less disturbed.

Horizontal Distribution Characteristics of PAHs

The PAHs content was relatively high at a depth of 0–1 m across the entire factory area. To facilitate comparison, samples at depths of 0–1 m and 1–3 m were selected from the plot to analyze the distribution characteristics of PAHs in different functional areas and depths. The average horizontal distributions of the 16 PAHs at the 0–1 and 1–3 m levels are shown in Fig. 5.

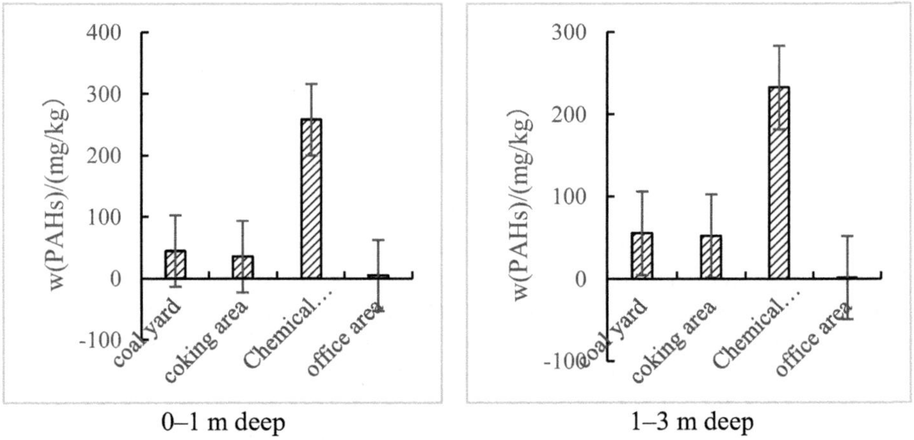

Fig. 5. Horizontal distribution of the 16 polycyclic aromatic hydrocarbons (PAHs)

Figure 5 shows that the content of the 16 types of PAHs was relatively high in chemical production areas, followed by coal yards. In the chemical production area, the point with a high content of coal in the gas tank pneumatic conveying workshop was HIX6-1, the content of 16 PAHs was 13218.7 mg/kg, and the content of phenanthrene monomer PAHs was 2786 mg/kg,The soil pollution in the chemical producing areas is the most severe. The reason for the high content of the 16 types of PAHs and phenanthrene in the gas tank coal pressure delivery workshop may be that phenanthrene exists in the coal tar and that the highest point of phenanthrene content (HIX6) was located on the west side of the coke tank area in the chemical production area. The groundwater flow direction in this area is from northeast to southwest, which may lead to a high content of PAHs (especially phenanthrene) in the soil of the chemical production area due to the phenomenon of "running, emitting, dripping, and leaking" in the long-term production process. Tar leakage in the tar storage tank area may be the primary reason for the high PAHs content. The production processes in chemical production areas have a long history, and environmental management is relatively extensive. A large amount of tar may leak into the soil, resulting in a high total amount of PAHs.

3.3 Source Analysis of 16 PAHs in Different Functional Areas

In existing studies, the proportion of medium and high-ring PAHs to low-ring PAHs can be used to indicate the source of the PAHs (Maliszewska-Kordybach et al.,1996). Four–six-ring (medium and high-ring) PAHs primarily originate from fossil combustion, while two–three-ring (low-ring) PAHs primarily originate from petroleum pollution or natural diagenesis. In general, when the content ratio of low rings to medium and high rings, that is, w (low-ring PAHs)/w (medium and high-ring PAHs), is greater than one, the PAHs primarily originate from oils. Soil PAHs often originate from various oil chemicals, such as tar, crude benzene, and diesel [22]. When w (low-ring PAHs)/w (medium and high-ring PAHs) is < 1, it indicates that PAHs are primarily composed of medium and high-ring PAHs and primarily originate from the high-temperature combustion of fossil fuels, such as coal [19]. Xiaofeng [23] analyzed particulate phase PAHs in the atmospheric environment of a coking plant area and revealed that, in the context of the proportion of medium to high cyclic components, the proportion of 16 PAHs ranged from 67.77% to 92.58%. Enke [24] used an experimental coke oven to measure the content of PAHs in the particle phase generated during the coking process and found that the proportion of components above the middle and high-ring PAHs exceeded 80%.

The compositions of the 16 types of PAHs at different soil depths are shown in Fig. 6 for the low-ring and medium and high-ring PAHs.

Figure 6 shows that the proportion of low-ring PAHs content in the entire study area to the total amount of the 16 types of PAHs was 78.07%, and the proportion of medium and high-ring PAHs content was 21.92%. Noticeably, w (low-ring PAHs)/w (medium and high-ring PAHs) = 3.56. Overall, the PAHs at the site were primarily low-ring, and the low-ring PAHs were primarily phenanthrene. Among the 16 types of monomer PAHs, the phenanthrene content accounted for the highest proportion of the total amount of the 16 types of PAHs, which was 15.54% (accounting for Σ, the proportion of low-ring PAHs reached 22.68%). The highest concentration of phenanthrene in the chemical production area was HIX6-1, with a maximum concentration of 2786 mg/kg.

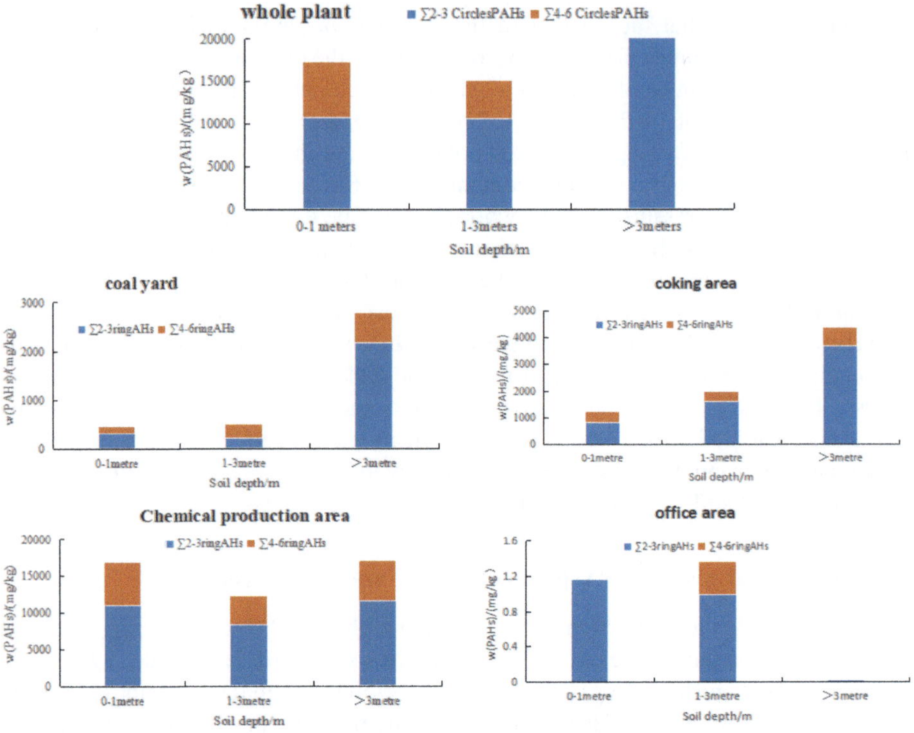

Fig. 6. Composition of the 16 PAHs at different soil depths

Monomer phenanthrene exists in coal tar, and the coke tank area is located in the chemical production area, which may cause pollution due to pipeline "running, emitting, dripping, and leaking."

The high proportion of low-ring PAHs in the coal fields, coke fields, and chemical production areas indicates that the composition of PAHs in the soil of these areas was significantly affected by atmospheric sedimentation. The sedimentation of coal smoke generated in coke fields often results in the soil being dominated by medium to high-ring PAHs. The concentrations of high and low-ring PAHs in the coking area were similar. This is because the production process is relatively extensive, and a large amount of coal tar leaks into the soil. These coal tars are the primary source of low-ring environmental PAHs, and the total petroleum hydrocarbon (C10–C40) content in the region was as high as 7630 mg/kg; In contrast, due to the low chimney and nearly no tail gas treatment equipment, coking generates a large number of atmospheric particles that settle near to the surface, and the middle and high-ring PAHs components are primarily enriched in these particles.

3.4 Ecological Risks in Different Functional Areas

PAHs in the soil of the coking plant may pose risks to the ecological environment and residents. Therefore, research was conducted on the ecological and human health risks of PAHs in soils of different functional areas of the plot. First, the Q_{ME} values of the soil in different functional areas were calculated according to Eq. (1), and the ecological risk levels of different functional areas were represented by the proportion of different pollution levels (Fig. 7).

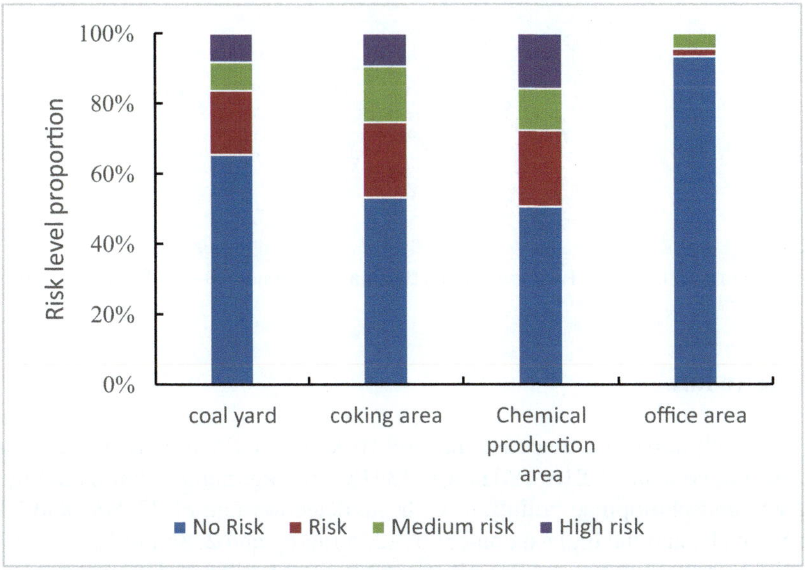

Fig. 7. Ecological risk in different functional areas Ecological risks: No Risk, $Q_{ME} \leq 0.1$; Low Risk, Q_{ME} 0.1–0.5; Medium Risk, Q_{ME} 0.5–1.5; High Risk, Q_{ME}>1.5

From the perspective of the proportion of high-risk ecological sites, the order of risk levels was chemical production areas > coke fields > coal fields > office areas. Chemical production areas had the highest ecological risk, with high ecological risk points accounting for over 15.8%.

According to the "Technical Guidelines for Soil Pollution Risk Assessment of Construction Land" (HJ25.3 2019), the risk of PAHs in soil with depths of < 3 m in different functional areas to human health was calculated (Fig. 8). The points with high human health risks were primarily located in chemical production areas, and the point with the highest risk of cancer and noncancer was HIX6-1 (located in chemical production areas).

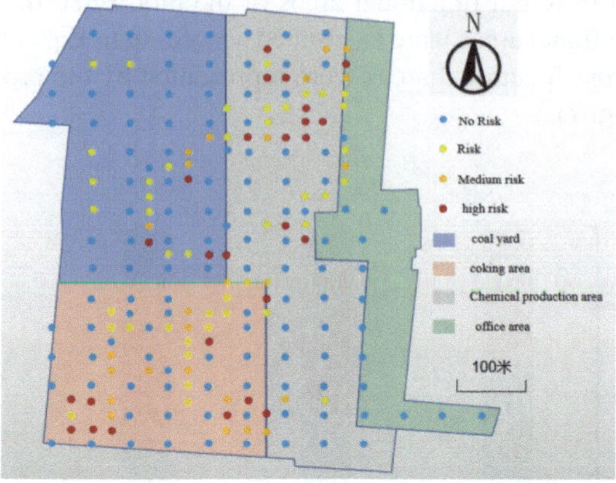

Fig. 8. Human health risks of PAHs in soil depths of approximately 0–3 m for different functional zones

4 Conclusion

(1) In the study area, the detection rate of 16 kinds of PAHs was 38.22%, and the maximum content of Σ16 PAHs was 13218.7 mg/kg, mainly dominated by naphthalene and philippinae pollution, with the detection rate of 27.24% and 23.29%, respectively, and the highest content of 1235 mg/kg and 2786 mg/kg.

(2) PAHs in the soil of the coking plant exhibited a clear spatial distribution pattern. At the horizontal level, the average content of the 16 PAHs was the highest in the production area, followed by the coal yard area, coking plant area, and office area; 52.64% of PAHs in the vertical soil layers were concentrated in the 0–3 m soil layer. From the analysis of the ratio of the contents of low-ring (two–three rings) to medium and high-ring(four–six rings) PAHs, it can be concluded that the PAHs in the soil of the chemical production area primarily originated from the leakage of tar during the production process, while the PAHs in the coke and coal fields primarily originated from the sedimentation of coal smoke produced by the coke oven. The ecological risk level analysis showed that the chemical production area had the highest risk level, with high-risk points accounting for over 15.8%, followed by coke fields, coal fields, and office areas.

Declaration Statement. All authors have read, understood, and have complied as applicable with the statement on "Ethical responsibilities of Authors" as found in the Instructions for Authors.

Declaration of Competing Interest. The authors declare that they have no known competing financial interests or personal relationships that could have appeared to influence the work reported in this paper.

Funding Statement. This study was supported by the Special Research Award Fund of Shandong Coalfield Geological Bureau.

References

1. Du, F.F., Yang, Y., Liu, M., et al.: Distribution and source apportionment of polycyclic aromatic hydrocarbons in surface soils in Shanghai. China Environ. Sci. **34**(4), 989–995 (2014)
2. Kim, K.H., Jahan, S.A., Kabir, E., et al.: A review of airborne polycyclic aromatic hydrocarbons (PAHs) and their human health effects. Environment International **60**, 71–80 (2013)
3. Manoli, E., Samara, C., Konstantinou, I., et al.: Polycyclic aromatic hydrocarbons in the bulk precipitation and surface waters of Northern Greece. Chemosphere **41**(12), 1845–1855 (2000)
4. Wang, P., Jiang, P., Zhang, H., et al.: Distribution characteristics and polluting process of PAHs in soil and groundwater at a coking plant site. Res. Environ. Sci. **28**(5), 752–759 (2015)
5. Wang, L., Zhang, T.F., Yang, S.C., et al.: Characteristics of microbial community diversity in PAHs contaminated soil of a coking plant. J. Environ. Eng. Technol. **11**(4), 720–726 (2021)
6. Ran, Z.X., Chen, J.Y., Wang, Y.T., et al.: Characteristics and Influencing Factors of Polycyclic Aromatic Hydrocarbons in Surface Soils from Typical Industrial Areas of Chengdu. Environmental Science **40**(10), 4594–4603 (2019)
7. Zhang, D., Cao, S.P., Sun, J.L., et al.: Occurrence and spatial differentiation of polycyclic aromatic hydrocarbons in surface soils from Shenzhen, China. Environmental Science **35**(2), 711–718 (2014)
8. Feng, Y., Lü, Y.-L., Jiao, W., et al.: Distribution and risk of polycyclic aromatic hydrocarbons in soils from different workshops of an abandoned coking factory in Beijing. Asian Journal of Ecotoxicology **4**(3), 399–407 (2009)
9. Kong, L.L., Shi, M.J., Liang, J.J., et al.: Concentration and origin of polycyclic aromatic hydrocarbons in the soil of Dagang oil field. Environmental Science & Technology **41**(5), 151–157 (2018)
10. Zhang, H.L., Wang, Y., Zhou, Y., et al.: Emission characteristics and environmental impact of PAHs in PM2.5 from one coking plant. Chinese Journal of Environmental Engineering **11**(10), 5571–5576 (2017)
11. Zhong, M.Y., Li, H.Y., Jia, X.Y., et al.: A comparative study on the pollution characteristics of polycyclic aromatic hydrocarbons (PAHs) in the soil of different coking plants. Journal of Ecology and Rural Environment **37**(5), 627–635 (2021)
12. USEPA: Guidelines for Ecological Risk Assessment. Washington DC: US Environmental Protection Agency, EPA/630/R-95/002F (1998)
13. Du, S.L., Ding, T.T., Dong, H.J., et al.: Pollution and risk assessment of polycyclic aromatic hydrocarbons in the water environment of the Shaying River Basin. J. Agro-Environ. Sci. **39**(3), 601–611 (2020)
14. Song, Y., Hou, D., Zhang, J., et al.: Environmental and socio-economic sustainability appraisal of contaminated land remediation strategies: a case study at a mega-site in China. Sci. Total. Environ. **610–611**, 391–401 (2018)
15. Akhbarizadeh, R., Moore, F., Keshavarzi, B., et al.: Aliphatic and polycyclic aromatic hydrocarbons risk assessment in coastal water and sediments of Khark island, SW Iran. Mar. Pollut. Bull. **108**(1/2), 33–45 (2016)
16. Peng, C., Liu, X., Zhou, Z.R., et al.: Characteristics and risk assessment of heavy metals in the soil around copper smelting sites. Environmental Science **44**(01), 367–375 (2023)
17. Li, Q., Li, M., Li, Y., et al.: Analysis on distribution characteristics of soil polycyclic aromatic hydrocarbons (PHAs) in a large coking plant site. J. Ecolo. Rural Environ. **37**(12), 1623–1632 (2021)
18. Maliszewska-Kordybach, B.: Polycyclic aromatic hydrocarbons in agricultural soils in Poland: preliminary proposals for criteria to evaluate the level of soil contamination. Applied Geochemistry **11**(1/2), 121–127 (1996)

19. Hao, L.H., Zhang, S.C., Wu, Z.H., et al.: Spatial distribution characteristics of PAHs in soil at hilly areal coking plant. China Environ. Sci. **38**(7), 2625–2631 (2018)
20. USEPA: Polycyclic Aromatic Hydrocarbons (PAHs): EPA Fact Sheet. Washington DC, USA: National Center for Environmental Assessment, Office of Research and Development (2008)
21. Richnow, H.H., Eschenbach, A., Mahro, B., et al.: The use of 13C-labelled polycyclic aromatic hydrocarbons for the analysis of their transformation in soil. Chemosphere **36**(10), 2211–2224 (1998)
22. Mai, B.X., Fu, J.M., Sheng, G.Y., et al.: Chlorinated and polycyclic aromatic hydrocarbons in riverine and estuarine sediments from pearl river delta, China. Environmental Pollution **117**(3), 457–474 (2002)
23. Liu, X.F.: Research on characteristics of polycyclic aromatic hydrocarbons and carbonaceous composition in surrounding of mechanical coke oven. Taiyuan University of Technology, Taiyuan (2014)
24. Li, E.K., Cheng, X.L., Cang, D.Q., et al.: Study of generation characteristics of polycyclic aromatic hydrocarbons in coking process. Modern Chemical Industry **30**(1), 74–77 (2010)

Research on Parameter Identification of a Normalized Creep Model for Clay in Environmental Engineering Applications

Changxi Yue[1,2](✉), Wei Pan[2], Yonghua Cao[1,2], Guoliang Ye[2], Liqiang Sun[3], and Changyi Yu[1,2]

[1] Tianjin Port Engineering Institute Co., Ltd. of CCCC First Harbor Engineering Co., Ltd., Key Laboratory of Port Geotechnical Engineering of Ministry of Communications, Key Laboratory of Port Geotechnical Engineering of Tianjin, Key Laboratory of Geotechnical Engineering, CCCC, Tianjin, China
yuechangxitab@163.com
[2] CCCC First Harbor Engineering Co., Ltd., Tianjin, China
[3] School of Architecture and Engineering, Tianjin University, Tianjin, China

Abstract. Understanding the creep behavior of soft soils is essential for evaluating the long-term stability of structures such as embankments, landfills, and other civil engineering projects that interact with the environment. By normalizing the creep parameters of the generalized Kelvin model, a comprehensive creep model was proposed, which takes into account soil consolidation load, bulk density, void ratio, liquid limit, depth, and other relevant indicators. An empirical formula for the creep parameters of the generalized Kelvin model was also established. The proposed model can aid in assessing the long-term stability and environmen-tal impact of such structures, ensuring compliance with sustainability standards and minimizing adverse effects on surrounding ecosystems. By integrating geotechnical analysis with environmental considerations, this research contributes to the development of more resilient and environmentally responsible engineering solutions.

Keywords: geo-environmental · soft soil creep · normalization · generalized Kelvin model · settlement prediction

1 Introduction

The study of creep behavior in soft soils is critical for the design and long-term performance of geotechnical structures. Accurate identification of creep model parameters is fundamental to predicting soil deformation under sustained loads. Recent advancements in computational techniques and experimental methodologies have significantly enhanced our understanding of soft soil creep behavior [1, 2]. However, the complexity of soil structure interactions and the inherent variability of soil properties continue to pose challenges in parameter identification. Several models, such as the Modified Cam-Clay model and the Burgers model, have been widely adopted to describe the creep behavior of soft soils [3–6]. Despite these efforts, discrepancies between predicted and observed behaviors necessitate further refinement of these models.

D. Han et al. (Eds.): ICEPG 2024, ESE, pp. 682–687, 2025.
https://doi.org/10.1007/978-3-031-86829-0_77

In the realm of environmental engineering and governance, understanding the creep behavior of soft soils is equally crucial. Soil deformation can significantly impact the stability of infrastructure, as well as influence the migration of contaminants and the behavior of underground water flow, thereby affecting environmental safety and sustainability. For instance, the long-term settlement of landfills and waste containment systems is directly influenced by the creep behavior of the underlying soft soils, which can lead to potential environmental hazards if not accurately predicted and managed [7, 8]. Therefore, developing robust creep models not only aids in geotechnical design but also serves as a vital tool for environmental monitoring and pollution control.

Based on the five-element generalized Kelvin model, this paper proposes a normalized creep model that comprehensively considers consolidation load, bulk density, void ratio, liquid limit, and depth of the soil. This enhanced model aims to bridge the gap between theoretical predictions and observed behaviors, offering a more reliable framework for both geotechnical and environmental applications.

2 Creep Test

The soil sample was collected from a land reclamation project, with sampling depths ranging from 0.5 m to 27.0 m below the ground surface, using a thin-walled sampler. The physical properties of the soil sample are presented in Table 1. Uniaxial creep tests were conducted on the aforementioned soil samples at loading levels of 25 kPa, 50 kPa, 80 kPa, 100 kPa, 200 kPa, 250 kPa, 400 kPa, and 800 kPa. The soil is considered to be in a stable creep stage when the deformation of the specimen is less than 0.005 mm within 24 h under each load level. Once stable creep conditions are achieved, lower load levels can be applied, provided the total test duration reaches at least 120 days.

Table 1. Soil physical indicators

No.	Depth(m)	Bulk density(kN/m^3)	Void ratio	Natural moisture content(%)	Liquid limit (%)
A1	6.0	17.6	1.236	43.6	48.6
A2	7.0	17.5	1.264	44.6	46.7
A3	8.0	17.2	1.317	46.0	47.0
A4	9.0	17.6	1.250	44.5	45.0
A5	11.0	17.9	1.140	39.8	42.8
A6	15.0	17.7	1.178	41.2	37.7
A7	18.0	16.6	1.587	55.6	53.3
A8	20.0	16.1	1.817	64.3	62.8
A9	24.0	16.1	1.791	62.8	60.7
A10	27.0	18.2	1.121	40.9	44.5

3 Creep Model and Parameter Identification

3.1 Creep Model

To describe the instantaneous elasticity and viscoelastic properties of soil and to characterize the attenuation stability characteristics of creep, this paper employs a five-element generalized Kelvin model to fit the creep behavior under different load levels. The five-element generalized Kelvin model consists of a Kelvin model and a Maxwell model arranged in series. The creep equations for the model are presented in Eq. (1) and Eq. (2).

$$\varepsilon(t) = J(t)\sigma \tag{1}$$

$$J(t) = \frac{1}{E_{\mathrm{H}}} + \frac{1}{E_{\mathrm{K1}}}\left(1 - e^{-\frac{E_{\mathrm{K1}}}{\eta_{\mathrm{K1}}}t}\right) + \frac{1}{E_{\mathrm{K2}}}\left(1 - e^{-\frac{E_{\mathrm{K2}}}{\eta_{\mathrm{K2}}}t}\right) \tag{2}$$

In the equation, J(t) is the creep softness, E_{H} is the elastic modulus of the Hooke body, η_{K1} and η_{K2} are the viscosity coefficients of the Kelvin body, E_{K1} and E_{K2} are the elastic modulus of the Kelvin body. By fitting the results of creep tests on samples A1 through A10 using Eq. (1) and Eq. (2). Overall, the fitting accuracy of the five-element generalized Kelvin model under various loading levels is approximately 0.99.

3.2 Parameter Identification

When fitting the creep characteristics of soil using the generalized Kelvin model, the five parameters of the model must vary with changes in the soil stress state. Specifically, there are five parameters for each of the samples A1 through A10 at each load level, resulting in a total of 400 creep parameters across 10 tests under load levels of 25 kPa, 50 kPa, 80 kPa, 100 kPa, 200 kPa, 250 kPa, 400 kPa, and 800 kPa. Given the substantial number of parameters, direct application to engineering practice is challenging. Therefore, a normalization analysis is con-ducted on the creep parameters to establish their relationship with soil physical properties and creep parameters.

Based on the fitting results of the generalized Kelvin model parameters, empirical formulas are established to relate consolidation pressure (σ), natural porosity (e), liquid limit porosity (e_{L}), bulk density (γ), and atmospheric pressure (P) of samples at different burial depths (h) to the corresponding model parameters E_{H}, E_{K1}, E_{K2}, η_{K1} and η_{K2}. The correlation coefficient (R^2) exceeds 95%.

$$E_{\mathrm{H}} = \sigma\left\{\begin{array}{l} 5.81 + 172.97\exp\left[3.65 - \dfrac{\sigma\left(1+\frac{e}{e_{\mathrm{L}}}\right)}{0.04\mathrm{P}\left(1+\exp(1-\frac{\sigma}{\gamma h})\right)}\right] + \\ 71.16\exp\left[0.3129 - \dfrac{\sigma\left(1+\frac{e}{e_{\mathrm{L}}}\right)}{0.495\mathrm{P}\left(1+\exp(1-\frac{\sigma}{\gamma h})\right)}\right] \end{array}\right\} \tag{3}$$

$$E_{\mathrm{K1}} = \gamma h\left[19.28 + \frac{6868.77}{1+\exp\left(\frac{E_{\mathrm{H}}}{2.21\sigma} - 0.472\right)}\right] \tag{4}$$

$$\frac{E_H}{E_{K2}} = 0.3065\left(\frac{E_H}{E_{K1}}\right)^4 - 2.0154\left(\frac{E_H}{E_{K1}}\right)^3 + 3.6878 \left(\frac{E_H}{E_{K1}}\right)^2 - 0.7871\left(\frac{E_H}{E_{K1}}\right) + 0.2145 \tag{5}$$

$$\eta_{K1} = \frac{E_H}{7.7291 + 0.01884e^{-28.1655\frac{\sigma}{\gamma h}}} \tag{6}$$

$$\eta_{K2} = \frac{E_H}{0.0031 + 0.97138e^{-21.4086\frac{\sigma}{\gamma h}}} \tag{7}$$

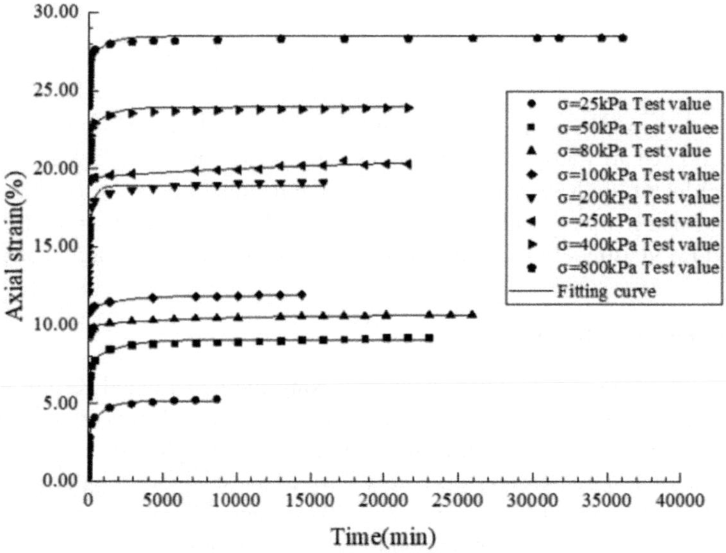

Fig. 1. Fitting curve based on normalized creep model

Using the A4 soil sample as an example, creep deformation was fitted under different burial depths and consolidation pressures using Eqs. (1) to (7), as illustrated in Fig. 1. The study demonstrates that the fitting results are robust, and the normalized five-element generalized Kelvin model exhibits high fitting accuracy. The experimental data points are uniformly distributed around the fitted curve, and the error can be controlled within 5%, meeting the requirements of practical engineering applications. This model can effectively predict one-dimensional consolidation creep deformation of soft soil under varying burial depths and consolidation pressures.

3.3 Numerical Simulation Calculation

Utilizing the self-developed UMAT subroutine, a finite element simulation of the creep test process was conducted on a uniaxial creep soil sample (A3) at a depth of 8 m.

The finite element analysis was conducted in three steps. First, the initial pre-stress field was established, and the Geostatic analysis step was configured to balance the initial stress state. In the second step, stress was applied to simulate instantaneous loading in one-dimensional consolidation creep tests. Finally, creep analysis was performed under constant load conditions to calculate the creep deformation. The simulation results are presented in Fig. 2. The error between the finite element calculations and experimental results is within 5%, and the creep trend is consistent with the experimental values, thereby validating the reasonableness of the subroutine development.

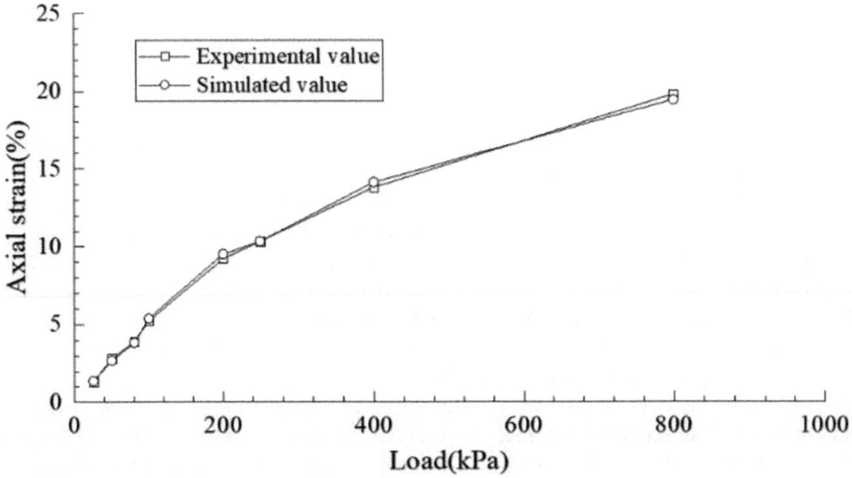

Fig. 2. Comparison curve of numerical simulation and experimental results

4 Conclusion

Through uniaxial creep tests, the creep characteristics of soft soil at different burial depths were investigated, and a normalized creep model for soft soil was established. The following conclusions were drawn:

1. Under the initial loading level, surface soft soil exhibited maximum compressive deformation in the samples. The sedimentation history and depositional environment of soft soil significantly impact its compressive deformation. At lower load levels, soft soil primarily exhibits attenuated stable-type creep characteristics; at higher load levels, creep deformation transitions to a constant-rate creep state, with both the rate and duration of constant-rate creep increasing with the load level.
2. Based on a five-element generalized Kelvin model, this study established a comprehensive creep model that considers consolidation load, bulk density, void ratio, liquid limit void ratio, burial depth, and other indicators. A normalized empirical formula for the five parameters in the model was derived. The fitting curves of this model align well with experimental values and accurately describe the creep characteristics of soft soil.

3. A UMAT subroutine was developed for the proposed normalized five-element generalized Kelvin creep model, which solves the problem of creep parameters changing with load. The reliability of the subroutine was validated through simulations of uniaxial creep tests, with an error margin within 5%.

Acknowledgment. This paper is funded by CCCC Key Program. (2023-ZJKJ-03) and CFEC Key Program. (2022-YHJJBGS-04).

References

1. Kaczmarek, Ł., Dobak, P.: Contemporary overview of soil creep phenomenon. Contemporary Trends in Geoscience, 6 (2017)
2. Xiang, G., Yin, D., Cao, C., et al.: Creep modelling of soft soil based on the fractional flow rule: Simulation and parameter study. Appl. Math. Comput. **403**, 126190 (2021)
3. Bacquaert, G., Raude, S., Alves-Fernandes, V., Voldoire, F., Kondo, D.: A standard thermodynamic-based extension of the Modified Cam-Clay soil model and its applications. European Journal of Mechanics-A/Solids **103**, 105122 (2024)
4. Sivasithamparam, N., Karstunen, M., Bonnier, P.: Modelling creep behaviour of anisotropic soft soils. Comput. Geotech. **69**, 46–57 (2015)
5. Dey, A., Basudhar, P.K.: Applicability of Burger model in predicting the response of viscoelastic soil beds. GeoFlorida 2010: Advances in Analysis, Modeling & Design, 2611–2620 (2010)
6. Kovačević, M.S., Bačić, M., Librić, L., Gavin, K.: Evaluation of creep behavior of soft soils by utilizing multisensor data combined with machine learning. Sensors **22**(8), 2888 (2022)
7. Koda, E., Podlasek, A., Osiński, P., Markiewicz, A., Vaverková, M.D.: Geotechnical and environmental aspects of landfill reclamation: a case study of an old municipal solid waste landfill in Poland. Japanese Geotechnical Society Special Publication **9**(4), 94–98 (2021)
8. Gao, W., Kavazanjian, E., Jr.: A constitutive model for municipal solid waste considering mechanical creep and biodegradation-induced compression. Acta Geotech. **17**(1), 37–63 (2022)

Study on Vegetation Community Characteristics and Soil Physical Properties of Artificial *Hippophae Rhamnoides* Forest in Shuanglonggou Abandoned Mining Area

Lide Wang[1,2,3,4], Guangwu Liu[1], Jing He[4](✉), Fanglin Wang[1,2], Dacheng Song[1,2,3], Jingyi Si[4], and Lianghaoran Wang[4]

[1] Gansu Institute of Desertification Control, Lanzhou 730070, China
wldzy69@163.net
[2] Minqin Desert Grassland Ecosystem National Field Scientific Observation and Research Station, Minqin 733300, Gansu, China
[3] Hexi Corridor Forest Ecosystem National Station, Wuwei 733000, Gansu, China
[4] College of Forestry, Gansu Agricultural University, Lanzhou 730070, China
hejing268@aliyun.com

Abstract. [Objective] In order to investigate the vegetation community characteristics and soil physical properties of artificial *Hippophae rhamnoides* forest with different restoration years (CK, 3a, 6a and 11a) in Shuanglonggou abandoned mining area, the artificial *Hippophae rhamnoides* forest was selected as the research object. [Methods] The spatial and temporal substitution method was used to analyze the characteristics of vegetation community and soil physical properties in different restoration years. [Results] The results showed that: (1) During the restoration of 11a, 65 species of plants appeared in 24 families, 48 genera, and 11 families were single families, single genera, and single species. The composition of vegetation species was as follows: most species belonged to minority family, and few species belonged to the majority families; (2) With the extension of repair years, the dominant species of the community gradually evolved from annual or perennial herbs to perennial herbs or bushes, and *Hippophae rhamnoides, Leontopodium ochroleucum, Potentilla fruticosa* and *Poa annua* became the dominant species of the community, and the sum of important values of the four plants reached 44.55; Simpson dominance index showed the trend of "first rise~then decline~then rise". Max reached 0.277 while fixing 3a. Shannon diversity index and Pielou evenness index showed a trend of "first decline~then rising~then decline". When 6a was repaired, Shannon diversity index reached the maximum value of 3.286, and Pielou evenness index reached the maximum value of 0.761. (3) With the extension of restoration years, soil water content showed an increasing trend, and reached the maximum value of 12.263, 9.798 and 8.797% at 11 years of restoration, respectively. Soil bulk density and particle size showed a trend of continuous decline, and soil particles gradually developed in the direction of thinning. (4) The number of plant individuals and Shannon diversity index were positively correlated with soil water content. Plant individual number, plant species number and Margalef richness index were negatively correlated with soil bulk density and particle size. There was a significant negative correlation between Pielou

D. Han et al. (Eds.): ICEPG 2024, ESE, pp. 688–705, 2025.
https://doi.org/10.1007/978-3-031-86829-0_78

evenness index and water content. [Conclusion] It can be seen that with the extension of restoration years, vegetation community characteristics and soil quality have been significantly optimized, which promotes the restoration of ecosystems in the abandoned mining areas.

Keywords: Shuanglonggou mining area · different restoration years · vegetation community characteristics · soil physical properties

1 Introduction

Shuanglonggou area is the main catchment area of the upper reaches of the Shiyang River, with vast forests and grasslands and rich mineral resources [1]. As we all know, large-scale mineral mining often leads to land excavation loss, collapse and occupation, leading to ecological and environmental problems such as vegetation degradation and even death [2]. The Shuanglonggou area has experienced 16 years of coal mining and gold mining activities, which seriously damaged the ecological environment and left a large area of bare gravel, pits and gangue hills [3]. Although the gold and coal mining activities in Shuanglonggou have long ceased, the natural recovery and renewal of vegetation is slow due to the harsh climatic conditions in the high altitude area, which greatly affects the sustainable development of the ecosystem in Shuanglonggou and the water security of the downstream people [4]. Therefore, how to scientifically control and restore the abandoned mining area of Shuanglonggou and maintain the stability of plant community structure and plant diversity are particularly important.

Vegetation restoration is an important symbol of ecological restoration in mining areas and one of the most commonly used means to prevent soil erosion and restore the integrity of the ecosystem. Among vegetation restoration measures, natural restoration takes a long time and cannot achieve rapid restoration of vegetation, while artificial restoration takes a short time, which is suitable for rapid restoration of large exposed wounds. Creating good external conditions for the cycle of mining ecosystem can accelerate vegetation restoration and is an effective way to promote the restoration of mining ecosystem [5, 6]. Plant species diversity is the expression of biodiversity at the level of vegetation species, reflecting the competition and coordination between species in the stand community and environmental resources, and also reflecting the stability, complexity and succession trend of plant community functions. Mining activities in mining areas will cause direct and indirect pressure on the protection of plant species diversity [7, 8]. In the process of ecological restoration in mining areas, Zhang Guanxiong [9] studied the maintenance mechanism of plant species diversity in new wetlands in coal-mining subsidence areas, and believed that vegetation restoration time was very important for land reclamation and ecosystem restoration. The research of Li Xinying et al. [10] on understory plant diversity in coal mining subsidence area shows that artificial vegetation restoration in coal mining subsidence area can increase soil water content and soil total nitrogen, and has positive significance for improving vegetation community structure. The research results of Cheng Wei et al. [11] on degraded soil in western mining areas show that vegetation restoration mode has an important impact on ecological restoration in mining areas, and soil humus layer reconstruction technology can effectively improve soil structure and enhance soil nutrients. Chen Jing et al. [12] studied the vegetation species diversity of Shendong mining area, and showed that the comprehensive evaluation index of biodiversity of artificial vegetation restoration community was higher than that of natural restoration community, and the protection effect of plant species diversity in mining area was remarkable. Fernandez et al. [13] studied the vegetation restoration technology in abandoned mining areas, indicating that the key technologies of vegetation restoration in abandoned mining areas include plant species selection, seedling planting, community construction and vegetation maintenance. Li Pengfei [14]'s research on the soil quality of Hedaigou Coal mine dump showed that the mixed planting mode of tree, shrub and grass had a good effect on improving the soil physical characteristics in this area. It can be seen that the study of vegetation community characteristics and soil physical properties has far-reaching significance for ecological restoration in mining areas.

At present, the research on Shuanglonggou mining area mainly involves the characteristics of soil enzyme activity, soil heavy metal pollution control and microbial characteristics [15–17], but there are few studies on the vegetation community characteristics and soil physical properties of artificial *Hippophae rhamnoides* forest. Therefore, based on the current situation of Shuanglonggou abandoned mining area, this paper adopts the spatio-temporal substitution method to analyze the vegetation community characteristics and soil physical properties and their changes in the artificial *Hippophae rhamnoides* forest with different restoration years in Shuanglonggou abandoned mining area. The purpose is to provide a theoretical basis for the restoration of the ecosystem in this area, and provide technical support for the ecological environment construction of Shuanglonggou abandoned mining area and its similar areas.

2 Overview of the Study Area and Research Methods

2.1 Overview of the Study Area

The study area is located in Shuanglonggou (37°22 ′24. 6″ N, 102°25′ 24. 8″ E), Haxi Town, Tianzhu Tibetan Autonomous County, Gansu Province. It belongs to the Lenglongling Mountain on the north slope of Qilian Mountains. The terrain is high in the south and low in the north, with rich plant resources. The altitude of Shuanglonggu area ranges from 2400 to 4800 m. The average annual evaporation is 1677 mm and the annual precipitation is more than 400 mm. The rainfall is mainly distributed in 8 September, accounting for 76% of the annual precipitation. The average annual temperature is 3. 3°C, the frost-free period is 110 days, and the average annual sunshine duration is 2547 h. In the Middle and late Pleistocene and early Holocene, under the influence of glacier (meltwater) and a series of complex chemical reactions, a typical ice water deposit - terrace sand gold deposit was formed in Shuanglonggou Basin, rich in sand (rock) gold, copper, iron, coal and other mineral resources; In this area, *Picea crassifolia* is the established species, and a small number of *Picea crassifolia*, *Betula albosinensis* and *Populus davidiana* are mixed. The main bushes are *Salix cupularis*, *Rhododendron simsii* and *Caragana sinica*. The plantations are mainly *Hippophae rhamnoides*, *Potentilla fruticosa* and *Potentilla glabra*. The main herbaceous plants are *Elymus nutans*, *Potentilla anserina*, *Potentilla multicaulis* and *Taraxacum mongolicum* [18].

2.2 Layout of Sample Plots and Investigation of Vegation

From March 28 to 30, 2022, the project team visited and investigated local herdsmen and consulted relevant data on the management of Haxi Town in Tianzhu County and the mining area of Haxi Protection Station in Qilian Mountain National Natural Park. By using the spatio-temporal substitution method, the project selected similar natural environments, small differences in micro-topography, and uniform plant growth in the centralized and contiguous-based management areas. The abandoned lands with treatment years of 3a, 6a, 11a and no treatment years were used as the control plots (CK). During the survey from August 31 to September 2, 2022, three 10 m × 10 m large samples were set in each plot, and a total of 12 large samples were used for the determination of bush species, height, coverage, crown width and other indicators. At the same time, a 1 m × 1 m small quadrat was set in the center and four corners of each large quadrat, and a total of 60 small quadrat squares were used for the determination of herb species, species number, height, coverage and other indicators. The dominant species, geographical coordinates and altitude of the community in the quadrat were measured and recorded in detail by GPS. (Table 1).

Table 1. Basic status of sample plots

Repair period	Altitude(m)	Latitude	Longitude	Repair years(a)	Community dominant species
CK	3110	N37°22′14″	E102°22′54 ″	0	*Artemisia hedinii, Ranunculus tanguticus, Elymus nutans*
3a	3050	N37°22′16″	E102°23′9 ″	3	*Artemisia scoparia, Potentilla anserina, Elymus nutans, Hippophae rhamnoide*
6a	3020	N37°22′6″	E102°23′45 ″	6	*Leontopodium ochroleucum, Poa pratensis, Potentilla chinensis, Hippophae rhamnoide*
11a	3000	N37°22′19″	E102°23′55 ″	11	*Elymus nutans, Potentilla fruticosa, Hippophae rhamnoide*)

2.3 Samples Were Collected

There are 4 plots, namely 3a, 6a, 11a and CK. 3 quadrates are set in each plot, and 2 soil sample collection points are arranged diagonally off each quadrate (to avoid damaging the plant sample square) Soil profile samples are collected, and soil samples are collected at 3 depths of 0~5 cm, 20~25 cm and 40~45 cm, respectively. Each layer was repeated 3 times and a total of 216 soil samples were collected. After the 6 soil samples were mixed evenly, 36 soil samples were obtained and brought back to the laboratory for analysis to measure the physical and chemical properties of the soil. Each soil sample was measured 3 times, and the average value was taken as the fractional data. The sampling depth and stratification depth of the soil measured were consistent.

2.4 Methond of Determination

Determination of Plant Community Characteristics
Plant species diversity plays an important role in maintaining the stability and function of forest ecosystems, and is one of the main indicators for evaluate the effectiveness of vegetation restoration [19]. Therefore, plant importance value, individual number, species number, Margalef richness index, Pielou evenness index, Simpson dominance index and Shannon diversity index were selected to determine the vegetation community. The calculation formulas are as follows [20]:

Important value: IV = (relative abundance + relative coverage + relative height) 100/3.

Simpson dominance index $C = \sum_{i=1}^{S} \frac{N_i(N_i-1)}{N(N-1)}$.

Shannon Diversity index $H = -\sum_{i=1}^{s} p_i 1np_i$.

Pielou evenness index $E = \frac{H}{1nS}$.

Margalef richness index $D_{MG} = \frac{S-1}{1nN}$.

Where Pi is frequency, Ni is the number of individuals in the i group, S is the number of groups, N is the total number of individuals of all classes.

Determination of Soil Physical Properties
Soil moisture content is determined by electric oven drying method; Soil bulk density was measured by ring knife method. Soil particle size was determined by Malvern laser particle size analyzer (mechanical composition of soil was determined by Malvern Mastersizer 2000 laser particle size analyzer).

2.5 Data Processing the Experimental Data

The experimental data were statistically sorted and plotted using Excel 2019 software, and the soil and vegetation data were analyzed by one-way ANOVA and correlation analysis by SPSS 26. 0 software. Multiple comparisons were performed by LSD under the conditions of significant differences.

3 Results and Analysis

3.1 Plant Community Species Composition of Artificial *Hippophae Rhamnoides* Forest in Shuanglonggou

It can be seen from Table 2 that 65 species of plants of 48 genera and 24 families appeared in the 4 plots (3a, 6a, 11a and CK) during the plant restoration of the artificial *Hippophae rhamnoides* forest in Shuanglonggou. Among them, Asteraceae (13 species in 10 genera), Rosaceae (10 species in 5 genera), Poaceae (6 species in 4 genera), Ranunculaceae (4 species in 3 genera) and Polygonaceae (3 species in 3 genera) accounted for 20. 83% of the total family and 55. 38% of the total species. There were 2 genera and 3 species in Brassicaceae family, 2 genera and 2 species in caryophyllidae and Lamiaceae, 2 genera and 3 species in Amaranthaceae. The 4 families accounted for 16. 66% of the total family and 15. 38% of the total species. Elaeagniaceae, Cyperaceae, Violaceae and Caprifoliaceae family belong to 1 genus and 2 species. The 4 species accounted for 16. 66% of the total family and 12. 30% of the total species. The other 11 families of plants are 1 genus 1 species, that is, single family, single genus and single species, accounting for 45. 83% of the total family number, accounting for 16. 92% of the total species number.

Table 2. Plant community species composition of artificial sea buckthorn forest in Shuanglonggou

ID	Family	Number of genera	Number of species
1	Asteraceae	10	13
2	Rosaceae	5	10
3	Poaceae	4	6
4	Ranunculaceae	3	4
5	Polygonaceae	3	3
6	Brassicaceae	2	3
7	Caryophyllaceae	2	2
8	Lamiaceae	2	2
9	Amaranthaceae	2	3
10	Elaeagnaceae	1	2
11	Cyperaceae	1	2
12	Violaceae	1	2
13	Caprifoliaceae	1	2
14	Plantaginaceae	1	1
15	Apiaceae	1	1
16	Geraniaceae	1	1
17	Iridaceae	1	1
18	Gentianaceae	1	1
19	Onagraceae	1	1
20	Rubiaceae	1	1
21	Mazaceae	1	1
22	Thymelaeaceae	1	1
23	Fabaceae	1	1
24	Urticaceae	1	1
total	24	48	65

3.2 Composition of Plant Life Forms in the Artificial *Hippophae Rhamnoides* Forest in Shuanglonggou

According to the characteristics of plant life forms in the artificial *Hippophae rhamnoides* forest in Shuanglonggou (see Table 3), there were 4 plant life forms in the survey sample, of which 16 species of annual herbs accounted for 24. 61% of the total species. The main ones are *Chenopodium album*, *Stellaria palustris* and Artemisia hedinii. Perennial herbs (41 species) accounted for 63. 07% of the total species, mainly *Leontopodium leontopodioides*, *Elymus nutans* and *Potentilla anserina*, etc. Bush (7 species) accounted for 10. 77% of the total species, mainly *Hippophae rhamnoides* and *Potentilla fruticosa*. Vine (1 species) accounted for 1. 53% of the total species.

Table 3. Plant life-type composition characteristics of artificial seabuckthorn forest in Shuanglonggou

Life form	Family		Genera		Species	
	Number	Account for(%)	Number	Account for(%)	Number	Account for(%)
Annual herb	9	37. 50	14	29. 16	16	24. 61
Perennial herb	18	75	31	64. 58	41	63. 07
Bush	4	16. 66	7	14. 58	7	10. 77
Vine	1	4. 16	1	2. 08	1	1. 53

3.3 Important Values of Plant Community Species of Artificial *Hippophae Rhamnoides* Forest in Shuanglonggou for Different Restoration years

As can be seen from Table 4, in the unrepaired CK plots, a total of 17 plants appeared, including 15 species of herbs and 2 species of bushes. *Artemisia hedinii, Ranunculus glareosus, Stellaria palustris* and *Plantago asiatica* were the most important herbs. They were 14. 763, 13. 372, 11. 912 and 9. 778, respectively. Annual herbs and perennial herbs were dominant.

After the restoration of 3a, the plant species decreased, and a total of 15 plants appeared, including 13 herbaceous plants and 2 bush plants. At this time, the dominant species of *Artemisia hedinii, Ranunculus glareosus* and *Stellaria palustris* in the plot CK disappeared, and the important value of *Plantago asiatica* decreased by 72. 54%. But plants such as *Potentilla anserina* and *Artemisia scoparia* appear in abundance. In restoration 3a, *Elymus nutans, Potentilla anserina, Hippophae rhamnoides* and *Elymus dahuricus* had the highest importance. They were 18. 548, 12. 921, 9. 028 and 7. 998, respectively, and the perennial herbs occupied the dominant position.

After 6a was repaired, there was a large increase in plant species, with a total of 28 plants appearing, of which 22 were herbaceous, an increase of 8. *Poa pratensis, Taraxacum mongolicum, Potentilla acaulis, Artemisia sieversiana* and other herbs appeared in large numbers, and 5 species of bush plants, Two species were added, *Potentilla glabra* and *Potentilla fruticosa*, while the vine *Lonicera japonica* appeared for the first time. The important values of *Leontopodium ochroleucum, Potentilla glabra, Hippophae rhamnoides* and *Poa pratensis* were the highest. 16. 296, 10. 501, 10. 303 and 9. 361, respectively, with perennial herbs and bushes occupying the dominant position.

After 11a restoration, the plant species reached the peak, and 34 plants appeared, of which 31 were herbaceous, 4 were increased, 3 were bushes, lianas disappeared, and all the plants in CK in the sample plot withdrew except a few plants. The most important values were *Hippophae rhamnoides*, *Leontopodium ochroleucum*, *Potentilla fruticosa* and *Poa annua*. They were 15. 163, 10. 810, 10. 422 and 8. 155, respectively, and perennial herbs and bushes still occupied the dominant position.

Table 4. Important values of plant community species in Shuanglonggou artificial sea *Hippophae rhamnoides* forest

ID	Species	Life form	Repair period			
			CK	3a	6a	11a
1	*Plantago asiatica*	Herbal	9. 778	2. 977	2. 355	1. 189
2	*Poa annua*	Herbal	2. 267	6. 612	2. 066	8. 155
3	*Elymus nutans*	Herbal		18. 548	1. 212	6. 789
4	*Saussurea japonica*	Herbal	7. 772	5. 831		
5	*Senecio thianschanicus*	Herbal	1. 609	4. 253	2. 143	0. 564
6	*Cerastium pusillum*	Herbal	7. 036	4. 433		
7	*Artemisia hedinii*	Herbal	14. 763			2. 180
8	*Urtica triangularis*	Herbal	4. 763			
9	*Poa pratensis*	Herbal			9. 361	
10	*Draba mongolica*	Herbal	4. 771			
11	*Axyris hybrida*	Herbal	2. 226			
12	*Ranunculus glareosus*	Herbal	13. 372			
13	*Galium spurium*	Herbal	2. 303			2. 466
14	*Stellaria palustris*	Herbal	11. 912			3. 132
15	*Chenopodium album*	Herbal	6. 579			
16	*Carduus crispus*	Herbal	0. 980			
17	*Lepidium alashanicum*	Herbal	0. 975			
18	*Potentilla anserina*	Herbal		12. 921	1. 727	0. 535
19	*Artemisia scoparia*	Herbal		3. 822		
20	*Potentilla virgata*	Herbal			2. 794	2. 800
21	*acobaea argunensis*	Herbal		4. 127		
22	*Crepis flexuosa*	Herbal		5. 144		
23	*Artemisia mongolica*	Herbal		4. 179		2. 103
24	*Ranunculus tanguticus*	Herbal		6. 732		1. 880
25	*Elymus dahuricus*	Herbal		7. 998		4. 242
26	*Taraxacum mongolicum*	Herbal			1. 214	1. 728
27	*Potentilla acaulis*	Herbal			2. 311	
28	*Artemisia sieversiana*	Herbal			1. 829	
29	*Draba nemorosa*	Herbal			0. 777	
30	*Polygonum aviculare*	Herbal			0. 975	
31	*Carex spp*	Herbal			3. 599	
32	*Polygonum viviparum*	Herbal			3. 311	

(*continued*)

Table 4. (*continued*)

ID	Species	Life form	Repair period			
			CK	3a	6a	11a
33	*Leontopodium ochroleucum*	Herbal			16. 296	10. 810
34	*Carum carvi*	Herbal			3. 699	
35	*Cyperaceae*	Herbal			5. 036	
36	*Fragaria ananassa*	Herbal			1. 880	
37	*Leymus secalinus*	Herbal			2. 309	
38	*Thalictrum alpinum*	Herbal			1. 561	
39	*Koenigia pilosa*	Herbal			1. 537	
40	*Viola prionantha*	Herbal			1. 409	
41	*Viola violacea*	Herbal				0. 140
42	*Geranium sibiricum*	Herbal				1. 409
43	*Atriplex centralasiatica*	Herbal				0. 933
44	*Elsholtzia densa*	Herbal				1. 668
45	*Epelobium hirsutum*	Herbal				0. 895
46	*Saussurea salicifolia*	Herbal				2. 865
47	*Iris lactea*	Herbal				2. 710
48	*Swertia tetraptera*	Herbal				0. 563
49	*Potentilla bifurca*	Herbal				1. 357
50	*Lancea tibetica*	Herbal				2. 715
51	*Artemisia gansuensis*	Herbal				0. 568
52	*Anemone rivularis*	Herbal				0. 957
53	*Cleistogenes chinensis*	Herbal				1. 176
54	*Stellera chamaejasme*	Herbal				1. 003
55	*Potentilla multicaulis*	Herbal				0. 968
56	*Stipa breviflora*	Herbal				0. 741
57	*Erigeron elongatus*	Herbal				0. 237
58	*Rubus irritans*	Bush	6. 102			
59	*Lonicera × brownii*	Bush	2. 792			
60	*Hippophae neurocarpa*	Bush			1. 110	
61	*Potentilla glabra*	Bush			10. 501	5. 745
62	*Potentilla fruticosa*	Bush			5. 849	10. 422
63	*Hippophae rhamnoides*	Bush		9. 028	10. 303	15. 163
64	*Comarum salesovianum*	Bush		3. 386	4. 131	
65	*Lonicera japonica*	Vine			2. 361	

3.4 Changes of Vegetation Community Species Diversity of Artificial *Hippophae Rhamnoides* Forest in Shuanglonggou with Different Restoration years

According to Table 5, the number of plant individuals, number of plant species and Margalef richness index showed an increasing trend with the extension of restoration years in the restoration process of 11a, and the maximum values were 1234, 34 and 3. 213 in plot 11a, respectively. Simpson dominance index showed a trend of first increasing (CK~3a), then decreasing (3a~6a) and then increasing (6a~11a). The Simpson dominance index showed an upward trend on the whole, and the maximum value (0. 277) was found in plot 3a. The Shannon diversity index and Pielou evenness index decreased first (CK~3a), then increased (3a~6a) and then decreased (6a~11a), showing a downward trend as a whole. The maximum Shannon diversity index (3. 286) was found in plot 6a. The Pielou evenness index (maximum 0. 761) appeared in the CK plot.

Table 5. Changes of plant species diversity with restoration years

Index	Repair period/a			
	Ck	3	6	11
Individual number (N)	156 c	751 b	1231 a	1234 a
Species number (S)	17 c	15 d	28 b	34 a
Shannon (H)	3. 111 b	2. 400 d	3. 286 a	2. 894 c
Pielou (E)	0. 761 a	0. 614 c	0. 683 b	0. 592 d
Simpson (C)	0. 162 c	0. 277 a	0. 150 d	0. 263 b
Margalef (D)	2. 196 c	1. 465 d	2. 630 b	3. 213 a

Note: Different letters in the same line indicate significant difference between different repair years ($p < 0.05$), the same below

3.5 Changes in Soil Physical Properties of *Hippophae Rhamnoides* Forest in Shuanglonggou with Different Restoration Years

Changes in Soil Water Content and Bulk Density of Hippophae Rhamnoides Forest with Different Restoration years as Shown
As can be seen from Fig. 1, with the continuous extension of restoration years, Soil water content in the three soil layers (0~5 cm, 20~25 cm, and 40~45 cm) showed a trend of first increasing (CK~3a), then decreasing (3a~6a) and then increasing (6a~11a), with the maximum value appearing in plot 11a. The soil water content was 12. 263, 9. 798 and 8. 797%, and the soil water content was increased by 9. 45, 5. 86 and 4. 96 times, respectively. In the vertical direction, with the deepening of soil layer, soil water content in plots CK and 3a showed an increasing trend, and the maximum value of soil water content appeared in deep soil (40~45 cm), while the maximum value of soil water content in plots 6a and 11a showed a decreasing trend, and the maximum value appeared in surface soil (0~20 cm).

With the extension of restoration years, the soil bulk density in the three soil layers showed a decreasing trend, which was manifested as CK > 3a > 6a > 11a in the sample plot. The maximum value appeared in the CK of the control plot (1. 487, 1. 513 and 1. 527 g/cm3, respectively), and the minimum value appeared in the sample plot 11a. They were 1. 133, 1. 387 and 1. 333 g/cm^3, respectively. In the vertical direction, with the deepening of soil layer, soil bulk density CK showed an increasing trend against the same plot, and the maximum value was found in deep soil (40~45 cm). The other three plots all showed a trend of first increasing (surface soil~middle soil) and then decreasing (middle soil~deep soil). The maximum value was found in mid-layer soil (20~25 cm).

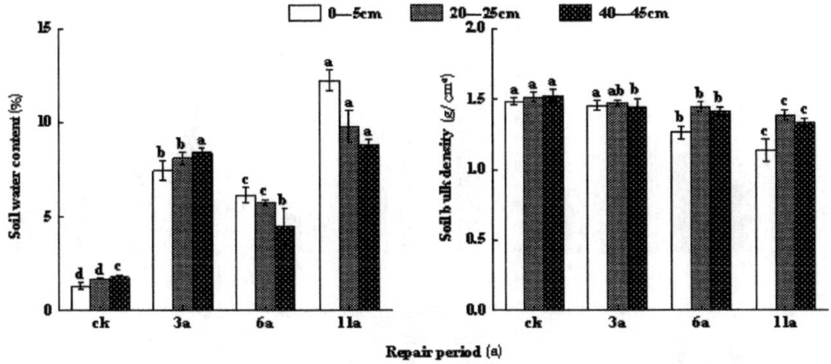

Note: Different lowercase letters indicate significant differences between different repair years (p <0. 05), the same below.

Fig. 1. Soil moisture content and soil capacity weight of artificial *Hippophae rhamnoides* forest in different restoration years.

Changes of Soil Particle Size in Artificial Hippophae Rhamnoides Forest with Different Restoration years.
Soil mechanical composition was divided into 6 grades according to particle size ≤ 5 mm, 5 mm~10 mm, 10 mm~25 mm, 25 mm~50 mm, 50 mm~80 mm and ≥ 80 mm (see Fig. 2 for details). With the extension of restoration years, the particle sizes of 5 mm~10 mm, 50 mm~80 mm and ≥ 80 mm showed a decreasing trend above the three soil layers (0~5 cm, 20~25 cm and 40~45 cm), which were shown as CK > 3a > 6a > 11a in the sample plots. The maximum values appear in plot CK and the minimum values appear in plot 11a. The Particle sizes ≤ 5 mm, 10 mm~25 mm and 25 mm~50 mm showed a trend of first increasing (CK~3a) and then continuously decreasing (3a~11a) with the increase of restoration years, with the maximum value appearing in plot 3a, and the particle sizes in plot 6a were lower than those in the corresponding CK. From the perspective of vertical direction, the grain size of different levels does not change with the depth of soil layer.

3.6 Correlation Between Plant Community Diversity and Soil Physical Properties as Shown

As can be seen from Table 6, the number of plant individuals was positively correlated with soil water content (P < 0. 01). There was a significant negative correlation with soil bulk density, particle size of 5~10 mm, 25~50 mm, 50~80 mm and ≥ 80 mm (P < 0. 01). Plant species number and Margalef richness index were negatively correlated with soil bulk density and soil particle size (P < 0. 01). There was a significant positive correlation between Shannon diversity index and soil water content (P < 0. 01). There was a significant negative correlation between Pielou evenness index and water content (P < 0. 01). There was no significant correlation between Simpson dominance index and soil physical properties (P > 0. 05).

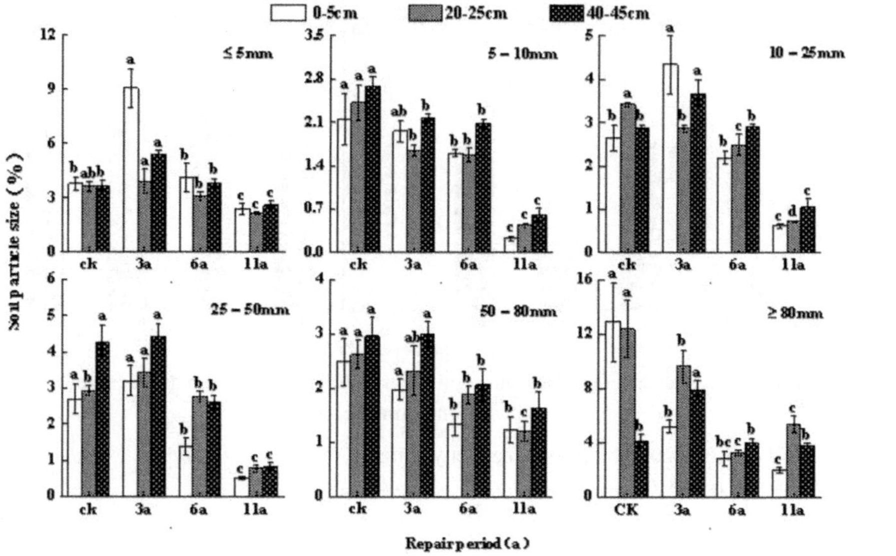

Fig. 2. Soil particle size of artificial *Hippophae rhamnoides* forest in different restoration years

Table 6. Correlation between vegetation community characteristics and soil physical properties

index	water content	unit weight	Particle size ≤ 5 mm	Particle size 5~10 mm	Particle size 10~25 mm	Particle size 25~50 mm	Particle size 50~80 mm	Particle size ≥ 80 mm
individual number(N)	0. 731**	−0. 699*	−0. 236	−0. 733**	−0. 552	−0. 748**	−0. 799**	−0. 742**
species number(S)	0. 520	−0. 740**	−0. 728**	−0. 823**	−0. 872**	−0. 886**	−0. 842**	−0. 765**
Shannon	0. 785**	−0. 211	−0. 510	−0. 066	−0. 426	−0. 424	−0. 306	−0. 268
Pielou	−0. 959**	0. 534	−0. 057	0. 596	0. 386	0. 424	0. 544	0. 434
Simpson	−0. 389	−0. 234	0. 226	−0. 508	−0. 155	−0. 132	−0. 175	−0. 054
Margalef	0. 285	−0. 702**	−0. 742**	−0. 710**	−0. 888**	−0. 860**	−0. 722**	−0. 709**

Note: * Significant correlation at 0. 05 level. ** Very significant correlation at 0. 01 level3

4 Discuss

Community characteristics are the basis of ecosystem restoration and reconstruction, as well as the a comprehensive embodiment of vegetation adaptation to ecosystem [21], and important values can show different status and roles of species in the community, and the role of species in the community is positively correlated with the important values. The greater the value of this index, the higher the status of species and the greater the role [22]. The results showed that in the restoration process of artificial *Hippophae rhamnoides* forest in Shuanglonggou, 5 families of plants, including Asteraceae, Rosaceae and Poaceae, accounted for 55. 38% of the total species, and Plantaginaceae. The plants in 11 families, including Apiaceae and Geraniaceae, accounted for 45. 83% of the total families, and they were single families, single genera and single species. This is consistent with the study of He Hongsheng et al. [23] on the characteristics of plant communities in Qingtu Lake area of Minqin. This indicates that the five families of plants, including Asteraceae, Rosaceae and Poaceae, play an important role in the restoration process of artificial *Hippophae rhamnoides* forest land. In order to adapt to different environments, plants constantly change in the composition of plant species in the process of succession, and the relationship between species is more complex, but they do not lose the species composition of "most species belong to the minority family, the minority species belong to the majority family," Most plant species are single family, single genus" [24]. According to the characteristics of life forms of the vegetation in Shuanglonggou, the number of perennial herbs is absolutely dominant, and they are the vegetation with the largest distribution range and the dominant life forms in the area. This is because the environment in Shuanglonggou area is harsh and the vegetation growth is limited, while the perennial herbs have good cold-tolerance, drought-resistance and disease-resistance, grow rapidly, have better durability and tenacious vitality, and can better adapt to the harsh environment in the abandoned mining area of Shuanglonggou [25]. This is consistent with the research results of Li Jiangwen et al. [26] on the diversity of vegetation species in desert steppe. In order to better adapt to the living environment, plants are constantly making adjustments conducive to survival. This is also different from the research results of Xin Yan 'an et al. [27] on the vegetation diversity of Fenhe Reservoir wetland, which is caused by the different test plots. The different plant growth environment (such as the influence of climate, soil, altitude and other factors) will lead to changes in the individual traits of plants and the symbiotic relationship between community species, and then lead to the different plant life forms. With the extension of restoration years, the succession of the plant community in Sanglongou was significantly progressive, which roughly experienced a succession process from annual herbs → annual herbs and perennial herbs → perennial herbs → perennial herbs and bushes, and finally formed a stable shrub-herbaceous complex system. It accords with the general characteristics of the flora in the high altitude cold region. This is consistent with the research results of Wang Lide et al. [28] on the secondary grassland in the Minqin farmland area, because plants will form community types with specific structures that adapt to the local ecological and geographical environment under long-term natural selection, and species with high adaptability and tolerance will gradually become the dominant species in the community. In addition, herbs such as *Plantago asiatica*, *Poa annua* and *Senecio Tianshanensis* were found in the four plots, indicating that vegetation succession in the artificial *Hippophae rhamnoides* forest has a certain continuity, and vegetation succession is a slow process. This is consistent with the research results of Geng Bingjin et al. [29] on the vegetation of reclamation land in open pit mining areas.

Plant community species diversity is not only an index reflecting the richness, variation degree or evenness of species in the community, but also the relationship between different physical and geographical conditions and the community [30]. The results showed that the number of plant individuals, the number of plant species and the Margalef richness index increased with the extension of restoration years during the 11a restoration in Shuanglonggou *Hippophae rhamnoides* forest. This is consistent with the research results of Wang Jia et al. [31] on the species diversity of retired arable land in Qingtu Lake area. This is because with the continuous extension of restoration years, the leaves of vegetation on the surface accumulate to form litter. There is no limit to the growth of herbs and bushes, and the variety and number of plants tend to be abundant. Simpson dominance index showed a trend of first increasing (CK~3a), then decreasing (3a~6a) and then increasing (6a~11a), and on the whole increased with the increase of repair years. This is consistent with the research results of Wang Lide et al. [32] on the vegetation species diversity in the process of gangue treatment in Shuanglonggou. This is because the plant community develops in a simple and stable direction in the succession process, and the dominant species of the community gradually changes from annual herbs and perennial herbs to perennial herbs and shrubs, and the status of the dominant species in the community is becoming more and more prominent. Vegetation species are not evenly distributed. Shannon diversity index and Pielou evenness index showed a change trend of first decreasing (CK~3a), then increasing (3a~6a) and then decreasing (6a~11a), and the overall trend was decreasing. This is consistent with the research results of Li Yongqiang et al. [33] on the species diversity of plant community in abandoned grassland areas. This is because the dominant species of *Hippophae rhamnoides* grows vigorously, occupies more resources, inhibits the growth of some plants, and some vegetation is gradually eliminated without sufficient resources. As a result, Shannon diversity index and Pielou evenness index show a downward trend [34]. This is also different from the research results of Zuo Liping et al. [35] on the species diversity of planted forest in Qingyang City, which is caused by the differences of the experimental plots. The experimental plots are different from the planted loess Plateau gully areas in Qingyang City in terms of hydrology, soil, topography and climate.

As the link and core of the air-vegetation-soil-groundwater system, soil water is also the basic condition for the survival of surface vegetation. Only when precipitation, irrigation water and groundwater are converted into soil water can they be absorbed and utilized by vegetation [36], and soil bulk density is one of the main physical property indexes reflecting the change of soil structure [37]. The results showed that soil water content increased and soil bulk density decreased with the extension of restoration years. This is consistent with the research results of Wu Hao et al. [38] on the physical and chemical properties of soil in the Minqin farmland area. This is because with the succession of vegetation, the transformation effect of plants on soil is continuously enhanced, the soil structure is optimized, while the soil bulk density is reduced, soil compacty and permeability are improved, soil porosity is increased, and the water retention effect of plant roots is continuously enhanced. At the same time, the litter layer on the surface of the soil also reduces the loss of water, and the water retention performance of the soil is constantly enhanced [39]. With the deepening of soil layer, the soil moisture layer showed an increasing trend for CK and 3a, and a decreasing trend for 6a and 11a. This is because in the early stage of restoration, plants were mainly annual herbs and perennial herbs with shallow roots, which mainly absorbed soil water in the surface layer, making the water content of deep soil higher than that of the surface soil. However, in restoration

6a and 11a, bushes appeared in large numbers with deep root influence, and consumed a lot of water in the deep soil. The water content of deep soil is lower than that of surface soil [40]. With the deepening of soil layer, the bulk density of soil increased first and then decreased in the three plots except CK. This is consistent with the research structure of Yu Hang et al. [41] on the physical and chemical properties of soil in the loess region of western Shanxi. Soil particle size composition can reflect the percentage content of different sediment fractions, which determines the structure and texture of the soil after restoration in the mining area [42]. Soil particle size showed an overall decreasing trend with the extension of restoration years, which was consistent with the research results of Su Tingting et al. [43] on the physical and chemical properties of soil in typical grasslands of loess hills, indicating that large plant growth effectively promoted the turnover of underground fine roots, enhanced the improvement effect of plant roots on soil, and enabled the accumulation of soil organic matter under the cementation of organic matter. Granular matter is accumulated, and the soil structure is greatly improved [44].

The number of plant individuals and Shannon diversity index were positively correlated with soil water content. This is consistent with the research results of Fengfei et al. [45] on plant community characteristics and soil physicochemical properties in Ningdong mining area, indicating that soil water has a significant impact on vegetation species diversity, and rich soil water provides a better environment for plant growth and development and promotes vegetation recovery and development [46]. Plant individual number, species number and Margalef richness index were negatively correlated with soil bulk density and particle size. This is consistent with the research results of Ma Rui et al. [47] on the vegetation community characteristics and soil properties of the alpine grassland in the eastern Qilian mountain and river Valley, indicating that the restoration of vegetation has extremely high requirements on soil bulk density and soil particle size, and inappropriate soil bulk density and particle size will hinder the restoration of plants. There was a significant negative correlation between Pielou evenness index and water content. This is consistent with the research results of Wang Mingtao et al. [48] on the vegetation community and soil characteristics of artificial grassland in northern Tibet, indicating that the increase of soil water content may not be conducive to the uniformity of vegetation distribution, and the increase of soil water content will promote the rapid growth and development of the dominant species in the community, which will become more prominent and significantly reduce the evenness index.

5 Conclusion

In the restoration process of artificial *Hippophae rhamnoides* forest in Shuanglonggou abandoned mining area, 65 species of plants appeared in 24 families, 48 genera, and the species composition showed that "most species belong to the minority family, and a few species belong to the majority family". Perennial herbs were the dominant life forms in the area. The dominant species of the community gradually evolved from annual herbs or perennial herbs to perennial herbs and bushes. With the extension of restoration years, the number of plant individuals, species and Margalef richness index showed an increasing trend, and the Simpson dominance index showed a trend of "first "increasing~then decreasing~then increasing", showing an overall increasing trend. The Shannon diversity index and Pielou evenness index showed a trend of "first decreasing~then increasing~then decreasing, and the overall trend was decreasing." Soil water content showed an increasing trend, while soil bulk density and soil particle size showed a decreasing trend. It can be seen that the restoration years have a great impact on vegetation species diversity and soil physical properties. The plant community gradually becomes single and stable, and the soil structure is optimized, which promotes the restoration of the ecosystem in the abandoned mining area to a certain extent.

Acknowledgment. National Natural Science Foundation of China: Evolution and driving Mechanism of soil system under gangue treatment and reconstruction in eastern Qilian Mountains (42167069); Key Talent Project of Gansu Province: Research and team Building on Key technologies of ecological restoration in Qilian Mountains and Shiyang River Basin based on the concept of "Two Mountains" (GZTZ20240415).

References

1. Chang, L., Wu, C.R., He, F.L., et al.: Changes of plant community composition and soil properties of Hippophae rhamnoides after artificial planting in abandoned gold mining area of Shuanglonggou. Chin. Agric. Sci. Bull. **35**(19), 83–90 (2019)
2. Zhenqi, H., Yuanyuan, L., Gensheng, L., et al.: Opportunities and challenges of land reclamation and ecological restoration in mining area under carbon neutrality. Coal Science and Technology **51**(01), 474–483 (2023)
3. Lu, H.Y.: Investigation and analysis of disturbance factors of vegetation restoration in Shuanglonggou abandoned mining area. Journal of Agriculture and Technology **35**(06), 7 (2015)
4. Wang, Y.C., Ma, J.B.: Causes and restoration measures of ecosystem degradation in Shuanglonggou Basin, Haxi. Chinese Forestry **19**, 61 (2009)
5. Chen, X.H., Duan, Z.H.: Changes in soil physical and chemical prop- erties during reversal of desertification in Yanchi County of Ningx- ia Hui autonomous region, China. Environ. Geol. **57**, 975–985 (2009)
6. Liutong, S., Changqing, S., Junjiao, Z., et al.: Restriction factors and countermeasures of ecological restoration in Helan Mountain open-pit Coal mine. Journal of Resources and Ecology **14**(04), 784–793 (2023)
7. Xue, R., Bihua, G., Jinping, L., et al.: The impact of soil desertification on the characteristics of regenerating plant communities and grassland productivity. Grassland and Turf **43**(02), 22–28 (2023)
8. Yuan, Z.J.: Effects of tending on understory vegetation biodiversity of Pinus tabulaeformis plantation on the Loess Plateau. Shaanxi Normal University (2018)
9. Zhang, G.X.: Study on biodiversity and spatial and temporal distribution characteristics of newborn wetland in coal mining subsidence area. Chongqing University (2020)
10. Li, X.Y., Zhang, M., Guo, Y.N., et al.: Relationship between understory plant diversity and soil factors in coal mining subsidence area. J. Soil Water Conserv. **36**(01), 268–276 (2022)
11. Cheng, W., Zhu, Z., Wang, W.J., et al.: Ecological effect of humus layer reconstruction on degraded soil in western mining area. J. China Coal Soc. **48**(07), 2850–2857 (2023)
12. Jing, C.: Protection and restoration of biodiversity in Shendong mining area. Coal Engineering **55**(S1), 182–187 (2023)

13. Fernandez, M.E., Passera, C.B., Cony, M.A.: Sapling growth, water status and survival of two natives hrubs from the Monte Desert, Mendoza, Argentina, under different preconditioning treatments. Revista De La Facultad De Ciencias Agrarias **48**(1), 33–47 (2015)
14. Li, P.: Study on the impact of vegetation restoration on soil properties and quality of the tailings pond in heidaigou coal mine. Northwest A&F University (2014)
15. Song, D., Wang, L.D., Wu, H., et al.: Soil enzyme activity and nutrient evolution characteristics of different artificial Hippophae rhamnoides forests in gangue hills. Research of Soil and Water Conservation **30**(05), 162–168+174 (2023)
16. Li, S.X., Yao, T., Wang, L.D., et al.: Changes of soil microbial biomass during restoration of degraded alpine grassland in eastern Qilian Mountains. Journal of Grassland Science **31**(12), 3668–3675 (2023)
17. Song, D., Wu, H., WL, D., et al.: Distribution of heavy metals in soil of artificial Hippophae rhamnoides forest with different planting years and its effect on enzyme activity in Shuanglonggou abandoned mining area. Journal of Pratacultural Science **32**(08), 61–70 (2023)
18. Zhou, H.Q.: Thoroughly remediate Shuanglonggou gold mining area in Wuwei area. Soil and Water Conservation in China **04**, 35–36 (1999)
19. Tilman, D.: The ecological consequences of changes in biodiversity: a search for general principles. Ecology **80**(5), 1455–1474 (1999)
20. Wang, Q., Wang, F., He, Y.T., et al.: Effects of stand factors on understory plant diversity of different forest communities in Shenmulei. Journal of Northwest Botany **43**(11), 1950–1958 (2023)
21. Zhang, Y., Asya, M., Xin, X., et al.: Effects of grazing prohibition and grazing on community structure, biomass and forage quality of warm steppe in Xinjiang. J. Grassland Sci. **28**(03), 815–821 (2020)
22. Liu, W., Jiang, F., Malili, et al.: Effects of organic fertilizer application rate and sowing method on plant community growth and soil nutrient in alpine mining area. Grassland and Turf **43**(02), 116–125 (2023)
23. He, H.S.: Study on vegetation succession and soil change in different years of reclaimed farmland in Qingtuhu Lake. Gansu Agricultural University (2021)
24. Qi, D.H., Yang, H.X., Lu, Q., et al.: Main types and characteristics of plant communities in Hunshandake sandy land. J. Desert Res. **41**(04), 23–33 (2021)
25. Shunxia, W., Jingjing, Z., Yunfei, L., et al.: Relationship between typical plant community diversity and soil factors in desert steppe of Ningxia. J. Lanzhou Univ. (Natural Science Edition) **59**(06), 711–719 (2023)
26. Jiangwen, L., Bangyin, H., Xiaoxi, Z., et al.: Effects of abnormal precipitation on species composition and diversity of plant communities with different stocking rates in desert steppe. Acta Ecol. Sin. **43**(15), 6433–6442 (2023)
27. Xin, Y., Li, S.: Floristic composition and species diversity of wetland in Fenhe Reservoir. Journal of Arid Land Resources and Environment **26**(12), 153–160 (2012)
28. Wang, L.D.: Study on vegetation and soil system evolution of secondary grassland in Minqin area. Gansu Agricultural University (2016)
29. Geng, B.J., Wang, S.F., Cao, Y.G., et al.: Comparative analysis of vegetation reconstruction characteristics in different years of reclamation land in Pingshuo open-pit mine, Shanxi Province. Acta Ecol. Sin. **42**(08), 3400–3419 (2022)
30. Zhang, L., Wang, J., Bai, Z., et al.: Effects of vegetation on runoff and soil erosion on reclaimed land in an opencast coal-mine dump in a loess area. CATENA **128**, 44–53 (2015)
31. Wang, J., Tian, Q., Wang, L.D., et al.: Effects of different years of farmland reclamation on soil moisture and species diversity in Qingtu Lake District of Minqin. Arid Zone Res. **39**(02), 605–614 (2022)

32. Wang, L.D., Song, D., Li, G.Y., et al.: Study on plant community succession and species diversity during gangue treatment in Shuanglonggou. Arid Zone Res. **40**(08), 1294–1303 (2023)
33. Li, Y., Xu, Z.: Species diversity changes during plant community succession in abandoned grassland area. J. Inner Mongolia Agricult. Univ. (Natural Science Edition) **04**, 26–31 (2002)
34. Zhou, S.P.: Study on understory vegetation diversity and soil physicochemical properties of teak plantation. Chinese Academy of Forestry (2017)
35. Zuo, L.P., Chang, J., Xi, Z.C., et al.: Effects of afforestation on species diversity in Qingyang City. J. Gansu Agricult. Univ. **57**(04), 147–155+162 (2022)
36. Wan, Y.X., Wang, Z.D., Chang, C.P.: Protective effect of cropland conver-sion with different years on soil in Bashang district, North China. J. Arid Land Resour. Environ. **32**(9), 84–89 (2018)
37. Jiao, F., Wen, Z.M., An, S.S.: Changes in soil properties across a chronosequence of vegetation restoration on the Loess Plateau of China. CATENA **86**(2), 11 (2011)
38. Wu, H., Wang, L.D., Song, D., et al.: Physical and chemical properties and enzyme activities of soil under different years in Minqin farmland. Agric. Res. Arid Areas **39**(01), 191–199 (2021)
39. Zhang, Y.F., Li, X.L., Jin, L.Q., et al.: Analysis of physical and chemical properties of reconstituted soil without guest soil based on mixed application of granular organic fertilizer and sheep manure in alpine mining area. Chinese Soil and Fertilizer **07**, 129–137 (2023)
40. Zhang, K., Zheng, Y., Liu, C.L., et al.: Effects of different vegetation patterns on surface soil properties of mine abandoned land. Science of Soil and Water Conservation in China, 1–13
41. Yu, H., Feng, T.J., Wei, W., et al.: Response of soil physicochemical characteristics to long-term vegetation restoration in loess region of western Shanxi Province. Acta Ecol. Sin. **07**, 1–13 (2024)
42. Zhu, Y., Jia, Z., Yang, X.: Resource-dependent water use strategy of two desert shrubs on interdune, Northwest China. J. Food Agric. Environ. **9**(3), 832–835 (2011)
43. Tingting, S., Hongbin, M., Yao, Z., et al.: Responses of soil physicochemical properties to ecological restoration measures in typical grassland of Loess Plateau. Acta Pratacul. Sin. **28**(04), 34–46 (2019)
44. Zhang, L.G., Cheng, L., Sun, Y.Y., et al.: Soil grain size characteristics of typical underlying surface in Mu Us Sandy Land. Chinese Journal of Ecology, 1–9 (2024)
45. Feng, F., Xu, L., Gu, Q.M., et al.: Characteristics of natural plant communities and their relationship with soil physicochemical properties in Ningdong mining area. Bull. Soil Water Conserv. **41**(06), 48–56 (2021)
46. Lu, N., Fu, B.J., Jin, T.T., et al.: Trade-off analyses of multiple ecosys-tem services by plantations along a precipitation gradient across Loess Plateau landscapes. Landscape Ecol. **29**(10), 1697–1708 (2014)
47. Ma, R., Zhao, J.M.: Vegetation community characteristics of alpine grassland in East Qilian Mountain Valley and its relationship with soil characteristics. Arid Zone Res. **37**(02), 374–381 (2020)
48. Wang, M.T., Fang, J.P., Wei, X.H., et al.: Effects of different years of abandonment on vegetation community and soil characteristics of annual artificial grassland in northern Tibet. Chinese Journal of Grassland **44**(02), 41–48 (2022)

Spatial Distribution and Correlation of Organic Matter and pH in Cultivated Land in Bei'an County, Heilongjiang Province

Youtao Xin[1,2], Yanda Lu[1,2](✉), Haonan Song[1,2], Mingxin Duan[1,2], Haicheng Zhang[1,2], Hongqiang Zhao[1,2], and Xidong Zhao[1,2](✉)

[1] Harbin Natural Resources Comprehensive Survey Center, China Geological Survey, Harbin 150086, China
zhaoxidong@mail.cgs.gov.cn

[2] Observation and Research Station of Earth Critical Zone in Black Soil, Harbin, Ministry of Natural Resources, Harbin 150086, China

Abstract. Soil organic matter is not only a key component of soil fertility, but also plays an important role in water retention, nutrient cycling, and biodiversity maintenance. It improves soil structure, enhances soil water-holding capacity, and promotes microbial activity, thereby affecting plant growth and development. Meanwhile, soil pH directly affects the availability of nutrients, the activity of microorganisms, and the growth of plants. Understanding the spatial distribution characteristics of soil organic matter and pH is of great theoretical and practical significance for agricultural management and ecological restoration. In this study, we used the inverse distance weighted spatial interpolation method in ArcGIS software and the correlation analysis method in SPSS to study the spatial distribution and correlation of soil organic matter and pH in 331 agricultural soil samples from Beian County, Heilongjiang Province. The results show that the average soil organic matter content in the study area is 56.60 g/kg, indicating a relatively high overall content and a relatively uniform distribution, showing good soil fertility potential. At the same time, the average pH of cultivated land is 6.04, which belongs to weak acidic soil, providing a suitable environment for crop growth. However, the study found that there was no significant correlation between soil organic matter content and pH in Beian County's cultivated land. This result is different from previous studies, indicating that there is no direct mutual influence between the changes of these two factors, which may also be affected by indirect factors of other variables. This suggests that in future studies, it is necessary to conduct in-depth analysis of indirect factors affecting soil organic matter and pH value changes in order to better understand the evolution process of soil fertility quality. This will provide more comprehensive theoretical support and practical guidance for agricultural sustainable development and ecosystem restoration.

Keywords: organic matter · soil PH · spatial distribution · cultivated land · Beian County · Heilongjiang Province

---

The original version of the chapter has been revised. The corresponding author indication in the chapter has been updated. A correction to this chapter can be found at https://doi.org/10.1007/978-3-031-86829-0_124

D. Han et al. (Eds.): ICEPG 2024, ESE, pp. 706–715, 2025.
https://doi.org/10.1007/978-3-031-86829-0_79

1 Introduction

Organic matter and pH are two important indicators in farmland soil [1], which directly affect soil fertility and crop yield. Organic matter is an important nutrient source in soil, which can not only improve soil structure and enhance soil water and fertilizer retention capacity, but also promote soil microbial activity, which is conducive to crop growth and development [2–4]. Soil pH directly affects the growth and reproduction of microorganisms in soil, the availability of nutrients in soil, and the absorption and utilization of nutrients by crops [5]. Therefore, studying the spatial distribution characteristics of organic matter and pH in cultivated land has important theoretical and practical significance for effectively utilizing soil resources and improving farmland productivity.

At present, studies on the spatial distribution of organic matter and pH in cultivated land have attracted extensive attention [6–9]. Through the investigation and analysis of soil organic matter and pH in cultivated land in different regions, different soil types and different management measures, the spatial distribution law can be better understood and the basis for scientific and reasonable soil management can be provided [10]. However, there are still some deficiencies in the current research on the spatial distribution characteristics of organic matter and pH in cultivated land, for example, the research scope is relatively limited and the research methods are not systematic and comprehensive enough [11, 12]. Therefore, this paper aims to deeply explore the spatial distribution characteristics of organic matter and pH in cultivated land, so as to provide scientific basis and technical support for optimizing soil management and improving farmland productivity.

2 Materials and Methods

2.1 Overview of the Study Area

Beian County belongs to Heihe City, Heilongjiang Province, is located in the northern regional center of Heilongjiang Province, is located in the northern Songnen plain. It has obvious monsoon climate characteristics, belongs to the temperate continental monsoon climate, and its geographical position is 126°16′ to 127°53′ E, 47°35′ –48°33′ N. Bei'an County arable land 4.39 million mu, per capita occupation of arable land 9.08 mu. The land is fertile, flat and open, suitable for agricultural mechanization continuous operation, rich in wheat, soybeans, corn, rice, grains and beans.

2.2 Sample Collection and Research Methods

According to terrain, geomorphology, geological conditions, soil type, land use type and other conditions, combined with equidistant grid and "map spot" to arrange sampling points, sampling density of 1 point /16 km^2, a total of 311 samples were collected. After sample collection, the samples were dried in a cool and ventilated place. In a clean, ventilated and pollution-free site, 1000 g of samples were intercepted by 2 mm (10 mesh) nylon sieve, bottled, labeled and sent to the laboratory for analysis. Organic carbon and PH were determined by potassium dichromate volumetric method and ion selective electrode method, respectively (Fig. 1).

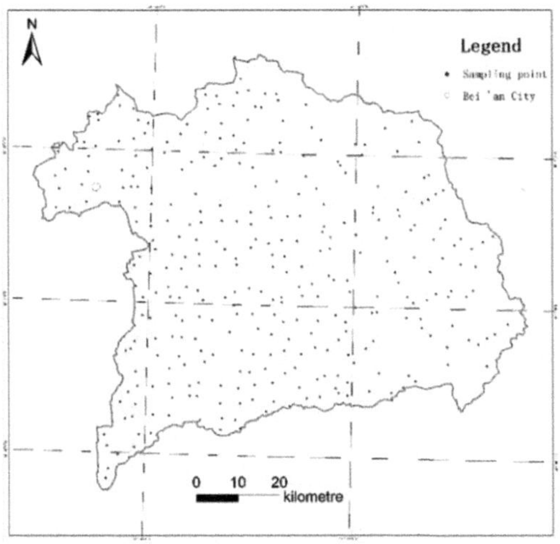

Fig. 1. Sampling point distribution map

2.3 Data Processing and Analysis

In this study, ArcGIS 10.8 software was used to perform spatial interpolation on soil organic matter and pH data of cultivated land in Bei'an County to generate spatial distribution maps of organic matter and pH. Using spatial interpolation technology, we were able to convert the discrete sampling point data into a continuous spatial distribution map to more clearly show the trend of soil organic matter and pH value across the region. In terms of data analysis, this study used IBM SPSS Statistics 19.0 software to conduct descriptive statistical analysis. After sorting and analyzing the collected soil sample data, Table 1 is generated, which lists the basic statistical information of organic matter and pH value in detail, including mean value, standard deviation, coefficient of variation, minimum value and maximum value. These statistical indicators provide basic data support for understanding the overall characteristics of soil. In addition, Pearson correlation analysis was performed to explore the relationship between organic matter and pH. This analysis method can effectively reveal the degree of linear correlation between the two variables, and help us understand whether there is a significant interaction between soil organic matter and pH value.

Table 1. Description statistics of organic matter and PH sampling points in cultivated land in Bei'an County

	N	Min	Max	Mean	Standard deviation	Variable coefficient
Organic matter	331	6.34	189.98	56.60	28.03	49.53%
PH	331	5.07	7.35	6.04	0.41	6.8%

The organic matter and PH classification standards of the third National Soil Census were used for classification, as shown in Table 2:

Table 2. Soil organic matter and PH grading standards

Ranks	organic matter		PH	
	Levels	Contents (g/kg)	Levels	Values
1	Very rich	>40.0	Strong acidity	<4.5
2	Abundant	30.0~40.0	acidity	4.5~5.5
3	Abundant	20.0~30.0	Faintly acidic	5.5~6.5
4	less	10.0~20.0	neutral	6.5~7.5
5	lack	6.0~10.0	alkalescence	7.5~8.5
6	Serious shortage	<6.0	alkalinity	8.5~9.5

3 Results and Analysis

3.1 Spatial Distribution of Organic Matter

The soil organic matter ranged from 6.34 to 189.98 g/kg, with an average content of 56.60 g/kg. It can be seen from Fig. 2 that the overall content of organic matter is high and the distribution is relatively uniform. High in the east and low in the west, gradually lacking. Areas with rich soil organic matter content (> 40.0 g/kg) are mainly distributed in the eastern part of Beian County, which is a forest farm and farm with fertile arable land and relatively high organic matter. The areas with abundant soil organic matter (30.0–40.0 g/kg) and medium areas (20.0–30.0 g/kg) are mainly distributed in and around densely populated areas such as Beian County, Zhaoguang Town and Tongbei Town in the west of Beian County.

The distribution area of cultivated land organic matter in Bei'an County is shown in Table 2. The level of cultivated land organic matter in the whole region ranges from 1 to 4. The area of cultivated land organic matter in the whole region is Grade 1 (> 40.0 g/kg) with an area of 6,948.57 km^2, accounting for 96.60% of the total area of cultivated land in Bei'an County, and the distribution area of the least is Grade 4 (10.0~20.0 g/kg). The area is 0.73 km^2, accounting for 0.01% of the total area. The distribution area of Grade 2 and 3 is 209.67 km^2 and 34.03 km^2 respectively, accounting for 2.92% and 0.47% of the total area.

3.2 Spatial Distribution of pH Value

Combined with the analysis results of Table 1 and Fig. 3, it can be clearly seen that the average pH value of cultivated land in Bei'an County is 6.04, showing the overall characteristics of weak acidity. This value indicates that the soil environment in the

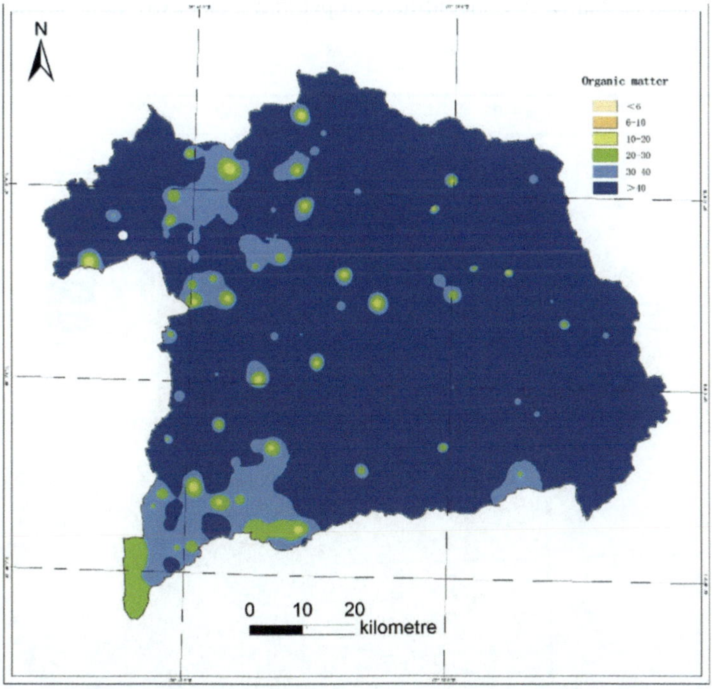

Fig. 2. The Spatial distribution map of organic matter content in Bei'an County

region is relatively suitable for the growth of most crops. Specifically, the maximum pH value is 7.35, while the minimum pH value is 5.07, and the change in this range shows the diversity of soil pH. In addition, the coefficient of variation was 6.8%, indicating that the distribution of soil pH value was relatively stable and the range of change was small.

From the perspective of spatial distribution, the pH value of Bei'an County showed a trend of increasing gradually from west to east. This phenomenon may be closely related to factors such as terrain, vegetation cover and human activities. Specifically, the pH value in the southern region is generally higher than that in the northern region, which may be related to the soil type and hydrological conditions in the south, which provide a more suitable growing environment for agricultural production.

Table 2 shows the distribution of soil pH in Bei'an County in detail. According to the data analysis, the largest proportion of pH value in the region is grade 3, that is, weakly acidic (pH 5.5~6.5), which covers an area of 6799.32 km^2, accounting for 94.53% of the total cultivated land area in Beian County. This result further confirms that the soil in Bei'an County is weakly acidic. At the same time, the area of soil with grade 2 acidity (pH 4.5 to 5.5) is 106.11 km^2, accounting for 1.5% of the total area, indicating the presence of a small amount of acidic soil in the region. In addition, the area of grade 4 neutral soil (pH 6.5~7.5) is 288.11 km^2, accounting for 4.0% of the total area. No cultivated soils of grade 1, 5 and 6 were found in the study area, which indicates that the soil in this area is relatively concentrated in pH and lacks extremely acidic or alkaline soil.

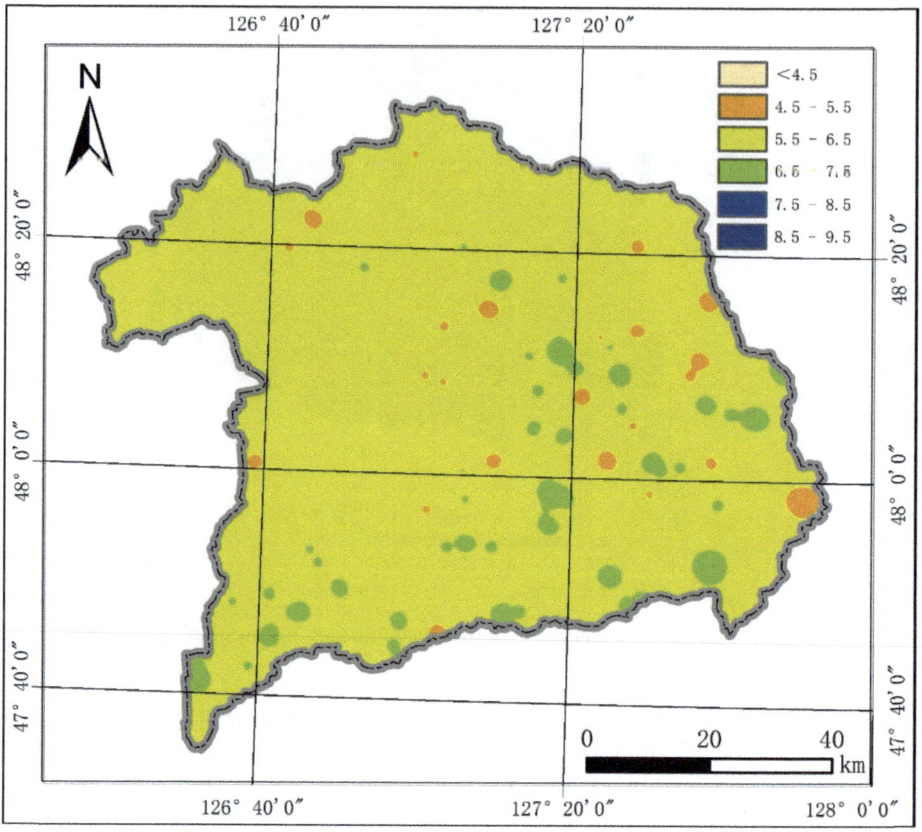

Fig. 3. The Spatial distribution map of PH in Bei'an County

3.3 Spatial Correlation Between Organic Matter and pH Value

According to the data in Table 3, the correlation analysis results between organic matter content of cultivated land and soil pH value in Bei'an County show that the significance level (Sig.) is 0.711, which is greater than 0.05, indicating that the degree of correlation between these two variables is not significant. Further analysis of the correlation coefficient value is −0.02, which also reflects the weak correlation between the two, almost can be considered as no linear relationship. This means that in the cultivated land of Beian County, there is no obvious interaction between organic matter content and soil pH value change, which may be due to the interference or influence of other factors.

Step 3 Discuss

Soil organic matter is an important index to measure soil fertility [13, 14] and one of the comprehensive indexes to weigh land quality [15]. It mainly comes from the residues

Table 3. Correlation between organic matter and PH in Bei'an County

	PH	Organic matter
PH	1	
Organic matter	−0.02	1

of animal and plant microorganisms [16–18]. Agricultural soil organic matter comes from a wide range of sources, including humus formed by animals and plants themselves and man-made fertilizers [19]. Studies have shown that the content of organic matter is not only significantly correlated with natural climatic factors such as temperature and rainfall, but also varies with the change of terrain height and geomorphic type [20, 21]. The overall content of organic matter in the cultivated land in Bei'an County is very high, and the reason why the organic matter content in the eastern part is abundant (> 40) and the organic matter content in the western part is reduced is largely related to human factors. The eastern part is mainly farms and forest farms with less human activities, and the land is more preserved in the original state. On the contrary, in the western part, cities and towns are distributed from south to north, which are greatly affected by urbanization and human activities, and the content of organic matter is correspondingly affected.

Soil PH plays an important role in the physical and chemical properties of soil, the form and availability of nutrients, microbial activity, plant growth, and biogeochemical cycles of carbon and nitrogen, and is a natural attribute generated by the comprehensive effects of many factors (soil parent material, topography, climate conditions, biological activities, etc.) in the process of soil formation [22, 23]. The study area is located in the southeast of Wudalianchi, and due to volcanic effects, the overall pH is weakly acidic [24].

Soil organic matter is an important part of soil, which is the nutrient source of plants and the energy source of soil microbial life activities. Soil pH plays an important role in soil non-circulation, energy flow and structural stability of regional ecosystem. Both are important factors in ensuring soil fertility and productivity. Some scholars have found that soil pH has an impact on the accumulation of organic matter [16], and pointed out that organic matter in cultivated land in Heilongjiang Province has a very significant negative correlation with pH, that is, soil pH increases, organic matter content significantly decreases [25, 26] (Dai Wanhong, Liu Yiying). In this study, the correlation between organic matter and pH in cultivated land in Bei'an County was not significant, which was inconsistent with previous research views, indicating that the changes of both were affected by other factors, either man-made or natural. Therefore, the factors affecting organic matter content and pH are relatively complex, and the relationship between the change difference levels in different regions may be inconsistent, which requires further research in order to better understand the evolution mechanism of soil environment, which is conducive to optimizing the management strategy of cultivated land and improving the quality of cultivated land.

4 Conclusion

In this study, soil organic matter and pH value of cultivated land in Beian County, Heilongjiang Province were analyzed in depth, and the following important conclusions were drawn:

(1) The overall content of organic matter in the cultivated land in Bei'an County was higher, with an average content of 56.60 g/kg, indicating good soil fertility. This result shows that the soil in the region is able to provide sufficient nutrients for crops to grow. At the same time, the spatial distribution of organic matter is relatively uniform, and the primary distribution area is the widest, accounting for 96.60% of the total area. However, due to the influence of human activities and natural environment, the content of soil organic matter shows some differences in space. Specifically, the content of soil organic matter in the eastern mountain area is generally higher than that in the western area due to better vegetation coverage and less human activities. This phenomenon reflects the important effects of land use and ecological environment on soil quality.
(2) In terms of pH value, the pH of cultivated land in Bei'an County ranges from 5.07 to 7.35, with an average value of 6.04, showing a weak acidic character as a whole. This weakly acidic soil is relatively suitable for the growth of most crops and can effectively promote the absorption of nutrients. The spatial distribution of pH value is mainly concentrated in grade 3 (pH 5.5~6.5), which has the largest area, accounting for 94.53% of the total area. However, the neutral soil (pH 6.5~7.5) only accounted for 4.0% of the total cultivated land in Beian County, and showed the characteristics of dot distribution. Further analysis showed that the pH value showed a trend of relatively high in the western region and low in the eastern region, while the pH value was high in the southern region and low in the northern region. This distribution pattern may be closely related to soil type, vegetation coverage and human activities.
(3) The results of this study show that there is no significant correlation between organic matter and pH value in Beian area. This finding suggests that in addition to organic matter and pH itself, there may be other factors influencing the changes in the local action relationship between the two. This suggests that we should not only rely on a single index when evaluating soil quality, but should carry out multi-factor coupling correlation analysis to fully understand the soil health status and change mechanism. Future research will need to combine climate change, land use, fertilization management and other factors, in order to provide a more scientific basis for cultivated land management and soil improvement in Beian County. Through such in-depth analysis, we can better maintain and improve soil quality and promote sustainable agriculture.

Acknowledgments. This study was supported collectively by the Geological survey project of China Geological Survey(No.DD20220852 and No.DD20230517). Thanks to other colleagues for their help and support in writing the paper.

References

1. Chen, Y., Huang, Y., Sun, W.: Using organic matter and pH to estimate the bulk density of afforested/reforested soils in northwest and northeast China. Pedosphere **27**(5), 890–900 (2017)
2. Qi, C., Zhou, Z., Cai, S., et al.: Spatial-temporal variation of soil organic matter decomposition potential in China. Soil and Tillage Research **235**, 105898 (2024)
3. Chenchen, Q., Ren, W., Li, X., et al.: Revisit soil organic matter. Chin. Sci. Bull. **67**(10), 913–923 (2022)
4. Huber, S., Bernardini, L.G., Bennett, et al.: Suitability of microbial and organic matter indicators for on-farm soil health monitoring. Soil Use and Management **40**(1), e12993 (2024)
5. Yun, L., Rillig, M.C., Chen, H.Y.H., Shan, R., Ma, Z.: Soil pH drives the relationship between the vertical distribution of soil microbial biomass and soil organic carbon across terrestrial ecosystems: A global synthesis. CATENA **238**, 107873 (2024)
6. Quaye, A.K., Doe, E.K., Attua, E.M., Yiran, G., et al.: Geospatial distribution of soil organic carbon and soil pH within the cocoa agroecological zones of Ghana. Geoderma **386**, 114921 (2021)
7. Ou, Y., Rousseau, A.N., Wang, L.X., et al.: Spatio-temporal patterns of soil organic carbon and pH in relation to environmental factors—A case study of the Black Soil Region of Northeastern China(Article). Agriculture, Ecosystems and Environment **245**, 22–31 (2017)
8. Huang, H.Q., Zhang, H., Hu, Z., et al.: Spatial distribution and influence between organic matter and ph of arable soil in mountainous areas of Guizhou. Southwest China Journal of Agricultural Sciences **36**(11), 2473–2479 (2023)
9. Bai, S., Pei, J., Li, S., et al.: Temporal and spatial dynamics of soil organic matter and pH in cultivated land of Liaoning province during the past 30 years. Chin. J. Soil Sci. **47**(3), 636–644 (2016)
10. Bárcenas-Moreno, G., Jiménez-Compán, E., San Emeterio, L.M., Jiménez-Morillo, N.T., González-Pérez, J.A.: Soil pH and soluble organic matter shifts exerted by heating affect microbial response. Int. J. Environ. Res. Pub. Health **19**(23), 15751 (2022)
11. Que, J., Tang, Z., Li, X., et al.: Analysis of Soil pH and organic matter of tobacco-planting soil in mile city in honghe prefecture. Southwest China Journal of Agricultural Sciences **32**(7), 1633–1638 (2019)
12. Shang, B., Li, D., Yimin, X., et al.: The characteristics of soil organic matter and pH in Wulingshan tobacco-growing areas. Chin. J. Soil Sci. **46**(3), 590–596 (2015)
13. Li, F., Li, X., Wu, P., et al.: Correlation Between Heavy Metal Pollution and Basic Properties of Agricultural Soils in Hainan Province. Soils **41**(1), 49–53 (2009)
14. Ping, W., Honglian, H., Zhiqiang, D., et al.: The effects of altitude and land use on organic matter and integrated fertility of soils in the northern tropics mountain. Tropical Geography **43**(1), 144–154 (2023)
15. Yeting, Z., Qingrui, C., Zhipeng, L., et al.: Spatial characteristics and changes of soil organic matter for cultivated land in suburban area of Xi'an from 1983 to 2 009. Transactions of the Chinese Society of Agricultural E ngineering **29**(2), 132–140 (2013)
16. Ćwieląg-Piasecka, I.: Soil organic matter composition and pH as factors affecting retention of carbaryl, carbofuran and metolachlor in soil. Molecules (Basel, Switzerland) **28**(14), 5552 (2023)
17. Kim, P.-G., Tarafdar, A., Kwon, J.-H.: Effect of soil pH on the sorption capacity of soil organic matter for polycyclic aromatic hydrocarbons in unsaturated soils. Pedosphere **33**(2), 365–371 (2023)
18. Chen, Y., Huang, Y., Sun, W.: Using Organic Matter and pH to Estimate the Bulk Density of Afforested/Reforested Soils in Northwest and Northeast China. Pedosphere **27**(5), 890–900 (2017)

19. Dongchu, L., Jing, H., Changbao, M., et al.: Soil organic matter content and its relationship with pH and bulk density in agricultural areas of China. J. Soil and Water Conservation **34**(6), 252–258 (2020)
20. Cen, Z., Wang, T., Wang, R., Feng, H.: Spatial and temporal variability of soil organic matters in se-rich areas of hainan island in recent 30 years. Agricultural Biotechnology **7**(4), 242–245 (2018)
21. Zhang, Y., Shen, H., Gao, Q., et al.: Estimating soil organic carbon and pH in Jilin Province using Landsat and ancillary data. Soil Science Society of America Journal **84**(2), 556–567 (2020)
22. Jiao, X., Chao, S., Yingchun, S., et al.: Spatial variation characteristics and influencing factors of soil pH in the Lu'an area of Anhui province. Chin. J. Soil Sci. **52**(1), 34–41 (2021)
23. Qingxia, C., Xiaohui, L., Chenglong, T.: Spatial variation and influencing factors of soil pH in Anshun City. Environmental Science **43**(4), 2124–2132 (2022)
24. Huang, Q.Y., Yang, F., Xie, L.H., et al.: Diversity and community structure of soil bacteria in different volcanoes, Wudalianchi. Acta Ecologics Sinica **41**(20), 8276–8284 (2021)
25. Wanhong, D., Yao, H., Li, W., et al.: Relationships between soil organic matter content(SOM) and pH in top soil of zonal soils in China. Acta Pedologica Sinica **46**(5), 851–860 (2009)
26. Liu, Y., Pei, J., Wang, J.: Spatial distribution and relationship between organic matter and pH in the typical black soil region of northeast China. J. Agricult. Resour. Environ. **36**(6), 738–743 (2019)

Spatial Distribution Characteristics of Land Mass Geochemistry in Zhaoguang Area, Beian County

Hongqiang Zhao[1,2], Mingxin Duan[1,2], Haonan Song[1,2], Haicheng Zhang[1,2], Youtao Xin[1,2(✉)], and Xidong Zhao[1,2(✉)]

[1] Harbin Natural Resources Comprehensive Survey Center, China Geological Survey, Harbin 150086, China
zhaoxidong@mail.cgs.gov.cn

[2] Observation and Research Station of Earth Critical Zone in Black Soil, Harbin, Ministry of Natural Resources, Harbin 150086, China

Abstract. Based on the data of 1:50,000 land quality geochemical survey and evaluation in Zhaoguang area of Beian County, soil pH, soil nutrient, soil environmental quality, land quality classification standard and spatial distribution characteristics were analyzed. According to the single index evaluation of soil biological essential elements, trace elements, health elements and comprehensive evaluation of soil environmental geochemistry, the farmland in Zhaoguang area was classified into grades. The comprehensive level of soil environment in Zhaoguang area was rated as first class, indicating that there was no risk of heavy metal pollution in the farmland soil.

Keywords: grading standard · Spatial distribution characteristics · Bei'an County

1 Introduction

The investigation area is located in the central and western part of Beian County,Binbei, Qibei, Beihei three railway trunk lines, Heida, Nianbei and other four national provincial highways, Jihei, Suibei and other five high-speed intersection; 47 km away from WudalianChidu Airport, it has the geographical location and transportation advantages of Nantong provincial capital, north border, east forest sea, west grain warehouse. The city is located in the northwest of the city, connected to Xunke County and Suilang County in the east, facing Hailun City by the Tongken River in the south, adjoining Baiquan County and Kedong County in the west, and bordering Wudalianchi City in the north [1–6].

The original version of the chapter has been revised. The corresponding author indication in the chapter has been updated. A correction to this chapter can be found at
https://doi.org/10.1007/978-3-031-86829-0_124

D. Han et al. (Eds.): ICEPG 2024, ESE, pp. 716–728, 2025.
https://doi.org/10.1007/978-3-031-86829-0_80

2 Review of Working Methods and Quality

2.1 Field Work Methods and Quality Review

Technical Standards
In this work, the field sample layout, sample collection, processing, analysis and testing were strictly implemented in accordance with the Code for Geochemical Evaluation of Land Quality (DZ/T0295-2016), and the sample analysis and testing was completed by the Laboratory of Harbin Natural Resources Comprehensive Center of China Geological Survey.

Field Work Methods
Layout Principle of sample points. This work adopts the principle of grid plus map spots to ensure that the samples are relatively uniform in space, and the sample points are mainly distributed in agricultural land. According to the requirements of the Code for Geochemical Evaluation of Land Quality (DZ/T0295-2016), the sampling density ranges from 4 points /km^2 to 16 points /km^2. In this work, the measuring network of farmland in Zhaoguang area is 500 m500 m, and the sampling density is 4 points /km^2, which can meet the sampling point layout standard and sampling density of 1:50,000.

Sampling Method. According to the requirements of the Code for Geochemical Evaluation of Land Quality (DZ/T0295-2016), technicians avoided ditches, forest belts, ridges, roadsides, old house foundations, manure piles and unrepresentative areas with unequal micro-topography during the sampling process according to the field conditions. The samples were collected from the soil of the tilling layer, the sampling depth was 0 ~ 20 cm, and the sampling was carried out by S-type sampling method, and three sub-samples were mixed in equal amounts to form a sample.

Sample Processing. The sample drying and processing sites of this work are clean and clean. After the soil samples collected from the field were cleaned and registered in time, they were placed in a clean and tidy ventilated area to dry. After the air-dried samples were passed through a 10-mesh nylon sample screen, some samples were packed into kraft paper bags and sent to the laboratory for analysis. The secondary samples were packed into clean 300 g plastic bottles and sent to the sample library for storage.

3 Soil Element Content Characteristics

This study carried out a land quality geochemical survey in Zhaoguang area of Beian County, covering an area of 23.55 km^2. 140 farmland soil samples were collected and 54 elements were analyzed.

3.1 Geochemical Characteristics and Classification of Soil Nutrient Elements

The geochemical parameters of nutrient elements in 140 farmland soil samples in Zhaoguang area were analyzed. The series of soil geochemical parameters mainly include the arithmetic mean value (□$\overline{X}$), standard deviation (S_0), coefficient of variation (CV), maximum value (Xmax), minimum value (Xmin), arithmetic mean value (□$\overline{X}_0$), standard deviation (S_0), coefficient of variation (CV_0) after iterative elimination by adding or subsuming the mean value by 3 times the standard deviation, etc. The calculation is as follows [7–9].

(1) arithmetic mean

In the statistical data, the arithmetic mean value of outlier data before and after elimination is expressed by □$\overline{X}$ and □$\overline{X}_0$, and the arithmetic mean value is $= \frac{1}{n}\sum_{i=1}^{n} xi$.

(2) Standard deviation

In the statistical data, the arithmetic standard deviation of outlier data before and after elimination is expressed by S and S0 respectively, and the arithmetic standard deviation is $= \sqrt{\frac{\sum_{i=1}^{n}(x_i - X)^2}{n}}$.

(3) Coefficient of variation

In the statistical data, the coefficient of variation before removing the outlier data was expressed by CV, and the coefficient of variation was = S/X × 100%.

(4) Maximum value

When no outlier data is removed, the largest value is the maximum value, expressed by Xmax.

(5) Minimum value

When no outlier data is removed, the smallest value is the smallest value, expressed in Xmin.

Geochemical Characteristics of Soil Nitrogen, Phosphorus, Potassium, Calcium, Magnesium, Sulfur, Organic Matter, etc.

Nitrogen, phosphorus, potassium, calcium, magnesium, sulfur, and organic matter are essential macronutrients for organisms. Overall, except for calcium, the content of other elements in the Zhaoguang area is higher than the Chinese soil background values, indicating a medium to rich state.

The geochemical parameters of each elemental indicator are listed in Table 1. As shown in Table 1, the enrichment coefficients of N, P, MgO, and S are between 1.44 and 4.82, indicating that most nutrient elements are enriched in the agricultural soil; the enrichment coefficient of CaO is 0.73, slightly lower than the Chinese soil background value, and the enrichment coefficient of Corg is 1.05, equivalent to the Chinese soil background value. From the coefficient of variation, the coefficients of variation of nutrient elements are between 0.06 and 0.56, with the largest being N, followed by Corg, S, and P, indicating that the distribution of Corg, N, S, and P is relatively uneven, while the coefficients of variation for K_2O and CaO are smaller, indicating a more uniform distribution. The relative enrichment of Corg in the agricultural fields of the Zhaoguang area is mainly due to the area being in the common influence zone of the Wuyuer River and Tongken River in Bei'an County, with relatively good soil aggregates and fertility, which is beneficial for agricultural production.

The enrichment coefficient = the average value after elimination / the background value of Chinese soil; The units of N, P, and S contents are mg/kg, and others are %.

Table 1. Statistical Table of Geochemical Characteristics of Nutrient Elements

Element	Xmax	Xmin	$\overline{X}$	S	CV	$\overline{X}_0$	S_0	CV_0	Chinese Soil Background Value	Enrichment Coefficient
N	6520	1090	3086	1721.39	0.56	3086	1721.39	0.56	640	4.82
P	2558	495	1342	410	0.31	1342	410.00	0.31	520	2.58
K_2O	2.68	1.98	2.25	0.12	0.06	2.25	0.12	0.06	1.86	1.21
CaO	2.85	0.66	1.13	0.26	0.23	1.13	0.26	0.23	1.54	0.73
MgO	1.53	0.67	1.12	0.12	0.10	1.12	0.12	0.10	0.78	1.44
S	1165	140	432.40	148	0.34	432.40	148	0.34	150.00	2.88
Corg	9.05	0.97	3.26	1.56	0.48	3.26	1.56	0.48	3.10	1.05

Nitrogen: The average nitrogen content of farmland soil samples in Zhaoguang area is much higher than the background value of soil in China. According to the grading evaluation standard of the Code for Geochemical Evaluation of Land Quality (DZ/T0295-2016), the grading evaluation standard of the Code for Geochemical Evaluation of Land Quality (DZ/T0295-2016), The nitrogen content of farmland in Zhaoguang area can be divided into first, second, third, fourth and fifth. On the whole, the nitrogen content of farmland soil in Zhaoguang area is higher. Lack of nitrogen in soil can lead to dwarf plants, leaf yellowing, delayed flower bud differentiation, small fruits and poor quality, while too much nitrogen will make plants grow barren, resulting in a large number of fallen flowers and fruit development stagnation. Therefore, nitrogen fertilizer application dose should be controlled in the land with excellent nitrogen content, and the use of nitrogen fertilizer should be increased appropriately in the land with poor nitrogen content.

Phosphorus: The average content of phosphorus in the samples is higher than the background value of soil in China. According to the grading evaluation standard of the Code for Geochemical Evaluation of Land Quality (DZ/T0295-2016), the content of phosphorus in farmland in Zhaoguang area can be divided into first class, second class, third class and fourth class. As a whole, the content of phosphorus in farmland in Zhaoguang area is mainly medium and above. The content of phosphorus is medium to abundant. Phosphorus deficiency can cause slow cell growth, short plants, dark green leaves, etc. Therefore, it is recommended to increase the use of phosphorus fertilizer in the land with phosphorus deficiency.

Potassium oxide: The average content of potassium in the samples is higher than the background value of Chinese soil. According to the grading evaluation standard of the Code for Geochemical Evaluation of Land Quality (DZ/T0295-2016), the content of potassium oxide in farmland in Zhaoguang area is first-class. Zhaogang area is rich in potassium and has high potassium content, which can promote the absorption of enzymes and enhance the resistance of crops to disease, drought, cold, lodge and salt. However, excessive potassium will cause the decrease of cations absorption of crops such as calcium and magnesium, which will weaken the production capacity of crops. Although the total potassium content in the farmland is first, the content of available potassium that can be directly used by the plants should also be paid attention to, and the application of potassium fertilizer should be determined after comprehensive consideration.

Magnesium oxide: The average content of samples is higher than the background value of soil in China. According to the grading evaluation criteria of the Code for Geochemical Evaluation of Land Quality (DZ/T0295-2016), the content of magnesium oxide in farmland in Zhaoguang area can be divided into grade 3, grade 4 and grade 5, and the soil magnesium in farmland in Zhaoguang area is in a medium-deficient state. This may be due to the low magnesium content of soil formed by Quaternary river deposits.

Calcium oxide: The average content of samples is lower than the background value of soil in China. According to the grading evaluation criteria of the Code for Geochemical Evaluation of Land Quality (DZ/T0295-2016), the content of phosphorus in farmland in Zhaoguang area can be divided into grade 3 and grade 4. The soil calcium in farmland in Zhaoguang area is in a moderate to relatively deficient state, and the lack of calcium will cause crop growth retardation. Yields and quality are reduced, and moderate calcium supplementation in calcium-deficient areas is recommended when growing crops.

Geochemical Characteristics of Trace Elements Such as Iron, Manganese, Zinc, Copper, Boron and Molybdenum in Soil

Iron, manganese, zinc, copper, boron and molybdenum elements in soil are essential micronutrients for crop growth. The classification standard of these micronutrients is calculated according to the surface soil analysis data obtained from the multi-target regional geochemical survey based on the element content of soil in Layer A of China. The specific data are shown in Table 2.

Table 2. Classification standards of soil nutrient indexes

Indicator	Indicator	First Class	Second Class	Third Class	Fourth Class
	Abundant	more abundant	medium	more lack-ing	lack
Fe（%）	＞5.30	4.60～5.30	4.15～4.60	3.40～4.15	≤3.40
Mn（mg/kg）	＞700	600～700	500～600	375～500	≤375
Zn（mg/kg）	＞84	71～84	62～71	50～62	≤50
Cu（mg/kg）	＞29	24～29	21～24	16～21	≤16
B（mg/kg）	＞65	55～65	45～55	30～45	≤30
Mu（mg/kg）	＞0.85	0.65～0.85	0.55～0.65	0.45～0.55	≤0.45

The geochemical parameters of trace elements were statistically analyzed, and the results are shown in Table 3. From the perspective of enrichment coefficient and variation coefficient, the enrichment coefficient of TFe_2O_3 and Mn is higher than 1.2, showing enrichment state, Zn, Cu and Cl are similar to the background values of Chinese soil, and the enrichment coefficient of Mo and B is only 0.45 and 0.63, showing strong depletion. From the perspective of the coefficient of variation, the value of CV ranges from 0.16 to 0.37, and the coefficient of variation of Mn is the largest, indicating that its spatial distribution is highly volatile; the coefficient of variation of Mo and Cl is relatively large, indicating that its spatial distribution is relatively volatile; the other elements are more evenly distributed.

Table 3. Statistical table of geochemical parameters of trace elements

Element	Xmax	Xmin	$\overline{X}$	S	CV	$\overline{X}_0$	S_0	CV_0	Chinese Soil Background Value	Enrichment Coefficient
TFe_2O_3	6.66	2.65	4.67	0.554	0.12	4.67	0.554	0.12	2.94	1.59
Mn	2955	381	1162	426.7	0.37	1162	426.7	0.37	583	1.99
Zn	89.3	48	67.5	6.79	0.10	67.5	6.79	0.10	74.2	0.91
Cu	31.4	13.5	22	3.08	0.14	22	3.08	0.14	22.6	0.97
B	47.9	22.3	30.2	4.91	0.16	30.2	4.91	0.16	47.8	0.63
Mo	1.67	0.48	0.89	0.207	0.23	0.89	0.207	0.23	2	0.45
Cl	130	31.2	67.8	19.5	0.29	67.8	19.5	0.29	68	1.00

Enrichment coefficient = average value after removal/Chinese soil background value; The unit of TFe_2O_3 is %, and the unit of other elements is mg/kg.

Total iron: The average content of the samples was higher than the Chinese soil background value, which may be related to the joint action of Wuyuer River and Tongken River. According to the "Land Quality Geochemical Evaluation Code" (DZ/T0295-2016), the Zhaoguang area can be divided into second class, third class, fourth class and fifth class farmland.

Manganese: The average content of the sample is higher than the Chinese soil background value. According to the "Code for Geochemical Evaluation of Land Quality" (DZ/T0295-2016), the Zhaoguang area can be divided into first, second, third, fourth and fifth grade farmland. The distribution pattern of low value region of Mn element is similar to that of iron element, which should be affected by the occurrence mode of the element.

Zinc element: The high value area is mainly distributed in two areas on the west side of Zhaoguang area. According to the Code for Geochemical Evaluation of Land Quality (DZ/T0295-2016), the Zhaoguang area can be divided into first, second, third, fourth and fifth grade farmland. The low value area is mainly distributed in the southwest of the east side.

Copper element: There is no obvious high value area in Zhaoguang area, only a small part of farmland in the southwest has moderate content, and the rest of the area is low value area. According to the "Land Quality Geochemical Evaluation Code" (DZ/T0295-2016), the Zhaoguang area can be divided into three, four and five grades of farmland.

Boron element: Low value area is widely distributed in Zhaoguang area, there is no high value area, and some farmland in the south has moderate content. According to the "Land Quality Geochemical Evaluation Code" (DZ/T0295-2016), the Zhaoguang area can be divided into three, four and five grades of farmland.

Molybdenum element: low value area is widely distributed in Zhaoguang area, and there is no high value area. According to the "Land Quality Geochemical Evaluation Code" (DZ/T0295-2016), the Zhaoguang area can be divided into first, second, third and fourth grade farmland.

Geochemical Characteristics of Trace Health Elements Such as Selenium, Iodine, and Fluorine in Soil

Trace health elements in soil include selenium, iodine, and fluoride, totaling three elements. The grading standards are shown in Table 4. The grading standards for soil selenium, iodine, and fluoride are given based on the statistical data of surface soil analysis obtained from multi-purpose regional geochemical surveys, referring to relevant domestic and international research results.

Table 4. Grading Standards for Soil Health Elements

Indicator	Indicator	First Class	Second Class	Third Class	Fourth Class
	Excess	High	Sufficient	Marginal	Deficient
Se(mg/kg)	>3.0	0.4~3.0	0.175~0.40	0.125~0.175	≤ 0.125
I(mg/kg)	>100	5~100	1.5~5	1~1.5	≤ 1
F(mg/kg)	>700	550~700	500~550	400~500	≤ 400

Geochemical Characteristics of Selenium Selenium is an important non-metal trace element in the ecological environment, with active chemical properties, but it is usually present in low concentrations in terrestrial ecosystems. Selenium has dual biological functions that are both beneficial and harmful to human health. An adequate intake of selenium can improve immunity and prevent cancer, while selenium deficiency can lead to diseases such as Keshan disease and kashin-beck disease. Excessive selenium intake can cause symptoms such as hair loss and nail shedding.

The Zhaoguang area is located on the selenium-deficient belt in China, with selenium content ranging from 0.1 to 0.4 mg/kg, an arithmetic mean of 0.24 mg/kg, a standard deviation of 0.0599, and a coefficient of variation of 0.25, indicating that the distribution of selenium is relatively uneven. The farmland in the Zhaoguang area is mainly classified as third, fourth, and fifth grades (Fig. 1).

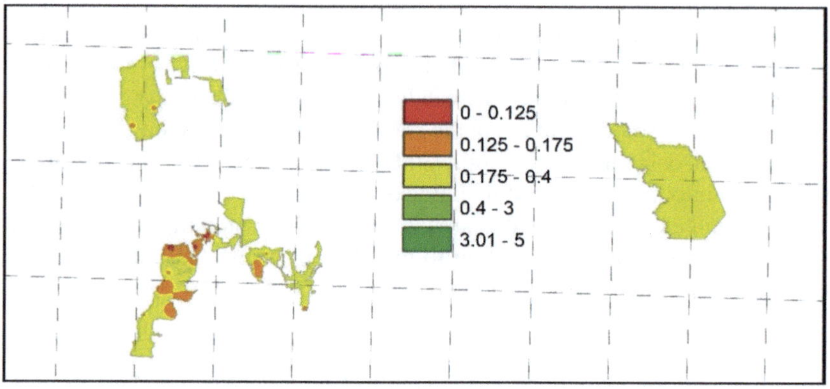

Fig. 1. Selenium geochemical single-factor evaluation map

Geochemical Characteristics of Iodine Iodine is known as the "intelligence element" and is an essential element for the synthesis of thyroid hormones, playing an extremely important role in the metabolism of humans and animals. Iodine deficiency can cause endemic goiter and endemic cretinism.

The content of iodine in the Zhaoguang area ranges from 0.73 to 4.9 mg/kg, with an arithmetic mean of 2.61 mg/kg, a standard deviation of 0.844 mg/kg, and a coefficient of variation of 0.32, indicating that the distribution of iodine is uneven.

The iodine in the farmland of the Zhaoguang area is mainly classified as deficient, marginal, and sufficient grades (Fig. 2), among which the area of iodine deficiency is 5.38 km^2, the area of marginal iodine is 0.21 km^2, and the area of sufficient iodine is 0.73 km^2.

Geochemical Characteristics of Fluorine Fluorine is an essential trace element closely related to human health. Both insufficient and excessive intake of fluorine can harm health. Insufficient intake can lead to dental caries, while excessive intake can cause skeletal fluorosis and dental fluorosis.

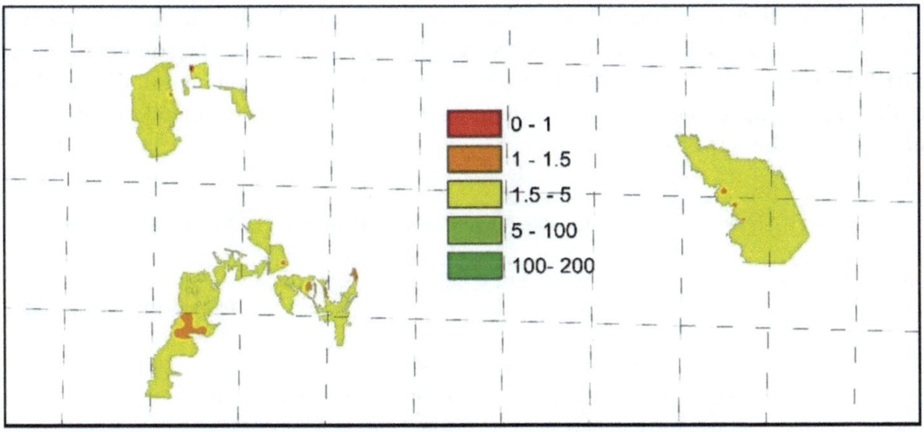

Fig. 2. Iodine geochemical single-factor evaluation map

The content of fluorine in the Zhaoguang area ranges from 264 to 632 mg/kg, with an arithmetic mean of 481.76 mg/kg, a standard deviation of 48.88, and a coefficient of variation of 0.1, indicating that the distribution of fluorine is relatively uneven (Fig. 3).

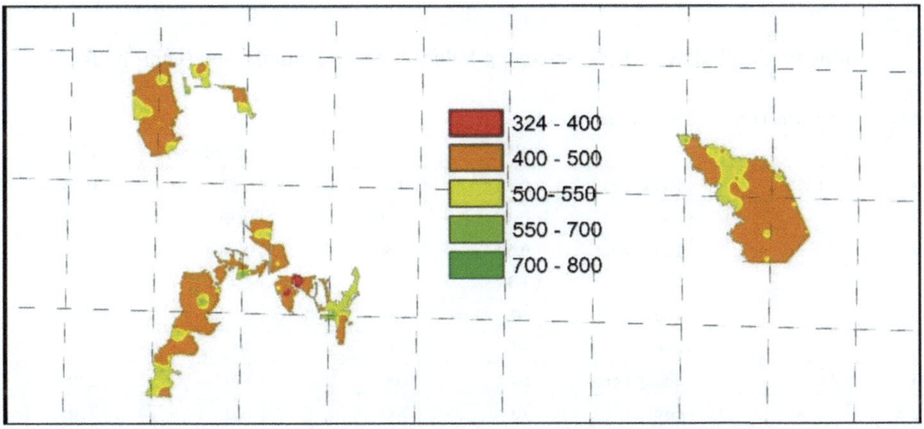

Fig. 3. Fluorine geochemical single-factor evaluation map

3.2 Soil Environmental Geochemical Characteristics and Grade Division

The geochemical parameters of eight heavy metal elements (As, Cd, Hg, Cr, Cu, Pb, Zn, Ni) and pH values in the Zhaoguang area were statistically analyzed, with results shown in Table 5. By comparing the average values of each heavy metal element with the risk screening values and control values of soil environmental pollution for agricultural land in Tables 6 and 7, it can be seen that the average values of the eight heavy metal elements are all less than the risk screening values and control values for soil pollution in agricultural land. As shown in Fig. 4, the comprehensive grade of the soil environment is risk-free. Therefore, the comprehensive grade of the soil environment in the farmland of the Zhaoguang area is first class, indicating that there is no heavy metal pollution risk in the soil of the farmland in this area.

Table 5. Statistical Table of Geochemical Parameters of Soil Heavy Metal Elements

Element	Xmax	Xmin	$\overline{X}$	S	CV	$\overline{X}_0$	S_0	CV_0	Chinese Soil Background Value	Enrichment Coefficient
As	24.2	4.97	11.92	2.227	0.19	11.92	2.227	0.19	11.2	1.06
Cd	0.27	0.033	0.109	0.047	0.43	0.109	0.047	0.43	0.097	1.12
Hg	0.088	0.017	0.04	0.013	0.33	0.04	0.013	0.33	0.065	0.62
Cr	83.5	57.4	70.9	5.57	0.08	70.9	5.57	0.08	61	1.16
Cu	31.4	13.5	22	3.08	0.14	22	3.08	0.14	22.6	0.97
Pb	34.1	19.2	25.03	3.25	0.13	25.03	3.25	0.13	26	0.96
Zn	89.3	48	67.5	6.79	0.10	67.5	6.79	0.10	74.2	0.91
Ni	35.4	18	27.92	3.19	0.11	27.92	3.19	0.11	26.9	1.04
PH	6.64	5.46	5.74	0.14	0.02	5.74	0.14	0.02	6.7	0.86

Table 6. Soil Pollution Risk Screening Values for Agricultural Land (Unit: mg/kg)

No.	Pollutant Item	Risk Screening Value				
			pH ≤ 5.5	5.5<pH ≤ 6.5	6.5<pH ≤ 7.5	pH>7.5
1	Cd	Paddy field	0.30	0.40	0.60	0.80
		else	0.30	0.30	0.30	0.60
2	Hg	Paddy field	0.50	0.50	0.60	1.0
		else	1.3	1.8	2.4	3.4
3	As	Paddy field	30	30	25	20
		else	40	40	30	25
4	Pb	Paddy field	80	100	140	240
		else	70	90	120	170
5	Cr	Paddy field	250	250	300	350
		else	150	150	200	250
6	Cu	Orchard land	150	150	200	200
		else	50	50	100	100
7	Ni		60	70	100	190
8	Zn		200	200	250	300

Table 7. Soil pollution risk control values of agricultural land (unit: mg/kg)

No.	Pollutant Item	Risk Screening Value			
		pH ≤ 5.5	5.5<pH ≤ 6.5	6.5<pH ≤ 7.5	pH>7.5
1	Cd	1.5	2.0	3.0	4.0
2	Hg	2.0	2.5	4.0	6.0
3	As	200	150	120	100
4	Pb	400	500	700	1000
5	Cr	800	850	1000	1300

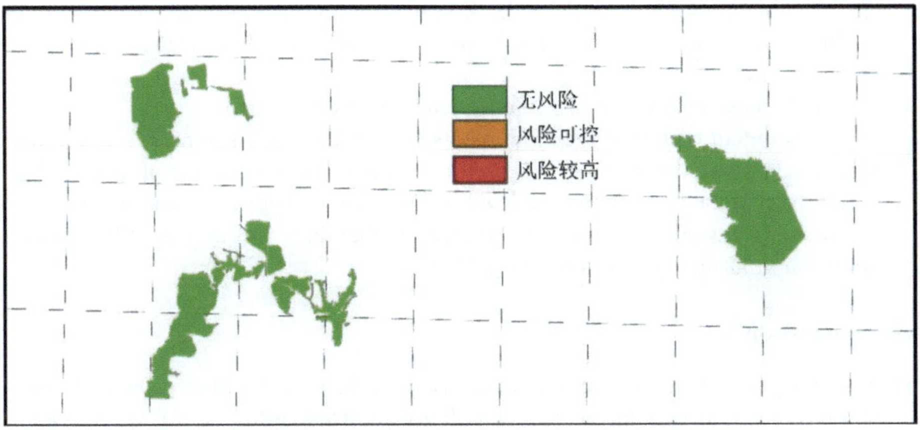

Fig. 4. Comprehensive grade of soil environment

4 Conclusions and Suggestions

4.1 Conclusion

(1) Single index evaluation of soil biological essential elements

According to the content of nitrogen in soil, farmland in Zhaoguang area can be divided into five grades: first class, second class, third class, fourth class and fifth class. According to the content of phosphorus in soil, the farmland in Zhaoguang area can be divided into four grades: first, second, third and fourth. According to the content of potassium in soil, the farmland in Zhaoguang area can be divided into first grade. According to the content of magnesium in soil, the farmland in Zhaoguang area can be divided into three grades: third grade, fourth grade and fifth grade. According to the content of calcium in soil, the farmland in Zhaoguang area can be divided into two grades, which are third grade and fourth grade [10, 11].

(2) Single index evaluation of soil trace elements

According to the content of iron in soil, farmland in Zhaoguang area can be divided into four grades: second class, third class, fourth class and fifth class. According to the content of manganese, the farmland in Zhaoguang area can be divided into five grades: first, second, third, fourth and fifth. According to the content of zinc in soil, the farmland in Zhaoguang area can be divided into five grades: first, second, third, fourth and fifth. According to the copper content in soil, the farmland in Zhaoguang area can be divided into three grades: third grade, fourth grade and fifth grade. According to the content of boron in soil, the farmland in Zhaoguang area can be divided into three grades: first, fourth and fifth. According to the content of molybdenum in soil, the farmland in Zhaoguang area can be divided into four grades: first class, second class, third class and fourth class [12, 13].

(3) Single index evaluation of soil health elements

According to the content of selenium in soil, the farmland in Zhaoguang area can be divided into three grades, which are deficiency, marginal and moderate. According to iodine content, farmland in Zhaoguang area can be divided into three grades, which are deficient, marginal and moderate. According to the fluorine content, the farmland in Zhaoguang area can be divided into three grades, which are deficient, marginal and high [14].

(4) Comprehensive evaluation of soil environmental geochemistry

The content of heavy metal elements in farmland soil samples in Zhaoguang area was compared with the risk screening value and control value of farmland soil pollution, and the comprehensive level of farmland soil environment in Zhaoguang area was concluded to be first-class, indicating that there was no risk of heavy metal pollution in farmland soil in this area [15].

4.2 Suggestions

(1) From the point of view of the content of essential elements for plant growth, the phosphorus content level in farmland soil in Zhaoguang area is low, so it is recommended to use phosphate fertilizer appropriately during the fertilization period.

(2) The content of trace health elements in farmland soil in Zhaoguang area is mostly in the deficient and marginal levels, and the lack of Se, I, F and other health elements in human body may cause endemic diseases. Therefore, it is recommended to use selenium fertilizer and iodine fertilizer in farmland lacking health elements.

References

1. Oyama, S., Oiwa, K., Nozawa, A.: Analysis of spatial distribution characteristics of facial skin temperature on stress coping. Artificial Life and Robotics **29**(2), 259–265 (2024). https://doi.org/10.1007/s10015-024-00942-x
2. Xuan, G., Qingwen, M., Wenjun, J.: Spatial distribution characteristics and differentiated management strategies of China nationally important agricultural heritage systems. J. Geog. Sci. **003**, 034 (2024)
3. Niu, Y., Wang, E., Li, Z., et al.: Study on electric potential response's spatial distribution characteristics to damaging localization evolution of gas? bearing coal under load. Bulletin of Engineering Geology and the Environment (1), 83 (2024)
4. Tian, H.-H., Xiao, T., Deng, M.M.: Temporal and spatial pattern analysis and susceptibility assessment of geological hazards in Hunan Province of China from 2015 to 2022. Stochastic Environmental Research and Risk Assessment **38**(4), 1453–1474 (2024)

5. Liu, J., Chen, Y., Shang, Y., et al.: Contamination Characteristics and Source Apportionment of Heavy Metal in the Topsoil of a Small Watershed in South Taihang (2024)
6. Spatial distribution of shallow landslides caused by Typhoon Lekima in 2019 in Zhejiang Province, China. Journal of Mountain Science **21**(5), 1564–1580 (2024). https://doi.org/10.1007/s11629-023-8377-y
7. Haixia, G., Shanqiang, F., Jiali, L.: Numerical and experimental study on the aerodynamic noise characteristics of 600-km/h high-speed maglev trains. Transportation Safety and Environment (2024). https://doi.org/10.1093/tse/tdae017
8. Zhang, Y., Niu, J., Wei, Z., et al.: Spatial distribution characteristics, source analysis and risk assessment of polycyclic aromatic hydrocarbons in topsoil of a typical chemical industry park. Front. Environ. Sci. (2023). https://doi.org/10.3389/fenvs.2023.1209137
9. Xiaxuan, H.E., Luxi, W., Minjia, Y.: Spatial distribution characteristics and influencing factors of traditional villages in northern Guangxi based on spatiotemporal big data and spatial syntax. Landsc. Res. **002**, 014 (2022)
10. Li, P., Shen, M., Perry, D.M., et al.: A comparative study on the spatial distribution characteristics and the driving factors of protected river systems between China and the United States of America. Ecological Indicators **135**, 108505 (2022). https://doi.org/10.1016/j.ecolind.2021.108505
11. Soejima, S.: Study on the spatial distribution characteristics of residents' consciousness for town development and urban infrastructure development. Proceedings of the City Planning Institute of Japan, Kansai Branch **20**, 93–96 (2022). https://doi.org/10.11361/cpijkansai.20.0_93
12. Xiao, J., Wang, B., Wei, R., et al.: Study on the geometry and spatial distribution characteristics of physical and mechanical indexes of soft clay. IOP Conference Series: Earth and Environmental Science (2021). https://doi.org/10.1088/1755-1315/861/4/042094
13. Shi, P., Jiao, Y., Diggle, P.J., et al.: Spatial distribution characteristics of stomata at the areole level in Michelia cavaleriei var. platypetala (Magnoliaceae). Annals of Botany (2021). https://doi.org/10.1093/aob/mcab106
14. Heyu, Z., Wenke, G., Zhipeng, L.I., et al.: Research on vegetation coverage and spatial distribution characteristics in Gobi region based on UAV image. Journal of Arid Land Resources and Environment (2020)
15. Fazhan, Z., Zhang, et al.: Spatial distribution characteristics of soil organic carbon in subtropical forests of mountain Lushan, China. Environmental monitoring and assessment (2018). https://doi.org/10.1007/s10661-018-6906-x

Study on the Structure Form and Parameter Selection of Multi-Soil-Layering

Jie Chen[1,2](✉), Jing Wang[1,2], Xiaojun Tang[1,2], and Shiming Bi[1,2]

[1] CCCC Fourth Harbor Engineering Institute Co., Ltd., Guangzhou 510230, China
83877340@qq.com

[2] Key Laboratory of Environment and Safety Technology of Transportation Infrastructure Engineering, CCCC, Guangzhou 510230, China

Abstract. Multi-Soil-Layering (MSL) is a kind of natural ecological and low-cost polluted water treatment technology, which shows the potential of removing many pollutants in polluted water with its unique structural design and treatment mechanism. This paper will evaluate the effects of different structural forms and operating parameters on the performance of MSL system, and deeply discuss the structural design and parameter optimization of MSL system. By comparing the effect of pollutant removal under different structures, the optimal MSL system configuration is determined to provide scientific basis and technical support for the actual water treatment project.

Keywords: Multilayer Percolation System · Structural Design · Parameter Optimization · System Configuration

1 Introduction

On the basis of the traditional soil percolation system, the Multi-Soil-Layering (MSL) system modulates the soil into a soil mixing module, which is alternately distributed through the soil mixing module and the water layer to set up multiple aerobic and anaerobic zones [1, 2]. The MSL system uses physical filtration, chemisorption and exchange, and biological decomposition to remove pollutants, especially organic matter and nitrogen and phosphorus [3–5]. A large number of practical applications have proved that MSL system can effectively remove pollutants in water bodies such as domestic sewage, river water and textile sewage [6, 7]. However, the structural design and parameter optimization of MSL system still need further research.

In this paper, the distance between the percolation module (SMBs) and the water passage layer (PLs), the specification of SMBs and the type and proportion of fillers in MSL system are determined by experiments, and the structural design and parameter optimization of MSL system are discussed. The optimal MSL system configuration is determined by comparing the contaminant removal effects under different structural forms.

D. Han et al. (Eds.): ICEPG 2024, ESE, pp. 729–735, 2025.
https://doi.org/10.1007/978-3-031-86829-0_81

2 Test of Packing Adsorption

2.1 The Purpose of the Test

Through the adsorption test of various fillers, the most suitable fillers were selected and applied in MSL system to improve the effect of nitrogen and phosphorus removal.

2.2 Test Method

Polyvinyl chloride plastic cylinder is used as the test device for the packed soil column test, as shown in Fig. 1. This structure not only successfully simulates the real operating environment of the MSL system, but also ensures the accuracy and reliability of the test results. In the test, a variety of fillers were selected, from left to right, vermiculite, biochar, iron carbon pellets, ceramic particles, red soil, blast furnace slag, wood chips, volcanic rock, rice husk, and the waste water was carried out in a top-down vertical flow.

Fig. 1. Test diagram of packed soil column

2.3 Results and Analysis of the Test

Laboratory configuration of simulated polluted water was used to evaluate the adsorption effect of different fillers. In water treatment engineering, chemical oxygen demand (COD), ammonia nitrogen (NH_4^+) and total phosphorus (TP) are important indicators to measure the pollution degree of polluted water. COD reflects the total amount of organic matter in water, while NH_4^+ and TP are the main nutrients that lead to eutrophication of water. Therefore, chemical oxygen demand (COD), ammonia nitrogen (NH_4^+) and total phosphorus (TP) were selected as the test analysis indexes.

As can be seen from Table 1, iron-carbon pellets, biochar, ceramite and volcanic rocks show excellent removal effects. The removal rates of COD and NH_4^+ of these fillers can reach more than 80%, which shows the strong organic matter and ammonia nitrogen adsorption capacity of these fillers. Moreover, these fillers were also able to remove more than 60% of TP, indicating that these fillers are equally effective in reducing phosphorus content in water bodies.

Table 1. Removal rates of COD, NH_4^+ and TP of different fillers

	COD (mg/L)			NH_4^+ (mg/L)			TP (mg/L)		
	Inlet	outlet	removal rate	Inlet	outlet	removal rate	Inlet	outlet	removal rate
vermiculite	301.73	101.45	66.38%	25.23	6.83	72.93%	5.1	2.84	44.31%
biochar	301.73	46.77	84.50%	25.23	3.88	84.62%	5.1	1.74	65.88%
iron carbon pellets	301.73	34.27	88.64%	25.23	3.74	85.18%	5.1	1.64	67.84%
ceramic particles	301.73	44.29	85.32%	25.23	4.01	84.11%	5.1	1.92	62.35%
red soil	301.73	164.65	45.43%	25.23	7.48	70.35%	5.1	2.54	50.20%
blast furnace slag	301.73	78.95	73.83%	25.23	10.2	59.57%	5.1	3.01	40.98%
wood chips	301.73	174.55	42.15%	25.23	13.24	47.52%	5.1	3.24	36.47%
volcanic rock	301.73	47.73	84.18%	25.23	4.27	83.08%	5.1	1.82	64.31%
rice husks	301.73	164.27	45.56%	25.23	10.21	59.53%	5.1	3.44	32.55%

In addition, the selection of fillers should also take into account cost-effectiveness and environmental impact. Low cost, easy to obtain and environmentally friendly fillers will be more conducive to promotion and application. In summary, by comparing the adsorption effects of different fillers on COD, NH_4^+ and TP in polluted water, this experiment selected red soil, biochar, iron carbon pellets, wood chips, rice husk, vermiculite and other fillers as the next step of the test.

3 Soil Column Adsorption Test

3.1 Purpose of the Test

In the process of research and development of MSL (multi-layer percolation) system, the selection of structural form and parameters is very important for the optimization of system performance. In order to evaluate the influence of different structural forms on the polluted water treatment effect, the soil column adsorption test as shown in Fig. 2 was carried out. The purpose of the experiment is to provide scientific basis for the design of MSL system.

3.2 Test Method

Three different structures were set up in the pre-test: soil column A, soil column B and soil column C.

The three types of soil columns are composed of SMBs (percolation module) and PLs (water passage layer). PLs (water layer) filler selection is made of zeolite, ceramic

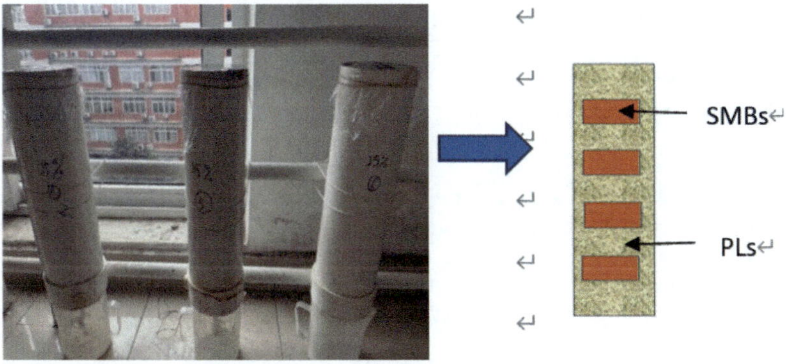

Soil column A Soil column B Soil column C.

Fig. 2. Soil column test for SMBs and PLs parameter selection

and volcanic rock mixture. SMBs (percolation module) is a square brick module made by mixing a variety of fillers selected by packing adsorption test according to a certain dry weight ratio.

The distance between SMBs (percolation module) and PLs (water vent layer) and the specifications of SMBs vary, as shown in Table 2. These structural forms are designed to explore optimal hydrodynamic conditions, the efficiency of pollutant adsorption and biodegradation, and the overall treatment capacity of the system.

Table 2. The table of MSL structure

Sort	Pacing of SMBs and PLs	Specifications of SMB (length: Width: height)
Soil column A	5 cm	20 cm × 10 cm × 5 cm
Soil column B	3 cm	20 cm × 10 cm × 5 cm
Soil column C	5 cm	15 cm × 10 cm × 3 cm

In the soil column adsorption test, a peristaltic pump is used to extract the simulated polluted supplied by the laboratory. The flow direction is set from top to bottom, which not only helps to simulate the flow state under natural conditions, but also facilitates the uniform distribution of pollutants in the filler layer and the full penetration of water flow. Polluted flows out through a hole in the bottom and is collected in a measuring cup placed below.

3.3 Test Results and Analysis

Through the detection of the concentration of key pollutants such as COD (chemical oxygen demand), TP (total phosphorus), TN (total nitrogen), NH_4^+ (ammonia nitrogen), NO_3^- (nitrate), NO_2^- (nitrite) in inlet and outlet water, the removal rate of various

Table 3. Pre-test results of soil columns with different structural forms

		Soil column A		Soil column B		Soil column C	
Index	Inlet (mg/L)	Outlet (mg/L)	Removal rate	Outlet (mg/L)	Removal rate	Outlet (mg/L)	Removal rate
COD	290–320	29.12–72.8	74.73%–89.06%	48.88–74.15	72.9%–81.6%	108.16–58.24	65.56%–81.11%
TP	4.9–5.1	1.7–2.03	64%–69%	1.62–2.28	51.7%–68.2%	1.92–2.115	47.67%–63.18%
NH_4^+	23–26	3.84–4.65	82.8%–86.92%	4.6–5.8	77.69%–84.32%	4.685–5.385	76.68%–82.67%
TN	25–30	9.23–14.9	49.45%–58.45%	8.3–14.08	49.01%–66.67%	8.7–15.63	45.21%–60.81%
NO_3^-	2.073–2.433	0.412–0.937	59.48%–81.24%	0.622–1.283	52.46%–67.32%	0.586–0.826	64.85%–80.12%
NO_2^-	0.031–0.04	0.006–0.012	58.24%–85%	0.008–0.011	70.27%77.76%	0.007–0.013	54.28%–82.5%

pollutants is calculated. Through the statistics and analysis of removal rate, the different structural forms of MSL systems were evaluated and compared.

As can be seen from Table 3, it can be seen from the data in the table that although the distance between SMBs and PLs and SMB specifications are changed, the removal effect is almost no big difference. However, when the structural form of SMBs and PLs is 5 cm apart, and the specification of SMB is 20 cm × 10 cm × 5 cm (length × width × height), the removal effect of the system on COD, TP, TN, NH_4^+, NO_3^- and NO_2^- in sewage is better than others. Therefore, the structural form of soil pillar A is selected for subsequent tests.

4 SMBs Packing Reference Test

In order to determine the SMBs packing reference, the packing combination adsorption test was carried out on the basis of the test conclusions in 2.3.

Through comprehensive and comparative analysis of the adsorption capacity, cost-effectiveness and promotion of the fillers on microbial activities, six fillers, red soil, biochar, iron carbon pellets, wood chips, rice husk and vermiculite, were selected in the test and mixed according to different dry weight ratios, as shown in Table 4, and made into a square brick module.

Table 4. Dry weight ratio of SMBs filler

Sort	Dry weight ratio of red soil, biochar, iron carbon pellets, wood chips, rice husk, vermiculite
Plan A	70: 5: 5: 5: 10: 5
Plan B	60: 5: 5: 15: 10: 5
Plan C	50: 5: 5: 20: 15: 5

According to three different ratio schemes, soil columns were prepared and adsorption tests were carried out, as shown in Table 5.

As can be seen from Table 5, after the structural form is determined through the structure selection test, the dry weight ratio of different SMB fillers has certain differences in the removal effects of COD, TP, TN, NH_4^+, NO_3^- and NO_2^- of MSL systems.

Table 5. Test results of dry weight ratio of different SMB fillers

		Soil column A		Soil column B		Soil column C	
Index	Inlet (mg/L)	Outlet (mg/L)	Removal rate	Outlet (mg/L)	Removal rate	Outlet (mg/L)	Removal rate
COD	290–320	29.12–72.8	74.73%–89.06%	30.88–74.23	65.8%–73.6%	31.38–68.53	63.56%–72.11%
TP	4.9–5.1	1.7–2.03	64%–69%	1.72–2.32	61.7%–65.2%	1.72–2.45	62.67%–66.18%
NH_4^+	23–26	3.84–4.65	82.8%–86.92%	4.0–4.8	78.69%–83.32%	4.22–4.95	72.68%–80.67%
TN	25–30	9.23–14.9	49.45%–64.45%	9.6–14.36	46.56%–63.47%	9.5–15.42	45.21%–64.81%
NO_3^-	2.073–2.433	0.412–0.937	59.48%–81.24%	0.425–1.043	57.46%–77.32%	0.456–0.928	57.85%–80.12%
NO_2^-	0.031–0.04	0.006–0.012	58.24%–85%	0.007–0.011	60.27%–83.76%	0.007–0.013	57.28%–83.5%

Plan A is superior to plan B and plan C, so plan A is chosen as the dry weight ratio of SMBs filler, that is, the dry weight ratio of red soil, biochar, iron carbon pellets, wood chips, rice husk and vermiculite is 70:5:5:5:10:5.

5 Conclusion

In this paper, through the adsorption test of different fillers, combined with the economy and universal availability of local materials, red soil, biochar, iron carbon pellets, wood chips, rice husk and vermiculite were selected as fillers for SMBs module. Through the soil column adsorption test, the optimal MSL system structure is obtained, that is, the distance between SMBs and PLs is 5 cm, and the size of SMB is 20 cm × 10 cm × 5 cm (length × width × height). The optimum dry weight ratio of red soil, biochar, iron carbon pellets, wood chips, rice husk and vermiculite was 70:5:5:10:5 by carrying out SMBs filler reference test.

Through this series of tests, we not only determined the optimal MSL system structure form, but also evaluated the effect of different filling ratios on the system performance. These results provide important reference for the design and optimization of MSL system.

In future studies, we will further explore the impact of different environmental conditions, characteristics of polluted water, and operating parameters on the performance of MSL systems, so as to achieve more efficient, economical, and sustainable polluted water treatment technologies.

References

1. Hong, Y., Huang, G., An, C., et al.: Enhanced nitrogen removal in the treatment of rural domestic sewage using vertical-flow multi-soil-layering systems: experimental and modeling insights. J. Environ. Manage. **2**(4), 273–284 (2019)
2. Lamzouri, K., Mahi, M., Ouatar, S., et al.: Application of multi-soil-layering technique for wastewater treatment in Moroccan rural areas: study of the operation process for an engineering design. J. Mat. Environ. Sci. **7**(2), 579–585 (2016)
3. Li, D., Wang, Z., Sheng, J., et al.: Nitrogen removal performances and mechanisms in a two-stage soil infiltration system by shunt distributing wastewater. Chinese Environmental Science **39**(2), 666–673 (2019)

4. Sbahi, S., Ouazzani, N., Hejjaj, A., et al.: Neural network and cubist algorithms to predict fecal coliform content in treated wastewater by multi-soil-layering system for potential reuse. J. Environ. Qual. **50**(1), 144–157 (2021)
5. Sbahi, S., Ouazzani, N., Latrach, L., et al.: Predicting the concentration of total coliforms in treatedrural domestic wastewater by multi-soil-layering (MSL) technology using artificial neural networks. Ecotoxicol. Environ. Saf. **20**(4), 111–118 (2020)
6. Zhu, L., Zhang, J., Mao, H.: Study on purification effect of a multi-layer composite infiltration system on initial rainwater runoff. Environ. Pollution and Prevention **45**(4), 506–510 (2023)
7. Kammoun, A., Ouazzani, N., El Fels, A.E.A., et al.: Enhancing pollutant removal efficiency in urban domestic wastewater treatment through the hybrid multi-soil-layering (MSL) system: A case study in Morocco. Water science and technology **89**(10), 2685–2702 (2024)

Evaluation of Geological Hazard Susceptibility Based on Analytic Hierarchy Process and Geographic Information System

Zhengzheng Mao(✉), Qiang Li, Hanwen Yang, Shouwan Zhong, and Yihua Zhang

Xi'an Mineral Resources Survey Center, China Geological Survey, Xi'an 710100, China
942124755@qq.com

Abstract. Based on the Analytic Hierarchy Process, six evaluation indicators, including stratum lithology, geological structure, rainfall, terrain slope, NDVI (Normalized Difference Vegetation Index), and human engineering activities, are selected to construct the geological disaster susceptibility evaluation index system in Cangshang Town, Baihe County. Then, the evaluation and zoning of geological disaster susceptibility are carried out.The results show that the extremely high - and high - susceptibility areas in geological disasters account for about 32.02%. They are mainly distributed in a band - like pattern along Leng - Hou Road, Lengshui River and its tributary gullies, and around human - populated areas on the slopes on both sides of Lengshui River, Peijia River and their tributary gullies. The medium - and low - susceptibility areas account for about 67.98%, mainly distributed in mountain valleys, areas with high vegetation coverage and less engineering activities. The evaluation results can provide references for geological disaster zoning and prevention in Cangshang Town, Baihe County.

Keywords: geological disasters · analytic hierarchy process · natural breaks method · susceptibility evaluation · disaster prevention

1 Introduction

Geological disasters are natural calamities that result from geological factors and severely impact human life, property, and the environment. China's unique tectonic environment, influenced by both the Pacific and Indian Ocean plates, leads to active block tectonic movements. Additionally, China spans multiple climatic zones, with significant temporal and spatial variations in rainfall, making it one of the countries most prone to geological disasters worldwide. According to the latest National Statistical Yearbook (2023), China experienced over 5,600 geological disasters in 2022, with landslides being the most frequent (over 3,900 cases), followed by collapses (over 1,300 cases), debris flows over 200 cases, ground collapses over 150 cases, ground subsidence 15 cases, and ground fissures 4 cases. These varying types and scales of geological disasters resulted in the deaths of over 90 people, 16 missing, 34 injured, and direct economic losses exceeding 1.5 billion yuan. It is evident that the geological disaster situation in China is quite severe.

D. Han et al. (Eds.): ICEPG 2024, ESE, pp. 736–748, 2025.
https://doi.org/10.1007/978-3-031-86829-0_82

Many scholars worldwide have conducted extensive in - depth research on geological disasters, such as susceptibility, hazard, and risk assessment, and have obtained numerous results.For example, Qian Lingling et al., [1] combined the fuzzy mathematics method with the Analytic Hierarchy Process to zone the geological disaster susceptibility in Ruili City; He et al., [2] used this method to assess landslide susceptibility in southern Anhui, China; Abay et al., [3] studied landslide susceptibility in Ethiopia's Tarmaber area based on the AHP method; Chen Fei et al., [4] selected eight indicators including rainfall to build a geological disaster hazard assessment index system and established a hierarchical structure model for geological disaster hazard assessment in Xiushui County, Jiujiang City, Jiangxi, using the Analytic Hierarchy Process; With the support of GIS, Sonker et al., [5] drew a landslide susceptibility map of the Sikkim - Himalayan region in India by using the Analytic Hierarchy Process (AHP); Zhou et al., [6] evaluated the landslide susceptibility of a photovoltaic farm in Yunxian County, western China, using this method; Wei Zupeng et al., [7] selected multiple factors influencing geological disaster development with the Analytic Hierarchy Process and zoned the geological disaster susceptibility in Tonghai County; Panchal and Shrivastava [8] used the AHP method for landslide hazard assessment on the slope of National Highway 5 in India; El Jazouli et al., [9] mapped landslide susceptibility in the Oum Er Rbia high basin in Morocco using analytic hierarchy process; Yuan Sufeng [10] took expressway linear projects as an example and used the fuzzy comprehensive evaluation method to assess the geological disaster hazard of expressway construction projects; Das et al., [11] zoned landslide susceptibility in the Himalayan region of Darjeeling with this method, achieving results similar to the actual situation; Xia Yifeng et al., [12] carried out the geological hazard risk assessment of the Yizhao Expressway slope based on fuzzy mathematics and analytic hierarchy process, and the evaluation results were close to the actual situation of the slope; Kohno et al., [13], with GIS support, used analytic hierarchy process (AHP) to study the landslide susceptibility of the entire slope of the Japan archipelago; Zangmen et al., [14] used this method to perform landslide - prone zoning in the Bafoussam - Dschang area of Western Cameroon; Nguyen et al., [15] evaluated the impact of landslide susceptibility in the mountainous areas of central Viet Nam under rainfall factors using this method.

Based on the research results of the above - mentioned scholars, this paper uses the Analytic Hierarchy Process under the support of the spatial analysis function of GIS to zone the geological disaster susceptibility in Cangshang Town, Baihe County, hoping to provide references for the zoning and prevention of geological disasters in Cangshang Town, Baihe County.

2 General Situation of the Research Area

Cangshang Town in Baihe County is located in the Qinba Mountains. The altitude within its territory ranges from 261 m to 1,148 m, and the total area is 107.29 square kilometers. It has jurisdiction over 1 urban community, 1 new - type rural community, 10 villages, 100 villagers' groups, and 9 community grids. The market town is surrounded by mountains on all sides, close to the mountains and by the water. The town is as beautiful as the mountains and has long been known as having the beauty of "Three Stars Surrounding the Moon". The landform is mainly low - mountain areas, accounting for 94.5% of the area.

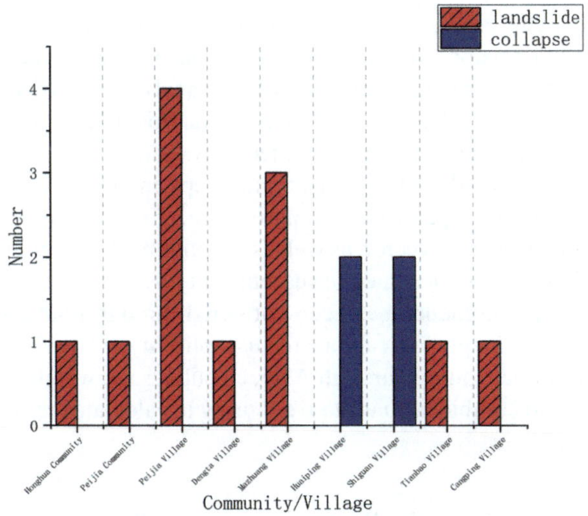

Fig. 1. Geological Hazard Statistics in Cangshang Town

Cangshang Town is located in the humid continental monsoon climate zone of the northern subtropics, with a mild climate and abundant rainfall. Due to the large topographic drop, the vertical change in temperature is significant, and it has the climate characteristics of "warm spring, summer drought, and autumn rainy days". The total precipitation amount exhibits a gradually diminishing tendency from the southeast towards the northwest, and the rainfall is between 750–800 mm. The water system in this area is mainly subordinate to the Lengshui River Basin. The Lengshui River originates from the Xiwuling Mountain in the county, flows through six townships including Xiying and Lengshui, and then joins the Han River at the Lengshui River estuary. The river is 65.5 km long, with a drainage area of 475 km^2, accounting for 32.6% of the whole area of the county. The average gradient is 7.84‰, mainly lateral erosion, and the erosion modulus is 2642t/y·km^2. There are currently 17 registered potential geological disaster sites inside the study area, including 13 landslides and 4 collapses. For the geological disaster statistics, see Fig. 1 [16].

Geological disasters in Cangshang Town are mainly distributed in the low - mountain areas spatially. Among the 17 geological disasters in this town, the largest number is distributed in the residential areas along the Leng - Hou Road, accounting for 70% of the total number in the whole town. The areas along rural roads and concentrated residential areas are also the places where geological disasters occur intensively. Temporally, the distribution of geological disasters has a concentrated pattern, occurring relatively intensively during periods of frequent human activities and the rainy season.

3 Research Approach

3.1 Brief Introduction to Analytic Hierarchy Process

The Analytic Hierarchy Process (AHP) is a decision - making approach. It decomposes elements relevant to decision - making into various levels, like objectives, criteria, and schemes. Subsequently, qualitative and quantitative analysis is carried out according to these levels. AHP has the advantages of clear hierarchical structure, flexibility, wide applicability, combining subjectivity and objectivity, and visualization, so it has been widely used in various fields. The fundamental steps involve creating a hierarchical structure model, building a pairwise comparison matrix, computing and performing consistency tests on the weight vector, as well as calculating the combined weight vector.

In the evaluation of geological hazard susceptibility, AHP has been applied for a long time, and its subjective influence has been corrected through multiple expert scoring. This article establishes the geological hazard susceptibility evaluation index system for Cangshang Town, Baihe County, through AHP, calculates the weight coefficients of the evaluation factors, and combines quantitative scoring to calculate the susceptibility value of geological hazards.

3.2 Steps of Analytic Hierarchy Process

Establish the Hierarchical Structure Model

Establishing a hierarchical structure model is a prerequisite for applying the Analytic Hierarchy Process. The hierarchical structure analysis model is presented in Fig. 2.

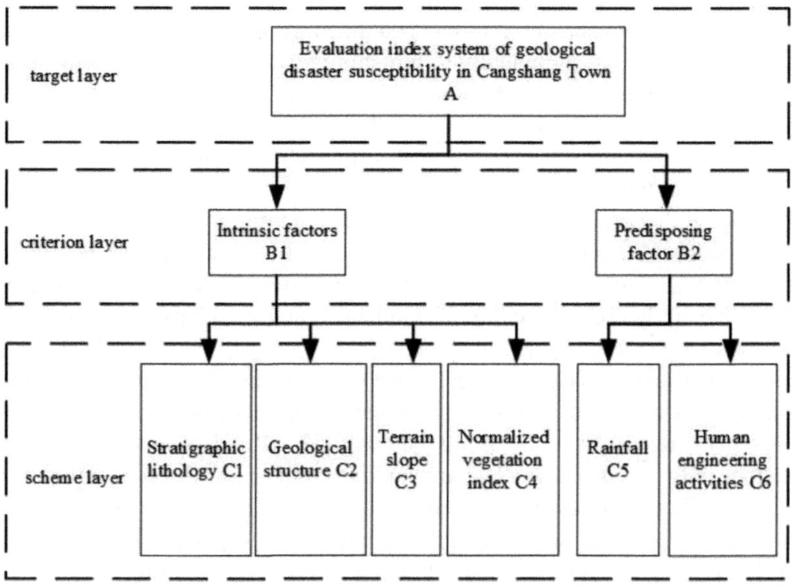

Fig. 2. Analysis Model of Hierarchical Structure

The target layer (A) is the constructed geological disaster susceptibility evaluation index system in Cangshang Town, Baihe County; the criterion layer (B) includes two aspects: internal factors and inducing factors; the scheme layer (C) consists of six indicators for this evaluation: stratum lithology, geological structure, terrain slope, NDVI, rainfall, and human engineering activities.

Construct the Judgment Matrix

One of the basic components of the AHP is to construct a judgment matrix, which is used to determine the relative importance relationships between different levels. The judgment matrix can provide decision - makers with a quantitative understanding of the relative importance between different levels in the hierarchical structure, and provide a basis for subsequent index weight calculation and comprehensive evaluation. The steps for constructing a judgment matrix are as follows:

(1) Determine the size of the judgment matrix: It depends on the number of different levels in the hierarchical structure. Since this article selects 6 evaluation indicators, the size of the judgment matrix is 6×6.
(2) Select an evaluation scale: According to the specific situation, select a suitable evaluation scale to describe the relative importance relationships. This article intends to select an evaluation scale of 1 - 9. The scale of the judgment matrix and its meanings are shown in Table 1 [17].

Table 1. Scale of Judgment Matrix (Importance Indicator) and Its Meanings

Serial Number	Meaning of Scale	Ratio
1	When the former element i is compared with the latter element j,i and j are of equal importance	$a_{ij} = 1$
2	When the former element i is compared with the latter element j, i is slightly more important than j	$a_{ij} = 3$
3	When the former element i is compared with the latter element j, i is significantly more important than j	$a_{ij} = 5$
4	When the former element i is compared with the latter element j, i is strongly more important than j	$a_{ij} = 7$
5	When the former element i is compared with the latter element j, i is absolutely more important than j	$a_{ij} = 9$
6	It indicates that the importance between element i and element j is in between the above judgments	$a_{ij} = 2,4,6,8$
7	If the relative importance scale between element i and element j is a_{ij}, then the relative importance scale between element j and i is $a_{ji} = 1/ a_{ij}$	reciprocal

(3) Conduct pairwise comparisons: For each level, pairwise comparisons are carried out. Decision - makers need to, based on their subjective judgments, use the selected

evaluation scales to give the comparison results of relative importance. The principle of comparison is to compare the contribution or influence of the two levels on the objective.

(4) Determine the judgment matrix: First, let 12 experts score, and then synthesize all the expert evaluation results to obtain the judgment matrices at each level. The synthesized evaluation results are then submitted to the 12 experts for review to determine the final evaluation results, as shown in Tables 2, 3, and 4.

Table 2. Judgment Matrix of A - B Layer

A	Internal factor B1	Inducing factor B2
Internal factor B1	1	3/4
Inducing factor B2	4/3	1

Table 3. Judgment Matrix of B1 - C Layer

A	Stratum lithology C1	Geological structure C2	VDVIC3	Topographic slope C4
Stratum lithology C1	1	2	4	3
Geological structure C2	1/2	1	3	2
VDVIC3	1/4	1/3	1	2/3
Topographic slope C4	1/3	1/2	3/2	1

Table 4. Judgment Matrix of B2 - C Layer

B2	Rainfall C5	Human engineering activities C6
Rainfall C5	1	2
Human engineering activities C6	1/2	1

(5) Consistency test: Once the judgment matrix is constructed, a consistency test must be carried out to verify its rationality and reliability. The eigenvector and the maximum eigenvalue can be calculated to perform this consistency test. If the consistency ratio CR is lower than 0.1, the judgment matrix is considered acceptable; otherwise, it needs to be adjusted.

$$CR = \frac{CI}{RI} \tag{1}$$

$$CI = \frac{\lambda_{max} - n}{n - 1} \quad (2)$$

$$\lambda_{max} = \sum\nolimits_{i=1}^{n} \frac{[\overline{A}W]_i}{nW_i}. \quad (3)$$

In the formula, CR represents the consistency index; n stands for the order of the judgment matrix. RI is the average random consistency index, with its values presented in Table 5, and CI is the random consistency ratio. λmax refers to the maximum eigenvalue of the judgment matrix. Here, is the judgment integration matrix, W is the weight vector, and $[\overline{A}\ W]_i$ is the i - th component of the matrix $[\overline{A}\ W]$.

Table 5. Random Consistency Table (RI) of Analytic Hierarchy Process

Matrix order(n)	1	2	3	4	5	6	7	8	9	10
RI	0	0	0.52	0.89	1.12	1.26	1.36	1.41	1.46	1.49

Calculation of Combined Weights

Based on the constructed judgment matrix, compute the relative weights of evaluation factors at each level. The calculation formula is:

$$W_i = \frac{(\prod_{j=1}^{n} a_{ij})^{\frac{1}{n}}}{\sum_{i=1}^{n} (\prod_{j=1}^{n} a_{ij})^{\frac{1}{n}}}, i = 1, 2, 3.., n \quad (4)$$

Calculation steps:

(1) Multiply the elements of each row of the matrix to obtain a new vector;

$$m_i = \prod_{j=1}^{n} a_{ij} \quad (5)$$

(2) Take the n - th root of each component of the obtained new vector;

$$x_i = \sqrt[n]{m_i} \quad (6)$$

(3) Perform standardized quantitative processing on the obtained vector, and form a matrix in sequence with the obtained vector and then carry out normalization processing.

$$w_i = x_i / \sum\nolimits_{i=1}^{n} x_i \quad (7)$$

The obtained w1, w2… wn are the weight coefficients of each evaluation index.

In this article's constructed hierarchical structure model, to obtain the weights of each index factor in layer C with respect to the overall goal, it is required to calculate the combined weight vector from the scheme layer (layer C) to the goal (layer A). Subsequently, a combined consistency test should be carried out according to the formula. If the test result satisfies the requirements, the combined weight vector can serve as a reference model for decision - making; otherwise, if the test fails, the judgment matrix needs to be readjusted. After normalizing the weight values calculated for criteria layers B1 and B2 and then conducting normalization processing under layer A, the weight values of the six indicators are obtained [18]. Table 6 shows the weights of each evaluation index.

Table 6. Results of combined weight calculation

Evaluation indicators	stratum lithology	geological structure	NDVI	terrain slope	rainfall	human engineering activities
Weight value	0.2009	0.1195	0.0441	0.0642	0.3810	0.1904

Substitute the maximum eigenvalues obtained from the three matrices into CI for inspection, and the consistency inspection is passed.

3.3 Quantitative Scoring by Analytic Hierarchy Process

The geological disaster susceptibility value can be comprehensively calculated based on the weight values and scores of the six factors. The process is as follows: Score each factor according to certain criteria, and then multiply the assigned value of each factor by its corresponding weight to calculate the geological disaster susceptibility value.

$$R = \sum RiWi \tag{8}$$

In the formula: Wi is the weight of each index factor, Ri is the weight assignment of each index factor, and R is the geological disaster susceptibility index. The scoring criteria are determined according to the size of the weight coefficients of each index factor. The quantitative scoring results are shown in Table 7.

Table 7. Weight Summary Table

Objective layer	criterion layer	relative weight	scheme layer	relative weight	combined weight
Susceptibility evaluation	Internal factors	0.4286	Stratum lithology	0.4687	0.2009
			Geological structure	0.2787	0.1195
			Terrain slope	0.1028	0.0441
			VDVI	0.1497	0.0642
	Precipitating factor	0.5714	Rainfall intensity	0.6667	0.3810
			human engineering activities	0.3333	0.1904

The quantitative scoring criteria for the six evaluation factors are as follows [19]:

(1) Scoring based on weights: First, sort the six evaluation factors in ascending order according to their weights. The factor with the smallest weight is scored as 1 point, the factor with the second - smallest weight is scored as 2 points, and so on until the factor with the largest weight is scored as 6 points.
(2) Scoring based on factor classification: If the measured value of a factor is at the lowest level, it is uniformly scored as 1 point; if it is at the next level up, it is scored according to its ranking position in terms of weight among the six factors and then added to this position score; if it is at an even higher level, the score is increased in an arithmetic progression.

According to the quantitative scoring results (Table 8), among the six selected evaluation indicators, rainfall is the most important factor inducing geological disasters. Among them, torrential rain has the largest susceptibility index and the highest scoring level; the secondary influencing factor is stratum lithology; the weight of NDVI is the smallest, and the corresponding susceptibility index is also the lowest.

Table 8. Quantitative scoring results of analytic hierarchy process

Evaluation factor	Weight	Factor classification	Factor assignment	Susceptibility index
stratum lithology	0.2009	better lithology	1	0.2009
		average lithology	5	1.0045
		poorer lithology	10	2.0090
geological structure	0.1195	(0, 200]	12	1.4340
		(200, 400]	9	1.0755
		(400, 600]	6	0.7170
		(600, 800]	3	0.3585
		(800, +∞]	1	0.1195
Terrain slope	0.0441	(0, 15°]	1	0.0441
		(15, 30°]	2	0.0882
		(30, 45°]	3	0.1323
		(45, 60°]	4	0.1764
		(60, 75°]	5	0.2205
		(75, 90°]	6	0.2646
VDVI	0.0642	[–0.1, –0.6]	1	0.0642
		(–0.6, –0.2]	2	0.1284
		(–0.2, 0.2]	4	0.2568
		(0.2, 0.6]	6	0.3852
		(0.6, 1.0]	8	0.5136
rainfall intensity	0.3810	light rain	1	0.3810
		moderate rain	6	2.2860
		heavy rain	12	4.5720
		torrential rain	18	6.8580
human engineering activities	0.1904	not severe	1	0.1904
		relatively severe	4	0.7616
		extremely severe	8	1.5232

3.4 Geological Disaster Susceptibility Zoning

According to the quantitative scoring results of the analytic hierarchy process, combined with the spatial analysis function of GIS [20], after processing the susceptibility indices of each grade of each evaluation index, import them, and use the natural break method to partition the geological disaster susceptibility [21]. The susceptibility in the study area is divided into four grades:High - susceptibility area: The susceptibility index range is [0.7, 1.0], accounting for about 17.29% of the total area. It is mainly distributed in a belt - like pattern along Leng - Hou Road, Lengshui River and its tributary gullies, and around human - inhabited areas on the slopes on both sides of Lengshui River, Peijia River and their tributary gullies; Medium - susceptibility area: The susceptibility index range is [0.45, 0.7), accounting for about 14.73% of the total area. It is mainly distributed in a strip - like pattern outside a certain distance around the high - susceptibility area of geological disasters, and some areas are connected in a planar pattern; Low - susceptibility area: The susceptibility index range is [0.2, 0.45), accounting for about 38.33% of the total area. It is mainly distributed in some mountain valleys, areas with high vegetation coverage and less engineering activities; Non - susceptibility area: The susceptibility index range is [0, 0.2), accounting for about 29.65% of the total area. See Fig. 3 for the schematic diagram of the geological disaster susceptibility zoning in Cangshang Town, Baihe County.

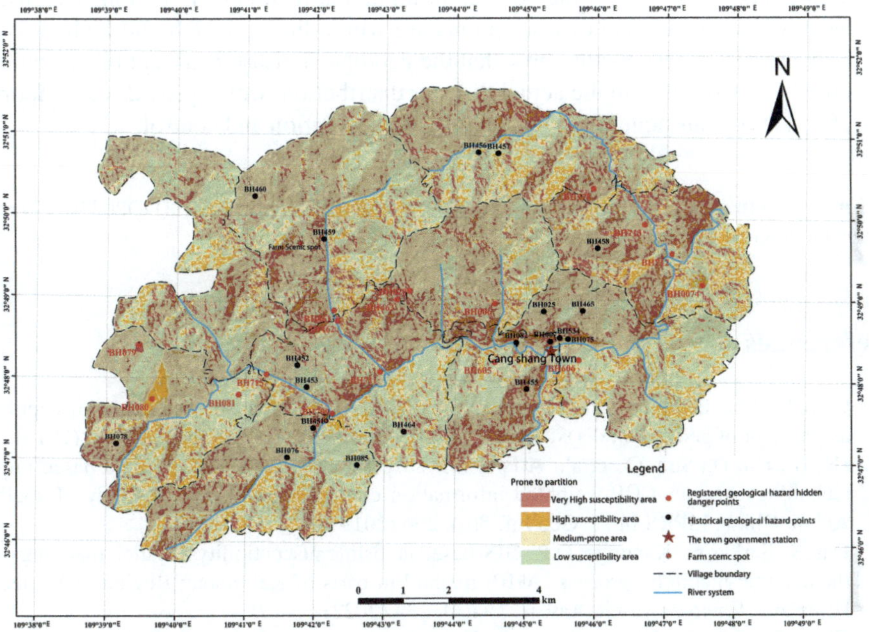

Fig. 3. Map of Geological Disaster Susceptibility Zoning

4 Conclusion

1. Based on the analytic hierarchy process, this paper establishes a geological disaster susceptibility evaluation index system for the study area in Cangshang Town, Baihe County, and determines the weights of each evaluation index (stratum lithology, geological structure, NDVI, topographic slope, rainfall, human engineering activities) as follows: 0.2009, 0.1195, 0.0441, 0.0642, 0.3810, 0.1904.
2. In combination with the spatial analysis function of GIS, the natural break method is employed to partition the geological disaster susceptibility within the study area. The study area is categorized into four grades, namely very high - susceptibility areas,

high - susceptibility areas, medium - susceptibility areas, and low - susceptibility areas.The low and medium susceptibility areas account for about 67.98% of the total area, mainly distributed in mountain valleys, areas with high vegetation coverage, and less engineering activities; the high and very high susceptibility areas have similar proportions, accounting for a total of 32.02%, mainly distributed in a belt-like pattern along Leng-Hou Road, Lengshui River and its tributary gullies, and around human-inhabited areas on the slopes on both sides of Lengshui River, Peijia River and their tributary gullies.

3. According to the evaluation results, the accuracy of geological disaster susceptibility area evaluation is tested and analyzed, using historical disaster statistics for verification. The results meet the following two indicators: (1) Over 90% of geological disaster points are within the very high, high, and medium susceptibility areas; (2) Over 60% of geological disaster points are within the very high and high susceptibility areas. The test results show that the geological disaster susceptibility zoning is roughly consistent with the actual disaster distribution, which provides an important reference for subsequent geological disaster prevention and control.

Acknowledgements. This work is funded by the geological survey project (DD20230522) of the China Geological Survey(CGS).

References

1. Qian, L., Xue, B., Fan, Z., et al.: Application of analytic hierarchy process in the susceptibility assessment of geological disasters in Ruili City. Mine Surveying **95**(04), 88–90 (2014)
2. He, H., Hu, D., Sun, Q., et al.: A landslide susceptibility assessment method based on GIS technology and an AHP-weighted information content method: A case study of southern Anhui, China. ISPRS Int. J. Geo Inf. **8**(6), 266 (2019)
3. Das, S., Sarkar, S., Kanungo, D.P.: GIS-based landslide susceptibility zonation mapping using the analytic hierarchy process (AHP) method in parts of Kalimpong Region of Darjeeling Himalaya. Environ. Monit. Assess. **194**(4), 234 (2022)
4. Chen, F., Guo, S., Xiong, R., et al.: Hazard assessment of geological disasters based on analytic hierarchy process. Nonferrous Met. Sci. Eng. **05**, 54–60 (2018)
5. Abay, A., Barbieri, G., Woldearegay, K.: GIS-based landslide susceptibility evaluation using analytical hierarchy process (AHP) approach: the case of Tarmaber District, Ethiopia. Momona Ethiop. J. Sci. **11**(1), 14–36 (2019)
6. Zhou, J., Tan, S., Li, J., et al.: Landslide susceptibility assessment using the analytic hierarchy process (AHP): a case study of a construction site for photovoltaic power generation in Yunxian County. Southwest China. Sustainability **15**(6), 5281 (2023)
7. Wei, Z., Sui, H., He, J., et al.: Application of analytic hierarchy process in geological disaster susceptibility zoning in Tonghai County. Soil Water Conserv. Chin. **05**, 28–31 (2019)
8. Sonker, I., Tripathi, J.N., Singh, A.K.: Landslide susceptibility zonation using geospatial technique and analytical hierarchy process in Sikkim Himalaya. Quat. Sci. Adv. **4**, 100039 (2021)
9. El Jazouli, A., Barakat, A., Khellouk, R.: GIS-multicriteria evaluation using AHP for landslide susceptibility mapping in Oum Er Rbia high basin (Morocco). Geoenvironmental Disasters **6**(1), 1–12 (2019)

10. Yuan, S.: Fuzzy comprehensive evaluation of the hazard of geological disasters in expressway construction projects. J. Catastrophol. **24**(02), 57–60 (2009)
11. Panchal, S., Shrivastava, A.K.: Landslide hazard assessment using analytic hierarchy process (AHP): a case study of National Highway 5 in India. Ain Shams Eng. J. **13**(3), 101626 (2022). https://doi.org/10.1016/j.asej.2021.10.021
12. Xia, Y., Yin, G., Lin, R., et al.: Slope geological disaster risk assessment method based on fuzzy comprehensive evaluation and its engineering application. Sci. Technol. Ind. **22**(10), 378–382 (2022)
13. Kohno, M., Higuchi, Y.: Landslide susceptibility assessment in the Japanese archipelago based on a landslide distribution map. ISPRS Int. J. Geo-Inf. **12**(2), 37 (2023). https://doi.org/10.3390/ijgi12020037
14. Zangmene, F.L., Ngapna, M.N., Ateba, M.C.B., et al.: Landslide susceptibility zonation using the analytical hierarchy process (AHP) in the Bafoussam-Dschang region (West Cameroon). Adv. Space Res. **71**(12), 5282–5301 (2023)
15. Nguyen, C.C., Vo, P., Doan, V.L., et al.: Assessment of the effects of rainfall frequency on landslide susceptibility mapping using AHP method: a case study for a mountainous region in Central Vietnam. Progress in Landslide Research and Technology, Volume 1 Issue 2, 2022. Cham: Springer International Publishing pp. 87–98 (2023)
16. Li, S.: Report on Geological Disaster Investigation and Zoning in Baihe County, Shaanxi Province: Northwest Geological Survey Institute of Nuclear Industry (2002)
17. Liu, W., Liu, S.: Construction of group decision - making judgment matrix in AHP. Syst. Eng. Electron. (2005)
18. Liu, Y., Deng, H., Xiong, Q.: Susceptibility assessment of slope geological disasters in Maoxian County based on analytic hierarchy process. J. Yangtze River Sci. Res. Inst. **34**(05), 31–35 (2017)
19. Tie, Y., Tang, C.: Application of AHP in single debris flow risk assessment. Chin. J. Geol. Hazard Control 79–84 (2006)
20. Zhang, Y., Hou, Y., Liu, W.: Application of analytic hierarchy process and GIS spatial analysis method in the susceptibility assessment of geological disasters in Lanzhou urban area. Gansu Geol. 18(04), 84–88 71 (2009)
21. Baocheng, S., Hao, Q., Yi, L., et al.: Geological disaster risk assessment in Tai'an City based on GIS analysis. Shandong Land Resour. **39**(06), 44–51 (2023)

Research on the Activation Method of Solid Waste from Oilfield Drilling

Jingang Zhao[1], Guanchao Wang[2], Yanbo Zhang[2], and Fang Liu[1(✉)]

[1] College of Chemistry and Chemical Engineering, China University of Petroleum, Qingdao 266580, China
liufangfw@upc.edu.cn

[2] Technology Inspection Center of Shengli Oilfield Branch, Dongying 257000, China

Abstract. Oil and natural gas are the main energy sources in the world today and the foundation industry that affects national economy and social development. As a major oil and gas consuming country, China's oil and gas extraction has been increasing in recent years, resulting in the generation of large amounts of water-based drilling mud. Water-based drilling mud has complex components and great harm, so the efficient resource utilization still faces huge challenges. The mud contains rich silicon and aluminum species, and transforming it into high-value ZSM-5 zeolite is a hot research topic. However, the silicon and aluminum components in the mud are usually inert, so the key to the resource utilization is to convert the inert natural minerals into highly reactive silicon and aluminum species. This study systematically investigated the activation methods such as acid-alkali activation at room temperature using water-based drilling mud as raw material. The results showed that the alkaline melting method could effectively activate the inert components in the mud, and the obtained precursors could successfully synthesize high-purity ZSM-5 zeolites with different Si/Al ratios (15, 25, 60, respectively).

Keywords: Water-based Drilling Mud · Acid and Alkali Activation · Alkali Melting Activation · ZSM-5 Zeolite

1 Introduction

Oil and natural gas play an important role in China's economic and social development. In the process of oil and gas exploration, drilling fluid is often used, which generates large amounts of drilling solid waste such as drilling cuttings and drilling mud that are difficult to effectively treat. These solid waste components are complex and contain oil, heavy metals, inorganic salts, and high molecular organic compounds, and if not treated or treated improperly, they can contaminate the soil and groundwater in the local area [1, 2]. Drilling mud utilization is an economical and environmentally friendly method of treating drilling mud and also meets the requirements of current ecological civilization construction. Drilling mud can be divided into oil-based drilling mud and water-based drilling mud according to its main components and the characteristics of the constituent substances [3]. Oil-based drilling mud is classified as hazardous waste, while water-based drilling mud is considered general industrial solid waste. At present, different

D. Han et al. (Eds.): ICEPG 2024, ESE, pp. 749–755, 2025.
https://doi.org/10.1007/978-3-031-86829-0_83

technologies have been adopted in the operation of drilling waste treatment, including solidification technology, solid-liquid separation technology, MTC technology, incineration technology, land cultivation, and biological remediation, in order to achieve the harmlessness, reduction, and resource utilization of drilling cuttings [4]. The products obtained after drilling mud treatment also need to be environmentally disposed of or efficiently utilized in resource recycling. Water based drilling mud is rich in silicon and aluminum elements, making it a good raw material for synthesizing zeolites because their composition is characterized by aluminosilicates and silicate phases, with SiO_2 and Al_2O_3 contents typically above 70%. However, SiO_2 and Al_2O_3 in water-based drilling mud mainly exist in the form of crystalline minerals such as quartz, mullite, and non crystalline mineral glass bodies. How to effectively dissolve silicon and aluminum from the mud is the key to activating and fully utilizing water-based drilling mud.

2 Materials and Method

2.1 Materials and Characterization

Water based drilling mud (solid waste from drilling in a certain oil field). The crystal phase structure of the synthesized sample was determined using a Bruker AXS D8 Avance X-ray diffractometer (XRD, Japanese Institute of Technology D/max-2400 X-ray diffractometer). The morphology of the synthesized sample was observed using a Hitachi S-4800 field emission environmental scanning electron microscope (SEM) produced by Hitachi, Japan. The chemical composition of the synthesized sample was determined using the Japanese Institute of Technology ZSX-100e 4580 X-ray fluorescence spectrometer (XRF).

2.2 Syntheses

Activation of Water-Based Drilling Mud Precursor
Acid activation: Place a solid-liquid ratio of 10 ml/g of mud precursor in a 2 mol/L hydrochloric acid solution, treat at 80 °C for 12 h, wash it to neutrality, and dry it. Alkali activation: Place 1 g of mud precursor in 10 ml, 1 mol/L, 2 mol/L, and 4 mol/L sodium hydroxide solutions, stir at 90 °C for 4 h, repeat three times, and finally wash it neutral and dry it. Alkali melting activation: Grind the mud and NaOH in a certain proportion evenly, transfer it to a 900 °C muffle furnace for high temperature treatment for 2 h, and remove and grind it for later use.

Synthesis of ZSM-5
After mixing the activated mud precursor, n-butylamine, seed crystals, and deionized water in a certain molar ratio, transfer them to a teflon reactor and hydrothermal crystallize at 170 °C for 48 h. After crystallization is complete, cool and filter off the mother liquor, wash the solid product to neutrality, dry at 100 °C, and finally place the solid product in a 550 °C muffle furnace for 4 h to obtain NaZSM-5 zeolite.

3 Results and Discussion

3.1 Activation of Water-Based Drilling Mud

Acid and Alkali Activated Mud Precursors

According to the chemical structure characteristics of the mud precursor, there are a large number of uncertain SiO_2 impurities and Fe impurities inside. The precursor is activated and modified with alkali to destroy the Si-O-Si and Si-O-Al structures of the crystalline SiO_2 in the mud, and to dissolve and remove impurities inside the mud. Acid modification changes the interaction forces between Al-O, Si-O-Si, and Si-O bonds in the mud, leading to expansion between crystal planes and promoting the dissolution of acidic substances [5].

Figure 1 shows the XRD spectra of the mud precursor and the products of acid and alkali activation treatment. It can be seen from the figure that both samples have obvious characteristic peaks of crystalline phase SiO_2 and Fe_2O_3 when compared with untreated mud precursor. However, compared to the precursor, the peak of SiO_2 impurity peak in the sample activated by acid and alkali has significantly decreased, but still exists clearly, indicating that this activation method cannot effectively activate the non oriented SiO_2 in the mud. And the characteristic peak of Fe_2O_3 in the sample activated by acid showed a slight decrease, while a large amount of Fe impurities remained in the sample.

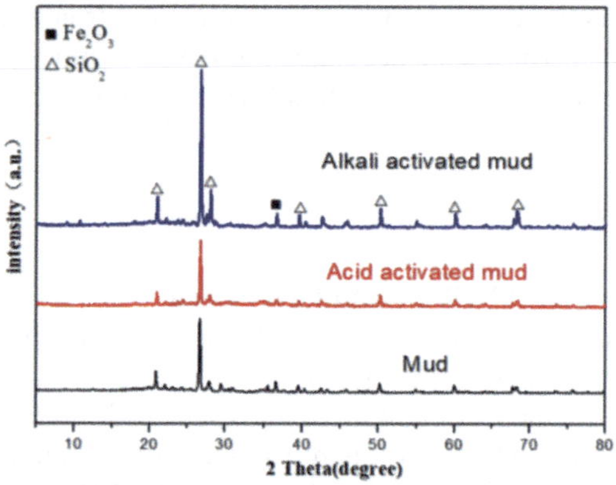

Fig. 1. XRD patterns of acid and alkali activated mud precursors

Alkali Melt Activated Mud Precursor

The essence of alkaline melting activation is to utilize the destructive effect of alkaline substances on silicon oxygen bonds and aluminum oxygen bonds to activate natural silicon aluminum minerals. Due to the action of chemical media, under certain conditions, high-temperature alkaline activation method can completely destroy the crystal phase structure of crystalline minerals and depolymerize them into low molecular weight aluminosilicates. Research has shown that high-temperature alkaline activation can more thoroughly and completely activate minerals [6].

Figure 2 shows the XRD spectra of the activated products from alkali melting of mud with different alkali contents. From Figure 2, it can be seen that when the mass ratio of sodium hydroxide to mud is less than 0.2, the characteristic peak of mud at 26.8 ° appears in the XRD spectrum, indicating incomplete activation. When the mass ratio is greater than 0.2, there are no obvious mud characteristic peaks in the XRD spectra of all samples.

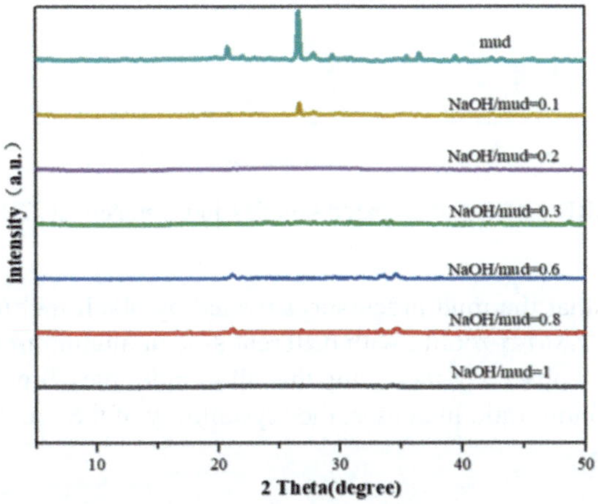

Fig. 2. XRD patterns of activated mud precursors with different alkaline concentrations

3.2 Characterization of Synthesized ZSM-5 Zeolite

Figure 3 shows the XRD patterns of ZSM-5 zeolite synthesized by crystallization of mud precursor after acid, alkali activation, and alkali melting activation treatment. It can be seen from the figure that all XRD patterns of the samples have typical diffraction peaks of MFI zeolite. ZSM-5 zeolite can be synthesized under this crystallization condition, but the amorphous SiO_2 impurity peak in the precursor is retained in both acid activated and alkali activated samples. This is because the steam generated at 170 °C during the crystallization process forms a large pressure under closed conditions. This pressure not only provides a driving force for the synthesis of zeolite, but also provides a crushing force for the SiO_2 impurity in the precursor, driving the amorphous SiO_2 to transform into a crystalline state. Due to the abundant presence of amorphous SiO_2 in the precursor, the pressure in the teflon reactor is limited and cannot break all SiO_2 impurities, resulting in the XRD spectrum showing that the synthesized ZSM-5 zeolite also retains the amorphous SiO_2 impurity peak in the precursor. However, no SiO_2 impurity peak was observed in the spectrum of the alkali melt activated sample, indicating that the product is high-purity ZSM-5 zeolite.

Figure 4 shows the XRD patterns of ZSM-5 zeolite with different silicon aluminum ratios prepared using mud precursors activated by alkali melting. From the figure, it can be seen that all XRD patterns of the samples have typical diffraction peaks of MFI type

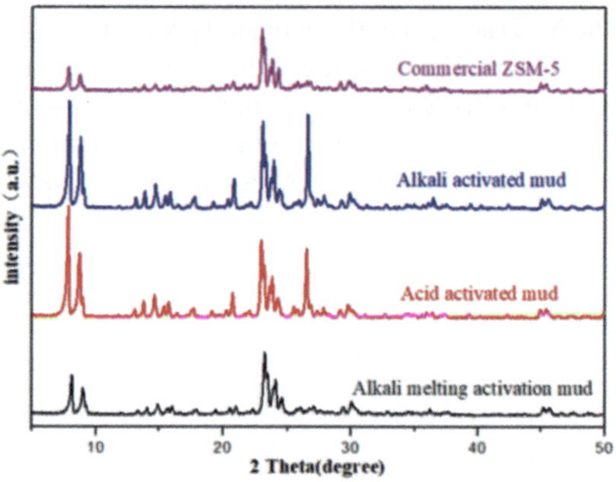

Fig. 3. XRD patterns of prepared samples and commercial ZSM-5 zeolite

zeolite, indicating that the mud precursor activated by alkali melting can successfully synthesize various ZSM-5 zeolite with different silicon aluminum ratios, and no other impurity peaks were observed, indicating that all samples have high purity. Meanwhile, as the silicon aluminum ratio increases, the crystallinity of the sample diffraction peaks gradually increases.

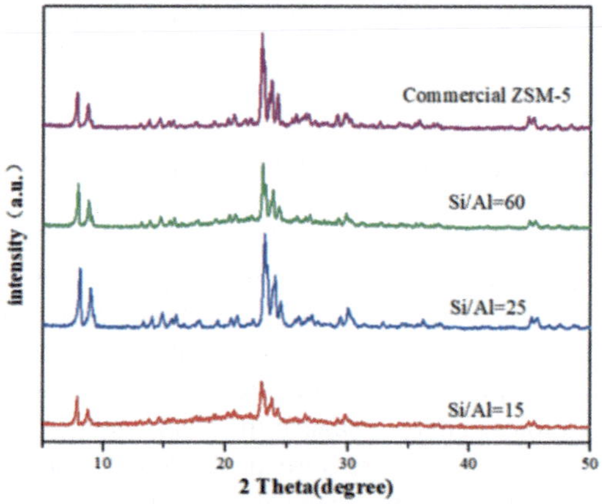

Fig. 4. XRD patterns of ZSM-5 zeolite with different silicon aluminum ratios prepared from mud precursors activated by alkali melting

Figure 5 shows SEM images of ZSM-5 zeolite with different silicon aluminum ratios prepared using mud precursors activated by alkali melting. Among them, A, B, and C are SEM images of ZSM-5 samples with silicon to aluminum ratios of 15, 25, and 60, respectively, and D is a SEM image of a commercial ZSM-5 zeolite. As can be seen from the figure, the morphology of all samples is not significantly different, with smooth surfaces, uniform particle size distribution, typical "coffin shape", and some agglomeration phenomena, which are consistent with the morphology characteristics of typical commercial ZSM-5 zeolite [7].

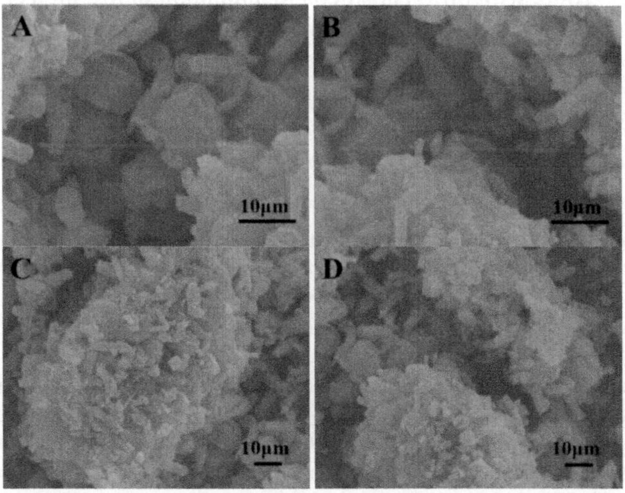

Fig. 5. SEM images of prepared samples and commercial ZSM-5 zeolite (A, B and C are ZSM-5 zeolite with silicon to aluminum ratios of 15, 25, and 60, respectively, B is a commercial ZSM-5)

Table 1 shows the chemical composition results of ZSM-5 zeolite with different silicon aluminum ratios prepared using mud precursors activated by alkali melting. As shown in the table, the main components of all synthesized ZSM-5 zeolite are SiO_2 and Al_2O_3. Through calculation, it can be seen that the bulk silicon aluminum molar ratios of the two are 15, 25, and 60, respectively, which are similar to the results obtained by XRF measurement.

Table 1. Chemical composition of synthesized samples

Sample	Component, wt%								Si/Al molar ratio
	Na_2O	Al_2O_3	SiO_2	MgO	K_2O	CaO	SO_3	Fe_2O_3	Bulk
Mud	5.5	13.4	41.8	2.8	0.5	9.0	1.5	4.7	5.3
Si/Al = 15	2.0	4.3	85.5	0.7	1.0	2.8	0.3	2.1	16.7
Si/Al = 25	1.2	2.3	91.8	0.5	0.4	1.6	0.1	1.3	33.0
Si/Al = 60	0.4	1.0	96.5	0.2	0.2	0.7	0.1	0.6	84

4 Conclusion

This article investigates the effects of different activation methods on the crystal structure of mud. ZSM-5 zeolite with different silicon to aluminum ratios, high crystallinity, and high purity can be directly synthesized from mud precursors activated by alkali melting. The optimal activation method for alkali melting activation is a sodium hydroxide to mud mass ratio of 0.2, activation temperature of 900 °C, and activation time of 4 h.

References

1. Liu, Y., Wang, Z., Chen, M., et al.: Potential of thermal desorption residue of oil-based drilling cuttings for use in subgrade materials. J. Environ. Eng. **149**, 04023061 (2023)
2. Yin, Q., Nie, H., Nie, M., et al.: Rapid effective treatment of waxy oily sludge using a method of dispersion combined with biodegradation in a semi-fluid state. Environ. Pollut. **319**, 120971 (2023)
3. Chen, H., Wang, X., Liang, H., et al.: Characterization and treatment of oily sludge: a systematic review. Environ. Pollut. **344**, 123245 (2024)
4. Siddioue, S., Yates, K., Matthews, K., et al.: Oil-based mud waste reclamation and utilisation in low-density polyethylene composites. Waste Manage. Res. **38**(12), 1331–44 (2020)
5. Shi, Y., Chen, Y., Zhang, Y., et al.: Low–temperature selective catalytic reduction of NO with NH_3 over fly ash loaded vanadium oxides. J. Northeast Dianli Univ. **35**(3), 59–63 (2015)
6. Belviso, C., Giannossa, L.C., Huertas, F.J., et al.: Synthesis of zeolites at low temperatures in fly ash-kaolinite mixtures. Micropor. Mesopor. Mat. **212**, 35–47 (2015)
7. Matthew, D., Jeffrey, D.: Synthesis of zeolites in the absence of organic structure-directing agents: factors governing crystal selection and polymorphism. Rev. Chem. Eng. **30**, 1–49 (2014)

Low Cost and Green Synthesis of ZSM-5 Zeolite Using Solid Waste of Silicon and Aluminum as Raw Materials

Guanchao Wang[1](✉), Zhaochuan Li[1], Yanbo Zhang[1], and Qingrun Meng[2](✉)

[1] Technology Inspection Center of Shengli Oilfield Branch, Dongying 257000, China
t-wangguanchao.slyt@sinopec.com

[2] Key Laboratory of Energy Chemical and Nano-Catalysis of Liaoning Province, School of Chemistry and Environmental Engineering, Liaoning University of Technology, Jinzhou 121000, Liaoning, China
meng081015@lnut.edu.cn

Abstract. With the development of the power industry, the amount of fly ash (FA) discharged from coal-fired power plants has increased year by year, becoming one of the largest industrial waste residues in China. FA has complex composition and great harmfulness, so the efficient resource utilization still faces huge challenges. FA contains rich silicon and aluminum species, and transforming it into high-value ZSM-5 zeolite is currently a hot research topic. However, the silicon and aluminum components in FA are usually inert, so the key to this strategy is to convert the inert natural minerals into silicon and aluminum species that can be extracted by acid/alkali solutions and have high reaction activity. In this study, FA was used as the raw material, and the systematic examination of the methods of acid activation at ambient temperature and sub-melting salt activation was conducted. Using these methods as raw materials, ZSM-5 zeolite was successfully synthesized. The results show that the sub-melting salt activation method can effectively activate the inert components and successfully synthesize ZSM-5 zeolite with a low Si/Al ratio (Si/Al = 40).

Keywords: Fly Ash · ZSM-5 Zeolite · Acid activation · Sub-molten salt activation

1 Introduction

In recent years, with the continuous advancement of China's industrialization process, industrial solid waste has been increasing year by year. As the main body of China's energy industry, power plants generate more than 4×107 tons of solid waste annually, which not only puts tremendous pressure on the surrounding environment but also leads to waste of resources. One of the largest volumes of waste produced is fly ash (FA), which is captured from the smoke of coal combustion [1]. If left untreated, large quantities of fly ash can cause dust pollution in the atmosphere, and if discharged into waterways, it can cause river sedimentation. The toxic chemicals contained in fly ash can also cause harm

D. Han et al. (Eds.): ICEPG 2024, ESE, pp. 756–762, 2025.
https://doi.org/10.1007/978-3-031-86829-0_84

to humans and living organisms. Zeolites have well-ordered pores, high crystallinity, diverse structures, and high specific surface areas, which make them widely used in catalysis, adsorption, and separation applications [2]. ZSM-5 zeolite has the characteristics of good thermal stability, acid resistance, hydrothermal stability, selective adsorption, and high adsorption efficiency, which can be used for the adsorption of phosphorus, ammonia nitrogen, heavy metal ions, and dyes in water. It occupies a pivotal position in the fields of petrochemical, fine chemical, environmental protection, etc. [3, 4] ZSM-5 zeolite is usually synthesized by the classical hydrothermal crystallization method using inorganic silicon-aluminum salt precursors such as water glass and aluminum sulfate in an alkaline environment. The inorganic silicon-aluminum salt precursors used in the synthesis of ZSM-5 zeolite are mainly obtained from natural minerals through complex reactions and separation processes, which is a typical high-energy, high-material, high-pollution emission process and therefore a high-cost, non-environmentally friendly process [5]. Using fly ash as part of the silicon source and the entire aluminum source not only properly disposes of the solid waste generated by thermal power generation, but also effectively alleviates the production cost and environmental pressure of zeolite materials, and further expands the raw material range of zeolite production, thereby developing a new low-cost, environmentally friendly technology route for zeolite material synthesis.

2 Materials and Method

2.1 Materials and Characterization

The FA contains 41.89% SiO_2 and 13.36% Al_2O_3.An X-ray diffraction analyzer (Bruker AXS D8 Avance) was used to identify crystal structure of FA and synthetic samples. Surface texture was observed using a scanning electron microscope (SEM, Hitachi S-4800). The chemical composition of the synthesized sample was determined using the X-ray fluorescence spectrometer (XRF, Japanese Institute of Technology ZSX-100e 4580). The surface areas were measured on a Micromeritics ASAP2020 instrument.

2.2 Syntheses

Activation of FA Precursor

Acid activation: Place the slurry precursor at a solid-to-liquid ratio of 10 ml/g into a 2 mol/L hydrochloric acid solution at 80 °C for 12 h, then wash it to neutrality and dry it. Sub-molten salt activation method: Mix NaOH and water in a certain mass ratio in a Teflon-lined container and dissolve it. Add 10 g of fly ash to the NaOH solution and mix it evenly. Heat the fly ash and the solution in a 240 °C oven for 2 h to allow the fly ash to react with the solution fully. Grind, wash, and dry the treated sample for later use.

Synthesis of ZSM-5

Mix the activated fly ash precursor, n-butylamine, seed, and deionized water in a certain mole ratio to form a homogeneous mixture. Transfer the mixture to a stainless steel crystallization autoclave lined with Teflon and heats it to 170 °C for static crystallization. After the crystallization is complete, cool, filter out the mother liquid, wash the solid product to neutrality, dry it at 100 °C, and finally bake it at 550 °C in a muffle furnace for 4 h to remove the template agent, thereby obtaining NaZSM-5 zeolite.

3 Results and Discussion

3.1 Different Activation Methods for FA

Acid Activation Treatment of FA

According to the chemical structural characteristics of FA precursor, it contains a large amount of SiO_2 impurity crystals and Fe impurities. By activating and modifying the precursor with acid, the Si-O-Si and Si-O-Al structures in the crystalline SiO_2 in the precursor are destroyed, and the impurities in the precursor are dissolved and removed [6].

The acid modification changes the Al-O, Si-O-Si, and Si-O bond interactions in the FA, causing the crystal planes to expand, increasing the surface porosity and specific surface area of the FA, and moderating the Si/Al ratio in the FA and removing iron, calcium, etc. metal impurities. Yu et al. showed that HCl can remove iron impurities and reduce the soluble calcium content in FA [7]. Figure 1 shows the XRD spectra of the FA precursor and the acid-activated product. From the Fig. 1, it can be seen that both samples have obvious crystalline SiO_2 and Fe_2O_3 characteristic peaks. However, compared with the precursor, the SiO_2 impurity crystal peak value of the acid-activated sample has a significant reduction, but it still exists, indicating that this activation method cannot effectively activate the slurry mud SiO_2 impurity crystals.

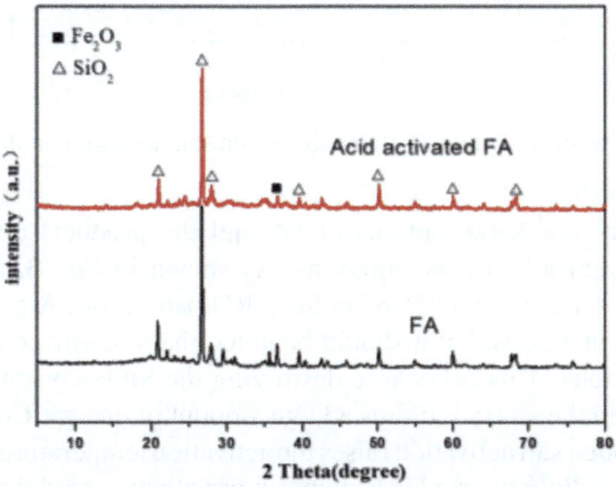

Fig. 1. XRD patterns of the FA and activated FA

Sub-Molten Salt Activation Method

Sub-molten salt is a non-conventional medium proposed by Professor Zhang Yi, with alkali metal salt concentration between molten salt and solution [8, 9]. Sub-molten salt not only can effectively destroy the crystal lattice structure of natural minerals, but also can completely disperse the layered silicon-aluminum species in natural minerals in high polymerization state into single polymerization state silicon-aluminum acid salt. Liu et al. found that the sub-molten salt activation method has good activation effects on minerals, i.e., it can completely disperse the silicon-aluminum species in natural minerals in high polymerization state into highly reactive single polymerization state silicon and aluminum acid salt species at lower activation temperature.

Figure 2 shows the XRD patterns of FA and the products after sub-molten salt activation at different temperatures. As shown in Fig. 2, the characteristic peaks at 20.6° and 26.8° in the original XRD pattern of FA at 180 °C are basically not visible in the XRD pattern of the activated product at the same temperature, indicating that the sub-molten salt activation method can completely destroy the crystal phase structure of the slurry at this temperature.

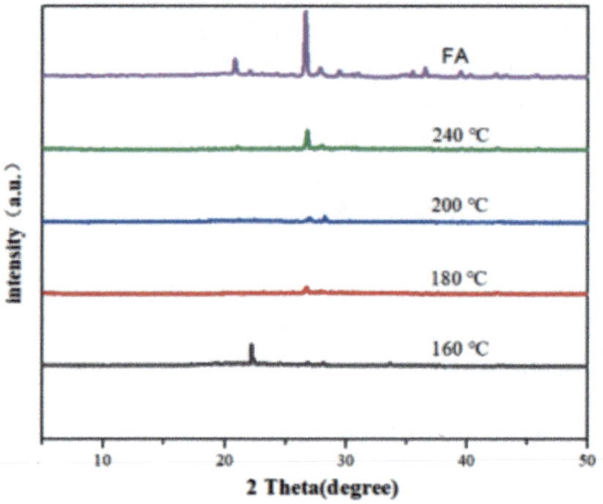

Fig. 2. XRD patterns of the products after sub-molten salt activation at different temperatures

Figure 3 shows the XRD patterns of FA and the products after sub-molten salt activation at different alkali concentrations. As shown in Fig. 3, the intensity of the characteristic peaks at 20.6° and 26.8° in the XRD pattern of FA gradually weakens as the alkali strength increases, but it should be noted that the diffraction peak at 26.8° is still relatively obvious. This is because destroying the SiO_2 crystal phase Si-O-Si and Si-O-Al structure in the slurry requires a large amount of energy. Compared with alkali activation, sub-molten salt activation raises the activation temperature in this basis. Under high temperature, sodium hydroxide becomes more alkaline, and the alkali provided by alkali activation at lower temperature is not enough to have sufficient alkalinity to break the covalent bond energy of amorphous SiO_2. When the mass ratio of FA to sodium hydroxide is 1, all impurity crystal characteristic peaks disappear, indicating that the alkaline submelting salt activation method can activate the slurry more thoroughly under this alkaline condition.

3.2 Characterization of Synthesized ZSM-5 Zeolite

Figure 4 shows the XRD pattern of the ZSM-5 zeolite synthesized from fly ash precursor after acid activation treatment. From the figure, it can be seen that the XRD pattern of the acid-activated sample has the typical diffraction peaks of the ZSM-5. This indicates that

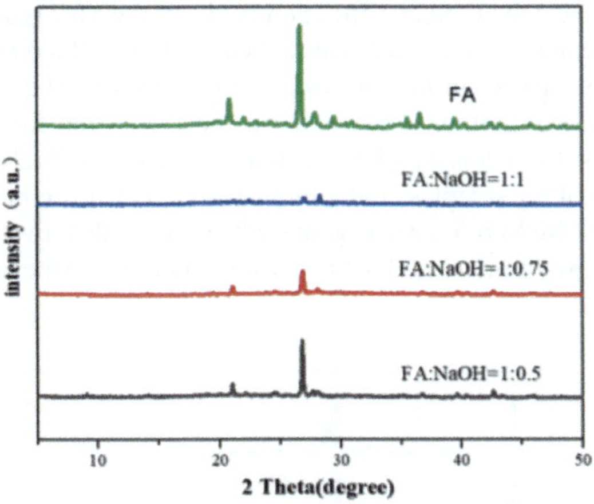

Fig. 3. XRD patterns of the products after sub-molten salt activation at different alkali concentrations

ZSM-5 zeolite can be synthesized under this crystallization condition, but the product retains the amorphous SiO_2 impurity peaks of the precursor because the steam generated at 170 °C in a closed condition forms a larger pressure. This pressure not only provides a driving force for the synthesis of zeolite, but also provides a crushing force for the amorphous SiO_2 impurities in the precursor, driving the amorphous SiO_2 to transform into crystalline state. Due to the large amount of amorphous SiO_2 in the precursor, the pressure in the Teflon-lined vessel is limited and cannot crush all the SiO_2 impurity peaks, so the XRD pattern of the acid-activated product also retains the amorphous SiO_2 impurity peaks of the precursor.

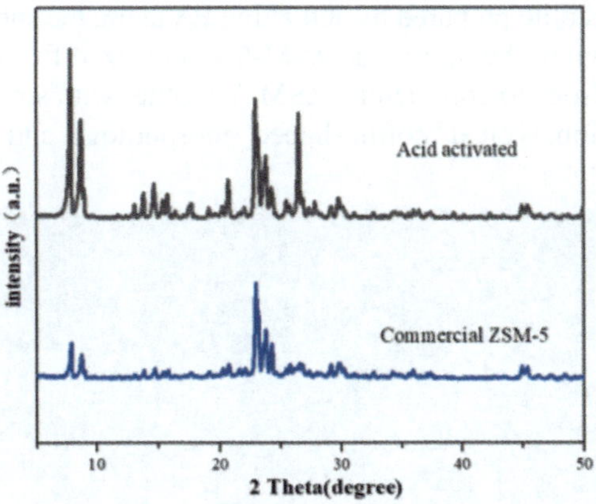

Fig. 4. XRD patterns of the sample prepared after acid activation

Figure 5 shows the XRD patterns of ZSM-5 zeolite synthesized by the sub-molten salt activation slurry method with different Si/Al ratios. From the figure, it can be seen that all the samples have the typical diffraction peaks of the MFI type zeolite, indicating that ZSM-5 zeolite can be synthesized within the range of the feed Si/Al ratio, and there are no other impurity peaks in the spectrum, indicating that the synthesized product is a high-purity ZSM-5 zeolite. It can also be seen from the figure that as the feed Si/Al ratio increases, the intensity of the diffraction peaks gradually increases and the relative crystallinity of the samples gradually increases. However, the crystallinity of the samples is low at low Si/Al ratio, because the sub-molten salt activation feed Si/Al ratio is lowered, the alkalinity is increased, which is not conducive to the growth of crystalline nuclei.

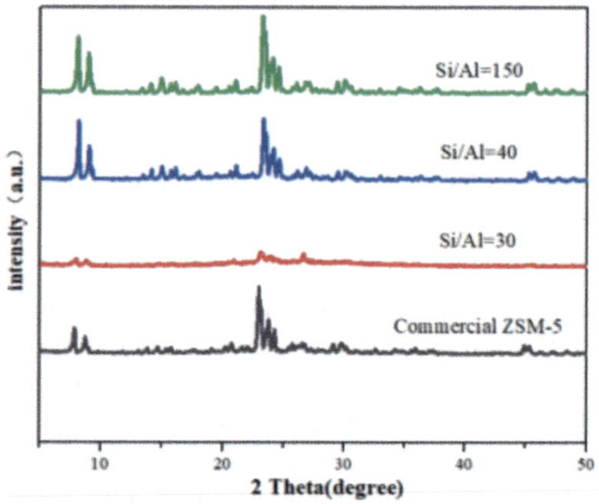

Fig. 5. XRD patterns of samples with different Si/Al ratios prepared after activation of sub-molten salt

Figure 6A and B show SEM images of ZSM-5 zeolite with Si/Al ratios of 40 and commercial ZSM-5 zeolite prepared by activating FA using the sub-molten salt method, respectively. As shown in the figure, the ZSM-5 synthesized from FA has similar morphological characteristics to commercial ZSM-5 zeolite, with smooth surface, uniform particle size distribution, typical "coffin shaped" morphology, and some agglomeration.

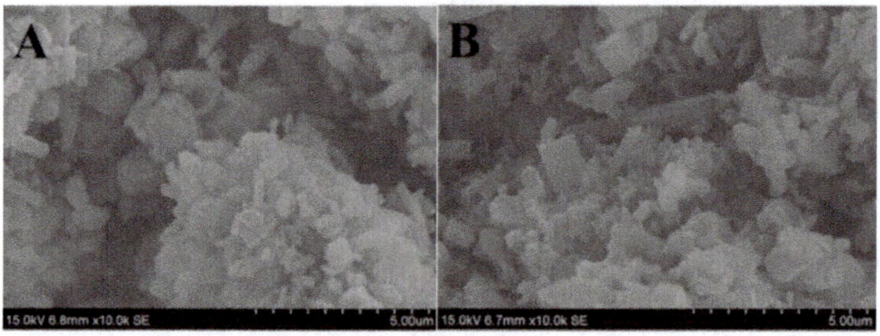

Fig. 6. SEM images of prepared samples and commercial ZSM-5 zeolite (A is a ZSM-5 sample with a Si/Al ratio of 40, B is a commercial ZSM-5)

4 Conclusion

This thesis uses FA as the raw material and activates it through sub-melting salt at a temperature of 200 °C. The main conclusions are as follows:

(1) The sub-molten salt activation method can activate most inert components, and the optimized optimal activation conditions are: a mass ratio of sodium hydroxide to FA of 1:1, and an activation temperature of 200 °C.
(2) Under these conditions, by adding part of the silicon source and using n-butylamine as the template agent, the ZSM-5 zeolite with a Si/Al ratio of 40 was successfully synthesized by seed-guided method. The XRD, SEM, and N_2 adsorption-desorption results show that the synthesized sample has a similar crystallinity, morphology, and pore structure to the commercial ZSM-5.

References

1. Jung-Hyun, K., et al.: An experimental study of chloride acceleration on the seawater resistance of fly ash antiwashout underwater concrete. J. Ocean Eng. Technol. **19**, 29–34 (2005)
2. Qi, G., et al.: Selective catalytic oxidation (SCO) of ammonia to nitrogen over Fe/ZSM-5 catalysts. Appl. Catal. A **287**, 25–33 (2005)
3. Corma, A.: From microporous to mesoporous molecular sieve materials and their use in catalysis. Chem. Rev. **97**, 2373–2420 (1997)
4. José, M.A., et al.: HZSM-5 zeolite as catalyst additive for residue cracking under FCC conditions. Energy Fuels **23**(9), 4215–4223 (2009). https://doi.org/10.1021/ef9002632
5. Irina, I., et al.: Micro–mesoporous materials obtained by zeolite recrystallization: synthesis, characterization and catalytic applications. Chem. Soc. Rev. **42**, 3671–3688 (2013)
6. Peng, P., et al.: Chemical modification of coal fly ash for the removal of phosphate from aqueous solution. Fuel **87**, 2469–2476 (2008)
7. Pizarro, J., et al.: Adsorption of Cu^{2+} on coal fly ash modified with functionalized mesoporous silica. Fuel **156**, 96–102 (2015)
8. Chen, H.D., et al.: Synthesis of cancrinite and zeolite W from K-feldspar by mixed sub-molten salt method. J. Synth. Cryst. **44**, 3158–3165 (2015)
9. Grant, S.H., et al.: The structure of silicate glasses and melts. Elements **2**, 269–73 (2006)

Distribution of Sediment Pollution and Analysis of Ecological Desilting Depth in Huizhou Jinshan Lake

Faming Huang(✉), Zongkai Ma, Cao Shen, Yulai Wang, Yuyang Liu, and Xu Xie

Chengdu Survey Design and Research Institute Co., Ltd., Chengdu 610000, China
1914753474@qq.com

Abstract. In order to study the relationship between sediment ecological desilting depth and pollutant content, the sediment in the Lake area of Huizhou was taken as the research object, the pollution risk of total nitrogen, total phosphorus and heavy metals in sediment and the effect of nitrogen and phosphorus adsorption-desorption equilibrium concentration on the dredging depth were studied. The results showed that the potential ecological harm index of heavy metals in the sediment of the lake area was at the low risk level. The comprehensive pollution index of TN and TP was in the heavy pollution level, with an average of 3.55 and 8.55, respectively. The average adsorption and apportionment equilibrium concentrations of NH_3-N and P in pollution layer, transition layer and normal lake mud layer are 1.41 mg/L, 1.14 mg/L and 0.86 mg/L, which have a great risk of pollution release. It was suggested that the pollution layer and transition layer should be dredged with an average dredging thickness of 54.41 cm.

Keywords: Huizhou · sediment · desilting thickness · pollution release · heavy pollution

1 Introduction

Due to the slow flow rate of urban river and lake water bodies, pollutants in the water continuously deposit at the bottom of the river, forming sediment. River and lake sediment was not only the central link of the material cycle, but also the enrichment site of nutrients, persistent organic pollutants, heavy metals, etc.. Pollutants in the sediment, under appropriate conditions (such as wind waves, waterways, river channel construction, etc.), become secondary pollution sources, with characteristics such as large release area, uncertain release time, pathways and release amounts [1, 2].

The water area of a lake in Huizhou was about 1.18km^2 with an average width of 150-250 m. The main pollutants were COD_{cr}, BOD_5, NH_4^+-N and TP, the comprehensive nutrient status index was 54.26–75.49, and the water quality was generally poor [3]. Combined with the development plan of Huizhou, comprehensive remediation of the lake water was carried out. Internal remediation was particularly critical. Chi Minghui [4] showed that the heavy metal pollution degree of sediment in the southwest gusher was Cd > Cu > Zn > Cr > Pb; The ecological hazards of heavy metals in the inland

D. Han et al. (Eds.): ICEPG 2024, ESE, pp. 763–769, 2025.
https://doi.org/10.1007/978-3-031-86829-0_85

waterway of Xiling Passage were Cd > Hg > Cu > Pb > As > Cr > Ni > Zn, and the ecological hazards of Cd were the strongest.

This paper takes the sediment of a lake in Huizhou as an example, through sampling and monitoring the sediment distribution sites in the lake area, adopts the organic index evaluation method and the potential ecological hazard index method to assess the ecological risk of the sediment, analyzed the equilibrium concentration of nitrogen and phosphorus release through adsorption-analytic test, and comprehensively determines the ecological dredging depth of the sediment, which effectively guides the dredging engineering design.

2 Materials and Methods

2.1 Distribution and Sampling

In order to better study the distribution of sediment pollution, 10 detection sections were set up at the sampling points in the lake district. Each section was provided with three detection points, one at the middle thread and one at each point 40 to 50 m away from the shoreline. A sample was taken from each point for detection.

The bottom mud in the lake area was sampled by a vessel equipped with an engineering drilling rig and a quick soil collection device. The samples were stored in polyethylene self-sealing bags.

2.2 Sample Determination and Data Processing

The samples were dried by removing gravel and plant residues and passed through a 100-mesh sieve. Acidity meter was used for pH, and atomic fluorescence spectrophotometer was used for As and Hg. Cr, Cu, Ni, Pb and Zn were plasma mass spectrometers, Cd was atomic absorption spectrometer, total nitrogen (TN) was determined by semi-micro Kelvin method, total phosphorus (TP) was determined by SMT total phosphorus method.

2.3 Sediment Pollution Risk Assessment and Adsorption Equilibrium Test

Comprehensive Pollution Index Evaluation Method

This paper intends to take the measured values of total nitrogen TN and total phosphorus TP of the sediment of Taihu Lake in 1960 in the Annals of Lakes of China as the evaluation standard values (background values) [5], and the formula for calculating the comprehensive pollution index evaluation method was as follows [6–8]:

$$\mathrm{S_i} = {}^{\mathrm{C_i}}/_{\mathrm{C_s}} \tag{1}$$

$$\mathrm{FF} = \sqrt{\frac{\mathrm{F}^2 + \mathrm{F}^2_{\max}}{2}} \tag{2}$$

Among them, S_i was a single evaluation index, and i represents TN and TP; If S_i was greater than 1, the pollutant content of the index sediment exceeds the standard value.

C_i was the measured value of evaluation index item i; Cs was the standard value of evaluation factor i, and the standard values for total nitrogen and total phosphorus were $C_{STN} = 0.067\%$ and $C_{STP} = 0.044\%$. F was the average value of the pollution index of n pollutants, and F_{max} was the largest single pollution index.

Potential Ecological Hazard Index Method
According to assessment of heavy metal pollution in sediment pollution assessment methods, the calculation formula of the potential ecological harm index method was as follows [9, 10]:

$$C_f^i = C_D^i / C_R^i \tag{3}$$

$$E_r^i = T_r^i \times C_f^i \tag{4}$$

$$RI = \sum_{i=1}^{n} E_r^i \tag{5}$$

where: C_f^i was the pollution coefficient of a single pollutant; C_D^i was the measured content of heavy metals in sediment. C_R^i was the parameter ratio required for calculation; Er^i was the potential ecological risk coefficient of a single pollutant. Toxicity response parameters of individual T_r^i pollutants; RI was the potential ecological risk index of many metals.

Adsorption-Analytical Thermodynamic Test
Weigh the drying sample of 0.5 g bottom mud and place it in a 100 mL centrifuge tube washed with hydrochloric acid. Add a certain proportion of ammonium chloride solution and potassium dihydrogen phosphate solution according to the water and soil ratio of 100:1, respectively. Oscillate at 25 °C for 2 h until adsorption equilibrium, and centrifuge (10000RPM, 10 min). The concentration of ammonia nitrogen and total phosphorus was determined after the supernatant was passed through 0.45 μm fiber filter membrane.

3 Results and Discussion

3.1 Sediment Thickness Distribution Characteristics

According to Fig. 1, there are certain differences in the thickness of the sediment pollution layer and transition layer in different sample sections. The pollution layer, transition layer and total thickness of the sediment in the lake area were 16.97–33.18 cm, 19.38–33.72 cm and 36.36–66.91 cm, respectively, and the average thickness is 28.18 cm, 26.22 cm and 54.41 cm, respectively.

The coefficients of variation were 5.00%–11.91%, 6.69%–15.08% and 5.00%–13.37%, respectively. The maximum and minimum values of pollution layer, transition layer and total thickness were 8# and 1#, 10# and 1#, 10# and 1#, respectively.

The polluted sediment was mainly distributed in the central and southwest regions of the lake district, which could be due to the construction of drainage pumping stations in the northeast of the lake district for water replenishment or drainage. The flow and velocity of the pumping stations vary greatly during operation, and the hydraulic erosion in this region was large, resulting in shallow distribution of sediment thickness in this region.

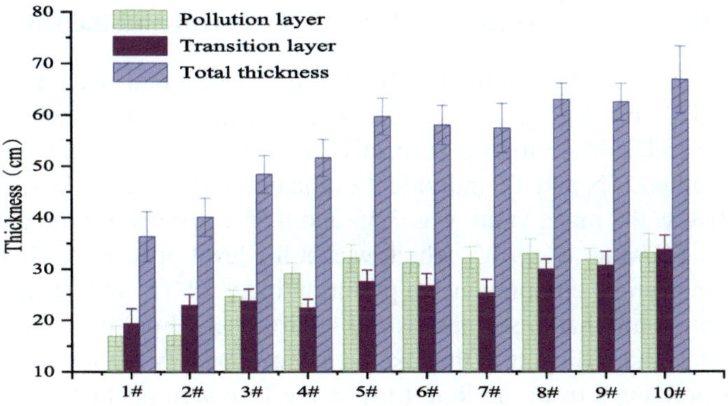

Fig. 1. Thickness map of contamination layer and transition layer in monitoring section

3.2 Distribution Characteristics of Heavy Metal Content and Pollution Evaluation

According to Table 1, The average content of Cr, Cd, As, Ni and Pb was lower than the background value. The average content of Cu and Zn were between 0.12 and 0.69, indicating that Cu and Zn are enriched, mainly in JSH-1 to 3#, but the enrichment degree was light. The over-standard rates of Cu and Zn were 57.14% and 71.42%, respectively, indicating that Cu and Zn exceeded the standard in most areas, and the over-standard rates of As and Ni were below 30%, and the overall pollution was small.

The risk of sediment contamination was assessed according to the potential ecological risk index method, the potential risk of individual heavy metals E_i And E_i Values between 0.13 and 7.51, and RI values between 12.34 and 17.15 are at the low risk level. It showed that the risk of heavy metal pollution in the lake sediment was low, basically no risk of heavy metal pollution, and the degree of heavy metal stress on the environment was small.

Table 1. Content of heavy metal in surface sediment in Lake

data type	Cu	Zn	Cd	Cr	As	Ni	Pb
Background value (mg / kg)	32	100	0.19	67	18	27.4	61
Maximum value (mg / kg)	293.75	162.55	0.00	45.60	28.75	43.50	42.40
Minimum value (mg / kg)	11.95	53.90	0.00	9.75	6.45	6.20	25.55
mean (mg/kg)	54.09	112.00	0.00	29.81	15.30	17.13	34.69
standard deviation SD	59.34	26.43	0.00	8.98	4.56	9.35	4.59
coefficient of variation (%)	109.72	23.60	0.00	30.14	29.82	54.58	13.23
over standard rate (%)	57.14	71.42	/	0	21.42	10.71	0
Average exceeds the multiple	0.69	0.12	/	/	/	/	/

3.3 Distribution Characteristics of TP and TN Content and the Risk of Pollution

According to Fig. 2, the content of TN and TP in different sections. TN and TP were 333.87–5237.62 mg/kg, 694.38–9146.61 mg/kg, and the mean values were 2442.65 mg/kg and 3815.45 mg/kg, respectively. According to the comprehensive pollution index method, TN and TP individual evaluation index were 1.39–5.85 and 3.79–12.55 respectively, the mean value was 3.55 and 8.55 respectively; the comprehensive pollution index FF value was 9.85, which was at the level of severe pollution. It could be seen that the average comprehensive pollution index of TN and TP was at the level of severe pollution and had a serious risk of environmental pollution. The overall TN and TP content of the sediment in the lake area was relatively high, which could be the deposition of pollutants from the long time in the lake area for historical reasons, the direct discharge of surrounding domestic sewage and the input of pollutants from fish ponds in the lake area.

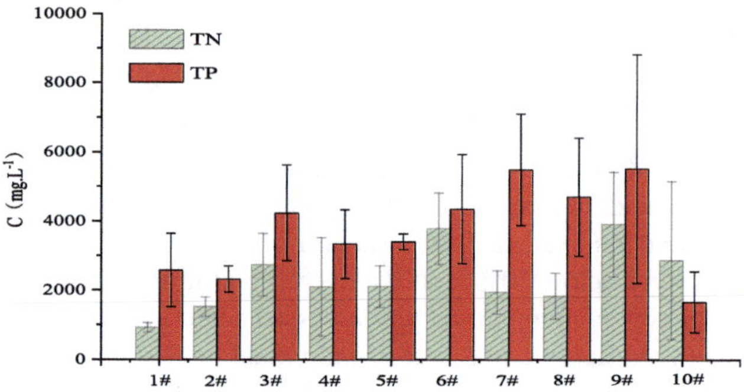

Fig. 2. Distribution of TN and TP in the lake sediment

The adsorption-analytical equilibrium concentration at different layered depths in the lake sediment was quite different, and the overall performance was top-down NH_3^-, The adsorption-analytical concentration of N and P was gradually decreasing. Sludge layer, transition layer, normal lake bottom layer NH_3^- The equilibrium equilibrium concentrations of N and P were 1.05–1.85 mg/L, 0.72–1.62 mg/L and 0.65–1.04 mg/L; 0.17–0.52 mg/L,0.08–0.36 mg/Land 0.01–0.09 mg/L, respectively; the average equilibrium concentrations were 1.41 mg/L, 1.14 mg/L and 0.86 mg/L;0.33 mg/L, 0.21 mg/L and 0.07 mg / L, respectively.

3.4 Dredging Thickness

Combined with the risk analysis of sediment pollution and the adsorption-analysis balance test, it could be seen that the sediment in a lake area of Huizhou was basically free of heavy metal pollution, the N and P comprehensive pollution indexes were in the level of severe pollution, the potential pollution risk was serious, and the pollution layer and transition layer were NH_3, The adsorption-analytical equilibrium concentration of N and P was higher than that of surface water class III, which had a high risk of pollution release. It was suggested to dredge the polluted layer and transition layer sediment in the lake area. The dredging depth of the lake area was 36.36–66.91 cm and the average dredging thickness was 54.41 cm. Reduce the risk of sediment pollution release, reduce the water pollution of internal source release, and ensure the water quality.

4 Conclusion

(1) The total thickness of the lake sediment contamination was 16.97–33.18 cm, 19.38–33 cm and 36.36–66.91 cm, and the average thickness were 28.18 cm, 26.22 cm and 54.41 cm, respectively.
(2) E in the potential ecological hazard index of heavy metals in the sedimenti As > Cd > Cu > Pb > Ni > Zn > Cr, between 0.13 and 7.51, and RI values between 12.34 and 17.15, all of which were low-risk levels and less stress on environmental heavy metal pollution.
(3) The total nitrogen and total phosphorus in content of sediment were 333.87–5237.62 mg/kg, 694.38–9146.61 mg/kg, and average 2442.65 mg/kg and 3815.45 mg/kg, respectively. TN and TP individual evaluation index were 1.39–5.85 and 3.79–12.55, respectively, with the average value of 3.55 and 8.55, respectively. The equilibrium equilibrium concentrations of N and P of subsumed contaminated layer, transition layer, normal lake mud layer NH_3^-were 1.05–1.85 mg/L, 0.72–1.62 mg/L and 0.65–1.04 mg/L; 0.17–0.52 mg/L, 0.08–0.36 mg/L and 0.01–0.09 mg/L, respectively; the average equilibrium concentrations were 1.41 mg/L, 1.14 mg/L and 0.86 mg/L; 0.33 mg/L, 0.21 mg/L and 0.07 mg/L, respectively.
(4) It was suggested that the polluted layer and the transition layer in the lake area, with the dredging depth of 36.36–66.91 cm and the average dredging depth of 54.41 cm, which is conducive to reducing the risk of pollution release.

References

1. Bai, J., Cui, B., Chen, B., et al.: Spatial distribution and ecologi-cal risk assessment of heavy metals in surface sediments from a typical plateau lake wetland, China. Ecol. Model. **222**, 301–306 (2011)
2. Dan, Z., Yong, Z., Yan, H., et al.: Progress in environmental protection dredging of river sediment. Water Purif. Technol **30**(1), 1–3,7 (2011)
3. Cheng, Z., Guolu, Y., Jing, L., et al.: Research progress of influencing factors affecting the release of sediment pollutants and sediment treatment. Environ. Eng. **34**(5), 113–117 (2016)
4. Minghui, C., Yingqun, M., Wei, C., et al.: Distribution characteristics of pollution from nitrogen, phosphorus and heavy metals in surface sediments in northern lake area of Jiaxing. Environ. Sci. Technol. **43**(3), 190–197 (2020)
5. Bhuyan, M., Bakar, M., Akhtar, A., et al.: Heavy metal contamination in surface water and sediment of the Meghna River, Bangladesh. Environ. Nanotechnol. Monit. Manag. **8**, 273–279 (2017). https://doi.org/10.1016/j.enmm.2017.10.003
6. Hakanson, L.: An ecological risk index for aquatic pollution control-a sedimentological approach. Water Res. **14**(8), 975–1001 (1980)

7. qiu, H.-r.: Study on heavy metal pollution characteristics and potential ecological risk of southwest Chung and tributary Dalong eddy. Guangdong: Guangdong University of Technology (2012)
8. Shuyun, S., Xiaofang, C., Guangbing, L.: Research on river silt dredging and sediment pollution assessment. Sichuan Environ. **43**(4), 33–40 (2024)
9. Jin, H., Van, L.C., Van, D.W., et al.: Impacts of sediment resuspension on phytoplankton biomass production and trophic transfer: implications for shallow lake restoration. Sci. Total Environ. **808**, 152–156 (2022)
10. Yin, H., Zhang, M., Yin, E., et al.: Characterization of internal phos-phorus loading in the sediment of a large eutrophic lake (Lake Taihu, China). Water Res. **225**, 119–125 (2022)

Screening and Solubilization Effect of Siloxane-Degrading Bacteria Enhanced by Surfactants

Fengle Sun[1], Fumiao Zhu[2], Jiangxue Long[1](✉), Panfeng Gao[1], and Shuaihao Liu[1]

[1] Xiamen University of Technology, Xiamen 361024, China
ljxczl@126.com
[2] China Mobile Tietong Co., Ltd. Taizhou Branch, Taizhou, China

Abstract. In this article, the gas-liquid mass transfer efficiency of hexamethyldisiloxane (L2) was improved by adding surfactants, and its removal rate was increased. Lauryl dimethyl amine oxide (OB-2), sodium dodecyl sulfate (SDS), Octylphenol polyoxyethylene ether (OPE) and Tween-80 surfactants were screened from three aspects: the solubilizing emulsification ability of Hexamethyldisiloxane, foam stability and toxicity to degrading bacteria. Finally, Tween-80, which has strong solubilizing emulsification ability, good foam stability and is non-toxic to the degrading bacteria, was selected to carry out follow-up enhanced degradation research. Tween-80 can be used as a carbon source for the growth and metabolism of degraded bacteria, but as a carbon source alone, it has no obvious effect on the growth of degraded bacteria. After the addition of Tween-80, the removal rate of the degraded bacteria in mixed culture was 86.31%, which was 5.69% higher than that in culture without surfactants.

Keywords: Surfactants · solubilization screening · hexamethyldisiloxane biodegradation

1 Introduction

In recent years, the extensive use of organosiloxane-containing materials has led to the introduction of siloxanes into the environment as pollutants, and the amount has increased year by year. Especially in biogas (biogas) produced during digestion in landfills and sewage treatment plants. Biogas is often used as a biofuel or injected directly into natural gas pipeline networks. However, biogas also contains other unwanted chemical components, especially the presence of siloxane, which poses the greatest threat to biogas use [1]. They will oxidize into white, fine SiO2 crystal particles during combustion, resulting in equipment corrosion [2]. SiO2 will produce deposition on the inner wall of the engine combustion chamber and the surface of the piston components, affecting heat conduction and component wear, improving the maintenance rate and operating costs, and seriously affecting the effective utilization of biogas [3]. In biogas boilers, SiO2 generated by combustion is deposited on the inner wall of the furnace, leading to blockage of

D. Han et al. (Eds.): ICEPG 2024, ESE, pp. 770–777, 2025.
https://doi.org/10.1007/978-3-031-86829-0_86

the combustion tube and reducing the boiler thermal efficiency [4]. Therefore, effective removal of VMS can avoid equipment wear and corrosion, extend equipment life, and reduce operating costs [5]. At present, the most widely used method for removing VMS is activated carbon adsorption. However, as biogas contains water vapor, VMS adsorption by activated carbon is not applicable under conditions of high moisture content [6]. The biological siloxane removal process is convenient, economical and environmentally friendly, and may overcome the shortcomings of physical and chemical processes [7]. Many microorganisms have been shown to degrade dimethylsilanediol, octamethylcyclotetrasiloxane (D4), and other organosilicon compounds, They include A. radiobacter, F. oxysporum, P. aeruginosa, P. fluorescens,P. putida and indigenous microorganisms [8]. Biotrickling filters are widely used in digestive biogas purification because they are easy to operate and have been identified as the lowest operating cost technology for removing siloxanes from biogas [9].

Hexamethyldisiloxane (L2) is a kind of siloxane which has high content in landfill biogas but is difficult to degrade. Hamethyldisiloxane (L2) has a low solubility in the organic phase due to its high vapor pressure. It resulted in a low mass transfer efficiency from the gas phase to different oil phases. In the biological process, surfactants, as an important treatment tool, have certain advantages in the treatment of hydrophobic waste gas. Therefore, in this study, the gas-liquid mass transfer efficiency of L2 was improved by adding surfactants, and its removal rate was further increased. In this article, the following four kinds of surfactants with different properties were screened: lauryl dimethyl amine oxide (amphoteric surfactant), sodium dodecyl sulfate (anionic surfactant), Octylphenol polyoxyethylene ether (non-ionic surfactant) and Tween-80 (non-ionic surfactant).

2 Materials and Methods

2.1 Strain Source

10 g of wet activated sludge obtained from the secondary sedimentation tank of a sewage treatment plant was inoculated into 200 mL of inorganic salt medium, and a certain amount of glucose carbon source was added, then placed into a constant temperature oscillator under oscillating conditions of 30 °C and 150 r/min for acclimatization culture. After many times of conversion and enrichment culture, the colonies with HMC degrading ability were successfully isolated on the inorganic salt medium with HMC disiloxane as the only carbon source, and finally four strains with degrading ability were purified. They were named JXL-1, JXL-2, FMZ-1 and FMZ-2 respectively. JXL-1 (GenBank: PQ198516) was identified by 16S rRNA gene sequencing as Alloalcanivorax xenomutans, belonging to Bacteroidota. JXL-2 (GenBank: PQ210955) is Lactococcus lactis, which belongs to Firmicutes. Sequence comparison of FMZ-1 (GenBank: PQ210956) and FMZ-2 (GenBank: PQ210970) showed that both were fungi, Apiotrichum mycotoxinivorans, belonging to Ascomycota.

2.2 Experimental Materials

Lauryl dimethyl amine oxide, Reagent grade; Sodium dodecyl sulfate, Reagent grade Octylphenol polyoxyethylene ether, Reagent grade; TweeN-80, Reagent grade.

2.3 Preliminary Screening of Surfactants

The solubilization of surfactants was determined by the determination of L2 content in surfactant, and then the surfactant with good solubilization was preliminarily selected [10].

First, five different surfactant solutions were prepared, each with a concentration of 1 g/L. Next, 3 mL of the surfactant solution and 1 mL of the L2 reagent are added to the 10 mL centrifuge tube. Process these mixtures for 5 min using a vortex oscillator to ensure adequate mixing. After that, place the centrifuge tube into the centrifuge, set the speed to 3000 r/min, and centrifuge for 10 min. After completing the above steps, remove 1 mL of the lower layer of aqueous solution from the centrifuge tube and transfer it to another 10 mL centrifuge tube containing 1 mL of n-hexane. Use the vortex oscillator again to oscillate for 1 min, then centrifuge at 2000 r/min for 10 min. After centrifugation, remove 1 mL of sample from the upper liquid and use a gas chromatograph to determine L2 content in it. Three parallel sets of each surfactant were made. The results were averaged and compared with the solubilization of water.

2.4 Determination of Critical Micelle Concentration of Surfactant After Initial Screening

CMC and the equilibrium surface tension (γcmc) at this concentration are the core parameters for evaluating the performance of surfactants [11]. In order to determine these parameters, the overdilution method was used with the critical micelle concentration measuring instrument: the surface tension value (γ) of a series of surfactants at different concentrations is first measured. These tension values are then plotted as Y-axis data and lgC as X- axis data for the surfactant concentration. The X coordinates corresponding to the turning points in the graph are the CMC of the surfactant.

2.5 Study on Foam Stability of Surfactant

Foam stability is one of the important indexes to evaluate the foam properties of surfactants. In this experiment, the stability of the foam was evaluated by measuring its half-life [12]. To do this, observe and record the time it takes for the foam to fall from its initial height to half. This time period is defined as the half-life of the foam. By this measurement, the stability of the surfactant foam can be quantified in this paper. This method is not only intuitive, but also can effectively reflect the changing characteristics of the foam with time.

2.6 Toxicity Test of Surfactants to Degrading Bacteria

In this experiment, four strains were selected for mixed culture. The suspensions of each strain were uniformly inoculated into 250 mL conical bottle, which contained 100 mL LB medium. The total inoculated volume was 1 mL, and the proportion of each strain was 25%. In the experiment, different types and concentrations of surfactants were accurately added to each oscillating bottle according to the predetermined setting: lauryl dimethyl amine oxide, octylphenol polyoxyethylene ether and Tween-80 at a concentration of 0.5

CMC. At the same time, a control group without surfactants was set up. All samples were cultured on a constant temperature oscillating shaker at 30 °C and 120 r/min. During the culture process, samples were taken at regular intervals, and the OD600 values of the samples were measured using an ultraviolet spectrophotometer. Based on these data, the growth curves of the strains were plotted to evaluate the effects of different surfactants on the degrading bacteria and their potential toxicity.

2.7 Experiment on L2 Degradation Performance of Biodegradable Bacteria Enhanced by Surfactants

(1)To determine whether the growth of bacteria will use surfactants as carbon source: In the blank group, 100 mL inorganic salt medium was added into 250 mL conical bottle, and a total of 1mL bacterial suspension of 4 strains was inoculated, with the content of 4 strains accounting for 25% each. Tween-80 was added to the control group on the basis of the blank group, so that its concentration was 0.5 CMC. After inoculation, the bacteria in both groups were cultured on a constant temperature shaking table at 30 °C and a rotating speed of 120 r/min for 4 days. After sampling, the OD600 value of the bacteria was determined by enzyme marker. Compared with the experimental group: the OD600 value of the experimental group affected by the addition of Tween-80 on the removal rate of degrading bacteria.

(2)Degradation performance experiment: 0.2 mL of L2 was added into 250 mL conical bottle containing 100 mL of inorganic salt nutrient solution, and then Tweene-80 was added to make its concentration 0.5 CMC. Finally, the bacterial suspension of 4 strains was inoculated with a total of 1mL, and the content of the four strains accounted for 25% each. After 4 days of culture in the shaking table, the concentration of the remaining L2 in the nutrient solution was determined by extraction, and the removal rate was calculated. The removal rate was compared with that of mixed culture of degradable bacteria without adding surfactants.

3 Results and Analysis

3.1 Preliminary Screening Results of Surfactants

The gas chromatographic test results are shown in Fig. 1. For hexamethyldisiloxane, Sodium dodecyl sulfate, Lauryl dimethyl amine oxide, Octylphenol polyoxyethylene ether and Tween 80 have good solubilizing and emulsifying ability. Therefore, these four surfactants were selected for further screening.

3.2 Critical Micelle Concentration of Surfactants

The X-coordinate (concentration value) corresponding to the turning point of the curve in the critical micelle concentration diagram of the surfactant is the CMC of the surfactant. Figure 2 shows the critical micelle concentration curve for four surfactants. The CMC concentration and surface tension of the four surfactants are shown in Table 1. The γcmc values of a single surfactant from small to large are: Lauryl dimethyl amine oxide, Sodium

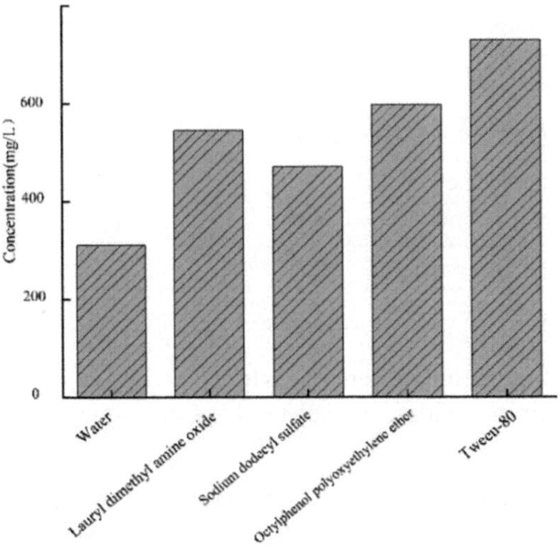

Fig. 1. Effect of surfactants on the solubility of L2

dodecyl sulfate, Octylphenol polyoxyethylene ether, Twain 80. In general, surfactants with lower CMC are more likely to form micelles, which means that their solubilization is usually more significant [13].

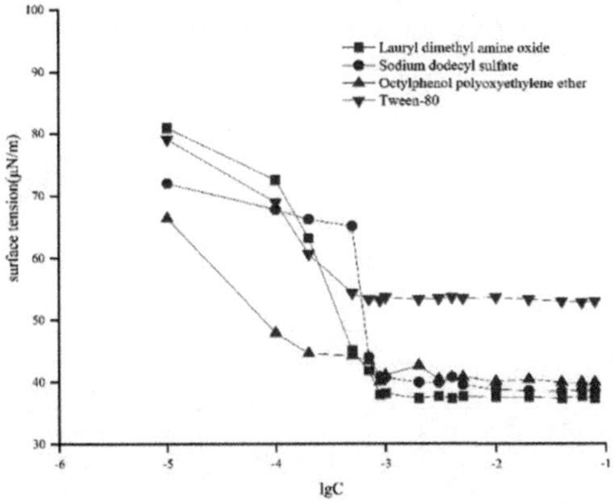

Fig. 2. Critical micelle concentration curves for four surfactants

3.3 Comparative Analysis of Foam Stability of Surfactant

As can be seen from Table 1, Octylphenol polyoxyethylene ether, Tween-80 and Lauryl dimethyl amine oxide foam amount is moderate and can quickly defoamer, while the foam half-life of sodium dodecyl sulfate is too long, which may cause energy consumption and cost of foam removal equipment, so sodium dodecyl sulfate as a solution-

Table 1. Four surfactants CMC and Foam Half-life

Kinds	CMC(mg/L)	Gamma (μN/m)	Half-life
Sodium dodecyl sulfate	800	38.2	76’ 00”
Octylphenol polyoxyethylene ether	400	40.7	1’ 03”
Tween-80	400	54.28	1’ 30”
Lauryl dimethyl amine oxide	800	37.99	1’ 32”

increasing surfactant is excluded here. Octylphenol polyoxyethylene ether, Tween-80 and lauryl dimethyl amine oxide were selected for subsequent experiments.

3.4 Strain Source Results of Toxicity Determination of Surfactants to Degrading Bacteria

The results are shown in Fig. 3. Lauryl dimethyl amine oxide is relatively toxic, which can inhibit the growth of degrading bacteria. Octylphenol polyoxyethylene ether inhibited the growth of degrading bacteria in the logarithmic growth stage of culture, but the OD600 value in the later stage was basically the same as that in the blank group, indicating that although Octylphenol polyoxyethylene ether inhibited the growth of degrading bacteria, its toxicity was less. Tween-80 had little effect on the growth of hexamethyldisiloxane degrading bacteria, but promoted the growth of degrading bacteria. Subsequently, Tween-80 was selected as a solubilizing surfactant to determine its ability to remove L2.

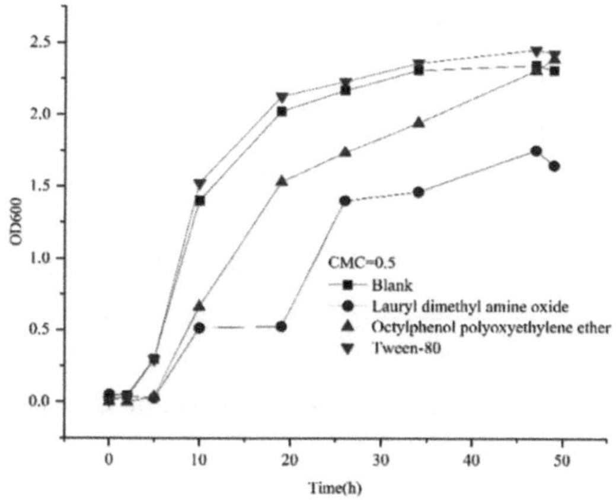

Fig. 3. Growth curve of degradation bacteria with addition of surfactants Toxicity curves of octylphenol polyoxyethylene ether

3.5 Analysis of L2 Degradation Performance of Bacteria Under the Action of Surfactants

As shown in Fig. 4, compared with the blank group, after 4 days of culture without additional carbon source, the OD600 value of the bacterial solution grew to a certain extent when Tween-80 surfactant was added only, but the increment was not large. However, when L2 carbon source was added for 4 days, the OD600 value was 6.2 times that of the group with only surfactant added. Therefore, Tween-80 may provide carbon source for degrading bacteria, but its growth promotion effect on degrading bacteria is not obvious, which is consistent with the research results of Zhou Xuexia [11].

Under the action of surfactant Tween-80, the removal rate of the mixed bacteria was 86.31%, which was 5.69% higher than that of the mixed bacteria without surfactant. The addition of surfactants can effectively promote the degradation of L2.

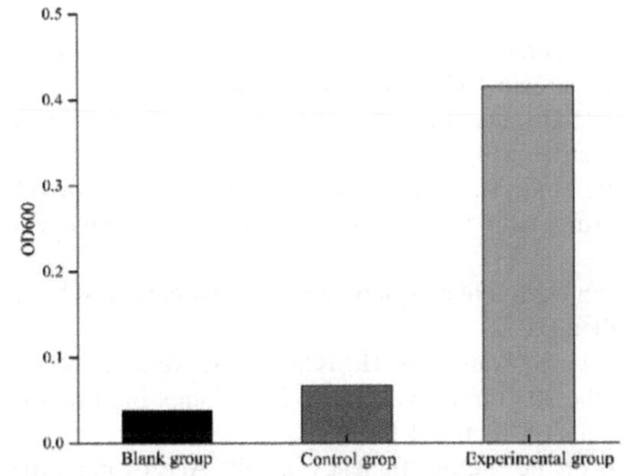

Fig. 4. Toxicity curves of Octylphenol polyoxyethylene ether

4 Conclusion

Four surfactants, were screened from three aspects: the solubilizing emulsification ability of hexamethyldisiloxane, foam stability and toxicity to degrading bacteria. Finally, Tween-80, which has strong solubilizing emulsification ability, good foam stability and is non-toxic to the degrading bacteria, was selected to carry out follow-up enhanced degradation research. After the addition of Tween-80, the removal rate of the degraded bacteria in mixed culture was 86.31%, which was 5.69% higher than that in culture without surfactants.

Foundation Item. Fujian Natural Science Foundation project (2023J011423); Fujian Natural Science Foundation project (2022J011230); Fuxiquan National Independent Innovation Demonstration Zone urban and rural ecological environment governance improvement key technology collaborative innovation platform project (2022-P-024); Xiamen Science and Technology Project (3502Z20226024).

References

1. Soreanu, G., Beland, M., Falletta, P., et al.: Approaches concerning siloxane removal from biogas–a review. Can. Biosyst. Eng. **53**(8), 1–8 (2011)
2. Muñoz, R., Meier, L., Diaz, I., et al.: A review on the state-of-the-art of physical/chemical and biological technologies for biogas upgrading. Rev. Environ. Sci. Bio/Technol. **14**, 727–759 (2015)
3. Accettola, F., Guebitz, G.M., Schoeftner, R.: Siloxane removal from biogas by biofiltration: biodegradation studies. Clean Technol. Environ. Policy **10**, 211–218 (2008)
4. Abatzoglou, N., Boivin, S.: A review of biogas purification processes. Biofuels Bioprod. Biorefin. **3**(1), 42–71 (2009)
5. Yunhui, L.: Basic research on Biodegradation of siloxane and hydrogen sulfide in energy gas. Tianjin University, Tianjin (2014)
6. Zheng, Y.H.: Construction of micro/narrow mesoporous graphene-based aerogel and its adsorption properties for methicone. Hebei Normal University, Shijiazhuang (2023). (in Chinese)
7. Syed, M., Soreanu, G., Falletta, P., et al.: Removal of hydrogen sulfide from gas streams using biological processes a review. Can. Biosyst. Eng. **48**, 2 (2006)
8. Shen, M., Zhang, Y., Hu, D., et al.: A review on removal of siloxanes from biogas: with a special focus on volatile methylsiloxanes. Environ. Sci. Pollut. Res. **25**, 30847–30862 (2018)
9. Soreanu, G.M., Be´land, P., Falletta, K., Edmonson, L., Svoboda, M. Al-jamal, S.P.: Approaches concerning siloxane removal from biogas -- a review. Canadian Soc. Bioeng (2011)
10. Xuexia, Z.: Study on treatment of xylene waste gas by enhanced biotrickling filter. Zhejiang University, Hangzhou (2012)
11. Dupont, C., Kulovesi, K., Vanasselt, H.: Editorial: Governing the EU's climate and energy transition through the 2030 framework. Rev. Eur. Comp. Int. Environ. Law **29**(2), 147–150 (2020). https://doi.org/10.1111/reel.12359
12. Zhang, Y.X.: Study on the elution efficiency of soil PAHs by microfoaming of biosurfactant. Shanxi University, Shanxi (2021)
13. Zabed, H.M., Akter, S., Yun, J., et al.: Biogas from microalgae: Technologies, challenges and opportunities. Renew. Sustain. Energy Rev. **117**, 109503 (2020)

Preparation of Bar-Shaped Catalysts from Alumina Sol and Study on Their Denitration Performance

Junmao Qie, Shuqi Zhang(✉), Qunying Wang, Zefeng Zhao, and Tong Xu

School of Energy and Environment, Inner Mongolia University of Science and Technology, Baotou 014010, Inner Mongolia, China
2074575452@qq.com

Abstract. Rare earth minerals have denitration activity, which can be used as the active component of catalyst. Various minerals in rare earth tailings, various minerals are embedded and wrapped with each other, which can play the role of combined denitration. In this article, the strip catalyst of nanoalumina and pseudobohemite were prepared as binders to explore the effect of the amount of binder on the performance of the strip catalyst. With the help of XRD, NH_3-TPD and H_2-TPR to analyze the performance of the molding catalyst. The denitrification efficiency of rare earth tailings is only 8.1% within the limits of 100 to 500 °C; the denitrification performance of the strip catalyst formulated with 15% nano aluminum gel exhibits an efficiency of 60.3%, while maintaining a longitudinal compressive strength of 1.14 Mpa. the catalyst surface is rich with nano alumina as binder, with Brønsted acidic site as the main adsorption site, and the stability and adsorption capacity of NH_3 species are improved. Upon the addition of aluminum sol, the denitration performance of the rare earth tailings underwent a significant enhancement, achieving a 52.2% increase.

Keywords: rare earth tailings · strength · denitration performance

1 Introduction

Nitrogen oxides (NO_x) have become the main pollutant that causes atmospheric environmental pollution and the important reason for the formation of haze in recent years, acid rain, and photochemical smog [1, 2]. Currently, the prevalent technology employed for NO_x treatment is the Selective Catalytic Reduction (SCR) method utilizing NH_3 (ammonia) as the reducing agent. This method is renowned for its exceptional purification efficiency, safe operation and no secondary pollution. NH_3 Selective reduction of NO_x to the non-toxic and harmless N_2 and H_2O [3]. Carbon monoxide selective catalytic reduction (CO-SCR) can remove CO and NO from industrial waste gas at the same time and avoid ammonium poisoning. It is considered a feasible technology with great application prospect [4]. Forming is a process, which involves the unit operation of various kinds of powder, particles, solutions or raw materials of molten state with each other under the action of a specific external force, and then the formation of solid

D. Han et al. (Eds.): ICEPG 2024, ESE, pp. 778–786, 2025.
https://doi.org/10.1007/978-3-031-86829-0_87

particles or products with a specific size, strength and shape. No matter what method is used to prepare a solid catalyst, it is finally used in the form of particles of different sizes and shapes in the catalytic reactor, and molding is an important process in the process of preparing the catalyst.

At present, vanadium-based catalyst [5] is the most widely used commercial NH3-SCR catalyst, but the difficult treatment of waste catalysts, complex preparation process, high cost, high temperature sintering, low temperature activity and biotoxic [6, 7] are the deficiencies of vanadium-based catalysts. Bayan Obo mine is the largest rare earth reserves in China, mainly rare earth, iron and niobium [8]. Bayan Obo rare earth tailings inherently contain iron, cerium, manganese, and other vital elements conducive to denitrification. However, when solely utilizing the tailings as a catalyst, the denitration efficiency is relatively low, so this paper uses the nano alumina and pseudobohemite as binder, so that the powder in rare earth tailings can be bonded together,in order to enhance the denitration performance of the catalyst, modifications or additional components may be introduced.

2 Materials and Methods

2.1 Rare Earth Tailings

In this experiment, rare earth tailings in Bayan Obo tailings dam in Baotou city were selected as raw material to prepare SCR (selective catalytic reduction) denitration catalyst. To understand the composition of the rare earth tailings, X-ray fluorescence (XRF) analysis was performed, as shown in Table 1. Rare earth tailings are rich in a variety of metal oxides, including manganese oxides (1.3%), iron oxides (28.60%) and rare earth metal oxides (2.47%). The study shows that these oxides, namely manganese oxide, iron oxide and rare earth oxide, each show a certain catalytic denitration efficiency. Based on the rare earth tailings as a polymetallic composite oxide mineral, its application prospect as a denitration catalyst is particularly broad and worthy of further study.

Table 1. Contents of main elements in rare earth tailings (wt%)

Analysis of the rare-earth tailings elements(%)									
Na_2O	K_2O	Al_2O_3	MgO	CaO	BaO	SrO	Fe_2O_3	TiO_2	MnO
2.26	0.636	1.71	5.29	21.87	1.62	0.136	28.60	0.557	1.30
ZnO	ZrO_2	CeO_2	PbO	Nb_2O_5	SiO_2	P_2O_5	SO_3	Others	
0.0422	0.0013	1.86	0.0092	0.600	19.25	2.25	2.14	9.49	

2.2 Preparation of the Catalyst

The rare earth tailings selected for this experiment originated from the Bayan Obo tailings dam in Baotou City. In the process of preparing strip catalyst, it usually needs to go through three key stages: first, the careful preparation of plastic sludge, followed by the extrusion process of the catalyst, and finally the drying and roasting of the catalyst.

In the first stage, the preparation process of plastic sludge is as follows: First, the rare earth tailings collected from Bayan Obo tailings dam in Baotou are put into the high speed ball mill. The speed of the ball mill was then set to 1100 rpm and continued for 2 h. After the ball grinding process, the screening equipment is used to screen for 200 tailings powder for subsequent processing. Add binder (such as nano-alumina or alumina) to the powder and then mixed until the mixture forms a uniform mud mass; then the mud-like catalyst body is transferred to the agate tank and then mixed in a planetary ball mill at 600 rpm for 2 h to ensure that all components are fully and evenly fused together. Subsequently, a plastic clay mass is both sticky and liquid. The mud is then tightly wrapped around the catalyst and let it through a 24-h stale process.

The second stage involves the extrusion molding step of the catalyst: the stale catalyst mud is placed in the extrusion molding machine and extruded into a strip by the machine. Subsequently, these formed catalyst embryos were cut precisely according to the required length of the experiment.

The third stage covers the drying and roasting process of the catalyst: First, the strip catalyst embryo body is placed in a drying box and dried at a constant temperature of 80 °C for 1 h. Immediately, it was transferred to a maver furnace and roasted at 600 °C for 2 h. After roasting, the strip catalyst is removed and cooled to room temperature. Thus, the finished strip catalyst is prepared.

2.3 Characterization Method of the Catalyst

X-ray Diffraction Analysis (XRD) Analysis

The X-ray diffraction instrument used in this experiment is from Brooke, Germany, model D8 ADVANCE. The test conditions are: Cu target (Kα-ray), the measurement mode is continuous scanning, the scanning angle range is 10–90, and the scanning speed is 4 / min.

Procedure Warming Off-Attachment (NH3-TPD) Analysis

The instrument used in this experiment is PCA-1200. The analytical method is continuous flow procedure, and the detector is gold-plated TCD detector. The measurement conditions are: the heating rate is 10 °C / min, first heat the furnace to 200 °C under N2 atmosphere for 30 min for dehydration, then NH3 under NH3 atmosphere for NH3 to 100 °C and NH3 atmosphere to 800 °C for NH3 temperature.

Mechanical Strength Analysis

The universal test machine produced by steel mining Research Testing Technology Co., Ltd., the model is GNT200. The prepared strip catalyst was placed on the universal test machine, and the maximum pressure was read when the pressure suddenly disappeared by continuous pressure, and multiple experiments were averaged. Thus, the transverse and radial compressive strength of the strip catalyst is measured.

Analysis of Program Warming Reduction (H2-TPR)

The instrument used in this experiment is produced in Beijing, the model is PCA-1200.

The analytical method is continuous flow procedure, and the detector is gold-plated TCD detector. The measurement conditions are: the heating rate is 10 °C/ min, the furnace is heated to 200 °C in N2 atmosphere and constant temperature for 30 min, and then heated to 800 °C in H2 atmosphere.

3 Results and Discussion

3.1 Denitration Performance

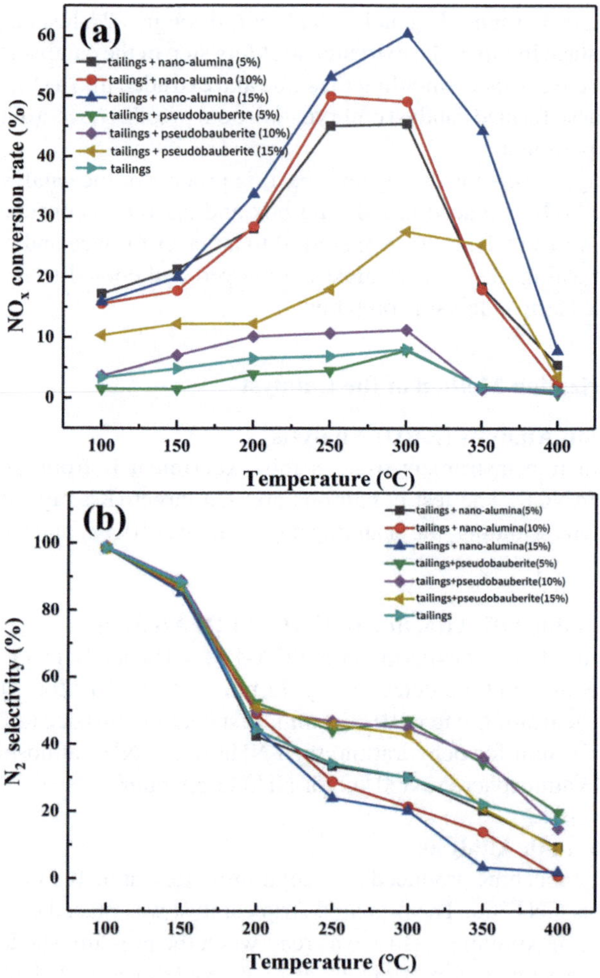

Fig. 1. Denitration activity of aluminum sol forming catalyst (a-NO_x conversion, b-N_2 selectivity), The reaction conditions: NO = 500 ppm, NH_3 = 500 ppm, O_2 = 3%, N_2 as balance gas, and airspeed 10 000 h^{-1}

To study the effect of the addition of aluminum sol on the denitrification efficiency of the strip SCR denitration catalyst, and to simulate the strip catalyst forming process. This experiment used two different aluminum of nano alumina gel and thin aluminum as binder to prepare strip catalyst, catalyst is tailings + nanoalumina (5%), tailings + nanoalumina (10%), tailings + nanoalumina (15%), tailings + pseudobohemite(5%), tailings + pseudobohemite (10%), tailings + pseudobohemite(15%), tailings, denitrification efficiency and N_2 selectivity of different binder strips shown in Fig. 1.

The denitration efficiency of a tailings was only 8.1%, in comparison to the tailings alone, the denitration efficiency of the strip catalyst significantly improves upon the addition of nanoalumina. As the quantity of nanoalumina incorporated increases, so does the denitration efficiency of the catalyst. Specifically, when the nanoalumina content reaches 15%, the denitrification efficiency of the catalyst surges to 60.3%. The strip catalyst prepared by the addition of pseudobohemite is compared with the tailings, when the pseudobohemite volume of 5%, the catalyst denitration rate decreased, when the pseudobohemite volume of 10%, the denitration efficiency of the catalyst is relatively modest when the volume of pseudoboehmite is 15%, achieving a maximum of 27.2% efficiency under these conditions. In general, the denitrification efficiency of the strip catalyst prepared by adding these two binders has been improved compared with the tailings, and the denitrification efficiency of the catalyst prepared by adding nano alumina has been more significantly improved. However, the catalysts prepared by the two binders still have disadvantages such as N_2 selectivity difference and narrow denitrification temperature interval.

3.2 Analysis of the Object-Phase and Crystal-Phase Structure

The phase and form of the catalyst has a certain influence on the catalytic activity of the catalyst, in order to determine the relationship between the phase and form and the catalytic activity of the catalyst, the catalyst XRD analysis, Fig. 2 is the XRD diagram of tailings + nanoalumina (15%), tailings + pseudobohemite (15%), tailings.

As can be seen from Fig. 2, the diffraction peak intensity of catalysts Fe_2O_3 and SiO_2 prepared by the addition of two different binders changed little, and the diffraction peak of CaF_2 decreased, indicating that the addition of binder had little effect on the crystal phase structure of the tailings itself, and the decrease of the diffraction peak of CaF_2 may be covered by γ-Al_2O_3 particles. The diffraction peak of γ-Al_2O_3 could not be found for either catalyst, indicating that the γ-Al_2O_3 generated by roasting of these two binders could be distributed evenly on the surface of the catalyst, which is more conducive to the adsorption of NH_3.

3.3 NH_3 Analysis of the Adsorption Characteristics

The main method for the catalyst acid site characterization in the laboratory is NH_3-TPD, which is the main method to study the catalyst reaction mechanism: the quantitative or qualitative calculation of the catalyst acid site. Therefore, for the effect of NH_3 adsorption capacity on the surface of the prepared strip catalyst, the NH_3-TPD characterization of these seven catalysts was analyzed, and the analysis results are shown in the following Fig. 3.

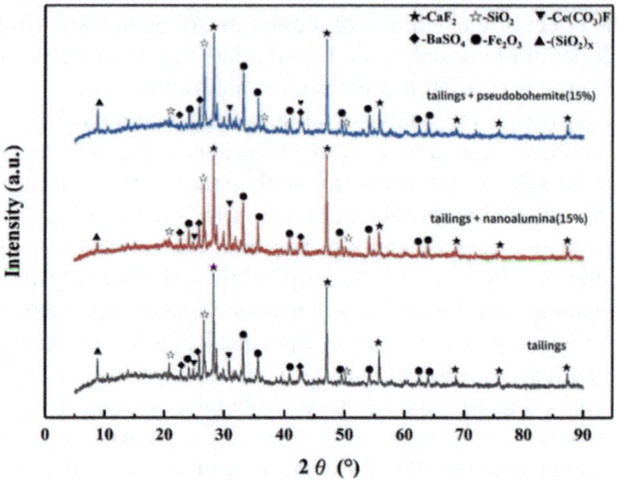

Fig. 2. XRD pattern of aluminum sol forming catalyst

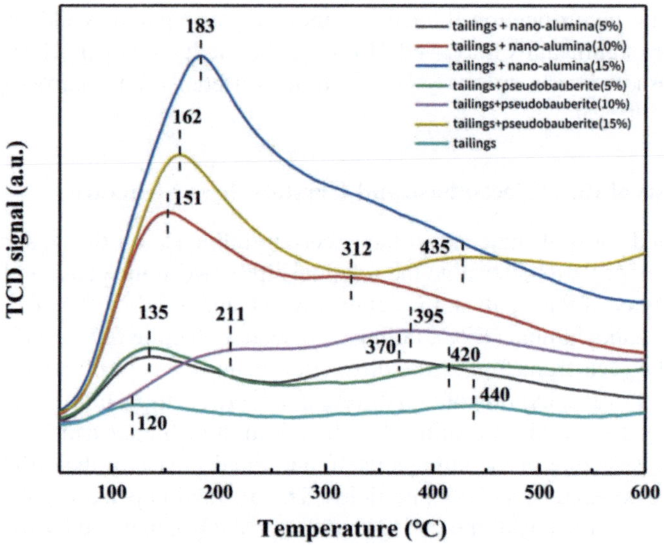

Fig. 3. NH_3-TPD diagram of aluminum sol forming catalyst

Figure 3 shows that the tailings + nanoalumina (5%) has two detachment peaks of 135 °C and 370 °C, Tailings + nano-alumina (10%) has two detachment peaks of 151 °C and 312 °C, Tailings + nano-alumina (15%) has a detachment peak of 183 °C, Tailings + (5%) of 135 °C and 420 °C, Tailings + thin cement (10%) has two detached peaks of 211°C and 395 °C, Tailings + thin (15%) of 162 °C and 435 °C, The tailings have two desorption peaks of 120 °C and 440 °C. Compared with the tailings, the more the tailings and ALCO 20. Because the prepared strip catalyst after the addition of nanoalumina has a larger NH3 adsorption, the catalyst has more surface acid, and the more the acid on the surface of the catalyst, the more smooth the denitration reaction.

3.4 Mechanical Strength Analysis

The mechanical strength of the strip catalyst prepared with different proportions of nanoalumina is shown in Table 2. As the quantity of nanoalumina was incrementally increased from 0 to 15%, the longitudinal strength of the material rose markedly from 0.28 MPa to 1.14 MPa, while the transverse strength also experienced a notable enhancement, climbing from 0.11 MPa to 0.33 MPa. Because the structure changes less, good adhesive performance and strong bonding after roasting, even after undergoing roasting, nano-alumina continues to function as an effective binder. Consequently, as the amount of nano-alumina added increases, the mechanical strength of the catalyst gradually strengthens. According to Table 3, illustrates the mechanical strength of catalysts prepared by incorporating varying amounts of pseudoboehmite. As the pseudoboehmite content rises from 0 to 10%, the longitudinal strength experiences a notable increase, from 0.28 MPa to 1.36 MPa, while the transverse strength also shows an upward trend, rising from 0.11 MPa to 0.44 MPa, however, it is noteworthy that when the addition of pseudoboehmite exceeds 10% and reaches 15%, the mechanical strength of the catalyst undergoes a decrease. Because of the good soluble performance of pseudobohemite, a large number of additives can lead to the increase of viscosity of plastic paste, serious particle aggregation, a large amount of water loss during roasting, the bonding performance decreases, resulting in the reduction of mechanical strength.

Table 2. Mechanical strength of nanoalumina forming catalyst

content	0%	5%	10%	15%
longitudinal strength MPa	0.28	0.53	0.77	1.14
transverse strength MPa	0.11	0.14	0.27	0.33

Table 3. Mechanical strength of pseudobohemite forming catalyst

content	0%	5%	10%	15%
longitudinal strength MPa	0.28	1.09	1.36	1.16
transverse strength MPa	0.11	0.32	0.44	0.36

3.5 Redox Activity Analysis

To verify the activity of the catalyst, the redox capability of the catalyst was investigated using H_2-TPR. From Fig. 4, tailings have two reduction peaks, respectively at 503 °C and 629 °C; 503 °C reduction peak belongs to Fe_2O_3 to Fe_3O_4, 629 °C reduction peak is Fe_3O_4 to FeO [9]; tailings + nanoalumina (5%) has two reduction peaks at 653 °C and 716 °C; tailings + nanoalumina (10%) has two reduction peaks at 615 °C and 757 °C respectively; tailings + nanoalumina (15%) has two reduction peaks at 586 °C and 714 °C respectively. Among them, the reduction peaks of 586 °C, 615°C and 653 °C are the reduction peaks of Fe_3O_4 to FeO, and 714°C, 716 °C and 757 °C are the reduction peaks of FeO to Fe [10]. Therefore, when compared with the tailings, different proportions of nanoalumina are added to the catalyst, and the reduction peak is shifted to the high temperature region. Moreover, with the gradual increase of the amount of nano-alumina added, the area of the reduction peak showed a trend of increasing first and then decreasing. 5% of thin aluminite + tailings have a reduction peak of 703 °C; 10% of thin aluminite + tailings have a reduction peak of 656 °C; 15% of thin aluminite + tailings have a reduction peak of 652 °C; the reduction peak from Fe_3O_4 to FeO and FeO to Fe is 652 °C, 6 and 656 °C and 703 °C respectively; when compared with the tailings, different proportions of thin aluminite are added to the catalyst, and the reduction peak shifts to the high temperature region, and the quantity is also decreasing. Because the number of reducing peaks of the catalyst decreases and the reducing peak area is reduced, this change leads to the denitrification efficiency of the strip catalyst prepared by thin alumina is lower compared with the strip catalyst prepared by nano-alumina.

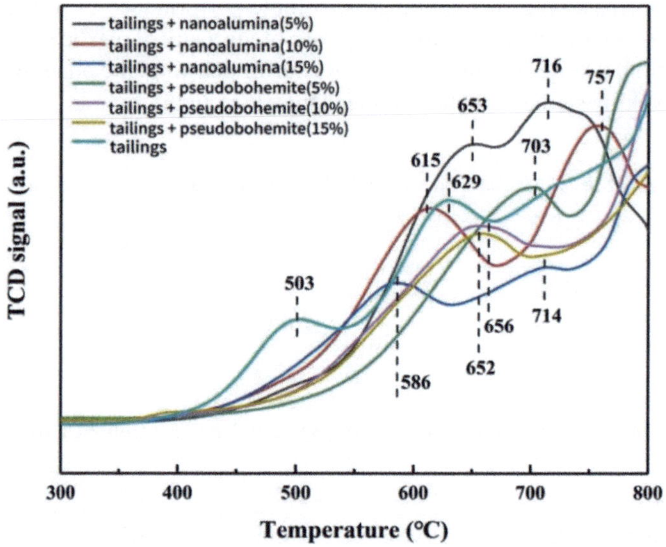

Fig. 4. H_2-TPR diagram of aluminum sol forming catalyst

4 Conclusion

This paper tests the denitration properties of two different strip catalysts prepared by nanoalumina and pseudobohemite, and analyzes the characterization of mechanical strength, XRD, NH_3-TPD H_2-TPR:

(1) As the amount of binder added increases, the mechanical strength of the strip catalyst produced using either binder undergoes significant enhancement. In general, the catalyst prepared with a binder exhibits superior mechanical strength compared to those without.
(2) There are Brønsted acidic sites and Lewis acidic sites on the surface of rare earth tailings, but the adsorption capacity of NH_3 species is weak; the catalyst surface is rich with nano alumina as binder, with Brønsted acidic site as the main adsorption site, and the stability and adsorption capacity of NH_3 species are improved.
(3) The denitrification efficiency of the tailings stands at a mere 8.1%, but substantial improvements are observed with the incorporation of nano alumina. Specifically, when 15% nano alumina is added, the maximum denitrification efficiency soars to 60.3%. However, using 15% thin alumina yields a maximum denitrification rate of 27.2%, which is still inferior. Despite these significant enhancements in denitration activity, challenges persist, including suboptimal N_2 selectivity and a narrow temperature range suitable for denitration.

References

1. Jabłońska, M.: Recent progress in the selective catalytic reduction of NO_x with NH_3 on Cu-SAPO-34 catalysts. Mol. Catal. **518**, 112111 (2022)
2. Mohajan, H.: Acid rain is a local environment pollution but global concern (2018)
3. Huang, L., Zeng, Y., Chang, Z., Zong, Y., Wang, H., Zhang, S., et al.: Promotional effect of phosphorus modification on improving the Na resistance of V_2O_5-MoO_3/TiO_2 catalyst for selective catalytic reduction of NO_x by NH_3. Mol. Catal. **506**, 111565 (2021)
4. Song, Y.J., Jesús, Y.M.L., Fanson, P.T., Williams, C.T.: Kinetic evaluation of direct NO decomposition and NO-CO reaction over dendrimer-derived bimetallic Ir-Au/Al_2O_3 catalysts. Appl. Catal. B-Environ. **154**, 62–72 (2014)
5. Meng, Z., Yan, X., Cui, M., ,et al.: Study on denitration performance of mineral catalyst prepared by high energy ball milling of rare earth concentrate supported Fe_2O_3 for NH_3-SCR. Mater. Chem. Phys. **276**, 125248 (2022). https://doi.org/10.1016/j.matchemphys.2021.125248
6. Kim, M., Yang, K.: The role of Fe_2O_3 species in depressing the formation of N_2O in the selective reduction of NO by NH_3 over V_2O_5/TiO_2-based catalysts. Catalysts **8**(4), 134 (2018). https://doi.org/10.3390/catal8040134
7. Liang, Q., Li, J., He, H., Liang, W., Zhang, T., Fan, X.: Effects of SO_2 on the low temperature selective catalytic reduction of NO by NH_3 over CeO_2-V_2O_5-WO_3/TiO_2 catalysts. Front. Environ. Sci. Eng. (2017). https://doi.org/10.1007/s11783-017-0926-6
8. Wenbo, Y.: Study on the physical separation of rare earth resources ofrare earth tailings in Baotou .Northeastern University (2013)
9. Apostolescu, N., Geiger, B., Hizbullah, K., et al.: Selective catalytic reduction of nitrogen oxides by ammonia on iron oxide catalysts. Appl. Catal. B **62**, 104–114 (2006)
10. Xiubing, H., Peng, W., Jinchang, T., et al.: CeO_2 modified Mn-Fe-O composites and their catalytic properties of NH_3-SCR denitrification. J. Inorg. Mater. **35**(5), 573–580 (2020)

Fast Pyrolysis of PET Plastic and Peanut Shell: Assessment of Blending Ratio Effect on Synergism

Shagali Abdulmajid Abdullahi, Haoran Qing, Kai Xu, Yi Wang, Sheng Su, Jun Xiang, and Song Hu(✉)

State Key Laboratory of Coal Combustion, School of Energy and Power Engineering, Huazhong University of Science and Technology, Wuhan 430074, China
husong_hust@hotmail.com

Abstract. The work presents a comprehensive thermal behavior analysis during fast pyrolysis of polyethylene terephthalate (PET) and peanut shell (PS), as the co-pyrolysis process represents a promising approach to producing valuable chemicals. The study further investigates the effect of blending ratio on kinetic parameters and bio-oil obtained under a higher heating rate of 500 °C/min. Thermogravimetry (TG) analysis proved the existence of synergistic effects. The mass loss rate presents a relative increase with PET ratio increase. The comprehensive pyrolysis index (CPI) change for PS/PET blends increased to ~320% when PET ratio increased from 25 wt% to 75 wt%. The average deviation ΔD value of –4.96% was observed for PS75PET25% blend, reflecting the optimal blend ratio during the co-pyrolysis process. The study also highlights the influence of varying blending ratios on bio-oil component selectivity. Furthermore, the presence of PET promoted the synthesis of aromatic oxygenated hydrocarbons in the liquid product while the content of derivatives such as terephthalic and benzoic acids was remarkably increased. Supervised machine learning model based on polynomial regression approach accurately predicted the experimental data and blending ratio influence on the activation energy and overall bio-oil selectivity.

Keywords: Pyrolysis · Synergistic Effect · Machine Learning

1 Introduction

Sustainable management and use of plastic and agricultural waste is a solution brought on by the increasing energy demand caused by population growth and tremendous global waste generation. The pyrolysis utilization of agricultural residue and plastic is considered an alternative to reducing CO_2 emissions and also reduction of process energy, which significantly lowers char yield, attributing to the H-abstraction in plastic polymers [1, 2]. Compared to traditional biomass pyrolysis, co-pyrolysis parameters like temperature, heating rate, blending ratio, etc., are important in synthesizing good-quality biofuels and recovering essential chemicals [3]. The thermogravimetric analysis (TGA) is used to

D. Han et al. (Eds.): ICEPG 2024, ESE, pp. 787–795, 2025.
https://doi.org/10.1007/978-3-031-86829-0_88

analyze the kinetic parameters and degradation properties of solid carbonaceous materials because of its ease of use and excellent precision [2]. TG analysis is further required to study the kinetic mechanism and pyrolysis characteristics of individual and blended materials [4].

The synergistic interaction of plastic and biomass is one of the main factors for improving the properties of co-pyrolysis by-products. During co-pyrolysis of biomass with plastics, Zhou et al. [5] reported an apparent synergistic effect leading to an increasing mass loss rate of the blends at high-temperature levels. Heating rate and mixing ratio are fundamental process parameters that dictate the overall pyrolysis kinetics characteristics and final product distribution [6–8]. The influence of varying blending ratios changes the reaction mechanism of the overall co-pyrolysis process [9]. Vo et al. [10] stated that blending plastic with biomass impacts the selectivity of aliphatics and aromatic hydrocarbons in the pyrolysis oil.

Furthermore, a strong synergistic effect was observed when the plastic ratio was reduced to 20%, and an apparent discrepancy in thermokinetic behavior was evident [10]. It has been reported that increasing the plastic blending ratio could promote secondary reactions of co-pyrolysis volatiles in hydrogen production and biomass tar cracking [6, 11]. In this study, pyrolysis kinetic behavior and synergistic effects of peanut shells (PS) and polyethylene terephthalate (PET) were systematically studied at different blending ratios (i.e., 25%, 50%, and 75%). A high pyrolysis heating rate heating of 500 °C/min was chosen, and the results of thermal characteristic mechanisms were reported.

2 Methodology

Peanut shell (PS) and commercially available polyethylene terephthalate (PET) were chosen for the co-pyrolysis study. Before blending the sample, the raw materials were ground to obtain about 200 mesh particle size and dried at 85 °C in an oven overnight. Detailed elemental and industrial analyses of individual feedstock were reported elsewhere [2]. TG experiments of PS with PET were carried out in a novel concentrated photothermal TG analyzer system (CP-TGA). A thorough operational explanation of the experimental setup has been published elsewhere [4, 12]. The complete process was conducted in an N_2 environment at a flow rate of 50 mL/min at a higher heating rate (500 °C/min) from ambient temperature to 900 °C and was selected for the co-pyrolysis of PS and PET. The experiment's reproducibility was confirmed. To comprehend how the blending ratio affects the co-pyrolysis samples, a crucible containing about 0.2 g sample is heated to 750 °C using a laboratory-scale photothermal pyrolysis reactor at a higher heating rate of 500 °C/min. Pyrolysis oil was trapped and collected in the U-tubes. Every experiment was repeated to guarantee data reliability.

The synergistic interactions effect during PS and PET co-pyrolysis of using the additive model is as follows (1) and (2):

$$D_{TM} = aW_{PS} + bW_{PET} \tag{1}$$

$$\Delta D = D_E - D_{TM} \tag{2}$$

where a and b stand for blended sample's mass ratio (a + b = 1). The discrepancy value relationship (ΔD) between the experiment value (D_E) and theoretical value (D_{TM}) are computed to determine the degree of synergy or inhibition [4].

3 Results and Discussion

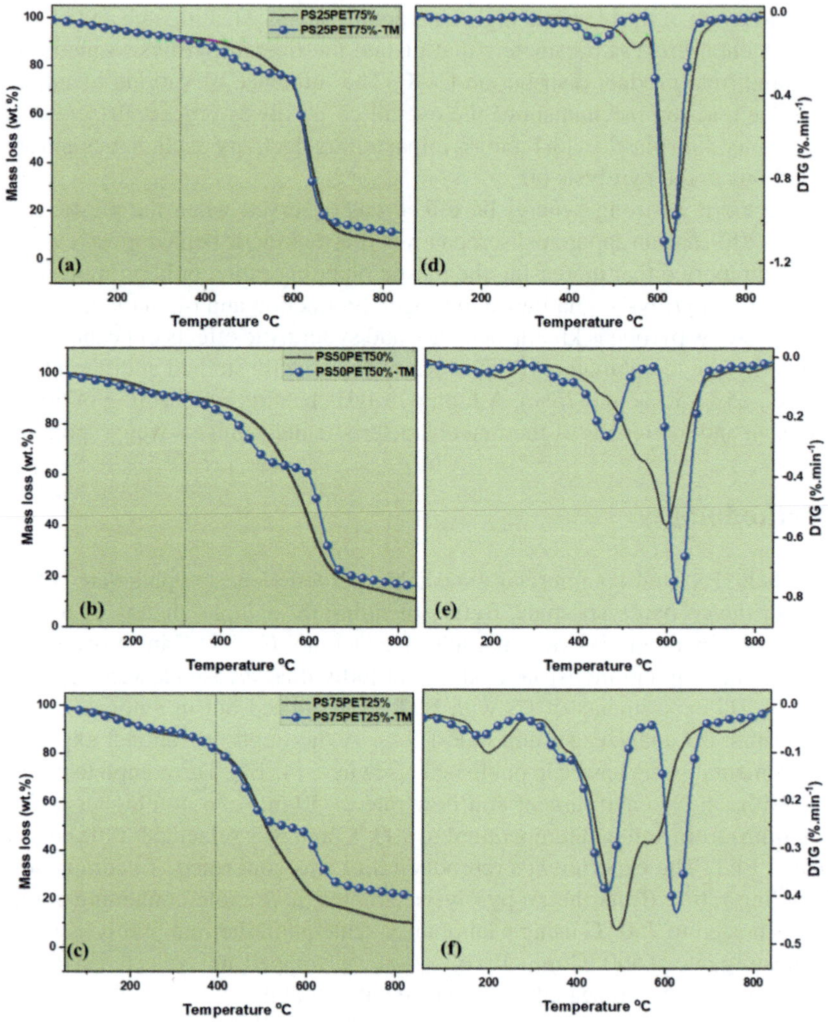

Fig. 1. Thermal analysis profiles of blend samples at 500 °C/min heating rate. (a)–(c) TG curves; (d)–(f) DTG curves.

The TG profiles of PS and PET mixtures are presented in Fig. 1a, b, c). In order to comparatively study the co-pyrolysis characteristics using varying approaches, normalized theoretical weight loss curves were obtained by multiplying the weight loss values of PS and PET by the corresponding mixing ratios. The calculated blends involve the theoretical addition of the TG values of PS and PET. The residual char variances between the experimented sample compared with PS25PET75%–TM, PS50PET50%–TM and PS75PET25%–TM were 5.5%, 6.02%; and 14.37%, respectively. As presented in Table 1, the initial temperature (T_{int}) and the maximum peak temperature (T_{max}) corresponding to DTG_{max} of PET and PS exhibited varying thermal behavior with each other. Contrarily to the thermal degradation of PS, which is mainly relatable to the behavior of biopolymer components via three significant degradation phases, i.e., dehydration, devolatilization, and char formation zones; the active pyrolysis phase of PET was a one-step reaction process between 581–684 °C at 500 °C/min heating rate. Even though some PS components are degraded during co-pyrolysis, especially when PS is 75%, PET melts without changing its composition and still mixes evenly with PS components [1]. It is worth noting that the differential thermogravimetry (DTG) profiles (Fig. 1) presented varying discrepancies, especially in Fig. 1e, f. It is evident from the figures that the DTG curves were not completely coinciding, indicating that there was a synergic interaction during PS and PET co-pyrolysis. Compared to PS25PET75% blend, the interaction between PS50PET50% and PS75PET25% was more pronounced, showing varying mass loss characteristics. The comprehensive pyrolysis index (CPI) is used to assess the performance and efficiency of pyrolysis, and a larger CPI value indicates the facile decomposition of the blended samples. In Table 1, it can be deduced that the CPI values for the theoretical model (TM) blends are 1.18, 3.16, and 1.77 that of experimented PS25PET75%, PS50PET50% and PS75PET25%, respectively. It is worth noting that the discrepancies in CPI value can be attributed to variability in the chemical composition.

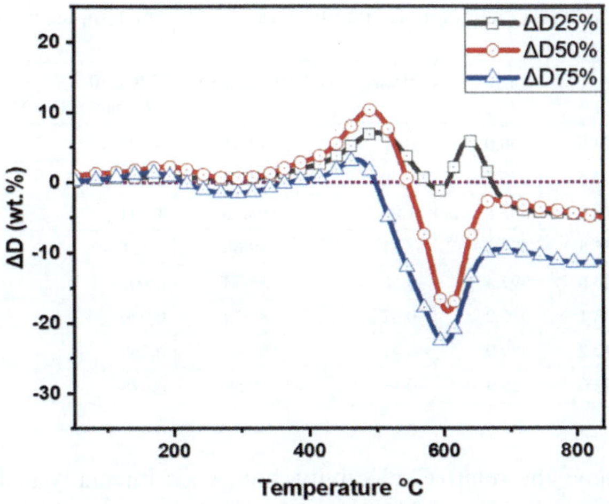

Fig. 2. Comparative TG mass loss deviation at different blending ratios.

For an in-depth understanding of the synergistic interactions during the co-pyrolysis process, the discrepancy ΔD between the TG experimental and theoretical TG mass loss data can be used to quantify the existing interactions. The comparative variations (ΔD) of the experimental against a theoretical model with temperature increase are shown in Fig. 2. For PS25PET75%, the value of ΔD almost remains unchanged at low temperatures up until around 380 °C, then, a sharp increasing trend was further observed. Positive ΔD values indicate an inhibitory effect due to the coating effect on PS by melting PET. As a result of the higher PET fraction (75 wt.%), the PET plasticity state hindered the association with PS and prevented volatile emission in this temperature range [13]. The ΔD of PS75PET25% is basically below 0, reflecting synergistic behavior between PET and PS during the co-pyrolysis process. The discrepancies in the synergistic effects could be attributed to the specific mixing ratio, which may represent a threshold for effective interaction since some of the components in the mixture begin to decompose at lower temperatures [13].

In the last column of Table 1 is the activation energy E of samples estimated using the Coats-Redfern method. The best-fitted results in accordance with a higher value of linear regression coefficient R^2 were chosen and reported. Notable, a higher value of R^2 does not guarantee that the chosen reaction mechanism of the pyrolysis process is suited the most. Since the activation energy is the least amount of energy needed to initiate the pyrolysis process, the energy needed to break the structure of PET materials was about six times higher than that of PS due to its complex aromatic ring structure. For the actual experimental blends, it was observed that the increase in PS ratio is directly proportional to the decrease in activation energy. Significant variance was observed between the theoretical and experimental activation energy. The variation is mainly attributed to the blending ratio and the chemical interactions of evolved volatile compounds driven by synergetic interactions [14]. All the experimentally obtained activation energy values for blends were higher than the theoretical values.

Table 1. Pyrolysis characteristic parameters of different samples at 500 °C/min

Sample	T_{int} (°C)	T_{max} (°C)	DTG_{max} (%/°C)	M_{loss} (%)	CPI × 10^{-6} ($\%^2.min^{-2} °C^{-3}$)	ΔD (%)	E (kJ/mol)
PS	304.0	468.0	–0.52	73.37	0.401	–	48.79
PET	570.6	626.2	–1.59	94.21	2.862	–	288.17
PS25PET75%	518.0	633.8	–1.06	93.86	1.294	–	134.98
PS25PET75%-TM	535.8	626.2	–1.21	89.00	1.521	0.65	46.12
PS50PET50%	404.6	597.4	–0.56	88.83	0.307	–	74.98
PS50PET50%-TM	368.1	626.2	–0.82	83.79	0.970	–0.58	28.87
PS75PET25%	312.2	489.0	–0.47	89.87	0.268	–	57.87
PS75PET25%-TM	295.0	625.3	–0.43	78.58	0.475	–4.96	31.47

Figure 3 displays the relative selectivity, both experimentally and theoretically of various chemical compositions of bio-oils produced by co-pyrolysis of PS, PET, and their blends. The family pyrolysis oil composition contains phenols, ethers, ketones, acids, aldehydes, amides, esters, etc. Several phenolic compound groups constituent the pyrolysis oil obtained from PS; similar studies have reported an excessive fraction of phenolic derivatives from PS bio-oil [15]. In contrast, the principal components in PET-derived oil were mainly terephthalic acid (~30%) and benzoic acid (~19%) due to high pyrolysis temperature. The decomposition of PET is initiated by β-scission, where the ester link is broken to form benzoic acid and other carboxylic derivatives. Considering the complex composition of co-pyrolysis oil, three classifications were adopted, i.e., acids, aromatics, and aliphatics derivatives. This approach enhances the understanding of the chemical composition and helps identify valuable compounds for fuels and to maximize desirable yields. It also makes comparisons with conventional fuels easier and highlights undesirable acids that may need to be reduced. Overall, this approach supports the development of efficient and valuable bio-oil production from biomass and plastic waste. About 12%, 14%, and 74% of acids, aromatic, and aliphatics derivatives were found in PS, respectively. Acidic and aliphatic components in PS bio-oil are primarily attributed to the thermal degradation of hemicellulose and cellulose, respectively [16]. Besides, the low selectivity of aliphatic derivatives in PET is due to its degradation into benzene components, and further decomposition into lighter compounds does not occur [17].

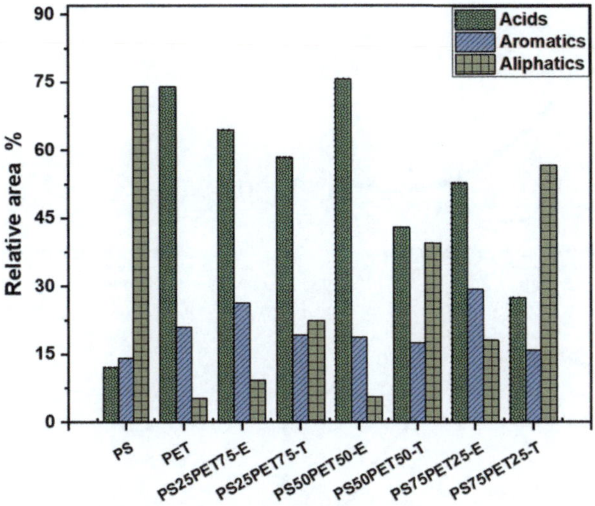

Fig. 3. GC–MS data obtained at 500 °C/min heating rate.

When comparing the influence of the blending rate of PET, it was found that the relative selectivity of acids increased while aliphatics derivatives significantly decreased with increasing PET blending ratio. Moreover, the interaction between PET and the bio-oil components of PS might cause the formation of new chemicals or the stabilization of certain fractions, thereby reducing the overall aliphatic content [18]. Accordingly, a decrease in PET blending ratio (i.e., an increase in the peanut shell ratio) effect on acids, aromatics, and aliphatics derivatives presented synergistic deviations of 9.4, 4.4, and 5.3, respectively. Thus, low or higher blending ratios (25, 50, and 75%) can hinder the formation and selectivity of various bio-oil components during the pyrolysis interaction of plastic and biomass.

Figure 4 presents a comparative predictive analysis using machine learning (ML) based on the polynomial regression model (PRM). ML has excellent prediction capability, and PRM analysis is a prediction model that works on statistical rules where the relationship between the independent and dependent variables is modeled to represent the relationship mathematically [19, 20]. The experiment datasets of activation energy and the selectivity of acids, aromatics, and aliphatics derivatives are utilized to predict the effect of the increase in blending ratio. Higher coefficient of determination R^2 values were observed for activation energy, showing that the model adequately captured the relationship between the variables.

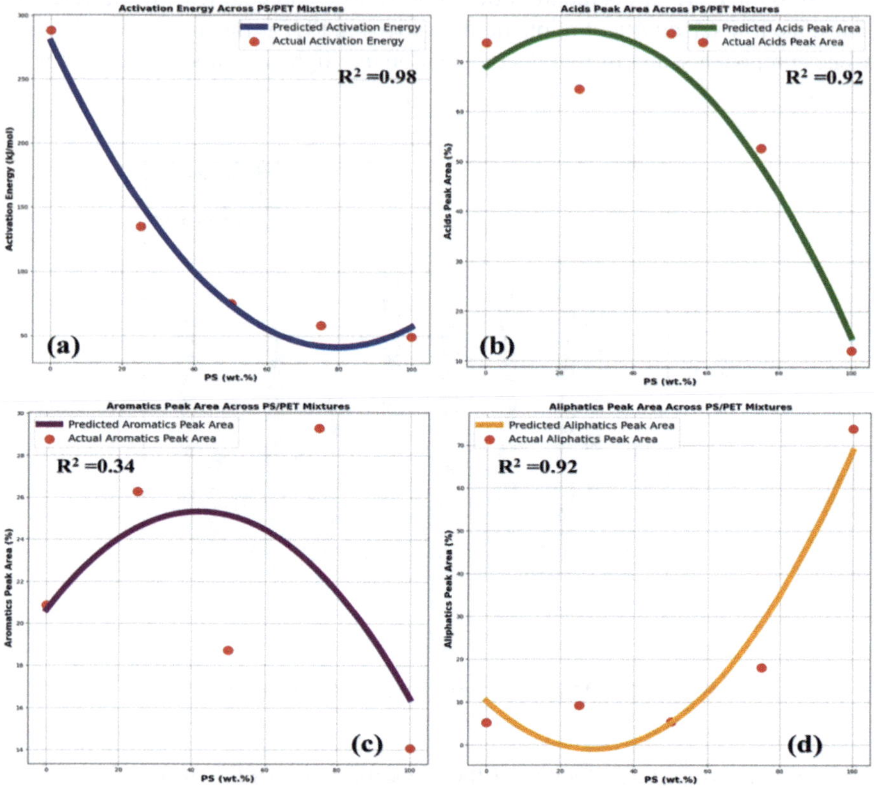

Fig. 4. Comparative analysis using machine learning approach.

4 Conclusion

The impact of blending ratio during co-pyrolysis of PET with peanut shell (PS) was examined in this study to understand the thermal kinetic behavior and synergistic interaction. TG/DTG profiles indicated that the co-pyrolysis properties of blends differ significantly from the theoretical combination of the individual samples, signifying the existence of synergism between PS and PET. Adding PS to PET rapidly decreased the activation energy to about 500% from 0% to 75% blending ratio. The PS/PET ratio played a crucial role in bio-oil component selectivity. The high selectivity of different acids and aromatic hydrocarbons was evident due to the interaction of PET. The developed machine learning model has revealed an outstanding prediction capability in relation to blending ratio with R^2 value relatively higher than 0.90 for activation energy, acids, and aliphatics derivatives selectivity.

Acknowledgement. The National Natural Science Foundation of China provided the research financial (NSFC No. 52076097).

References

1. Burra, K.G., Gupta, A.K.: Kinetics of synergistic effects in co-pyrolysis of biomass with plastic wastes. Appl. Energy **220**, 408–418 (2018)
2. Shagali, A.A., et al.: Thermal behavior, synergistic effect and thermodynamic parameter evaluations of biomass/plastics co–pyrolysis in a concentrating photothermal TGA. Fuel **331**, 125724 (2023)
3. Seah, C.C., et al.: Co-pyrolysis of biomass and plastic: circularity of wastes and comprehensive review of synergistic mechanism. Results in Engineering **17**, 100989 (2023)
4. Abdullahi Shagali, A., et al.: Synergistic interactions and co-pyrolysis characteristics of lignocellulosic biomass components and plastic using a fast heating concentrating photothermal TGA system. Renewable Energy **215**, 118936 (2023)
5. Zhou, L., Wang, Y., Huang, Q., Cai, J.: Thermogravimetric characteristics and kinetic of plastic and biomass blends co-pyrolysis. Fuel Process. Technol. **87**, 963–969 (2006)
6. Engamba Esso, S.B., et al.: Review on synergistic effects during co-pyrolysis of biomass and plastic waste: significance of operating conditions and interaction mechanism. Biomass Bioenerg. **159**, 106415 (2022)
7. Shagali, A.A., et al.: Pyrolysis characteristics and kinetic parameters assessment of typical agricultural residues using high heating photothermal TGA. J. Anal. Appl. Pyrol. **174**, 106109 (2023)
8. Biagini, E., Fantei, A., Tognotti, L.: Effect of the heating rate on the devolatilization of biomass residues. Thermochim. Acta **472**, 55–63 (2008)
9. Ullah, F., et al.: Assessing pyrolysis performance and product evolution of various medical wastes based on model-free and TG-FTIR-MS methods. Chem. Eng. J. **473**, 145300 (2023)
10. Vo, T.A., et al.: Co-pyrolysis of lignocellulosic biomass and plastics: a comprehensive study on pyrolysis kinetics and characteristics. J. Anal. Appl. Pyrol. **163**, 105464 (2022)
11. Dewangan, A., Pradhan, D., Singh, R.K.: Co-pyrolysis of sugarcane bagasse and low-density polyethylene: influence of plastic on pyrolysis product yield. Fuel **185**, 508–516 (2016)
12. Abdullahi Shagali, A., et al.: Fast co-pyrolysis of corncob with plastics: evaluation of thermal behavior using deconvolution procedure, kinetic analysis and product characterization. Fuel **381**, 133300 (2025)
13. Najafi, H., Laye, Z.R., Sobati, M.A.: Deep insights on the co-pyrolysis of tea stem and polyethylene terephthalate (PET): unveiling synergistic effects and detailed kinetic modeling. J. Environ. Chem. Eng. **12**(5), 113906 (2024). https://doi.org/10.1016/j.jece.2024.113906
14. Söyler, N., Ceylan, S.: Thermokinetic analysis and product characterization of waste tire-hazelnut shell co-pyrolysis: TG-FTIR and fixed bed reactor study. J. Environ. Chem. Eng. **9**, 106165 (2021)

15. Bu, Q., Cao, M., Wang, M., Vasudevan, S.V., Mao, H.: Enhancement of bio-oil quality over self-derived bio-char catalyst via microwave catalytic pyrolysis of peanut shell. J. Anal. Appl. Pyrol. **164**, 105534 (2022)
16. Kumar, M., Rai, D., Bhardwaj, G., Upadhyay, S.N., Mishra, P.K.: Pyrolysis of peanut shell: kinetic analysis and optimization of thermal degradation process. Ind. Crops Prod. **174**, 114128 (2021)
17. Suriapparao, D.V., Kumar, D.A., Vinu, R.: Microwave co-pyrolysis of PET bottle waste and rice husk: effect of plastic waste loading on product formation. Sustainable Energy Technol. Assess. **49**, 101781 (2022)
18. Yamoum, C., Magaraphan, R.: Effect of peanut shell content on mechanical, thermal, and biodegradable properties of peanut shell/polylactic acid biocomposites. Polym. Compos. **38**, 682–690 (2017)
19. Gong, Y., Zhang, P.: Predictive Analysis and research of python usage rate based on polynomial regression model. In: 2021 3rd International Conference on Artificial Intelligence and Advanced Manufacture (AIAM) pp. 266–270 (2021). https://doi.org/10.1109/AIAM54119.2021.00061
20. Huang, Z., et al.: Effects of waste-based pyrolysis as heating source: meta-analyze of char yield and machine learning analysis. Fuel **318**, 123578 (2022)

Feasibility of Recycling Waste Lime Soil in Road Subgrade

Yongjian Chang[1], Leilei Gu[1], Kefei Fan[2], Peng Zhang[3](✉), and Shengnian Wang[3]

[1] China First Highway Engineering Co., Ltd., Beijing 100024, China
[2] Nanjing Jiangbei New Area Public Utilities Holding Group Co.Ltd., Nanjing 225009, China
[3] College of Transportation Engineering, Nanjing Tech University, Nanjing 211816, China
zhangpeng-mail@njtech.edu.cn

Abstract. The cementitious internal structure of waste lime soils has been destroyed after subjecting to years of traffic loadings and their recycling often requires secondary technical treatment. This study summarized the advantages and disadvantages of traditional treatment technologies, including lime and new agent stabilization. The feasibility of recycling waste lime soils reused in road subgrades was evaluated. Their restriction issues in engineering practices were discussed. The challenges of new low-carbon agent stabilization for recycling waste lime soils in road subgrade were analyzed. The outcomes of this study could be helpful technical support for recycling wasted lime soils.

Keywords: waste lime soil · feasibility · recycling treatment · road subgrade

1 Introduction

With the rapid development of the national economy, China's public transportation industry has flourished. And because of increasing People's demand for transportation, old roads are unable to withstand the growing traffic loadings and meet common driving requirements. When expanding or rebuilding these old roads, their construction activities will be bound to lead to issues such as how to stack these demolished old roadbed filling materials and how to obtain new high-quality roadbed filling materials. Particularly, the current policies have issued many tightening restrictions on natural soil and rock resource exploitation, resource recycling problems with the increasing amount of construction solid wastes cannot be ignored anymore. At present, there are three main methods for solid waste treatment in China. The one is nearby dumping. However, due to the lack of planning and unreasonable distribution of solid waste disposal sites in various cities in China, which are far from the source of solid waste generation, great burdens are caused by transportation costs and time consumption. As a result, the method of nearby dumping is usually chosen in engineering practices despite it being certainly the most wasteful, environmentally polluting, and least desirable approach. Another way of solid waste treatment is simple landfills. It is the most important solid waste treatment method in China at present. However, land occupation and secondary environmental issues are troublesome problems engineers have to face. The third way of solid waste

D. Han et al. (Eds.): ICEPG 2024, ESE, pp. 796–801, 2025.
https://doi.org/10.1007/978-3-031-86829-0_89

treatment is comprehensive treatment. It follows the principles of disposal reduction, resource utilization, and harmlessness for the recycling of solid waste. These principles can effectively achieve the recycling of solid waste, reduce environmental pollution, and have good economic and environmental benefits. Nevertheless, although China has made certain achievements in the disposal of solid waste, the research results and implementation levels are still far from enough, especially in the face of the reconstruction of a large number of old roads. There is still no clear and suitable method for the disposal of waste lime soils. Conducting secondary recycling research on waste lime soils is thus of significance in road engineering practices.

2 Characteristics of Waste Lime Soils

It is well known that lime soil subgrade and pavement have good mechanical properties and stability, and are characterized by sufficient material sources and easy construction. Especially in the southeastern region of China, the common engineering soils are mainly composed of fine-grained soils. The method of lime stabilization is a highly popular way of increasing soil strength, as shown in Fig. 1. Some scholars have conducted much research on lime-stabilized soils. Liu et al. [1] found that the application of lime and fly ash in expansive soils could lead to a decrease in their plasticity, swelling, and shrinkage index and a mechanical performance increase; Wang et al. [2] showed that the stress-strain relationship of lime-stabilized clayey soil under freeze-thaw cycles belonged to the mechanical hardening type with a plastic failure as the main failure mode; Zeng [3] pointed out that as the mixing ratio of lime increased in red clay, the maximum dry density of these soils gradually decreased, the optimal moisture content increased, and their compressive strength showed a trend of first increasing and then decreasing with an optimal mass mixing ratio of 8%; Zhang et al. [4] studied the permeability changes of lime-stabilized loess and found that the addition of lime significantly changed the permeability of fine-grained soil. The permeability of the soil changed significantly within 90 days of curing time, and the permeability coefficient decreased most significantly within 30 days; Zhao et al. [5] reported that lime could effectively inhibit the development of cracks in improved weathered phyllite and red clay, and their best stabilization effect could be achieved when the lime mixing ratio was 0% ~ 3%. However, when it exceeded 5%, an unexpected counterproductive effect would be triggered seriously.

Fig. 1. Technical scheme of traditional lime-stabilized soil

3 Recycling of Waste Lime Soils by Lime

Engineering wastes, as the main part of construction waste in China, have the characteristics of a large base, diverse components, and difficult treatment. Their disposal seriously affects the development of cities. At the same time, with the rapid development of the transportation industry, the volume of highway traffic has increased significantly. Most of the arteries in China have been overloaded, resulting in accelerated deterioration of the overall strength of many road subgrades, and some main sections even urgently need maintenance and reconstruction. However, owing to factors such as difficulty in obtaining materials in certain areas, it is challenging to use conventional road reinforcement schemes to complete roadbase construction. To fully utilize the original pavement materials and reduce reconstruction investment, the recycling of waste lime soils can not only turn wastes into treasure, save engineering costs, and shorten the earthwork engineering cycle, but also improve the economic benefits of the project. Therefore, some scholars have conducted a series of studies on their disposal and recycling, as shown in Fig. 2. Shen et al. [6] utilized four different types of engineering waste soils to produce recycled bricks with different proportions and found that the strength of these recycled bricks could be increased by 26.7%. Qin et al. [7] addressed the problems of sludge disposal by simultaneously incorporating engineering waste soil and cement for improvement and pointed out that both of them could improve the physical and mechanical properties of the soil and meet the requirements for earthwork. Scholars have also provided some other methods for the treatment of waste lime soils. For example, Zhang et al. [8] conducted experiments on cement cold recycled waste lime soil drawing on the experience of cold recycled old asphalt concrete materials and showed that the compressive strength, splitting strength, and frost resistance of waste lime soils after adding cement were significantly improved. Good road performance can thus be achieved.

Fig. 2. Characteristics of waste lime soils restabilized with lime

4 Advantages of New Agents for Waste Lime Soil Stabilization

In recent years, geopolymer materials have attracted widespread attention from researchers around the world due to their high strength, durability, frost resistance, locally sourced materials, more sustainable development, and less environmental pollution than traditional cement. These new agents are essentially a network of dense cementitious mineral aggregates formed by the depolymerization and geopolymerization of active aluminosilicates (such as blast furnace slag, fly ash, etc.) in alkaline environments. If they were used for the improvement of roadbed fillers, their high strength and durability can increase the bonding force between filler particles, making them less likely to take place significant deformation due to the transmission of upper structures and long-term traffic loads. However, the current research on geopolymers at home and abroad mostly focuses on the types and proportions of their raw materials and activators, as well as the alkaline excitation environment. Granizo et al. [9] found that the main products of metakaolin activated by calcium hydroxide are sodium aluminosilicate and calcium silicate hydrate gel. Hana et al. [10] studied the effects of temperatures and curing times on the strength of geopolymers using fly ash and alkali activators. Feng et al. [11] pointed out that the strength of geopolymer binder could reach the maximum when the mass ratio of fly ash to slag was 1, the water glass modulus was 1.0, the sodium peroxide content was 6%, and the water-binder ratio was 0.5. Chu et al. [12] reported the optimal mixing ratio of metakaolin for improving the corrosion resistance of cement soil. Zhu [13] and Peng et al. [14] applied geopolymer to repair concrete pavement by comparing the mechanical performance differences between geopolymer and cement stabilization. Although the mechanical properties of geopolymer do have significant advantages compared to ordinary Portland cement, these properties are not inherent to geopolymer raw materials but can be obtained through reasonable design of the ratio between raw materials and alkali activators, as well as correct processing technology. Particularly, geopolymer reactions require alkaline environments and higher temperatures, which poses challenges to environmental safety and troubles in construction conditions. There is lacking more detailed research on how to stably excite aluminosilicates in a weak alkaline environment at room temperature. Since the residual alkalinity of waste lime soil itself is certain, there is currently a lack of relevant research on how to economically and effectively restabilize waste lime soils by geopolymers to recycle roadbed filling. Great corresponding work thus needs to be launched to provide theoretical and technical support for the reuse of waste lime soils.

In addition, the metal oxide hydration products present in the active materials or reinforcement binders in the reuse of waste lime soil can spontaneously undergo thermodynamic and chemical reactions with CO_2 in the air, forming carbonates. This technology is expected to achieve low energy consumption and large-scale CO_2 storage. Especially, most industrial solid wastes, such as fly ash, blast furnace slag, steel slag, and waste cement, contain many active mineral components like CaO, MgO, etc. Therefore, their raw materials can be pretreated through thermal activation or alkali activation to effectively increase their reaction activity, thereby achieving CO_2 capture and mineralization. Therefore, it is not only beneficial for solving the problem of waste disposal and improving engineering efficiency but also for achieving negative carbon sequestration and serving the national dual carbon goals if they were used in the restabilization of waste lime soils as shown in Fig. 3.

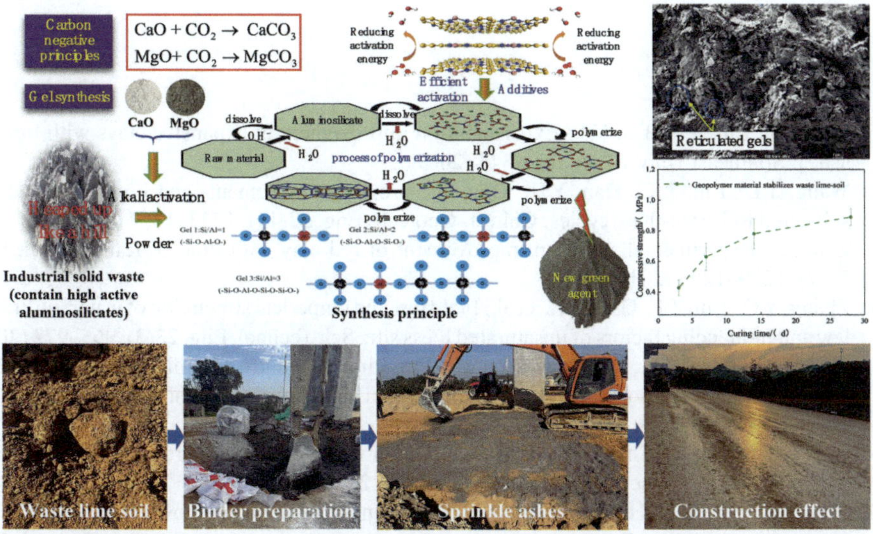

Fig. 3. Application of new geopolymer agents in the restabilization of waste lime soils

5 Conclusion

This study summarized the advantages and disadvantages of traditional and new treatment technologies for recycling waste lime soils in road engineering practices. Some main conclusions are obtained as follows.

(1) The chemical reaction of waste lime soils has already entered the diagenetic stage after subjecting to years of traffic loadings. When their internal cementation was destroyed, it was impossible to restore them as before. Conducting recycling research on waste lime soils is of significance in road engineering practices.
(2) Due to differences in the original process parameters and service life of waste lime soil, the residual activity content in waste lime soil varies significantly. If lime is reused to stabilize waste lime soils and its usage amount is not properly controlled, the drying and thermal shrinkage shall be significant in the restabilized soils.
(3) The restabilization of waste lime soils with geopolymer can avoid drying shrinkage problem. However, their raw material composition design and engineering performance evolution of restabilized waste lime soil are unclear at present.
(4) The physical and mechanical properties of geopolymer-restabilized waste lime soils have significantly changed compared with that of lime soil. These changes may affect the construction process of subgrade filling to some extent. More research needs to be carried out to clarify its construction quality control standards.

Acknowledgment. This study was supported by the Key Program of the National Natural Science Foundation of China (42330704), the Science and Technology Planning Project of Jiangsu Province(No.BE2022605), the National Natural Science Foundation of China (41902282), and the Science and Technology Development Planning Project of Nanjing, China (202211011).

References

1. Zha, F.S., Liu, S.Y., Du, Y.J.: Experiment on improvement of expansive clays with lime-fly ash. J. Southeast Univ. **37**(2), 339–344 (2007)
2. Wang, T.L., Liu, J.K., Tian, Y.H.: Dynamic properties of cement- and lime-improved soil subjected to freeze-thaw cycles. Chin. J. Geotech. Eng. **32**(11), 1733–1737 (2010)
3. Zeng, J.: Experimental study on improvement of red clay with lime. J. Railway Sci. Eng. **13**(7), 1289–1293 (2016)
4. Zhang, Y.C., Liu, Z.P., Gao, X.L., et al.: Field test and numerical simulation of water infiltration law and influencing factors of unsaturated loess site. Sci. Technol. Eng. **23**(3), 965–972 (2023)
5. Zhao, X.S., Rao, J.L., Chen, Z.X., et al.: Experimental study on fully weathered phyllite composite improved soil and numerical analysis of subgrade settlement. J. Railway Sci. Eng. **20**(2), 554–564 (2023)
6. Gao, Q., Xiao, J.Z., Shen, J.Y.: Effects of garden waste on the properties of sintered bricks prepared with construction spoil. J. Build. Mater. **25**(11), 1195–1202 (2022)
7. Qin, Y.Y., Qin, Z.P., Huang, H., et al.: Experimental study on composite solidification and improvement of silt soil by cement and coarse particle engineering spoil. Sci. Technol. Eng. **24**(12), 5076–5087 (2024)
8. Zhang, Q., Wei, H.W., Wang, Y.B., et al.: Study on cement stabled cold recycling lime-treated soil. J. Xi'an Univ. Archit. Technol. **44**(5), 632–636 (2012)
9. Granizo, L.M., Alonso, S., Blanco-Varela, T.M., et al.: Alkaline activation of metakaolin: effect of calcium hydroxide in the products of reaction. J. Am. Ceram. Soc. **85**(1), 225–231 (2002)
10. Hana, O., Jonáš, T., Petra, M., et al.: Effect of mass ratio and milling on compressive strength and corrosion resistance of blast-furnace slag/fly ash geopolymer activated by solid alkali activator. J. Sustain. Metall. **8**(4), 1961–1974 (2022)
11. Feng, H., Wen, J., Shao, Q., et al.: Carbonation resistance of fly ash/slag based engineering geopolymer composites. Constr. Build. Mater. **449**, 138471 (2024)
12. Chu, C.F., Wang, Y.H., Zong, W.Q.: Effect of aggregate grading on durability of expansive soil modified by solid waste. J. Build. Mater. **27**(3), 237–244 (2024)
13. Zhu, B.Q.: Application of fly ash geopolymer in cement concrete pavement repair. Railway Constr. Technol. **11**, 95–97 (2015)
14. Sun, K.K., Peng, X.Q., Ran, P., et al.: The Influence factor of the anti-freeze of geopolymer concrete. Mater. Rep. **35**(24), 24095–24100 (2021)

Study on Shell-Wrapped Granulation of Demolished Foam Blocks

Hongling Pan, Yuting Xiang, Shengnian Wang(✉), Peng Zhang, Jue Gao, and Xingyi Luo

School of Transportation Engineering, Nanjing Tech University, Nanjing 211816, China
shengnian.wang@njtech.edu.cn

Abstract. This study innovatively improved the performance of demolished foam blocks by way of shell-wrapped granulation. Their regeneration process was investigated by a series of indoor tests, including the water-cement ratio for shell-wrapped granulation, the air blast temperature, and the granulating rotation speed. The results showed that these three technological parameters were crucial for the shell-wrapped granulation rate of demolished foam blocks. The recommended value of the water-binder ratio should be 0.2–0.3. The ideal air blast temperature lay between 60 °C and 70 °C. There should be a threshold for the rotation speed to control the shell-wrapped granulation rate, which was suggested to be 35 r/min. This study can provide a practical reference for the recycling of construction and demolition wastes.

Keywords: construction and demolition waste · shell-wrapped granulation · regeneration process · technological parameter

1 Introduction

With the acceleration of urbanization, the amount of construction and demolition wastes has risen sharply [1], causing problems such as land occupation and environmental pollution [2]. Fortunately, these construction and demolition wastes contain more than 90% of waste concrete, crushed stone, bricks, and other pollution-free inorganic materials. If they were converted into recycled aggregates by some specific technologies to partially or completely replace natural aggregate [3], it would be very beneficial to effectively reduce resource consumption and environmental impact induced by the open-air storage of these construction and demolition wastes.

Recycled aggregates from construction and demolition wastes are different from natural aggregates in that they have insufficient initial strength, adhered residual mortar, developed internal cracks, and strong water absorption. How to eliminate such impact of these influencing factors, minimizing the inherent pores and micro-cracks in recycled aggregates is always the key to comprehensive utilization. Up to now, many scholars have tried to strengthen recycled aggregates by reinforcing or removing adhered residual mortar on the surface of aggregates. For example, Jiang et al. [4] and Li et al. [5]

D. Han et al. (Eds.): ICEPG 2024, ESE, pp. 802–808, 2025.
https://doi.org/10.1007/978-3-031-86829-0_90

adopted the volcanic mortar immersion method to improve the performance of recycled aggregates. However, the existence of residual micro volcanic ash particles might reduce the workability of concrete. Santha [6] attempted to remove these adhered residual mortar by combining heat and chemical treatment. Their shortcoming was a large amount of wastewater containing chemicals generated due to multiple cleaning. Liu et al. [7] pointed out that the strength of recycled aggregates could meet national standards. However, for the study of shell-wrapped granulation, there are a few research reports on construction and demolition wastes up to now, no matter of theoretical research and regeneration technologies. Since the recycled aggregates from the shell-wrapped granulation of construction and demolition wastes possess a composite structure aggregate with strong strength in external shells and weak strength in inner cores, their compressive performances do not only depend on the shell-wrapped materials but are also significantly related to granulation technological parameters which might cause problems like multiple core molding, easy crushing of shell-wrapped materials, and uneven shell-wrapped thickness. Therefore, further study on the shell-wrapped granulation of recycled aggregates from construction and demolition waste is of great significance for improving the performance of recycled aggregates and the resource utilization of construction and demolition wastes.

This study innovatively improved the performance of demolished foam blocks by way of shell-wrapped granulation. The influence of the water-cement ratio, the air blast temperature, and the granulating rotation speed on the granulation rate and mechanical performance of demolished foam blocks were investigated, and their suggested regeneration technological parameters were summarized.

2 Materials and Methods

2.1 Testing Materials

The P.O. 42.5 ordinary Portland cement produced by Jiangxi Xinyu Fenyi Hailuo Cement Co., Ltd., China was employed as the shell-wrapped material. The recycled demolished foam blocks collected from Nanjing Kangheng Renewable Resources Utilization Co., Ltd., China (Fig. 1) were selected to conduct experimental research by a disc granulator (Fig. 2). Their basic physical properties are listed in Table 1.

Table 1. Physical and mechanical properties of demolished foam blocks

Index	Apparent density (kg/m^3)	Water absorption rate (%)	Crushing value (%)
Values	956.3	43.5	42.3

2.2 Testing Scheme

The water-cement ratio, the air blast temperature, and the granulating rotation speed could have a significant impact on the shell-wrapped granulation process. So this experimental study just mainly focused on the influence of these factors. To quantitatively

Fig. 1. Foam blocks

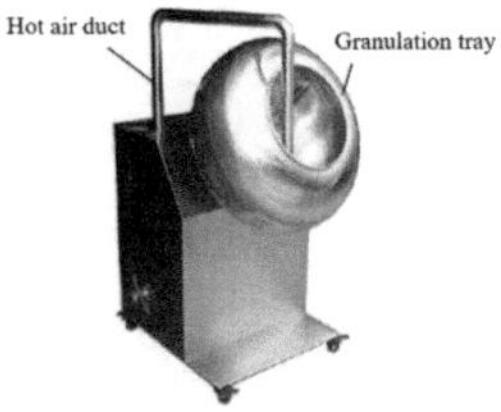

Fig. 2. BY-400 disc granulator

characterize the effect of granulation, the molding rate refers to the ratio of the quantity of qualified shell-wrapped recycled aggregates to the initial recycled aggregates. These qualified recycled aggregates refer to hard shell-wrapped aggregates that have a uniform surface, no sharp protrusions, and no cracks. The crushing values of these recycled aggregates were used to evaluate their mechanical performance.

(1) Water-cement ratio

The water-cement ratio has a significant impact on the formation and strength improvement of shell-wrapped recycled aggregates. Considering that cement slurry shall be wrapped on the recycled aggregates by way of thermal spray. The water-cement ratios of 0.1, 0.2, 0.3, 0.4, and 0.5 were designed in the tests. The air blast temperature was 40 °C, and the granulating rotation speed was kept at 20 r/min.

(2) Air blast temperature

The influence of air blast temperature on the granulation rate was designed with a temperature program from 40 °C to 100 °C at an interval of 10 °C. The water-cement ratio was fixed at 0.2, and the granulating rotation speed was kept at 20 r/min.

(3) Granulating rotation speed

Some studies have shown that the diameter, inclination angle, and rotational speed of the disc granulator have a significant impact on the compactness of recycled aggregates [8], and follow a critical rotation speed (Fig. 3)

$$N_c = 42.28 \times \sqrt{D^{-1} \cdot \sin\theta} \quad (1)$$

where Nc is the critical rotation speed, rad/s; θ is the dip angle of the disc granulator; D is the size of the disc granulator. According to the size and design of the disc granulator used in this study, the critical rotation speed should be Nc = 54.68 r/min.

The practical research showed that the actual granulating rotation speed should be controlled between 0.5Nc-0.75Nc. So the theoretical granulating rotational speed during operation should be 27.34–41.01 r/min. Following these suggestions, the initial granulating rotational speed was set to 25 r/min and was gradually increased from 5 r/min until it reached 50 r/min. The water-cement ratio was fixed at 0.2, and the air blast temperature was 40 °C. The failure proportion of recycled aggregates can be used to characterize the granulation effect. So it was used for regeneration evaluation.

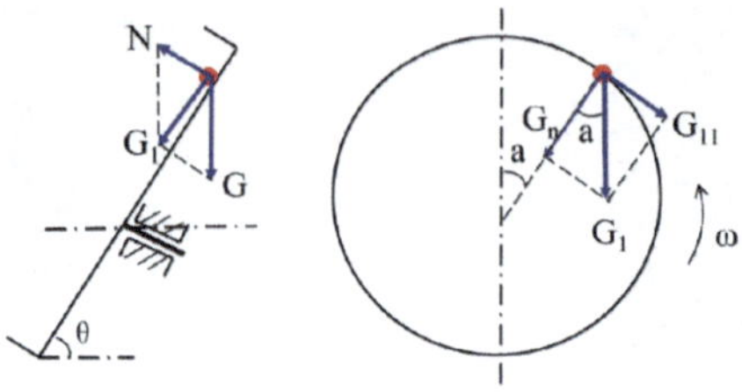

Fig. 3. Force diagram of shell-wrapped recycled aggregates in the granulator

2.3 Process of Shell-Wrapping

According to the experimental scheme, the recycled aggregates were firstly presoaked with sodium silicate solution, then they were placed in a disk granulator and the cement slurries were sprayed onto the surface of the aggregates by a gun connected to an air compressor as the granulator rotated. Since the cement slurries were solidified quickly due to the action of hot air and the mutual interaction forces between the aggregates could dense the wrapped shell, these recycled aggregates with a hard denser external shell could be generated rapidly as shown in Fig. 4. After granulation, these regenerated aggregates were cured for seven days in an environment with a room temperature of 20 °C ± 3 °C and a relative humidity of no less than 90%. Then their molding rate and failure proportion were calculated.

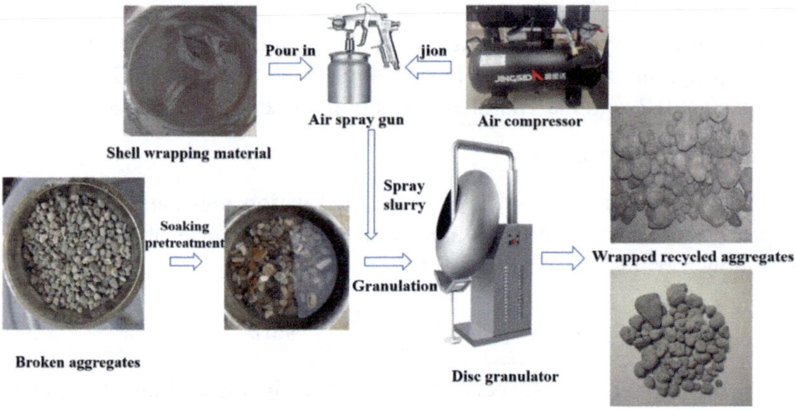

Fig. 4. Shell granulation process

3 Results and Analysis

3.1 Influence of Water-Cement Ratio on Molding Rate

Figure 5 illustrates the molding rate variation of recycled aggregates with different water-cement ratios. It could be observed that the molding rate of recycled aggregates exhibited a sharp increase followed by a gradual decline as the water-cement ratio increased. When the water-cement ratio reached 0.2, the molding rate achieved the best value. The reason why the molding rate rose significantly might be that the lower water-cement ratios could not easily adhere cement to the surface of recycled aggregates, and cement might form agglomerates itself, leading to a lower molding rate. When the water-cement ratio increased, the adherence of the cement increased, and the molding rate was improved. But this improvement would also cause adjacent aggregates prone to agglomeration and their wrapped thickness might decrease if the water-cement ratio was much higher. Therefore, the water-cement ratio by multiple tries was suggested within 0.2–0.3 for ensuring the molding rate higher than 70%.

3.2 Influence of Air Blast Temperature on Molding Rate

Figure 6 illustrates the molding rate variation of recycled aggregates with different air blast temperatures. It could be observed that the molding rate increased first and then decreased with the air blast temperature. When the temperature reached 70 °C, the molding rate achieved its best value. When the temperature exceeded 70 °C, the shell-wrapped recycled aggregate would crack due to the fast evaporation of moisture and the solidification of cement. Therefore, the recommended air blast temperature for shell-wrapped granulation was 60–70 °C.

3.3 Influence of Granulating Rotation Speed on Failure Proportion

Figure 7 shows the failure proportion of recycled aggregate with different granulating rotation speeds. It could be found that when the rotation speed was lower than 35 r/min, the failure proportion of shell-wrapped recycled aggregates was less than 5%. When the rotation speed continued to increase, the failure proportion rose rapidly. Therefore, the working rotation speed of the granulator should be controlled below 40 r/min. Following the suggested value of theoretical working speed given in Sect. 2.2, the granulating rotation speed of 35 r/min should be the optional value, which could ensure a relatively high molding rate and a lower failure proportion of shell-wrapped recycled aggregates in the actual granulation.

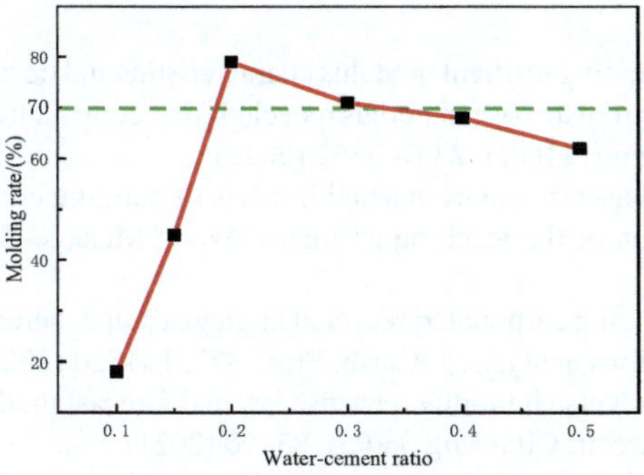

Fig. 5. Influence of water-cement ratio

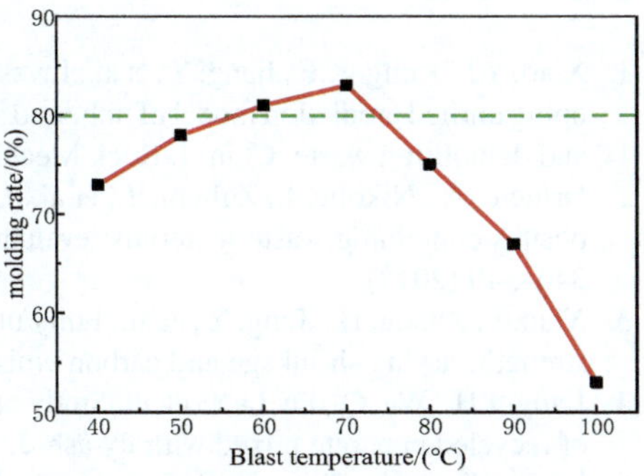

Fig. 6. Influence of air blast temperature

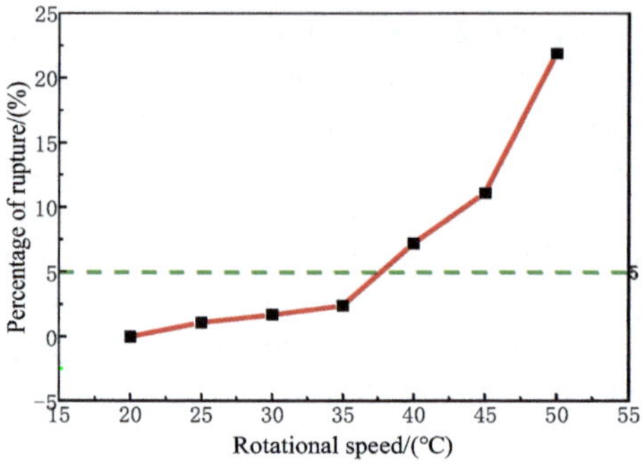

Fig. 7. Influence of granulating rotation speed

4 Conclusions

This study innovatively improved the performance of demolished foam blocks by way of shell-wrapped granulation. Their regeneration process controlled by some main influencing factors was explored. Some main conclusions were obtained as follows.

(1) The water-cement ratio could cause the molding rate to increase first and then decrease. Its suggested value was 0.2–0.3.
(2) The suggested air blast temperature for shell-wrapped granulation was 60–70 °C since high temperature would result in fast moisture evaporation and solidification.
(3) The optional granulating rotation speed was 35 r/min, which could ensure a relatively high molding rate and a lower failure proportion of aggregate regeneration.

Acknowledgment. This study was supported by the Key Program of the National Natural Science Foundation of China (42330704), the Science and Technology Planning Project of Jiangsu Province(No.BE2022605), the National Natural Science Foundation of China (41902282), and the Science and Technology Development Planning Project of Nanjing, China (202211011).

References

1. Xiao, Y.J., Kong, K.F., Jiang, Y., et al.: Investigating resilient modulus characteristics and developing unified prediction model of unbound aggregate base materials recycled from construction and demolition waste. Chin. J. Rock Mech. Eng. **41**(11), 2337–2352 (2022)
2. Drinčić, A., Nikolić, I., Zuliani, T., et al.: Long-term environmental impacts of building composites containing waste materials: evaluation of the leaching protocols. Waste Manage. **59**, 340–349 (2017)
3. Xiao, J., Zhang, H., Tang, Y., et al.: Fully utilizing carbonated recycled aggregates in concrete: strength, drying shrinkage and carbon emissions analysis. J. Clean. Prod. **377**, 134520 (2022)
4. Jiang, J.H., Wu, Q., Fu, Q.Y., et al.: Study on reverse humidity response law and forecast model of recyeled concrete mixed with fly ash. J. Archit. Civil Eng **38**(05), 83–90 (2021)
5. Li, J., Xiao, H., Zhou, Y.: Influence of coating recycled aggregate surface with pozzolanic powder on properties of recycled aggregate concrete. Constr. Build. Mater. **23**(3), 1287–1291 (2009)

6. Santha Kumar, G., Minocha, A.K.: Studies on thermo-chemical treatment of recycled concrete fine aggregates for use in concrete. J. Mater. Cycles Waste Manage. **20**(1), 469–480 (2017). https://doi.org/10.1007/s10163-017-0604-6
7. Liu, Q.D., Zhang, X.L., Qin, W.P., et al.: Study on strengthening and application of recycled aggregate of discarded brick. Concrete **2**, 42–45 (2018)
8. Xie, M.D.: Study on Strengthening Recycled Coarse Aggregate by Microbial Carbon Fixation. Southeast University, Nanjing (2021)

Soil Moisture Retrieval of Tea Plantations Based on Sentinel Data

Yan Li[1,2](✉) and Jing Zhang[1,2]

[1] Key Laboratory of Estuarine Ecological Security and Environmental Health of Fujian Province, Zhangzhou 363105, Fujian, China
{yanzi16,jingz421}@xujc.com

[2] School of Environmental Science and Engineering, Xiamen University Tan Kah Kee College, Zhangzhou 363000, Fujian, China

Abstract. The water cloud model has a good retrieval effect for low and sparse vegetation. This paper proposes an improved water cloud model for the retrieval of soil moisture in tea plantations. Based on the classic model, this improved approach takes into account the scattering effects of crop stems, incorporating two additional parameters: vegetation height (h) and radar backscattering cross-section of vegetation leaves and stems per unit volume (σ_{pq}). By using Sentinel-1A data and utilizing the AIEM model to establish the relationship between soil moisture content and radar backscattering coefficients on the soil surface, a direct connection between radar backscattering coefficients and soil moisture status under vegetation cover is established. The results show that the improved model significantly enhances the precision of soil moisture retrieval compared to the classic model. Particularly, the VV polarization shows better performance, with the R^2 increasing from 0.6009 to 0.8177, and the RMSE decreasing from 0.0954 to 0.0708. The results satisfied the requirement for SSM retrieval in the study region.

Keywords: Soil moisture · Water cloud model · Microwave remote sensing · Sentinel-1 · Sentinel-2

1 Introduction

Soil moisture is an important factor in the soil-plant-atmosphere continuum [1, 2]. Accurate soil moisture can provide support for crop water requirements estimation, essential support for estimating crop water requirements, formulating irrigation plans for crop growth, monitoring irrigation water usage, and assessing irrigation effectiveness, as well as offering fundamental information for drought prevention and mitigation efforts [3]. However, the complex nature of soil conditions and meteorological factors, as well as the strong spatiotemporal heterogeneity of soil moisture, make accurate acquisition of soil moisture extremely challenging.

In recent years, the dynamic and real-time characteristics of remote sensing technology have significantly improved the accuracy and coverage of soil moisture retrieval [4, 5]. While optical remote sensing technology has achieved maturity in theoretical research

D. Han et al. (Eds.): ICEPG 2024, ESE, pp. 809–815, 2025.
https://doi.org/10.1007/978-3-031-86829-0_91

and applications related to soil moisture, research results often fail to achieve sufficient sensitivity due to the significant influence of cloud cover, solar radiation conditions, and vegetation coverage on satellite-based optical remote sensors [6]. Numerous studies have shown that active microwave remote sensing, represented by Synthetic Aperture Radar (SAR), with its all-day, all-weather, multi-polarization capabilities, certain penetration ability, and high sensitivity to soil moisture, provides a powerful means for soil moisture research [7, 8]. The water cloud model(WCM), proposed by Attema and Ulaby in 1978, treats vegetation as a layer of uniformly distributed scatterers of similar shape and size floating on the soil surface, making it suitable for retrieving soil moisture in low-height vegetables, especially crops [9, 10]. However, this model only considers single scattering and ignores multiple scattering between the canopy and the ground, rendering it unsuitable for taller vegetables.

Based on the WCM, this paper introduces parameters, such as vegetation height and radar backscatter cross-section of vegetation leaves and stems per unit volume. First, vegetation cover is incorporated. NDVI and NDWI are derived from Sentinel-2 to calculate vegetation cover and water content. Then, an improved WCM is proposed based on Sentinel-1A multi-polarization data and the AIEM model. Finally, the relationship between soil moisture and radar backscattering coefficients is established, thereby obtaining soil moisture information for tea plantations.

2 Study Area and Data

2.1 Study Area

Xiamen City (24°23′–24°54′ N, 117°53′–118°26′ E) is located in the northeastern part of Fujian Province, with a total area of 1700.61 square kilometers. The terrain is primarily composed of coastal plains, terraces, and hills, with more mid to low-altitude mountain ranges in the northwest, sloping from the northwest to the southeast. The study area is situated in the northwestern part of Xiamen's Tong'an District, specifically in Lianhua Town. The region reaches an elevation of nearly 1 km and is characterized by a relatively high overall vegetation coverage as it is predominantly a mountainous area. Tea trees are the primary economic crop in this locality. The location map of the study area is shown in Fig. 1.

2.2 Remote Sensing Data

The data used in this paper are derived from the European Space Agency's (ESA) Sentinel satellite series, including Sentinel-1 and Sentinel-2.

Sentinel-1 carries a C-band SAR system with a revisit period of 6 days, featuring all-day, all-weather, wide coverage, multiple modes, and various applications [11]. The Sentinel-1A SAR data selected for this study is the TOPS-mode SLC data from May 19, 2024, operating in the interferometric wide swath mode with a swath width of 250 km and a spatial resolution of 5 m × 20 m. The radar polarization modes used are VV and VH.

Sentinel-2 is equipped with a high-resolution Multi-Spectral Imager (MSI), covering 13 spectral bands from visible light to shortwave infrared. It has a high spatial resolution of up to 10 m and includes three bands in the red edge range [12]. The Sentinel-2B MSIL2A data from May 17 was selected for this study.

The preprocessing of images was performed using the European Space Agency's SNAP software.

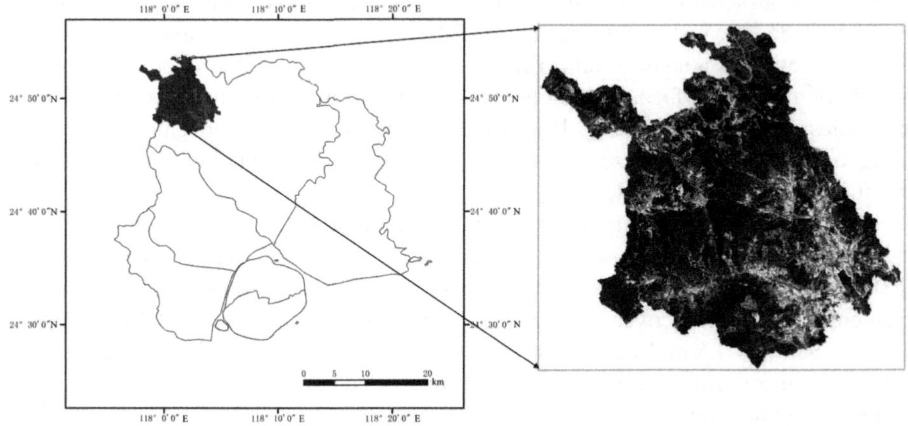

Fig. 1. The location map of the study area (Sentinel-2 image)

2.3 Ground Data

In the study area, a total of 20 ground points were established for synchronous observation of surface parameters during satellite overpasses. Soil moisture was measured using the ring knife method, with soil samples collected at each observation point using a soil auger from a depth of 0–20 cm. The gravimetric water content of the soil was determined by drying and weighing the samples.

3 Methodology

The WCM proposed by Attema and Ulaby in 1978 treats vegetation as a layer of uniformly distributed scatterers, uniformly shaped and sized, floating above the soil surface [9]. This model can directly utilize the back scattering coefficients for soil moisture retrieval. The model is composed of the volume scattering of the vegetation and the soil surface scattering after the two-way attenuation through the vegetation. The specific mathematical expression of this model is as follows:

$$\sigma_c^0(\theta) = \sigma_v^0(\theta) + \gamma^2(\theta)\sigma_s^0(\theta) \tag{1}$$

$$\sigma_v^0(\theta) = A\cos(\theta)\left(1 - \gamma^2(\theta)\right) \tag{2}$$

$$\gamma^2(\theta) = \exp(-2\tau \cdot \sec(\theta)) \tag{3}$$

where, $\sigma_c^0(\theta)$ is the total radar backscattering coefficient under the surface covered by vegetation; $\sigma_v^0(\theta)$ is the backscattering coefficient of the direct vegetation layer; $\sigma_s^0(\theta)$ is the backscattering coefficient of the direct surface; $\gamma^2(\theta)$ is the two-layer attenuation factor of radar penetration into the crop layer; the value of A depends on the different types of crops and the frequency of the incident electromagnetic wave; θ is the radar angle of incidence; and τ is represented by the water content of the vegetation, which is expressed by the formula as follows:

$$\tau = b \cdot m_v \tag{4}$$

where the value of b depends on the different crop types and the frequency of the incident electromagnetic wave.

In order to simulate the backscattering coefficient from bare soil to the surface under full vegetation cover, the vegetation coverage (f_v) is added to the WCM:

$$\sigma_c^0(\theta) = f_v(\sigma_v^0 + \gamma^2(\theta)\sigma_s^0) + (1 - f_v)\sigma_s^0(\theta) \tag{5}$$

$$f_v = (NDVI - NDVI_{min})/(NDVI_{max} - NDVI_{min}) \tag{6}$$

where $NDVI_{max}$ and $NDVI_{min}$ are the normalized vegetation indices under complete vegetation cover and completely bare soil, respectively.

The MIMICS model divides vegetation cover into three parts based on microwave scattering characteristics, especially refining the vegetation layer into vegetation canopy and vegetation stem parts. According to the MIMICS model [13], it can be concluded that:

$$\sigma_v^0(\theta) = \frac{\sigma_{pq1}\cos\theta}{2k_{epq}}\left(1 - \gamma^2(\theta)\right) \tag{7}$$

$$\gamma^2(\theta) = \exp(-2\tau \cdot \sec(\theta)) \tag{8}$$

where, σ_{pq1} is the radar backscattering cross section of the vegetation leaves and stems per unit volume, k_{epq} is the canopy extinction coefficient, $k_{epq} = \tau/h$.

The Eq. (1) is changed to:

$$\sigma_c^0(\theta) = \frac{a\cos\theta}{2bm_v}h(1 - \exp(-2bm_v\sec(\theta)))f_v + \sigma_s^0\exp(-2bm_v sec(\theta))(1 - f_v) \tag{9}$$

The active radar is more sensitive to surface roughness and soil moisture, the relationship model of radar incidence angle, combined roughness, soil moisture content and backward scattering coefficient of soil surface is established according to the AIEM model, which is expressed as follows:

$$\sigma_{Spq}^o = A_v(\theta)ln(m_v) + B_v(\theta)ln(Z_s) + C_v(\theta) \tag{10}$$

where, σ_{Spq}^o is the surface back scattering coefficient, pq is the polarization mode; $A_v(\theta)$, $B_v(\theta)$ and $C_v(\theta)$ are all parameters in terms of variables of the radar incidence angle;

m_v is the soil moisture; Z_s is the combined roughness, $Z_s = s^2/l$, where s is the root-mean-square height, and l denotes the correlation length.

Bringing Eq. (10) into Eq. (9) yields the final equation for soil moisture:

$$\sigma_c^0(\theta) = \frac{a\cos\theta}{2bV} h(1 - \exp(-2bV\sec(\theta)))f_v +$$

$$\left[\exp(-2bVsec(\theta))(1 - f_v)\right]A_v(\theta)\ln(m_v) + B_v(\theta)\ln(Z_s) + C_v(\theta) \tag{11}$$

This formula is the corrected water cloud model in this paper, and the corrected water cloud model can directly inverse perform the water content of the soil based on the radar back scattering coefficient.

4 Results and Analysis

4.1 Comparison of the Results of Improved and WCM

In this study, soil moisture in the 0–20 cm depth soil layer was measured and recorded during the same period at the monitoring station. The soil moisture calculated from the improved WCM was verified for accuracy using measured soil moisture, and the test data are shown in Table 1. The RMSE represents the root mean square error, which is a measure of the difference between the calculated and actual values; the R^2 evaluates the degree of correlation between the calculated and actual values.

Table 1. Evaluation of different model accuracies

models	VV/VH	RMSE	R^2
Improved WCM	VV	0.0328	0.8177
	VH	0.0708	0.4027
WCM	VV	0.0424	0.6009
	VH	0.0954	0.3443

To further test the applicability and reliability of the method, the improved WCM was compared with the WCM using accuracy validation samples, the results of which are represented in Fig. 2 and Fig. 3:

As can be seen in Fig. 2 and Fig. 3, most of the scatters in the improved WCM under VV polarization are located near the 1:1 line, while the correlation coefficient squared is 0.8177, while most of the scatters in the WCM are located in the part of the 1:1 line, and at the same time, the correlation coefficient squared is only 0.6009. It is not difficult to see that the VH polarization is relatively ineffective in both methods. In summary, the improved WCM can more accurately retrieval soil moisture under VV polarization mode.

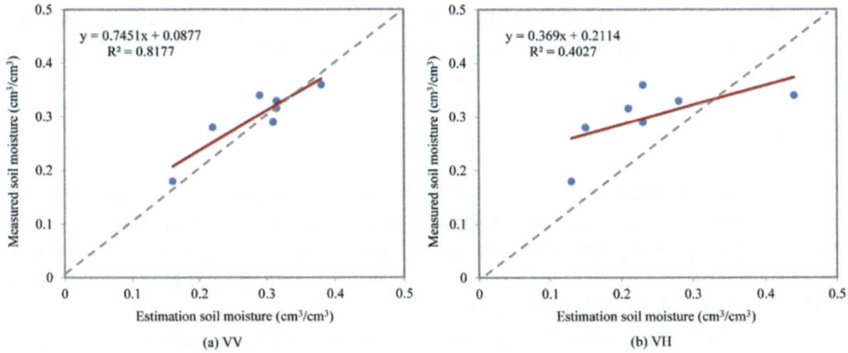

Fig. 2. Scatter plots of measured and estimated improved WCM

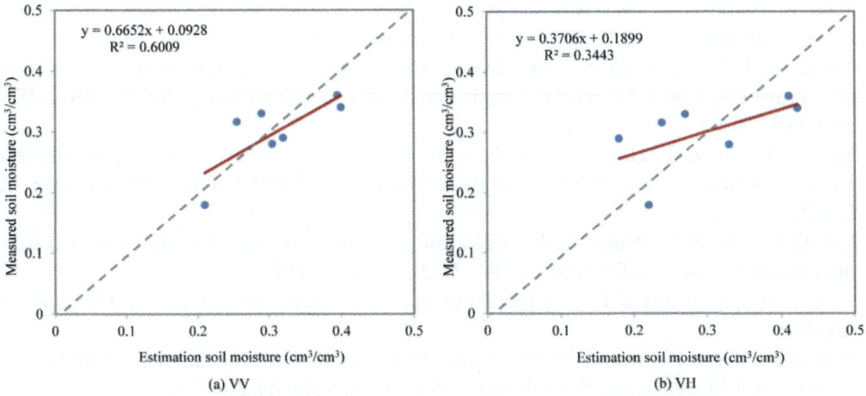

Fig. 3. Scatter plots of measured and estimated WCM

5 Conclusions

In this paper, a soil moisture retrieval method is proposed for tea plantations. The method utilizes the developed WCM, which combines Sentinel-1 SAR and Sentinel-2 optical data.

In the improved WCM, two parameters were added: vegetation height (h) and radar backscattering cross-section of vegetation leaves and stems per unit volume (σ_{pq}). Compared with the measured results, the improved method can effectively retrieval the soil moisture of tea plantations. However, the retrieval results demonstrated varying accuracy of soil moisture retrieval using different polarization modes. Specifically, under VV polarization, the RMSE of 0.0328 and an R^2 of 0.8177, while under VH polarization, the RMSE is 0.0708 and an R^2 is only 0.4027.

The proposed improved WCM was based on the theoretical model of first-order scattering. The study aims to further research using multi-order scattering theoretical models to correct for vegetation scattering effects and achieve vegetation scattering correction.

Funding. This work was supported by the School-level scientific research incubation project of Tan Kah Kee College, Xiamen University (Grant No. YY2022L05).

References

1. Zhang, L., Zhou, W., Li, D.X.: Research progress in irrigation mode selection of high-efficiency water-saving agriculture. J. Drain. Irrig. Mach. Eng. **37**(5), 447–453 (2019)
2. Wu, D.L., Liu, C., Guo, F.C., et al.: Regional monitoring of soil moisture content in winter wheat field based on multi-source remote sensing data and optimal model selection. Chin. Agric. Sci. Bull. **40**(25), 147–154 (2024)
3. Shao, C.L., Wu, D.L.: Review of applicability of soil moisture measurement techniques. Meteorol. Sci. Technol. **47**(1), 1–9 (2019)
4. Li, Y., Zhang, C.C., Heng, W.D.: Soil moisture retrieving method based on depth learning by multi-source data. Water Saving Irrig. **02**, 57–64 (2023)
5. Gao, N., Zhang, A.Q., Mei, H.B., et al.: Current status and development trends of soil moisture monitoring technologies. J. Intell. Agric. Mech. **5**(3), 51–62 (2024)
6. Yang, J.J., Deng, Q.H., Li, S., et al.: Synergistic inversion of soil in vegetation covered areas of the lower Yellow River based on optical and SAR Remote Sensing. Yellow River **45**(11), 106–110 (2023)
7. Yang, L.P., Su, Z.Q., Hou, C.L., et al.: Soil moisture content retrieval in arid area based on random forest using polarimetric SAR data. J. Jilin Univ. (Earth Sci. Ed.) **52**(04), 1255–1264 (2022)
8. Liu, Q.X., Gu, X.F., Wang, C.M., et al.: Soil moisture retrieval on both active and passive microwave data scales. Earth Sci. Front. **31**(2), 42–53 (2024)
9. Attema, E.P.W., Ulaby, F.T.: Vegetation modeled as a water cloud. Radio Sci. **13**(2), 357–364 (1978)
10. Sun, X.X., Hou, D., Bai, Y.F.: Study on soil moisture retrieval based on Sentinel-1 and Landsat8-OLI data. Gansu Water Resour. Hydropower **60**(01), 29–32 (2024)
11. Ramon, T., Paul, S., Dirk, G.,et al.: GMES Sentinel-1 mission. Remote Sens. Environ. **120**(6), 9–24 (2012)
12. Su, W., Hou, N., Li, Q., et al.: Retrieving leaf area index of corn canopy based on Sentinel-2 remote sensing image. Trans. Chin. Soc. Agric. Mach. **49**(01), 151–156 (2018)
13. Ulaby, F.T., Sarabandi, K., Mcdonald, K., et al.: Michigan microwave canopy scattering model. Int. J. Remote Sens. **11**(7), 1223–1253 (1990)

Assessment of Heavy Metal Pollution and Potential Risks in the Soil of Bayan County, Heilongjiang Province

Youtao Xin[1,2], Tao Liu[1,2(✉)], Xidong Zhao[1,2(✉)], Haonan Song[1,2], Haicheng Zhang[1,2], and Hongqiang Zhao[1,2]

[1] Harbin Center for Integrated Natural Resources Survey, China Geological Survey, Harbin 150086, China
zhaoxidong@mail.cgs.gov.cn

[2] Observation and Research Station of Earth Critical Zone in Black Soil, Ministry of Natural Resources, Harbin 150086, China

Abstract. In order to comprehensively understand the current status of heavy metal pollution in Bayan County and the potential harm it may cause to the ecosystem, this study collected surface and deep soil samples from 22 locations and conducted area analysis. The contents of five heavy metals were determined, and the samples were evaluated using single factor index, Nemerow index, and potential ecological risk index. The results showed that the contents of As, Cd, Cr, Hg, and Pb in the surface soil of Bayan County were 9.86 mg/kg, 0.11 mg/kg, 61.35 mg/kg, 0.060 mg/kg, and 23.76 mg/kg, respectively, which were all below the screening values specified in the "Soil Environmental Quality Standards for Agricultural Land Pollution Risk Control" (GB 15618-2018). Among them, As, Cd, Cr, and Pb mainly came from parent materials, while the coefficient of variation of Hg was 1.6, indicating that it was more affected by human activities, and there was significant spatial distribution difference among different regions. According to the Nemerow comprehensive pollution grade evaluation, Hg was classified as heavy pollution, while Cd and Cr were at the light pollution level. The potential ecological hazard degree of various heavy metals in order from high to low was: Hg > Cd > As > Pb > Cr, and they were all within the low potential ecological hazard grade range. The comprehensive ecological risk index (RI) ranged from 44.75 to 345.41, with an average of 87.04, within the low ecological risk region.

Keywords: Soil heavy metals · distribution characteristics · ecological risk · Harbin

1 Introduction

On a worldwide scale, soil contamination has turned into a severe environmental problem, presenting a threat to human well-being and the stability of ecological systems [1–5]. Heavy metal pollution, in particular, has a more profound impact on the environment and organisms due to its persistence and bioaccumulation. As an important

The original version of the chapter has been revised. The corresponding author indication in the chapter has been updated. A correction to this chapter can be found at https://doi.org/10.1007/978-3-031-86829-0_124

D. Han et al. (Eds.): ICEPG 2024, ESE, pp. 816–826, 2025.
https://doi.org/10.1007/978-3-031-86829-0_92

agricultural area in Heilongjiang Province, Bayan County has attracted wide attention for soil heavy metal pollution [6, 7].

Bayan County is located in the west of Harbin City, close to the upper reaches of Songhua River, with rich agricultural resources. However, in the process of urbanization, due to many factors such as agricultural production and industrial waste treatment, soil heavy metal pollution is increasingly aggravated.

This paper aims to draw the attention of society to the problem of soil heavy metal pollution, promote the formulation and implementation of relevant policies, and protect the land resources and ecological environment of our country. At the same time, we explored how to use scientific and technological progress and management innovation to realize the sustainable utilization and management of soil resources and contribute to China's green development.

2 Experimental Part

2.1 Study Area

Bayan County is situated to the south of the Songhua River on the north bank of the middle reaches of Heilongjiang Province, close to the eastern edge of the Songnen Plain. The south of the county is adjacent to the Songhua River, across the river to see Bin County; West of the Piao River, bordering Hulan County; North of the Nihe River, adjacent to Suihua City and Qing'an County; The east is connected with Camel Stone Mountain, Huangni River and Mulan County. It is 97 km away from the provincial capital Harbin and is a suburban county of the city. It is located in the gentle slope area bordering Songnen Plain and the southwest foot of Lesser Hinggan Mountains, and Lesser Hinggan Mountains pass through the eastern half of the county. It belongs to the hilly terrain and is relatively flat in the Shaoling River, the middle and lower reaches of the Piao River, the Nihe River and the Songhua River. Bayan County has a terrain that is higher in the east and lower in the west. It is distributed in the north and south directions. The middle area shows a hilly and undulating state. The yearly average temperature is 2.6 °C (declining from southwest to northeast). The main precipitation is concentrated between June and August [8].

2.2 Sample Collection and Determination

In June 2022, a backpack drill was used in the study area to select an area with a soil layer thickness greater than 5 m for sampling, and the sampling density was 1 point/64 km^2 (Fig. 1). Take soil surface and deep samples, the deep sample collection location is the same as the surface sample, and the control depth is 5 m. During the sample collection process, the handheld computer combined with the electronic map was used to accurately position the soil type, soil thickness, sampling depth, soil color and other contents of the sampling point were recorded in detail. The original sample of no less than 400 g was bagged as the sample to be processed to ensure that the weight of the sample to be measured after fine processing was above 150 g. After natural drying, the collected soil samples were processed in a clean, ventilated and pollution-free site, and 100 g of 2 mm

(10 mesh) grain samples were intercepted and be delivered to the laboratory for analysis and examination.

In laboratory, the content of each element was determined by the following methods: The As element content was determined by atomic fluorescence spectrophotometry; The contents of Cr and Pb were determined using an X-ray fluorescence spectrometer. The concentration of Cd and Hg in soil was determined by automatic atomic fluorescence spectrophotometer AFS-230E [9–13]. The detection limits of As, Cd, Cr, Hg and Pb were 0.2, 0.02, 5, 0.005 and 2 mg/kg, respectively. And the relative error (RE) and relative standard deviation (RSD) of each element were less than 15% and less than 10%. The testing unit is the laboratory of Harbin Natural Resources Comprehensive Survey Center of China Geological Survey.

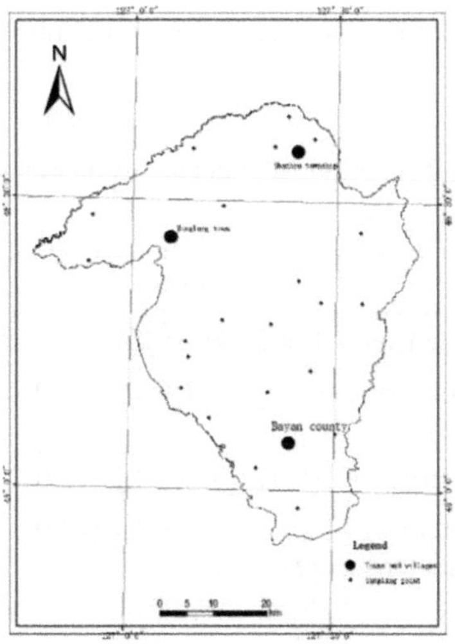

Fig. 1. Distribution of sampling points

2.3 Evaluation Method

Nemerow index method and potential ecological risk assessment method [14–17] are effective means to assess the level of soil heavy metal pollution, which are favored and widely used at home and abroad.

Nemerow's Index Method

Single factor index approach: Utilizing the single factor evaluation method to evaluate the degree of pollution by heavy metal pollutants in the study area is highly significant:

$$P_i = \frac{C_i}{S_i} \quad (1)$$

In Eq. 1, P_i is a single pollution parameter. C_i is heavy metal in surface soil.

Concentration measured value. S_i is the background reference value of soil. A lower P_i value indicates a lower risk value of agricultural soil pollution.

The Nemerow index approach thoroughly considers the average and maximum values. It is capable of highlighting the impact of heavy metal elements on the comprehensive pollution degree of the study area. Hence, it has been widely employed in the calculation of comprehensive pollution indices both domestically and internationally [11].

$$P_n = \sqrt{(P_a^2 + P_{max}^2)/2} \tag{2}$$

In Eq. 2, P_n is the comprehensive pollution index of soil heavy metal elements; P_a represents the average value of each single factor pollution index of heavy metal elements in soil.

The classification standards of single factor and comprehensive pollution index are shown in Table 1.

Table 1. Soil environmental quality evaluation classification

Individual pollution index	class of pollution	Composite pollution index	pollution level
Pi ≤ 1	Grade I soil (clean)	$P_Z \le 0.7$	safety
$1 < P_i \le 2$	Secondary soil (safe)	$0.7 < P_Z \le 1.0$	warning line
2 < Pi ≤ 3	Tertiary soil (Alert)	$1.0 < P_Z \le 2.0$	mild pollution
Pi > 3	Quaternary soil(contaminate)	$2.0 < P_Z \le 3.0$	Moderate pollution
		$P_Z > 3$	heavy pollution

Potential Ecological Risk Assessment Methods

The potential ecological risk assessment approach was proposed by Swedish geochemist Lars Hakanson in 1980. An integrated approach to assessing the extent to which pollutants in the environment may cause damage to ecosystems. This method can thoroughly reflect the comprehensive impacts of various heavy metal pollutants. Thus, it is widely used in addressing current heavy metal pollution problems.

$$E_r^i = \frac{T_r^i \times C_s^i}{C_n^i} \tag{3}$$

In Eq. 3, E_r^i represents the potential ecological risk index of heavy metal element. T_r^i is the toxicity coefficient of heavy metal. C_s^i is the measured value of heavy metal element. C_n^i is the background value of heavy metal element in soil. The toxicity effects of As, Cd, Cr, Hg and Pb were 10, 30, 2, 40 and 5 mg/kg, respectively.

$$RI = \sum\nolimits_{i=0}^{n} E_r^i = \sum T_r^i \times P_r^i \tag{4}$$

RI represents the comprehensive potential ecological risk index; P_r^i is the single pollution coefficient of the measured heavy metal elements. The classification of the potential ecological risks of individual heavy metal elements and comprehensive potential ecological risks is shown in Table 2.

Table 2. Classification of pollution degree by potential ecological hazard index method

E^i_r	pollution level	RI	Hazard level
$E^i_r < 40$	slight	RI < 150	slight
$40 \geq E^i_r < 80$	medium	$150 \leq RI < 300$	medium
$80 \geq E^i_r < 160$	Strong degree	$300 \leq RI < 600$	Strong degree
$160 \geq E^i_r < 320$	Stronger degree	$RI \geq 600$	Stronger degree
$E^i_r \geq 320$	Extremely strong		

2.4 Data Processing

In this research, Microsoft Excel 2016 software was utilized for conducting statistical analysis on soil heavy metal content data. ArcGIS 10.8 software was applied to make the sampling point distribution map. Inverse distance weight interpolation was used to handle geographical data. The geostatistical analysis module of ArcGIS 10.8 software was employed to generate the spatial distribution map of heavy metals and assess heavy metal pollution [18–20].

3 Results, Analysis

3.1 Descriptive Analysis

In Bayan County of Harbin City, the average concentrations of heavy metals As, Cd, Cr, Hg, and Pb in the surface soil were 9.86 mg/kg, 0.11 mg/kg, 61.35 mg/kg, 0.060 mg/kg, and 23.76 mg/kg respectively (see Table 3). All these values were lower than the screening values stipulated in the Soil Environmental Quality Management and Control Standard (GB15618-2018). According to the background value of Chinese soil, Cd and Cr are relatively high, while As, Hg, and Pb are relatively low, belonging to a relatively clean area in the country. In comparison with deep soil, the content of heavy metal Cr in surface soil is lower, and the contents of other heavy metal elements are slightly higher. As, Cd, Hg, and Pb were 1.05, 1.09, 2.45, and 1.02 times the corresponding heavy metals in deep soil respectively. The coefficient of variation reflects the degree of variation of variables in the sample. The greater the coefficient of variation, the stronger the influence of human activities on its content and the greater the difference in spatial distribution. The variation coefficients of the five heavy metals in surface soil were Hg > Cd > Cr > As > Pb. The coefficient of variation of Hg was relatively large. This indicates that the content of Hg underwent significant changes and was likely to be greatly influenced by human activities. Regarding skewness and kurtosis, As, Cd, and Pb are all nearly equal to 0 and had a normal distribution, suggesting that the content of these metal elements is mainly influenced by the parent material.

Table 3. Results of soil heavy metal analysis in Bayan County (mg/kg)

Element	min	Max	AVG	standard deviation	variable coefficient	skewness	kurtosis	background value	Average heavy metal content in deep soil
As	6.47	12.1	9.86	1.24	0.13	−0.58	1.58	11.2	9.38
Cd	0.044	0.19	0.112	0.038	0.34	0.55	−0.05	0.097	0.103
Cr	28.9	71.5	61.35	9.16	0.15	−2.29	7.10	61	63.42
Hg	0.019	0.47	0.06	0.097	1.62	4.00	16.77	0.065	0.024
Pb	19.6	25.3	23.76	1.40	0.06	−1.34	2.24	26	23.4

3.2 Spatial Distribution Characteristics of Heavy Metals

By making use of the geostatistical spatial analysis function of the ArcGIS 10.8 software system, the heavy metal contents in the surface soil of Bayan County were interpolated through inverse distance weighting, and the content distribution map of each element was created (Fig. 2).

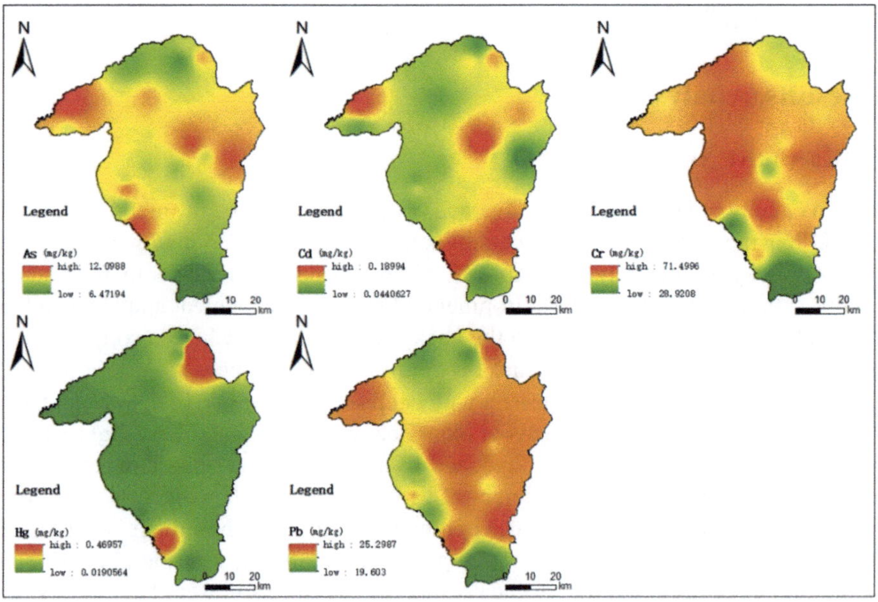

Fig. 2. The spatial layout of heavy metals in the soil of Bayan County

The As content varied from 6.47 to 12.1 mg/kg. The high-value region was mainly distributed from east to west of the study area, and the low-value region was in the south and north. The Cd content was 0.044–0.19 mg/kg. The high-value area was mainly in the middle and southern part of the study area. The Cr content was 28.9–71.5 mg/kg. The high-value area was extensive, and the south was a low-value area. The Hg content ranged from 0.019 to 0.47 mg/kg. High values were distributed in the south and north, and low values were elsewhere. The Pb content ranged from 19.6 mg/kg to 25.3 mg/kg. It was generally lower than the background value of Chinese soil. The relatively high-value area was located in the middle.

Heavy metal elements As, Cd, Cr, Hg, and Pb are basically distributed in a dot pattern. Except for Hg, the rest of the heavy metal elements have a characteristic of being high in the middle and low in the north and south.

3.3 Evaluation of Soil Heavy Metal Pollution

Single Factor Pollution Index Evaluation

The single factor index method was employed to assess the soil environmental quality indexes in Bayan County and the analysis results were shown in Table 4. According to the evaluation results, the average single factor index of soil heavy metals in Bayan County was Cd(1.16), Cr(1.01), Hg(0.92), Pb(0.91) and As(0.88) from high to low. The pollution levels of the two sampling points of Hg element are in alert and polluted state, and other heavy metal elements are in clean and safe state, among which As and Hg are mainly in clean state, Cr and Cd are mainly in safe state, and Pb is completely clean state. It indicates that Bayan County as a whole is in a state of light pollution.

Table 4. Results of evaluating the single-factor index of soil heavy metals in Bayan County.

Element	min	Max	AVG	Each pollution level %				comprehensive index
				$P \leq 1$	$1 < P \leq 2$	$2 < P \leq 3$	$P > 3$	
As	1.08	0.58	0.88	81.82	18.18	0	0	0.99
Cd	1.96	0.45	1.16	36.36	63.64	0	0	1.61
Cr	1.17	0.47	1.01	31.82	68.18	0	0	1.09
Hg	7.23	0.29	0.92	90.91	0	4.55	4.55	5.15
Pb	0.97	0.75	0.91	100	0	0	0	0.94

Evaluation of Nemero Pollution Index

As evident from the results in Table 4, when taking the background value of soil in China as a reference, the comprehensive pollution index of Hg is greater than 3, signifying a serious degree of pollution. The comprehensive pollution index of Cd and Cr lies between 1.0 and 2.0, which indicates light pollution. The comprehensive pollution index of As and Pb is within the range of 0.7 to 1.0, which is at the warning stage. The average composite pollution index of soil heavy metals in the study area is 1.96, presenting the general traits of light pollution. The heavy metal element Hg contributes the most to the Nemerow pollution index, and the pollution is relatively more severe in the south of Bayan County and the north of Shanhou Township.

Assessment of Ecological Risk from Soil Heavy Metal Pollution

As is evident from Table 5, the average Ei value of heavy metals in soil in Bayan County is Hg > Cd > As > Pb > Cr, indicating a slight risk level on the whole. The average Ei of Hg was 36.95, of which the severe ecological risk samples accounted for 9%(2), the average Ei of Cd was 34.71, and the moderate risk sample sites accounted for 12% (4), The results showed that mercury and cadmium were the main influencing factors of soil pollution in Bayan County. As shown in Fig. 3.

After calculation, the comprehensive ecological risk index RI ranges from 44.75 to 345.41, with an average of 87.04, and the comprehensive ecological risk grade is mild (<150). In addition to the moderate ecological risk level of Hg element in two samples, The ecological risk of other heavy metal elements is comparatively light, and there is essentially no potential ecological hazard. The maximum RI value is 345.41, which is located near Houshan Township, Bayan County. The supply rate of Hg is 79.43%, and the supply rate of other elements is less than 15%. The RI value of the sample located near Bayan County was 184.63, and the delivery rates of Hg and Cd were 52.94% and 37.45%, respectively. On the whole, the soil environmental quality of Bayan County is good, and it belongs to the low ecological risk area (Fig. 4).

Table 5. Outcomes of the potential ecological risk coefficient for soil heavy metals in Bayan County

Element	Each pollution level %			Percentage of the number of sampling points (%)				
	min	Max	AVG	slight	medium	Strong degree	Stronger degree	Extremely strong
As(Ei)	10.80	5.78	8.80	100.00	0	0	0	0
Cd(Ei)	58.76	13.61	34.71	88.00	12.00	0	0	0
Cr(Ei)	2.34	0.95	2.01	100.00	0	0	0	0
Hg(Ei)	289.23	11.69	36.95	91.00	0	0	9	0
Pb(Ei)	4.87	3.77	4.57	100.00	0	0	0	0

4 Discussion

Soil heavy metal pollution not only influences the growth and quality of crops but also may enter the human body through the food chain, resulting in health risks. Therefore, the academic world, the government, and the public have all begun to pay attention to this problem.

In the academic domain, soil heavy metal pollution has become a research focus in the fields of environmental science, ecology, and agricultural science. A large number of

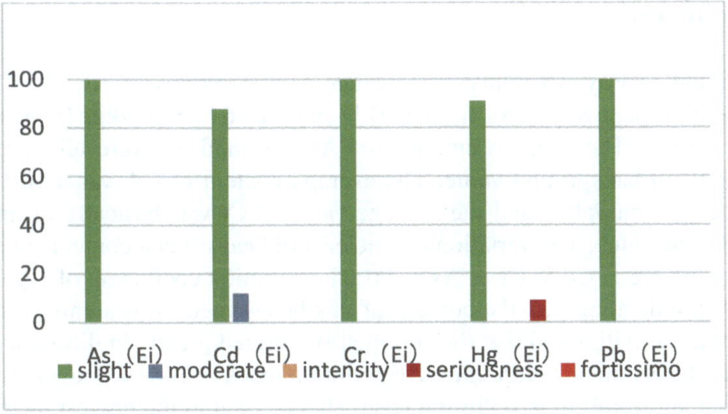

Fig. 3. Distribution of frequency of potential ecological hazards of heavy metals in the soil of Bayan County

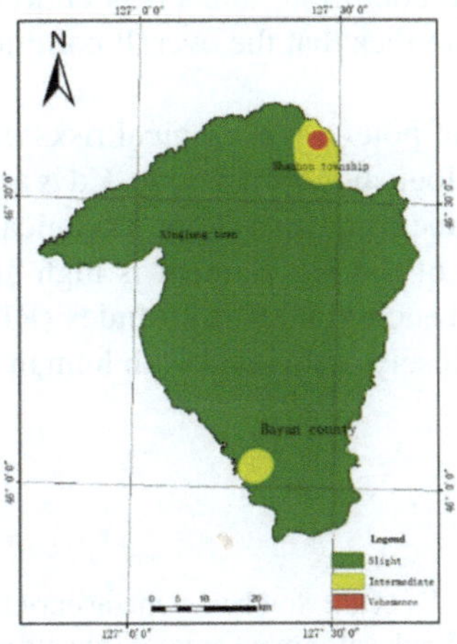

Fig. 4. Spatial distribution of soil heavy metal potential ecological hazard index in Bayan County

researchers have been engaged in exploring the sources, migration and transformation rules, influencing factors, and remediation methods of heavy metal pollution. In recent years, based on the traditional assessment of heavy metals in soil, the combination with Arcg is geostatistical analysis provides a powerful spatial analysis and visualization tool for the research and management of soil pollution in the study area. It can perform a more comprehensive and intuitive evaluation and analysis of soil heavy metal pollution and offer powerful spatial analysis and decision-making support tools for related research and management.

5 Conclusion

(1) In the soil of Bayan County, the average concentrations of As, Cd, Cr, Hg, and Pb were respectively 9.86 mg/kg, 0.11 mg/kg, 61.35 mg/kg, 0.06 mg/kg, and 23.76 mg/kg. The average amounts of As, Hg, and Pb were all lower than the national soil background value. The average content of Cd was a bit higher than the background value, and the average value of Cr was basically in line with the background value. The variation coefficients of heavy metal contents in descending order were Hg > Cd > Cr > As > Pb. The variation coefficient of Hg was greater than 1.0, indicating that the content of this heavy metal was significantly affected by human activities and that the soil quality differed greatly in different areas.

(2) The soil heavy metal comprehensive pollution index of Bayan County stood at 1.96, showing an overall light pollution level. Hg emerged as the biggest factor affecting the assessment of the Nemero pollution index. From the single-factor pollution evaluation outcomes, the order of heavy metal pollution degree from high to low is Cd, Cr, Hg, Pb, and As. In this regard, As and Hg are basically free from pollution, Cd and Cr are in a safe condition, and Pb is entirely clean. The comprehensive pollution index of 1.96 implies that the overall pollution degree of Bayan County is relatively slight.

(3) The assessment results of potential ecological risks of heavy metals reveal that Hg has attained a strong ecological hazard degree, Cd is at a moderate ecological hazard degree, and the other elements are at a slight ecological hazard degree. The potential ecological harm degree of the Hg element is high and must not be ignored. The comprehensive Potential ecological Hazard Index (RI) is rated as a mild ecological hazard, which may be closely associated with human activities.

References

1. Wu, W.W., Shen, C., Sha, C.Y., et al.: Soil heavy metal enrichment characteristics, risk assessment, and source analysis in redevelopment areas during urban industrial plots. Ecol. Environ. Sci. **33**(5), 791–801 (2024)
2. Du, J., Lu, X., Cai, K., et al.: Cadmium pollution and remediation technology in soil and plant system: status and prospects. Geology in China
3. Wang, S.Y., Hu, Z.X., Wang, E.W., et al.: Research hotspot and evolution of global soil heavy metal contamination assessment from 1991 to 2022. Environ. Pollut. Control **46**(04), 589–595 (2024)
4. Yu, Q.Y., Li, W.Y., Zhang, N., et al.: Present pollution status and safe utilization technologies of heavy metal-polluted farmland soil: a review. Soils **56**(02), 229–241 (2024)
5. Zhao, B., Wang, L., Wei, Y.Q., et al.: Current status and prospects of agricultural soil heavy metal contamination prevention and control standard systems in China. Res. Environ. Sci. **37**(05), 1092–1103 (2024)
6. Xin, Y.T., Li, Y.F., Chen, Z., et al.: Evaluation and analysis of soil heavy metal pollution based on Nemerow index and Geoaccumulation index methods: a case study of Songling area in Daxinganling mountains. Geol. Resour. **33**(02), 196–201+221 (2024)
7. Bao, L.X., Jiang, J., Sun, G.X.: Soil heavy metal pollution and risk evaluation in black soil areas of Heilongjiang province. Environ. Sci. Technol. **47**(04), 205–216 (2024)

8. Li, T.J., Zheng, Y.S., Wang, Y.: Soil Geography, 2nd edn., pp. 42–55. Higher Education Press, Beijing (1984)
9. Zhu, L.M., Fu, K.B., Zhong, Q.H., et al.: Study on the remediation of composite heavy metal pollution of slag in a typical copper mine in Sichuan using sulfate-reducing bacteria. Metal Mine **03**, 278–283 (2024)
10. Zhang, D.F., Sun, J.W.: Water body and sediment of Xiaohe Watershed in Xiongershan ore concentration area heavy metal pollution assessment. Multipurp. Util. Miner. Resour. **45**(1), 109–119 (2024)
11. Ma, T.L., Dong, Y.B., Lin, H., et al.: Acid production law and heavy metal release behavior of waste rock from a copper mine in Jiangxi province. Metal Mine **01**, 276–284 (2024)
12. Zhang, Z.G., Tan, Y.N., Zheng, Y.H., et al.: Effect of exogenous organic acids on activation of heavy metal Cr in reclaimed soil. Metal Mine **01**, 285–292 (2024)
13. Wang, C.G., Peng, J.S., Li, W.Y.: The analysis and evaluation on the features of the heavy metals in coal mining soil in the middle reaches of the Yellow River. Min. Saf. Environ. Protect. **49**(05), 124–130 (2022)
14. Zhang, D.Q., Wang, Y.N., Qiao, W., et al.: Evaluation of heavy metal pollution in soils from a mining area in Southern Jiangxi. Multipurp. Util. Miner. Resour. (03), 181–191 (2023)
15. Zhang, Y.K., Feng, N.Q., Liu, Y., et al.: Speciation analysis and risk assessment of heavy metals in the soil of a lead-zinc mining area. Multipurp. Util. Miner. Resour. (03), 199–204+210 (2023)
16. Wang, Z.B., Ren, D.J., Xiao, Y.L., et al.: Assessment of heavy metal pollution and potential ecological risk of farmland soil around ancient copper mine site in Daye city. Metal Mine **11**, 290–298 (2023)
17. Shi, S.H., Bai, J.L.: Heavy metal migration and soil pollution assessment in an intensive mining area in the southwest China. Metal Mine **02**, 194–200 (2022)
18. Lan, F.N., Zhao, Y., Li, J., et al.: Health risk assessment of heavy metal pollution in groundwater of a karst basin, SW China. J. Groundw. Sci. Eng. **12**(1), 49–61 (2024)
19. Zhang, W.Y., Liu, T.Y., Wang, W.Z., et al.: Source analysis of dissolved heavy metals in the Shaying River Basin, China. J. Groundw. Sci. Eng. **11**(4), 408–421 (2023)
20. Shuang, Y., Yang, Z.H., Hu, F.: Migration characteristics of heavy metal elements between soil and rape in the high geological background area of three gorges reservoir region. Ecol. Environ. Monit. Three Gorges **8**(04), 56–65 (2023)

Ecological Communities and Restoration Technologies

The Innovative IMS-MBBR Integrated Treatment Process Operation Performance Study

Jialiang Liu[1,2], Zongkai Ma[1,2](✉), Binbin Qiu[1,2], and Yawen Luo[1,2]

[1] PowerChina Chengdu Engineering Corporation Limited, Chengdu 611130, Sichuan, China
MZK_PowerChina@163.com

[2] Sichuan Municipal Water Environment Treatment Engineering Technology Research Center, Chengdu 611130, Sichuan, China

Abstract. Relying on the Integrated Modular Self-reflow Moving Bed Biofilm Reactor (IMS-MBBR), this article elaborates on the operational effect of this device in a rural domestic sewage treatment plant, combined with actual operating conditions. After 90 days of operation, the results show that the device is simple to operate, stable in operation, and effective in treatment. Under the impact of low-temperature conditions and mutations in water quality and quantity, the average effluent concentrations of COD, total nitrogen, ammonia nitrogen, and total phosphorus are 24.58 mg/L, 10.92 mg/L, 2.49 mg/L, and 0.26 mg/L, respectively, with stable effluent quality reaching Class 1A standards. Meanwhile, the device adopts a prefabricated reinforced concrete construction method, which effectively shortens the construction period and enhances the service life. The innovative integrated layout and pipeline design greatly reduce the number of equipment, enhance the stability and operability of the device, and lower the operation and maintenance costs. Overall, the IMS-MBBR facility provides a new solution for subsequent rural sewage treatment.

Keywords: Rural Township Sewage Treatment · Innovative IMS-MBBR Treatment Process · Integrated Treatment Process · Treatment Effectiveness

1 Introduction

Since The 12th Five-Year Plan was implemented, China has made significant progress in the field of rural sewage treatment, with a large number of integrated sewage treatment facilities constructed and operated nationwide. However, serious equipment aging and unsatisfactory actual operation effects have emerged after more than ten years of operation. Meanwhile, due to the economy level constraint and insufficient policy implementation in some regions, the sewage collection rate in rural towns is generally low, and most of the effluent cannot meet the required discharge standard, resulting in prominent water environmental pollution in rural towns. In addition, the widespread issues of significant fluctuations in influent water quality and quantity in rural sewage treatment,

D. Han et al. (Eds.): ICEPG 2024, ESE, pp. 829–839, 2025.
https://doi.org/10.1007/978-3-031-86829-0_93

coupled with the adoption of traditional large-scale plant design concepts for most existing integrated sewage treatment facilities in rural towns, have led to many problems, including numerous equipment facilities, complex processes, high difficulties in operation and maintenance, high energy consumption, big chemical and labor consumption, and high maintenance costs. The above problems on the existing rural sewage treatment process have made it difficult to achieve stable and compliant effluent quality.

In order to improve the rural water environment for boosting the rural revitalization strategy, it is quite essential to improve the reliability, stability and durability of rural sewage treatment facilities and simplify the operation and maintenance for them.

2 Overview of Rural Sewage Treatment Study Progress

The characteristics of rural sewage are closely related to local environment, economic development level, regional lifestyle habits, etc. [1], typically manifesting in wide fluctuations in water quality and quantity [2]. The small scale of individual sewage treatment stations in rural towns leads to vulnerability to big fluctuations. In response to these characteristics, the main research directions of integrated sewage treatment units in recent years include: 1) combining different processes based on various influent and effluent quality parameter; 2) conventional processes improvement; 3) optimization of operating parameters; 4) simplification and smartlization of operation and maintenance; 5) exploring the research, development, and application of novel technologies and materials; 6) upgrading technology towards carbon neutrality, etc. [3–5].

Liu Xudong et al. [6] proposed a new integrated sewage treatment facility tailored to the water quality and quantity characteristics of villages and towns. The average removal rates of COD, TN, and TP were 86.1%, 84.3%, and 77.4%, respectively.

Zheng Jun et al. [7] developed an integrated mud-film coupled sewage treatment facility suitable for rural areas. Based on the AO process, the facility adds fillers to the aerobic and combined aerobic-anaerobic zone. The removal rates of COD, ammonia nitrogen, TN, and TP can reach 94.96%, 95.03%, 81.82%, and 91.17%, respectively.

Liu Tingzhuo et al. [8] optimized the structure of the treatment tank, stirring, and sludge return methods based on the traditional AO process, reducing the use of kinetic energy equipment to achieve energy-saving purpose.

In order to overcome the difficulties in treating rural sewage and excess sludge, Jiang Weiqing et al. [9] conducted wastewater treatment experiments using a micro-pressure cyclone reactor (MPSR) under five different biomass concentrations (MLSS concentrations). The results showed that MPSR is suitable for operation at high biomass concentrations, with the biomass concentration maintained between 7.98 and 11.45 g/L. When the biomass concentration increased from 4.14 to 7.98 and 11.45 g/L, the amount of excess sludge decreased by 27% and 21%, respectively.

Han Yunping et al. [10] have developed a new piece of equipment based on the New Integrated Self-Reflux Rotating Biological Contactor (NISRRBC) for rural sewage treatment. The operating results of the pilot-scale NISRRBC indicated that the effluent meets the Class B Discharge Standard for Urban Wastewater Treatment Plants (GB 18918–2002). This facility features low sludge production, integrated design concepts, and simple operation modes.

Currently, most research still remains at the laboratory stage, lacking practical application and verification under complex and variable actual operating conditions. The water quality standards and stability of the effluent still need to be further improved.

3 Overview of the IMS-MBBR Facility Design

3.1 Introduction of the Facility

The IMS-MBBR facility was innovatively developed by the PowerChina Chengdu Engineering Corporation Limited based on the traditional Moving Bed Biofilm Reactor (MBBR) process. It integrates the biochemical treatment and sedimentation into a compact reactor, resulting in a small footprint. Prefabricated reinforced concrete blocks were used for constructing the facility, which not only reduces the construction period but also enhances the service life. The facility polishes the pipe design and considers a jet aeration system, achieving stable suspension of fillers, sufficient and uniform aeration, and sludge return by gravity while minimizing the number of equipment required. The facility is primarily composed of an anoxic zone, a multifunctional zone, a sedimentation zone, and an aerobic zone, as illustrated in Fig. 1.

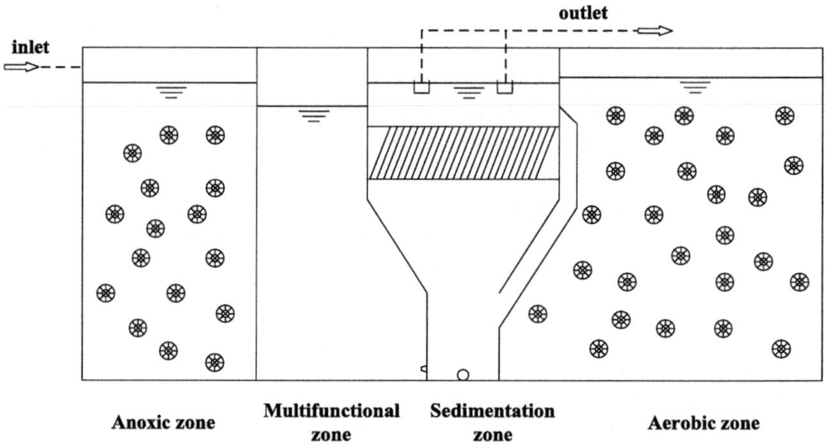

Fig. 1. Schematic Diagram of the Zones in the IMS-MBBR Facility.

By integrating submerged sewage pumps and jet aerators within the multifunctional zone, efficient aeration and suspension mixing of the media in both aerobic and facultative zones are achieved. The operation of submerged sewage pumps elevates the water level in the aerobic zone, facilitating a favorable condition for gravity-driven recirculation. Innovatively, customized diffusers are installed at the bottom of both the facultative and aerobic zones, ensuring comprehensive and uniform suspension of the media within these zones, thereby significantly enhancing the biological treatment efficacy and reaction rates.

To achieve precise control over the treatment process, regulating valves are installed on the pipelines connecting the multi-functional zone to the facultative zone. Manipulation of these valves allows for flexible distribution of flow between the facultative and aerobic zones, adapting to varying treatment requirements and operational conditions. Furthermore, the jet aerator inlet pipes are equipped with regulating valves and flowmeters, enabling fine-tuning of the aeration volume to ensure it meets the operational demands of the system.

Gravity internal return pipes are established from the aerobic zone to the multi-functional zone, while gravity external return pipes are set up from the sedimentation zone to the multi-functional zone. Valves and flowmeters are installed on these return pipes to control and monitor the return flow rates, facilitating adjustments to match operational needs. The design and calculation of gravity return flow rates refer to fluid mechanics formulas, as detailed in Eq. 1.

$$Q = uA\sqrt{2gH} \tag{1}$$

where:

Q represents the gravity return flow rate,
u is the discharge coefficient,
A is the cross-sectional area of the outflow,
g is the gravitational acceleration,
H is the effective head.

Additionally, all valves and monitoring instruments installed on the system are integrated into the Internet of Things platform and the operation and maintenance management platform. This integration allows for real-time monitoring and adjustment of operational conditions, enabling unattended operation and enhancing overall system efficiency and reliability. The schematic diagram of pipeline layout is shown in Fig. 2. The on-site picture of the constructed IMS-MBBR facility is shown in Fig. 3.

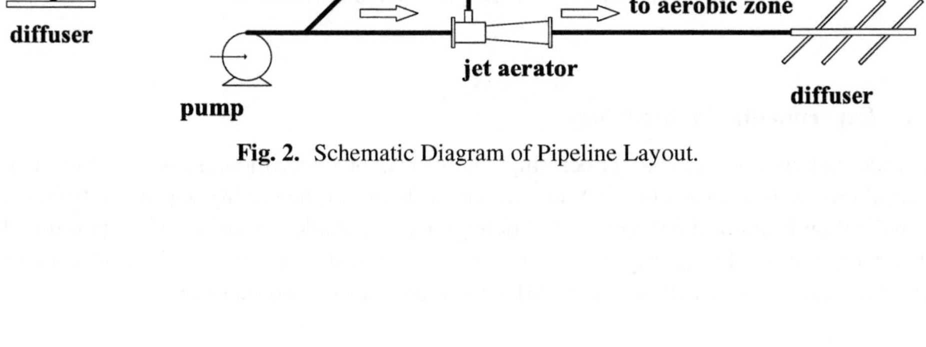

Fig. 2. Schematic Diagram of Pipeline Layout.

Fig. 3. Picture of the Constructed IMS-MBBR Facility at the Site.

3.2 Process Flow

The sewage first flows through the screen channel to remove large particles, and then flows into the equalization basin before entering the IMS-MBBR wastewater treatment device. In the big container the sewage successively enters the anoxic zone, the multi-functional zone, the aerobic zone, and the sedimentation zone. The treated water is then disinfected and discharged to outside with the effluent parameter met the required value. The process flow diagram is illustrated in Fig. 4.

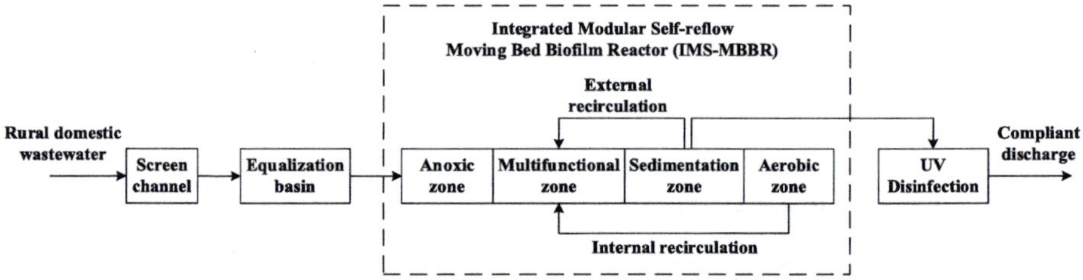

Fig. 4. Process Flow Diagram of the IMS-MBBR.

3.3 Experimental Methodology

In order to better reflect actual operating conditions, the experimental site was chosen at a rural sewage treatment plant in Sichuan. The influent for the facility was taken from the equalization basin of the device. After debugging and stable operation, The operational data were recorded regularly every day. The experimental period was from December 2023 to March 2024, with a total of 90 days of data tested and recorded.

The design parameters of the device are as follows: design treatment capacity Q = 250 m^3/d, hydraulic retention time T = 12.6 h, media filling rate of 25%, internal return ratio ranging from 200% to 400%, external return ratio of 100%. The design inlet and outlet water quality are detailed in Table 1.

Table 1. Water Quality Standards for Inlet and Outlet of Equipment Design.

Item	COD_{Cr}	BOD_5	SS	T-N	NH_3-N	T-P	pH
Influent	300	150	200	65	55	4	6–9
Effluent	≤50	≤10	≤10	≤15	≤5(8)[a]	≤0.5	6–9

[a] The numerical value outside the brackets is the control index when the water temperature is above 12 °C, while the numerical value inside the brackets is the control index when the water temperature is at or below 12 °C

4 Analysis of Operation Performance

4.1 Analysis of Influent Flow Capacity, Temperature, and pH

During the experimental period, the curve of influent capacity is shown in Fig. 5. The influent flow rate varied significantly between 202 and 288 m^3/d. During the initial stable operation period (days 1 to 15), the average influent flow capacity was approximately 92.26% of design capacity. As the system ran, the flow capacity gradually increased and fluctuated around 100%, where there were approximately 36 days of operation exceeding 100% load. The average daily influent flow rate was 247.92 m^3/d, and the peak flow rate reached 287.32 m^3/d, exceeding the designed influent flow rate by 14.93%. Throughout the experimental period, the influent flow rate effectively matched the designed influent flow rate requirements of the system, and also appropriately validated the treatment capacity under overload conditions.

The variation curves of water temperature and pH value are presented in Fig. 6. The water temperature ranged from 11.3 to 18.1 °C, with an average temperature of 13.7 °C, indicating an overall low temperature condition. Notably, during the periods of days 42 to 46 and 60 to 66, the water temperature dropped below 12 °C. Low temperatures can inhibit microbial activity, thereby affecting the efficiency of wastewater treatment. According to the "Discharge Standard of Pollutants for Municipal Wastewater Treatment Plant" (GB18918-2002), when the water temperature is below 12 °C, a lower standard for the effluent ammonia nitrogen concentration may be applicable. This temperature range is thus representative of the operational conditions for wastewater treatment during low temperatures in winter.

The pH value varied within a narrow range of 6.80 to 7.93, with an average of 7.16. Extremely low or high pH values can diminish or even inactivate enzyme activity, leading to reduced treatment efficiency. Typically, the pH value should be maintained within 6 to 9. During this experimental period, the pH values of the wastewater remained within the acceptable range and showed minimal variation. Therefore, the impact of pH on the treatment performance of the system will not be analyzed separately.

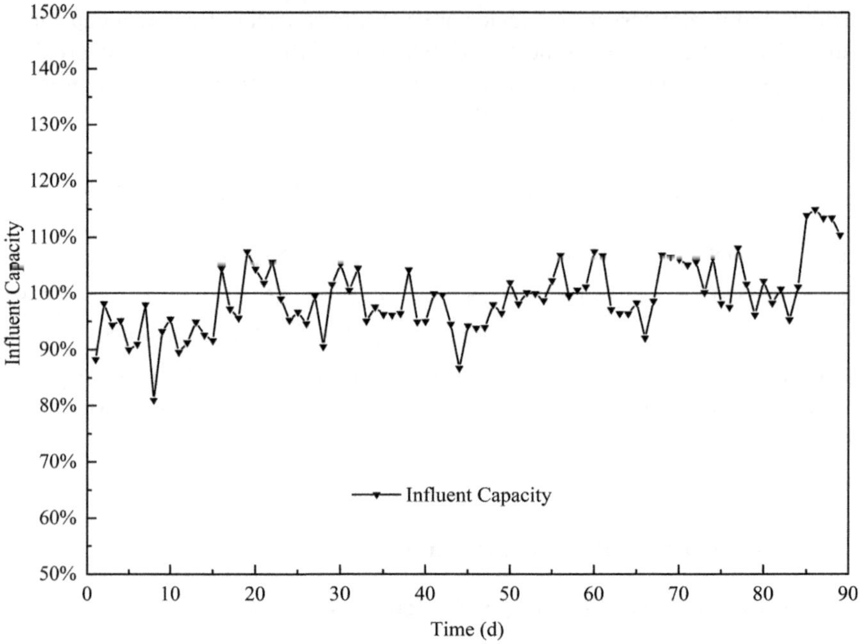

Fig. 5. The Curve of Influent Capacity for Facility.

4.2 Analysis of COD Removal Efficiency

As shown in Fig. 7, the mostly range of influent COD concentrations was approximately 120 mg/L to 350 mg/L, with an average concentration of 225.51 mg/L. Notably, during days 39 to 45, the influent COD concentration increased significantly, averaging 483.21 mg/L, exceeding the designed influent COD concentration by 61.07%, and peaking at 652.64 mg/L on day 44. The effluent COD concentration ranged from 17.78 mg/L to 38.44 mg/L, and its maximum value (38.44 mg/L) occurred on day 44 due to the influence of influent concentration variations. The average effluent COD concentration was 24.58 mg/L, with an average removal rate of 87.72%, indicating stable effluent COD concentrations that met the Class 1A Discharge Standard.

During the experimental period, the effluent COD concentration showed a certain positive correlation with the influent COD concentration. When the influent COD concentration was higher from day 39 to day 45, the average effluent COD concentration was 29.23 mg/L, which was 18.92% higher than the average (24.58 mg/L). Temperature variations during the experiment did not significantly impact the COD removal efficiency. Remarkably, from day 42 to day 45, under the dual stress of low temperature conditions (<12 °C) and high influent concentrations (>400 mg/L), the COD removal rate exceeded the average (87.72%) to reach 93.82%, with effluent meeting the Class 1A Discharge Standard. This demonstrates the excellent COD removal performance and strong resilience of the system to shocks.

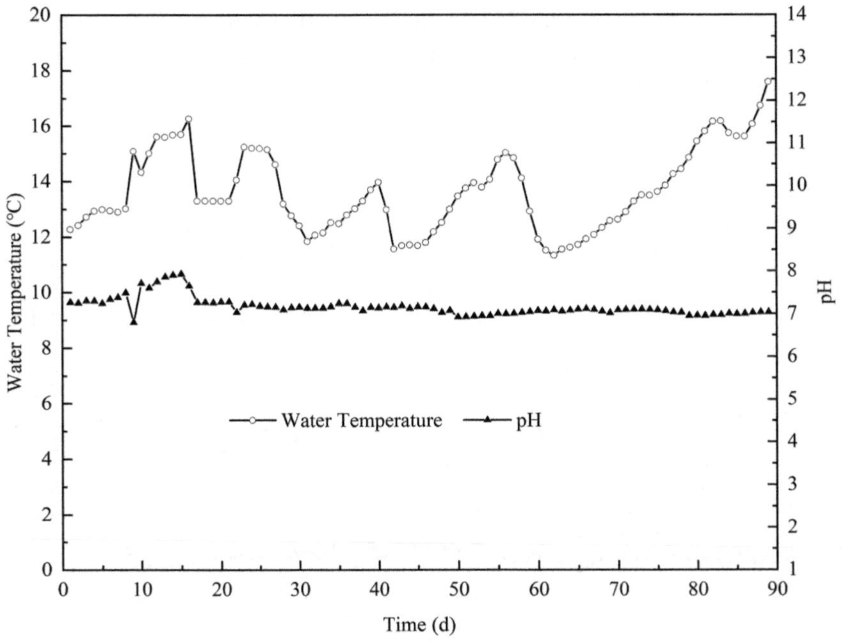

Fig. 6. Temperature and pH Change Curve Diagram.

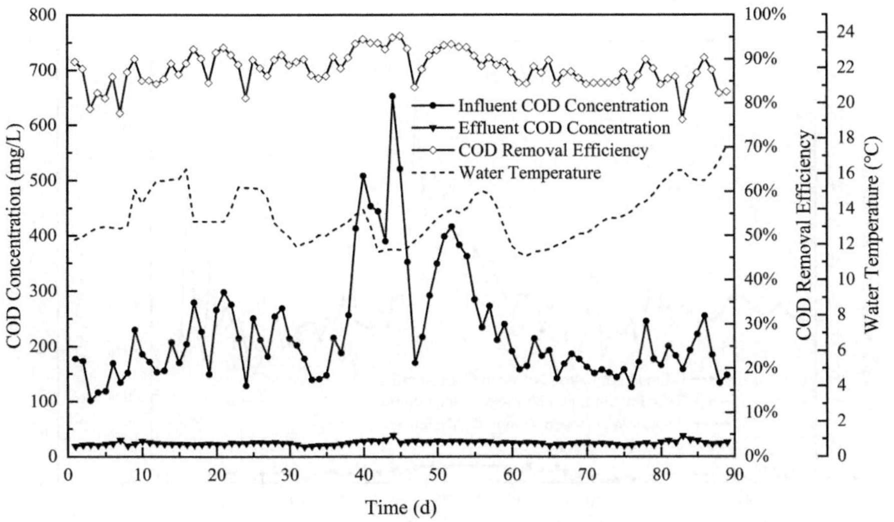

Fig. 7. COD Removal Efficiency.

4.3 Analysis of Total Nitrogen and Ammonia Nitrogen Removal Efficiency

The removal efficiencies of total nitrogen and ammonia nitrogen are presented in Fig. 8 and Fig. 9, respectively. The influent total nitrogen concentration ranged from 34.57 mg/L to 106.64 mg/L, with an average of 54.28 mg/L. The effluent total nitrogen concentration varied between 6.50 mg/L and 14.75 mg/L, averaging at 10.92 mg/L, resulting in an average removal rate of approximately 79.36%.

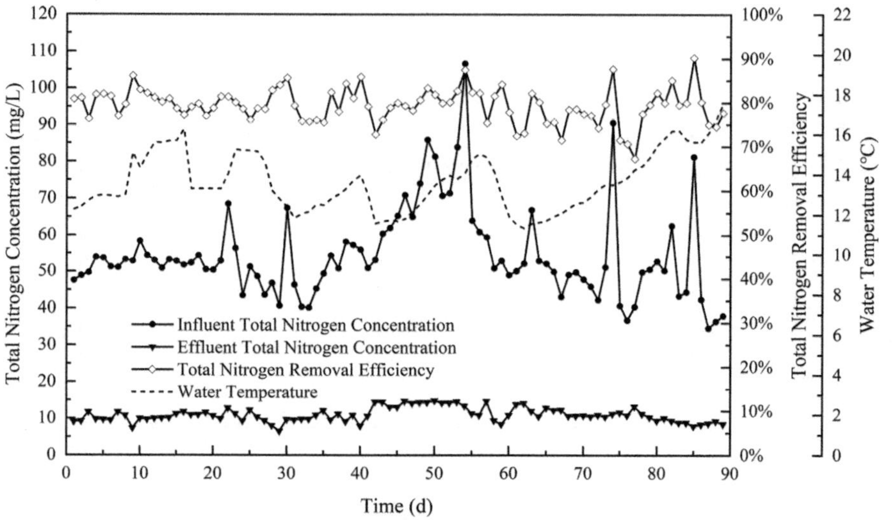

Fig. 8. Total Nitrogen Removal Efficiency.

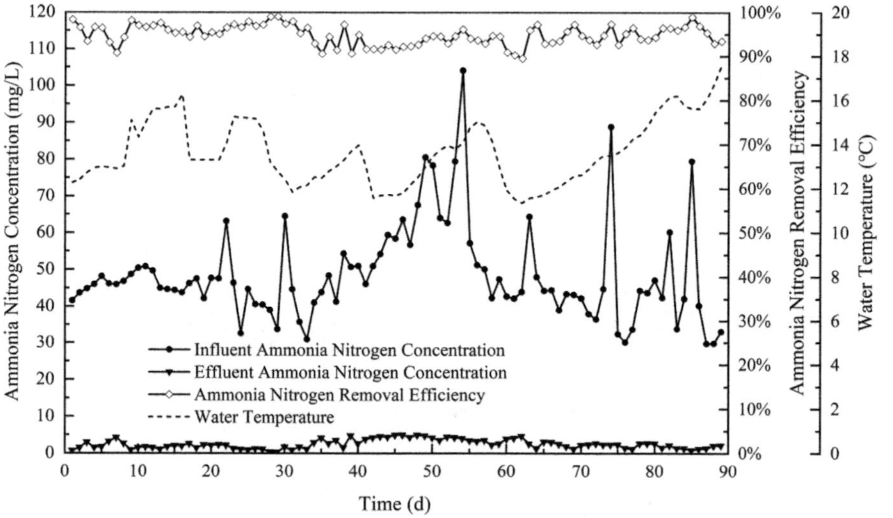

Fig. 9. Ammonia Nitrogen Removal Efficiency.

The influent ammonia nitrogen concentration was within the range of 29.75 mg/L to 104.07 mg/L, with an average of 48.39 mg/L. The effluent ammonia nitrogen concentration ranged from 0.33 mg/L to 4.97 mg/L, averaging at 2.49 mg/L, yielding an average removal rate of approximately 94.82%. Both the effluent total nitrogen and ammonia nitrogen concentrations remained stable and met the Class 1A Discharge Standard.

Overall, the removal of total nitrogen and ammonia nitrogen was effective. Between days 42 and 53, minor increases in the effluent total nitrogen and ammonia nitrogen concentrations were observed, averaging at 14.16 mg/L and 4.47 mg/L, respectively. Comparative analysis with influent concentrations and water temperature revealed that the water temperature fell below 12 °C from days 42 to 46, and the influent total nitrogen and ammonia nitrogen concentrations significantly increased from days 43 to 56, averaging at 72.97 mg/L and 66.94 mg/L, respectively. Despite the combined effects of decreased water temperature and increased influent concentrations, the overall effluent quality remained stable, meeting the Class 1A Discharge Standard, demonstrating the system's strong resilience against shocks.

4.4 Analysis of Total Phosphorus Removal Efficiency

As shown in Fig. 10, during the experimental period, the influent total phosphorus concentration ranged from 2.43 mg/L to 3.84 mg/L, with an average of 2.85 mg/L. The effluent total phosphorus concentration remained stable, ranging approximately from 0.18 mg/L to 0.37 mg/L, averaging at 0.26 mg/L, achieving an average removal rate of 90.96%. This stable effluent total phosphorus concentration met the Class 1A Discharge Standard. Notably, under low-temperature conditions, the removal of total phosphorus was still effective, and temperature variations did not significantly impact the effluent total phosphorus concentration.

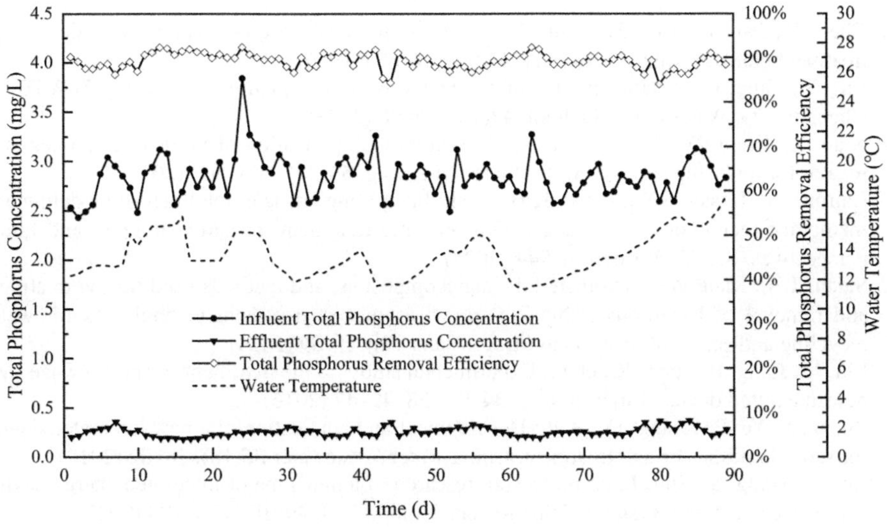

Fig. 10. Total Phosphorus Removal Efficiency.

5 Conclusions

Based on the experimental data recorded over a period of 90 days, the following conclusions can be drawn.

(1) Stable Operation and Excellent Effluent Water Quality
Through continuous monitoring of the IMS-MBBR wastewater treatment facility for 3 months, stable compliance of the effluent was achieved without the addition of external phosphorus removal agents. The effluent parameters were more excellent than the required value of Class 1A Discharge Standard.

(2) Good and Stable Nitrogen Removal Efficiency during Low-Temperature Conditions and Significant Influent Quality Fluctuations
The IMS-MBBR wastewater treatment facility demonstrated excellent performance in low-temperature environments, even when the influent temperature drops below 12 °C, ensuring stable effluent compliance. Moreover, the effluent quality remained stable despite significant fluctuations in influent water quality in rural areas, showing good adaptability during such variations.

(3) Obvious advantages and bright future for application in rural sewage treatment
The IMS-MBBR integrated wastewater treatment facility was demonstrated with excellent treatment efficiency, good flexibility and adaptability, and simple operation and maintenance. It can effectively address the current issues faced by rural wastewater treatment and has broad application prospects in the field of rural wastewater treatment.

References

1. Chen, Y., Fan, R., Liu, Z., et al.: Research progress of integrated rural domestic sewage treatment plant. J. Anhui Agric. Sci. **44**(09), 84–88 (2016)
2. Liu, Y., Tang, Q.: Characteristics of influent wastewater quality in township WWTPs in Chengdu city. Water Purif. Technol. **42**(10), 76–86 (2023)
3. Xiao, X., Zhou, W., Chen, X., et al.: Research and application of vertical flow integrated reactor for township sewage. Water Wastewater Eng. **56**(S1), 500–503 (2020)
4. Cairone, S., Hasan, S.W., Choo, K.H., et al.: Integrating artificial intelligence modeling and membrane technologies for advanced wastewater treatment: research progress and future perspectives. Sci. Total Environ. **944** (2024)
5. Saleh, T.A.: Materials, nanomaterials, nanocomposites, and methods used for the treatment and removal of hazardous pollutants from wastewater: treatment technologies for water recycling and sustainability. Nano-Struct. Nano-Objects **39** (2024)
6. Liu, X., Jiang, F., Feng, X., et al.: Experimental study on the treatment of rural sewage by a new integrated device. Environ. Eng. **34**(12), 38–42, 47 (2016)
7. Zheng, J., Yu, P., Zhang, D., et al.: Development and application of a new integrated device for rural domestic sewage treatment. China Water Wastewater **35**(18), 78–82 (2019)
8. Liu, T., Wang, S., Sun, J., et al.: Design research of a new type of integrated energy-saving wastewater treatment system. China Resour. Compr. Util. **40**(01), 186–189 (2022)
9. Jiang, W., Ma, Y., Nie, Z., et al.: Improving nitrogen and phosphorus removal and sludge reduction in new integrated sewage treatment facility by adjusting biomass concentration. J. Water Process Eng. **50** (2022)
10. Han, Y., Ma, J., Xiao, B., et al.: New integrated self-refluxing rotating biological contactor for rural sewage treatment. J. Clean. Prod. **217**, 324–334 (2019)

Study on the Coupling and Coordinated Development of Water-Food-Energy-Ecosystem in Shiyang River Basin

Wanzhen Wang[1] and Haihua Ma[2(✉)]

[1] Water Resources Department, Water Resources Utilization Center of Shiyang River Basin, Wuwei 733000, China
[2] Gansu Forestry Vocational and Technical University, Lanzhou 730050, China
2820915635@qq.com

Abstract. To explore the coordinated and sustainable development of the Shiyang River Basin water-food-energy-ecosystem (WFEE) to avoid the harm caused by other resource departments due to the development of a single resource sector, the northwest inland The WFEE system of the Shiyang River Basin, one of the river regions is a research object, builds the Basin WFEE system coupling and coordinated evaluation index system. And the characteristics of space-time evolution were studied, and the coupling collaboration of the Shiyang River Basin from 2020–2030. The results showed that: the comprehensive development index of the WFEE system in the Shiyang River Basin in 2000–2020 is at a medium level, and the comprehensive development index of the energy subsystem grows negative; The upward trend has gone through six stages of mild disorders to good coordination; the prediction results show that the coupling coordination of the basin is in a good coordinated transition stage during the predicted period. The research results can provide scientific basis for the sustainable management and efficient development and utilization of the Shiyang River Basin.

Keywords: Water-Food-energy-ecosystem · CRITIC weighting method · Comprehensive development index · Coupling coordination degree · Shiyang River basin

1 Introduction

Water, food and energy are essential resources for human social development. For a long time, the problems of water resource shortage, serious soil erosion, weak resource and environmental carrying capacity and ecological fragility have seriously restricted the high-quality development of the river basin. Social and economic activities such as population change, economic growth, agricultural development and urbanization promote the growth of human demand for water, food and energy. However, the unreasonable resource utilization methods in the river basin will further lead to conflicts and threaten the realization of high-quality development goals. In November 2011 in Bonn, Germany, water-food-energy (WFE) association meeting [1] first proposed water, food, energy ties,

D. Han et al. (Eds.): ICEPG 2024, ESE, pp. 840–854, 2025.
https://doi.org/10.1007/978-3-031-86829-0_94

points out that three subsystems is not separate development but interdependent, complementary correlation, and ecosystem as water, food, energy resources important carrier, and WFE mutual dependence, mutual influence, ecosystem and water, food and energy matching and coupling relationship between the regional development and stability, so it is necessary to add ecological factors into the ties.

At present, experts and scholars in this field have gradually introduced economic, production, social and other factors, and carried out a lot of research work on the relationship of water resources, food, energy and ecological link. For example, Zhang [2] and Zhang et al. [3] conducted the bibliometric analysis of the water-food-energy correlation system from 1985 to 2020, reviewed the common research methods, and sorted out the research history and current situation of WEF correlation system. Wang et al. [4] have joined the ecological subsystem on the basis of the water-food-energy system, and found that it is necessary to coordinate the social governance process of the water, food, energy and other resource sectors from the perspective of ecosystem services. Mi et al. [5] used system dynamics to predict China's population, economic development level and demand scale for food, fresh water and energy. Deng Peng et al. [6] constructed the coupled coordination degree model of WEF system in Jiangsu Province, quantitatively evaluated the coordinated development level of Jiangsu Province, and found that the level of water resources and energy was continuously lower than the grain level. Peng et al. [7] constructed the WEF research framework in the Yellow River Basin based on the synergy theory, and optimized and solved the collaborative optimization model of grain production, energy development and water resources allocation in the Yellow River Basin. Zhou et al. [8] based on the life cycle theory, we studied the mutual demand relationship in the process of resource flow. Shi et al. [9] used the Bayesian network to study the WFEE bond in the Aral Sea basin. Sun et al. [10] used SBM-DEA model, network DEA model and bivariate spatial autocorrelation method to study the efficiency and change characteristics of WF ties in 30 provinces in China from 1997 to 2017. Lin et al. [11] have systematically sorted out the concept and connotation of WFE bond relationship, and found that the substantive concept and general framework of the bond relationship have not been unified. Through the improved GWAS model, Chang [12] and Zhao [13] constructed a water-food-energy-production cooperative regulation model to realize the transmission and mutual feeding of key elements of each subsystem, providing effective suggestions for solving the destructive water shortage in the Beijing-Tianjin-Hebei region with the most significant link.

Shiyang River basin is not only an important commodity grain base in Gansu, but also an "energy river basin" in Gansu. The resource endowment and development status of the Shiyang River Basin determine that the Shiyang River Basin is a typical area with prominent and concentrated contradictions in the WFEE system. Promoting the coordinated management of ties and ensuring the security of water resources, food, energy and ecological security in Shiyang River Basin not only have great strategic significance to the implementation of ecological protection and high-quality development of Shiyang River Basin, but also has an important and far-reaching impact on the realization of the long-term goal of social and economic development in Gansu. Therefore, Taking the Shiyang River Basin, one of the three major inland river basins in northwest China, Combined with the differences in the spatial distribution of natural resources in the river basin, And the diversified characteristics of economic structure development, Construct the evaluation index system of the WFEE system, And according to the different site conditions and ecological positioning of each county, In combination with the 14th Five-Year Development Plan, Different weights of the WFEE system of each county-level administrative region, Using the coupled coordination model, Quantitative study of the coupling characteristics of the WFEE system from 2000–2020 from time and spatial scale, respectively, In order to reveal the coordinated development mechanism of WFEE system in Shiyang River Basin, It also provides a reference basis for sustainable economic and social development and local government decision-making.

2 Research Area Overview

Shiyang River Basin (37°02′–39°17′N, 100°57′–104°12′E) located in the Hexi Corridor in Gansu Province, it is one of the three major inland river basins in China. The Shiyang River basin has eight main water systems, Flow through Wuwei, Jinchang, Zhangye, Baiyin four cities, nine counties and districts, The river basin area is 41,900 km^2, Accounting for 9.84% of the land area of Gansu Province; Shiyang River basin belongs to a continental temperate arid climate, Rainfall decreases gradually from south to north, The area with the least precipitation is located to the north at the intersection of the Taklimakan and Tengger deserts, Annual precipitation is only 35.0 mm, Rainfall is extremely uneven in its spatial and temporal distribution, The annual allocation is mainly concentrated in July, August and September, The average annual evaporation within the basin is three to four times the amount of rainfall, Annual average annual runoff of 1.306 billion m^3, Accounting for 5.03% of Gansu Province, The per capita water resources in the basin is 725.11 m^3, For 44.15% of the per capita water resources in Gansu Province, There is a large gap in the per capita water resources and water resources utilization rate in the upper, middle and lower reaches. In 2020, the surface water resources were 1.119 billion m^3, the underground water resources were 1.092 billion m^3, and the total water resources accounted for 3.25% of Gansu Province. The Shiyang River Basin has great grain output potential. In 2020, the total farmland area of the whole basin was 8.1678 million mu, the effective irrigation area of farmland was 4.6012 million mu, and the total grain output was 1.4718 million tons, accounting for 12.24% of the total grain output of Gansu Province. The energy in Shiyang River Basin is mainly purchased from foreign sources. In 2020, the total energy consumption is 4.319 million tons of standard coal, including 4.2788 million tons of raw coal, 167 tons of gasoline and 4,622 tons of diesel. Since the comprehensive management of Shiyang River Basin in 2005, the groundwater level has recovered and the ecological environment has been greatly improved.

3 Research Method

3.1 Selection of Indicators and Data Sources

In order to accurately grasp the shiyang river basin water, food, energy and ecological coupling coordinated development level, on the basis of the existing literature, follow the scientific, systematic, comprehensive and authenticity principle, 2000–2020 selected representative four subsystems, a total of 32 indicators, build the shiyang river basin WFEE system coupling coordinated development evaluation index system, data mainly from the Gansu province water resources bulletin Zhangye statistical yearbook, the Jinchang statistical yearbook the Wuwei statistical yearbook the Baiyin statistical yearbook, etc.

3.2 Data Standardization and Weight Determination

Data Standardization

The positive and negative indexes are calculated as follows:

$$M_{ij} = \frac{h_{ij} - minh_{ij}}{maxh_{ij} - minh_{ij}} \tag{1}$$

$$M_{ij} = \frac{maxh_{ij} - h_{ij}}{maxh_{ij} - minh_{ij}} \tag{2}$$

Formula: M_{ij} is normalized data; a_{ij} is raw data; $maxh_{ij}$ is the maximum value; $maxh_{ij}$ is the minimum value.

Index Weight Determination

Some indicators in the index system constructed in this paper may have a correlation relationship, and the index weight is determined by CRITIC empowerment method [14]. The calculation formula is as follows:

$$r_{ij} = \frac{\sum\left(x_{ij} - \overline{x_{ij}}\right)\left(y_{ij} - \overline{y_{ij}}\right)}{\sqrt{\sum\left(x_{ij} - \overline{x_{ij}}\right)^2\left(y_{ij} - \overline{y_{ij}}\right)^2}} \tag{3}$$

$$c_j = \sigma_j \sum\nolimits_{i=1}^{n} \left(1 - r_{ij}\right) \quad j = 1, 2, 3 \cdots n \tag{4}$$

$$V_j = \frac{c_j}{\sum_{j=1}^{n} c_j} \quad j = 1, 2, 3 \cdots n \tag{5}$$

Formula: r_{ij} is the correlation coefficient of column j in row i of R matrix; c_j is the information contained by the j index; j is the standard deviation of each indicator; V_j is the j index weight.

The coupling coordination evaluation index system of WFEE system in Shiyang River basin is shown in Table 1.

Table 1. Shiyang River Basin WFEE system coupling coordination evaluation index system

Subsystem	Index	Unit	Weight	Style
Water resources subsystem	1. Surface water resources	100 million m^3	0.0570	Positive
	2. Groundwater resources	100 million m^3	0.0522	Positive
	3. Total water resources	100 million m^3	0.0545	Positive
	4. Water resources per capita	m^3/person	0.0502	Positive
	5. Water consumption per capita	m^3/person	0.1196	Negative
	6. Water permeability coefficient	/	0.0673	Positive
	7. Modulus of water production	/	0.0535	Positive
	8. Rainfall	100 million m^3	0.0833	Positive
	9. Proportion of groundwater supply	%	0.1053	Negative
	10. Proportion of agricultural water consumption	%	0.0850	Negative
	11. Proportion of industrial water consumption	%	0.0951	Negative
	12. Proportion of ecological water consumption	%	0.0699	Positive
	13. Proportion of domestic water consumption	%	0.1069	Negative
Food subsystem	1. Total grain production	10000 tons	0.0935	Positive
	2. Grain output per unit area	kg/acre	0.1154	Positive
	3. Per capita grain production	kg/person	0.0931	Positive
	4. Grain planting area	10000 acres	0.1834	Positive
	5. Effective irrigation area	10000 acres	0.2022	Positive
	6. Fluctuations in total grain production	/	0.1794	Negative
	7. Area affected	10000 acres	0.1330	Negative

(*continued*)

Table 1. (*continued*)

Subsystem	Index	Unit	Weight	Style
Energy subsystem	1. Raw coal consumption	ton	0.1576	Negative
	2. Gasoline consumption	ton	0.2273	Negative
	3. Diesel consumption	ton	0.1171	Negative
	4. Total energy consumption	10000 tons of standard coal	0.2105	Negative
	5. Energy consumption of industrial added value	tons of standard coal/10000 yuan	0.1406	Negative
	6. Energy consumption per unit of GDP	tons of standard coal/10000 yuan	0.1468	Positive
Ecological subsystem	1. Forest coverage	%	0.1822	Positive
	2. Afforestation area	10000 acres	0.1626	Positive
	3. The area of mountain closure and afforestation	10000 acres	0.1659	Positive
	4. Ecological land area	km^2	0.2071	Positive
	5. Desert area	km^2	0.1397	Negative
	6. Population density	10000 people	0.1425	Negative

3.3 Water-Food-Energy-Ecosystem Composite Index

The comprehensive index method of the weighted sum of each index is used to calculate the comprehensive development index of the water, energy and food security subsystem, indicating the development degree of each subsystem. The calculation formula is as follows:

$$\begin{cases} F_w = \sum_{j=1}^{n} V_j W_{wj}\prime \\ F_f = \sum_{j=1}^{n} V_j X_{fj}\prime \\ F_{e1} = \sum_{j=1}^{n} V_j Y_{e1j}\prime \\ F_{e2} = \sum_{j=1}^{n} V_j Z_{e2j}\prime \end{cases} \tag{6}$$

Formula: F_w, F_f, F_{e1}, F_{e2} are the comprehensive development indexes of water, food, energy and ecological subsystem respectively; V_w, V_f, V_{e1}, V_{e2} are the index weights of water, food, energy and ecology respectively; $W_{wj}\prime$, $X_{fj}\prime$, $Y_{e1j}\prime$, $Z_{e2j}\prime$ are the standardized data of each index.

3.4 Coupled Coordination Degree Model

The coupling degree is used to measure the strength of the interactions between the subsystems. The coupling degree C_i between the subsystems is calculated as follows:

$$C_i = \frac{4\sqrt[4]{F_w F_f F_{e1} F_{e2}}}{F_w + F_f + F_{e1} + F_{e2}} \tag{7}$$

Formula: the value range of the coupling degree is [0,1], and the size of the value is determined by the quality of the sustainable subsystem. The larger the value of C_i, the interaction between the subsystems are, the stronger. When $C_i = 1$, the subsystems are in the best coordination state.

In order to more accurately represent the coupling coordination degree between the subsystems, construct the coupling coordination degree model [15]:

$$D_i = \sqrt{C_i T_i} \tag{8}$$

$$T_i = aF_w + bF_f + cF_{e1} + dF_{e2} \tag{9}$$

Formula: D_i is the coupling coordination degree, $D_i \in [0, 1]$; T_i is the comprehensive evaluation index of WFEE system; *a, b, c, d* are the importance weight of each subsystem. According to the different positioning of land planning and ecological function areas in the counties of counties, the weighted average. Specific divisions are shown in Table 2.

Table 2. Shiyang River Basin WFEE system subsystem subsystem weight division

Area/district (county)	Weight division				Area/km^2
	Water	Food	Energy	Ecology	
jinchuan	0.30	0.15	0.40	0.15	3060.26
yongchang	0.20	0.25	0.30	0.25	4509.19
liangzhou	0.20	0.40	0.20	0.20	4907.96
minqin	0.30	0.20	0.20	0.30	15900.00
gulang	0.20	0.20	0.30	0.30	5103.00
tianzhu	0.20	0.30	0.15	0.35	7150.00
sunan	0.20	0.20	0.20	0.40	20177.00
shandan	0.20	0.40	0.20	0.20	5402.43
jingtai	0.20	0.30	0.20	0.30	5483.00
Shiyang River Basin	0.22	0.25	0.22	0.31	/

According to the existing literature, the water-food-energy-ecosystem coupling coordination degree is divided, as shown in Table 3:

Table 3. Coupling degree grade classification standard

Coupling level	Data range	Coordination type	Grade	Range of coupling coordination values
Low level coupling	[0, 0.3]	Disorder decline	Extreme disorder	[0, 0.1)
			severe disorder	[0.1, 0.2)
			Moderate disorder	[0.2, 0.3)
Antagonistic stage	(0.3, 0.5]		mild disorder	[0.3, 0.4)
			near the disorder	[0.4, 0.5)
Run-in stage	(0.5, 0.8]	Transition	Barely coordination	[0.5, 0.6)
			primary coordination	[0.6, 0.7)
		Harmonious development	intermediate coordination	[0.7, 0.8)
High level coupling	(0.8, 1]		Good coordination	[0.8, 0.9)
			high-quality coordination	[0.9, 1]

4 Results and Analysis

4.1 Comprehensive Level Analysis of the WFEE System

Firstly, calculate the Eq. (1)-(7) of the evaluation index, comprehensive development index, coupling degree and coupling coordination degree of the WFEE system from 2000 to 2020 (Table 4); draw the timing evolution diagram of the comprehensive development index of Shiyang River Basin (Fig. 1). As can be seen from Fig. 1, the comprehensive development index of water resources subsystem fluctuates greatly in the early stage and basically tends to be stable in the later period, with an average annual growth rate of 0.84% and the highest value of 0.6981 in 2003. The comprehensive development index of grain subsystem is relatively stable compared with other subsystems, which is due to the developed agriculture in Shiyang River Basin, large planting scale and high grain yield, with an average annual growth rate of 1.73%. The comprehensive development index of energy subsystem changed greatly from 2002 to 2006, and then it tended to be stable, basically above 0.5, with an average annual growth rate of-0.54%. The comprehensive evaluation index of ecological subsystem fluctuated between 0.0426 and 0.6959, growing slowly after 2005, with the largest growth rate from 2015 to 2016 (53.2%), which is closely related to the comprehensive management of Shiyang River basin implemented since 2005.

On the whole, the comprehensive development index of water and energy subsystems in Shiyang River Basin from 2000 to 2020 shows a "wave" distribution pattern. Water-food-energy-ecosystem index overall on the rise, the average annual growth rate of 2.25%, before 2008 the total system and the subsystem index variation is larger, after change, visible Shiyang river basin water-food-energy-ecosystem internal coupling correlation, four subsystems affect the stable development of the total system.

Table 4. The comprehensive index, coupling degree and coupling coordination degree of WFEE system in Shiyang River basin

Year	Water subsystem F_w	Food subsystem F_f	Energy subsystem F_{el}	Ecological subsystem F_{e2}	Comprehensive evaluation index T	Coupling degree C	Coupling coordination degree D	Coupling level	Coupling coordination type
2000	0.3578	0.3784	0.7707	0.0406	0.306	0.426	0.361	Antagonistic stage	Mild imbalance
2001	0.3047	0.3769	0.6508	0.0852	0.175	0.478	0.289	Antagonistic stage	Mild imbalance
2002	0.3381	0.3555	0.7189	0.2206	0.341	0.796	0.521	Run-in stage	Barely coordinating
2003	0.6981	0.2472	0.8204	0.3287	0.575	0.422	0.492	Antagonistic stage	Barely coordinating
2004	0.5361	0.2940	0.5913	0.2932	0.384	0.849	0.571	High level coupling	Barely coordinating
2005	0.4374	0.3588	0.7589	0.2673	0.438	0.895	0.626	High level coupling	Primary Coordination
2006	0.4295	0.3947	0.4425	0.2903	0.306	0.93	0.534	High level coupling	Barely coordinating
2007	0.4635	0.4001	0.4106	0.3006	0.316	0.802	0.503	High level coupling	On the brink of imbalance
2008	0.2945	0.3748	0.4464	0.3336	0.245	0.525	0.359	Run-in stage	Mild imbalance
2009	0.4685	0.4780	0.3868	0.3423	0.369	0.505	0.431	Run-in stage	On the brink of imbalance
2010	0.5250	0.5467	0.5260	0.3790	0.522	0.968	0.711	High level coupling	Intermediate coordination
2011	0.4843	0.3895	0.5351	0.4000	0.43	0.976	0.647	High level coupling	Primary Coordination
2012	0.5455	0.4112	0.6699	0.3839	0.534	0.977	0.722	High level coupling	Intermediate coordination
2013	0.3764	0.4538	0.5161	0.4755	0.432	0.912	0.628	High level coupling	Primary Coordination
2014	0.5533	0.5143	0.6096	0.4335	0.587	0.997	0.765	High level coupling	Intermediate coordination
2015	0.5136	0.5428	0.6251	0.4542	0.598	0.997	0.772	High level coupling	Intermediate coordination
2016	0.5929	0.4995	0.6257	0.6959	0.729	0.971	0.841	High level coupling	Good coordination

(continued)

Table 4. *(continued)*

Year	Water subsystem F_w	Food subsystem F_f	Energy subsystem F_{e1}	Ecological subsystem F_{e2}	Comprehensive evaluation index T	Coupling degree C	Coupling coordination degree D	Coupling level	Coupling coordination type
2017	0.6088	0.5185	0.6183	0.4847	0.647	0.991	0.8	High level coupling	Intermediate coordination
2018	0.5686	0.6752	0.5492	0.4398	0.655	0.95	0.789	High level coupling	Intermediate coordination
2019	0.5662	0.6975	0.5898	0.5395	0.733	0.966	0.841	High level coupling	Good coordination
2020	0.4985	0.6969	0.5732	0.6469	0.737	0.938	0.832	High level coupling	Good coordination

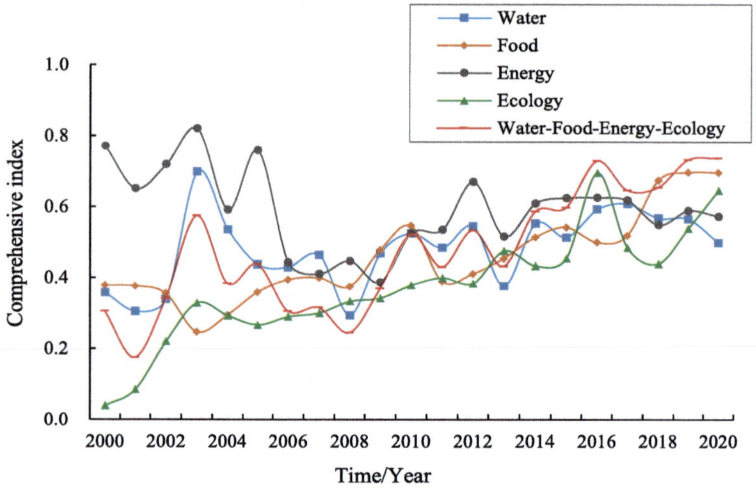

Fig. 1. 2000–2020 Shiyang River Basin WFEE Comprehensive Evaluation Index

In this paper, the distribution pattern of the comprehensive development index of water-food-energy-ecosystem in four cities and nine counties covered by the Shiyang River Basin from 2000 to 2020 (see Fig. 2) is further studied. As can be seen from Fig. 2, the spatial distribution of water resources in Shiyang River basin is uneven, and the rainfall in central China is low, mainly due to the great influence of Tengger Desert and Badain Jaran Desert. The average comprehensive development index of water resources in Jingtai County is the highest (0.642), due to the relatively large underground water resources in this area. Grain subsystem average comprehensive development index area mainly concentrated in jinchuan district and yongchang county, the main reason is that the effective irrigation area and grain planting area is high, the rest of the grain subsystem development index gradually from west to east, jingtai county grain subsystem average development index (0.395), the reason is that the region grain planting area is small. The average comprehensive development index of the energy subsystem in Wuwei region is relatively low, mainly due to the developed agriculture and the slow development of the energy industry. Jingtai has the highest average energy comprehensive development index, which is mainly affected by large gasoline consumption. The average ecological comprehensive development index of each district and county is between 0.4 and 0.53, which is not much different. The high ecological comprehensive development index in Jinchuan and Yongchang areas is mainly affected by the large area of mountain closure and forest cultivation. The low ecological average comprehensive index in Sunan county and Shandan counties is due to the low forest coverage rate, small afforestation area and low population density.

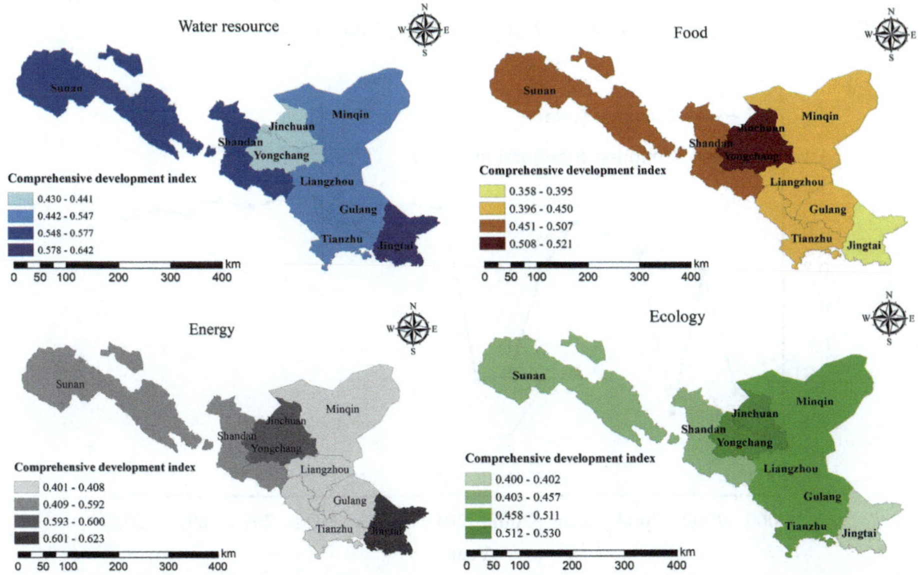

Fig. 2. 2000–2020 Shiyang River Basin water-food-energy-energy-comprehensive development index of ecological development index

4.2 Analysis of Spatiotemporal Dynamics of WFEE System

Time Evolution Analysis

Draw the coupling degree and the coupling coordination degree of the WFEE system in the Shiyang River Basin from 2000 to 2020 (Fig. 3). As can be seen from Fig. 3, the coupling degree of Shiyang River basin was between 0.422 and 0.997, which mainly experienced three stages: jie resistance, run-in and high level coupling. The average coupling degree of WFEE system in Shiyang River basin was high level. In 2003, the coupling degree suddenly decreased to the lowest value (C = 0.422), mainly due to the lower per capita water consumption in 2003 and the low volatility of total grain output. In 2008 and 2009, the coupling degree decreased sharply, mainly due to the lower grain output per unit area, per capita water resources per unit area and energy consumption than usual. After 2010, the coupling degree was higher than 0.9, showing a steady state fluctuation, indicating that each subsystem influenced and promotes each other.

From Fig. 3, 2000–2020 Shiyang river basin coupling coordination experienced mild disorder, almost disorder, barely coordination, primary coordination, intermediate coordination, good coordination, the change trend and coupling degree, including 2007–2009 coupling coordination is lower than the medium level, the main reason is that water resources, afforestation area is lower than the perennial level. In 2011 and 2013, the coupling coordination degree decreased significantly, because the total water resources in Shiyang River Basin decreased in 2011 and 2013, and the grain planting area decreased in 2011. In general, the coupling degree and coupling coordination degree of Shiyang River basin are high, indicating that the water, food, energy and ecosystem are closely related, with strong coordination and good development level.

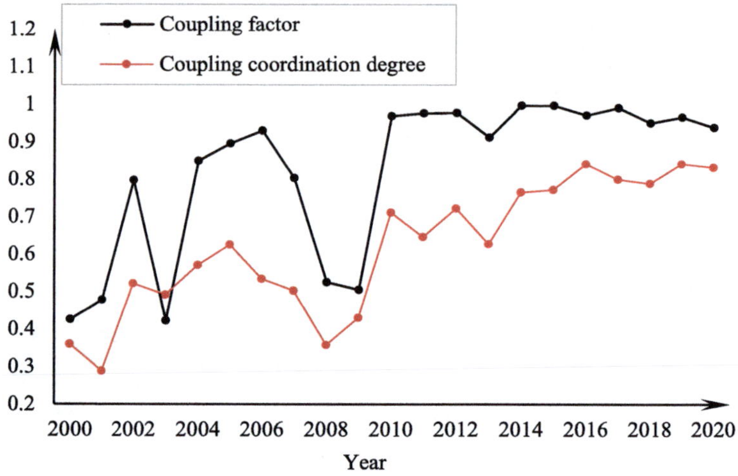

Fig. 3. 2000–2020 Shiyang River Basin WFEE system Coupled and Cou at the Coordination of Coupling

Characteristic Analysis of the Spatial Evolution

Draw the coupling degree and the coupling coordination degree of the WFEE system in the Shiyang River Basin from 2000 to 2020 (Fig. 4). As can be seen from Fig. 4, the average coupling degree level of WFEE from 2000 to 2020 gradually decreases from west to east, and the coupling degree of WFEE system in Jingtai County is the lowest compared with that of other regions (C = 0.949). On the whole, the coupling degree of each district (county) is higher than 0.94, which is in a high level coupling state. The coupling coordination degree of WFEE is between 0.68–0.73, among which the coupling coordination degree of Minqin, Liangzhou, Gulang, Tianzhu and Jingtai is in the primary coordination stage, while Sunan, Shandan, Jinchuan and Yongchang are in the intermediate coordination state. It shows that the development of WFEE coupling system in 2000–2020 is coordinated, and the coupling degree of WFEE system has regions is between regions.

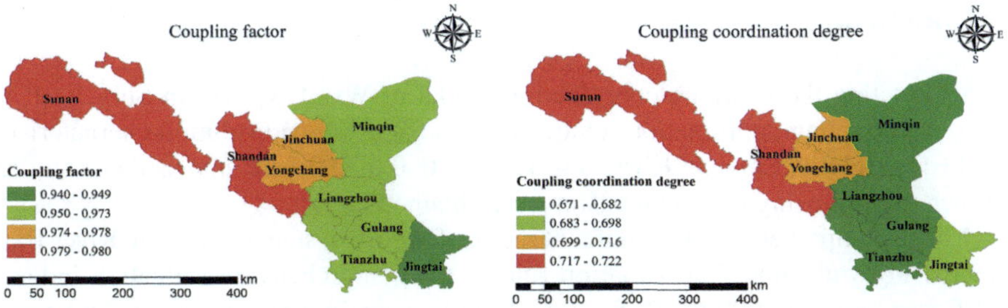

Fig. 4. 2000–2020 Shiyang River Basin's water-food-energy-energy-ecosystem coupling and coupling coordination

4.3 Coupled Coordination Degree Prediction

The ARIMA prediction model was constructed to predict the development trend of WFEE system coupling coordination in Shiyang River basin from 2021 to 2030 (Fig. 5). The difference order is 1, combined with the ACF and PACF plots, the autoregressive order p and the moving mean order q were finally determined to be 0. The mean squared error was 0.01 and the mean absolute error was 0.084. All the indicators showed a good fit of the model. As can be seen from Fig. 5, the coordination degree of water-food-energy-ecosystem coupling in Shiyang River Basin from 2021 to 2030 is above 0.85, showing a linear growth trend. From 2021 to 2023, the high-quality coordination state is from 2024 to 2030. It shows that the correlation between the subsystems of Shiyang River basin is stronger in the future.

In the future, we should strengthen the coordinated management of the subsystems, coordinate the economical and intensive use of water resources and pollution prevention, accelerate the low-carbon transformation, pay attention to ecological protection and restoration, promote the energy transformation and the high-quality development of the energy industry; and strengthen the overall management of water-food-energy-ecological ties through the spatial planning of the river basin.

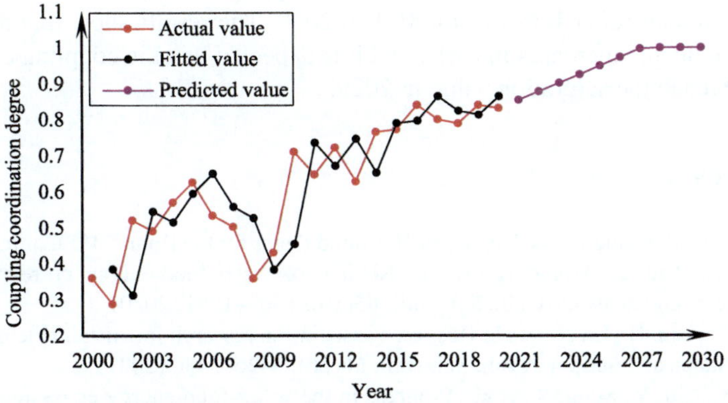

Fig. 5. Shiyang River Basin 2021–2030 coupling collaborative scheduling forecast

5 Conclusion

By constructing the coupling coordination model of WFEE system in Shiyang River Basin, the coupling coordination relationship and space-time evolution characteristics of WFEE system in Shiyang River Basin from 2000 to 2020 are analyzed and studied, and then the following main conclusions are obtained:

(1) The comprehensive development index of WFEE system in Shiyang River Basin is increasing with time. The variation rule of the comprehensive evaluation index of WFEE system from 2000 to 2015 is mainly influenced by water resources and energy subsystem. The comprehensive development index of each subsystem showed a "wave" trend in time distribution, while the comprehensive evaluation index of grain and ecological subsystem changed relatively little.

(2) The coupling degree of WFEE system in Shiyang River Basin basically maintains a high level of coupling, and the subsystems have close connection and strong interaction. The degree of coupling coordination fluctuated slightly in time distribution. From 2005 to 2008, primary coordination to mild coordinated development in 2008, to intermediate coordinated development in 2008 to 2010, and to good coordinated development in the next 10 years.

(3) Jinchuan District and Yongchang County should continue to follow the idea of "water saving first", "four water and four determination", Further implement the principle of "saving urban and rural domestic water use, ensuring water use for industrial production, gradually increasing basic ecological water use, and rationally controlling agricultural water use"; The Jingtai County takes the agricultural supply-side structural reform as the main line, Strictly implement the strategy of "storing grain on the ground and using technology", Ensuring food security, Increasing the overall grain production capacity, Developing ecological and circular agriculture, Optimize the layout structure of agricultural industry; Continuous optimization and upgrading of the energy structure in Wuwei region, Strengthening the energy supply capacity, To build a green, low-carbon, safe and efficient energy supply system, Match the energy demand of the four cities in the Shiyang River Basin.

(4) Predict the development trend of the coupling coordination degree of WFEE system in Shiyang River Basin from 2021 to 2030. The results show that the coupling coordination degree increases linearly, and it is expected that the coupling coordination degree will reach the saturation value in 2029.

References

1. Hoff, H.: Understanding the Nexus: Background Paper for the Bonn 2011 Conference (2011)
2. Zhang, Z., Liu, J., Wang, K., et al.: Review of water-food-energy correlation system: bibliometry and analysis. Chin. Sci. Bull. **65**(16), 1569–1581 (2020)
3. Zhang, C., Liu, B., Li, N., et al.: Resource association research for sustainable development: current status and outlook. Chin. Sci. Bull. **66**(26), 3426–3440 (2021)
4. Wang, Y., Liu, Y., Song, S., et al.: Progress in the water-food-energy-ecosystem association study. Adv. Earth Sci. **36**(07), 684–693 (2021)
5. Mi, H., Zhou, W.: System simulation of China's grain, fresh water and energy demand in China in the next 30 years. Popul. Econ. **1**, 1–7 (2010)

6. Deng, P., Chen, J., Chen, D., et al.: Study on the coordinated evolution characteristics of regional water-energy-food coupling—Take Jiangsu province as an example. J. Water Resour. Water Eng. **28**(6), 232–238 (2017)
7. Peng, S., Zheng, X., Wang, Y., et al.: Co-optimization of water resources, energy and food in the Yellow River Basin. Adv. Water Sci. **28**(5), 681–690 (2017)
8. Zhou, L.: Research on the water-energy-food link in water resources management. Shandong University of Science and Technology (2019)
9. Shi, H., Luo, G., Zheng, H., et al.: Analysis of water use in the Sr River basin based on "water – energy – food – ecology" link causality and Bayesian network. Acta Geogr. Sin. **75**(5), 1036–1052 (2020)
10. Sun, C., Yan, X.: Safety evaluation and Spatial Association analysis of water resources-energy-food coupling system in China. Water Resour. Protect. **34**(5), 1–8 (2018)
11. Lin, Z., Liu, X., Chen, Y., et al.: Research progress and prospect of the link between water-food-energy. Acta Geogr. Sin. **76**(7), 1591–1604 (2021)
12. Chang, H., Zhao, Y., Sang, X., et al.: Study on co-regulation of water resources-food-energy-ecology in Beijing-Tianjin-Hebei I: methods and models. J. Hydraul. Eng. **53**(6), 655–665 (2022)
13. Zhao, Y., Chang, H., Sang, X., et al.: Research on the coordinated regulation of water resources-food-energy-ecology in Beijing, Tianjin and Hebei: application. J. Hydraul. Eng. **53**(10), 1251–1261 (2022)
14. Niu, L., Hu, T.: Evaluation of water resource carrying capacity in Liaoning Province based on the extended cloud model. J. Yangtze River Sci. Res. Inst. **40**(10), 22–29 (2023)
15. Qian, F., Wang, W., Wang, Q.: Quantifying the synergistic relationship between the natural quality of cultivated land and site conditions based on the coupled coordination degree model. Trans. Chin. Soc. Agric. Eng. **34**(18), 284–291 (2018)

Research Progress of EPS Recovery: Component Identification and Content Regulation

Jialu Tang[1], Zhiguo Li[1], Xueting Wang[1], Xiaoling Ma[1], Heng Yuan[1], and Junfeng Wan[1,2(✉)]

[1] School of Ecology and Environment, Zhengzhou University, Zhengzhou 450001, People's Republic of China
wanjunfeng@zzu.edu.cn

[2] Henan International Joint Laboratory of Environment and Resources, Zhengzhou University, Zhengzhou 450001, People's Republic of China

Abstract. Recently, there has been a growing interest in the recovery of extracellular polymeric substances (EPS) from traditional and cutting-edge biological wastewater treatment systems in academia and industry, especially aerobic granular sludge (AGS), which has attracted wide attention due to its rich and diverse EPS components. Numerous studies have shown that EPS extracted from AGS has broad application potential. However, the conversion of EPS from laboratory research to practical engineering applications still faces many challenges and a long way to go. In order to bring EPS to the market, it is necessary to accurately identify its components and regulate its content. This paper aims to introduce the latest progress in the identification and content control of EPS components. In this paper, the research trends of EPS recovery from aerobic granular sludge are reviewed by combining qualitative and quantitative analysis methods. Firstly, the existing identification and characterization techniques of EPS and the currently known EPS components are summarized. Subsequently, the regulatory factors affecting the production of EPS are discussed, and an outlook on the future research direction of EPS resourcing is presented, aiming to provide recommendations for the transition of wastewater treatment plants to resourcing.

Keywords: EPS · component identification · regulation · resource utilization

1 Introduction

At present, more research on wastewater treatment plants has gradually shifted from pollutant removal to resource recovery, especially for extracellular polymeric substances (EPS). EPS is a 'sticky' substance produced by the flora in the process of cell metabolism. It contains many substances such as protein (PN), polysaccharide (PS), humus and DNA. It is a biopolymer with complex components [1]. Among them, proteins and polysaccharides can account for more than 90% of the total. Compared with activated sludge, aerobic granular sludge produces a large amount of EPS, which is several times higher than that of ordinary activated sludge, and the maximum content can reach 40% of the

D. Han et al. (Eds.): ICEPG 2024, ESE, pp. 855–861, 2025.
https://doi.org/10.1007/978-3-031-86829-0_95

biomass of aerobic granular sludge [2]. It has been reported that EPS can be used in many fields, including water treatment, agricultural use and as a flame retardant and waterproof coating. Recovery of EPS from aerobic granular sludge is very attractive to the market, EPS for the market must be considered both qualitatively and quantitatively, to understand the specific composition of EPS and the amount of EPS extracted is the first step in its resourcefulness. Therefore, the identification and regulation of extracellular polysaccharide (EPS) components in aerobic granular sludge is very important for its resource recovery. Due to the complex composition of EPS, it is difficult to separate and purify it, which hinders our in-depth understanding of it and increases the difficulty of commercial utilization. However, a number of recent studies have revealed the composition of EPS using a variety of techniques. By understanding the effects of physical and chemical factors on the expected EPS production, it is expected to improve the resource recovery of EPS. This paper mainly summarizes the identification and characterization methods of EPS in aerobic granular sludge, and expounds the internal mechanism of regulating EPS secretion in AGS and the influencing factors of external environment.

2 The Identification and Characterization of EPS Components

The extraction of EPS from aerobic granular sludge is a prerequisite for the identification of EPS components. The composition and characterization of EPS largely depend on the extraction method. Different extraction methods will lead to differences in the quantity and quality of EPS [3]. At present, many reports have elucidated different extraction methods, including physical, chemical or biological methods. Physical methods mainly use external forces to destroy and dissolve EPS; chemical methods mainly use extraction techniques; and biological methods are mainly enzymatic treatments. In general, the extraction amount of EPS by physical method is less, and the chemical method can produce a higher amount of EPS. The existing extraction methods rarely consider the practical application of EPS. Most of them are crudely extracted by the above methods, and EPS is simply divided into protein (PN), polysaccharide (PS) and humus. The rough classification of EPS components ignores the diversity of their components, which is a challenge to the identification of special components in aerobic granular sludge. Despite many difficulties, there have been many recent attempts to identify specific components of EPS, particularly with regard to PN, PS and heteropolymers with specific/hypothetical functions.

2.1 Identification Method

Regardless of the concentration level of the target component, detailed quantitative analysis can be performed using a range of powerful tools to establish reliable data over a wide linear range. However, these methods are used to analyze known substances with unknown concentrations. As a quantitative analysis method, colorimetric method determines the content of the component to be tested by comparing or measuring the color depth of the colored substance solution. As shown in Table 1, the components of EPS (proteins, carbohydrates, uronic acids, humic substances) can be characterized by colorimetry.

Colorimetric method is fast, simple and cheap. But colorimetric method also has some defects. On the one hand, the determination of a single target component is susceptible to other factors. The determination of protein is affected by the selection of standards and methods. The quantification of sugar is sensitive to the selection of standard and wavelength. In the phenol sulfuric acid method, different sugars lead to different maximum absorbance values. On the other hand, there is cross-interference between EPS components. EPS contains proteins, neutral sugars, aldehyde acids and phenolic compounds, and the colorimetric quantification of one component may be interfered by the presence of other components. It has been reported that protein quantification may be overestimated in the presence of phenolic compounds, glucose, galacturonic acid and glucosamine; humic acid and galacturonic acid can lead to overestimation of the measured sugar content. Therefore, it is necessary to develop new quantitative measurement methods to increase the accuracy of quantitative analysis.

2.2 Specific Components

Particle size exclusion chromatography (SEC) allows further separation of crude extracted EPS to obtain different types of polymers. Some studies have used this method to isolate ovalbumin, aldolase, ferritin, thyroglobulin and blue dextran [4]. The extracted EPS can be analysed for total protein and polysaccharide analysis, elemental analysis, chemical structure analysis and electrophoretic protein analysis using advanced techniques. Several specific EPS components were initially identified using advanced analytical techniques, including: amino acids, glycosylated amyloid, sialic acid, hyaluronic acid-like, alginate-like extracellular polysaccharides, sulfonated glycosaminoglycans (GAGs) and sulfated polysaccharides. By isotope dilution mass spectrometry, 14 different amino acids (glycine, leucine, alanine, isoleucine, etc.) were detected in the EPS of AGS. Through the analysis of the secondary structure of the protein, the secondary structure of PN also contributes to the granulation process and the stability of the particles. Two components of exopolysaccharides (PS) have been identified: alginate exopolysaccharides and glycosaminoglycans. Alginate exopolysaccharide is a mixture of neutral polysaccharide and aldehydic polysaccharide, which helps AGS to form a physical structure in the form of gel. Felz et al. [5] found glycosaminoglycans (GAGs) in the EPS of AGS, namely hyaluronic acid-like and sulfated GAGs-like biopolymers. GAGs-like polymers and sialic acid may be widely distributed in the particles and play an important role in performance stability. Glycosaminoglycan is a linear heteropolysaccharide of EPS, which is easily ignored by the commonly used colorimetric method. The presence of glycosaminoglycans in AGS and AnGS was confirmed by FTIR and staining. It is speculated that they play a role in attracting cations, forming hydrogels, binding proteins, and signal transduction in biological processes.

Some of the identified extracellular polymer (EPS) components have been found to have significantly high added value, as listed in Table 2. This augurs well for a wide range of potential applications of EPS, but the complexity of the EPS components prevents some of them from being efficiently extracted and purified according to conventional extraction methods. Through comprehensive multidisciplinary analyses, in-depth research and development of these high value-added EPS components, we expect to witness the birth of a series of innovative products and solutions.

Table 1. Colorimetric Analysis of Different EPS Components

Target component	Methods	Principle	Advantages / Disadvantages
Protein	Lowry	Peptide bond chelates with Cu^{2+} to form a complex, which reduces phenolic phosphomolybdic acid to produce blue compound	Poor specificity, many interfering substances
	BCA	Proteins reduce Cu^{2+} to Cu^{+} and BCA chelating produces a blue-violet colour	There was a large deviation in the determination of different proteins
	Bradford	Coomassie Brilliant Blue G-250 binds to proteins in blue	Affected by chelating agent, high concentration of reducing agent
Polysaccharides	Phenol-sulfuric acid	Hydrolysis, dehydration to furfural derivatives, condensation with phenol to form coloured compounds	The operation is simple and rapid, and the color is stable, but the results are easy to be high for colored samples
	Anthrone	Hydrolyzed, dehydrated, and then combined with anthrone to form a colored compound	Poor stability; some interference with tryptophan
aldehyde acid	carbazole-sulfuric acid	Hydrolysis produces galacturonic acid, which reacts with carbazole reagent in sulfuric acid to form a purple red compound	The operation is simple and the stability is good

3 Regulation of EPS Content

From the composition, structure and physicochemical properties of EPS extractable products, the microbial community and its selection environment are the key factors in regulating EPS. The factors regulating EPS content can be divided into two categories: internal factors and external factors.

Table 2. High value-added materials and their potential applications

High value-added materials	Category	Property and function	Potential application
Amyloid-like proteins	Proteins	High mechanical strength, good chemical resistance	Biosensor nanomaterials
Alginate-like extracellular polymers	polysaccharides	Good gel forming properties	Flame retardants, seed coating and adsorbents etc
Sialic acids	polysaccharides	Molecular and membrane stabilisation	Targets for the development of new drugs for the treatment of various cancers and inflammatory processes
Sulfated polysaccharides	polysaccharides	Antiviral, antioxidant and immunomodulatory	Drug carriers
Hyaluronic acid-like and sulfated glycosaminoglycans like polymers	polysaccharides	Components of the mammalian cell ECM	Tissue engineering materials

3.1 External Influences

Qualitative and quantitative EPS yields are strongly dependent on the carbon supply present in the bacterial medium. Several studies have shown that microbial EPS is associated with the depletion of carbon sources that serve as substrates for bacterial growth, but that bacterial enzymes have a substantial impact on the breakdown of complex carbon sources. The findings suggest that organic invasion dramatically alters EPS propagation and bacterial conformation. The addition of glucose improved the EPS distribution of polysaccharides and proteins, with a large rise in protein content. More specifically, by increasing the COD concentration from 0 to 200 mg/L, the total EPS concentration increased from 33 mg/gVSS to 181 mg/gVSS, while increasing the COD to 300 mg/L resulted in a decrease in EPS concentration to 155 mg/gVSS [6].

Similar to carbon sources, nitrogen sources present in bacterial culture media have been shown to play an important role in the assembly and improvement of bacterial EPS. The change of COD/N ratio leads to a great change in the functional groups and composition of EPS. Shi and Liu [7] fed wastewater with COD/N ratios of 100: 5 and 100: 10 in two reactors, respectively. At low COD/N, the granulation speed was fast and the granules were stable. The characteristics of EPS showed that the higher the content of polysaccharide, the higher the content of β-sheet in protein, the lower the adhesion and viscoelasticity of mineral surface, which was beneficial to granulation and particle maintenance of aerobic sludge.

The production of EPS may be affected by the presence of pollutants (including micro-pollutants) in wastewater. It has been reported that polystyrene nanoplastics inhibit the total production of EPS and PNs after 12 h of exposure [8]. Exposure to saline/hypersaline conditions leads to changes in EPS composition/structure and particle surface properties (e.g., increased hydrophobicity) [9]. Many studies have shown that increased salinity leads to increased PN secretion. This behaviour may be attributed to microbial osmotic pressure regulation and protective mechanisms of cation-π interactions of monovalent ions with extracellular PN. It is important to emphasise that the trend of increased EPS production/fraction is not always proportional to the elevated salt concentration: some studies have even reached controversial conclusions. For example, an increase in salt concentration leads to a decrease in the PN content of AGS-EPS.

3.2 Internal Factors

Various factors such as wastewater composition, temperature, F/M, exogenous chemicals have been reported to affect the synthesis and secretion of EPS, which may be attributed to the fact that these factors affect microbial growth, which, in turn, alters the activity of EPS synthase and adenosine triphosphate (ATP). In the final analysis, EPS is synthesised by microorganisms, and the synthesis and secretion of EPS is fundamentally regulated by microorganisms. Quorum sensing (QS) is a system that uses signalling molecules as carriers to transmit information between cells, and the high concentration of signalling molecules secreted by cells will control group behaviour. QS has been shown to directly affect EPS synthesis and secretion by regulating microorganisms.

Community sensing (QS) is a system by which microorganisms communicate with each other. QS signalling molecules mainly include intraspecific signalling molecules represented by N-acyl homoserine lactones (AHLs) and interspecific signalling molecules represented by Autoinducer 2 (AI-2). Most of the current studies have focused on the effects of AHLs on EPS. Microorganisms with EPS secretion function are regulated by AHLs, which affect the content and chemical composition of EPS. Wei Gao [10] added exogenous AHLs to AGS and analysed the changes in EPS subcomponents and found that the addition of 50 nm exogenous AHLs significantly increased the TB-EPS production by 22.07% after 36 days compared to the control reactor, leading to an increase in the PN/PS index. Similarly, in addition to AHL, other signalling molecules secreted by QS were detected to regulate EPS secretion. The concentration of the signalling molecule cyclic diguanylate (c-di-GMP) was positively correlated with the synthesis of EPS in AGS within a certain range.

4 Conclusions and Outlook

The complexity and diversity of EPS constitute an under-explored treasure trove of great potential and value. Although preliminary studies have revealed its potential exploitation value, the conversion of EPS into commercially available products in the market is still in the early stages of our exploration. In order to advance the process of EPS resourcing in AGS (aerobic granular sludge), future research could focus on the following key directions: (1) Technological innovation with multidisciplinary integration: developing new technological tools to identify and characterise EPS from a multidisciplinary perspective. At present, the identification of biomolecules is limited, and their molecular structure and distribution characteristics still need to be further clarified. (2) In-depth exploration of commercial value: Explore EPS components with commercial potential and study effective methods for their isolation and purification. (3) Exploration of targeted regulatory mechanisms: Finding effective ways to regulate and optimise the generation of specific EPS components. (4) Expansion of application fields: Based on the characteristics and functions of EPS, promote its application in different fields, such as biomaterials and bioenergy. (5) Comparative study of resourcefulness modes: Comparative analysis of the feasibility and efficiency of EPS resourcefulness in wastewater treatment plant systems and pure culture systems. Through these research directions, substantial progress can be expected in the field of EPS resourcing, paving the way for the realisation of its commercial application.

References

1. Huang, L., Jin, Y., Zhou, D., et al.: A review of the role of extracellular polymeric substances (EPS) in wastewater treatment systems. Int. J. Environ. Res. Public Health **19**, 12191 (2022)
2. Liu, Y., Lv, W.Z., Zhang, Z.Q., et al.: Influencing characteristics of short-time aerobic digestion on spatial distribution and adsorption capacity of extracellular polymeric substances in waste activated sludge. RSC Adv. **8**, 32172–32177 (2018)
3. Cydzik-Kwiatkowska, A.: Biopolymers in aerobic granular sludge—Their role in wastewater treatment and possibilities of re-use in line with circular economy. Energies **14**, 7219 (2021)
4. Görner, T., de Donato, P., Ameil, M., et al.: Activated sludge exopolymers: separation and identification using size exclusion chromatography and infrared micro-spectroscopy. Water Res. **37**, 2388–2393 (2003)
5. Felz, S., Neu, T.R., van Loosdrecht, M.C.M., et al.: Aerobic granular sludge contains Hyaluronic acid-like and sulfated glycosaminoglycans-like polymers. Water Res. **169**, 115291 (2020)
6. Li, X., Luo, J., Guo, G., et al.: Seawater-based wastewater accelerates development of aerobic granular sludge: a laboratory proof of-concept. Water Res. **115**, 210–219 (2017)
7. Shi, Y., Liu, Y.: Evolution of extracellular polymeric substances (EPS) in aerobic sludge granulation: composition, adherence and viscoelastic properties. Chemosphere **262**, 128033 (2021)
8. Qian, J., He, X., Wang, P., et al.: Effects of polystyrene nanoplastics on extracellular polymeric substance composition of activated sludge: the role of surface functional groups. Environ. Pollut. **279**, 116904 (2021)
9. Corsino, S.F., Capodici, M., Torregrossa, M., et al.: Physical properties and extracellular polymeric substances pattern of aerobic granular sludge treating hypersaline wastewater. Bioresour. Technol. **229**, 152–159 (2017)
10. Gao, W., Hu, Y., Jiao, X.: Recovery of structure and activity of disintegrated aerobic granular sludge after long-term storage: effect of exogenous N-acyl-homoserine lactones. Chemosphere **281**, 130894 (2021)

Carbon Dynamics of Farmland Ecosystem in Hailun Typical Black Soil Region Based on DNDC Model

Xidong Zhao[1,2], Yanda Lu[1,2](✉), Youtao Xin[1,2], Hongqiang Zhao[1,2], Haicheng Zhang[1,2], Zhuang Li[1,2], and Zongyue Lu[3,4](✉)

[1] Harbin Natural Resources Comprehensive Survey Center, China Geological Survey, Harbin 150086, China
zhaoxidong@mail.cgs.gov.cn
[2] Observation and Research Station of Earth Critical Zone in Black Soil, Harbin, Ministry of Natural Resources, Harbin 150086, China
[3] Center for Geophysical Survey, China Geological Survey, Langfang, China
[4] Technology Innovation Center for Earth Near Surface Detection, China Geological Survey, Langfang, China

Abstract. Based on the DNDC model, the carbon storage of farmland ecosystem in Hailun City during 2010–2060 was simulated and predicted, and the interannual dynamics of soil carbon source/sink in the farmland ecosystem in Hailun City were analyzed. The results showed as follows: (1) The SOC content of soil simulated by DNDC model in 2017 was compared with the soil organic matter survey results of Hailun Agricultural and Rural Bureau in 2017. The root-mean square error is 0.009 kg·C/kg, and the average absolute percentage error is 28.49%. It can be seen that the DNDC model can simulate soil SOC content change in Hailun City well. (2) According to the simulation results, after excluding the years affected by bad weather, it was found that planting corn was conducive to SOC accumulation in the comparison of multiple planting systems. (3) If the soybean recropping system continues to be implemented, the surface SOC in the study area will remain in the state of carbon source, which is not conducive to the implementation of the national carbon peak strategy in 2030. (4) According to the simulation results, a reasonable arrangement to implement alternating planting systems of corn-soybean rotation and corn recropping during 2022–2060 can ensure that the typical black soil area of Hailun can reach the state of carbon sequestering in 2030 and in the future (2030–2060) for a long time, supporting the action plan of carbon peak and carbon neutrality. It has a good guiding effect on agricultural planting in the study area.

Keywords: SOC · Typical black soil area · Carbon source/sink · DNDC model

The original version of the chapter has been revised. The corresponding author indication in the chapter has been updated. A correction to this chapter can be found at https://doi.org/10.1007/978-3-031-86829-0_124

D. Han et al. (Eds.): ICEPG 2024, ESE, pp. 862–871, 2025.
https://doi.org/10.1007/978-3-031-86829-0_96

1 Introduction

This conclusion makes the protection of black soil appear in the public's vision again, and arouses the attention of the whole society to black soil.

Black soil in China is mainly distributed in Heilongjiang, Jilin, Liaoning provinces and the Eastern Four League Areas of Inner Mongolia Autonomous Region, with a total area of more than 1 million km^2. The farmland ecosystem of black soil area is mainly concentrated in the Sanjiang Plain and Songnen Plain [1, 2]. Hailun City of Heilongjiang Province is located in the transition zone from the eastern edge of Songnen Plain to the Lesser Khingan Mountains [3], with an area of 4667 km^2, of which the cultivated area is about 2940 km^2, accounting for 63% of the total area. On September 22, 2020, China proposed that before 2030, CO2 emissions should strive to reach the peak, and in 2060, China should strive to achieve the artificial removal of CO2 in the prescribed period to offset CO2 emissions globally, which can achieve net zero CO2 emissions, namely carbon neutrality. This marks the first time that China has made a formal commitment to carbon neutrality in front of the world. China will face great challenges and invest more energy. As a typical black soil area, the research on the carbon storage of farmland ecosystem and organic carbon pool becomes more realistic.

As a complex ecosystem, the changes in carbon storage and existence forms of farmland ecosystem are affected by multiple factors such as soil physical and chemical properties, field management, climate conditions, vegetation conditions, etc. [4], so the research methods often adopt a comprehensive model simulation instead of a single conventional test. In 1992, Professor Li Changsheng's research group first published the DNDC model and triggered a strong response. Subsequently, the biogeochemical model developed rapidly, providing new research ideas and research methods for the study of soil carbon sequestration, nitrate leaching, and the emission process of carbon and nitrogen in forest, grassland, and farmland ecosystems [5]. DNDC is the abbreviation of Denitrification-Decomposition. Denitrification and organic matter decomposition are the main biogeochemical processes that lead to the loss of nitrogen and carbon from soil and transfer to the atmosphere. Li Changsheng used the DNDC model to study the greenhouse gas emissions of the whole farmland system in China [6], He Mei et al. used the DNDC model to analyze the evolution law of black soil organic carbon in Northeast China and the synergistic relationship between the yield of other crops [7]. In this study, the DNDC model was used to study the interannual changes of carbon source/sink of farmland ecosystem in the typical black soil area of Hailun City from 2010 to 2060, and a systematic and in-depth modeling study was conducted on carbon cycle of farmland ecosystem, in order to provide a theoretical basis and important basis for the analysis of carbon source/sink function of farmland ecosystem.

The DNDC model has a good predictive ability for soil organic carbon in the study area after the simulation verification of "point mode", and is suitable for promotion in Songnen Plain [8].Through information integration processing and field measured data, the accuracy of the DNDC model and the adaptability of the model to the study area were verified, and the changes of soil organic carbon storage and crop yield were simulated and analyzed to explore the changes and influencing factors of soil organic carbon storage in the study area. This study provides a small-scale basic example for the global application of the DNDC model and the change of global soil carbon flux, and has reference significance for the scientific management of regional agricultural ecosystem [8].

2 Materials and Methods

2.1 Overview of the Study Area

The research area is located at the northeast end of Songnen Plain and the west foot of Xiaoxing'an Mountains. The terrain is mainly hilly and unhilly, with an average elevation of 239 m. Belongs to the temperate continental climate, frost-free period of about 120 days, effective accumulated temperature 2200–2400 °C [9], annual precipitation 500–600 mm. Within the territory of Tongken River, Zayin River, Ke Yin River, Helen River, three Oolong Valley five main rivers.

Hailun City is located in the middle of Heilongjiang Province, latitude 46.58 to 47.52° north, longitude 126.14 to 127.45° east, located in the northeast end of Songnen Plain, the west foot of the Lesser Hinggan Mountains. The hinterland of Songnen Plain, one of the three major black soil plots in the world, is a typical representative of the farmland ecosystem in the northeast black soil region. The average frost-free period of the year is 120 days, the average precipitation is 500–600 mm [10], the effective accumulated temperature is 2300–2600°C, and the sunshine is 2600–2800 h. The soil type is 82% black soil, and the suitable climate and soil conditions are very suitable for the growth of soybean. The annual soybean planting area is more than 1 million mu, and soybean planting has become an important source of economic income for farmers in Hailun City.

2.2 Basic Principle and Parameter Determination of DNDC Model

DNDC (denitritide-Decomposition Model) is designed to simulate the carbon and nitrogen cycles and water dynamics in terrestrial ecosystems. A biogeochemical process model that combines ecological drivers (e.g., climate, soil, vegetation type, and human management practices), environmental factors (e.g., temperature, humidity, pH, Eh, etc.), and related physicochemical processes [11]. DNDC model simulates carbon and nitrogen cycling dynamics in terrestrial ecosystems by simulating denitrification and decomposition, two of the most important biogeochemical processes.

The meteorological data of the study area were from the National Meteorological Science Data Center, the soil data were from the land quality geochemical survey and literature, the grain yield data, field management data and crop rotation data were from the Hailun Municipal Bureau of Agriculture and Rural Affairs, and the slope data were from the DEM data [12]. The soil data for 2021 are derived from the average measured data of the black soil surface matrix layer survey project in Hailun Area, Songnen Plain, Harbin Comprehensive Survey Center of China Geological Survey, as shown in Table 1.

Table 1. The required parameters for DNDC model

Entscheidender Faktor	Input parameters	Value (average)	Sources
Weather	Years of simulation	81	-
	Daily maximum temperature (year-round model)	-	China Meteorological Data Network
	Daily minimum temperature (year-round model)	-	China Meteorological Data Network
	Rainfall (full year model)	-	China Meteorological Data Network
	Wind speed (all year model)	-	China Meteorological Data Network
Soil	Land use type for the simulated site	Arable dry land	Geochemical survey of land quality
	2010 Soil bulk density (g/cm^2)	1.38	Geochemical survey of land mass
	Soil pH in 2010	6.65	Geochemical survey of land quality
	2010 Soil SOC (kg·C/kg)	0.028	Geochemical survey of land quality
	2017 Soil SOC (kg·C/kg)	0.0275	Hailun Agriculture and Rural Bureau
	Slope (%)	0 to 5	DEM data
	2021 Soil bulk density (g/cm^2)		Measured value
	Soil pH for 2021		Measured values
	Soil SOC in 2021 (kg·C/kg)		Measured value
Crops	Corn yield (jin/mu)	1009.38	Hailun City Agriculture and rural Bureau
	Soybean yield (jin/mu)	285.64	Hailun Agriculture and Rural Bureau
	Planting system	1980-present	Hailun Agriculture and Rural Bureau

2.3 Sample Point Settings

In this paper, 16 survey sites were selected from the soil organic carbon database in 2010 and 2017, which were located in 16 townships in the central and western part of Hailun City [13] (one sample site for each township, excluding eastern farms). The sample sites were located in dry farmland. The geographic coordinates of sampling points, fertilization (fertilization frequency, quality, time and mode), tillage (tillage frequency, time and mode), irrigation (irrigation frequency, time and mode), straw return ratio and other original information data were obtained by various methods.

2.4 Model Verification and Data Processing

The DNDC model has been verified by many research projects and personnel, but for the farmland ecosystem in typical black soil area of Hailun area [14], the research and application of this model are few. Therefore, the accuracy analysis of the simulated value of the DNDC model and the real value measured should be done, so as to prove that the DNDC model can be applied to this region [15].

In this paper, correlation coefficient r, mean absolute percentage error (MAPE) and root mean square error (RMSE) were used to verify the adaptability of the model:

The correlation coefficient r is as follows:

$$r = \frac{\sum(X - \overline{X})(Y - \overline{Y})}{\sqrt{\sum(X - \overline{X})^2 \sum(Y - \overline{Y})^2}} \tag{1}$$

The average absolute percentage error formula is as follows:

$$MAPE = \frac{100}{n}\sum\nolimits_{i=1}^{n} \frac{|Y_i - X_i|}{X} \tag{2}$$

The root mean square error formula is as follows:

$$RMSE = \sqrt{\frac{1}{n}\sum\nolimits_{i=1}^{n} (Y_i - X_i)^2} \tag{3}$$

In this study, Excel2019 was used for statistical analysis and CorelDRAW X8 (64-Bit) was used for image processing.

3 Results and Analysis

3.1 Change Trend of Soil Organic Carbon at Different Monitoring Sites

The changes of soil organic carbon in 2010 and 2017 at 23 sampling points in the central and western part of Hailun City are shown in Fig. 1. As shown in the figure, from 2010 to 2017, the soil organic carbon content in the central and western parts of Hailun City was in a state of reduction, with a small reduction.

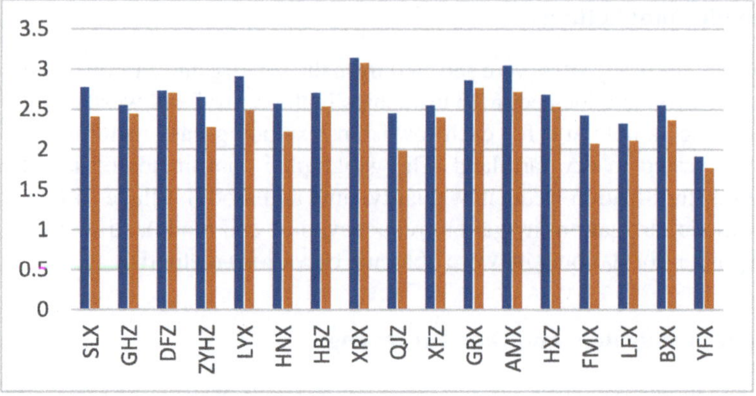

Fig. 1. Soil organic carbon change map of 23 towns in central and western Helene County from 2010 to 2017

3.2 DNDC Model Simulation and Verification

The simulation results of SOC at different sampling points are shown in Fig. 2. It can be seen from the figure that the correlation trend of the data simulated by DNDC model is relatively good, the correlation coefficient r is above 0.95, the root mean square error is 0.009 kg·C/kg, and the average absolute percentage error is 28.49%. It can be seen that the DNDC model can better simulate the change of soil SOC content in Hailun City.

In this paper, soil surface SOC from the results of the 2010 land quality geochemical survey was used as the initial value for simulation. The simulation period was from 2010 to 2060, and the results of the 2017 survey of soil organic matter by the Hailun Bureau of Agriculture and Rural Affairs were used as the data verification. Five groups of samples with the same or close locations (the distance between the sites is less than 500 m) were selected (Fig. 1). In this paper, root-mean-square error and mean absolute percentage error are used to evaluate the model.

The soil SOC content simulated by DNDC model in 2017 was compared with the soil organic matter survey results of Hailun Agriculture and Rural Bureau in 2017. The root-mean square error is 0.009 kg·C/kg, and the average absolute percentage error is 28.49%. It can be seen that the DNDC model can simulate soil SOC content change in Gonghe Town, Hailun City.

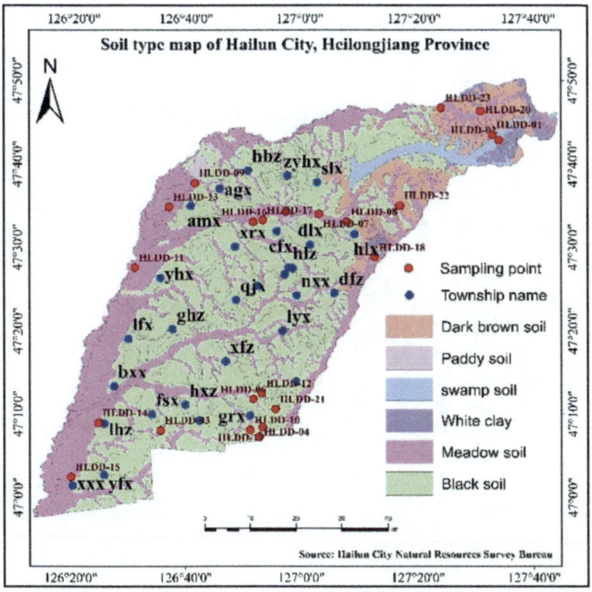

Fig. 2. Distribution of sample points

3.3 Prediction of Future Change Trend of Soil Organic Carbon

In this paper, the sampling point of Gonghe Town (the largest dry field planting area) was selected as an example, and the planting system, the proportion of straw returned to the field and the use of organic fertilizer were respectively tested by the control variable method, and the change trend of SOC content in the next 20 years was predicted under different test conditions. Experiment 1: Based on soil data, meteorological data and tillage system data in 2010, four different straw-returning ratios of 0, 20%, 40% and 60% were used for simulation, and the results were shown in Fig. 3.

According to the simulation results (Fig. 3), it can be seen that a reasonable arrangement of the 2021–2060 planting system and the implementation of the two planting systems of corn-soybean rotation and corn recropping can ensure that the study area can reach the carbon sink state in 2030 (that is, the soil SOC change is positive) and remain in the carbon sink state for a long time in the future (2030–2060). Support the Carbon Peak and carbon Neutrality action Plan.

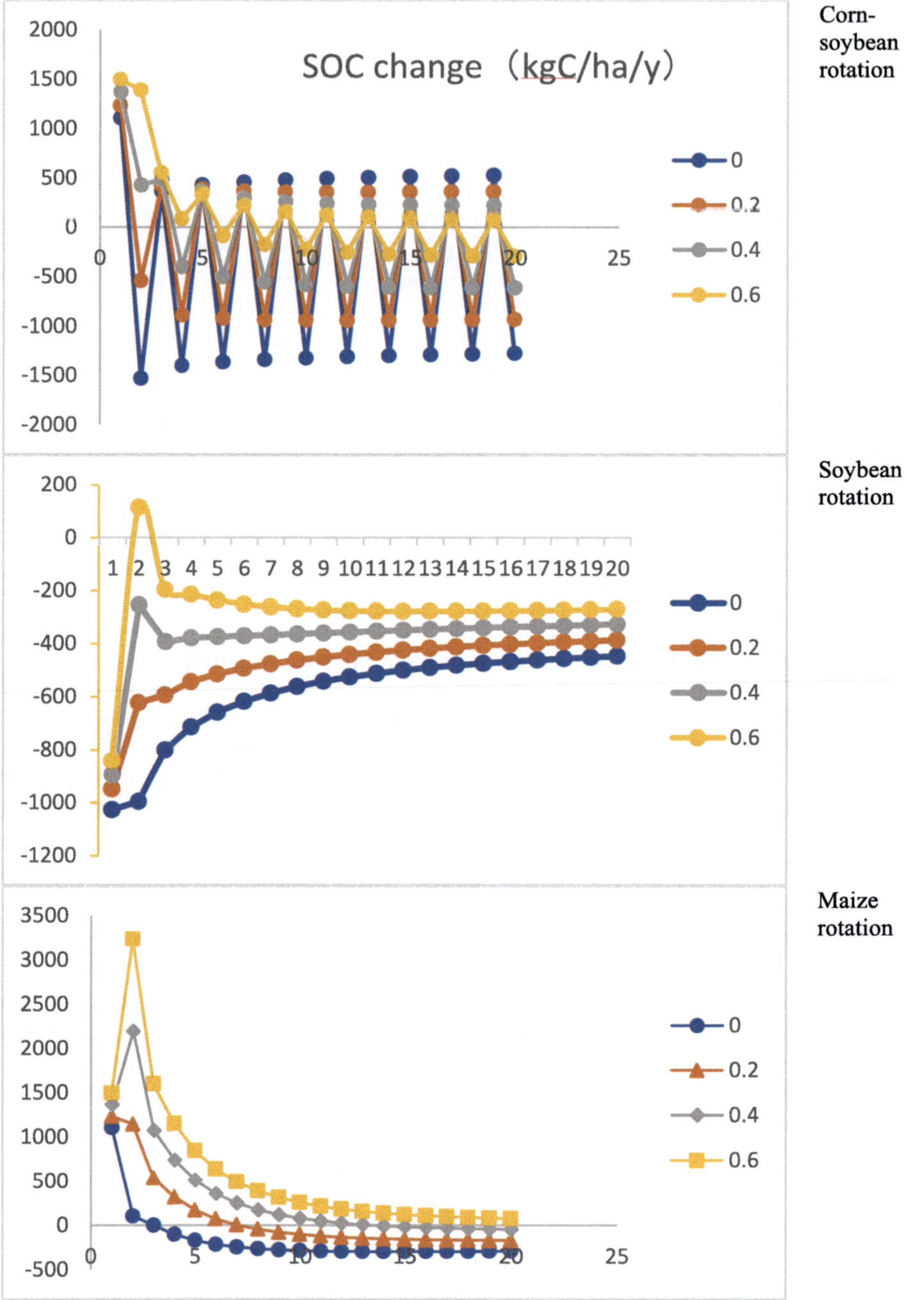

Fig. 3. SOC content trends under different conditions over the next 20 years

4 Discussion

1. The amount of organic carbon in soil is the balance result of the amount of plant residues entering the soil and the amount of decomposition and loss under the action of soil microorganisms. Under natural conditions, the amount of plant residues is determined by the type of vegetation.
2. The overall SOC consumption capacity of soybean root microorganisms was stronger than that of corn root.
3. The amount of plant residues planted in corn is more than that of soybeans.

Conclusion: Planting corn is more conducive to SOC accumulation than planting soybean.

5 Conclusion

(1) According to the comparison of the survey values in 2017, the DNDC model can simulate the carbon dynamics of the farmland ecosystem in the typical black soil area of Hailun City more accurately.
(2) According to the simulation, after excluding the years affected by bad weather, it is found that under the corn-soybean rotation planting system, planting corn contributes to SOC accumulation.
(3) If the soybean recropping system continues to be implemented, the SOC in Helene will remain in a state of carbon source, which is not conducive to the implementation of the national carbon peak strategy in 2030.
(4) The implementation of alternating planting systems of corn-soybean rotation and corn recropping can ensure that the study area will reach the carbon sink status in 2030 and in the future (2030–2060) for a long time, which has a guiding effect on agricultural planting in the study area.

References

1. Wang, G., Li, Y., Fan, L., et al.: The response of soil organic carbon content of grasslands in Northern Xinjiang to future climate change. Phys. Chem. Earth (2024). 134.https://doi.org/10.1016/j.pce.2024.103576
2. Singh, J., Ale, S., Delaune, P.B., et al. Modeling the impacts of cover crops and no-tillage on soil health and cotton yield in an irrigated cropping system of the Texas Rolling Plains. Field Crops Res. **287**, 108661 (2022)
3. Zhang, H., Hobbie, E.A., Feng, P., et al.: Can conservation agriculture mitigate climate change and reduce environmental impacts for intensive cropping systems in North China Plain?. Sci. Total Environ. **806**, 151194 (2022). https://doi.org/10.1016/j.scitotenv.2021.151194
4. Jiang, Q., Madramootoo, C.A., Qi, Z.: Soil carbon and nitrous oxide dynamics in corn (Zea mays L.) production under different nitrogen, tillage and residue management practices. Field Crops Res. **277**, 108421 (2022).https://doi.org/10.1016/j.fcr.2021.108421
5. Yu, S., Yue, W.: Analysis of reactive nitrogen emissions from maize ethanol production based on the DNDC model. IOP Conf. Ser.: Earth Environ. Sci. **1011** (2022). https://doi.org/10.1088/1755-1315/1011/1/012002

6. Wang, L., Zhang, K., Guo, K., et al.: Multifactor effects on the N2O emissions and yield of potato fields based on the DNDC model. Environ. Sci. Pollut. Res. **29**(17), 25448–25460 (2022).https://doi.org/10.1007/s11356-021-17700-2
7. Shaukat, M, Muhammad, S., Maas, E.D.V.L., et al.: Predicting methane emissions from paddy rice soils under biochar and nitrogen addition using DNDC model. Ecol. Model. **466** (2022).https://doi.org/10.1016/j.ecolmodel.2022.109896
8. Yang, Y., Zong, Y., Qin, R.,et al.: Impact of planting Lycium barbarum L. on the carbon–land–water nexus of a Haloxylon ammodendron nature reserve in the Qaidam basin, China. Land Degrad. Dev. **33**(12) (2022). https://doi.org/10.1002/ldr.4312
9. Sereni, L., Guenet, B., Blasi, C., et al.: To what extent can soil moisture and soil Cu contamination stresses affect nitrous species emissions? Estimation through calibration of a nitrification–denitrification model. Biogeosciences (2022). https://doi.org/10.5194/bg-19-2953-2022
10. Singh, J., Ale, S., Delaune, P.B., et al.: Simulated effects of cover crops with no-tillage on soil and crop productivity in rainfed semi-arid cotton production systems. Soil Tillage Res. **230**, 105709 (2023)
11. Song, J., Bao, S., Bai, J., et al.: Modelling future climate effects on N2O emission and soil carbon storage in maize fields under controlled-release urea and straw incorporation. J. Environ. Manag. **351** (2024). https://doi.org/10.1016/j.jenvman.2023.119854
12. Yang, J., Qin, R., Shi, X., et al.: The effects of plastic film mulching and straw mulching on licorice root yield and soil organic carbon content in a dryland farming. Sci. Total Environ. **826** (2022). https://doi.org/10.1016/j.scitotenv.2022.154113
13. Chu, Q, Xing, Y, He, W, et al.: Modeling ammonia emissions and abatement potential from the rice-wheat rotation fields using the calibrated DNDC model: a case study in Shanghai, China. Atmos. Environ. **305**, 119782 (2023)
14. Gheisari, A, Asgharipour, M.R., Mousavi-Nik, M., et al.: Utilization of the DNDC model to estimate yield and CO2 and CH4 emissions in a cotton-wheat rotation under the influence of various tillage treatments. Ecol. Model. **481**, 110357 (2023)
15. Dai, Z., Trettin, C.C., Burton, A.J., et al.: Estimated mangrove carbon stocks and fluxes to inform MRV for REDD plus using a process-based model. Estuar. Coast. Shelf Sci. **294**, 108512 (2023)

Research on Evaluation and Prediction of Mine Ecological Restoration Effectiveness Based on PSR-Bootstrap-LSTM

Daiying Bi(✉)

School of Environmental Science and Engineering, Liaoning Technical University, Fuxin 123000, Liaoning, China
3040760058@qq.com

Abstract. The study focused on the Haizhou open-pit Coal Mine in Fuxin City, Liaoning Province. The PSR model was used to create an evaluation system to measure the effectiveness of ecological restoration efforts in the mine. The mining area's field investigation data from 2015 to 2021 was used to evaluate the positive and negative grey correlation degree between each scheme and the best and worst solution. This evaluation was done using the Topsis integrated grey correlation analysis method. Subsequently, the comprehensive score was calculated. Furthermore, the evaluation is based on scores that are categorized into distinct intervals. The score data was reconstructed using an LSTM neural network model for rolling prediction. The error distribution of the model prediction results was analyzed using the Bootstrap method to determine the interval range of the model prediction output. Subsequently, the future restoration effectiveness was predicted within different confidence intervals. The experimental results served as valuable references for assessing the effectiveness of future ecological restoration efforts.

Keywords: Mine ecological restoration · Effectiveness evaluation system · Bootstrap · LSTM neural network

1 Introduction

1.1 A Subsection Sample

For sustainable development, mine ecological restoration effect evaluation and prediction are crucial. Using land use change and normalised vegetation index, literature [1] creates an evaluation system. Literature [2, 3] predicted ecological restoration potential using two primary ecological quality and ecosystem service indicators. Spatial principal component analysis (SPCA) quantified the research area's ecological security [4].

This study introduces the PSR model for a quantitative assessment of current mine ecological restoration impacts and employs the Bootstrap-LSTM neural network to predict future trends. The model includes a prediction interval via Bootstrap to illustrate varying outcomes[5], providing a scientific basis for ecological restoration decision-making in mining areas.

D. Han et al. (Eds.): ICEPG 2024, ESE, pp. 872–878, 2025.
https://doi.org/10.1007/978-3-031-86829-0_97

2 Based on Topsis Comprehensive Evaluation Model

2.1 Evaluation System

The Pressure-State-Response (PSR) model is an invaluable framework for evaluating environmental stress and the subsequent reaction of the system egarding the ecological restoration of the Haizhou open-pit coal mine [6]:

Pressure (P) encompasses various elements that impact the ecological conditions of the mine, including noise pollution (x_{11}), surface collapse resulting from mining activities (x_{12}), and the discharge of waste from coal combustion (x_{13}).

State (S) evaluates the present environmental condition, reflecting the influence of various pressures. The process entails observing fluctuations in air quality (x_{21}), the extent of vegetation (x_{22}), and the amount of arable land per person (x_{23}).

Response (R) pertains to the policies and managerial measures implemented in reaction to alterations in the environment. The main criteria to consider are the rate at which vegetation is restored (x_{31}), the changes in investment in environmental protection (x_{32}), the amount of money the government spends on environmental initiatives (x_{33}), the efficient use of waste from combustion (x_{34}).

Table 1 provides a comprehensive overview of the assessment method and the various data kinds.

Table 1. Evaluation system of mine ecological restoration effect

Target layer	Criterion layer	Decision-making level	Index type	Nature
Evaluation of mine ecological restoration effect	Pressure	x_{11}	Quantify	+
		x_{12}	Quantify	+
		x_{13}	Quantify	+
	State	x_{21}	Quantify	–
		x_{22}	Quantify	–
		x_{23}	Quantify	–
	Response	x_{31}	Quantify	–
		x_{32}	Quantify	+
		x_{33}	Quantify	+
		x_{34}	Quantify	–

2.2 Comprehensive Evaluation Model Based on GR-Topsis

The grey relational degree (GR) is used to measure the distance between positive and negative ideal solutions by combining subjective judgement and objective analysis. This method takes into account the ambiguity and fuzziness of the data. The ideal solution

V^+ and the negative ideal solution V^- are respectively:

$$\begin{aligned} V^+ &= \{\max(d_{ij} \big| j \in J^+) \cup \min(d_{ij} \big| j \in J^-)\} \\ V^- &= \{\min(d_{ij} \big| j \in J^+) \cup \max(d_{ij} \big| j \in J^-)\} \end{aligned} \tag{1}$$

where J^+ is the benefit-based indicator set, and J^- is the cost-based indicator set.

Grey correlation analysis not only focuses on static differences between data, but also reflects dynamic trends in data over time or other factors. For each scheme i and each indicator *j*, calculate the grey correlation degrees:

$$\begin{aligned} \gamma_{ij}^+ &= \frac{\min_{ij} \min \left| d_{ij} - d_j^+ \right| + \rho \max_{ij} \max \left| d_{ij} - d_j^+ \right|}{\left| d_{ij} - d_j^+ \right| + \rho \max_{ij} \max \left| d_{ij} - d_j^+ \right|} \\ \gamma_{ij}^- &= \frac{\min_{ij} \min \left| d_{ij} - d_j^- \right| + \rho \max_{ij} \max \left| d_{ij} - d_j^- \right|}{\left| d_{ij} - d_j^- \right| + \rho \max_{ij} \max \left| d_{ij} - d_j^- \right|} \end{aligned} \tag{2}$$

where ρ is the resolution coefficient, usually 0.5.

3 First Section

3.1 LSTM Neural Network Rolling Prediction Model

Long short term memory (LSTM) uses a gating mechanism to control the rate of information accumulation, which includes selectively adding new information and selectively forgetting previously accumulated information, thereby improving the long-term dependence of RNNs or recurrent neural networks The neural network has three gates, which are "forgetting gate", "output gate" and "output gate".

3.2 Bootstrap Model

Bootstrap estimates the uncertainty of the statistic by resampling to approximate the sampling distribution of the population. There is no need to make any assumptions about the overall distribution of the data [7], and the implementation process is as follows.

The same number of observations are extracted from a raw sample $Y = \{y1, y2, \ldots, yn\}$ *containing* n observations to form a new Bootstrap sample. This process is repeated B times, and each draw is independent.

Where the confidence interval can be expressed as:

$$CI = (\overline{T} - Z_{\frac{\alpha}{2}} \times SE(\overline{T}), \overline{T} + Z_{\frac{\alpha}{2}} \times SE(\overline{T})) \tag{3}$$

where $\overline{T}$ is the mean of the Bootstrap statistic, $SE(\overline{T})$ is its standard error, and $Z_{\frac{\alpha}{2}}$ is the $\frac{\alpha}{2}$ quantile of the standard normal distribution [8].

4 First Section Results and Analysis

4.1 Full-Text Modeling Process

In order to improve the accuracy of the PSR model and make more accurate predictions about future restoration outcomes, the specific process is shown in Fig. 1, the model implementation includes the incorporation of variable probability intervals.

Step 1. Data Preprocessing: Remove missing values and outliers from the experimental data, and thereafter normalise PSR indicators to mitigate the impact of different dimensions.

Step 2. Topsis Analysis: Apply the Topsis method to the annual sample data by calculating the grey relation degree with respect to ideal solutions instead of using Euclidean distance. This will result in a comprehensive score.

Step 3. LSTM Prediction: Utilise LSTM to reconstruct score data for rolling 5-year forecasts, while modifying model parameters to enhance generalisation. Evaluate the performance of the model using Mean Squared Error (MSE) and Root Mean Squared Error (RMSE).

Step 4. Confidence Interval Estimation: Calculate intervals based on the features of the Bootstrap distribution, by re-evaluating scores for various confidence levels.

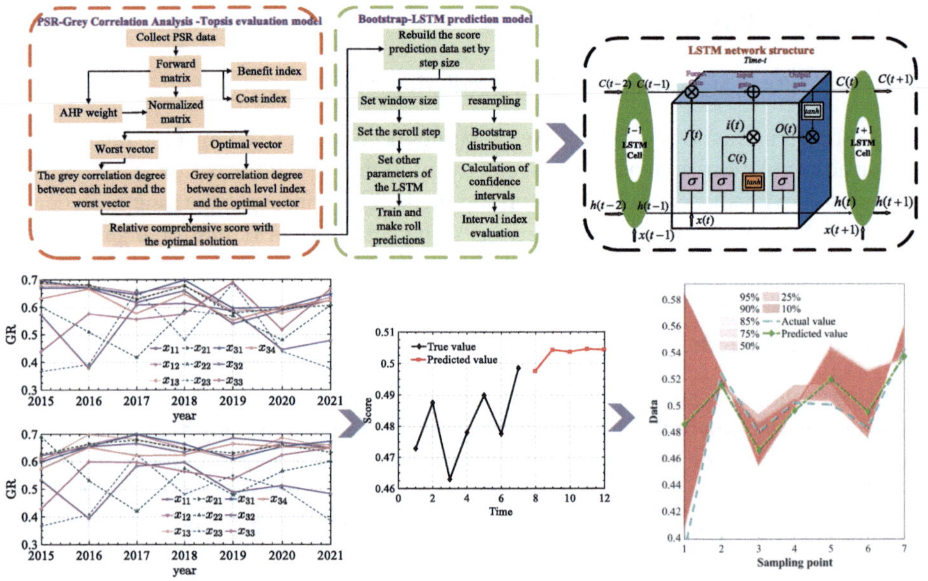

Fig. 1. Schematic diagram of PSR-Bootstrap-LSTM neural network evaluation and prediction model

4.2 Result Analysis

According to the comprehensive score solved by fusion Topsis, different ecological restoration levels were divided, and the corresponding four intervals were excellent, good, average and bad, as shown in the Table 2:

Table 2. According to the scoring interval, the ecological restoration grade evaluation was divided

Ecological restoration level	section
excellent	(0, 0.4527]
good	(0.4527, 0.4695]
average	(0.4695, 0.5369]
bad	(0.5369, 1]

Then, the performance of LSTM rolling prediction model was tested to obtain the performance index of the prediction model, and the evaluation index of Bootstrap confidence interval prediction was shown in the Fig. 2:

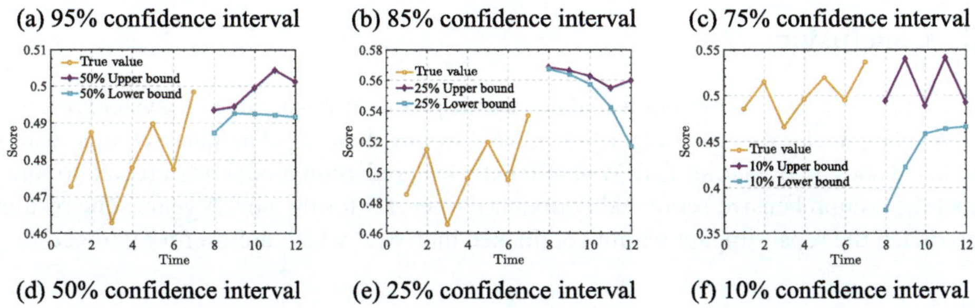

Fig. 2. Interval prediction rank distribution under different confidence intervals

Through the different intervals divided, we can intuitively see the interval of the comprehensive score of ecological restoration in the next five years, and make different degrees of preparation through the corresponding grades of the intervals, so as to provide references for enterprises. Finally, different confidence intervals were statistically analyzed, and the upper and lower limits of confidence intervals and the range of scores were observed. The results are shown in the Table 3.

Table 3. Prediction intervals of ecological restoration grade evaluation under different confidence intervals

prediction interval	2022	2023	2024	2025	2026
10% prediction interval	[0.3758, 0.4943]	[0.4225, 0.5406]	[0.4586, 0.4892]	[0.4641, 0.5417]	[0.4663, 0.4927]
25% prediction interval	[0.5677, 0.5687]	[0.5639, 0.5666]	[0.5574, 0.5627]	[0.5424, 0.5550]	[0.5169, 0.5600]
50% prediction interval	[0.5272, 0.5983]	[0.5254, 0.5950]	[0.5252, 0.5953]	[0.5245, 0.5992]	[0.5235, 0.6228]
75% prediction interval	[0.5137, 0.5149]	[0.4842, 0.5290]	[0.5043, 0.5392]	[0.5142, 0.5489]	[0.5242, 0.5426]
85% prediction interval	[0.4965, 0.4612]	[0.4733, 0.4672]	[0.4988, 0.4627]	[0.5102, 0.4578]	[0.5027, 0.4527]
90% prediction interval	[0.4659, 0.5175]	[0.4730, 0.5157]	[0.4691, 0.5191]	[0.4716, 0.5159]	[0.4695, 0.5256]
95% prediction interval	[0.6027, 0.6152]	[0.5747, 0.6204]	[0.5636, 0.6052]	[0.5591, 0.5679]	[0.5569, 0.6158]

It can be concluded from the above conclusions that PICP at 85% and 50% confidence levels is closest to the confidence level, and the interval prediction model has high reliability under different confidence levels, which not only ensures the accuracy of prediction, but also does not lose practical significance due to high values. Therefore, this model can be used as a reference for future ecological restoration assessment.

5 Conclusion

This study used grey correlation analysis and Topsis to evaluate mine ecosystem restoration. The panel data-based algorithm gave a thorough score. The data set was reconstructed and the Bootstrap-LSTM neural network prediction model was created to forecast the comprehensive score and confidence intervals for the next 5 years. The results predicted the repair impact within confidence intervals, which are useful references.

References

1. Wang, L., Li, Z., Wang, D., et al.: Unbalanced social-ecological development within the Dongting Lake basin: inspiration from evaluation of ecological restoration projects. J. Clean. Prod. **315**, 128161 (2021)
2. Niu, L.N., Shao, Q.Q., Ning, J., et al.: Evaluation on the degree and potential of ecological restoration in Loess Plateau. J. Nat. Resour. **38**(3), 779–794 (2023)
3. Lv, T., Zeng, C., Lin, C., Liu, W., Cheng, Y., Li, Y.: Towards an integrated approach for land spatial ecological restoration zoning based on ecosystem health assessment. Ecol. Indic. **147**, 110016 (2023)

4. Zhang, J., Chen, H., Ma, Y., et al.: Identification of priority areas for ecological restoration based on ecological security and landscape elements. Environ. Sci. Pollut. Res. **30**, 35307–35325 (2023)
5. Song, S., Luo, S., Li, G., et al.: Short-term PV power interval prediction based on MPA-LSTM model and Bootstrap method. J. Guangxi Univ. Nat. Sci. Ed. **47**(4), 986–997 (2022)
6. Wang, J., Lu, C., Li, F., et al.: Study on ecological security evaluation of open-pit mining area based on PSR and GA-Elman models. China Min. Mag. **29**(02), 65–71 (2019)
7. Mishra, A.P., Singh, S., Jani, M., Singh, K.A., Pande, C.B., Varade, A.M.: Assessment of water quality index using Analytic Hierarchy Process (AHP) and GIS: a case study of a struggling Asan River. Int. J. Environ. Anal. Chem. **104**(5), 1159–1171 (2024)
8. Venskus, J., Treigys, P., Markevičiūtė, J.: Unsupervised marine vessel trajectory prediction using LSTM network and wild bootstrap** techniques. Nonlinear Anal. Model. Control **26**(4), 718–737 (2021)

Research on the Mechanism of Spontaneous Combustion of Minerals and Prevention Technology in Funshun Open-Pit Mines

Wei Zhu[1](✉), Xuwen Zhu[2], Yue Li[1], Xing Li[1], Lixia Zhang[1], Yingzhe Wei[1], and Lina Yang[1]

[1] Ecological Environment Technology Co. Ltd., CCTEG, Beijing 100013, China
63399926@qq.com

[2] Shandong Shenbo Roadway Support Technology Co. Ltd., Jinan 250000, Shandong, China

Abstract. In response to the frequent occurrence of spontaneous combustion disasters in Fushun West Open Pit Mine, which has led to environmental pollution and production safety issues, coal samples were taken on site for laboratory industrial analysis and high-temperature programmed heating experiments, and the results of coal quality analysis and the laws of gas evolution during the spontaneous combustion process were obtained. The specific surface area and pore size distribution show that the minerals are prone to oxidation and adsorption, and the apparent structural characteristics explain why the minerals are prone to oxidation and spontaneous combustion. The oxidation heating process releases huge amounts of heat, which promotes the further development of self-heating to spontaneous combustion. Spontaneous combustion in open-pit mines not only affects the ecological environment, but also destroys high-quality coal resources, reduces slope strength, and may further trigger geological disasters such as landslides. An infrared thermal imaging system can be used to accurately delineate high-temperature self-heating and spontaneous combustion areas by combining auxiliary isotope radon measurement and borehole temperature measurement, so as to select appropriate prevention and control measures for precise management, effectively controlling the development of spontaneous combustion in the mine.

Keywords: Spontaneous Combustion Characteristics · Oxidation · Gas Precipitation · Monitoring and Prevention

1 Introduction

In the process of open-pit coal mining, residual coal is exposed to the air, and after weathering and fragmentation, it is easy to trigger oxidative spontaneous combustion and release gaseous products after prolonged and extensive contact with fresh air. Spontaneous combustion is one of the common disasters in surface coal mining, releasing toxic and harmful gases and heat, polluting the environment, eroding resources, reducing the strength of the rib gating-side, and leading to a series of safety and environmental

D. Han et al. (Eds.): ICEPG 2024, ESE, pp. 879–888, 2025.
https://doi.org/10.1007/978-3-031-86829-0_98

problems [1–4]. The spontaneous combustion and oxidization characteristics of different kinds of coals in different regions were experimented by temperature programmed experiments, on-site measurements and thermogravimetry, respectively, to obtain the spontaneous combustion characteristic parameters of coals under specific conditions, which provide a variety of iconic parameters for the monitoring of spontaneous combustion and the prevention and control of spontaneous combustion [5–9]. For the monitoring of spontaneous combustion in open-pit mines, the paper [10] researched the spontaneous combustion mechanism of open-pit mine and gangue mountain and the prevention and control technology of natural ignition respectively, and gained the practical experience of spontaneous combustion disposal and treatment, which provided reference for the prevention and control of similar disasters.

The Fushun West Open-Pit Mine has a mining history of more than 100 years, during which numerous spontaneous combustion accidents have occurred in the pit, and natural fires still occur from time to time, if not controlled in a timely manner, can cause serious impacts on the mine's production and the surrounding ecological environment. This paper takes the spontaneous combustion of coal in Fushun West Open-Pit Mine as the research object, obtains the physical properties of coal and the characteristic parameters of spontaneous combustion through physical tests and high-temperature programmed temperature-raising experiments, explains the mechanism that the coal wall of the open-pit mine is prone to spontaneous combustion, puts forward the method of monitoring spontaneous combustion of the open-pit mine and proposes the spontaneous combustion prevention and control technological methods for the characteristics of open-pit mines, which effectively controls the development of spontaneous combustion in the pits, and provides the reference experience for the precise prevention and control of spontaneous combustion of the similar mines.

2 The Current Situation of Spontaneous Combustion Disaster

Fushun West Open-Pit Mine is with a long history and a grand scale, with 109 years of mining history since the mine was established in 1914, and the pit site is shown in Fig. 1(a and b). With the increase of mining depth and internal soil discharge, the ventilation conditions are getting much worse, and it is difficult to discharge a large amount of harmful gases produced by spontaneous combustion of minerals in the pit, and there is an inversion layer in the upper part of the pit, and the harmful gases accumulate at the bottom of the pit, which affects the working environment. The spontaneous combustion of the mixed rows of residual coal and residue in the pit is shown in Fig. 1(c and d).

It can be seen that the spontaneous combustion of minerals and slag mixtures in the mine pit has seriously affected the normal production of the mine and the regional ecological environment, therefore, it is necessary to formulate effective and appropriate prevention and control methods for the spontaneous combustion of minerals in the open pit.

Fig. 1. General situation of western open-pit mine and spontaneous combustion in the mine

3 Spontaneous Combustion Characterization Experimental Test Analysis

The physical and spontaneous combustion experiments were conducted to analyze the physical properties of West Open-Pit coal such as component and elemental composition, specific surface area and pore size, etc. The high-temperature program heating system was used to test the characteristic parameters of high-temperature oxidation and spontaneous combustion of long-flame coal in West Open-Pit, including the change rule of precipitated gas concentration.

(1) Coal quality analysis

The coal seam extracted from West Open-Pit Mine is Jurassic Cretaceous coal, and the coal type is long-flame coal, and the industrial analysis of the coal quality is shown in Table 1, in which M_{ad} is the moisture; A_{ad} is the ash; V_{ad} is the volatile matter; and FC_{ad} is the fixed carbon. It can be seen that the volatile matter as well as the fixed carbon content are high, and the moisture of the experimental raw coal samples is <5%. Volatile fractions as high as 42% can greatly contribute to the development process of oxidative spontaneous combustion.

(2) Specific surface area and pore size analysis

The specific surface area and pore size distribution of the experimental coal samples, are shown in Table 2. According to the classification standard of pore structure of IUPAC organization: microporous (<2 nm), mesoporous (2~50 nm), macroporous (>50 nm), the macroporous and mesoporous of coal samples accounted for more than 95% of

Table 1. Coal analysis result (%)

Coal samples	Mine pit
FC_{ad}	46.93
V_{ad}	42.01
A_{ad}	8.12
M_{ad}	2.94

the total, which increased the chances of reacting with the contact of oxygen, and the specific surface area of the original coal reached 39.38 m^2/g, which was easy to oxidize and adsorb, and explained the reason why the coal seam was easy to be oxidized and spontaneous combusted to a certain degree from the apparent structural characteristics.

Table 2. Specific surface area and pore size distribution of sample

Coal samples	Raw coal
Mesoporous (%)	56.15
Macroporous (%)	43.23
Specific surface area (m^2/g)	39.38
Microporous (%)	0.62

(3) Analysis of gases precipitated by spontaneous combustion

Coal releases carbon oxides and hydrocarbons during low-temperature oxidative development to high-temperature spontaneous combustion, and the composition of the gases and their concentrations are extremely closely related to experimental temperature correlations. In this experiment, the concentrations of CO, CO_2, CH_4, C_2H_4, C_2H_6, and C_3H_8 produced by the coal samples in the process of warming and their changing rules were determined.

(a) Carbon-oxygen compounds

During the high-temperature programmed warming experiments, CO and CO_2 gases were detected from the beginning with low concentration levels, and the concentrations of the two gases increased more slowly when the coal temperature was less than 200 °C; after 200 °C, the concentrations of CO and CO_2 showed an overall sharp increase with the increase in coal temperature, as shown in Fig. 2 CO concentration at 350 °C reached the 1st peak point, in this stage of the coal oxygenation and synthesis reaction is intense, resulting in a sharp rise in the rate of oxygen consumption, the oxygen concentration decreased rapidly, and then due to the carboxyl groups, aliphatic hydrocarbons, and other functional groups to reduce the content, resulting in the concentration of CO gas fluctuates back to the point of the sample as the oxidation of the reaction is further aggravated, pyrolysis reaction occurs, so that the amount of CO production increased again, and reached a peak of 60,760 ppm at 555 °C, and then began to decline, until the end of the experiment.

In the low-temperature oxidation stage, CO_2 gas is subject to van der Waals force adsorption, the concentration change is relatively flat, with the sample temperature will gradually desorption of CO_2 gas, the gas concentration in the rapid heating stage rises rapidly into the high-temperature combustion stage after the rate of growth slows down a little bit, and after the rapid increase in 465 °C to reach a peak value of 168,704 ppm, after the CO_2 concentration of the slight fluctuations, which is due to the oxidation of aromatics, aliphatic hydrocarbons and other oxidizing, and produce part of CO_2 gas, resulting in the concentration of the phenomenon of the rise of the magnitude of the phenomenon again.

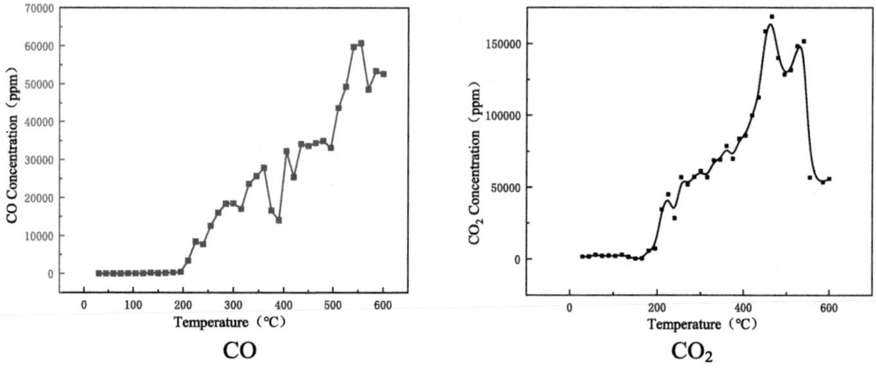

Fig. 2. Chang of gas concentration and growth rate with temperature

(b) Hydrocarbons

CH_4, C_2H_6, C_3H_8, C_2H_4 concentration with coal temperature change rule is generally similar, as shown in Fig. 3, the concentration of these four gases in 300 °C before the precipitation of the amount is relatively stable, but with the increase in coal temperature is also gradually increased. As CH_4 in coal samples is easier to desorb, it was detected at lower temperatures at the beginning of the experiment, and C_2H_6, C_3H_8, and C_2H_4 appeared at 120 °C, 195 °C, and 135 °C, respectively, with relatively small concentrations, and the concentration increased slowly with the increase of coal temperature. As the temperature rises, the oxidation reaction intensifies, and the aliphatic side chains on the aromatic rings within the coal macromolecule and the aliphatic hydrocarbons in the free phase are gradually cleaved through free radicals to form C_2H_4 and C_2H_6 and C_3H_8. The concentrations of CH_4, C_2H_6, C_3H_8, and C_2H_4 increased sharply from 340 °C, 330 °C, 350 °C, and 345 °C, respectively, and reached the maximum values of their respective gas concentrations at 480 °C, 430 °C, 435 °C, and 435 °C, respectively. In addition, a certain amount of fulvic acid contained in the coal undergoes thermal decomposition, resulting in the production of various hydrocarbon gases. After that, the cracking of the active functional groups in the coal molecule ends gradually, the CO concentration is in a high growth state, which inhibits the precipitation of hydrocarbon gases, and the CH_4 concentration shows a decreasing trend after 480 °C until the end of the combustion, and the concentration of precipitation is at a high level. C_2H_6 and C_3H_8, C_2H_4 show a sharp decrease after 435 °C, 480 °C, 480 °C, respectively, and all of them maintain the concentration level as in the low-temperature oxidation phase after 530 °C.

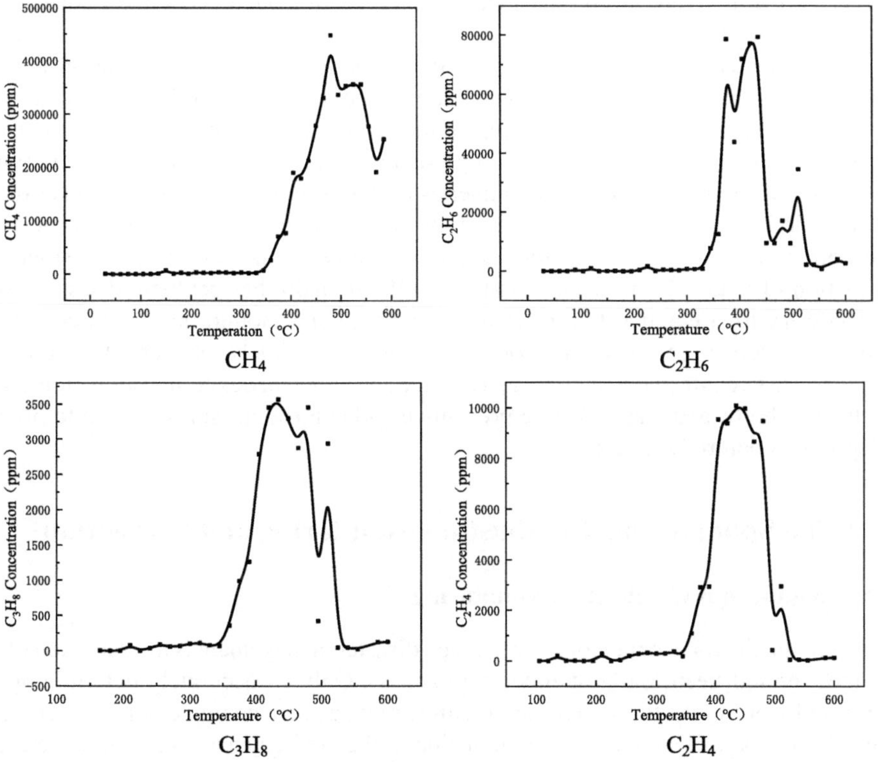

Fig. 3. Change of concentration with temperature

4 Analysis of Spontaneous Combustion Mechanism and Influence

Open pit mine slopes are exposed for a long time, affected by blasting vibration, mine shock, wind and rain, etc., the weathering zone fissure development, full contact with the air, the area of oxygen absorption increases, which provides the necessary conditions for the occurrence of spontaneous combustion and the continuous development of spontaneous combustion. At the low temperature stage, oxidation proceeds slowly, generating unstable oxidation products, while releasing heat, prompting the oxidation rate to accelerate, and the chemical activity of the minerals is enhanced, and this stage is the self-heating latent period. With the oxidation rate further increased, when the oxidation of the heat released too late to release, then the temperature inside the heap rises, the active functional groups in the minerals to further participate in the oxidation reaction, while precipitating a variety of indicators of the gas, this stage for the self-heating period. Further accumulation of heat in the heap, the temperature rises to above the critical temperature of spontaneous combustion of minerals, the rate of oxidation increases exponentially, this stage is accelerated oxidation period. When the temperature in the pile continues to increase to reach the ignition point and the oxygen supply is sufficient, the minerals will produce external combustion phenomenon, this stage is the spontaneous combustion period.

Spontaneous combustion disaster on the working environment of the pit and the regional ecological environment caused serious pollution impact, is not conducive to the safe production of mine operations, at the same time, the spontaneous combustion of the coal seam erosion damage to high-quality coal resources, reducing the purity and value of the resources, the spontaneous combustion of the waste rock slag of the internal drainage field and the bottom of the slope residual coal fire, easy to cause the rock body loose strength reduction, is not conducive to the stability of the slope, the formation of landslides may be landslides collapse and other geologic disasters, so it is necessary to strengthen the internal heat storage and monitoring of the coal body, and take effective preventive and control measures in a timely manner. Therefore, it is necessary to strengthen the internal heat storage and spontaneous combustion monitoring of the coal body, and take effective preventive and control measures in time to prevent the problem before it occurs.

5 Mine Spontaneous Combustion Monitoring and Prevention

5.1 Spontaneous Combustion Monitoring

The prerequisite for spontaneous combustion disaster management is the detection of fire sources (or self-heating high-temperature zones). Only by accurately determining the depth and scope of the fire source can scientific management be implemented. Generally, the primary stage of spontaneous combustion of the ore body is not easy to be detected, and the appearance of obvious smoke or fire is already a factual stage, so it is necessary to monitor the early warning of spontaneous combustion hazardous areas in a timely manner. Infrared thermal imaging monitoring system in the mine pile, gangue mountain, open pit mine spontaneous combustion monitoring plays an important role, as shown in Fig. 4, not only can detect the spontaneous combustion of the parts, but also can detect the severity of spontaneous combustion, is a non-contact, long-distance, wide-range monitoring method, and the operation is simple and convenient, can be fixed-point monitoring or can be equipped with drones scanning and monitoring.

For the deep self-heating and spontaneous combustion area can be assisted by combining the method of drilling temperature measurement, thermometer and thermocouples can be used to detect the temperature at different depths of the drill holes, mark the high temperature drill holes, and circle the self-heating and spontaneous combustion area, and at the same time, isotope radon measurement can be adopted to assist in the monitoring method. By testing and analyzing the temperature distribution of drill holes at different

a

b

Fig. 4. Infrared thermal imaging scanning

depths, the depth and range of the fire zone can be grasped to provide basic data for the precise prevention and control of spontaneous combustion.

5.2 Spontaneous Combustion Disaster Disposal and Treatment

(1) Water Pouring

Water into the fire area can play the role of cooling, suitable for eliminating open flames or fire roots within 2-m of the fire area, as shown in Fig. 5. The fire extinguishing cost is low, the effect is fast, applicable to the fire area range is small, the water consumption is less.

(2) Overlay method

After crushing the soft rock layer and the inner row of soil and other fine materials using construction machinery to cover the spontaneous combustion area, to achieve the purpose of isolation of oxygen, fire extinguishing. It is suitable for small fire areas with flat terrain and fast construction speed, but not for areas with complicated terrain and inconvenient transportation.

Fig. 5. Water pouring treatment

Fig. 6. Removal of fire roots

(3) Remove the fire root

The use of engineered equipment to excavate the fire root zone clear, as shown in Fig. 6, is the most direct and quite effective method. The method is relatively simple to implement when the fire is small and on the surface. It is suitable for fire areas with shallow roots and small range.

(4) Grouting method

Grouting fire extinguishing method is mainly for the underground area of large removal difficulties in the fire area for drilling and filling slurry, as shown in Fig. 7(a), until the high temperature or spontaneous combustion target area, so that the slurry is filled to the voids within the combustible accumulation body, so that the combustible material is isolated from the air, to achieve the purpose of prevention and control of spontaneous combustion. The slurry is mainly selected from fly ash slurry, clay slurry or fly ash clay composite slurry or yellow mud slurry, while the lime material can be added, as shown in Fig. 7(b). In recent years, the use of drilled holes to infuse the new extinguishing material of three-phase foam, as shown in Fig. 7(c). By circulating pressurized irrigation, a large amount of mud and three-phase foam in the deep high temperature area to form a pile of coverage and wrapping, can effectively eliminate the high temperature hazard and spontaneous combustion state.

a b c

Fig. 7. Borehole grouting to eliminate underground spontaneous ignition source and high temperature area

6 Conclusions

(1) The coal type of the West Open Pit Mine is long-flame coal, and the industrial analysis results show high volatile content and large specific surface area. The pore size distribution shows that the proportion of macropores and mesopores in the samples is more than 95%, and the minerals are prone to adsorption and oxidation reactions, which explains why the mixed-row minerals are easy to oxidize and spontaneous combustion.

(2) High-temperature program heating experiments show that the samples in the oxidative heating process is easy to precipitate CO, CO2 and other carbon-oxygen compounds, as well as CH4, C2H4, C2H6 and other hydrocarbons, the composition and concentration of the gas is closely related to the temperature. The oxidative heating process releases heat, which promotes the further development of self-heating to spontaneous combustion, in which the coal sample generates the greater amount of heat.

(3) In view of the spontaneous combustion characteristics of the West Open Pit Mine, an infrared thermal imaging system can be adopted to accurately define the high-temperature self-heating zone and the spontaneous combustion zone with the aid of drilling temperature measurement and isotope measurement of radon, so as to take preventive and curative measures for accurate management.
(4) Fushun West Open Pit Mine is currently used to dispose of the self-heating and spontaneous combustion zone by water pouring method, covering method, fire root removal method, and drilling and grouting method, and better prevention and control effects can be achieved by choosing the appropriate disposal method according to the specific development of the fire zone.

References

1. Qin, B., Zhong, X., Wang, D., et al.: Research progress of coal spontaneous combustion process characteristics and prevention technology. Coal Sci. Technol. **49**(1), 66–99 (2021)
2. Mursito, A.T., Hirajima, T., Listiyowati, L.N., Sudarsono, S.: Surface physicochemical properties of semi-anthracitic coal from Painan-Sumatra during air oxidation. Int. J. Coal Sci. Technol. **5**, 156–166 (2018)
3. Wang, D., Xin, H., Qi, X., et al.: Mechanism and relationships of elementary reactions in spontaneous combustion of coal: the coal oxidation kinetics theory and application. J. Coal Sci. **39**(8), 1667–1674 (2014)
4. Onifade, M., Genc, B.: A review of research on spontaneous combustion of coal. Int. J. Min. Sci. Technol. **30**(3), 303–311 (2020)
5. Külaots, I., Goldfarb, J.L., Suuberg, E.M.: Characterization of Chinese, American and Estonian oil shale semicokes and their sorptive potential. Fuel **89**, 3300–3306 (2010)
6. Onifade, M., Genc, B.: Comparative analysis of coal and coal-shale intrinsic factors affecting spontaneous combustion. Int. J. Coal Sci. Technol. **5**, 282–294 (2018)
7. Deng, J., Liu, L., Lei, C., et al.: Spatiotemporal distributions of the temperature and index gases during the dynamic evolution of coal spontaneous combustion. Combust. Sci. Technol. **193**(10), 1679–1695 (2021)
8. Onifade, M., Genc, B.: Spontaneous combustion liability of coal and coal-shale: a review of prediction methods. Int. J. Coal Sci. Technol. **6**, 151–168 (2019)
9. Lu, X.X., Wang, D.M., Qin, B.T., et al.: Novel approach for extinguishing large-scale coal fires using gas-liquid foams in open pit mines. Environ. Sci. Pollut. Res. **22**(23), 18363–18371 (2015)
10. Yang, Z., Xiao, P.: Research and application of fire prevention and extinguishing technology in the waste dump of fushun west open pit mine. Mine Saf. **33**(7), 33–34 (2002)

Effects of Co-composting of Gentamicin Mycelia Residues with Different Auxiliaries on Gentamicin Residue, Enzyme Activity and Microbial Community Dynamics

Wenjing Bu[1], Zhining Li[2,3], Xingtao Ren[4], Jinyu Du[1], Shuqing Zhang[1], and Junfeng Wan[3](✉)

[1] Henan University of Animal Husbandry and Economy, Zhengzhou 450001, People's Republic of China

[2] Henan Academy of Sciences, Zhengzhou 450000, People's Republic of China

[3] Zhengzhou University, Zhengzhou 450001, People's Republic of China
wanjunfeng@zzu.edu.cn

[4] Henan Normal University, Xinxiang 453007, Henan, People's Republic of China

Abstract. This research aims to explore how incorporating various materials influences gentamicin residue, enzyme activity, and microbial populations during the composting of gentamicin mycelia residues (GMRs). In the study, rice chaff, corn cobs, mushroom residue, and peanut shells were added to GMRs in specific ratios for co-composting. The results showed that the addition of rice chaff, corn cobs and mushroom residue led to a rise in composting temperature, prolonged the thermophilic phase, and enhanced gentamicin degradation. The elevated temperatures also induced shifts in bacterial community composition and fostered greater diversity. Throughout composting, gentamicin suppressed the activities of cellulase, urease, phosphatase, and catalase.

Keywords: Gentamicin · Aerobic Composting · Gentamicin Mycelia Residues · Enzyme Activity

1 Introduction

Gentamicin mycelia residues (GMRs) from spent gentamicin production cultures is a hazardous solid waste characterized by high concentrations of antibiotic residues. Aerobic composting, an efficient and cost-effective compost management and reutilization method, has gained widespread application globally, particularly in the treatment of organic solid waste.

The use of conditioning agents in the composting process can enhance microbial diversity, promote microbial community succession, effectively shorten composting time, and improve the fertility of compost products [1]. Biomass resources such as mushroom residues, corn cobs, and peanut shells are commonly utilized as compost auxiliaries due to their wide availability, low cost, and rich organic content. Due to the

D. Han et al. (Eds.): ICEPG 2024, ESE, pp. 889–895, 2025.
https://doi.org/10.1007/978-3-031-86829-0_99

high complexity of organic substrates, their complete biological transformation during the composting process requires the collaborative action of various enzymes [2]. Therefore, studying key enzymes-such as cellulase, urease, phosphatase, and catalase-provides a theoretical basis for accurately evaluating composting effectiveness.

We investigated the changes in gentamicin residue, enzyme activity, and microbial communities during the aerobic composting process with various additives. The main objectives were as follows: (1) To determine the most effective types of auxiliary materials for achieving compost maturation, and (2) to examine how various additives influence the enhancement of aerobic composting efficiency for gentamicin residues.

2 Materials and Methods

2.1 Experimental Material

GMRs was supplied by Henan Renhua Biotechnology Co., LTD, while rice chaff, mushroom residue, corn cobs, and peanut shells were sourced from local farms in Ruzhou. The main physical and chemical properties of compost accessories are shown in Table 1.

Table 1. Key physicochemical parameters of the feedstock. The values are presented as means ± SDs (n = 3).

Parameter	Rice chaff	mushroom residue	corn cobs	peanut shells
pH	6.05 ± 0.11	8.36 ± 0.13	6.49 ± 0.21	6.40 ± 0.26
EC(mS/cm)	0.69 ± 0.09	1.62 ± 0.14	0.56 ± 0.24	0.59 ± 0.20
MC[a](%)	10.55 ± 0.05	63.55 ± 0.34	21.64 ± 0.26	19.15 ± 0.31
OM[b](%)	78.28 ± 0.19	71.92 ± 0.24	91.30 ± 0.25	90.30 ± 0.30
TN[c](%)	0.34 ± 0.02	1.69 ± 0.04	2.56 ± 0.06	3.12 ± 0.08
C/N ratio	116.18 ± 7.66	24.68 ± 6.26	20.69 ± 4.56	16.79 ± 4.12

[a]means moistrue content; [b]means organic matter; [c]means total nitrogen

2.2 Experiment Design and Sampling

Three composting treatments were set up as follows: (1) GMRs + rice chaff (CK); (2) GMRs + rice chaff + 6% corn cobs + 4% mushroom residue (T1); (3) GMRs + rice chaff + 10% peanut shells (T2). To ensure uniformity in pile size, each compost pile treatment contained approximately 1000 kg of GMRs.

The collected samples were split into two portions: one was freeze-dried and stored at −20 °C for 16S rDNA diversity sequencing and gentamicin content analysis, while the other was kept at 4 °C for microbial enzyme activity tests. The samples from the three treatments were named as follows: T1_0, T1_6, T1_64, T2_0, T2_6, T2_64, CK_0, CK_6, and CK_64.

2.3 Physicochemical Parameters

The gentamicin degradation percentage (GDP) was calculated following a previously described method [3]. The enzyme activity was determined using colorimetric methods and titration methods [2].

2.4 Data Processing and Statistical Evaluation

Spearman correlation analysis was conducted to identify the relationships between the top 16 bacterial genera and environmental factors, and enzyme activities. Heatmaps were generated using the "pheatmap" package in R, and other graphics were created with Origin 9.0.

3 Results and Discussion

3.1 Effect of Different Additives on Gentamicin Residue Degradation

Figure 1 illustrates the variations in gentamicin residue across the three compost piles during the composting process. Among them, the T1 treatment showed the fastest degradation rate, which may be related to the addition of additives such as rice chaff, corn cobs and mushroom residue. The richness of raw materials can increase microbial diversity, which may contribute to a higher gentamicin degradation rate. Mushroom residue, being a byproduct after cultivating edible fungi, contains pre-degraded cellulose, hemicellulose, lignin, and abundant amino acids and trace elements (such as iron, calcium, zinc, magnesium, etc.), all of which promote microbial growth, reproduction, and metabolism, and are beneficial for gentamicin degradation.

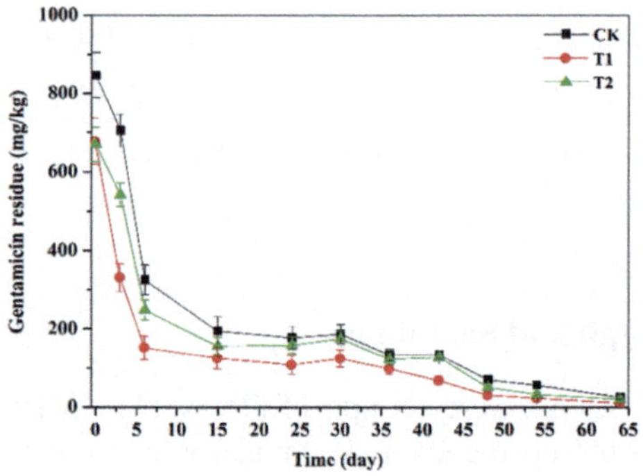

Fig. 1. Degradation of gentamicin in three treatments during composting.

3.2 Changes in Enzyme Activity During the Composting Process

Figure 2a presents the changes in cellulase activity across the three treatments during the composting process. During the composting process, the T1 treatment consistently showed higher cellulase activity than the other two treatments. This suggests that the addition of mushroom residue and corn cobs enhanced the lignocellulose content, leading to increased cellulase activity. Cellulase activity reached its peak at the high-temperature phase of composting, as the optimal temperature for cellulose degradation was approximately 65 °C, and the enzyme's activity was strongly influenced by temperature. As the compost temperature gradually decreased, cellulase activity also declined.

Figure 2b illustrates the trends in urease activity across the three treatments during composting. Throughout the composting process, T1 treatment consistently exhibited lower urease activity compared to the other two treatments, indicating that the addition of mushroom residue and corn cobs can reduce the hydrolysis of urea, thereby helping to inhibit NH_3 volatilization and reduce nitrogen loss [4].

Figure 2c shows the trends in phosphatase activity during composting for different treatments. Throughout the composting process, T1 treatment consistently exhibited higher phosphatase activity compared to the other two treatments, suggesting that the addition of mushroom residue and corn cobs in the compost promoted the conversion of organic phosphorus to inorganic phosphorus [5].

Figure 2d presents he trends in catalase activity for the three treatments during the composting process. Compared to the initial enzyme activity, all three treatments showed an increase. Among them, T1 and T2 treatments had higher catalase activity than CK, which may be related to the higher OM content in the added mushroom residue, corn cobs, and peanut shells.

3.3 Evolution of Bacterial Communities During Co-composting

Figure 3 depicts the dynamic shifts in bacterial communities at the genus level during composting under different treatments of GMRs. In the mesophilic phase, the bacterial community structure across the three treatments was relatively simple. *Aerococcus, Enterococcus, Lactococcu, Desemzia, Psychrobacter* and *Staphylococcus* were the dominat genera in the three treatments; however their relative abundance were varied. This indicated that adding compost additives can introduce inherent species and has a certain inoculation effect, increasing bacterial diversity during composting. In the high-temperature phase, the dominant genera across the three treatments were not clearly defined, with numerous species exhibiting relative abundances between 1.30% and 16.63%, indicating a more balanced distribution among genera. In the cooling phase, like in the high-temperature phase, the dominant genera across the three treatments were more evenly distributed, with relative abundances ranging from 3.38% to 15.70%. This distribution was significantly different from that observed in the mesophilic and thermophilic phases. This indicated that high-temperature composting can alter the bacterial community structure and increase bacterial diversity [6].

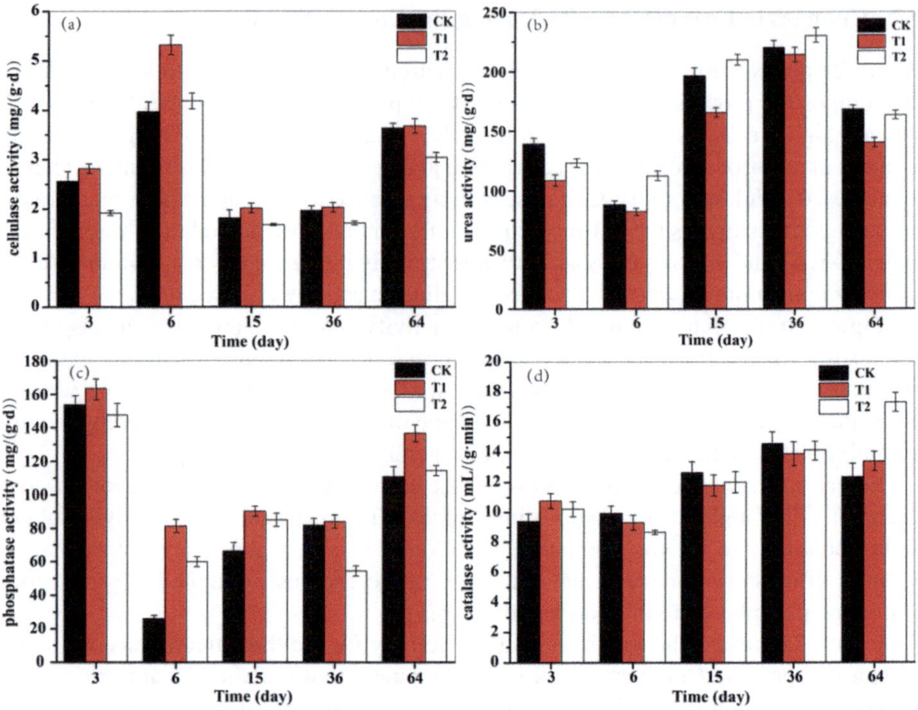

Fig. 2. Changes in the enzyme activities during the composting process for the three treatments include: (a) cellulase activity, (b) urea activity, (c) phosphatase activity, and (d) catalase activity.

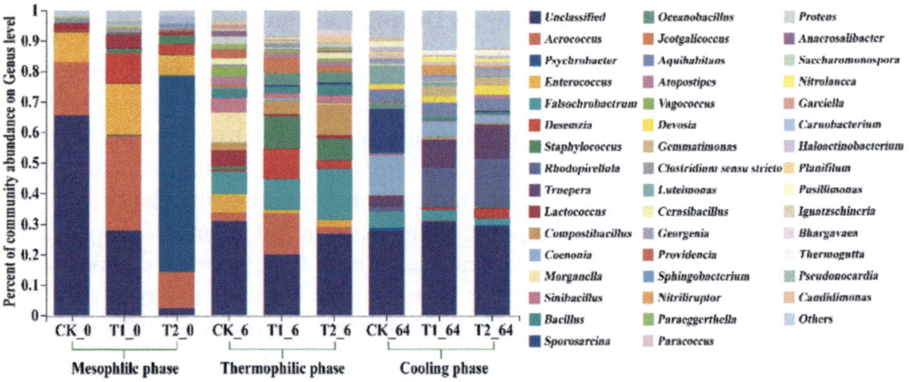

Fig. 3. Changes in the relative abundance of bacterial genera across three treatments during composting.

3.4 Analysis of the Relationship Between Environmental Factors, Enzyme Activity, and Microbial Communities During Composting

Figure 4 demonstrates a significant positive correlation between gentamicin residue and the bacterial genera *Aerococcus*, *Enterococcus*, and *Lactococcus*, suggesting that these microorganisms may play a major role in gentamicin degradation. This study also found that microorganisms negatively (or positively) correlated with cellulase, urease, phosphatase, and catalase activities were positively (or negatively) correlated with gentamicin residue. The main bacterial genera involved were *Aerococcus*, *Enterococcus*, *Falsocrobactrum*, *Lactococcus*, *Compostibacillus*, and *Coenonia*. This suggested that gentamicin residue may influence microbial metabolic functions, thereby inhibiting the activity of cellulase, urease, phosphatase, and catalase.

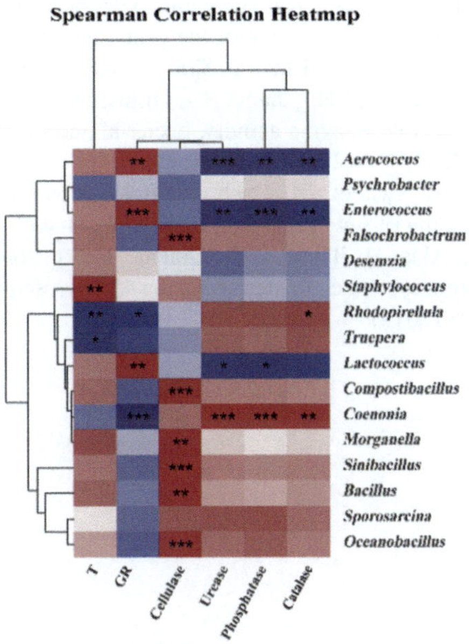

Fig. 4. Correlation heatmap of the top 16 genera with environmental factors and enzyme activity.

4 Conclusion

Rice chaff, mushroom residue, corn cobs, and peanut shells were selected as additives for co-composting with GMRs. The results indicated that the incorporating mushroom residue, corn cobs, and rice chaff (T1) shortened the composting duration to 64 days and enhanced the gentamicin degradation rate. The bacterial genera *Aerococcus*, *Enterococcus*, and *Lactococcus* may be the primary genera involved in gentamicin degradation. During composting, gentamicin inhibited the activity of cellulase, urease, phosphatase, and catalase.

Acknowledgments. The authors express their gratitude for the financial support provided by the Henan University of Animal Husbandry and Economy Doctoral Research Start-up Fund Project (Grant No. 2023HNUAHEDF012).

References

1. Jiang, Z.W., Lu, Y.Y., Xu, J.Q., Li, M.Q., Shan, G.C., Li, Q.L.: Exploring the characteristics of dissolved organic matter and succession of bacterial community during composting. Bioresour. Technol. **292**, 121942 (2019)
2. Wang, B., Liu, H.L., Cai, C., Thabit, M.G., Wang, P., Duan, Z.: Effect of dry mycelium of Penicillium chrysogenum fertilizer on soil microbial community composition, enzyme activities and snap bean growth. Environ. Sci. Pollut. R **23**(20), 20728–20738 (2016)
3. Bu, W.J., et al.: Metagenomics insights into gentamicin degradation and the dynamic of antibiotic resistance genes during co-composting of the gentamicin myelial residues with addition of various organic wastes. Sci. Total. Environ. **885**, 163848 (2023)
4. Ma, C., Hu, B., Wei, M.B., Zhao, J.H., Zhang, H.Z.: Influence of matured compost inoculation on sewage sludge composting: enzyme activity, bacterial and fungal community succession. Bioresour. Technol. **294**, 122165 (2019)
5. Malińska, K., Zabochnicka-Świątek, M., Dach, J.: Effects of biochar amendment on ammonia emission during composting of sewage sludge. Ecol. Eng. **71**, 474–478 (2014)
6. Adrian, L.F., Urooj, Z., Alan, H., Thomas, B., Graeme, F., Robson, G.D.: Fungal succession in an in-vessel composting system characterized using 454 pyrosequencing. FEMS Microbiol. Ecology **88**(2), 296–308 (2014)

Evolve to the Best: Parsing Community Succession

Yilin Rao(✉), Mengdie Hu, and Haochuan Lin

Department of Civil Engineering, Tongji University, 1239 Siping Road, Shanghai 200092, China
tjtongzhou10@163.com

Abstract. Plant communities are increasingly vulnerable to ecological threats, and they are employing species-to-species facilitation as a strategy to combat drought and safeguard biodiversity. To explore effective defense mechanisms, we analyzed species interactions and developed a model to understand how plant communities evolve and stabilize when facing drought stress, in which validation of existing studies accompanied. Our approach involved several critical steps. Firstly, we selected data from five cities with typical climates to accurately describe drought conditions, including adjusted precipitation equivalents. Secondly, we optimized the Lotka-Volterra competition model and introduced the Competition-Facilitation Model (CFM) with a differential system to unveil the intricate interplay of competition and facilitation among plant species. This included fitting the data and deriving quantitative parameters for the differential equations, thus creating a model that can describe changes in biomass within multi-species communities. We also defined three fundamental roles and a foundational combination (ERC) within simple communities to identify the number and types of high-performing species during successional processes. Thirdly, we conducted an in-depth analysis of biomass fluctuations in response to drought and wet conditions. By artificially intensifying drought frequency and intensity, we unraveled the mechanisms through which plant communities establish stability when facing drought and how species diversity affects different environments.

Keywords: Competition-Facilitation · Difference equation · Biomass diversity · Simulation experiments

1 Introduction

In contrast to the more focused conservation of animal ecology, only a minority of plant community biodiversity has received attention and protection, and the importance of plant biodiversity has been greatly underestimated, making the degree of risk to plants even greater [1]. At the same time, plant biodiversity in arid regions is under great threat and pressure [2]. Early results suggest that plant communities with large biodiversity are more resistant to drought in biodiversity-poor communities [3]. In a plant community, different species of plants have different ways of coping with drought stress, and there are complex interactions between species, with some plants overlapping ecological niches (The position occupied by a population in an ecosystem, in time and space, and

D. Han et al. (Eds.): ICEPG 2024, ESE, pp. 896–904, 2025.
https://doi.org/10.1007/978-3-031-86829-0_100

its functional relationships and roles with related populations.) and competing fiercely with each other, while others can maximize resource utilization through root dislocation. Callaway and Walker [4] indicated that plant communities succeed in a dynamic equilibrium of competition and facilitation between species under different environmental pressures. Miller [5] took an early old-field plant community as an example and found that moderately good competitors can work together to defeat both good and bad competitors.

2 Construction of the Model

2.1 Data Preparation

- Precipitation and SPEI index: We classify areas into different drought degrees using precipitation and obtain related data from *wheata.cn*. And we believe temperature and other factors, affecting plant water content, might also influence biomass. To better describe drought severity, we introduced the SPEI index as a correction factor, obtained from the Global SPEI database *Global SPEI database (csic.es)*. It enables us to assess the combined effects of other factors on different species at a smaller time scale. The SPEI index introduces potential evapotranspiration (PET), incorporating temperature via the Thornthwaite formula: $PET = 16K\left(\frac{10T}{I}\right)^m$. The factors K and I represent the impacts of insolation and heat.
- Data Analysis and Pre-processing: we selected five representative cities with four different climate classifications based on annual precipitation levels. We collected monthly precipitation and SPEI data from 2010 to 2020 for Las Vegas (arid), Los Angeles (semi-arid), Xining (semi-arid), London (semi-wet), and Crawford (wet). The effect of precipitation needs to be modified to corrected precipitation equivalent (P_{cor}) to explore the relationship between biomass and the degree of drought. Defining the drought correction factor (f_{cor}): $f_{cor} = \left(1 + \frac{SPEI}{2\max\{abs(SPEI)\}}\right) \times 100\%$, then $P_{cor} = P_i \times f_{cor}$.

3 Competition-Facilitation Model for Prediction

3.1 The Difference Equation for Plant Biomass

Based on the theory of biological populations, *the Lotka-Volterra equation* is commonly used to quantify the competitive relationships between species [6]. After enhancing the equation *with a facilitation factor,* we established *a plant community species intercropping system* and derived ***the Competition-Facilitation equation*** (C-F equation):

$$\frac{dBi}{dt} = r_i(P_{cor}(t))B_i\left(1 - \sum_{j=1}^{n} C_{ij}\frac{B_j}{K_i(P_{cor}(t), t)}\right) \tag{1}$$

where B_i is the biomass of the i-th plant; $r_i(\cdot)$ is the endogenous growth rate under certain environmental resources of the i-th plant, related to P_{cor}; C_{ij} is Competition-Facilitating Factor (CFF) of the j-th plant to the i-th plant; $K_i(\cdot, \cdot)$ is environmental holding capacity of the i-th plant, related to P_{cor} and t. r_i, Cij, K_i are all climate and species followers. We use numerical iterative algorithms on time series and MATLAB ode solver in solving the model.

3.2 The Parameters of the Difference Equation

The Endogenous Growth Rate ri.
When environmental resources (especially water) are insufficient, r_i ($r_i = f_w \times r_{i\max}$) can be considered as the result of the relative water utility factor (f_w) acting on the maximum growth rate ($r_{i\max}$). We defined three plant categories: *arid plants* (Arid.), *mesophytes* (Me.), and *wet plants* (Wet.) classified by the optimal water demand (P_{best}) and set f_w at optimal water demand to 1. Since f_w always tends to decrease when P_{cor} deviates from P_{best}, we define the possible relative curve of $f_w - P_{cor}$ mainly in the form of quadratic functions. We set r_i to *0* when P_{cor} deviates by one σ_p' from P_{best}, so the curve has characteristic points $(\mu_p', 1)$, $(\mu_p' - \sigma_p', 0)$, $(\mu_p' + \sigma_p', 0)$. The linear functions were used to succeed in the part where r_i is less than *0* to obtain the whole $f_w - P_{cor}$ curves (Fig. 1).

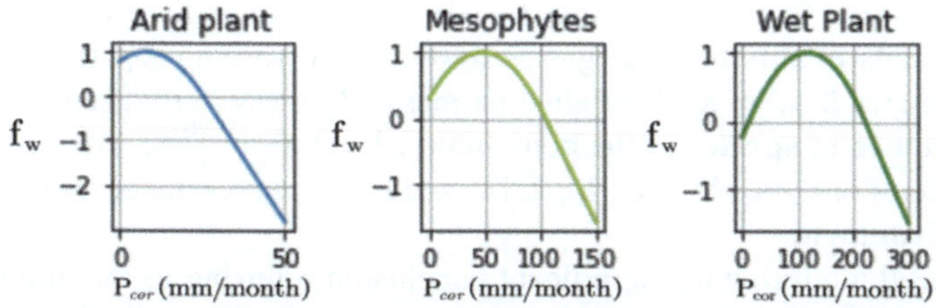

Fig. 1. $f_w - P_{cor}$ curves for the three types of plants.

The Competition-Facilitating Factor Cij (CFF) and the Environment Holding Capacity Ki.
C_{ij}(CFF), or Competition-Facilitation Factor quantifies interactions containing competition and facilitation between plant species. They are primarily influenced by the location of the root system structure. *Herbs* (Herb.), with shallow roots, tend to compete more in the surface soil, *Shrubs* (Shrub.) intermediate, and *Macrophanerophytes* (Macro.) access both deep and shallow water. Differences in root structure allow plants to fully utilize water from various soil layers, acting as a *facilitation factor. Parasitic or Symbiotic Relationships* between plants and *Drought Stress* can also lead to different C_{ij}(CFF). We establish 'CASE' Library (interaction coefficients are related to Categories, moisture Adaptations, Special Effect between species) by Monte Carlo stochastic simulations (for simulation) or expert scoring (for research) to comprehensively calculate and analyze C_{ij}(CFF), and constructed a Competition-Facilitating Factor Transformation Matrix covering vertical mechanism, horizontal mechanism, biochemical mechanism.

For specific species populations, the maximum population biomass is limited by environmental resources reserved. The ranges of K_i value for these plants are shown in Table 1. When the **corrected precipitation equivalent** is insufficient, the K_i value decreases; when the living environment is improved, the K_i value increases. When drought stress is detected, it can be assumed that the change in K_i value occurs with

Table 1. The approximate high and low values of K_i of different categories.

Category	High (g/m^2)	Low (g/m^2)
Herbs	4000	1500
Shrubs	150	100
Macrophanerophytes	15000	10000

a lag for the presence of water resource reserves, but changes within a short period for the great destructiveness.

3.3 Implementation

We set the r_i of the five plants in the 'CASE' library to simulate with the initial values. Since plants of different categories have widely varying K_i, the environmental control parameter $K_{cor} = K_i/2$, relative biomass (B_{re}) was introduced to describe **the self-performance of species** in the community: $B_{re}(t, i) = \frac{B(t)}{K_{cor}}$. After 120 months of differential iterations, we obtained B_{re}(t, i)-t and $B(t, i)$ curves under three conditions in the **arid environment.**

We deduced the following significant conclusion referring to the graphs in Fig. 2: when under *Drought Stress*, two dry plants have evolved facilitation mechanisms to make each other the dominant species, which would confirm Miller's conclusions.

4 Community Succession Model

4.1 Combination Effect: 1 + 1 > 3

As for long-term community succession processes under wet (occasional drought) conditions, evolving from short-term *dominant herbs* (HW) to long-term *dominant macro plants* (MW), Swanson et al. [7] pointed out that herbaceous and shrub species in early forest ecosystems had high productivity, large biomass, and complex structure. It has important value in maintaining community biodiversity.

So, we modeled the growth of Example (Herbs. Wet.1, Herbs. Wet.2, Macro. Wet., Macro. Me., Shrubs. Me.) with weather data in Crawford, and the shortens of the longer-term response was achieved by increasing the growth rate. In Fig. 3, by comparing the two graphs (a) and (b), we proved Swanson's theory and got the following conclusions: 1) Herbs plants (HW) with low demand are more suitable to promote community succession than macro plants (MW) in the early stage, while in the long term, macro plants (MW) have more succession potential. 2) The facilitation mechanisms of *dominant* + *companion* species still work and act as a community stabilizer in the early stages. 3) Mesophytes can only grow for a short time in this community and are quickly eliminated due to poor root respiration.

As for community succession subjected to sudden disasters (e.g., high-frequency droughts *see* Sect. 4.2), Dios et al. [8] found that when there is still an active competition (grasses & shrubs) under drought conditions, the death of grasses can lead to strong

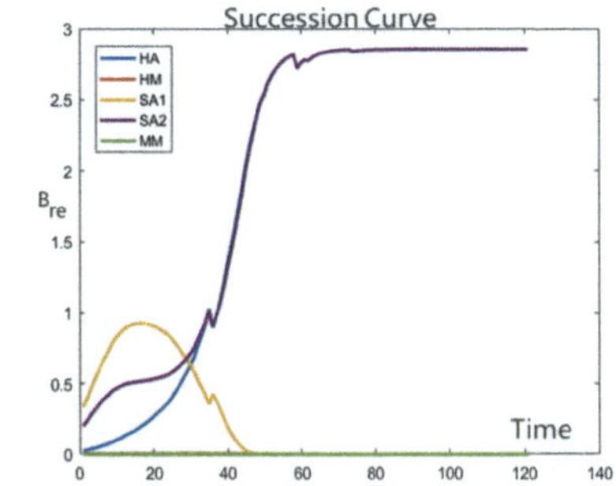

(a) B_{re} with Strong Facilitation

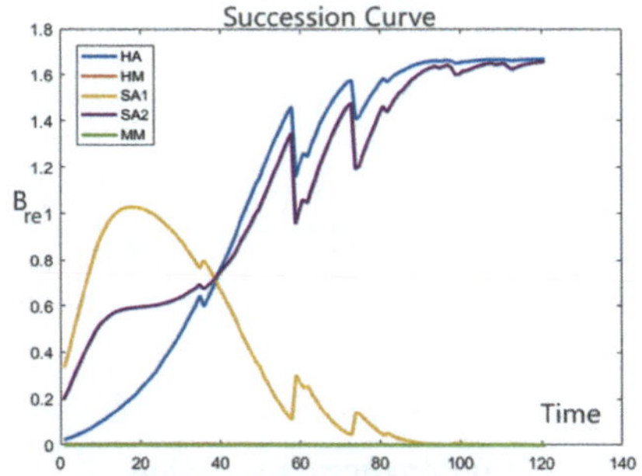

(b) B_{re} with Mild Facilitation

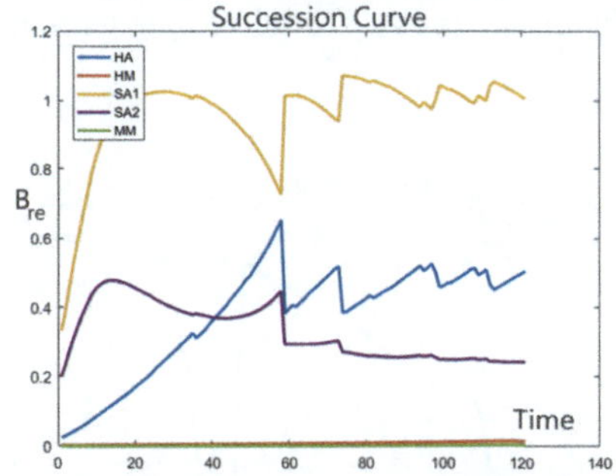

(c) B_{re} with Strong Competition

Fig. 2. Under an arid environment, different succession mechanisms between species. (a) and (b): species with facilitation (Herb. Arid. & Shrub. Arid. 2) become dominant concurrently with only differences in time. (c): occupying the favorable resources earlier, species with higher starting biomass will win in strong competition relationships.

passive facilitation for shrubs to go through the sudden drought, which informed that not only facilitation mechanisms with companion species are needed, but also strong competition mechanisms. In Fig. 4, we set up a sudden disaster around Time = 70, and there was a worrying sudden drop in HW1 which was previously the dominant specie, that is to say, a single population is likely to be unable to adapt to a variable environment. So, the introduction of HW3 was necessary to act as an Evolutionary Risk

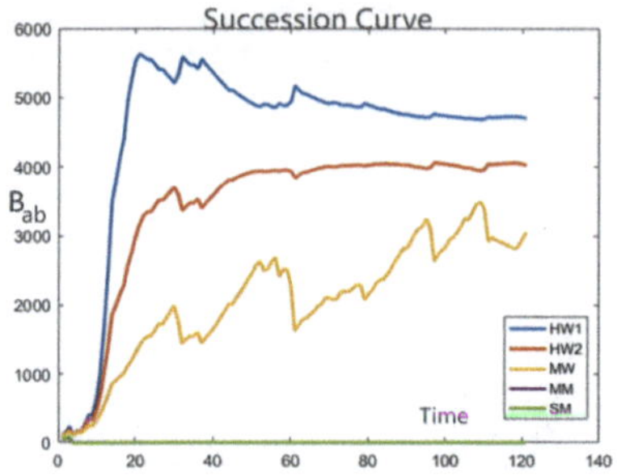

(a) Short-term succession

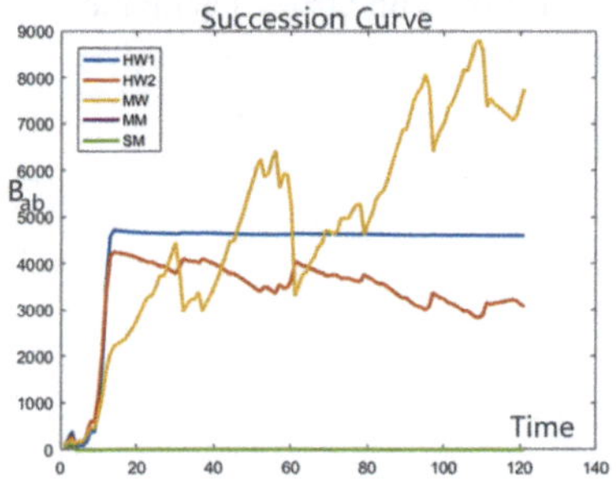

(b) Long-term succession

Fig. 3. The succession of Example under wet weather. (a) In the short term, herb plants (HW) with companions become dominant species. (b) In the long term, macro plants (MW) become more competitive.

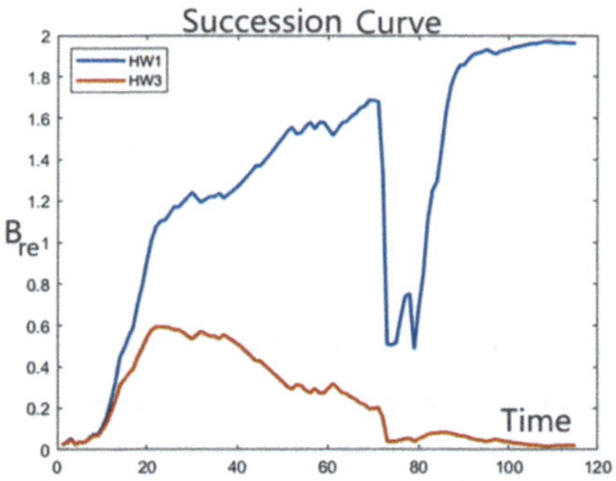

Fig. 4. B_{re} change in strong competition of two species. (a sudden disaster around Time = 70)

Combination (ERC) to safeguard the community's stability just like what Dios said. Then, we introduce the Shannon-Weiner index $H = -\frac{1}{B}\sum B_i \log B_i$ to describe the species diversity to reflect *the disaster-stability* of the community.

Based on the above characteristics of the two community successions, it only makes sense to discuss a pair of ERC with several companion species, since the addition of competing species will bring about elimination risk, and inferior species do not play a role in the community succession. The experimental results are shown in Table 2:

Table 2. Experimental results of the Shannon-Weiner index **H** and **Biomass** in various combinations. (HW_4 is facilitative with HW_3).

	Combination	H	Total biomass		Combination	H	Total biomass
1	1 ERC	0.06	7939.6	4	1 ERC + HW_1' 3 companion	1.26	9413.6
2	1 ERC + HW_1' 1 companion	0.69	9412.4	5	1 ERC + HW1' 1 companion + HW4	1.24	8602.5
3	1 ERC + HW_1' 2 companion	1.05	9433.3	6	1 ERC + MW	0.43	8354.2

Combining all results of experiments, there are the following conclusions: 1) Comparing the experiments of group 1–4, for this case, the best number of companion species for one dominant specie is 2. As this number continues to increase, the total biomass may decrease due to over-dominance of the dominant species. 2) The higher H of group 5 experiments suggests that the one-to-one pairing of "dominant + companion species" may provide better long-term stability, which has been suggested in Schnabel's research [9] to be a stress resistance mechanism brought by specific plants diversity and non-synchronization of production, but the "dominant + single dominant species companion species" has more overall biomass in short-term community succession. 3) Group 6 experiments suggest that long-term dominant species (MW) is not a substitute for the role of companion species. In summary, the key to benefiting the entire community lies in the combined effect of *"dominant species + companion species"* and *ERC*, typically involving about four species. Adding more species leads to strong competition, elimination, and excessive dominance of one specie, which reduces diversity.

4.2 Plant Community Response Under Controlled Drought Conditions

For the Higher intensity of drought, we artificially increased the frequency and extent of drought occurrence. Considering that plant communities are more regulated in *moderately humid environments*, to obtain significant effects, we conducted simulation experiments just in *arid and wet environments*.

By observing and comparing the graphs in Fig. 5, two effects are listed: 1) Regardless of arid environments or wet environments, the exposure to more frequent and prolonged droughts decreased the B_{re} of important species. At the same time, the reduction of B_{re} under arid environments was greater than that under wet conditions. 2) In response to stress in drought environments, plants have evolved the community stability.

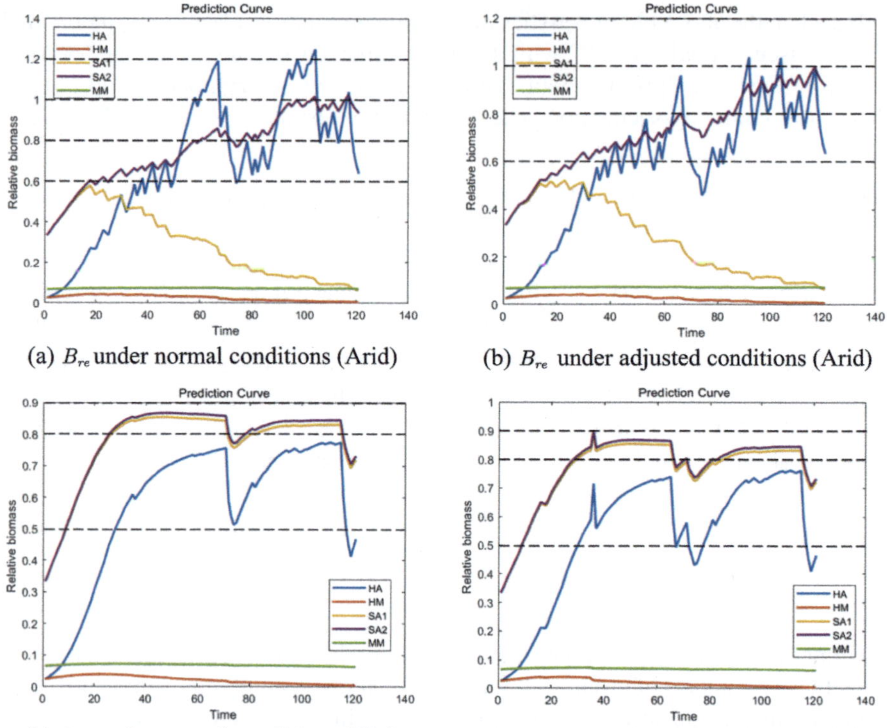

(a) B_{re} under normal conditions (Arid)

(b) B_{re} under adjusted conditions (Arid)

(c) B_{re} under normal conditions (Wet)

(d) B_{re} under adjusted conditions (Wet)

Fig. 5. B_{re} curves before and after drought frequency increase with isometric lines of 0.6 to 1.2 in (a) and (b), and isometric lines of 0.5 to 1.2 in (c) and (d). The fluctuation range of B_{re} in the arid plant community was unchanged and became larger in the wet plant community.

For the lower intensity of drought, we reduced the number of droughts for Los Angeles (semi-arid) data. To explore the effects of lower-frequency drought on community successions, we adopted the experimental protocol in Sect. 4.1. The experimental results are shown in Table 3.

Table 3. Experimental species assemblages and results.

Combination	H		Total biomass	
	Normal	Less- drought	Normal	Less- drought
1 ERC + HW_1' 1 companion	0.70	0.70	4191.7	5951.2
1 ERC + HW_1' 2 companion	1.08	1.06	4195.1	6019.4
1 ERC + HW_1' 3 companion	1.31	1.27	4196.3	6019.9

Overall, when the environment is at a moderately humid climate, the biomass can be enhanced while the index H decreased due to the sufficient water as the frequency of drought decreases, which can validate the conclusion that droughts can cause the evolution of more robust plant combinations and more diverse species, which is similar to the general patterns and consequences of positive interactions between plant community species in the face of frequent drought mentioned in Ploughe's study [10].

5 Conclusion

We constructed a numerical simulator of plant community biomass responses that can vary with environmental drought levels which will provide a modelling tool for the mechanisms of temporal evolution of community drought responses. Here are the basic conclusions we verified: In the long run, the direction of community succession is to increase the number of woody plants (although the growth rate is small, the environmental accommodation is large), but in the early stage, different species combinations which facilitation mechanisms from ecological niche complementarity and competition mechanisms from ecological niche overlap are needed to favor successions of the plant community. In more intense droughts, the plant combination described above evolved community stability, as evidenced by the stable magnitude of changes in plant diversity and biomass when subjected to drought.

References

1. Corlett, R.T.: Plant diversity in a changing world: status, trends, and conservation needs. Plant Divers. **01**, 11–18 (2016)
2. Goettsch, B., Hilton-Taylor, C., CruzPinon, G., et al.: High proportion of cactus species threatened with extinction. Nat. Plants **1**, 15142 (2015)
3. Isbell, F., Craven, D., Connolly, J., Loreau, M., Schmid, B., Beierkuhnlein, C., et al.: Biodiversity increases the resistance of ecosystem productivity to climate extremes. Nature **526**(7574), 574–577 (2015)
4. Callaway, R.M., Walker, L.R.: Competition and facilitation: a synthetic approach to interactions in plant communities. Ecology **78**(7), 1958–1965 (1997)
5. Miller, T.E.: Direct and indirect species interactions in an early old-field plant community. Am. Nat. **143**(6), 1007–1025 (1994)
6. Davis, J.D., Olivença, D.V., Brown, S.P., et al.: Methods of quantifying interactions among populations using Lotka-Volterra models. Front. Syst. Biol. **2**, 1021897 (2022)
7. Hanusch, M., He, X., Ruiz-Hernández, V., et al.: Succession comprises a sequence of threshold-induced community assembly processes towards multidiversity. Commun. Biol. **5**(1), 424 (2022)
8. Banerjee, S., Das, D., Zhang, H., et al.: Grassland-woodland transitions over decadal timescales in the Terai-Duar savanna and grasslands of the Indian subcontinent. For. Ecol. Manag. **530**, 120764 (2023)
9. Schnabel, F., Liu, X., Kunz, M., et al.: Species richness stabilizes productivity via asynchrony and drought-tolerance diversity in a large-scale tree biodiversity experiment. Sci. Adv. **7**(51), eabk1643 (2021)
10. Ploughe, L.W., Jacobs, E.M., Frank, G.S., et al.: Community Response to Extreme Drought (CRED): a framework for drought-induced shifts in plant–plant interactions. New Phytol. **222**(1), 52–69 (2019)

Ecosystem Health Assessment of the Yellow River Estuary Based on the DPSIRM Model

Jiarong Li, Min Ji(✉), and Fengxiang Jin

Shandong University of Science and Technology, Qingdao 266590, China
345812294@qq.com

Abstract. Dongying City is located at the mouth of the Yellow River, with a long coastline, and its urban ecosystem and climate are significantly influenced by the marine environment. Considering the intense development activities in the Yellow River estuary area and their profound impact on the ecological environment, this study aims to comprehensively assess the ecological health status of the Yellow River estuary, with Dongying City as the study area. Based on the literature analysis method, key indicator factors affecting the ecology of the Yellow River estuary were screened using Pearson correlation analysis and principal component analysis, and 23 indicator factors were selected based on the DPSIRM model. By combining the improved AHP-CRITIC-Entropy weight method and the comprehensive index method, an integrated ecosystem health evaluation system was constructed, and the ecological health status of the Yellow River estuary from 2010 to 2022 was analyzed. The study found that during this period, the ecosystem at the Yellow River estuary generally maintained a sub-healthy to healthy state, with an overall trend of steady improvement. Particularly since 2020, the ecosystem has entered a relatively stable and healthy state. In 2022, the ecological health index reached 0.7077, reflecting a trend towards a healthy development of the Yellow River estuary ecosystem.

Keywords: Ecosystem · Health Assessment · DPSIRM Model · Dongying City

1 Introduction

Dongying City is a crucial development area in the Yellow River Delta. The city not only boasts an extensive coastline but also possesses abundant wetland resources. The wetland ecosystem can respond swiftly to external changes and plays a critical role in ensuring global ecological security [1, 2]. However, in recent years, due to human activities and the development and utilization of water resources [3], the problem of wetland degradation has become increasingly severe [4, 5]. The wetland area has been shrinking, and its ecological functions have declined, leading to ecosystem fragility, threats to biodiversity, and disruptions in ecological balance [6, 7, 13].

Therefore, conducting a scientific ecological health assessment of Dongying City is of significant importance for protecting the coastal wetland environment and for the scientific planning of land use and development. Implementing measures to protect the

D. Han et al. (Eds.): ICEPG 2024, ESE, pp. 905–923, 2025.
https://doi.org/10.1007/978-3-031-86829-0_101

ecological environment is essential to ensure the health and sustainable development of ecosystems. Consequently, many scholars in China have been paying long-term attention to this issue and have conducted corresponding research [8–12]. Li Dongying used remote sensing technology to study the dynamic changes in major ecological factors in the Yellow River Estuary [14]. Wang Wei [15] and others selected 27 indicators from various perspectives, such as ecological characteristics, functional integration, and social environment, to evaluate the health of the wetland ecosystem, concluding that the Yellow River Delta wetlands are in a fragile state. Niu Mingxiang's [16] research, which focused on pressure, state, and response, showed that the health of the ecosystem in the Yellow River Estuary area had significantly declined in 2013 compared to 1991. Wang Guangzhou [17] evaluated the health and trends of the ecosystem in the Yellow River Estuary and its adjacent marine areas, finding that the ecosystem was generally in a sub-healthy to healthy state. These studies highlight the timeliness and dynamism of the ecosystem's health status, demonstrating continuous improvements in the environment of the Yellow River Estuary and the potential positive impacts of Dongying City's environmental protection policies on ecological health [18]. Given that the health status of the ecosystem changes over time and that its pressure and response states vary across different periods, there is an urgent need for a new ecological health assessment to account for the latest developments in Dongying City's environmental protection policies.

However, in the existing ecological health evaluations, the selection of indicator factors is often derived from literature analysis, which tends to be highly subjective. Additionally, there is a high degree of correlation between the indicators, increasing the workload and leading to inaccurate evaluation results. The issues of strong subjectivity and high correlation among indicators complicate the evaluation process and may result in inaccurate assessments.

In recent years, the Driver-Pressure-State-Impact-Response (DPSIR) model, an improvement of the PSR model, has become an important tool for analyzing and describing the relationship between environmental issues and social development. However, in the process of ecological vulnerability assessment, this model does not sufficiently reflect human subjectivity, neglecting the role of human activities in system evolution. Therefore, researchers have added a "Management" factor to the DPSIR model, forming the DPSIRM model [19]. The DPSIRM model has been successfully applied in evaluating watershed ecological vulnerability [20], habitat safety [21], grassland ecological security [22], water resource safety [23], and water resource carrying capacity [24].

Given the aforementioned background and existing problems, this study combines the specific ecological environment of Dongying City with literature analysis, Pearson correlation analysis, and principal component analysis techniques to select key indicators. Based on the DPSIRM model, the study employs the AHP-CRITIC-entropy method for weight determination to construct an ecological health assessment model specific to the Yellow River Estuary. The model evaluates the ecological health status of Dongying City and identifies the most representative indicators for predicting ecological health levels. The aim of this study is to provide Dongying City with a scientific and rational framework for ecosystem health assessment, promoting the formulation and implementation of ecological protection and sustainable development strategies.

2 Materials and Methods

2.1 Overview of the Study Area

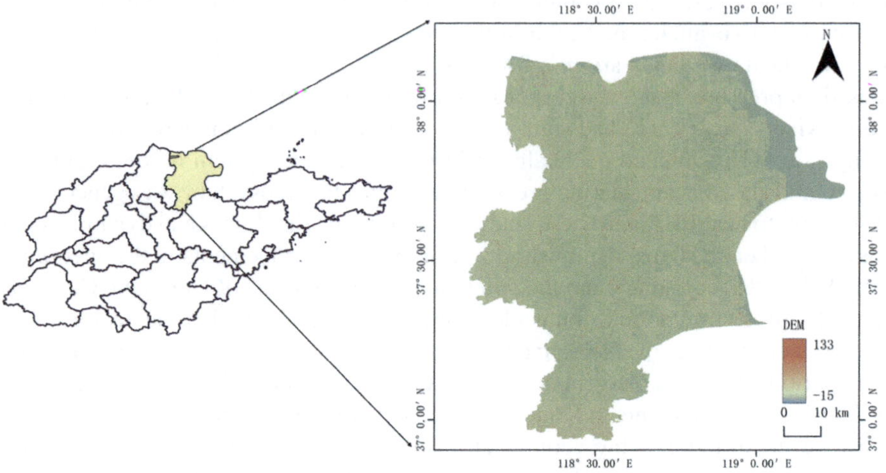

Fig. 1. Map of the Study Area in Dongying City.

Dongying City is located in the northeastern part of Shandong Province, Fig. 1 adjacent to the Bohai Sea and the mouth of the Yellow River. It is the core of the Jiaodong Economic Zone. This city is renowned not only for its rich natural resources and geographical advantages but also as an important base for petrochemical industries and port logistics, while maintaining a solid foundation for agricultural development. However, the rapid industrialization and urbanization of Dongying City have placed unprecedented pressure on its ecological environment, leading to increasingly severe issues such as water pollution, soil degradation, and loss of biodiversity.

Over the past decade, the wetland area in the region has shrunk by approximately 20%, and heavy metal concentrations in the soil in certain areas have exceeded safe limits, primarily including cadmium, chromium, and lead. Concurrently, climate change and human activities have further exacerbated these challenges, resulting in mild to moderate pollution in some water bodies.

To comprehensively assess the ecological health of Dongying City's ecosystem and effectively address related issues, it is essential to conduct an in-depth study of the ecological environment characteristics, ecosystem structure, and functions of the region. This study aims to explore the key factors influencing ecosystem health and their interrelationships. The findings of this research not only provide scientific recommendations for ecological management and sustainable development strategies for Dongying City but also offer valuable references and insights for other regions facing similar challenges.

2.2 Data Sources

The indicator factors summarized by the literature analysis method were subjected to data screening. Statistical data were sourced from various materials such as the Dongying Statistical Yearbook (2010–2021), Yellow River Water Resources Bulletin, Dongying Yearbook, China Marine Ecological Environment Bulletin, and Dongying City National Economic and Social Development Statistical Bulletin. The vegetation cover index research utilized vegetation NDVI data products from the National Aeronautics and Space Administration (NASA) dataset, with batch projection, transformation, and NDVI extraction implemented using the MRT tool in Python. Batch SG filtering and NDVI quality control were conducted using Matlab. The maximum value synthesis method was used to generate NDVI datasets for 2010–2023 based on the maximum NDVI values for each month (January-December) of each year, with a spatial resolution of 250 m. Land use data were sourced from Wuhan University's 30 m land use data, which were cropped, projected, and reclassified into six primary categories: arable land, forest land, grassland, construction land, water bodies, and unused land, using ArcGIS software. Grid-to-area conversion was performed to obtain their respective areas, and the saved TIFF data were calculated in Fragstats.

2.3 Data Preparation and Research Methods

Indicator Selection

In the process of scientific research or policy formulation, it is crucial to select and analyze research indicators correctly to derive valuable conclusions. In existing ecological health assessments, the selection of indicator factors is often based on literature analysis, which is subjective, and the indicators are highly correlated, leading to increased workload, inaccurate assessment results, and difficulty in assessment. This study extensively collected 255 potential indicator factors through literature analysis, covering various aspects such as environmental health, socio-economics, and ecological balance. In the preliminary screening process, considering the specific ecological environmental conditions of Dongying City and the accessibility of data in practical operation, the indicators were screened, and ultimately 60 indicator factors closely related to the ecological health status of Dongying City were determined. Due to the high correlation between indicators, which may affect the research results, in order to ensure that the selected indicators accurately reflect the ecological health status of the Yellow River Estuary and provide strong data support for subsequent analysis, Pearson correlation analysis was conducted on these 60 indicator factors [25]. Some of the results are shown in Table 1:

Table 1. Partial Results of Pearson Correlation Analysis.

		GDP	Per capita GDP	proportion of the tertiary industry	general public budget revenue	per capita fixed asset investment	average wage	disposable income of urban residents	agricultural output value	educational investment
GDP	Pearson correlation	1	.999**	.986**	.940**	−0.321	.972**	.986**	.843**	.918**
	Sig.		0.000	0.000	0.000	0.310	0.000	0.000	0.001	0.000
Per capita GDP	Pearson correlation	.999**	1	.980**	.942**	−0.306	.963**	.981**	.855**	.920**
	Sig.	0.000		0.000	0.000	0.334	0.000	0.000	0.000	0.000
proportion of the tertiary industry	Pearson correlation	.986**	.980**	1	.939**	−0.339	.993**	.998**	.800**	.908**
	Sig.	0.000	0.000		0.000	0.281	0.000	0.000	0.002	0.000
general public budget revenue	Pearson correlation	.940**	.942**	.939**	1	−0.062	.906**	.956**	.907**	.934**
	Sig.	0.000	0.000	0.000		0.847	0.000	0.000	0.000	0.000
per capita fixed asset investment	Pearson correlation	−0.321	−0.306	−0.339	−0.062	1	−0.407	−0.301	0.037	−0.225
	Sig.	0.310	0.334	0.281	0.847		0.189	0.342	0.909	0.482
average wage	Pearson correlation	.972**	.963**	.993**	.906**	−0.407	1	.987**	.761**	.888**
	Sig.	0.000	0.000	0.000	0.000	0.189		0.000	0.004	0.000

According to the chart, for example, the Sig. (two-tailed) value of GDP and per capita GDP is less than 0.001, thus removing an indicator with a relatively small contribution to the ecological health of the Yellow River Estuary. This analysis aims to assess the relationship between various indicators and identify groups of indicators that are strongly correlated with each other. Through this analysis, indicators that are highly correlated and have a Sig. (two-tailed) value less than 0.001 are selected, thereby identifying key indicators.

To further refine the set of indicators and ensure that the selected indicators effectively reflect the core elements of the ecological health of the Yellow River Estuary, we used Principal Component Analysis (PCA). Principal Component Analysis [26] is a statistical technique that simplifies the data structure by transforming potentially correlated variables into a set of linearly uncorrelated variables, known as principal components. In the Principal Component Analysis process, the number of indicators is required to be less than the number of samples, so we classified the indicator factors into positive and negative indicators and analyzed them separately. The principal components with eigenvalues greater than 1 were extracted, and their component matrices and rotated component matrices are shown in Tables 2, 3, and 4:

Table 2. Rotated Component Matrix of Principal Component Analysis for Negative Indicators.

	component	
	1	2
Human Interference Index	−.839	−.059
Population Growth Rate	.801	.511
Total Tourism Revenue	−.172	.965
Population Density	−.793	.488
Cultivation Area	.893	−.014
Patch Density	−.048	.915
Annual Average Nitrogen Dioxide	.870	−.197
Annual Average Sulfur Dioxide	.757	−.577
pm2.5	.839	−.440
Economically Active Population	.833	−.342

As shown in the figure, for negative indicators, they are divided into two principal components. The first principal component extracts the human interference index, population growth rate, economic activity population, aquaculture area, annual average sulfur dioxide, annual average nitrogen dioxide, and PM2.5. The second principal component extracts total tourism revenue, port cargo throughput, and patch density.

Table 3. Rotated Component Matrix of Principal Component Analysis for Positive Indicators.

	component		
	1	2	3
Sediment Transport Volume	.165	.964	.041
River Discharge into the Sea	.255	.867	.333
NPP	.756	.475	.166
NDVI	−.064	.166	.604
Proportion of Days with Good Air Quality	.912	.273	.013
Number of Students	.101	−.118	.937
GDP	.923	.310	.125
Agricultural Output Value	.976	−.015	−.085
Green Coverage Rate in Built-up Areas	.809	−.220	−.415
Water Supplement from the Yellow River	−.152	−.467	−.738
Environmental Protection Investment	.652	.456	−.427
Afforestation Area	−.790	−.243	−.325
Education Investment	.921	.168	.133

As shown in the figure, four principal components were extracted for positive indicators, and all indicators can be retained.

Based on the above method, a total of 23 indicator factors closely related to the ecological health status of the Yellow River Estuary were finally determined. This not only enhances the accuracy and effectiveness of our research but also provides a scientific and reasonable indicator system for the assessment and monitoring of the ecological health status of the Yellow River Estuary. The determination of these indicators provides a solid foundation for further in-depth analysis of the ecological health status of the Yellow River Estuary and also provides strong data support for relevant policy formulation.

Evaluation Indicator System Construction

(1) Evaluation Model DPSIRM Model:

The DPSIRM model is used as a framework, following the principles of systematicness, representativeness, quantifiability, and accessibility. It consists of six subsystems: Driving force, Pressure, State, Impact, Response, and Management. Based on the ecological environment of the Yellow River Estuary and following the principles of scientificity, systematicness, and standardization in the construction of indicator systems, a total of 23 indicators were selected to construct the evaluation indicator system for the ecological health of the Yellow River Estuary, as shown in Table 4.

Table 4. Ecosystem Health Evaluation Indicator System for Yellow River Delta.

Goal Level	Criterion Level	Indicator Level	Indicator Attributes	Comprehensive Weight
Ecosystem Health Assessment	Driving Force	Human Interference Index	–	0.0536
		Population Growth Rate	–	0.0318
		Total Tourism Revenue	–	0.0258
		Population Density	–	0.0363
	Pressure	Cultivation Area	–	0.03
		Sediment Transport Volume	+	0.044
		River Discharge into the Sea	+	0.0648
	State	NPP	+	0.0568
		NDVI	+	0.0585
		Proportion of Days with Good Air Quality	+	0.057
	Impact	Patch Density	–	0.0633
		Annual Average Nitrogen Dioxide	–	0.0296
		Annual Average Sulfur Dioxide	–	0.0338
		pm2.5	–	0.0386
	Response	Economically Active Population	+	0.0314
		Number of Students	+	0.0287
		GDP	+	0.0547
		Agricultural Output Value	+	0.0397

(*continued*)

Table 4. (*continued*)

Goal Level	Criterion Level	Indicator Level	Indicator Attributes	Comprehensive Weight
		Green Coverage Rate in Built-up Areas	+	0.0506
	Management	Water Supplement from the Yellow River	+	0.0383
		Environmental Protection Investment	+	0.0537
		Afforestation Area	+	0.0395
		Education Investment	+	0.0395

Weight Determination

Existing methods for assigning weights to indicators have limitations. The CRITIC method [27] is an objective weighting method that considers the influence of indicator correlations but cannot reflect the degree of dispersion of the indicators. In contrast, the entropy weighting method can determine weights based on the dispersion between indicators. Therefore, by combining the CRITIC method with the entropy weighting method, which has a strong theoretical basis, the combined method can overcome the limitations of considering only the influence of subjective factors. Furthermore, the Analytic Hierarchy Process (AHP) is integrated to further improve the weight determination method. The final weights are determined by synthesizing the three methods. This approach effectively overcomes the limitations of assigning weights to indicators based solely on subjective ideas [29]. The steps are as follows:

Indicator Standardization Use the Min-max normalization method [28] to normalize and standardize each evaluation indicator. The processed indicator values range from 0 to 1, with values closer to 1 indicating stronger health and values closer to 0 indicating weaker health.

Positive Indicators: $r_{ij} = \frac{(x_{ij} - x_{min})}{(x_{max} - x_{min})}$

Negative Indicators: $r_{ij} = \frac{(x_{max} - x_{ij})}{(x_{max} - x_{min})}$

In the formula, r_{ij} represents the standardized value, x_{ij} represents the actual value of the indicator, and x_{max} and x_{min} represent the maximum and minimum values of the indicator in the study area, respectively.

Calculation of the entropy for ej the j-th indicator according to the entropy weighting method is given by:

$$e_j = -h\sum_{i=1}^{m} f_{ij}\ln(f_{ij}) \quad (1)$$

$$f_{ij} = \frac{c_{ij}}{\sum_{i=1}^{m} c_{ij}} \quad h = 1/\ln m \quad (2)$$

In the formula, c_{ij} represents the standardized value, and m represents the number of indicators.

3. The formula to calculate the standard deviation of CRITIC is:

$$\delta_j = \sqrt{1/N \sum_{i=1}^{n} (f_i - \mu)^2} \tag{3}$$

δ_j is the standard deviation, f_i is the value of indicator j, μ is the arithmetic mean of f_i, N is the total number of f_i.

4. Calculate the weights of the indicator layer.

$$p_j = (e_j + \delta_j) \sum_{i=1}^{n} (1 - \rho_{ij}) \; w_{j2} = p_j / \sum_{j=1}^{n} p_j \tag{4}$$

p_j is the information content of the j indicator, ρ_{ij} is the correlation coefficient obtained from Pearson correlation analysis, and w_{j2} is the weight of the j indicator.

5. Combination weighting

$$w_z = (1 - \alpha) w_{j1} - w_{j2} \tag{5}$$

w_{j1} is the indicator weight obtained from the Analytic Hierarchy Process (AHP), w_{j2} is the indicator weight obtained from the CRITIC-entropy method.

$$\alpha = \frac{n}{n-1} [\frac{2}{n} (w_{q1} + 2w_{q2} + \ldots nw_{qn}) - \frac{n+1}{n}] \tag{6}$$

$w_{q1}, w_{q2} \ldots w_{qn}$ is the weights of the evaluation indicators determined by the Analytic Hierarchy Process (AHP), rearranged in ascending order;n is the number of evaluation indicators in the AHP.

The Analytic Hierarchy Process (AHP) analysis is conducted using the Yaahp software. A scale of 1 to 9 is used to pairwise compare the importance of elements, which helps construct the element layer judgment matrix. Consistency tests are then performed. Finally, the entropy weighting method is applied to obtain the weights of various indicators.

Ecological System Health Assessment Method

In this study, the expert scoring method is used, and relevant research results are referred to construct the judgment matrix. The Comprehensive Ecological Health Evaluation Index (CEI) of the Yellow River Estuary is constructed through weighted summation [30], and the ecological system health assessment results are obtained.

$$CEI = \sum_{i=1}^{n} E_i \times W_i \tag{7}$$

In the formula, CEI represents the Comprehensive Ecological Health Evaluation Index, with values ranging from 0 to 1.nis the number of evaluation indicators,, Wi is the weight of the i indicator, E i is the standardized value of the i indicator.

The classification of ecological system health status does not have a unified or recognized standard. In this study, reference is made to relevant literature [31], and the ecological health status of the Yellow River Estuary is divided into five levels (Table 5).

Table 5. Ecosystem Health Evaluation Grades for Yellow River Delta.

Evaluation Index	(0, 0.2]	(0.2, 0.4]	(0.4, 0.6]	(0.6, 0.8]	(0.8, 1]
Health Status	Pathological	Unhealthy	Subhealthy	Healthy	Very Healthy

3 Results and Analysis

3.1 Driving Force Analysis

Driving forces are the primary drivers of ecological vulnerability, divided into natural factor driving forces and human activity driving forces. In general, natural and anthropogenic factors are in balance and do not lead to the unhealthy development of the ecosystem. The study focuses on human activity factors, including unsustainable development, uncontrolled population growth, resource extraction, and environmental pollution. These activities not only disrupt the existing ecosystem structure but also increase the burden on ecological carrying capacity, leading to the degradation of regional ecosystems in the short term. The selected driving force indicators and weights are shown in Table 6:

Table 6. Weights of Driving Force Indicators for the Yellow River Estuary.

Criterion Level	Indicator Level	Weight
Driving Force	Human Interference Index	0.363
	Population Growth Rate	0.216
	Total Tourism Revenue	0.1749
	Population Density	0.2461

From the perspective of the pressure index, the Human Interference Index has a significant weight in the driving forces. This index is calculated as the proportion of the sum of arable land area and urban land area to the total area, revealing the impact of human activities on the environment. The data shows that this index initially decreased and then fluctuated, reflecting that urban expansion has led to a continuous reduction in arable land area, causing environmental degradation.

Data from 2010 to 2022 shows that the driving force index in the study area fluctuated between 0.1983 and 0.7736, showing a "high-low" trend. In recent years, with population growth and intensified human development activities, especially the rise of tourism, although it has promoted stable growth in per capita GDP, it has also significantly increased pressure on the ecosystem. The ecological driving forces in the Yellow River Estuary are in a sub-healthy or even unhealthy state.

3.2 Pressure Analysis

Pressure is the manifestation of driving forces and is the direct factor affecting the ecosystem, including the extent of the impact of both natural and human activities. In areas with high population density and rapid economic development, the cultivation of forests, grasslands, and agricultural activities, as well as urban construction, reduce the service value and carrying capacity of the ecosystem. This exacerbates negative succession processes and increases unhealthy conditions. The selected indicators and their weights are shown in Table 7:

Table 7. Weight of State Indicators for Yellow River Delta.

Criterion Level	Indicator Level	Weight
Pressure	Cultivation Area	0.216
	Sediment Transport Volume	0.317
	River Discharge into the Sea	0.467

Within the pressure indicators, the contribution of the river runoff into the sea to the assessment results is the highest, at 0.403. The Yellow River annually transports a substantial amount of freshwater, sediment, and nutrients to the Bohai Sea, providing an excellent environment for the growth and reproduction of marine life in the estuary and its adjacent waters. The runoff into the sea from the Yellow River shows a "high-low-high" trend, with the lowest value recorded in 2016 at 74.88 cubic meters, and the highest value in 2021 at 434.1 cubic meters. Additionally, as sediment is a crucial adsorption carrier for pollutants, fluctuations in sediment transport inevitably lead to changes in water quality, thereby affecting the ecological health of the Yellow River estuary and adjacent sea areas.

Data from 2010 to 2022 shows that the pressure index in the study area fluctuated between 0.00975 and 0.9049. The decrease in aquaculture area, the increase in sediment transport, and the peak value of river runoff into the sea in 2021 made the pressure health level reach its highest, while 2017 had the lowest health level with a sediment transport of only 0.07 cubic meters. The increase in the pressure index indicates that the ecological pressure in the Yellow River estuary area has decreased, gradually shifting to a healthier state.

3.3 Status Analysis

The status reflects the actual performance or changes in the natural environment system under the combined effects of driving forces and pressure. Additionally, the structure, function, and value of the ecosystem also reflect the system's status. The better the natural environment status under the coupling of driving forces and pressure, the higher the degree of ecological health. The selected response indicators and weights are shown in Table 8:

Table 8. Weights of State Indicators for the Yellow River Estuary.

Criterion Level	Indicator Level	Weight
State	Net Primary Productivity (NPP)	0.3297
	Normalized Difference Vegetation Index (NDVI)	0.3395
	Proportion of Days with Good Air Quality	0.3308

In the status indicators, we selected the Normalized Difference Vegetation Index (NDVI), Net Primary Productivity (NPP), and the proportion of days with good air quality. These indicators collectively reflect the health status of the ecosystem. An increase in NDVI and NPP signifies an improvement in ecosystem productivity and stability, while a higher proportion of days with good air quality indicates a healthier atmospheric ecosystem. The weights of these three indicators are relatively similar and jointly influence the overall state of the ecosystem.

From 2010 to 2022, data show that the status index in the study area fluctuated between 0.1771 and 0.9946. The status index was at its lowest value in 2010 and reached its highest value in 2022, showing an overall upward trend. This trend indicates that Dongying City has been committed to environmental protection and has continuously improved the health level of the ecosystem through a series of measures. For example, the implementation of vegetation restoration projects and air quality management policies has significantly increased the NDVI and the proportion of days with good air quality. The improvement in NPP indicates that the biological productivity of the ecosystem has gradually strengthened.

3.4 Impact Analysis

Impact factors are the results of the interaction between driving forces and pressure factors on status factors, reflected through both natural elements and socio-economic elements (Table 9).

Table 9. Weights of Impact Indicators for the Yellow River Estuary.

Criterion Level	Indicator Level	Weight
Impact	Patch Density	0.3829
	Annual Average Nitrogen Dioxide	0.1791
	Annual Average Sulfur Dioxide	0.2045
	pm2.5	0.2335

Among all status indicators, patch density contributes the most to the evaluation results, highlighting the importance of terrestrial productivity and ecosystem complexity. An increase in patch density indicates improvements and stabilization in the ecological environment, reflecting the degree of landscape fragmentation, which refers to the number and distribution of different types of patches within an area. Higher patch density typically implies greater ecological fragmentation, which negatively affects the ecosystem's status. Additionally, sulfur dioxide (SO_2), nitrogen dioxide (NO_2), and particulate matter (PM2.5) are three common air pollutants that can affect the local ecological environment through vegetation, rainfall, and other pathways.

Data from 2010 to 2022 show that the status index in the study area fluctuated between 0.2717 and 0.7260. The lowest value appeared in 2016, and the highest value was in 2021, presenting an overall "high-low-high" trend. This fluctuation is mainly due to the rapid development of the industrial economy, which emitted a large amount of pollutants, causing damage to the ecological environment. However, with industrial transformation, economic system reform, and the implementation of the "Dongying City Blue Sky Defense Action Plan" by the Dongying Municipal Government, the air quality has significantly improved, and pollutant emissions have decreased. This further demonstrates the continuous improvement in the natural environment of the Yellow River estuary.

3.5 Response Analysis

Response refers to the proactive measures taken by human activities to influence the development of the ecosystem, aimed at preventing, mitigating, or eliminating negative impacts. For example, to reduce pollution of the natural environment, humans can actively change development methods and adjust economic and industrial structures. For severely degraded ecosystems, technical means can be used for restoration and management, guiding the ecosystem towards positive succession. However, the response phase is usually reactive and targeted, with a certain degree of lag (Table 10).

Table 10. Weights of Indicators for the Yellow River Estuary.

Criterion Level	Indicator Level	Weight
Response	Economically Active Population	0.1531
	Number of Students	0.1399
	GDP	0.2667
	Agricultural Output Value	0.1936
	Green Coverage Rate in Built-up Areas	0.2467

Within the response indicators, GDP holds the largest weight, reflecting the close relationship between economic development and environmental improvement: economic growth promotes investment in ecological system enhancement. Additionally, the expansion of urban green areas has further driven improvements in the ecological environment, highlighting the impact and response of human activities on ecological health.

Data from 2010 to 2022 show that the response index in the study area fluctuated between 0.2356 and 0.9948, indicating a steady upward trend. In recent years, Dongying

City has actively responded to national guidance on the "three rural issues" (agriculture, rural areas, and farmers), advancing agricultural development strategies. For three consecutive years, grain production has been on the rise, with the total grain output surpassing 2.9 billion kilograms in 2022, leading to a steady increase in agricultural value. Agricultural output, to some extent, reflects the conditions of soil, water quality, biodiversity, and climate, and its continuous improvement is a key factor in the gradual rise of the ecological response index. Moreover, as social development has deepened, especially with the promotion of education and the increasing number of students, public awareness of environmental protection and understanding of ecological regulations have significantly increased. Education has enhanced the public's environmental consciousness, encouraging more people to participate in environmental protection actions, thereby promoting the continuous improvement of the ecological environment.

3.6 Management Analysis

Management is proposed based on response factors and represents the elevation and sublimation of human society's understanding of the development of the ecological vulnerability system. Similar to response factors, management factors refer to a series of measures taken by human society to improve ecological vulnerability. However, management differs in that it controls and improves ecological vulnerability throughout the entire process from the source and fundamentally, providing overall guidance through the formation of development concepts and management systems. Management involves conducting vulnerability assessments at various stages, making precise response measures, and comprehensively enhancing the ecosystem's ability to recover and stabilize (Table 11).

Table 11. Weights of Management Indicators for the Yellow River Estuary.

Criterion Level	Indicator Level	Weight
Management	Water Supplement from the Yellow River	0.2239
	Environmental Protection Investment	0.314
	Afforestation Area	0.231
	Education Investment	0.2311

Within the management index, afforestation area has the largest weight, reflecting the direct impact of reforestation projects on environmental quality. These projects contribute to environmental improvement by absorbing carbon dioxide, filtering harmful substances and dust from the air, showcasing efforts to combat environmental degradation.

Data from 2010 to 2022 indicate that the management index in the study area fluctuated between 0.2315 and 0.6645. Increased environmental investments and growth in educational expenditures illustrate the proactive measures Dongying City has undertaken for environmental health. The Yellow River water diversion alleviated water scarcity, improved wetland ecosystems, enhanced biodiversity, regulated regional climate, prevented land salinization, and supported agricultural irrigation. Additionally, through the Yellow River water diversion, Dongying City achieved comprehensive benefits such as flood prevention and disaster reduction, soil structure improvement, landscape value enhancement, promotion of ecological restoration projects, and support for sustainable economic and social development. As environmental issues become increasingly severe, Dongying City has increased funding for environmental projects, covering aspects such as air quality improvement, water pollution control, and solid waste treatment. These investments not only improved environmental quality but also provided a healthier living environment for urban residents. Moreover, by increasing educational funding, Dongying City has improved educational infrastructure, enhanced teaching quality, and strengthened environmental education. By disseminating environmental protection knowledge in schools and communities, public environmental awareness has been raised, encouraging more people to actively participate in environmental protection actions.

3.7 Comprehensive Analysis of Ecological Health

Based on the analysis of regional characteristics and relevant literature, the weights for the driving force, pressure, state, impact, response, and management are 0.1475, 0.1388, 0.1723, 0.1653, 0.2051, and 0.171, respectively. The comprehensive weights of each indicator are listed in Table 5. Using the comprehensive index model for health evaluation and based on the aforementioned analysis and weight values, the comprehensive health index for the Yellow River estuary is obtained as follows (Table 12):

Table 12. Ecosystem Health Index for Yellow River Delta.

	Driving Force	Pressure	State	Impact	Response	Management	Ecosystem Health Assessment
2010	0.773612551	0.359471017	0.177131022	0.610101196	0.235636769	0.231526513	0.383291966
2011	0.732497587	0.308736806	0.450479327	0.62458527	0.280363515	0.311766214	0.442572176
2012	0.6914705	0.538696523	0.582396609	0.557761948	0.272615054	0.448681236	0.501945801
2013	0.755267955	0.458311196	0.483948663	0.307826625	0.35560244	0.490460443	0.466086509
2014	0.672135867	0.130891089	0.311791789	0.291790442	0.497126697	0.538353757	0.413281587
2015	0.688322767	0.151477879	0.391101104	0.301063795	0.545558947	0.526560073	0.441641216
2016	0.604439325	0.072399354	0.42644517	0.271732614	0.497012377	0.544617734	0.412664606
2017	0.506100516	0.009750292	0.723234867	0.420399496	0.4749569	0.628734131	0.475035768
2018	0.439885592	0.785460628	0.664954777	0.533579181	0.553307702	0.65232653	0.601708653
2019	0.469892663	0.722099916	0.517608593	0.46569261	0.586073076	0.519484428	0.54473501
2020	0.267614818	0.836260699	0.737545357	0.644277676	0.630042655	0.664489703	0.631973823
2021	0.19831531	0.904914054	0.976806156	0.726034906	0.832640311	0.451449805	0.691143294
2022	0.417406426	0.579230899	0.994587921	0.700906191	0.994760354	0.435830078	0.707744281

Between 2010 and 2022, the ecological health index of the Yellow River estuary improved from 0.3829 to 0.7077, indicating a transition from a sub-healthy to a healthy state (Fig. 2). The overall trend shows a consistent rise in the ecological health index, with a significant milestone achieved in 2020 when the index reached 0.63197, signifying that the estuary had entered a healthy state.

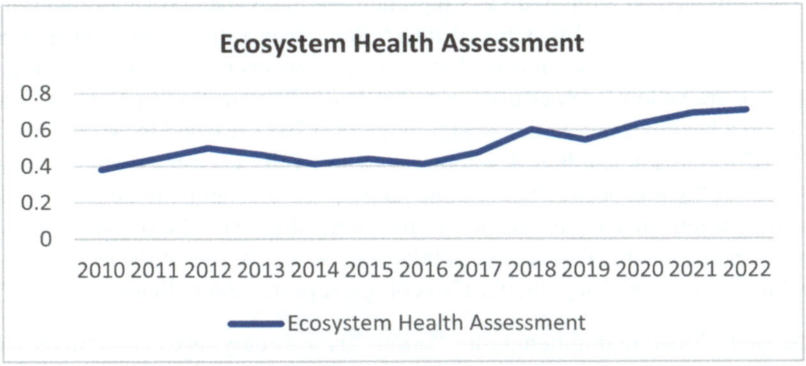

Fig. 2. Trend Chart of Ecological Health Index.

In recent years, Dongying City has focused on the goals of ecological protection and high-quality development in the Yellow River Basin, comprehensively promoting critical tasks such as structural adjustment, pollution control, ecological protection, and climate change mitigation. The city is committed to accelerating the improvement of ecological environment quality and enhancing levels of ecological protection and green development. As part of the national strategy for ecological protection and high-quality development in the Yellow River Basin, Dongying City is implementing the construction of an ecological corridor along the Yellow River. This project aims to promote cultural tourism along the Yellow River, align with the construction of international wetland cities, and serve as a key project for ecological revitalization along the Yellow River. Specifically, ecological construction projects on both sides of the Yellow River, focusing on greening enhancements and wetland protection and restoration, have successfully established Dongying City's ecological corridor along the Yellow River. This not only contributes to the formation of the ecological "backbone" of the Yellow River Delta but also provides an important model for the ecological protection and construction of the lower reaches of the Yellow River, forming an overall framework for river basin ecological protection. Therefore, it is anticipated that the ecosystem of Dongying City will remain stable and continue to develop positively, laying a solid foundation for Dongying to become a distinctive ecological security barrier in the Yellow River Basin.

4 Conclusion

(1) This study conducted an in-depth exploration of key indicators affecting ecosystem health through a literature review. Taking into account the actual situation and data availability in Dongying City, 58 preliminary indicators were selected. Through

Pearson correlation analysis and principal component analysis coupling techniques, the indicators were further screened, ultimately identifying 23 core indicators that have significant impacts on the health of the Yellow River estuary ecosystem and exhibit low correlation. This refined screening process aims to improve the accuracy and practicality of the evaluation model.

(2) Based on the DPSIRM model and incorporating an improved AHP-CRITIC-entropy weighting method along with a comprehensive index method, an evaluation model for the ecological health status of the Yellow River estuary was constructed. The application results of this model show a steady improvement in the ecological health index of the Yellow River estuary from 2010 to 2022, transitioning from a sub-healthy state to a healthy state, with a health index of 0.7077 reached in 2022. This reflects a trend towards a healthier development of the Yellow River estuary ecosystem. It is noteworthy that despite fluctuations in pressure indicators in the early stages, the positive trends in state and response indicators align with the overall improvement in ecological health. This further validates the effectiveness of the evaluation model and the success of Dongying City's ecological protection policies.

The study found that although the Yellow River estuary faces continuously changing ecological pressures, effective environmental policies and management measures, as well as increased societal awareness of ecological protection, have led to positive trends in ecological state and response measures. Particularly, the sustainable development efforts of Dongying City and the Yellow River estuary nature reserve have made significant contributions to improving and protecting the ecological environment. This study not only provides Dongying City with a scientific tool for evaluating and monitoring ecological health but also offers valuable experience and reference for ecosystem management and protection efforts in other regions. In the future, continuous monitoring, evaluation, and corresponding environmental protection measures will be key to ensuring the sustained health of the ecosystem in Dongying City and the surrounding areas.

References

1. Chapman, V.J.: Wet Coastal Ecosystems. Scientific Pub.Co, Amsterdam (1977)
2. Ancog, R., Ruzol, C.: Urbanization adjacent to a wetland of international importance: the case of Olango Island Wildlife Sanctuary, Metro Cebu, Philippines. Habitat Int. **49**, 325–332 (2015)
3. Gu, R., Duo, L., Zou, Z., et al.: Spatial and temporal evolution characteristics of ecosystem health in urban agglomerations in the lower yellow river. Bull. Soil Water Conserv. **42**(06), 285–292 (2022). https://doi.org/10.13961/j.cnki.stbctb.20221114.001
4. Wang, X., Zhi, X., Wan, P., et al.: Analysis of wetland utilization and influencing factors in the yellow river delta. People's Yellow River **10**, 10–11 (2004)
5. Jiang, Z., Shi, B., Cheng, Y., et al.: Characteristics of landscape changes in the modern yellow river delta wetlands. People's Yellow River (11), 12–13+46 (2008)
6. Zhang, X., Li, S.: Monitoring and analysis of ecological environment in the nearshore area of the yellow river estuary from 2015 to 2020. Bull. Soil Water Conserv. **42**(03), 139–147 (2022). https://doi.org/10.13961/j.cnki.stbctb.2022.03.019

technology (Air sparging (AS) [4], in-situ thermal desorption (ISTD) [5], etc.), chemical remediation technology (Permeable reactive barrier [6], in-situ chemical oxidation technology [7, 8], etc.), and biological remediation technology (microbial remediation [9, 10], phytoremediation [11–13], etc.) [14] and the combined use of multiple technologies. P&T is now one of the most widely used ex-situ remediation technologies, which reduces the concentration of pollutants and effectively controls the scope of the pollution plume by extracting contaminated groundwater for treatment at the surface, thus achieving the purpose of removing groundwater pollution at low cost and high efficiency [15, 16].

Benzene, toluene and 1, 2-dichloroethane (1, 2-DCA) are widely used as organic solvents in industrial production, but inevitable leakage during transportation, storage and other processes seriously jeopardizes the environment and human health and safety [17, 18]. Benzene and toluene have the characteristics of easy dissolution, volatility and strong migration ability. Within a short time of leakage, they can penetrate into the soil and groundwater and migrate and spread, both of which can damage liver and kidney functions and paralyze the central nervous system even at low concentrations [19, 20]. 1,2-Dichloroethane is a highly toxic volatile organic pollutant that is easily carcinogenic, teratogenic and mutagenic and has been listed as one of the pollutants to be prioritized for control in China's water environment. Its high solubility and difficult degradation make it difficult for it to be adsorbed and degraded after contaminating the groundwater, and the formation of a source of contamination continues to damage the groundwater environment [17, 20].

This study was based on the former site of a pesticide factory in Jiangsu Province, China. The previous site pollution investigation showed that the content of benzene, toluene and 1,2-DCA in the soil and groundwater of this site exceeded the Class IV water in the "Standard for groundwater quality (GB/T 14848–2017) [21]" seriously. An area with serious groundwater pollution located in the west of the site was selected for experiments to study the remediation effect of P&T on groundwater contaminated with benzene, toluene and 1,2-DCA. The paper can provide some scientific basis for the remediation of organic pollution in relevant contaminated sites.

2 Material and Methods

2.1 Overview of the Study Area

The study area was located in the west of the former site of a pesticide factory in Jiangsu Province, China, on the alluvial plain of the Yangtze River Delta and the south bank of the Yangtze River estuary. The terrain is mainly plain, flat and slightly sloping from northeast to southwest. The study area was the key pollution area of the plant, the thickness of the aquifer is 5.9–6.2 m and the lithology is clayey silt. The groundwater flow direction is south-east-north-west, the hydraulic gradient is 2‰ and the groundwater flow rate is very slow.

Based on the previous environmental investigation, benzene, toluene and 1,2-DCA were detected to varying degrees, with the highest concentrations at 6.56 mg/L, 23.5 mg/L and 23.7 mg/L.

Study on the Demonstration Effect of Pump-And-Treat to Repair Volatile Organic Compounds

Xue Xia[1,2,5], Zhihui Qu[1,2,5](✉), Biao Yu[1,2], Kaixuan Zheng[3], Yan Su[4], Shengdong Zhang[1,2], and Jiayu Xiao[1,2]

[1] Institute of Disaster Prevention, Sanhe 065201, China
qzh-19831001@163.com
[2] Hebei Key Laboratory of Resource and Environmental Disaster Mechanism and Risk Monitoring, Sanhe 065201, China
[3] State Environmental Protection Key Laboratory of Environmental Pollution Health Risk Assessment, South China Institute of Environmental Science, Ministry of Ecological Environment, Guangzhou 510655, People's Republic of China
[4] Shenyang Academy of Environmental Sciences, Shenyang, Liaoning 110000, China
[5] Beijing Disaster Prevention Science and Technology Co., Ltd., Beijing, China

Abstract. In the process of the increasingly prominent problem of groundwater remediation in organic contaminated sites, this study conducted pump-and-treat experiment on the basis of an former pesticide factory site in Jiangsu Province, China, to rehabilitate groundwater contaminated by benzene, toluene and 1,2-dichloroethane. Groundwater reached steady state when pumping at 29 L/h; Within the radius of influence of the pumping, concentrations of benzene, toluene and 1,2-dichloroethane in groundwater were reduced to fell below the Class IV remediation target values in the "Standard for groundwater quality". The study showed that pump-and-treat had good effects on the remediation of groundwater at sites contaminated with benzene, toluene and 1,2-dichloroethane and provided a basis for pump-and-treat to remediate groundwater at relevant contaminated sites.

Keywords: Pump-and-treat · volatile organic compounds · groundwater remediation

1 Introduction

With the accelerated industrialization of society, the problem of groundwater pollution has become prominent in China, and many old urban factories and chemical enterprises have relocated or ceased production, with a large number of pollutants left behind or leaking from industrial sites during the producing process, causing serious pollution of soil and groundwater [1, 2]. The Chinese government has made some regulations for managing contaminated sites, environmental investigation or remediation is required before the change of use of the polluted sites [3]. At present, the commonly used groundwater pollution control and remediation technologies mainly include physical remediation

D. Han et al. (Eds.): ICEPG 2024, ESE, pp. 924–931, 2025.
https://doi.org/10.1007/978-3-031-86829-0_102

7. Wolters, M.L., Sun, Z.C., Huang, C., Kuenzer, C.: Environmental awareness and vulnerability in the yellow river delta: results based on a comprehensive household survey. Ocean Coast. Manag. **120**, 1–10 (2016)
8. Mackintosh, T.J., Davis, J.A., Thompson, R.M.: The influence of urbanisation on macroinvertebrate biodiversity in constructed stormwater wetlands. Sci. Total. Environ. **536**, 527–537 (2015)
9. Wang, Q., Zhang, H., Zhang, T., et al.: Research on geological and environmental assessment of wetland resources in the yellow river delta and its conservation measures. Shandong Land Resour. **34**(5), 81–86 (2018)
10. Zhao, X., Dong, P., Feng, K., et al.: Evaluation of the ecological health of Zhaolanxin River and improvement strategies. J. Environ. Eng. Technol. **14**(03), 856–866 (2024)
11. Wang, M., Xu, Z., Yin, X.: Study on fish diversity and aquatic ecosystem health assessment in Dian Nong River Basin, Yinchuan City. Yellow River **46**(06), 85–89+102 (2024)
12. Yi, F.: Ecological vulnerability assessment of the yellow river delta wetland. [Dissertation]. China University of Geosciences, Beijing (2011)
13. Wang, J., Hu, P., Gong, J.: Implementing the great protection of the yellow river estuary promotes the construction of ecological civilization in the yellow river basin. People's Yellow River **41**(10), 7–10 (2019)
14. Li, D.: Study on the dynamic changes of major ecological factors in the yellow river estuary based on remote sensing. Shanghai Ocean University, Shanghai (2011)
15. Wang, W., Chen, W., Li, Q., et al.: Evaluation index system of wetland ecological system health in the yellow river delta. Water Resour. Protect. **28**(1), 13–16 (2021)
16. Niu, M., Wang, J., Xu, B.: Ecological system health assessment of the yellow river estuary based on PSR. Acta Ecol. Sin. **27**(3), 943–952 (2017)
17. Stuary and adjacent sea area based on DPSIR. Yellow River **45**(6), 92–97, 162 (2023)
18. Zhang, C., Wang, C., Meng, Q., et al.: Analysis of the health status and biodiversity of typical marine ecosystems in China. Mar. Environ. Sci. **41**(03), 423–429 (2022). https://doi.org/10.13634/j.cnki.mes.2022.03.013
19. Zhang, F.: Ecological vulnerability assessment of the Yinma river basin based on SD and DPSIRM models. Northeast Normal University (2019)
20. Xiang, L., Zhou, W., Ren, J., et al.: Ecological security assessment of plateau city wetlands based on DPSIRM model—taking the Xining section of the Huangshui river basin as an example. J. Ecol. **41**(10), 2064–2071 (2022). https://doi.org/10.13292/j.1000-4890.202209.003
21. Lv, T., Qiu, R., Zhao, Q., et al.: Temporal and spatial characteristics of ecological security and obstacle factors of urban agglomerations in the middle reaches of the Yangtze river based on DPSIRM model. Res. Soil Water Conserv. **31**(02), 379–388 (2024). https://doi.org/10.13869/j.cnki.rswc.2024.02.017
22. Du, H.: Evaluation and predictive analysis of grassland ecological security in Gannan prefecture. [Dissertation]. Beijing Forestry University (2022). https://doi.org/10.26949/d.cnki.gblyu.2022.000459
23. Xu, S., Chen, G., Zhang, Y., et al.: Water Environment security evaluation of hebei province based on DPSIRM model and improved TOPSIS method. Water Resour. Power **40**(05), 50–54 (2022)
24. Wang, J., Zhan, W.: Research on water resources carrying capacity in Guangxi based on DPSIRM model. People's Yellow River **45**(S2), 61–63 (2023)
25. Chen, M., Huang, W.: Analysis of meteorological factors affecting flood losses in Sichuan province. Northeast Water Resour. Hydropower **40**(12), 46–50+72 (2022). https://doi.org/10.14124/j.cnki.dbslsd22-1097.2022.12.011
26. Zhang, P.: Comprehensive evaluation study based on principal component analysis. Nanjing University of Science and Technology, Nanjing (2004)

27. Tan, X., Shi, Y., Yang, Q., et al.: Optimization of processing technology for rehmannia glutinosa charcoal based on AHP-CRITIC composite entropy weight method and grey relational degree analysis. Asia-Pac. Tradit. Med. **20**(05), 23–30 (2024)
28. Peng, T., Chen, X.: Ecological system health evaluation of typical estuaries in the Haihe River Basin. J. Wuhan Univ. Technol.: Eng. Edn. **42**(5), 631–634, 639–639 (2009)
29. Xu, M., Li, J., Peng, J., Niu, J., Cao, L.: Ecological system health assessment based on RS and GIS. J. Ecol. Environ. Sci. **19**(8), 1809–1814 (2010)
30. Department of Water Affairs and Forestry. Water Resource Protection Policy Implementation. Resource Directed Measures for Protection of Water Resources: Estuarine Ecosystems. Version 1.0.Department of Water Affairs and Forestry, Pretoria (1999). http://www.africanwater.org/dwaf_rdm_page.htm
31. Han, Y., Wang, X., Ge, L.: Assessment of the impact of 10 years of yellow river water discharge on the ecological environment. Water Resour. Protect. **29**(2), 76–81 (2013)

The concentration changes of benzene, toluene and 1, 2-DCA in the pumping well are shown in Fig. 2. Benzene concentration declined rapidly within an hour of the start of pumping, after which the rate fluctuated downward. With the progress of pumping, the concentration of benzene entered the trailing phase at the 50th hour. The concentration of toluene decreased significantly within 17 h after the start of pumping, and then the decline rate slowed down, with a tailing phenomenon at the 100th hour. The 1,2-DCA contaminant concentration declined at a relatively small rate over 2 h, with the rate of decline increasing from the 16th through the 22nd hour, followed by a trailing phase at the 30th hour. With the continuous pumping, the pollutant concentration of benzene and toluene still decreased slightly after entering the trailing phase, and remained relatively stable after 84.67 h and 144.67 h, until it was below the detection limit. Concentration of 1,2-DCA remained relatively stable until they were below the detection limit after entering the trailing phase. The average concentrations of benzene, toluene and 1,2-DCA during the trailing phase were 0.002 mg/L, 0.001 mg/L and 0.001 mg/L.

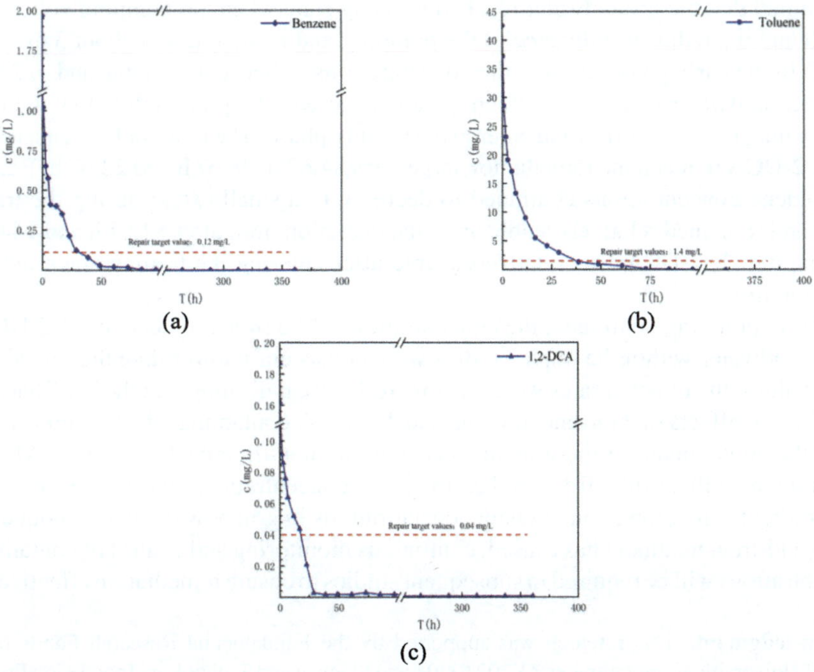

Fig. 2. (a) Benzene concentration curve with time in pumping well. (b) Toluene concentration curve with time in pumping well. (c) 1,2-DCA concentration curve with time in pumping Well.

The removal of contaminants from groundwater in the pumping well is shown in Table 3. P&T can greatly reduce the concentration of pollutants, the reduction rate is close to 100% and the concentrations of three pollutants were far lower than the remediation target value. The time for benzene, toluene and 1, 2-DCA to reach the remediation target was 84.67 h, 38.67 h and 22.67 h.

Table 3. Removal of contaminants in groundwater from pumping well.

No.	Pollutants	Initial concentration (mg/L)	Final concentration (mg/L)	Remediation target (mg/L)	Time to reach remediation goals (h)	Removal rate (%)
1	Benzene	1.97	0.00	0.12	84.67	100.00
2	Toluene	42.59	0.001	1.40	38.67	99.99
3	1,2-DCA	0.19	0.00	0.04	22.67	100.00

4 Conclusions

(1) This study was based on the former site of a pesticide factory in Jiangsu Province, China. According to the hydrogeological conditions of the site, it was experimentally determined that the groundwater reached a steady state when the pumping volume was 29 L/h and the radius of influence of the pumping and treatment was about 3 m.

(2) In the early phase of P&T, the concentrations of benzene, toluene and 1, 2-DCA in the groundwater within the influence radius decreased rapidly, fell below the repair target value in a short period and entered the trailing phase. The time for benzene, toluene and 1, 2-DCA to reach the remediation target was 84.67 h, 38.67 h and 22.67 h. Benzene and toluene concentrations continued to decrease to a small extent during the trailing phase and remained relatively stable up to the detection limit after 84.67 h and 144.67 h of pumping. 1, 2-DCA was relatively stable after entering the trailing stage until the detection limit.

(3) As pumping continued, the concentrations of benzene, toluene and 1,2-DCA in the groundwater within the impact radius were consistently lower than the remediation target values, the removal rates were all close to 100%, indicating that the P&T had good remediation effects on benzene, toluene and 1,2-DCA contaminated site groundwater.

If the contaminant in the groundwater is in the non-dissolved phase or NAPL, the contaminant will diffuse from higher to lower concentrations due to the concentration gradient. Therefore, contaminant concentrations in groundwater may rebound after pump-and-treat treatment has ceased, continuous monitoring and control of contaminant concentrations will be required in subsequent studies to ensure remediation effectiveness.

Acknowledgment. This research was supported by the Fundamental Research Funds for the Central Universities, grant number ZY20210304 and Science and Technology Innovation Program for Postgraduate students in IDP subsidized by Fundamental Research Funds for the Central Universities, grant number ZY20240307 and ZY20240309.

References

1. Peng, J.Q., Li, L., Zheng, C., et al.: Practice and consideration for environmental investigation of chemical enterprises' relocated sites. Environ. Sci. Survey **30**(6), 71–73 (2011). https://doi.org/10.13623/j.cnki.hkdk.2011.06.022

2. Yang, B., Li, H.Y., Wu, B., et al.: Engineering remediation techniques and its application for volatile organic compounds contaminated sites. J. Environ. Eng. Technol. **3**(1), 78–84 (2013). https://doi.org/10.3969/j.issn.1674-991X.2013.01.014
3. Wang, P., Han, Z.T., Zhang, H.L., et al.: Simulation and optimization of a pumping and treating system for the remediation of ammonia polluted groundwater. Hydrogeol. Eng. Geol **47**(3), 34–43 (2020). https://doi.org/10.16030/j.cnki.issn.1000-3665.201907053
4. Xu, L., Hu, X.Y., Zha, F.S., et al.: Mass transfer enhancement of air sparging on VOCs contaminated low-permeability soil by establishing pressure gradient. Chemosphere (313), 137416–137416 (2023). https://doi.org/10.1016/J.CHEMOSPHERE.2022.137416
5. Chen, Z.F., Chen, Z.G., Li, Y., et al.: A review on remediation of chlorinated organic contaminants in soils by thermal desorption. J. Ind. Eng. Chem. (133), 112–121 (2024). https://doi.org/10.1016/J.JIEC.2023.12.022
6. Li, C.W., Ding, H.R., Chen, C., et al.: Application of permeable reactive barrier technology in treatment of solid waste landfill leachate. Chin. J. Environ. Eng. **17**(02), 625–634 (2023). https://doi.org/10.12030/j.cjee.202209023
7. Wei, K.H., Ma, J., Xi, B.D., et al.: Recent progress on in-situ chemical oxidation for the remediation of petroleum contaminated soil and groundwater. J. Hazard. Mater. **432**, 128738 (2022). https://doi.org/10.1016/J.JHAZMAT.2022.128738
8. Liu, H.Z., Bruton, T.A., Doyle, F.M., et al.: In situ chemical oxidation of contaminated groundwater by persulfate: decomposition by Fe(III)- and Mn(IV)-containing oxides and aquifer materials. Environ Sci. Technol. **48**(17), 10330–6 (2014). https://doi.org/10.1021/es502056d
9. Zheng, X.H., Tafa, B.O., Shen, C.B., et al.: Effect of the bacterial community assembly process on the microbial remediation of petroleum hydrocarbon-contaminated soil. Front Microbiol (141), 196610–1196610 (2023). https://doi.org/10.3389/FMICB.2023.1196610
10. Li, J., Liang, Y.P., Zou, S.Z., et al.: Research progress on application of microorganism in groundwater pollution remediation. Environ. Pollut. Control **43**(05), 638–643 (2021). https://doi.org/10.15985/j.cnki.1001-3865.2021.05.020
11. Qian, X.R., Dong, M.M.: Phytoremediation of PDCB contaminated soil. Environ. Pollut. Control **31**(08), 67–70 (2009). https://doi.org/10.15985/j.cnki.1001-3865.2009.08.01
12. Kaur, H., Kumar, A., Bindra, S., et al.: Phytoremediation: an emerging green technology for dissipation of PAHs from soil. J. Geochem Explor. (259), 107426 (2024). https://doi.org/10.1016/J.GEXPLO.2024.107426
13. Xue, J.P., Cai, J.X., Li, Y.W., et al.: Research progress on application of phytoremediation technology for environmental pollution. North. Hortic **02**, 131–138 (2024). https://doi.org/10.11937/bfyy.20232463
14. Pu, M.: Reviews on groundwater contaminants control and remediation technology: pump and treat. Environ. Eng. **35**(04), 6 (2017). https://doi.org/10.13205/j.hjgc.201704002
15. Zhao, Y.S.: Control and Remediation of Groundwater Contaminated Site, 1st ed., pp. 150–151. Science Press, Beijing, China (2015). ISBN 978-7-03-043632-0
16. Fan, C., Shi, Z.M., Jia, Y.F., et al.: Polluted groundwater pump and treat system optimization based on weighted optimization method on the area. Hydrogeol. Eng. Geol **51**(1), 201–214 (2024). https://doi.org/10.16030/j.cnki.issn.1000-3665.202212029
17. Yan, Y., Wang, M.Y., Chen, J.P., et al.: Lab experiments and numerical simulations of 1,2-DCA remediation in groundwater at contaminated sites. Earth Environ **49**(03), 250–259 (2021). https://doi.org/10.14050/j.cnki.1672-9250.2020.48.115
18. Nagarajan, K., Loh, K.C.: Formulation of microbial cocktails for BTEX biodegradation. Biodegradation **26**, 51–63 (2015). https://doi.org/10.1007/s10532-014-9715-0
19. Miseon, J., Eldon, R.R., Soohong, K., et al.: Removal of BTEX compounds by industrial sludge microbes in batch systems: statistical analysis of main and interaction effects. World J. Microbiol. Biotechnol. **24**(1), 73–78 (2008). https://doi.org/10.1007/s11274-007-9441-4

20. Xie, T.: Application research of 1,2-dichloroethane contaminated groundwater remediation through in situ chemical oxidation enhanced by hydraulic regulation. Doctor. South China University of Technology, Guangzhou (2022). https://doi.org/10.27151/d.cnki.ghnlu.2022.003834
21. Standard for groundwater quality (GB/T 14848–2017)
22. Water quality-Determination of volatile organic compounds-Purge and trap/gas chromatography-mass spectrometer (HJ 639-2012)

Carbon Emission Prediction Study Based on Improved Adaptive Empirical Modal Decomposition ICEEMDAN and Improved Marine Predator IMPA

Zhengyi Zhang(✉)

College of Letters and Science, University of California, Davis, CA 95616, USA
eyzhang@ucdavis.edu

Abstract. In the face of the increasing global challenge of climate change, accurate prediction of carbon emissions is crucial for the development of effective mitigation strategies. To this end, this paper proposes a novel carbon emission prediction model that integrates the Improved Adaptive Empirical Modal Decomposition (ICEEMDAN) and the Improved Marine Predator Algorithm (IMPA). ICEEMDAN can efficiently decompose the raw carbon emission data into a series of easier-to-handle IMFs, whereas the IMPA, through an intelligent optimization mechanism, finely tunes the parameters of the BiLSTM network to achieve the optimal prediction effect. The results show that the proposed model exhibits excellent performance in five indicators, including MSE, and provides a powerful decision support tool for future carbon emission management.

Keywords: Carbon emission projections · BiLSTM · ICEEMDAN · IMPA

1 Introduction

Against the backdrop of global warming, research on carbon emission forecasting has become a key link in environmental science and policymaking. With the acceleration of industrialization and the continuous growth of energy consumption, the escalation of carbon emissions has had a significant impact on global climate change, exacerbated the greenhouse effect, triggered the frequent occurrence of extreme weather events and posed a serious threat to the ecosystem and human society. Therefore, accurate prediction of carbon emission trends is not only crucial for understanding future climate change, but also the basis for formulating effective emission reduction policies and realizing sustainable development goals [1]. Carbon emission forecasting research can help governments and enterprises to plan ahead, take targeted measures to optimize the energy structure, improve energy efficiency, and promote green and low-carbon transformation, thus providing a scientific basis for building a clean, safe and efficient modern energy system [2].

In recent years, scholars at home and abroad have carried out extensive and in-depth research in the field of carbon emission forecasting, utilizing a variety of forecasting

D. Han et al. (Eds.): ICEPG 2024, ESE, pp. 932–938, 2025.
https://doi.org/10.1007/978-3-031-86829-0_103

models and techniques, and exploring the dynamic characteristics and influencing factors of carbon emissions from different perspectives. In China, scholars have mainly focused on carbon emission forecasting at the regional and industry levels [3], for example, based on gray system theory, time series analysis, support vector machine and other methods, they have studied the trend of carbon emission changes in various provinces or key industries in China, and revealed the influence of the level of economic development, industrial structure, energy consumption and other factors on carbon emission. However, some of the studies have problems such as strong data dependence and limited prediction accuracy, especially the ability to deal with nonlinear and non-stationary signals needs to be improved.

The researches of foreign scholars in the field of carbon emission prediction have widely adopted machine learning and deep learning techniques, such as random forest and LSTM, to capture complex data patterns, but they face the challenges of large data demand and weak model interpretability.

2 Method

2.1 BiLSTM

In BiLSTM, the forward LSTM layer processes inputs from the beginning of the sequence toward the end, while the reverse LSTM layer does the opposite, processing from the end of the sequence toward the beginning. This bi-directional processing ensures that the hidden state at each moment not only contains past contextual information, but also incorporates future contextual information, which improves the expressive and generalization ability of the model [4].

The basic procedure of BiLSTM can be summarized as follows:

Putting the input sequence $X = [x_1,x_2,\ldots,x_T]$ is transformed into a form suitable for model inputs, where x_t denotes the number in the sequence t input vector at moment t.

The forward LSTM layer is processed step by step from the beginning $t = 1$ of the sequence to the end $t = \mathrm{T}$.

At each time step t, the LSTM unit receives the input vector x_t along with the previous hidden state h_{t-1} and cell state c_{t-1}. It then proceeds to calculate the new hidden state h_t and cell state c_t according to the following equations:

$$i_t = \sigma\left(W_i \cdot \left[h_{t-1}, x_t\right] + b_i\right) \tag{1}$$

$$f_t = \sigma\left(W_f \cdot \left[h_{t-1}, x_t\right] + b_f\right) \tag{2}$$

$$c_t = f_t \odot c_{t-1} + i_t \odot tanh\left(W_c \cdot \left[h_{t-1}, x_t\right] + b_c\right) \tag{3}$$

$$o_t = \sigma\left(W_o \cdot \left[h_{t-1}, x_t\right] + b_o\right) \tag{4}$$

$$h_t = o_t \odot tanh(c_t) \tag{5}$$

where i_t,f_t,o_t are the activation values of the input gate, oblivion gate, and output gate, respectively, σ denotes the Sigmoid function, and ⊙ denotes the element product.

The reverse LSTM layer progressively processes from the end $\mathrm{t} = \mathrm{T}$ of the sequence to the beginning $t = 1$, similarly computing the hidden state and cellular states [5].

2.2 ICEEMDAN

Conventional EMD is prone to modal aliasing, i.e., different frequency components are incorrectly assigned to the same IMF, which limits the decomposition effect to a large extent. To overcome this problem, ICEEMDAN adds a certain amount of white noise in the decomposition process, which effectively suppresses modal aliasing and improves the stability and accuracy of the decomposition by repeating the decomposition several times and averaging the results [6].

The ICEEMDAN implementation process is broadly divided into the following key steps:

Determine the signal *x(t)* as input, set the noise level during decomposition, usually a scaling factor based on signal energy, and initialize the loop counter and IMF list [7].

White noise *n(t)* is added to the signal *x(t)* to form the added noise signal $x_n(t) = x(t) + n(t)$. The noisy signal $x_n(t)$ is decomposed using a conventional EMD algorithm to obtain a series of IMFs and residual terms $r_n(t)$. The above process was repeated M times, each time using a different random noise sequence to collect all IMFs and residual terms.

The IMFs with the same serial number obtained from each decomposition are averaged to obtain the final IMF sequence. Similarly, all residual terms are averaged to obtain the final residual term *r(t)*.

2.3 IMPA

The Improved Marine Predator Algorithm (IMPA) is an advanced metaheuristic optimization technique inspired by the predatory behaviors of marine animals. It builds upon the foundational concepts of the original Marine Predator Algorithm (MPA) by introducing innovative mechanisms to enhance its exploration and exploitation capabilities [8].

The flow of the IMPA algorithm is as follows:

IMPA begins by initializing a population of predators, each represented as a potential solution in the search space [9].

The algorithm then simulates the predatory behaviors of marine animals by defining rules for the movement of predators towards prey. Prey, in this context, represents the optimal solution or a set of promising solutions [10].

The position update equation for each predator *i* at iteration *t* is given by:

$$P_i(t+1) = P_i(t) + \alpha \cdot (\Pr eyBest(t) - P_i(t)) \tag{6}$$

where $P_i(t)$ is the current position of predator *i*, *PreyBest*(t) is the position of the best prey (solution) found so far, and α is a parameter controlling the step size towards the prey.

3 Experiments

3.1 Experimental Environment Setup

The hardware configuration details are as follows: an Intel ® Core TM i7-7500U CPU, which runs between 2.70 GHz and 2.90 GHz, with 12 GB RAM; and the operating system is selected to be Microsoft Windows 10 (64-bit version). MATLAB R2023a was utilized as the simulation tool in the experiments.

In this study, LSTM, BiLSTM, BiLSTM-ICEEMDAN-IMPA model are placed in the same arena. To ensure the consistency of the experimental conditions, all the algorithms involved in the comparison are set to have a population size of 10 and an upper limit of 300 iterations [11].

3.2 Experimental Evaluation Indicators

Under the research framework of this paper, we adopt a series of evaluation criteria to construct a comprehensive model evaluation system, specifically including: mean square error (MSE), root mean square error (RMSE), mean absolute error (MAE), mean absolute percentage error (MPAE), and coefficient of determination (R^2). Below are the formulas for these key evaluation metrics [12].

3.3 Analysis of Results

In this paper, ablation comparison experiments are conducted based on the collected carbon emissions of a Chinese city in a specific time period, respectively.

The results of testing the effect of the dataset on the three models are given below as follows:

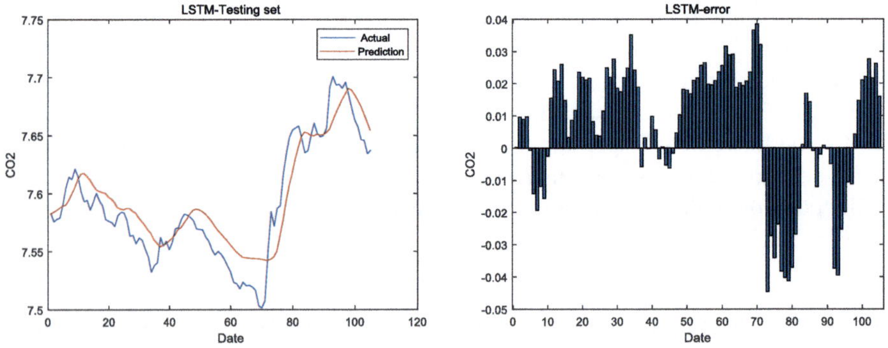

Fig. 1. LSTM testing set

Comparison of Figs. 1, 2 and 3 shows that the model fit superiority of BiLSTM-ICEEMDAN-IMPA is significantly better than that of LSTM and BiLSTM models.

According to Figs. 1, 2 and 3, it can be found that in the error map, the number of the BiLSTM-ICEEMDAN-IMPA model proposed in this paper with error more than

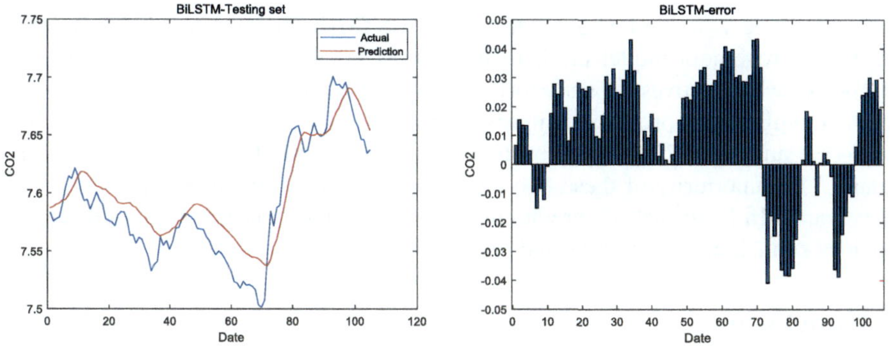

Fig. 2. BiLSTM testing set

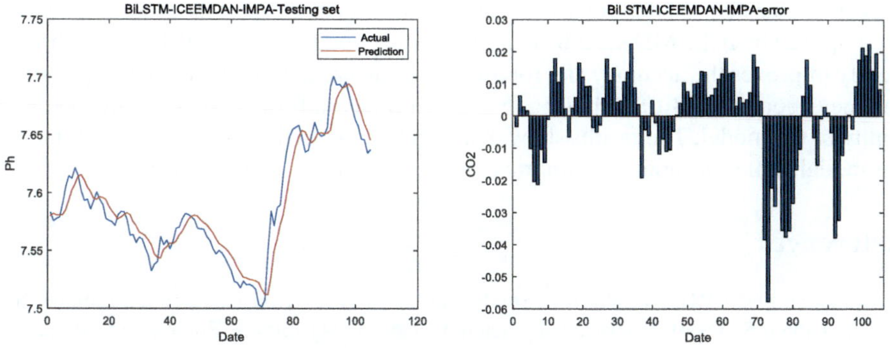

Fig. 3. BiLSTM-ICEEMDAN-IMPA testing set

3% in the predicted results is much less than that of the pre-improvement LSTM model and BiLSTM model. Meanwhile, in the interval range of 10–70 where the prediction errors of LSTM and BiLSTM models are large, the prediction errors of the improved BiLSTM-ICEEMDAN-IMPA model are almost always within 2%.

In order to evaluate the models more intuitively, the evaluation metrics for each model are calculated as shown in the table below:

Table 1. Evaluation of ablation experiments

Model	MSE	RMSE	MAE	MAPE	R^2
LSTM	0.000506	0.02251	0.019466	0.25673%	79.6582%
BiLSTM	0.000451	0.021255	0.017987	0.23713%	81.8631%
BiLSTM-ICEEMDAN-IMPA	0.000234	0.015324	0.011883	0.15641%	90.5727%

From Table 1, it can be concluded that BiLSTM-ICEEMDAN-IMPA outperforms the pre-improvement model in all metrics, most notably the R^2, which improves by 10.9145%, which proves the superiority of the proposed model in this paper.

Meanwhile, the proposed improved BiLSTM-BiLSTM-IMPA model achieves 0.15641% and 90.5727% in the evaluation metrics of MAPE and goodness-of-fit R-square. The superiority of these two metrics indicates that the model proposed in this paper can fulfill the task of predicting carbon dioxide concentration well, and also provides some help for other related prediction tasks.

4 Conclusion

In the study, this paper presents an innovative carbon emission prediction method that combines the improved adaptive empirical mode decomposition (ICEEMDAN) with the improved marine predator algorithm (IMPA). Complex time series data are effectively decomposed by ICEEMDAN into a set of intrinsic mode functions (IMFs), which significantly improves the accuracy and robustness of the prediction model. Subsequently, the parameter search of BiLSTM using IMPA optimization further enhances the prediction ability of the model. The results show that compared with the pre-improvement method, our model achieves significant improvement in various evaluation indexe.

References

1. Chukwunonso, P.B., Wesabi, A.I., Shixiang, L., et al.: Predicting carbon dioxide emissions in the United States of America using machine learning algorithms. Environ. Sci. Pollut. Res. Int. **31**(23), 33685–33707 (2024)
2. Khalil, H., Wainer, G.: A framework for modeling, generating, simulating, and predicting carbon dioxide dispersion indoors using cell-DEVS and deep learning. Simulation **100**(5), 421–434 (2024)
3. Moore, T. R., Damon Matthews, H., Yann, C.: Linking historical and projected trends in extreme precipitation with cumulative carbon dioxide emissions. Atmosphere-Ocean **62**(2), 165–182 (2024)
4. Zhang, X., Luo, Z., Xiao, W.: CNN-BiLSTM-DNN-based modulation recognition algorithm at low SNR. Applied Sciences **14**(13), 5879–5879 (2024)
5. Kim, H., Kim, M.: Malware detection and classification system based on CNN-BiLSTM. Electronics **13**(13), 2539–2539 (2024)
6. Lu, G., YanLing, G., Changzheng, C., et al.: Wind turbine gearbox bearing fault diagnosis method based on ICEEMDAN and flexible wavelet threshold. Journal of Failure Analysis and Prevention **24**(3), 1181–1198 (2024)
7. Yun, P., Zhou, Y., Liu, C., et al.: Forecasting China carbon price using an error-corrected secondary decomposition hybrid model integrated fuzzy dispersion entropy and deep learning paradigm. Environ. Sci. Pollut. Res. Int. **31**(11), 16530–16553 (2024)
8. Zaidi, M.T.M., Ashraf, M.A., Helmi, M.S., et al.: An improved marine predators algorithm tuned data-driven multiple-node hormone regulation neuroendocrine-PID controller for multi-input–multi-output gantry crane system. J. Low Frequency Noise, Vibration and Active Control **42**(4), 1666–1698 (2023)
9. Shuhan, Z., Shengsheng, W., Ruyi, D., et al.: A multi-strategy improved outpost and differential evolution mutation marine predators algorithm for global optimization. Arabian J. Sci. Eng. **48**(8), 21–24 (2023)

10. Xiangqi, L., Donghe, Y.: Color constancy computation for dyed fabrics via improved marine predators algorithm optimized random vector functional-link network. Color Research & Application **46**(5), 1066–1078 (2021)
11. Liu, G., Zhong, K., Li, H., et al.: A state of art review on time series forecasting with machine learning for environmental parameters in agricultural greenhouses. Information Processing in Agriculture **11**(2), 143–162 (2024)
12. Hossein, A., Tianfeng, Z., Mohamed, E.A., et al.: Predicting crystallite size of Mg-Ti-SiC nanocomposites using an adaptive neuro-fuzzy inference system model modified by termite life cycle optimizer. Alexandria Engineering Journal **84**, 285–300 (2023)

Ecological Security Pattern Construction and Ecological Restoration Zoning Based on Ecological Protection Red Line

Daohan Zha[1] and Qingda Duan[2(✉)]

[1] Hunan Vocational College of Engineering, Changsha 410151, China
[2] College of Forestry, Xinyang Agriculture and Forestry University, Xinyang 464000, China
duanqingda@126.com

Abstract. This study takes Yueyang County as the study area, adopts the optimization method of ecological protection red line to identify the ecological source in all directions, and applies the minimum cumulative resistance surface to identify the ecological corridor and ecological strategic points, so as to construct the ecological security pattern of Yueyang County, which is "one screen, two corridors" and "two nuclei and multiple points". The ecological security pattern of Yueyang County will be constructed. Combined with the ecological importance evaluation, ecological sensitivity evaluation and ecological security classification, ecological restoration zoning was carried out, and corresponding ecological restoration and protection suggestions were made. Results: 1. Yueyang County has a high ecological importance rating, with 61.45% of the area of high importance and above; the ecological sensitivity rating of Yueyang County is low sensitivity, with 92.62% of the area being insensitive. 2. A total of 32 ecological corridors and 42 ecological strategic points have been extracted. 3. The overall ecological security level of Yueyang County is high, but the corridors in the southeastern part of the county are dense but the strategic points are sparse, and the network connectivity is not high.3. Yueyang County is divided into four major ecological restoration zones, namely, the East Dongting Lake Biodiversity and Hydro-ecological Restoration Zone, the Habitat Comprehensive Improvement and Ecological Restoration Zone, the Agricultural Ecological Improvement and Hydro-ecological Restoration Zone, and the Mountainous Water Source Nourishment and Soil and Water Conservation Restoration Zone. On this basis, ecological protection is proposed to realize the landscape and idyllic urban and rural spatial structure, and to establish a new land-use pattern with a beautiful habitat.

Keywords: yueyang county · ecological protection red line · ecological security pattern · ecological restoration zoning

1 Introduction

Along with the continuous advancement of social and economic development process, human's predatory demand for nature has made the ecological carrying capacity of most of the natural resources close to or reached its upper limit, and the natural ecosystems in

D. Han et al. (Eds.): ICEPG 2024, ESE, pp. 939–953, 2025.
https://doi.org/10.1007/978-3-031-86829-0_104

Fig. 1. Geographic location of Yueyang County

the red line of ecological protection, the ecological sensitivity evaluation of the national land space of Yueyang County is carried out to clarify the ecological protection objectives and priorities, and to spatially identify ecologically sensitive and fragile areas.

Evaluation of erosion sensitivity.

(1) Evaluation model

$$[\text{Erosion sensitivity}] = \sqrt[4]{R \times K \times LS \times C} \tag{1}$$

where R is the rainfall erosivity factor, K is the soil erodibility factor, LS is the topographic relief factor, and C is the vegetation cover factor.

Optimization of the Ecological Protection Red Line

On the basis of the evaluation of ecological importance and ecological sensitivity of the national territory, a proposal for the delineation of the ecological red line of ecological protection in sensitive areas of ecologically important zones is formed. According to the relevant regulations, the spatial distribution of ecologically important areas is delineated, and expert knowledge is combined to comprehensively judge the consistency of the assessment results with the actual ecological conditions. For assessment results that do not meet the actual situation, on-site verification and adjustment are carried out to optimize the ecological protection red line delineation plan. The optimization of the red line of ecological protection mainly includes the deduction of fragmented patches, the exclusion of construction land and basic farmland, the approval of the boundaries of the red line of ecological protection for prohibited development zones and other specific areas, supplementary mapping, and up and down docking, and other steps.

Ecological Source Extraction

Ecological source extraction was mainly realized through four steps: ① using morphological spatial pattern analysis method to classify landscape types; ② comparing the changes in the number of patches of different sizes within the ecological protection red line to roughly determine the location of ecological sources; ③ applying Conefor 2.6

to evaluate the degree of landscape connectivity; and ④ determining ecological sources through the index of possible connectivity.

3.2 Ecological Resistance Analysis Methods

Selection of Resistance Factors

The selection of resistance factors is mainly considered from five aspects: topographic location index, land use type, distance from roads, rivers and settlements. Among them, in order to comprehensively reflect the resistance effect of topographic conditions on the spread of ecological source sites, the resistance generated by the expansion of ecological source sites in Yueyang County's topography was described by combining elevation and slope into a topographic location index through geographic information modeling. The formula is as follows:

$$T = Log\left[\left(\frac{E}{\overline{E}} + 1\right) \times \left(\frac{S}{\overline{S}} + 1\right)\right] \tag{2}$$

where T represents the topographic position index, the elevation at E represents the elevation at any point in the county, $\overline{E}$ represents the average of the elevations in the county, and S represents the slope at any point in the county, $\overline{S}$ represents the average of the slopes in the county.

Establishment of a Minimum Cumulative Resistance Surface

The Minimum Cumulative Resistance (MCR) model analyzes the minimum resistance that an ecological source site needs to overcome as it expands, by first calculating the cumulative distance that the source site must travel to other source sites, and then calculating the minimum resistance that the source site must overcome as it expands toward the outside.The MCR formula is as follows:

$$MCR = f_{\min} \sum_{j=n}^{i=m} D_{ij} \times R_i \tag{3}$$

where MCR is the cumulative resistance value of source spreading to any point; f is the positive correlation between MCR and landscape ecological processes; min represents the minimum cumulative resistance value; D_{ij} represents the distance from source j to a unit i in space; and R_i represents the resistance coefficient of space unit i to ecological expansion.

Identification of Ecological Corridors

The least-cost pathway tool in ArcGIS 10.2 was used to identify and obtain the least-cost pathway, or ecological corridor, starting from each source site to any of the remaining source sites. Since there are two corridors and inconsistent paths from a source to a source b and from a source b to a source a obtained through the calculation, and since some of the least-cost paths between sources roughly overlap with the paths of the intermediate pathway sources, some of the corridors are eliminated through quantitative analysis, and the number and length of ecological corridors can be obtained in the end.

some areas have shown different degrees of degradation, with ecological and environmental problems becoming more and more serious, and the ecological security of the national territory space is facing a great challenge, which has seriously constrained the sustainable economic and social development [1, 2]. As most of the past management of the ecological environment stays in the local area of water, soil and other single ecological elements and land sand, soil erosion and other single ecological processes, so that the top-level design of ecological restoration is often a lack of holistic and systematic considerations, and the arrangement of ecological restoration projects are also a lack of articulation, coordination, and often appear ecological restoration of the local area of the effect of a good but the overall effect of the outstanding problems, which led to ecosystems The improvement of the overall structure and function of the ecosystem is not significant [3, 4]. The continuous degradation of natural ecosystems and the increasing aggravation of ecological and environmental problems have prompted people to reflect deeply on the traditional industrial civilization, and the concept of ecological civilization has emerged in this context [5, 6]. Nowadays, it is of great significance to explore how to scientifically and effectively construct the technical paradigm of ecological restoration zoning, strengthen ecological protection and enhance ecosystem service function, in order to enrich the theory and practice of ecological restoration planning and improve the quality of the regional ecological environment.

At present, scholars have carried out a lot of useful explorations and achieved certain results in the study of ecological restoration zoning, and the research mode of "source identification - resistance surface evaluation - corridor extraction - pattern construction" has been formed [7]. For source identification, common methods include: (1) directly designating a nature reserve as the source [8]; (2) determining the source through quantitative evaluation methods (e.g., ecological importance evaluation, ecological sensitivity evaluation) [9]. However, in terms of coverage, the results of ecological importance and ecological sensitivity evaluation are only a part of the scope of the ecological protection red line, and rarely the ecological protection red line is included in the identification of ecological source areas. In terms of resistance surface evaluation, the more common way of selecting resistance factors when constructing the minimum cumulative resistance surface includes both anthropogenic and natural aspects, because incomplete considerations lead to the selection of indicators that are too homogeneous, making the construction of the later ecological security pattern lack of practicality and feasibility [10]. Ecological restoration zoning methods include the construction of ecological safety pattern and comprehensive evaluation methods, of which the comprehensive evaluation methods are mainly ecological safety evaluation and ecosystem health evaluation [11]. Ecosystem health evaluation and ecological safety evaluation have the disadvantages of over-reliance on landscape and geomorphology data, and no consideration of corridors and strategic points compared to the construction of ecological safety pattern [12]. In recent years, the construction of ecological security pattern has been widely used in ecological restoration zoning, and the main zoning methods include the ecological restoration zoning under the perspectives of "importance-vulnerability-service value", "natural background-dominant ecological function-ecological stress problem", etc. The ecological restoration zoning is a combination of the ecological restoration zoning and the ecological restoration zoning. The coupling of ecological restoration zoning with the construction of ecological security pattern is a key research direction in the field of ecological restoration [13].

Yueyang County is characterized by complex topography and geomorphology, fragile and sensitive ecological environment. The innate fragile natural background conditions coupled with irrational human development activities have jointly led to the continuous decline of ecological service functions, and there is an urgent need to take the necessary measures to repair the damaged territorial spatial functions of the region. Based on this, this study proposes the construction of an ecological security pattern and ecological restoration zoning based on the ecological red line, identifying the ecological source on the basis of the circled ecological red line, using it as the starting point of the diffusion of the ecological process, and then determining the ecological corridors and strategic points after the establishment of the minimum cumulative resistance surface, so as to ultimately build up a scientific and reliable ecological security pattern.

2 Overview of the Study Area

Yueyang County is located in the northeastern part of Hunan Province, between latitude 28°57′11″ ~ 29°38′41″ north and longitude 112°44′14″ ~ 113°43′35″ east, with a total area of 2,809 square kilometers. As shown in Fig. 1, the landform of the county is tilted in a step-like manner from the remnants of Mufu Mountain in the northeast to Dongting Lake in the southwest, forming the geographic environment of the eastern mountainous areas, the hilly and granular areas in the middle and southern part of the county, and the lake plains area in the west. Climate is subtropical monsoon humid climate, with abundant rainfall, mild climate, abundant sunshine, four distinct seasons, winter and changeable, long hot period, short cold period. Yueyang County has a relatively complete range of mineral resources and abundant energy reserves, but due to factors such as the expansion of the scope of urban development and construction, mining activities and the natural terrain dominated by sloping arable land, soil erosion is serious, landslides, avalanches and other geologic hazards are frequent, and the protection and restoration of national land space is imminent.

3 Research Methodology

3.1 Ecological Source Area Identification

Ecological Importance Assessment

Ecological importance assessment can be used to measure the service functions of various ecosystems, including forests, grasslands and deserts, in terms of water conservation and soil conservation. The ecological importance indicator can characterize the structure and function of ecosystems, and is an important indicator for regional main function zoning, which is important for the delineation of restricted development zones and prohibited development zones.

Ecological Sensitivity Assessment

Based on the relevant normative documents and technical methods for the delineation of

Eco-Strategic Point Identification

The tangent point of the equal resistance line between two neighboring ecological source areas and the intersection point of the corridor with the main road and the equal resistance line were selected as the ecological strategic points. Based on the established ecological corridor network and the minimum cumulative resistance surface for identification, points with spatial distance less than 2 km were excluded, and the number of ecological strategic points was extracted.

3.3 Ecological Security Pattern Zoning and Ecological Restoration Zoning

The ecological network of Yueyang County was constructed through the identification of ecological sources, ecological corridors and ecological strategic points. According to the resistance value after the reconstruction of ecological resistance evaluation, it is divided into 4 levels from low to high using the natural breakpoint method, which corresponds to the division of Yueyang County into 4 types of ecological security patterns, which are high level security pattern, higher level security pattern, lower level security pattern, and low level security pattern.

Combining the ideas of "top-down" and "bottom-up" zoning, we have constructed a framework for ecological restoration zoning based on "natural land conditions - dominant ecological functions - ecological stress problems". The "top-down" approach grasps the overall pattern of the region and carries out ecological restoration zoning under the principle of guaranteeing territorial integrity, while the "bottom-up" approach identifies key ecological restoration areas based on the dominant ecological functions of the region, combined with the main ecological sensitivities and characteristics of anthropogenic disturbances.

4 Analysis of Results

4.1 Ecological Security Patterns

Ecological Sources

(1) Evaluation of ecological importance

The ecological protection importance evaluation results of Yueyang County were derived by synthesizing the water source containment function importance grading and biodiversity maintenance function importance grading. As can be seen from Fig. 2, overall, Yueyang County has a high percentage of area with ecological importance grading of extremely important, which is mainly distributed on the east and west sides of Yueyang County, with specific distribution of Dongdongting Lake Nature Reserve, Xinwuhe National Wetland Park, Dayunshan National Forest Park, and specific townships of Zhangguying Township, Yuetian Township, and so on. The remaining grades are scattered within Yueyang County.

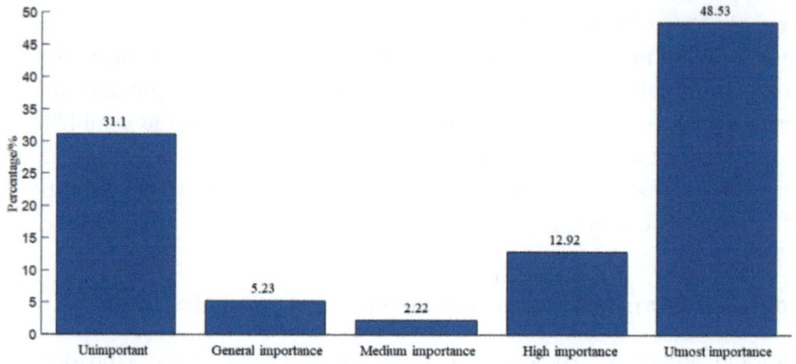

Fig. 2. Percentage of area evaluated for ecological importance in Yueyang County

(2) Evaluation of ecological sensitivity

Most of the major rivers in Yueyang County originate in the eastern mountains, the river source area has a high terrain, the surface vegetation is dominated by shrub forests and grasslands, the soil layer is thin, the sand content is high, the slope drop of the surface is steep, the wind speed is high, and it is difficult to recover from the destruction of vegetation, so that the risk of soil erosion is greater. As can be seen from Fig. 3, within the county, the vast majority of geographical soil erosion sensitivity is insensitive, with an area of 260,313. 15 hm^2 of insensitive area, 19,061. 25 hm^2 of generally sensitive area, mainly distributed in the central and eastern part of Yueyang County, and 541. 42 hm^2 of moderately sensitive area, distributed in a point-like manner, and mainly located in the eastern mountainous areas. The highly sensitive and extremely sensitive areas are 16. 99 hm^2 and 100. 63 hm^2 respectively, which are mainly distributed in a point-like manner and are located in the towns of Yuetian, Zhangguying, Maotian, Gongtian, Xinkai and Rongjiawan.

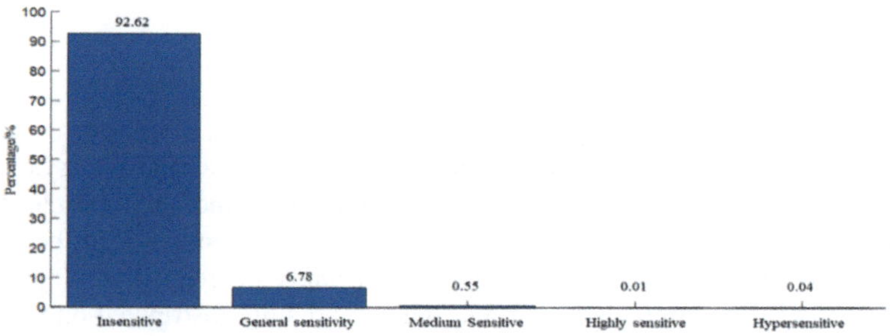

Fig. 3. Percentage of area evaluated for ecological sensitivity in Yueyang County

(3) Optimization of the red line for ecological protection

As can be seen from Fig. 4, the ecological protection red line in Yueyang County shows a spatially concentrated distribution pattern, mainly in the Dongting Lake area in the east of Yueyang County and the Tieshan Reservoir area in the west, with almost no distribution in the center. In terms of resource type, it is mainly water resources, which is an important water source conservation area, as well as biodiversity and soil and water conservation area.

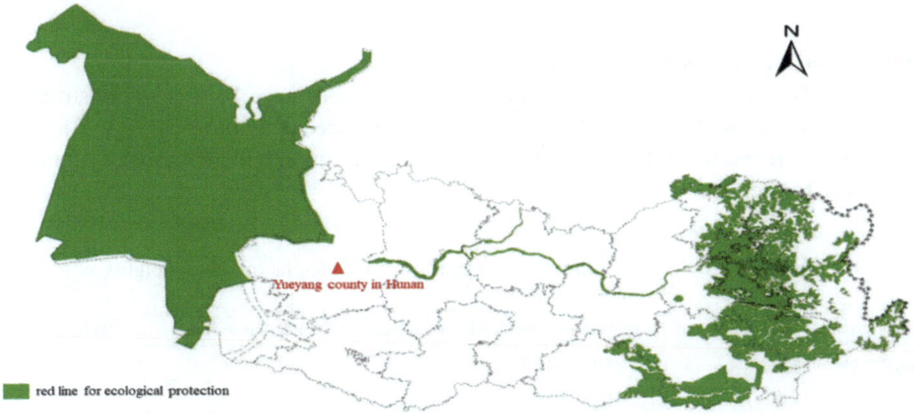

Fig. 4. Yueyang County Ecological Protection Red Line Optimization and Adjustment Map

(4) Ecological source extraction

Based on the evaluation of ecological importance, evaluation of ecological sensitivity and the determination of optimization of the ecological protection red line, the core area of ecological space and the important area of ecological space are classified as the ecological source area. This area has a high ecosystem service level and high ecological sensitivity, and the land cover type is dominated by forest land and cultivated land, with an area of about 1018.51 km^2, accounting for 36.24% of the ecological space area in Yueyang County.

4.2 Ecological Resistance

Selection of Resistance Factors

According to the actual situation in Yueyang County, five aspects are considered: topographic location index, land use type, distance from roads, rivers and settlements.

(1) Terrain position index. According to the DEM digital elevation data of Yueyang County, the slope was calculated, and the slope and elevation were converted to the topographic position index and then categorized into five classes by the natural intermittent point method.
(2) Land use types. The land use types in Yueyang County were divided into eight types, and the resistance surface was calculated by assigning resistance values.
(3) Distance from settlements. The distance from a settlement is categorized into five levels: less than 200 m, 200 m–500 m, 500 m^{-1} km, 1–2 km, and more than 2 km.

(4) Distance from roads. By extracting the roads above the county level in Yueyang County, the distance from the road is divided into five classes of less than 1 km, 1–2 km, 2–3 km, 3–5 km, and more than 5 km according to the class of the road after vectorization.
(5) Distance from rivers. Distance from rivers is categorized into five classes: less than 1 km, 1–2 km, 2–3 km, 3–5 km, and more than 5 km.

Determination of Resistance Values

In this study, we refer to the assignment criteria of resistance factors in constructing resistance surfaces in existing studies, set the resistance value interval as 1–500, and then determine the weights of each resistance factor by hierarchical analysis, and refer to Table 1 for details.

Table 1. Ecological source land expansion resistance factors and weights

Drag factor	Classification of resistance factors	Resistance value	Weights
Topographic index	0.694—1.019	30	0.3
	0.608—0.693	10	
	0.512—0.607	150	
	0.396—0.511	200	
	0.108—0.395	300	
Land use type	Woodland	1	0.3
	Grasslands	10	
	Garden area	50	
	Streams	50	
	Plow land	100	
	Village site	400	
	Urban building land	500	
	Other sites	450	
Distance from populated areas	>2 km	50	0.2
	1—2 km	100	
	500 m—1 km	200	
	200—500 m	300	
	<200 m	400	
Distance from road	>5 km	50	0.1
	3—5 km	100	
	2—3 km	200	
	1—2 km	250	
Distance from river	<1 km	300	0.1
	<1 km	1	
	1—2 km	50	

(*continued*)

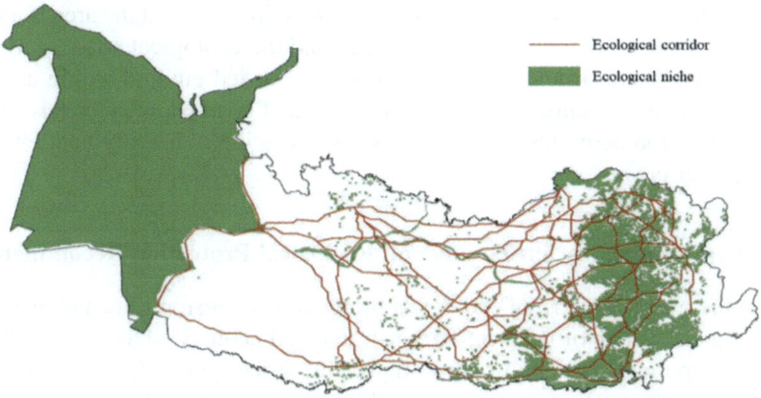

Fig. 8. Schematic diagram of the ecological corridor

with a spatial distance of less than 2 km. By superimposing the strategic points with the ecological corridor layer, an ecological security pattern of "one screen and two corridors" and "two cores and multiple points" is finally formed in Yueyang County, which is shown in Fig. 9.

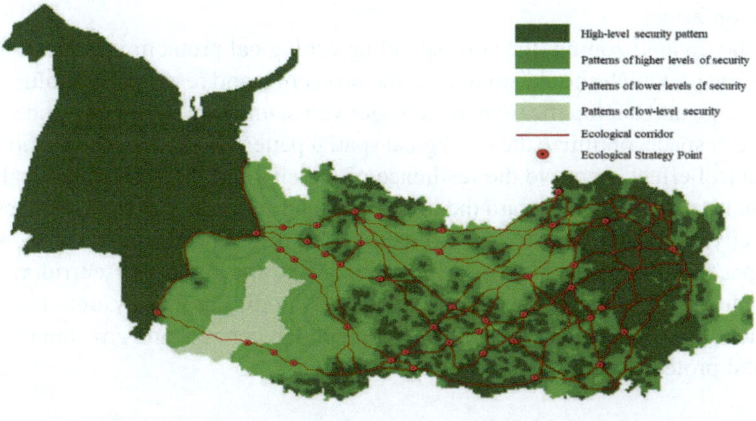

Fig. 9. Schematic diagram of ecological security patterns

The results show that the high-level security pattern covers an area of 1,482.78 km^2, with a range that basically overlaps with the ecological source area, which is important for maintaining the balance of the ecological environment within the region and providing diversified ecosystem services, etc., and it is strictly prohibited to carry out any construction activities that undermine the dominant ecological functions; The area of the higher level security pattern is 606.66 km^2, which is slightly larger than the ecological source area and spreads outward from the source area, and is an important transition zone connecting the urbanized area and the ecological source area, and urban construction activities should be strictly controlled; The lower level security pattern covers an area of 617.89 km^2, which basically covers all the township seats, and the area is a transition zone between the production and living space and the ecological space, which should be developed and constructed activities should be carried out and within a reasonable range; The low-level security pattern covers an area of 102.56 km^2, which is suitable for urban construction activities because of its low elevation, relatively gentle terrain and rich land cover types.

4.3 Ecological Restoration Zoning and Ecological Protection Recommendations

Based on the overall pattern of ecological protection and restoration, and in accordance with the logical system of "climate-terrain-watershed-ecology", based on the key watersheds and important mountain ranges, highlighting the types of ecological problems, combined with the evaluation of ecological importance and ecological sensitivity, etc., it divides the ecological protection and restoration zones of Yueyang County, and guides the landing of the major engineering projects of ecological protection and restoration.. According to the integrity and connectivity of natural geography and ecosystems, this study divides the county's land space into four major ecological restoration zones with full coverage, no intersection and no overlap: East Dongting Lake Biodiversity and Water Ecological Restoration Zone, Habitat Comprehensive Improvement and Ecological Restoration Zone, Farmland Ecological Improvement and Water Ecological Restoration Zone, and Mountainous Water Source Nourishment and Soil and Water Conservation Restoration Zone.

For the divided zoning, the corresponding ecological protection suggestions should be put forward, which should strengthen the protection and restoration of blue and green space, dredge and unclog the revitalized water veins, improve the quality and function of urban green space, optimize the ecological spatial pattern, reduce the urban flooding and flood control effect, improve the resilience of the city and the ecological quality of the human habitat, and push forward the integration and symbiosis of the natural ecosystem and the city, and deploy ecological protection and restoration in terms of the structural green space, the water system of rivers and lakes, the important corridor, the flood control and the high risk areas of flood control and flooding as the key areas for ecological restoration, focusing on water management, greening, corridor and governance to deploy ecological protection and restoration.

5 Discussion

Based on the construction of a scientific and reliable ecological security pattern, this study considers the national land space in an integrated manner, carries out ecological restoration zoning and puts forward protection and restoration suggestions. In order to solve the deficiencies of the current ecological source extraction method, this study adopts the optimization of ecological protection red line to identify ecological source in all directions; establishes the minimum cumulative resistance surface to identify ecological corridors and ecological strategic points, and finally divides the county land space into four ecological restoration zones and puts forward corresponding ecological restoration suggestions. However, because the ecological corridors obtained in this paper are

Table 1. (*continued*)

Drag factor	Classification of resistance factors	Resistance value	Weights
	2—3 km	100	
	3—5 km	200	
	>5 km	300	

Based on the assignment of each resistance factor type in the above table, the single factor resistance value of each resistance factor was obtained (Fig. 5). According to the weights, the single-factor resistance values were weighted and superimposed to form a comprehensive resistance value (Fig. 6). According to the calculation results, the high value area of diffusion resistance of ecological source in Yueyang County is mainly concentrated in the county city and the seat of each town government, and the low value area is mainly concentrated in the area with better ecological environment and higher ecosystem service level.

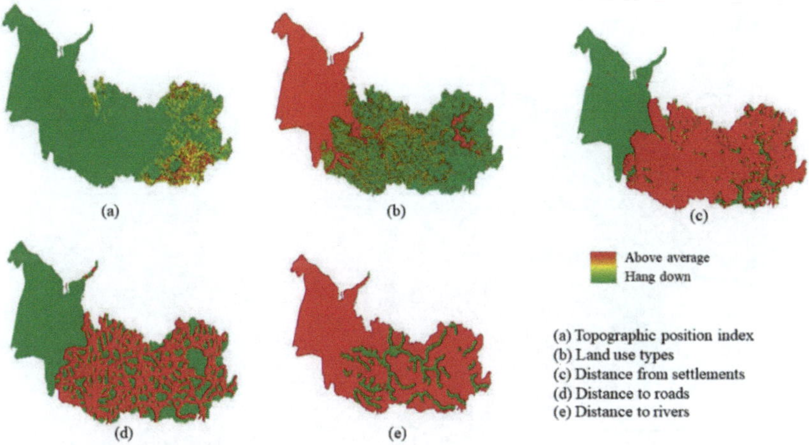

Fig. 5. One-factor resistance values for ecological source expansion

Establishment of the Minimum Cumulative Resistance Surface

Applying the cost-distance module of ArcGIS 10.2, the ecological source site was used as the elemental source data to calculate the minimum cumulative resistance surface (Fig. 7). From the figure, it can be seen that the resistance is increasing outward with the ecological source as the center, and the closer to the ecological source, the smaller the resistance value is. Overall, the high value area of the minimum cumulative resistance value is mainly concentrated in the western part of the town of Rongjiawan, Xinkai Town, Xinwall Town, Huangshajie Town town with better economic development, and these areas are largely affected by the urbanization factor, and the land use type is more complex and fragmented, so that the process of ecological expansion is more difficult in this type of area.

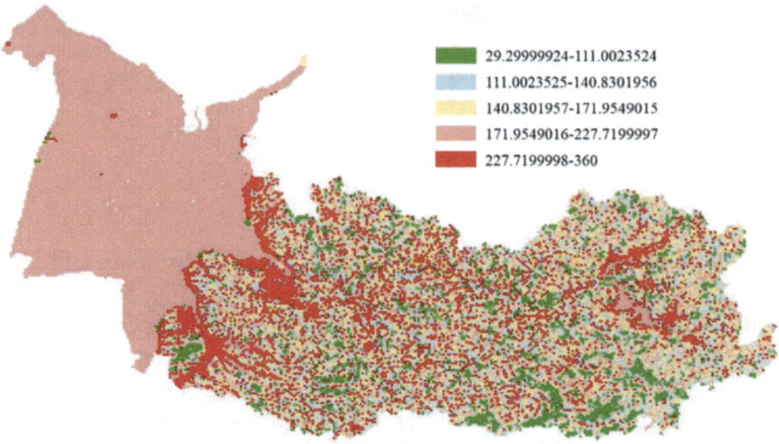

Fig. 6. Combined resistance values to ecological source expansion

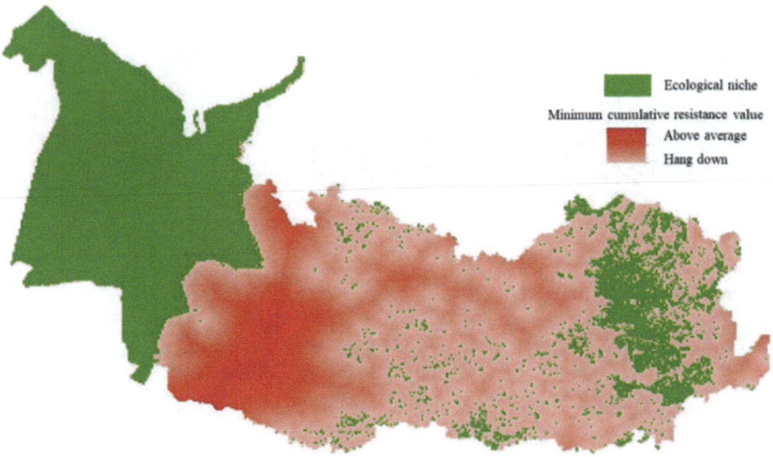

Fig. 7. Schematic diagram of minimum cumulative resistance surface

Identification of Ecological Corridors

In this paper, based on the results of ecological source identification and minimum cumulative resistance calculation, the ecological source is used as the source data, and the minimum cost path tool in ArcGIS10.2 is used to identify the ecological corridors, finally obtaining a total of 32 ecological corridors in Yueyang County, with the length of a total of 787.93 km, and the corridors are distributed as a whole densely in the east and sparsely in the west, which is shown in Fig. 8.

Eco-Strategic Point Identification

Based on the established ecological corridor network and the minimum cumulative resistance surface, 42 ecological strategic points were extracted by eliminating points only identified by the least-cost path tool, and the width of the corridors is not rigorously considered (with the different widths of the corridors, the land use types, landscape heterogeneity, etc. also change accordingly), so this study lacks the discussion on the setting of the widths of ecological corridors, and in the future, the author will focus on the study of the widths of the ecological corridors of artificial ecosystems, the migratory corridors required by different animals, and so on. The future author will focus on the setting of the width of ecological corridors in artificial ecosystems, migration corridors needed by different animals, etc. In addition, the level of ecological security is not static, it is always changing with the intensity of human activities, land use, etc. In the next step, the author needs to consider monitoring the ecological security for a certain period of time around the clock and predicting the development trend of different types of trends, in order to adapt to the current needs of ecological restoration of the national territory as much as possible.

6 Conclusions

Taking Yueyang County as the study area, this study proposes the construction of ecological security pattern and ecological restoration zoning based on the ecological protection red line, which is a reference for the study of related fields (ecological protection engineering, etc.). The conclusions of the study mainly include: the results of ecological protection importance evaluation in Yueyang County (31.1% unimportant, 5.23% generally important, 2.22% moderately important, 12.92% highly important, and 48.53% extremely important), of which the area with the ecological importance level of extremely important accounts for a high percentage of the total area, which is mainly distributed in the east and west of Yueyang County; the ecological sensitivity of Yueyang County is low, and the vast majority of the geographic area ecological nature is insensitive (92.62% insensitive, 6.78% generally sensitive, 0.55% moderately sensitive, 0.01% highly sensitive, 0.04% extremely sensitive); a total of 32 ecological corridors are obtained in Yueyang County, and the corridors are distributed as a whole to the east dense and west sparse; 42 ecological strategic points are extracted, which are mainly distributed in the east and the central region of the transportation network dense area; Yueyang County ecological security pattern is better as a whole, the The area of higher level safety pattern and above accounts for 74.36% of the area, but there are still 3.65% of the area with ecological safety grade in the low level safety pattern; combined with the analysis of ecological source, ecological resistance surface, ecological corridor and ecological strategic points, Yueyang County is divided into 4 large ecological restoration sub-areas.

Acknowledgment. Project of Hunan Provincial Department of Natural Resources "Research on the Measurement of Ecological Security and Protection of Dongting Lake Ecological and Economic Zone" (Foundation number: 2022G04); Joint Fund of Natural Science Foundation of Hunan Province (Foundation number: 2023JJ60190).

References

1. Gao, Q., Han, G.: Policy connotation, strategic focus and practical progress of modernized agricultural development. Zhongzhou Journal **07**, 38–46 (2024)

2. Liu, Y., Qin, M., Ouyang, H., et al.: Optimization of spatial resource allocation in the national territory based on the balance of supply and demand of ecosystem services. J. Nat. Resour. **39**(06), 1358–1383 (2024)
3. Dai, L., Ding, Z., Ma, H., et al.: Spatial correlation and endogenous dynamics of technology transfer in Guangdong-Hong Kong-Macao Greater Bay Area. J. Geogr. **79**(06), 1503–1520 (2024)
4. Wang, T., Yue, W.: Optimization of national spatial pattern for carbon sinks: theoretical framework and action logic. J. Nat. Resour. **39**(05), 1008–1021 (2024)
5. Xu, Y., Gao, M., Zhang, Z.: Study on habitat quality change in the middle and lower reaches of the Yangtze River based on the InVEST model. Research on Soil and Water Conservation **31**(05), 355–364 (2024)
6. Xiao, R., Huang, W., Li, Q., et al.: Optimization and evolution analysis of ecologically most relevant hydrological indicators in the lower reaches of Fuhe River based on PCA-RF. Yangtze River Basin Resources and Environment **33**(07), 1563–1575 (2024)
7. An, T., Wang, T., Gao, J.: Ecosystem-based zoning method and application for coastal zone management. Ecological Economy **40**(07), 159–165 (2024)
8. Xia, F., Guo, Z., Zhang, M., et al.: Exploration and progress of urban biodiversity conservation management. World Forestry Research **37**(03), 33–39 (2024)
9. Ma, Y., Liu, X., Zhang, L., et al.: Characteristics of soil desiccation response to 40 years of vegetation restoration in plantation forests in Baiyu Mountain Area. Soil and Water Conservation Research **31**(05), 192–198 (2024)
10. Liu, Y., Gao, Y., Chen, M., et al.: Construction and analysis of ecological security pattern of land space in Dongguan City, Guangdong Province. Remote Sensing of Natural Resources **36**(02), 126–134 (2024)
11. Lu, Z., Gao, Q., Yuan, J., et al.: Status quo of environmental geological problems and ecological restoration program of abandoned mines in arid areas: a case study of Baotou City. China Mining Industry **33**(S1), 111–115 (2024)
12. Xu, Q., Yan, F., Ding, Z., et al.: Dynamic evaluation of ecological security in the Yangtze River Economic Zone based on three-dimensional ecological footprint. Progress in Geoscience **43**(06), 1184–1202 (2024)
13. Jia, H., Ma, X., Wang, W., et al.: Spatial identification of ecological protection and restoration in typical karst areas-taking Shilin Yi autonomous county as an example. J. Yunnan Agricult. Univ. (Natural Science) **39**(02), 145–154 (2024)

Light Pollution Control Model Based on Grading and Optimization

Yihao Wu[1(✉)], Chengqi Dong[2], Zefeng Wu[1], Junhao Long[1], and Chenyue Ji[1]

[1] School of Information Science and Engineering, Dalian Polytechnic University, Dalian 116034, China
2085008383@qq.com

[2] School of Environmental Science and Engineering, Nanjing Tech University, Nanjing 211816, China

Abstract. The spread of light pollution engulfing the night sky also has adverse effects on organisms and the environment. In order to alleviate the problem of light pollution, this paper established a comprehensive evaluation model of light pollution degree by optimizing TOPSIS model based on the combination of analytic hierarchy process (AHP) and EWM. In order to measure the level of regional light pollution widely, six effective indicators were selected to evaluate the level of light pollution in different regions, namely: development level, population, air quality, night light intensity, climate, and geography. The weights of selected indicators were calculated by Matlab, and the regional light pollution was evaluated by TOPSIS score.

Ahp and EWM method are used to solve the weight, which makes the model more convincing. Finally, R2 test and f test were carried out on the model, and the model fit reached more than 95%.

Keywords: Evaluation model · Matlab · TOPSIS · AHP

1 Introduction

As early as the 1930s, the international astronomical community already suggested that "it is artificial lighting that causes the fading of the beautiful night sky". In a clear night sky without light pollution, the naked eye can see up to 7,000 stars, but under the influence of city light, only less than 30 can be seen. The vastness and brilliance of the Milky Way galaxy has given mankind much room for imagination, but the overuse of artificial light sources has caused light pollution phenomena such as light invasion, excessive illumination and light clutter [1].

Light pollution has a negative impact on astronomical observations and also produces a waste of resources: all outdoor lighting in the U.S. consumes 30% of the electricity scattered out each year, causing a loss of about $4.5 billion. Light pollution also affects urban environments, plant and animal growth rhythms, and human health to varying degrees, and light pollution is a growing concern for organizations around the world,

D. Han et al. (Eds.): ICEPG 2024, ESE, pp. 954–960, 2025.
https://doi.org/10.1007/978-3-031-86829-0_105

along with water and air pollution. There is an urgent need for good visibility and a safe night environment [2].

Taking into account the background information and constraints identified in the problem statement, the main task of this paper is to determine a measure of the light pollution risk level. We need to consider the influencing factors of the light pollution risk level and establish a measure that can fully reflect the influencing factors [3].

2 Notations

The team selected the level of development, population level, air quality, and the night light intensity as the measurement indicators based on the important factors affecting the level of light pollution, both directly and indirectly, and their relationship with light pollution and data selection are shown in the following Table 1.

Table 1. Influencing factors of light pollution level and data selection

Metrics	Relationship with light pollution	Data Selection
Level of development	Lighting demand is proportional to the level of urban development, which can indirectly reflect the level of light pollution in the city	Annual data of GNP by prefecture-level cities
Population	Light pollution is generated by human activities, so it can indirectly reflect the level of light pollution	Annual data on total population by prefecture-level city
Air quality	Air quality affects the propagation of light through the air and thus the propagation and distribution of light pollution	Annual data of Pm2.5 by prefecture-level cities
Night light intensity	It can reflect the impact of human activities on the natural environment at night, and thus express the luminous intensity of light sources generated by light pollution	Annual data of nighttime lighting level in each prefecture-level city
Climate	Areas with relatively high average annual temperatures are more suitable for human habitation and use more artificial light sources, thus increasing light pollution	Annual average temperature data by prefecture-level city
Geography	Ports usually have a lot of artificial lighting and ship lights, these light sources will emit light pollution to the surrounding environment; the closer the distance to the nearest port means that the more developed the area is, the more serious the light pollution	Distance of each prefecture-level city from the nearest port

3 Application of TOPSIS

In this section, we use TOPSIS to form a set of scoring rules by which to measure light pollution levels.

- Indicator forward.

Among the indicators selected by the team, the geographical location is selected as the distance data from the nearest port, which is inversely proportional to the light pollution level, and the other indicators are all extreme value indicators, so the indicator indicator needs to be positive to make it an extreme value indicator [4].

- Index standardization.

Normalize the normalized matrix and construct the normalized matrix.

$$X\overset{*}{ij} = \frac{X_{ij} - \min(X_j)}{\max(X_j) - \min(X_j)} \tag{1}$$

- Standardization Matrix Construction.

$$Xij* = \begin{bmatrix} X_{11}^* & X_{12}^* & \cdots & X_{1j}^* & \cdots & X_{1m}^* \\ X_{21}^* & X_{22}^* & \cdots & X_{2j}^* & \cdots & X_{2m}^* \\ \vdots & \vdots & \ddots & \vdots & \ddots & \vdots \\ X_{i1}^* & X_{i2}^* & \cdots & X_{ij}^* & \cdots & X_{im}^* \\ \vdots & \vdots & \ddots & \vdots & \ddots & \vdots \\ X_{n1}^* & X_{n2}^* & \cdots & X_{nj}^* & \cdots & X_{nm}^* \end{bmatrix} \tag{2}$$

- Weighted normalization of the matrix.

$$r_{ij} = W_i X_{ij} \tag{3}$$

- The distance to the maximum value and the distance to the minimum value of each indicator are calculated according to the definition of the Euclidean distance formula D_i^+, D_i^-.

$$D_i = \sqrt{\sum_{j=1}^{m} \left(r_{ij} - s_j\right)^2} \tag{4}$$

Finally, the composite evaluation value of each indicator is calculated.

$$C_i = \frac{D_i^-}{D_I^+ + D_i^-} \tag{5}$$

Weights ωj will be explained in detail in the next section

4 Weight Solving Model Based on Hierarchical Analysis and Entropy Weight Method

Both hierarchical analysis and entropy method are common methods to determine weights, and each has its own advantages and disadvantages. The weights derived from the hierarchical analysis method are more dependent on the subjective judgment of people, which can better reflect the judgment criteria of contemporary people combined with the current level of economic and technological development, but lack a certain degree of objectivity; the entropy method can objectively analyze the weights of indicators by analyzing data, but lack the judgment of people's subjective consciousness. By combining the two methods, we can establish a weight solving model based on Analytic Hierarchy Process(AHP) and entropy method to obtain more accurate weighting results [5].

4.1 Analytic Hierarchy Process (AHP) to Determine the Weights

- Judgment matrix was constructed by comparing the importance between two two light pollution measures

$$a_{ij} = \begin{bmatrix} 1 & a_{12} & \cdots & a_{1n} \\ \frac{1}{a_{12}} & 1 & \cdots & a_{2n} \\ \vdots & \vdots & \ddots & \vdots \\ \frac{1}{a_{1n}} & \frac{1}{a_{2n}} & \cdots & 1 \end{bmatrix} \tag{6}$$

- Normalizing this matrix yields the matrix

$$b_{ij} = \frac{\alpha_{ij}}{\sqrt{\sum_{j=1}^{m} \alpha_{ij}^2}} \tag{7}$$

- The consistency index CR (consistency ratio) of the eigenvectors of the judgment matrix is calculated to check the consistency, and if $CR < 0.1$, the judgment matrix passes the consistency test.
- Calculate the weights of each indicator

$$w_{ij} = \frac{\sum_{k=1}^{n} b_{ik}}{\sum_{l=1}^{n} \sum_{k=1}^{n} b_{ik}/b_{il}} \tag{8}$$

- Creation of hierarchical models, the structure diagram is shown in Fig. 1.

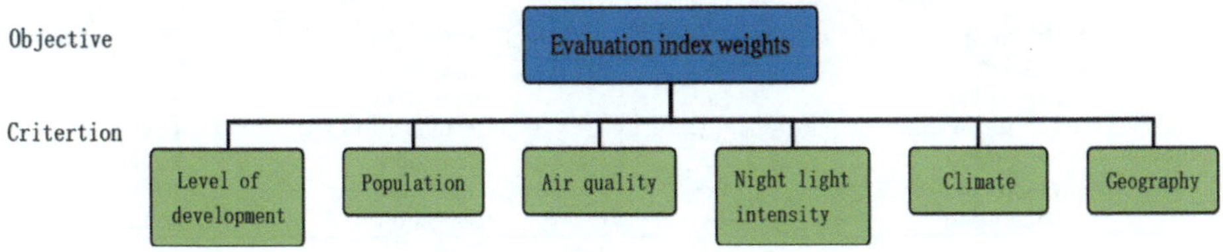

Fig. 1. Hierarchy Chart

4.2 Entropy Weighting Method to Find the Weights

The constructed matrix using the entropy weighting formula (where m is the number of samples) to obtain the weights of the normalized light pollution indicators

$$w_i = \frac{1 - \sum_{j=1}^{m}\left(\frac{x_j}{\sum_{k=1}^{m} x_{kj}} \ln \frac{x_{ij}}{\sum_{k=1}^{m} x_{kj}}\right)}{m-1} \tag{9}$$

4.3 Integrated Weight Solving

The weights of the two methods for each indicator are averaged to find the final weights considering subjective and objective

$$w_i = \frac{w_i + \sum_{j=1}^{Ni} w_{ij}}{2} \tag{10}$$

Solving for the final weights Wi After that, continuing the subsequent steps of topsis, the risk level of light pollution in each city can be measured [6].

5 Application Measuring Risk for Different Location Types

5.1 Apply TOPSIS Model for Scoring Light Pollution Levels

- Collect the corresponding index data, clean the data and then normalize it
- The indicator weights were calculated by the weight solution model based on AHP analysis and entropy weight method, and the results are shown in Fig. 2.

- A comprehensive evaluation model of light pollution risk level was applied to score the light pollution level of these areas, so as to evaluate and compare their light pollution levels, and the light pollution level scores are shown in Fig. 3.

Entropy method	
Projects	Weights (%)
GDP	17.66
Population	34.399
Average temperature	3.712
Air quality	9.523
Night light intensity	32.205
Geography	2.501

AHP	
Projects	Weights(%)
GDP	14.291
Population	28.496
Average temperature	4.689
Air quality	8.523
Night light intensity	40.628
Geography	3.374

Projects	Average weight (%)
GDP	15.9755
Population	31.4475
Average temperature	4.2005
Air quality	9.023
Night light intensity	36.4165
Geography	2.9375

Fig. 2. Weight calculation process chart

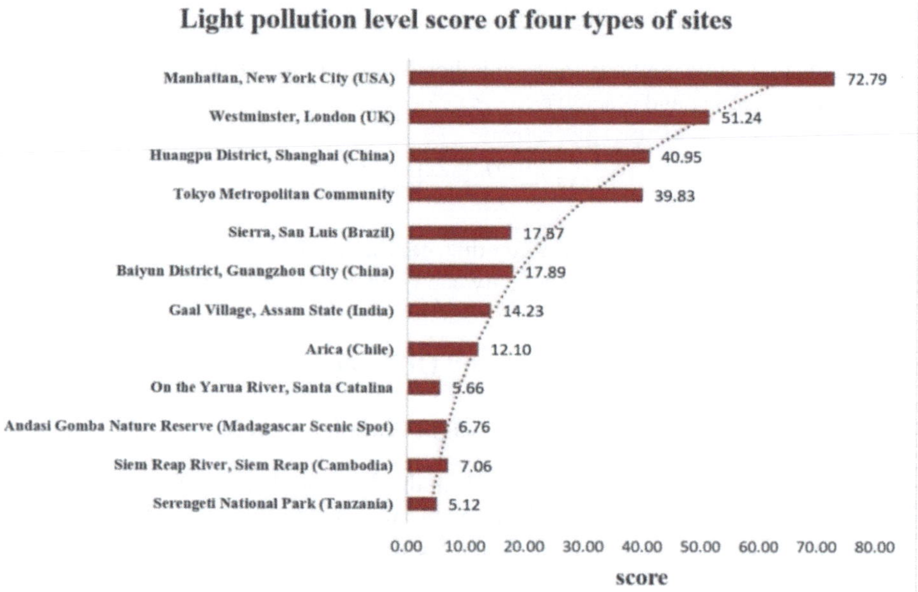

Fig. 3. Light pollution risk level score chart

5.2 Analysis of Model Implementation Effects

The light pollution level scores of the 12 sites selected by the team showed a logarithmic distribution from low to high, with the four types of sites showing light pollution levels of: protected land < rural communities < suburban communities < urban communities.

Protected lands are usually located in remote, inaccessible areas with low human activities, relatively far from cities. Protected objects are also more sensitive to light pollution at night. Therefore, the level of light pollution is relatively low.

Rural communities are relatively less densely populated than urban communities, have relatively less lighting needs, and are not as busy as urban areas, resulting in relatively low levels of light pollution.

Suburban communities are usually located between urban and rural areas. The population density of these communities is relatively low, but with the continuous urbanization, the level of light pollution in suburban communities is gradually increasing. And therefore at slightly higher levels of light pollution relative to rural communities.

Urban communities are usually economic, cultural and political centers with dense populations and high demand for lighting, and therefore have the highest levels of light pollution.

The results obtained from the model application are consistent with the above common sense and the model validation passes.

6 Conclusion

In this paper, TOPSIS and other methods are used to establish a comprehensive evaluation model of light pollution, and the results are as follows: The population density of rural communities is relatively lower than that of urban communities, the demand for lighting is relatively less, and the light pollution level is relatively low. Urban communities, often economic, cultural and political centers, are densely populated and have high demand for lighting, hence the highest levels of light pollution. We combined the analytic hierarchy process (AHP) and entropy method to calculate the weights, and the evaluation model combining subjective and objective made our results more supportive. Our research objects were all regions of China, the data samples were large and reliable, and the CR value of model test was less than 0.1, indicating that our model was objective and accurate, and the model we proposed could score any region. The level of light pollution in the area is derived.

References

1. Megha, K., et al.: Spatial-temporal expansion and determinants of light pollution in India's riparian habitats. Environmental Impact Assessment Review, 98 (2023)
2. Monika, C., Anna, K.: How light pollution can affect spring development of urban trees and shrubs. Urban Forestry & Urban Greening, 77 (2022)
3. Chris, S.-W.: Caterpillars hit by light pollution. New Scientist **251**(3350) (2021)
4. Herdiwijaya, D.: Light pollution at Bosscha Observatory, Indonesia. Journal of Physics: Conference Series **1153**(1) (2019)
5. Lim, H.S., Ngarambe, J., Kim, J.T., Kim, G.: The reality of light pollution: a field survey for the determination of lighting environmental management zones in South Korea. Sustainability **10**(2) (2018)
6. Gun, C.Y.: The Sunlight Reflection Lighting Cases and Light Pollution Possibility. KIEAE Journal **18**(1) (2018)

A Review of the Research and Application of Automotive Remote Supervision Technology Both Domestically and Internationally

Zihan Yin[1,2], Hao Zhang[1], Yanyan Yang[3], Jingyuan Li[1], Hanzhengnan Yu[1], Yu Liu[1](✉), Zhixin Liu[1], and Xionghui Zou[1]

[1] China Automotive Technology Research Center Co., Ltd., Tianjin, China
liuyu2016@catarc.ac.cn
[2] Zhejiang University, Hangzhou, Zhejiang, China
[3] Beijing Municipal Ecological and Environmental Supervision Center, Beijing, China

Abstract. Strengthening remote supervision for heavy-duty vehicles is a crucial measure for pollution prevention. It supports the development of an automotive powerhouse and contributes to building a beautiful China. To address the lack of comparative analysis between domestic and international emission remote supervision technology, a study was conducted. This research also aimed to clarify future development directions. The study found that remote supervision technology is now widely accepted as a key method for online and real-time supervision of vehicle emissions in leading automotive countries. Currently, China's technology has gradually caught up with those of the United States, Europe, and Japan. The implementation of this technology has effectively regulated heavy-duty vehicle emissions and enhanced the enforcement capabilities of regulatory authorities. However, future efforts must focus on improving remote supervision for heavy-duty vehicles. (1) a data quality control system should be established to address data quality issues; (2) methods and devices must be developed to detect and prevent data manipulation by users; (3) further research is needed to advance the application of remote supervision technology in the fields of energy consumption and emissions.

Keywords: remote supervision technology · heavy-duty vehicles · policies and regulations · emission control

1 Introduction

The global environmental pollution issue is becoming increasingly severe. According to relevant statistics [1], the total emissions from vehicles in China reached 14.66 million tons in 2022. Of these emissions, over 80% of nitrogen oxides (NOx) and more than 90% of particulate matter (PM) were attributed to heavy-duty vehicles. The importance and urgency of preventing pollution from vehicles, especially heavy-duty vehicles, are increasingly prominent. To intensify efforts in pollution control, the State Council of China proposed in 2018 to accelerate the implementation of special emission control

D. Han et al. (Eds.): ICEPG 2024, ESE, pp. 961–972, 2025.
https://doi.org/10.1007/978-3-031-86829-0_106

measures for heavy-duty vehicles. Strengthening online supervision is a key measure for controlling vehicle emissions [2]. It helps address the challenges of supervision, enforcement, and evidence collection in the process of vehicle emission control.

To achieve online supervision of vehicle emissions and enhance emission control capabilities, extensive research on remote supervision technology for vehicles has been conducted both domestically and internationally. For instance, the European Union has developed the On-Board Fuel Consumption Supervision (OBFCM) system, the United States has implemented the OBD-II system, and Japan has promoted an enhanced version of OBFCM.

Specifically, the European Union has mandated the installation of OBFCM in vehicles and has established a regulatory framework for data collection and public disclosure. The United States has required vehicles to be equipped with OBD-II systems, which are monitored in real-time by operators. Japan has adopted an enhanced OBFCM system to improve the accuracy of fuel and energy consumption supervision. Simultaneously, China has also launched initiatives for the remote supervision of heavy-duty vehicles. In 2018, the Ministry of Ecology and Environment and the State Administration for Market Regulation jointly issued the national standard "Limits and Measurement Methods for Pollutant Emissions from Heavy-Duty Diesel Vehicles (GB17691-2018)" [3]. This standard formally required the installation of on-board supervision terminals in heavy-duty vehicles to achieve online remote supervision of emissions, providing strong support for the intelligent visualization and dynamic control of vehicle emissions.

In summary, remote supervision of vehicles will be the most crucial regulatory tool for controlling vehicle emissions in the future. Effective emission supervision will significantly enhance the precision management of pollution control for heavy-duty vehicles. Although China has made some progress in remote supervision technology for vehicles, continuous efforts are needed to keep pace with the advancements and applications seen in other countries. Due to the lack of comparative analysis of remote supervision technology between China and other nations, it is challenging to clearly identify the strengths and weaknesses of China's approach, which hinders further development. Therefore, this paper aims to investigate and analyze the policies, regulations, and current research on remote supervision technology for vehicles both domestically and internationally. By exploring the strengths and weaknesses of China's remote supervision technology, this paper seeks to identify the future development directions, thereby supporting China's efforts to surpass global standards and contribute to the development of a powerful automotive industry and a beautiful China [4].

2 Research on International Remote Supervision Technology for Vehicles

2.1 Remote Supervision Technology in Europe

In 2017, the European Union officially issued Regulation (EU) 2017/1151, mandating the installation of On-Board Fuel and/or Energy Consumption Monitoring (OBFCM) devices in light-duty vehicles [5]. These devices recorded fuel consumption and mileage data, which were transmitted wirelessly to a centralized management platform. This system enabled closed-loop supervision and management of carbon emissions from light-duty vehicles within the region.

In November 2022, the European Union proposed the Euro 7 emission standards, which required motor vehicles to be equipped with On-Board Monitoring (OBM) system to monitor NOx and PM emissions. Additionally, heavy-duty vehicles were required to monitor NH3 emissions. The OBM system also had the capability to record instances of excessive emissions and the duration of these exceedances. Furthermore, it issued warnings to the driver when emissions exceeded the set limits, thereby ensuring continuous supervision of vehicle pollutant emissions. The Euro 7 standards are scheduled to be implemented for light-duty vehicles in July 2025, extended to heavy-duty vehicles by 2027, and fully enforced by 2031.

Following the implementation of remote supervision policies, the European Union has made significant progress in reducing vehicle carbon emissions. Continuous carbon emission data from 2021 to 2022 indicate that remote supervision technology effectively tracks the actual emissions of light-duty vehicles in Europe. This has provided substantial data support for the EU's goal of achieving carbon neutrality [6].

2.2 Remote Supervision Technology in the United States

The United States was an early pioneer in the field of remote supervision for vehicles. In 1990, the U.S. implemented the Federal Clean Air Act Amendments, requiring all vehicles to be equipped with On-Board Diagnostics (OBD) system to monitor vehicle emissions. The widely used OBD-II system was effective in detecting emission issues, but it did not provide drivers with immediate remote warnings. The ongoing development of the OBD-III system aims to address this gap by enabling real-time supervision of vehicle operational data and using on-board wireless communication technology for data transmission and aggregation. These data include vehicle identification numbers, fault codes, and positioning information. Through data mining and analysis, the system can effectively assess vehicle emissions, performance, and fault reports, facilitating remote diagnosis of vehicle emissions. When a vehicle's emissions significantly exceed the legal limits or when non-compliance occurs, the system can issue warnings to the driver and, in dangerous situations, take mandatory actions. Additionally, the OBD-III system can provide drivers with vehicle repair recommendations and deadlines for resolving emission issues, thereby ensuring vehicle emission performance and offering strong technical support for sustainable development.

2.3 Remote Supervision Technology in Japan

In 2023, Japan revised the Road Transport Vehicle Act, mandating the installation of an enhanced version of the On-Board Fuel and/or Energy Consumption Monitoring (OBFCM) system [7]. This device was capable of real-time supervision of fuel consumption in traditional internal combustion engine vehicles and collecting data on energy consumption and battery status in new energy vehicles. The regulation required that all new vehicles must be equipped with the enhanced OBFCM starting in October 2023, with existing models required to upgrade their OBFCM devices by September 2025.

The installation of the enhanced OBFCM standardized the collection of fuel and energy consumption data, significantly improving the consistency and accuracy of supervision across different vehicle models. This measure effectively curbed the practice of some companies misreporting vehicle performance parameters [8]. Additionally, the collection of energy consumption and battery status data supports the long-term development of new energy vehicles in Japan, aiding the achievement of carbon neutrality goals.

3 Remote Supervision Technology for Vehicles in China

3.1 Policies and Regulations

During the 14th Five-Year Plan period, China's ecological civilization construction has focused on synergistic efforts to reduce pollution and carbon emissions [9]. Addressing the emissions from heavy-duty vehicles, the "Opinions of the Central Committee of the Communist Party of China and the State Council on Comprehensively Strengthening Ecological and Environmental Protection and Resolutely Fighting the Battle against Pollution" explicitly called for the establishment of an "emission remote supervision platform of heavy-duty vehicle" to achieve real-time supervision and precise control of emissions from heavy-duty vehicles.

The importance of heavy-duty vehicles pollution control has been repeatedly emphasized by the central government. This emphasis was evident from the inclusion of the "special emission control of heavy-duty diesel vehicle" in the State Council's Government Work Report and the Central Financial and Economic Affairs Commission's directive to "fight the tough battle of diesel vehicles pollution control." In June 2018, the State Council issued the "three-year plan of action for winning the war to protect blue skies," which proposed the establishment of an "integrated supervision system for sky, land, vehicles, and people." Here, "vehicles" specifically referred to the remote supervision platform for vehicles. In December of the same year, the Ministry of Ecology and Environment, along with 11 other ministries, jointly issued the "Action Plan for the Battle against Diesel Vehicles Pollution," further underscoring the importance of constructing a remote supervision platform for heavy-duty vehicles.

Against this backdrop, the Ministry of Ecology and Environment organized the development of a comprehensive series of standards titled "Technical Specifications for Remote Supervision of Heavy-Duty Vehicle Emissions" in 2021 [10–12]. These standards defined the technical requirements and implementation plans for remote supervision technology. In addition, the Stage 6b of China's National VI emission standards, issued in 2023, mandated that new vehicles must be equipped with on-board terminals to collect emission data and upload it to the national platform. This ensured the online and remote supervision of heavy-duty vehicle operations and emissions [13, 14]. Subsequently, China officially commenced preliminary research on the National VII emission standards in 2023. The new standards will place significantly greater emphasis on the advancement and widespread application of remote supervision technology for heavy-duty vehicles [15].

3.2 Remote Supervision Platform

In 2019, China officially established and successfully launched a multi-tiered remote supervision platform for heavy-duty vehicles, encompassing the "Vehicle—Enterprise Platform—National Platform—Local Platform" structure. This platform relied on advanced technologies such as intelligent vehicle networks and big data analytics.

Through on-board terminals, the platform collected real-time data on vehicle parameters, including speed, NOx emissions, engine parameters, and vehicle identification number, as detailed in Table 1. The collected data was transmitted via wireless networks to the national platform. By systematically analyzing this emission data, the platform facilitated real-time online remote supervision of vehicle operation and emission statuses.

Benefiting from national policy support, China's heavy-duty vehicle remote supervision platform have made significant progress. Many regions have mandated the use of remote supervision platforms for emission control. Cities such as Beijing, Shanghai, and Zibo have introduced local standards requiring that vehicles sold or used must be equipped with remote supervision technology that complies with these standards.

Table 1. Remote supervision data parameters

No.	*Data item*	*No.*	*Data item*
1	Speed	12	SCR outlet temperature
2	Atmospheric pressure	13	DPF differential pressure
3	Engine net output torque	14	Engine coolant temperature
4	Engine friction torque	15	Fuel tank level
5	revolutions per minute	16	Positioning status
6	Engine fuel flow	17	Longitude
7	Output value of SCR upstream NOx sensor	18	Latitude
8	Output value of SCR downstream NOx sensor	19	Accumulated mileage
9	Excess reactant	20	Time information
10	Air inflow	21	Vehicle identification number
11	SCR inlet temperature	/	/

In Beijing, the efficient regulation enabled by the heavy-duty vehicle remote supervision platform led to a 43% reduction in NOx emissions from heavy-duty vehicles in 2020 compared to 2017. This reduction surpassed that achieved in California during the same period [2]. In Shanghai, the remote supervision platform has integrated over 100,000 heavy-duty vehicles. By dynamically analyzing large-scale emission data, the platform tracked and assessed high-emission vehicles across different models and seasons, supporting the development of vehicle exemption programs [16]. In Zibo, Shandong Province, the remote supervision platform's emission data has facilitated the creation of a blacklist and whitelist for heavy-duty vehicles. The "whitelist" management approach benefited the public, while the "blacklist" management focused on stringent inspections, continuously enhancing remote supervision for heavy-duty vehicles [17].

3.3 Comparison of Domestic and International Remote Supervision Technology

Research on the emission remote supervision technology policies and current practices in China, Europe, the United States, and Japan revealed that using the technology for real-time emission control has become a consensus among most major automotive nations. However, the focus of remote supervision varied across countries.

Internationally, Europe placed more emphasis on supervision light-duty vehicle fuel consumption and carbon emissions, with NOx and PM emission supervision still in the planning stages. The United States relied on the iterative upgrade of OBD systems for supervision, without establishing separate remote supervision platform. Japan has upgraded the European OBFCM model to an enhanced OBFCM system, which not only monitors fuel consumption and carbon emissions but also includes supervision for electric vehicles' energy consumption and battery health. However, the supervision of vehicle pollutant emissions has not yet been incorporated into the plan.

In contrast, China has developed remote supervision for both light-duty and heavy-duty vehicles, as well as new energy vehicles. The domestic remote supervision platform primarily focuses on heavy-duty vehicles, while new energy vehicle supervision is managed by a dedicated national platform. Light-duty vehicles are also being monitored based on European OBFCM practices.

Currently, China's remote supervision technology (the main subject of this study) has largely caught up with the automotive powers of Europe, the United States, and Japan. With the implementation of remote supervision, China has effectively regulated and controlled heavy-duty vehicle emissions [18], and the administrative enforcement capabilities of government regulatory bodies regarding emission violations have been significantly strengthened.

4 Application of Heavy-Duty Vehicle Remote Supervision Platforms

4.1 Accuracy of Remote Supervision Data

As a new method for supervision heavy-duty vehicle emissions, the accuracy of remote supervision data needs further validation. Comparing remote supervision data with dynamometer data from laboratory tests and Portable Emissions Measurement Systems (PEMS) data from actual road conditions is significant [19]. Studies [20–24] have compared remote supervision data with dynamometer data and PEMS data. Results showed that the difference in fuel consumption calculated from remote supervision data and dynamometer data was within 10%. The accuracy of CO_2 emission data compared to PEMS data averaged 2.06%. The correlation coefficient (R^2) for vehicle speed information from remote supervision data and PEMS data exceeded 50% with R^2 greater than

0.99. NOx emission data also showed a high level of consistency with R^2 greater than 0.80, especially at higher NOx concentrations. After correction using machine learning algorithms, the cumulative NOx emission error was 2.4%, with a Pearson correlation coefficient (R) of 0.92. For NOx critical and severe exceedance conditions, relative errors were 4.7% and −2.1%, respectively. Additionally, article [25] demonstrated the use of remote supervision data to construct NOx emission factor models, with R^2 reaching 0.97. These studies indicated a high degree of consistency between remote supervision data and actual measurements for heavy-duty vehicle speed, fuel consumption, and NOx emissions, confirming the reliability and credibility of remote supervision data [26, 27].

4.2 Study of Heavy-Duty Vehicle Fuel Consumption and Emission Characteristics

Fuel consumption and emissions are critical performance indicators for heavy-duty vehicles. The remote supervision technology enables online and remote supervision of heavy-duty vehicle fuel consumption and emissions, helping to identify characteristic patterns and support vehicle performance optimization. It also aids in calculating emission factors and constructing emission inventories [28].

Zhang et al. [29] found that NOx conversion efficiency significantly improved when exhaust temperature exceeded 180 °C. NOx emission rates initially increased with vehicle speed and then decreased. Zhao et al. [30] discovered that heavy-duty vehicle fuel consumption and emissions showed significant seasonal variations. Cold temperatures in winter negatively impacted NOx emissions, and fuel consumption peaked in the first and fourth quarters, with the lowest levels occurring in the second quarter. Zhou et al. [31] examined fuel consumption data under different temperatures and altitudes, finding the highest fuel consumption at temperatures between 10–20 °C and an increase in fuel consumption with higher altitudes. Feng et al. [20] used micro-operation modes and traffic flow methods to analyze transient fuel consumption data from the remote supervision system, observing that relative fuel consumption factors decreased rapidly within the speed range of 0–30 km/h and then declined more slowly. Additionally, works [32–34] provided quantitative analyses of how vehicle speed, acceleration, and torque affected fuel consumption and emissions, showing that low speeds and aggressive driving behaviors increased emissions. These studies highlighted the broad application potential of remote supervision data in researching vehicle fuel consumption and emission characteristics.

Emission factors and inventories are crucial quantitative tools for vehicle emission control. Remote supervision technology offers a new approach for obtaining these metrics. Researchers in works [34–37] have used remote supervision data to accurately calculate CO2, NOx, and energy consumption emission factors for heavy-duty vehicles. Liu et al. [35] verified the correlation between NOx emission factors and average NOx concentration using remote supervision data, achieving an R^2 above 0.9. Ge et al. [38] compared CO2 emissions for heavy-duty vehicles under China VI and China V standards, finding that CO2 emissions under the China VI standard were approximately 10% higher than under the China V standard, demonstrating the effectiveness of new emission technology under China VI standards. Liu et al. [39] used remote supervision data to calculate localized carbon emission factors and build a carbon emission measurement system for urban roads. As a powerful tool for supervision heavy-duty vehicle fuel consumption and emissions, the current application of remote supervision technology in emission factors and inventories has not yet reached its full potential.

4.3 Online Identification of High Emission Vehicles

Supervision and identifying emission violations in heavy-duty vehicles has been challenging due to weak regulatory enforcement and difficulties in detecting violations through periodic inspections and roadside checks. With the development of remote supervision technology, online supervision and identification of high-emission heavy-duty vehicles have become feasible. Xu et al. [40] used the shortest distance method for clustering and developed a back-propagation neural network model to effectively identify daily NOx emission level changes in heavy-duty vehicles, providing technical support for preliminary screening of high-emission vehicles. Wang et al. [41] developed a large-scale screening method based on neural networks, utilizing remote supervision data such as vehicle speed, acceleration, and NOx emissions to calculate NOx emission intensity and identify high-emission vehicles effectively. Wu et al. [24] proposed a short-trip reconstruction PEMS method based on remote supervision data, creating a new vehicle operating condition curve reconstruction method and emission evaluation model to effectively identify changes and violations in vehicle emissions. Additionally, works [42–44] employed grid and zonal methods to calculate NOx-specific emissions, further supporting emission assessment methods in heavy-duty vehicle remote supervision. These studies demonstrated the application potential of remote supervision data in identifying high-emission heavy-duty vehicles, enhancing regulatory efficiency, vehicle emission control, and urban environmental protection. However, the accuracy of identifying high-emission vehicles using remote supervision data still needs further improvement.

4.4 Supervision of Heavy-Duty Vehicle SCR System

Selective Catalytic Reduction (SCR) technology is crucial for effectively reducing NOx emissions from heavy-duty vehicles. Real-time supervision of urea consumption in the SCR system and feedback to drivers and national supervision platform is essential for ensuring emission performance. Zheng, Wang, and others [45–48] calculated the actual urea consumption in the SCR system using parameters such as urea injection volume and urea level, and compared it with engine NOx emissions to effectively determine SCR urea consumption. Additionally, Pei et al. [46] assessed the reasonableness of SCR urea consumption by calculating the urea-fuel ratio. However, limitations in remote supervision for SCR system resulted in the acquisition of low-frequency resolution urea level data, making these methods challenging to apply effectively. Therefore, the SCR system supervision methods based on heavy-duty vehicle remote supervision require further research.

5 Conclusion

This paper focuses on emission remote supervision technology for heavy-duty vehicles, conducting a thorough investigation of both domestic and international technology and their applications. The research findings highlight a global consensus among major automotive nations on the importance of using remote supervision technology as a primary method for real-time, online supervision of vehicle exhaust emissions. While China's emission remote supervision technology for heavy-duty vehicles has gradually caught up with leading automotive nations such as the United States, Europe, and Japan, there is still a need for continued efforts to advance this technology. Accordingly, this paper offers a detailed examination of current challenges faced by China in this domain and presents future development recommendations for emission remote supervision technology of heavy-duty vehicles in China.

5.1 Constructing a Quality Control System for Heavy-Duty Vehicle Remote Supervision Data

In the application of emission remote supervision technology for heavy-duty vehicles, significant data quality issues have been identified, including data loss, data outliers, data anomalies, and data invalidity. For instance, in April 2022, the average data unavailability rate for SCR downstream NOx sensor outputs for all Euro VI vehicles was as high as 37.67%, with approximately 26.52% of vehicles having a data unavailability rate exceeding 60%. The persistence of these data quality issues undermines the effectiveness of online remote supervision of heavy-duty vehicle emissions. To address this, it is essential to develop methods for identifying data quality issues, correcting abnormal data, and re-integrating clean data. Establishing a robust quality control system for remote supervision data will ensure data quality and consistency, thereby significantly enhancing the regulatory capabilities of emission remote supervision technology.

5.2 Developing Methods and Devices for Detecting Data Falsification by Heavy-Duty Vehicle Users

Data falsification by heavy-duty vehicle users presents a critical challenge to the accuracy of emission remote supervision platform. For example, some users may reduce or cut off urea solution consumption to lower operating costs [49], while simultaneously falsifying NOx emission and urea level data transmitted from on-board terminals. This deceptive behavior undermines the remote supervision platform's ability to accurately detect emissions exceedance, thereby compromising the overall effectiveness of emission supervision and regulatory enforcement. Therefore, there is an urgent need to develop methods and devices for detecting data falsification. This will enable accurate online detection of data manipulation by heavy-duty vehicle users, thereby enhancing the integrity and reliability of emission remote supervision platform. Addressing this issue is crucial for maintaining the effectiveness and credibility of emission control measures and advancing remote supervision platform for heavy-duty vehicles.

5.3 Addressing the Slow Progress in Data Application by Further Developing Research Potential

Currently, while the emission remote supervision technology has made notable progress in terms of basic functionality and development, its application in several critical areas remains limited. These areas include research on fuel consumption and emission characteristics, calculation of emission factors, construction of emission inventories, identification of high-emission vehicles, and supervision of SCR system. To address these gaps, it is essential to leverage advanced technologies such as the Internet of Things (IoT), big data analytics, and sophisticated machine learning models. Further theoretical research and practical application of remote supervision data are crucial. This will support fundamental research in heavy-duty vehicle energy consumption and emissions, and provide valuable recommendations for the control of motor vehicle pollution in China. Enhancing these aspects is a key priority for the future development of emission remote supervision technology of heavy-duty vehicles.

References

1. Ministry of Ecology and Environment: China mobile source environmental management annual report. Environ. Prot. **52**(02), 48–62 (2023)
2. School of Environment, Tsinghua University: "The need for synergistic efforts to reduce pollution and carbon emissions from heavy-duty vehicles under the 'dual carbon' strategy (2020). https://www.env.tsinghua.edu.cn/info/1251/7809.htm
3. Ministry of Ecology and Environment: Pollutant emission limits and measurement methods for heavy-duty vehicles (China Stage VI): GB 17691–2018. China Environmental Science Press, Beijing (2018)
4. Central Committee of the Communist Party of China and the State Council: Opinions on comprehensively promoting the construction of Beautiful China (2024). https://qstheory.cn
5. Capelec: OBFCM: transmission of new data becomes mandatory from 2023 (2022). https://www.capelec.com/en-GB/blog/obfcm-transmission-new-data-becomes-mandatory-2023
6. European Parliament and Council: Regulation (EU) 2023/851 of the european parliament and of the council of 19 April 2023 amending regulation (EU) 2019/631 as regards strengthening the CO2 emission performance standards for new passenger cars and new light commercial vehicles in line with the Union's increased climate ambition (text with EEA relevance), pp. 5–20 (2023)
7. Japanese Industrial Standards Committee: Product safety guidelines for non-electrical products, pp. 3–7 (2023)
8. Ministry of Health, Labour and Welfare, Japan: The 4th national health survey analysis report, pp. 1–8 (2022)
9. Chinese Government Website: Xi Jinping: Striving to build a modernization where man and nature coexist harmoniously (2024). https://www.gov.cn
10. Ministry of Ecology and Environment: Technical specification for remote supervision of heavy-duty vehicle emissions part 1: on-board terminal: HJ 1239.1-2021 (2022). https://www.mee.gov.cn/ywgz/fgbz/bz/bzwb/other/hjbhgc/202201/t20220119_967608.shtml
11. Ministry of Ecology and Environment: Technical specification for remote supervision of heavy-duty vehicle emissions part 2: enterprise platform: HJ 1239.2-2021 (2022). https://www.mee.gov.cn/ywgz/fgbz/bz/bzwb/other/hjbhgc/202201/t20220119_967610.shtml

12. Ministry of Ecology and Environment: "Technical specification for remote supervision of heavy-duty vehicle emissions part 3: communication protocol and data format: HJ 1239.3-2021 (2022). https://www.mee.gov.cn/ywgz/fgbz/bz/bzwb/other/hjbhgc/202201/t20220119_967611.shtml
13. Ministry of Ecology and Environment: "Announcement on the implementation of the national VI emission standards for vehicles (2020)
14. Ministry of Industry and Information Technology: Announcement on the implementation of the national VI emission standards for heavy-duty diesel vehicles by three ministries (2021)
15. State Council: Notice on the issuance of the action plan for continuous improvement of air quality (2021)
16. Shanghai Municipal Housing and Urban-Rural Development Commission: Notice on strengthening the planning and construction of electric vehicle charging facilities (2022). https://sthj.sh.gov.cn/hbzhywpt1272/hbzhywpt1023/20221201/8f7c8c91c8cb4ec0b2a61fcc3f5bd2ff.html
17. Qian, F., Liu, M.J., Wang, M.D., Wang, J., Yang, D., Hu, D.: Monitoring platform of heavy-duty vehicle pollutant emission based on distributed micro-service architecture. Comp. Sys. Appl. **31**(05), 94–101 (2022)
18. China Science and Technology Website: The first domestic report on green and low-carbon development of mobile sources released (2024). https://www.stdaily.com/index/kejixinwen/202407/360e085561ff48a4bf87e143c0849177.shtml
19. Society of Automotive Engineers of China: Technical specification for consistency testing of remote supervision data for heavy-duty vehicles: T/CSAE 258-2022 (2022)
20. Feng, Q., et al.: Analysis of real-world fuel consumption characteristics and high fuel consumption conditions of heavy-duty commercial diesel vehicles on OBD method. In: China Society of Automotive Engineers, Proceedings of the 2019 Annual Meeting of the China Society of Automotive Engineers (4). China Automotive Technology & Research Center Co., Ltd., p. 7 (2019)
21. Ren, S.J., Tong, C., Li, G., Zhang, C., Li, T.T., Liu, D.: A comparison study between heavy-duty remote monitoring and on-road experimental data. Journal of Automotive Safety and Energy **14**(02), 239–248 (2023)
22. Liu, B.L., Zhen, K., Bao, J.J., Ren, S.J., Yang, X.Z., Liu, L.: Research on the accuracy of remote monitoring of NOx emission of heavy duty diesel vehicle. Automotive Practical Technology **46**(01), 113–116 (2021)
23. Wu, C.L., et al.: Research on NOx Emission Calculation and Evaluation Method of Heavy-Duty Diesel Vehicles Based on On-Board Remote Emission Supervision. In: Proceedings of the 26th Atmospheric Pollution Control Technology Seminar, pp. 194–201 (2022)
24. Wu, C.L., Bai, X.X., Li, X., Liu, W.L., Jing, X.J., Pei, Y.X.: Study on evaluation of NOx emission from heavy-duty diesel based on vehicle remote monitoring. Environ. Superv. Manage. Technol. **35**(03), 49–52 (2023)
25. Hu, Q.Y., Xu, W.B., Huang, C., Ren, H.J.: Research on NOx emission characteristics of China V heavy-duty diesel vehicles based on on-board remote emission monitoring system. In: Proceedings of the 24th Atmospheric Pollution Control Technology Seminar (2021)
26. Zhang, S.J., Zhao, C., Wu, Y.: Method for determining the quality of OBD remote online supervision data. Beijing: CN201911075574.4 (2020)
27. Yang, L.H.Z.: Evaluating vehicle fuel consumption and nitrogen oxides emission characteristics based on on-board diagnostic approach, pp. 44–61. Tsinghua University, Beijing (2016)
28. Liu, H.Y., Zhang, X.W., Ren, S.J., Wang, T.Y.: Review on remote OBD system data analysis of heavy-duty diesel vehicle. Energy Environmental Protection **37**(01), 58–64 (2023)
29. Zhang, C.: Analysis of NOX emission characteristics of heavy-duty vehicles based on monitoring platform. Hebei University of Technology (2020)

30. Zhao, M.: Analysis and research on operating characteristics of heavy diesel vehicle based on remote transmission technology. Tianjin University of Technology (2023)
31. Hua, Z., Zhao, H.W., Feng, Q.: Effects of environmental parameters on real-world NOx emissions and fuel consumption for heavy-duty diesel trucks using an OBD approach. In: International Powertrains, Fuels & Lubricants Meeting (2018)
32. Yang, H.Y.Z., Zhang, S.J., Wu, Y.: Evaluating real-world CO2 and NOx emissions for public transit buses using a remote wireless on-board diagnostic (OBD) approach. Environ. Pollut. **218**, 453–462 (2016)
33. Yang, Q., Hu, Q.Y., Huang, C., Chen, Y., Liu, J.D., Li, L.: Hybrid electric bus SCR system operation and NOx emission characteristics based on remote communication technology. Environmental Science **39**(10), 4463–4471 (2018)
34. Zhang, S.J., He, L.Q., Wu, Y.: Method for acquiring NOx emission factors from heavy-duty vehicle exhaust based on OBD remote emission supervision data. Beijing, CN201911075598.X (2020)
35. Liu, S.C., Zhang, X., Liu, J.A., Yang, B.H., Ji, S.B., Pu, Y.X.: Research on correlation between NOx emission factors and concentrations of heavy-duty diesel vehicles. Environ. Sci. **37**(03), 545–553 (2024)
36. Zhang, X.W., Li, J.Y., Yang, Z.W., Xie, Z.K., Li, T.T.: Accuracy analysis of carbon emissions measurement of heavy heavy-duty diesel vehicles based on remote data. Environmental Science **42**(10), 4565–4570 (2022)
37. Xu, H.C., et al.: Research on accounting methods for road traffic carbon emissions. J. Automot. Eng. **13**(04), 496–505 (2023)
38. Ge, Z.H., Yin, H., Xu, L., Yang, Y., Ji, Z., Huang, Y.: Emissions of nitrogen oxides and carbon dioxide from heavy-duty diesel vehicles based on remote monitoring. China Environ. Sci. **44**(02), 646–653 (2024)
39. Liu, Y., Zhou, T., Sun, Q.M.: Study on urban road vehicle carbon emission calculation integrating multi-source data. In: Resilient Transportation: Quality and Service—Proceedings of the 2023 Annual Meeting of the China Urban Transportation Planning Association. Chongqing Institute of Transportation Planning, 11 (2023)
40. Xu, W.B., Huang, C., Ren, H.J., Hu, X.Y.: Diagnosis of NOx emission level of diesel vehicles based on remote online monitoring terminal. Acta Sci. Circum. **41**(06), 2329–2339 (2021)
41. Wang, J.G., et al.: Method for large-scale screening of high NOx emissions in heavy-duty diesel vehicles based on operating conditions. Tianjin, CN202011568576.X (2022)
42. Li, J.Y., Li, Y., Fu, T.Q.: Emission calculation method based on remote supervision data. Hubei, CN202011017250.8 (2021)
43. Feng, Q., et al.: Evaluation method for fuel consumption and NOx emissions of heavy-duty vehicles under real-world operating conditions based on vehicle networking. Tianjin, CN202110814296.0 (2024)
44. Liu, B.X., Feng, Q., Yang, Y.Y.: Remote online supervision method for NOx and CO2 emissions of heavy-duty vehicles. Beijing, CN202210618781.5 (2022)
45. Zheng, P., et al.: Statistical analysis system and method for urea consumption of commercial vehicles. Hubei, CN202210832957.7 (2022)
46. Pei, Y.Q., et al.: Diesel engine urea-to-fuel ratio calculation method based on NOx sensor values. Tianjin, CN202211262078.1 (2023)
47. Lu, H.L., Zhang, P., Meng, D.C., Wang, R., Gu, W.W.: Method and device for determining urea consumption based on vehicle networking. Beijing, CN202010469032.1 (2021)
48. Wang, T.T.: A method of remotely diagnosing the rationality of urea consumption in a diesel vehicle SCR system. Jiangxi, CN201910953378.6 (2020)
49. Li, X., Wu, C.L., Liu, W.L., Bai, X.X., Yang, Y.Z.: Research overview on the application of the emission remote monitoring platform for China VI heavy-duty diesel vehicles. Internal Combustion Engines **40**(03), 19–25 (2024)

Research on the Leaching Effect of Simulated Neutral and Sulfuric Acid Rain on Ni-Contaminated Fine Sand

Qiang Shan[1,2], Xizhao Tian[1,2], Yanling Lai[1,2], Ying Wang[1,2], Bingtai Ma[1,2], and Weiyan Chen[3,4,5](✉)

[1] Hebei Key Laboratory of Geological Resources and Environment Monitoring and Protection, Shijiazhuang 050021, China

[2] Hebei Geological Environment Monitoring Institute, Shijiazhuang 050021, China

[3] Institute of Disaster Prevention, Sanhe 065201, China
1311164560@qq.com

[4] Hebei Key Laboratory of Resource and Environmental Disaster Mechanism and Risk Monitoring, Sanhe 065201, China

[5] Beijing Disaster Prevention Technology Co., Ltd., Beijing 101199, China

Abstract. These experiments simulate the neutral precipitation and sulfuric acid precipitation leaching nickel polluted fine sand under diverse precipitation rate, explore the removal effect of leaching process and to provide basic research data support for the study of groundwater pollution. The research shows, when choose neutral water for leaching, the removal rate of nickel were 7.57%, 2.81%, and 0.67% under violent rain, very heavy rain, and heavy rain; when choose sulfuric acid water with pH 5.02, the removal rate were 17.03%, 12.89%, and 2.31%. Acidic rain increased the Ni removal rate by 1.64% to 10.09% compared to neutral rain (pH 7.0), with an average increase of 7.06%. The most significant improvement was seen in the violent rain group. Nickel was more easily leached form fine sand under acid environment.

Keywords: Simulated Natural Precipitation · Sulfuric Acid-type Acid Rain · Leaching · Fine Sand · Nickel Pollution

1 Introduction

According to the 2014 China Ecological Environment Status Bulletin, heavy metals are the primary pollutants affecting the soil environmental quality of agricultural land in China, with a nationwide total exceeding standard rate of 16.1%. The rate of heavy metal contamination in cultivated land can reach up to 19.4%, with cadmium (Cd), mercury (Hg), arsenic (As), copper (Cu), lead (Pb), chromium (Cr), and nickel (Ni) being the most prevalent [1]. The natural background levels of heavy metals in soil are generally higher in southern agricultural regions compared to northern areas, with the natural background concentration of nickel typically ranging from 20 to 30 mg/kg [2].

D. Han et al. (Eds.): ICEPG 2024, ESE, pp. 973–979, 2025.
https://doi.org/10.1007/978-3-031-86829-0_107

Ni, due to its stable chemical properties and desirable characteristics such as good plasticity, corrosion resistance, and magnetic properties, finds wide applications in steel smelting, electroplating processing, battery manufacturing, alloy production, and Ni-based catalyst production [3]. The distribution of nickel-contaminated soil is strongly associated with population density and industrial activities, predominantly located in Southwest China, East China, and South China. Pollution levels in these regions typically range from 90 to 300 mg/kg [4]. These areas typically coincide with regions experiencing frequent rainfall and significant acid rain deposition [5, 6], where the magnitude of natural rainfall directly influences the leaching speed of Ni from soil into groundwater environments.

This study aims to investigate different leaching effects under varying rainfall conditions by simulating natural rainfall and sulfuric acid-type acid rain leaching on Ni-contaminated fine sand. It seeks to provide fundamental research data support for preventing groundwater pollution.

2 Experimental Devices

Leaching apparatus construction, from top to bottom, is equipped with a water supply, a water distributor, a leaching column, and a leaching collection system (see Fig. 1). The leaching column is an acrylic cylinder with a height of 50 cm and a diameter of 10 cm, equipped with an outlet at every 10 cm interval to collect leachate. The column is filled with layers of quartz sand, polluted sand, a filter screen, and quartz sand. The quartz sand layer is 5 cm in height, while the polluted sand layer is 30 cm in height.

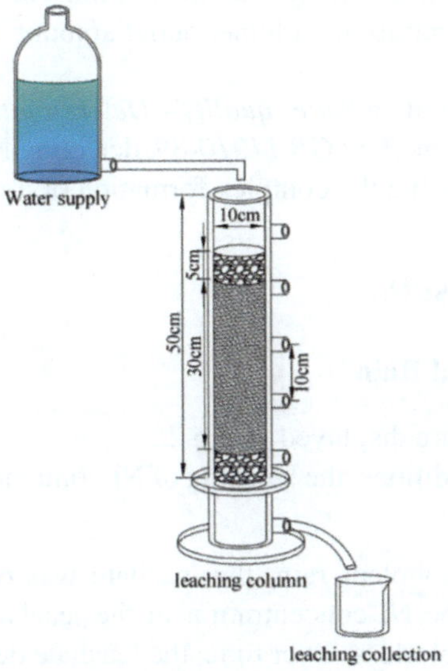

Fig. 1. Leaching Apparatus

Fine sand was purchased from a construction site which is located in Yancang Road, Yaojiao Economic and Technological Development Area, Sanhe City, Hebei Province, and was then ground to pass through 0.10 mm and 0.25 mm sieves. Each column contains 3.5 kg of fine sand, with Ni artificially added using nickel sulfate at a concentration of 600 mg/kg.

3 Experimental Methods

3.1 Calculate the Amount of Leachate

The precipitation rate is determined based on the *Grade of Precipitation GB/T 28592–2012*, with the precipitation volume calculated over 24-h periods. The precipitation rate for heavy rain, defined as between 25.00 and 49.90 mm over a 24-h period, is categorized as very heavy rain for 50.00 to 99.90 mm, and as violent rain for 100.00 to 249.90 mm, both over a 24-h period, respectively [7].

The mid-value of each precipitation grade was selected, resulting in precipitation amounts of 37.45 mm, 74.95 mm, and 174.95 mm per unit area for heavy rain, very heavy rain, and violent rain, respectively. The diameter of leaching column is 10cm, the cross-section area is 7.85×10^{-3} m^3. The 24-h addition volumes for each group are 293.98 ml, 588.36 ml, and 1373.36 ml, respectively.

3.2 Process of Experiment

Neutral water and a sulfuric acid solution with a pH of 5.02 were used as leaching agents. The leaching column was continuously washed with precipitation of varying intensities, including heavy rain, very heavy rain, and violent rain. The leachate was then collected and analyzed for Ni concentration, volume, outlet amount, and the residual Ni content in the soil.

The testing method based on *Water quality—Determination of nickel—Dimethylglyoxime spectrophotometric method GB 11910-89*, detecting Ni content by measuring the difference in absorbance using the complex formation of dimethylglyoxime and Ni [8].

4 Experimental Results

4.1 Leaching by Neutral Rain

The experimental results are displayed in Fig. 2.

Under neutral rain conditions, the leaching of Ni-contaminated soil exhibits different behaviors.

(1) During leaching with violent rain, the leachate was obtained during the second round of leaching. The Ni concentration in the leachate was 28.13 mg/L, and it was transparent and colorless. Over time, the leachate developed a yellow color and showed sediment accumulation. After 108 h of leaching, the Ni concentration could not be accurately measured, as it dropped below 0.13 mg/L and began to tailing, which is beyond the detection limit of our instrument. The residual Ni content in the soil is 554.57 mg/kg.

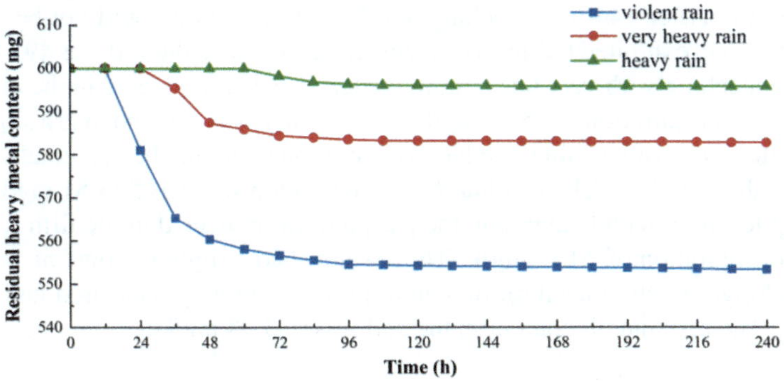

Fig. 2. Variation in Ni Concentration Over Time in Fine Sand Columns Leached with Neutral Rain of Different Grades

(2) During leaching with very heavy rain, the leachate was collected in the third round, with a Ni concentration of 31.47 mg/L. The leachate was clear but had a slight yellow tint. After 120 h of leaching, the Ni concentration could not be accurately measured and exhibited tailing. The residual Ni content in the soil was 583.16 mg/kg.
(3) During leaching with heavy rain, the leachate was collected in the sixth round, with a Ni concentration of 11.69 mg/L. The leachate had a slight yellow tint. After 156 h of leaching, the Ni concentration could not be accurately measured and exhibited tailing. The residual Ni content in the soil was 596.00 mg/kg.

4.2 Leaching by Sulfuric Acid Rain

The experimental results are displayed in Fig. 3.

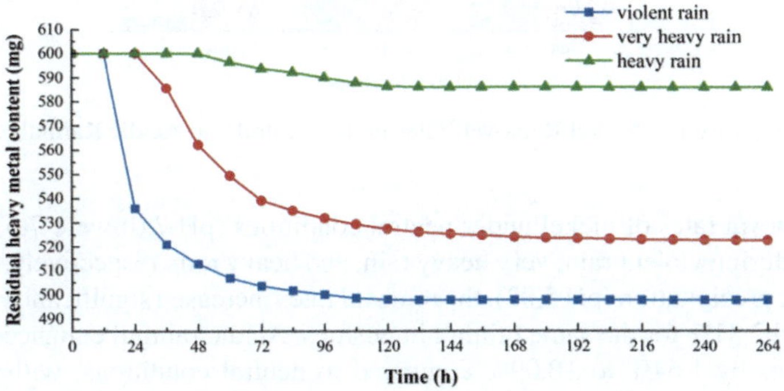

Fig. 3. Variation in Ni Concentration Over Time in Fine Sand Columns Leached with Acid Rain of Different Grades

Under acid rain conditions, the leaching of Ni-contaminated soil exhibits different behaviors.

(1) During leaching with violent rain, the leachate was collected in the second round, with a Ni concentration of 101.47 mg/L. The leachate was clear but had a slight

yellow tint. After 144 h of leaching, the Ni concentration could not be accurately measured and exhibited tailing. The residual Ni content in the soil was 497.86 mg/kg.

(2) During leaching with very heavy rain, the leachate was collected in the third round, with a Ni concentration of 85.58 mg/L. The leachate was clear but had a slight yellow tint. After 192 h of leaching, the Ni concentration could not be accurately measured and exhibited tailing. The residual Ni content in the soil was 523.51 mg/kg.

(3) During leaching with heavy rain, the leachate was collected in the fifth round, with a Ni concentration of 20.13 mg/L. The leachate had a slight yellow tint. After 180 h of leaching, the Ni concentration could not be accurately measured and exhibited tailing. The residual Ni content in the soil was 585.98 mg/kg.

4.3 Leaching by Sulfuric Acid Rain

The removal rate of Ni varies under neutral and acidic rainfall conditions, as illustrated in Fig. 4.

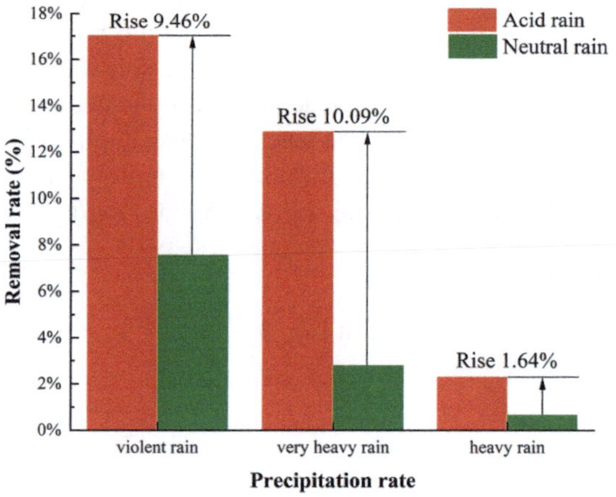

Fig. 4. Comparison of Nickel Removal Rates under Neutral and Acidic Rainfall Conditions

The removal rates of nickel under neutral conditions (pH 7.0) were 7.57%, 2.81%, and 0.67% during violent rain, very heavy rain, and heavy rain, respectively. In contrast, under acidic precipitation (pH 5.02), the removal rates increased significantly to 17.03%, 12.89%, and 2.31% for the same rainfall intensities. Acidic rainfall enhanced the nickel removal rates by 1.64% to 10.09% compared to neutral conditions, with an average increase of 7.06%. This phenomenon may be attributed to the high concentration of hydrogen ions in acidic rainfall, which can interact with soil particles. The hydrogen ions facilitate the exchange of nickel ions adsorbed on the particles with other cations, thereby promoting the migration of nickel into the water.

5 Conclusion

1. Natural precipitation has a limited ability to wash Ni-contaminated soil, with removal rates of 7.57%, 2.81%, and 0.67% under neutral conditions during violent rain, very heavy rain, and heavy rain, respectively. Under acidic precipitation (pH 5.02), the rates improved to 17.03%, 12.89%, and 2.31%.
2. Precipitation pH is the key factor in Ni leaching. Acidic rain increased the Ni removal rate by 1.64% to 10.09% compared to neutral rain (pH 7.0), with an average increase of 7.06%. The most significant improvement was seen in the violent rain group.
3. Precipitation grade is a key factor in Ni leaching. The removal rate of Ni increased by 14.72% under acidic conditions and by 6.90% under neutral conditions in the heavy rain group.

Acknowledgements. Funding Projects: This work was supported by Open Fund for Hebei Key Laboratory of Geological Resources and Environment Monitoring and Protection "Research on electric and chemical oxidation combined remediation technology for coking polluted soil" (JCYKT202005) and "Research on the strength deterioration of refuse soil and slope stability under the coupling of temperature and leachate" of the Natural Science Foundation of Hebei Province (E2021210007).

References

1. Ministry of Ecology and Environment of the People's Republic of China: https://www.mee.gov.cn/hjzl/sthjzk/zghjzkgb/201605/P020160526564730573906.pdf. 27 July 2015
2. Li, B.: Background Values and Baselines of Zu, Cu, Ni, Pb in Main Farmland Area of China. Doctor's thesis, China University of Geosciences (Beijing) (2023). https://kns.cnki.net/kcms2/article/abstract?v=691tpyMQYm09R4xEO_gmvN61_q8OMd-XXR_RtYWPGrQKwgmHFXdlOI79937CkqsOw52znmnWAAO-alpVJ4YRIVusZHWc8KCMrAkX1QMU26fx3cMvFOCN3Tn3QMWXKPea7AmZqV3ty9f8GXicz1Dl_JO6RrRdrqMd1viUUNrj8qRQw_iUEuh27270kmJoRGjfubEoMOcPBFJ5GLFmXLqPcvz9raoNkf6yBvojWBn4Hk3fU8kwBtF9hA==&uniplatform=NZKPT&language=CHS
3. Asia Metal: http://baike.asianmetal.cn/metal/ni/application.shtml. 1 Sep 2024
4. Fei, X., Lou, Z., Xiao, R., Ren, Z., Lv, X.: Contamination assessment and source apportionment of heavy metals in agricultural soil through the synthesis of PMF and GeogDetector models. Sci. Total Environ. **747**, 141–293 (2020). https://www.sciencedirect.com/science/article/abs/pii/S0048969720348221
5. National Meteorological Center: http://cmdp.ncc-cma.net/download/wise.ncc/QGNJSLFBT/2023/CMDP_DSTR_ACHN_L88_DATA_ELEMENT_YEAR_Y_107541_2023_00000000.png. 1 Sep 2024
6. Zhang, T., Xu, R., Zhao, Y., et al.: Interannual variation characteristics of acid rain from 2006–2021 in China. Environ. Pollut. Control **43**(06), 849–854 (2023). https://kns.cnki.net/kcms2/article/abstract?v=691tpyMQYm21y3a6VUJzeL9QAXnNX3lDz1MOgEhCQ2YBddLq7rzsy-kiDWJtiVBIN3XdRLtJdc7-PCCP4vYyd11P2zWMzvEOPAUpi8aYXt_3a2mUSSj1-Jolr3PuN-zff1_Y5BiE5pDu0f0fUYqsI1594Sca4LzKSUi6XyWqz6jBscJ2IIhFdBRYoNEcZIEpWmrUJGU8Wuc7N_k9WdXNwa-GB2qCT1jcZkC7MokBUDVdpPAxUk4D28qLkZ-GMxoZyBwFbKstFF046xLSuwKGXmwzkJsJzFBeE9e0GJYf5daIUwWWU0MVsM2-DS8Q8dX4mfvKfqPTEL7vN0uP6NRx7hliuvbkCiLvIG2xfoe9JF64orYWSAZBxA==&uniplatform=NZKPT&language=CHS

7. GB/T 28592-2012, Grade of Precipitation: General Administration of Quality Supervision. Inspection and Quarantine of the People's Republic of China (2012). https://www.mee.gov.cn/image20010518/3652.pdf
8. GB 11910-89, Water quality—Determination of nickel—Dimethylglyoxime spectrophotometric method. State Environmental Protection Agency (1989). http://c.gb688.cn/bzgk/gb/showGb?type=online&hcno=B4A00E4ABCF80F8C6A048C1D0121A97D

Agricultural Non-point Source Pollution Control by Using a Fuzzy-Boundary Interval Double-Side Chance-Constrained Programming

Yao Ji[1], Yu Fu[1,2], Jing Su[1], Dan Li[1], Chang Su[1,3], Yujiao Geng[1,2], Jinding Ha[1,4], Yanmei Yang[2], and Chi Zhang[1](✉)

[1] State Laboratory of Environmental Criteria and Risk Assessment, Chinese Research Academy of Environmental Science, Beijing 100012, China
{sujing,zhangchi}@craes.org.cn
[2] Hehai College, Chongqing Jiaotong University, Chongqing 400074, China
[3] College of Chemical Engineering, Beijing University of Chemical Technology, Beijing 100029, China
[4] College of Chemistry and Chemical Engineering, Xi'an Shiyou University, Xi'an 710065, China

Abstract. Non-point source pollution (NSP) has become the bottleneck in the sustainable development of agricultural. In this study, a fuzzy-boundary interval double-side chance-constrained programming (FBIDCCP) is established for agricultural non-point source pollution control, in order to coordinated the relationship between agricultural development and environment protection. To overcome the uncertainty of randomness and fuzziness in the agricultural system, the developed method integrates the concept of fuzzy boundary intervals (FBI) into the framework of inexact double-hand-side chance-constrained programming (IDCCP). Based on probability density functions (PDFs) and fuzzy set theory, this method introduces the default risk level (p_k) and the confidence level (α-cut) to transform random parameters appearing on both sides of constraints, as well as certain interval parameters with fuzzy boundaries on the upper and lower bound. The solution is achieved based on interactive algorithms and the fuzzy vertex analysis method. A case study is applied to verify the effectiveness of this method, and 36 decision scenarios is set, considering different combinations of p_k and α-cut. The results indicate that, the agricultural system obtains the maximum net economic benefits under the most loose environmental emission standard ($p_k = 0.1$) and the highest confidence level ($\alpha = 0.8$). FBIDCCP extends the applicability of traditional methods, and provides effective and flexible decision support for agricultural non-point source pollution control, supporting the sustainable development of agriculture.

Keywords: Fuzzy-boundary interval double-side chance-constrained programming · Agricultural syste · Water quality management · Nonpoint source pollution

D. Han et al. (Eds.): ICEPG 2024, ESE, pp. 980–998, 2025.
https://doi.org/10.1007/978-3-031-86829-0_108

1 Introduction

With the expansion of global agricultural scale and the widespread application of modern pesticide technologies, 30% to 50% of surface water is impacted by agricultural non-point source (NPS) pollution [1, 2]. It is necessary to coordinate the relationship between agricultural economy development and environmental protection requirements by implementing comprehensive agricultural management, including fertilizer application, irrigation water allocation, coordination of crop cultivation with livestock farming, etc. [3]. However, there are many uncertainties in agricultural systems due to the combined effects of market conditions, environment changes, and policy guidance, including parameter uncertainties, randomness in environmental process, and subjective uncertainties among decision-makers [4, 5]. The interaction among these uncertainties further complicates the uncertainty within agricultural systems, bringing challenges for agricultural non-point source pollution management.

Mathematical programming is an effective method for addressing the uncertainty in agricultural systems. For example, when the prices of agricultural products remain within a stable range, it can be expressed as interval parameters and addressed by Interval Mathematical Programming (IMP) effectively. The pollution loss process caused by rainfall runoff is often stochastic, and Stochastic Mathematical Programming (SMP) can characterize this type of random uncertainty. When it is necessary to account for manager's cognition or subjective judgment of fuzzy uncertainty, Fuzzy Mathematical Programming (FMP) is the optimal choice [6–8]. However, in practical management problems, a single planning method is insufficient to handle the multiple uncertainties within a system. Therefore, it is often necessary to couple or enhance the aforementioned methods to be applicable in scenarios involving multiple uncertainties. For example, developments such as ICCP (Interval Chance-Constrained Programming), ILCCP (Interval Linear Chance-Constrained Programming), IDCCP (Interval Dual Chance-Constrained Programming), and similar approaches have been built upon IMP and SMP. Some scholars have proposed combining chance-constrained programming with fuzzy mathematical programming, as seen in methods like MFSP (Mixed Fuzzy Stochastic Programming) and DFPSD (Dual Fuzzy Probabilistic Semi-Definite programming), which are designed to handle uncertainties represented in the form of fuzzy sets and probability distributions. From previous research, it appears that scholars have been focused on enhancing traditional methods to better and more comprehensively characterize and manage the complex multiple uncertainties in agricultural systems [9–11]. However, most of the mentioned methods combine stochastic parameters or fuzzy uncertain parameters with interval parameters, overlooking the fuzzy uncertainty inherent in the interval parameters themselves. In other words, the upper and lower bounds of these interval parameters may also be fuzzy.

Therefore, a Fuzzy-Bounded Interval Dual Chance-Constrained Programming (FBIDCCP) method is developed to deal with the above challenges, which integrates the concept of Fuzzy-Bounded Interval (FBI) into the framework of Uncertain Dual Chance-Constrained Programming (IDCCP), and is capable to address the stochastic parameters on both sides of the constraint and interval parameters with fuzzy upper and lower bounds, simultaneously. The decision management solutions provided can be flexibly adjusted based on environmental risk and subjective judgments of decision makers at different levels. This method offers a scientific representation and handling of uncertainty in complex systems, meeting the management requirements under multiple uncertainty conditions.

2 Methods

According to Huang, fuzzy boundary interval program model can be expressed as [11]:

$$\max \tilde{f}^{\pm} = \sum_{j}^{n} \left(\tilde{e}_j^{\pm} + c^{\pm}\right) x_j^{\pm} \tag{1}$$

subject to:

$$\sum_{j}^{n} a_{ij}^{\pm} x_j^{\pm} \leq b_i^{\pm}, i = 1, 2, 3, \ldots\ldots., m_1 \tag{2}$$

$$\sum_{j}^{n} \tilde{a}_{rj}^{\pm} x_j^{\pm} \leq \tilde{b}_r^{\pm}, r = 1, 2, 3, \ldots\ldots., m_2 \tag{3}$$

$$x_j^{\pm} \geq 0 \, j = 1, 2, 3, \ldots\ldots, n \tag{4}$$

where $\tilde{e}_j^{\pm}$, $\tilde{a}_{rj}^{\pm}$ and $\tilde{b}_i^{\pm}$ is fuzzy boundary interval parameter with fuzzy lower and upper bounds; m_1 and m_2 are the number of interval constraints and fuzzy constraints, respectively. The fuzzy vertex analysis (FVA) method can effectively deal with fuzzy sets [12]. By defining the decision boundaries of the fuzzy parameters using the α-cut level, input the combination of decision boundaries for each fuzzy variable at each α-cut level, and output the range of maximum and minimum values for the model at each α-cut level. Therefore, by introducing the FVA approach, the above model can be converted into as follows:

$$\max \tilde{f}^{\pm\alpha} = \sum_{j}^{n} \left(\left[\underline{\tilde{e}_j}^{\pm}, \overline{\tilde{e}_j}^{\pm}\right]^{\alpha} + c_j^{\pm}\right) x_j^{\pm} \tag{5}$$

subject to:

$$\sum_{j}^{n} a_{ij}^{\pm} x_j^{\pm} \leq b_i^{\pm}, i = 1, 2, 3, \ldots\ldots., m_1 \tag{6}$$

$$\sum_{j}^{n}\left[\underline{\widetilde{a_{rj}}}^{\pm},\overline{\widetilde{a_{rj}}}^{\pm}\right]^{\alpha}x_j^{\pm}\le\left[\underline{\widetilde{b_r}}^{\pm},\overline{\widetilde{b_r}}^{\pm}\right]^{\alpha}, r=1,2,3,\ldots\ldots., m_2 \tag{7}$$

$$x_j^{\pm}\ge 0\ j=1,2,3,\ldots\ldots,n \tag{8}$$

where $\left[\underline{\widetilde{c_j}}^{\pm},\overline{\widetilde{c_j}}^{\pm}\right]^{\alpha}$, $\left[\underline{\widetilde{a_{rj}}}^{\pm},\overline{\widetilde{a_{rj}}}^{\pm}\right]^{\alpha}$, $\left[\underline{\widetilde{b_r}}^{\pm},\overline{\widetilde{b_r}}^{\pm}\right]^{\alpha}$ and $\left[\underline{\widetilde{a_{kj}}}^{\pm},\overline{\widetilde{a_{kj}}}^{\pm}\right]^{\alpha}$ represents the fuzzy interval at the α-cut level. $\underline{\widetilde{c_j}}^{\pm}$, $\underline{\widetilde{a_{rj}}}^{\pm}$ and $\underline{\widetilde{b_r}}^{\pm}$ represents the lower bound of the fuzzy interval; $\overline{\widetilde{c_j}}^{\pm}$ $\overline{\widetilde{a_{rj}}}^{\pm}$ and $\overline{\widetilde{b_r}}^{\pm}$ represents the up bound of the fuzzy interval. The variation of α-cut level for fuzzy parameters affected by decision makers considering different preferences on tradeoff between environmental risks and economic benefit. The boundary intervals of the corresponding fuzzy sets can be obtained under different α-cut levels, and different values of the objective function will be acquired under different combination of boundary intervals.

$$\max[\tilde{f}^{\pm}]^{\alpha}=\sum_{j}^{n}\left(\left[\underline{\widetilde{e_j}}^{\pm},\overline{\widetilde{e_j}}^{\pm}\right]^{\alpha}+c_j^{\pm}\right)x_j^{\pm} \tag{9}$$

subject to:

$$\sum_{j}^{n}a_{ij}^{\pm}x_j^{\pm}\le b_i^{\pm}, i=1,2,3,\ldots\ldots., m_1 \tag{10}$$

$$\sum_{j}^{n}\left[\underline{\widetilde{a_{rj}}}^{\pm},\overline{\widetilde{a_{rj}}}^{\pm}\right]^{\alpha}x_j^{\pm}\le\left[\underline{\widetilde{b_r}}^{\pm},\overline{\widetilde{b_r}}^{\pm}\right]^{\alpha}, r=1,2,3,\ldots\ldots., m_2 \tag{11}$$

$$\Pr\left\{\sum_{j=1}^{n}a_{kj}\omega\left[\underline{\widetilde{c_{kj}}}^{\pm},\overline{\widetilde{c_{kj}}}^{\pm}\right]^{\alpha}x_j^{\pm}\le b_{kj}(\xi)\right\}\ge 1-p_k, k=1,2,\ldots\ldots,m_3 \tag{12}$$

$$a_{kj}(\omega)\sim N\left(\mu_{a,kj},\sigma_{a,kj}^2\right);\ b_{kj}(\xi)\sim N\left(\mu_{b,kj},\sigma_{b,kj}^2\right) \tag{13}$$

$$x_j^{\pm}\ge 0\ j=1,2,3,\ldots\ldots,n \tag{14}$$

where $a_{kj}(\omega)$ and $b_{kj}(\omega)$ are independently normally distributed random variables; p_k ($p_k \in [0, 1]$) is the level of default risk and $1 - p_k$ is the probability of constraint satisfaction; m_3 is the number of stochastic constraints. According to Ji's [13] linear transformation of the stochastic parameters, the linear inequality for different violation levels p_k can be obtained, and (12) can be transformed into:

$$\sum_{j=1}^{n}\left[\underline{\widetilde{a_{rj}}}^{\pm},\overline{\widetilde{a_{rj}}}^{\pm}\right]^{\alpha}\left[\mu_{a,kj}+\Phi^{-1}(1-p_k)\sigma_{a,kj}\right]x_j^{\pm}\le[\mu_{\mu_{b,kj}}-\sigma_{\mu_{b,kj}}\Phi^{-1}(1-p_k)] \tag{15}$$

where ϕ^{-1} is the inverse cumulative distribution function of a standard normal random variable. Based on the interactive algorithms and fuzzy vertex analysis method, upper and lower bound sub-models corresponding to the objective function are constructed. For each α-cut level, the interval endpoints of each fuzzy parameter are selected, and n sets of fuzzy boundary interval sets are constructed, resulting in 2^n combinations. A set of upper bound sub-models corresponding to $\tilde{f}^+$ is then constructed. By solving the sub-models, a set of values $(f_1^+, f_2^+, \ldots\ldots, f_{2^n}^+)$ can be obtained. The optimal upper bound interval values for the objective function $\tilde{f}^+$ are expressed as follows:

$$\left[\underline{f}_{opt}^{+}, \overline{f}_{opt}^{+}\right]^{\alpha} = \left[\min(f_1^-, f_2^-, \ldots, f_{2^n}^-), \min(f_1^-, f_2^-, \ldots, f_{2^n}^-)\right]^{\alpha} \tag{16}$$

Similarly, the optimal lower bound interval values for the objective function $\tilde{f}^-$ corresponding to each α-cut level are obtained as follows:

$$\left[\underline{f}_{opt}^{-}, \overline{f}_{opt}^{-}\right]^{\alpha} = \left[\min(f_1^-, f_2^-, \ldots, f_{2^n}^-), \min(f_1^-, f_2^-, \ldots, f_{2^n}^-)\right]^{\alpha} \tag{17}$$

Finally, the solutions obtained from the two sets of sub-models are integrated to obtain the solution of the objective function for a given α-cut level.

$$\tilde{f}^{\pm\alpha} = \left[\left[\underline{f}_{opt}^{-}, \overline{f}_{opt}^{-}\right], \left[\underline{f}_{opt}^{+}, \overline{f}_{opt}^{+}\right]\right]^{\alpha} \tag{18}$$

3 Case Study

The feasibility of FBIDCCP approach was validated through the utilization of an agricultural water quality management case [10]. The data involved are taken from the relevant literature. The total area is 143.7 ha, with a cultivable land area of 124 ha, all of the parameters in this case are listed in Appendix A. The tillable area is divided into three sub-areas (42, 42 and 39 ha, respectively) by three tributaries of a river that crosses the area. Crop cultivation and livestock breeding are the main agricultural economic sources. The main crops are wheat, vegetables and potatoes, while the main livestock are cattle, poultry and swine. The irrigation water for agriculture comes from various tributaries within each subarea. While generating economic benefits, crop cultivation and livestock breeding also contribute to agricultural NPS pollution. Unreasonable farming practices and land management lead to soil erosion, and excessive use of fertilizers and pesticides can cause water pollution through rainfall and irrigation water. Manure and urine produced by livestock contain large amounts of organic matter and nutrients, the indiscriminate discharge can lead to soil pollution and eutrophication of water bodies. The crop cultivation, livestock breeding and irrigation water, exert synergistic influence on both agricultural economic benefits and environmental pollution. Therefore, the core issue is to design agricultural production model, that take into account different economic objectives and environment protection requirements, by adjusting the relationships among crop cultivation, livestock breeding and irrigation water. The goal, which is to maximize the net economic benefits of the agricultural system, is pursued on the premise of meeting the requirement of water environmental quality. Parameters of the study case are shown in Tables 1, 2 and 3.

Furthermore, influenced by the complexity of the agricultural system and various uncertainty factors, the issue of water quality management becomes even more intricate. These uncertainty parameters are represented through different types of parameters. For example, the price trends of agricultural products follow the market's supply and demand dynamics, which generally remain within a reasonable range over the long term. The soil erosion and nitrogen and phosphorus content of soil are influenced by fertilizer application amount, rainfall and topography, which bring in both stochastic and fuzzy uncertainties. To tackle the above multiple uncertainties, a water quality management model is developed to account for various uncertainties within the system. The objective is to maximize the economic benefits of the agricultural system, and decision variables include the planting area of crops, the scale of livestock breeding, the amount of fertilizer/manure application, and the constraints are cropping-livestock cycle constraints, water environment quality constraints and irrigation water constraints. The specific FBIDCCP model can be expressed as follows:

3.1 Objective Function

$$\begin{aligned} \mathrm{Max} f^{\pm} = {} & \sum_{t=1}^{3}\sum_{i=1}^{3} \delta_i^{\pm} Y_i^{\pm} S_{it}^{\pm} + \sum_{j=1}^{3} \sigma_j^{\pm} T_j^{\pm} - \sum_{t=1}^{3}\sum_{i=1}^{3} P_i^{\pm} S_{it}^{\pm} - C_m^{\pm} \sum_{i=1}^{3} F_i^{\pm} \\ & - C_f^{\pm} \sum_{i=1}^{3} H_i^{\pm} - \sum_{t=1}^{3}\sum_{i=1}^{3} C_{it}^{\pm} W_{it}^{\pm} S_{it}^{\pm} \end{aligned} \tag{19}$$

The income from crop and livestock sales constitutes the main benefits. The main costs of crop production include crop cultivation, fertilizer usage and irrigation water. The cost of livestock breeding is the cost of manure disposal.

3.2 Constraints

Cropping-Livestock Cycle Constraints

(a) Manure Mass Balance Constraint

$$\sum_{i=1}^{3} F_i^{\pm} - \sum_{j=1}^{3} M_j^{\pm} T_j^{\pm} \le 0 \tag{20}$$

The above constraint ensures that the livestock farming industry can provide sufficient manure for the planting industry.

(b) Crop nutrient balance constraints

$$\sum_{t=1}^{3} q_i^{\pm} S_{it}^{\pm} - (1 - p_m^{\pm}) g^{\pm} F_i^{\pm} - (1 - p_f^{\pm}) H_i^{\pm} \leq 0, \forall i \tag{21}$$

The above constraint ensures that crops can obtain sufficient nitrogen from both manure and fertilizer.

(c) Livestock energy and digestible protein balance constraints.

$$\sum_{j=1}^{3} E_j^{\pm} T_j^{\pm} - \sum_{t=1}^{3} \sum_{i=1}^{3} Y_i^{\pm} \alpha_i^{\pm} S_{it}^{\pm} \leq 0 \tag{22}$$

$$\sum_{j=1}^{3} D_j^{\pm} T_j^{\pm} - \sum_{t=1}^{3} \sum_{i=1}^{3} Y_i^{\pm} \beta_i^{\pm} S_{it}^{\pm} \leq 0 \tag{23}$$

The above constraints ensure that farming can provide sufficient nutrients for livestock farming.

Water Environment Quality Constraints.

(d) Total nitrogen loss constraint

$$\sum_{i=1}^{3} \left(g^{\pm} F_i^{\pm} + H_i^{\pm} - q_i^{\pm} \sum_{t=1}^{3} S_{it}^{\pm} \right) \leq N_L^{\pm} \sum_{t=1}^{3} K_t \tag{24}$$

The above constraints require surplus nitrogen emissions to be below locally permitted emissions.

(e) Dissolved nitrogen and phosphorus loss constraints

$$\sum_{t=1}^{3} \sum_{i=1}^{3} (Q_{wi}^{\pm} N_{wi}^{\pm} + Q_{di}^{\pm} N_{di}^{\pm}) S_{it}^{\pm} \leq a_1^{\pm} \sum_{t=1}^{3} K_t \tag{25}$$

$$\sum_{t=1}^{3} \sum_{i=1}^{3} (Q_{wi}^{\pm} P_{wi}^{\pm} + Q_{di}^{\pm} P_{di}^{\pm}) S_{it}^{\pm} \leq a_2^{\pm} \sum_{t=1}^{3} K_t \tag{26}$$

The above constraints require that losses of dissolved nitrogen and phosphorus remain below locally allowable levels. The estimation of dissolved nitrogen and phosphorus depends on runoff and the concentrations of dissolved nutrient in runoff.

(f) Solid phase nitrogen and phosphorus loss constraints

$$\mathrm{pr}\left\{\sum_{t=1}^{3}\sum_{i=1}^{3} u_1\omega \widetilde{L_i} S_{it}^{\pm} \leq a_1(\xi)\sum_{t=1}^{r} K_t\right\} \geq 1 - p_k \tag{27}$$

$$\mathrm{pr}\left\{\sum_{t=1}^{3}\sum_{i=1}^{3} u_2\omega \widetilde{L_i} S_{it}^{\pm} \leq a_2(\xi)\sum_{t=1}^{r} K_t\right\} \geq 1 - p_k \tag{28}$$

$$u_1\omega \sim N\left(\mu_{u1}, \sigma_{u1}^2\right) \tag{29}$$

$$u_2\omega \sim N\left(\mu_{u2}, \sigma_{u2}^2\right) \tag{30}$$

$$a_1(\xi) \sim N\left(\mu_{a1}, \sigma_{a1}^2\right) \tag{31}$$

$$a_2(\xi) \sim N\left(\mu_{a2}, \sigma_{a2}^2\right) \tag{32}$$

$u_1\omega$ and $u_2\omega$ are the soil nitrogen and phosphorus contents (%). $a_1(\xi)$ and $a_2(\xi)$ are the maximum allowable solid-phase nitrogen and phosphorus losses (kg/ha) $1 - p_k$ represents the confidence level that the constraint is satisfied. A higher confidence level $(1 - p_k)$ corresponds to stricter environmental protection requirements, resulting in lower emissions of nitrogen and phosphorus. $\widetilde{L_i}$ is soil loss from land under crop i (kg/ha), representation in the form of triangular fuzzy numbers, As the α-cut level increases, the range of intervals gets closer to the "most probable" value.

(g) Soil erosion constraints

$$\sum_{t=1}^{r}\sum_{i=1}^{m} \widetilde{L_i} S_{it}^{\pm} \leq \widetilde{S_L}^{\pm}\sum_{t=1}^{r} K_t \tag{33}$$

$$\sum_{t=1}^{r} S_{it}^{\pm} \leq K_t, \forall t \tag{34}$$

The symbol $\widetilde{S_L}$ represents the maximum permissible soil erosion. $\widetilde{S_L}$ is the maximum allowable soil loss.

Irrigation Water Constraints

$$\sum_{i=1}^{m} W_{it}^{\pm} S_{it}^{\pm} \leq Q_t^{\pm} \tag{35}$$

The above constraints guarantee the amount of water used for irrigation of crops in each sub-zone.

Nonnegative Constraints

$$S_{it}^{\pm}, F_{i}^{\pm}, H_{i}^{\pm}, T_{j}^{\pm} \geq 0, \forall i, j, t \tag{36}$$

In an agricultural water quality management model, all decision variables (acreage, manure application, fertilizer application and livestock farming) must be non-negative.

3.3 Constraint Transformation

To ascertain the viability of FBIDCCP method, a scenario was created in which the soil nitrogen and phosphorus contents, along with the allowable losses of solid-phase nitrogen and phosphorus, were set as random parameters. Set the amount of soil loss caused by planting crops and the maximum allowable soil loss as fuzzy interval parameters. The superiority of FBIDCCP model is demonstrated by dealing simultaneously with the presence of random and fuzzy parameters in the constraints. According to (5a), the linear transformation form of inequalities (27), (28) and (33) can be transformed into:

$$\left[\underline{\widetilde{L_i}}^{\pm}, \overline{\widetilde{L_i}}^{\pm}\right]^{\alpha}\left[\mu_{u1} + \Phi^{-1}(1 - p_k)\sigma_{u1}\right]\sum_{j=1}^{n} S_{it}^{\pm} \leq [\mu_{a1} - \sigma_{a1}\Phi^{-1}(1 - p_k)]\sum_{t=1}^{r} K_t \tag{37}$$

$$\left[\underline{\widetilde{L_i}}^{\pm}, \overline{\widetilde{L_i}}^{\pm}\right]^{\alpha}\left[\mu_{u2} + \Phi^{-1}(1 - p_k)\sigma_{u2}\right]\sum_{j=1}^{n} S_{it}^{\pm} \leq [\mu_{a2} - \sigma_{a2}\Phi^{-1}(1 - p_k)]\sum_{t=1}^{r} K_t \tag{38}$$

$$\sum_{t=1}^{r}\sum_{i=1}^{m}\left[\underline{\widetilde{L_i}}^{\pm}, \overline{\widetilde{L_i}}^{\pm}\right]^{\alpha} S_{it}^{\pm} \leq \left[\underline{\widetilde{S_L}}^{\pm}, \overline{\widetilde{S_L}}^{\pm}\right]^{\alpha}\sum_{t=1}^{r} K_t \tag{39}$$

4 Results Analysis

4.1 System Economy Composition and Analysis

Figure 1 presents the economic benefit obtained by FBIDCCP under 36 scenarios. The results indicate that the economic benefits vary from \$1.92 million to \$2.34 million. The upper bound of the objective function refers to the optimal decision schemes with the highest net benefit in advantageous conditions, including the loosest pollutant emission standards, highest product unit price and yield. While lower bound refers to the highest pollutant losses in demanding conditions, respectively. For both upper and lower bounds, the net benefit declines with the confidence coefficient ($1 - p_i$) increases, and the tendency demonstrates that stricter environmental protection requirements demand less non-point source pollution, which downscales the crop planting and livestock breeding. Specifically, the highest net economic benefits obtained under the scenario $p_k = 0.1$ and $\alpha = 0.2$, while the lowest net economic benefits obtained under $p_k = 0.01$ and $\alpha =$

0.8. Under stricter environment requirements, the economic benefits are limit, and the boundary benefits decrease with the decrease in default risk levels. Indeed, confidence levels do not directly impact the system economic benefits, higher confidence levels provide much stricter decision-making. For example, at the same default risk levels (p_k = 0.1), the upper bounds for α = 0.2 and α = 0.8 are $ [2.30, 2.34] million and $ [2.31, 2.32] million. The solution for α = 0.2 includes the solution for α = 0.8, and as the confidence level increases, the decision space becomes much stricter. Decision makers can adjust default risk levels and confidence levels through comprehensive considerations to achieve a balance between system benefits and environment protection.

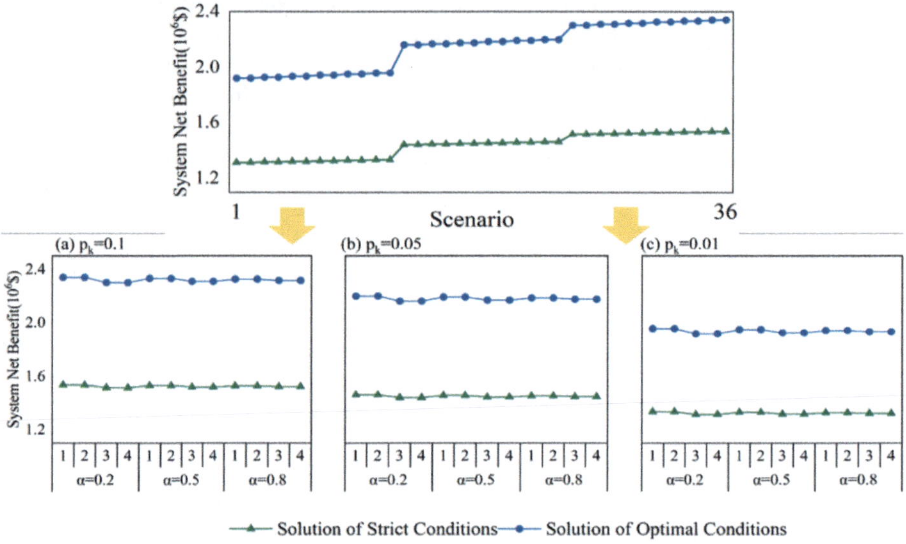

Fig. 1. System boundary benefits under different scenarios.

Figure 2 illustrates the composition of net economic benefits under different scenarios. Under strict conditions, livestock farming contributes the main economic benefits (57% to 73.74%) and increases as default risk level decreases. The economic benefits of crop cultivation gradually decrease as default risk level decreases (from 43% to 26.91%). Under optimal conditions, crop cultivation contributes the main economic benefits (53.04% to 58%), due to the loose emission standards. As the default risk levels decrease, livestock farming gradually becomes the primary economic source (42% to 55.23%). The solution indicates that, FBIDCCP has chosen to reduce the economics of crop cultivation to meet stricter environmental requirements, while maximizing the economics of livestock farming, besides, providing more manure to the crops, and effectively mitigate non-point source pollution caused by fertilizer application.

4.2 Crop Cultivation Planning Solution

Figure 3 shows that the crop area is primarily correlated with the yield, cost, and the nitrogen-phosphorus loss during the growth process. Among the three crops, vegetables

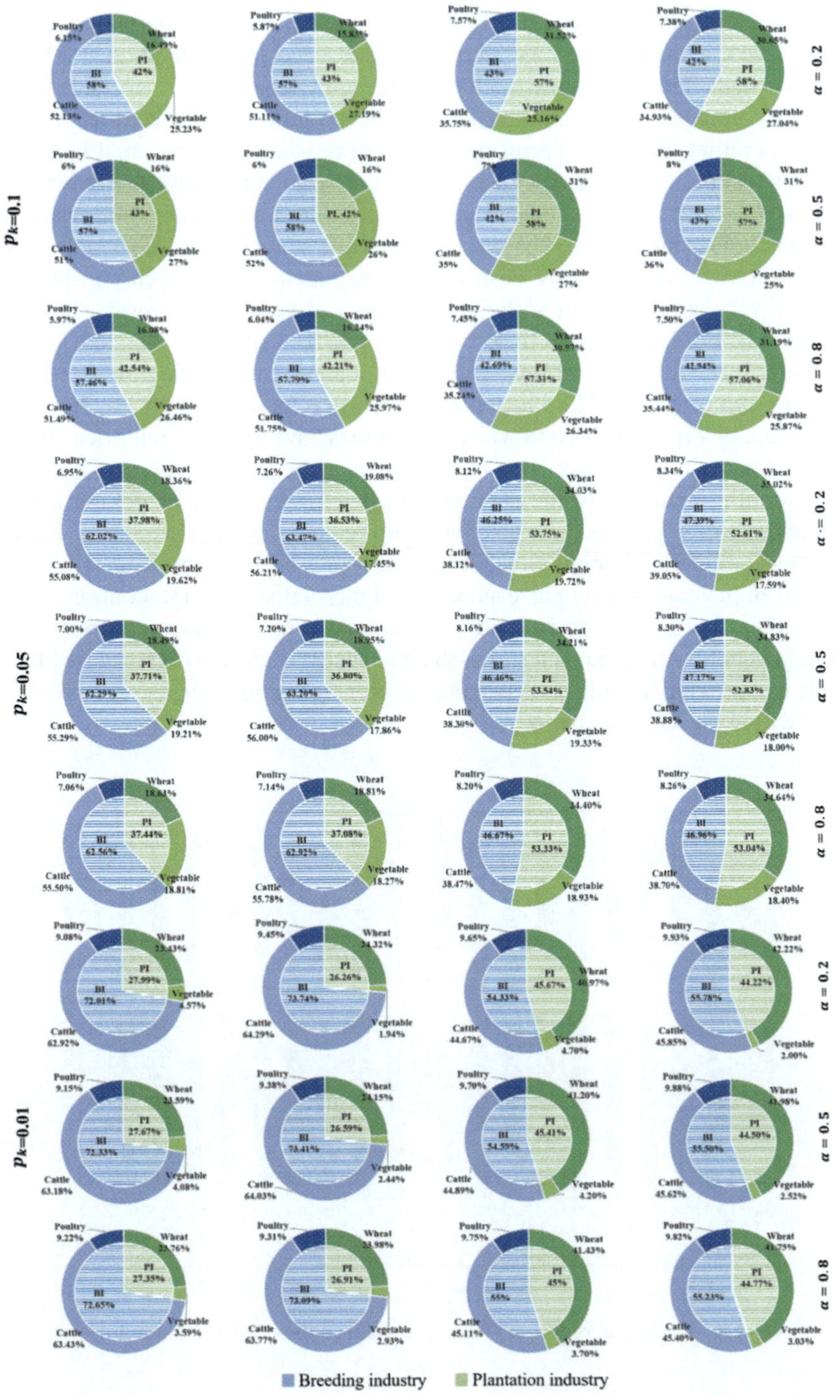

Fig. 2. Economic benefits composition under different scenarios.

have the highest planting benefits. To maximize economic benefits, it is necessary to cultivate a certain quantity of vegetables. However, vegetable cultivation also brings the highest nitrogen and phosphorus pollution, the cultivation area of vegetable should be limited. In comparison, the economic benefits and pollution of potatoes are both unspectacular. Therefore, FBIDCCP prioritizes cultivating wheat with minimal pollution and the highest energy content to provide maximum energy support for the growth of livestock. Most crop areas are allocated in subarea 1 and 2, since irrigation water cost in region 3 is higher than other subareas, and wheat is suggested in subarea 3 for its low water requirement and planting cost. Due to the maximum nitrogen-phosphorus loss caused by vegetable cultivation, the planting area of vegetable is further restricted as default risk levels decreases, from $p_k = 0.1$ to $p_k = 0.01$ ([85.47%, 93.34%]). With the reduction in vegetables cultivation, wheat can obtain more planting area and environmental emission capacity. The wheat planting area has increased by [[28.27%, 29.11%], [57.54%, 57.49%]]. Compared to wheat, vegetables have a higher unit price. Therefore, the reduction in the cultivation area of vegetables is the primary reason for the decline in the overall profitability of the planting industry. Under the same default risk levels, cropping areas of vegetable show an obvious uptrend as the α-cut increase in the scenarios aiming at minimum benefit. Since vegetable brings high economic benefit than other crops and higher α-cut level corresponds to higher values. On the contrary, cropping areas of vegetable show a downtrend as the α-cut decrease in the scenarios aiming at maximum benefit. And the result shows that the decision scheme obtained by FBIDCCP not only can provide flexible choice under different scenarios, but also can offer various options within the same scenario.

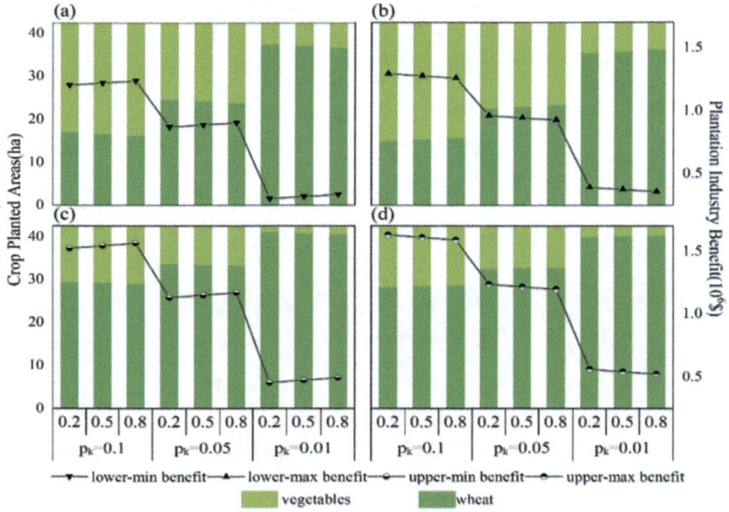

Fig. 3. Crop acreage and benefits (Taking sub-area 1 for example).

4.3 Livestock Farming Planning Solution

Figure 4 illustrate the scale and the economic benefits of the breeding industry. Livestock obtain the necessary energy and protein for their growth from the crops, the scale of livestock farming is influenced by the types and acreage of crops planted. Since wheat has higher protein and energy content than vegetable, the scale of livestock farming increases with the expansion of wheat cultivation scale. The breeding quantities of poultry and cattle have increased by [[8.54%, 8.61%], [24.99% 25.62%]] and [6.45%, 6.49%]. Although cattle generated the largest economy, accounting for 34.93% to 63.43% of the total economy in all scenarios. However, cattle farming requires more energy and protein, the scale of cattle farming is subject to strict limitations. In the same default risk levels, both poultry and cattle show a downtrend and the decision scope narrow as α-cut increase, demonstrating that the variation of livestock farming in different scenarios has the same tendency as the net benefit.

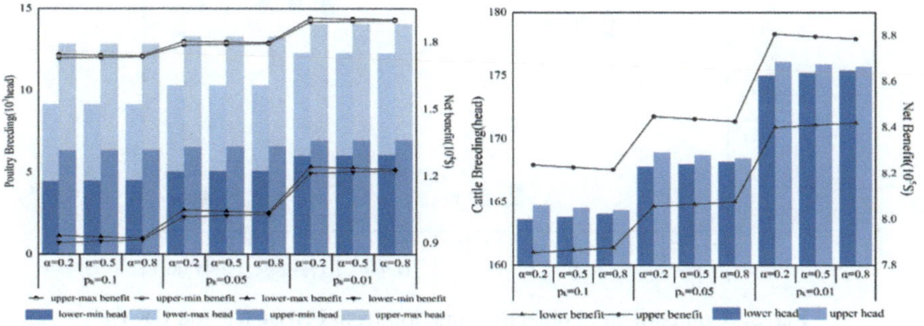

Fig. 4. Livestock breeding and benefits

4.4 Manure/Fertilizer Application Solution

Table 2 illustrates the solutions of Manure/Fertilizer application under different scenarios. Manure application is significantly higher than fertilizer application. For example, at the same scenario, the applied manure amount for wheat is [1125.04, 1198.78]t, while the applied fertilizer amount is [0.3408,0.3425]t, representing only three-thousandths of the manure amount. Manure derived from livestock is limited, and fertilizers are purchased externally as supplements. For example, with the decrease in default risk level (from 0.1 to 0.01), the cultivation area of wheat increased and demand for nitrogen and phosphorus was greater. Consequently, the fertilizer application has increased but only from [0, [0.2137, 0.2306]] tons to [0, [0.3408, 0.3425]] tons, owing to stringent environmental requirements and cost control. The results also show that, due to the particular structure and composition of the agricultural system in this case study, there is no competition between manure and chemical fertilizers. As the default risk level decreases, the reduction in vegetable acreage leads to a reduction in manure and chemical fertilizer application. When the area under vegetable cultivation reaches a minimum, the amount of chemical fertilizer applied drops to zero and only manure is applied.

5 Discussion

To further demonstrate the feasibility of FBIDCCP, a comparison was conducted by introducing inexact double-hand-side chance-constrained programming (IDCCP) into the same agricultural non-point pollution control management problem. To ensure the comparability of IDCCP, the lower bound minimum and upper bound maximum values of the fuzzy boundary interval parameters were utilized as discrete inputs for IDCCP, and other parameters remained consistent.

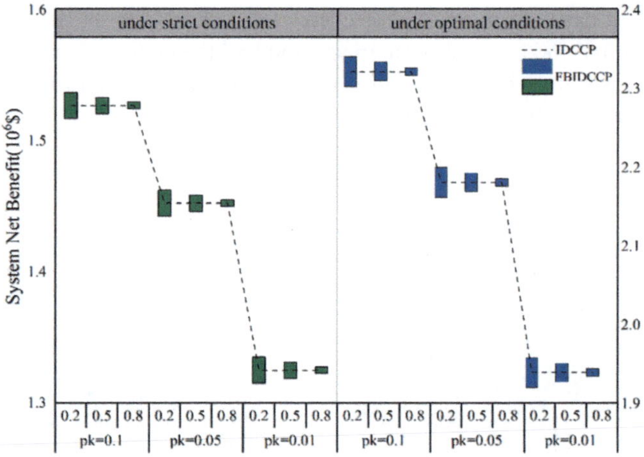

Fig. 5. Comparison of objective function from FBIDCCP and IDCCP.

Figure 5 illustrates the objective function values for IDCCP and FBIDCCP. As the default risk level decreases, the maximum systemic benefits of IDCCP decreased from 2.32×10^5 US$ to 1.94×10^5 US$. The decisions made by IDCCP and FBIDCCP are aligned. As the default risk level increases, the vegetable cultivation area of IDCCP decreases from 13.05 hectares to 1.41 hectares. FBIDCCP and IDCCP are chosen to reduce the vegetable cultivation area to meet the emission standards, and to offset the loss in economic benefits from vegetables, the cultivation area of wheat and livestock breeding has increased. As the environment constraints become more stringent, the decision-making processes of FBIDCCP and IDCCP are consistent. This In addition, the advantages of FBIDCCP can be further demonstrated by comparing the processing of decision variables between IDCCP and FBIDCCP. For the solution of IDCCP, in the case of the cultivation area of vegetables, IDCCP model provides a deterministic value, while FBIDCCP provides a range interval. On one hand, vegetables are the main economic and pollutant source, their fluctuations directly impact the entire agricultural system. Thus, under the constraints of multiple conditions, for the decision variables that have a significant impact (such as the cultivation area of vegetables and the number of cattle breeding), the decision space is also constrained within a relatively narrow range. On the other hand, IDCCP lost the information regarding fuzzy boundary distribution, which it can only consider decision options for each level of default risk. FBIDCCP provides a more precise characterization of the system's uncertain information. At the same default risk level, three confidence levels for the fuzzy boundary parameters are set, by inputting different combinations of fuzzy parameter boundaries corresponding to each α-cut level, various combinations of system net economic benefits can be obtained. The obtained decision solutions can be adjusted by choosing different confidence levels. FBIDCCP provides decision-makers with more accurate, stable, and feasible decisions.

In previous studies, related methods have been proposed to deal with the fuzzy uncertainty in interval parameters. Liu [14] developed a fuzzy-boundary interval programming (FBIP) to handle dual uncertainties expressed as crisp intervals and fuzzy-boundary intervals. The FBIP transforms the fuzzy-boundary interval parameters into definite interval by fuzzy membership function, and obtains the target values under different attitudes of decision makers. However, the FBIP does not consider random uncertainties in water quality management. In comparison, within the framework of FBIP, FBIDCCP integrates IDCCP to deal multiple uncertainties of fuzzy, random and interval. Meanwhile, FBIDCCP can provide reference of different scenarios based on the conditions and decision maker's judgment, and provide more alternatives. For example, under the lower levels of default risk and higher confidence level, the system can obtain greater benefits but withstand greater pollution emission pressures. In summary, FBIDCCP enables an in-depth analysis of the relationship between p_k and α-cut, weigh the relationship between environment pollution risk and system benefits, and consider more flexible and effective decision-making management programs.

6 Conclusion

A FBIDCCP method has been developed for managing agricultural non-point pollution control. The method can effectively handle various uncertainties presented in the form of fuzzy boundary intervals and stochastic distributions. It can also simultaneously reflect the impact of fuzzy boundary parameters on the system's economic benefits under a certain level of default risk and different confidence levels. A case of agricultural water quality management was used to demonstrate the feasibility of the method, and scenarios for all default risk levels and confidence levels are explored for pursuing the optimal agricultural production mode and achieving maximum economic benefits while meeting environmental protection requirements.

FBIDCCP results show that vegetables and cattle are the main economic sources of the entire agricultural system. At the highest level of default risk (loosest pollutant emission standards), the system achieves the highest economic benefits, which decrease as the default risk level decreases. As the primary source of pollutants in the system, the cultivation area of vegetables decreases as the default risk level decreases (due to stricter emission standards). Furthermore, FBIDCCP considers the impact of fuzzy boundary interval parameters with different confidence levels on the system's boundary benefits. At the same default risk level, obtaining the maximum boundary benefits at the lowest confidence level, the decision space obtained by FBIDCCP contracts as the confidence level increases. In the real world, the delineation of fuzzy boundary parameters primarily relies on the subjective judgment of decision-makers, who establish the boundary range of fuzzy parameters through varying confidence levels. FBIDCCP provides system decision solutions under different default levels and confidence levels, decision-makers can choose the optimal decision strategy based on a comprehensive consideration of various trade-offs.

Compared to traditional methods, FBIDCCP offers a more precise characterization of system uncertainties, and can adjust default risk levels and confidence levels to accommodate system uncertainty. The model can be widely applicable to resource management and environmental planning in both agricultural and non-agricultural sectors across different regions. Furthermore, future research can consider incorporating multi-objective planning problems, such as the protection of arable land and the acceptance level of residents, to address environmental issues of varying complexities in the actual cases.

Appendix A

Table 1. Plantation-related parameters.

Symbol		Corp		
		wheat (i = 1)	Vegetable (i = 2)	Potato (i = 3)
$Y_i^{\pm}$	Yield of crop i kg/ha)	[5500,6000]	[20000,21000]	[14500,15000]
$P_i^{\pm}$	Farming cost for crop i ($/ha)	[90, 110]	[5300, 5500]	[1400, 1600]
$L_i^{\pm}$	Soil loss from land planted to crop i (kg/ha)	[{2560,2600, 2640}, {2760,2800,2840}]	[{8960,9000, 9040}, {9160,9200,9240}]	[{3860,3900,3940}, {4060,4100,4140}]
$N_{wi}^{\pm}$	Dissolved nitrogen concentration in wet season runoff from land planted to crop i (mg/l)	[1.5,1.8]	[2.7,3]	[2.2,2.5]
$N_{di}^{\pm}$	Dissolved nitrogen concentration in dry season runoff from land planted to crop i (mg/l)	[0.8,1.1]	[1.7,2]	[1.4,1.7]
$p_{1i}^{\pm}$	Dissolved phosphorus concentration in wet season runoff from land planted to crop i (mg/l)	[0.15,0.19]	[0.26,0.3]	[0.23,0.27]
$p_{2i}^{\pm}$	Dissolved phosphorus cocentration in dry season runoff from land planted to crop i (mg/l)	[0.08,0.12]	[0.22,0.26]	[0.21,0.25]
$q_i^{\pm}$	Nitrogen requirement of crop i (kg/ha)	[90,110]	[140,155]	[80,120]
$Q_{wi}^{\pm}$	Wet season runoff from land planted to crop i (mm)	[65,80]	[82,96]	[86,102]
$Q_{di}^{\pm}$	Dry season runoff from land planted to crop i (mm)	[45,50]	[51,58]	[55,62]

(continued)

Table 1. (*continued*)

Symbol		Corp		
		wheat (i = 1)	Vegetable (i = 2)	Potato (i = 3)
$W_{it}^{\pm}$	Quantity of irrigation water required by crop i in subarea t (m^3/ha-s)			
	Subarea 1	[0.032,0.041]	[0.051,0.072]	[0.043,0.055]
	Subarea 2	[0.031,0.04]	[0.052,0.071]	[0.042,0.054]
	Subarea 3	[0.033,0.042]	[0.052,0.073]	[0.044,0.056]
$C_{it}^{\pm}$	Cost to deliver water to Sit (\$/(m^3/s))			
	Subarea 1	[1350,1400]	[1400,1450]	[1375,1425]
	Subarea 2	[2020,2070]	[2070,2120]	[2050,2100]
	Subarea 3	[2630,2680]	[2680,2730]	[2650,2700]
$\alpha_i^{\pm}$	Net energy content (Mcal/kg)	[3.3,3.5]	[0.22,0.32]	[0.8,1]
$\beta_i^{\pm}$	Digestible protein content of crop i (%)	[0.11,0.14]	[0.01,0.03]	[0.018,0.036]
$\delta_i^{\pm}$	Price of crop i (\$/kg)	[0.11,0.15]	[0.44,0.48]	[0.18,0.22]

Table 2. Farming related parameters.

Symbol		Livestock		
		Cattle (j = 1)	Swine(j = 2)	Poultry(j = 3)
$M_j^{\pm}$	Amount of manure generated by livestock j (t/unit)	[18,19]	[1.8,1.9]	[0.04,0.06]
$D_j^{\pm}$	Digestible protein requirement of livestock j (kg/unit)	[330,350]	[30,35]	[1.6,1.8]
$E_j^{\pm}$	Net energy requirement of livestock j (Mcal/unit)	[4800,5200]	[472,515]	[175,188]
$\sigma_j^{\pm}$	Average return from livestock j (\$/unit)	[480,500]	[35,40]	[2,2.7]

Table 3. Farming related parameters.

Symbol		
$N_L^{\pm}$	Maximum allowable total nitrogen losses (kg/ha)	[38,40]
$S_L^{\pm}$	Maximum allowable soil loss (kg/ha)	[{3700,3750,3800}, {5450,5500,5550}]
$a_1^{\pm}$	Maximum allowable solid-phase nitrogen loss (kg/ha)	$\eta_{a1} = 10$, $\tau_{a1}^2 = 1$
$a_2^{\pm}$	Maximum allowable solid-phase phosphorus loss (kg/ha)	$\eta_{a2} = 4$, $\tau_{a2}^2 = 0.4$
$g^{\pm}$	Nitrogen content of manure (kg/t)	[10, 13]
$C_m^{\pm}$	Cost of manure collection/disposal (\$/t)	[4, 5]
$C_f^{\pm}$	Cost of fertilizer application (\$/kg)	[0.7,0.9]
$u_1^{\pm}$	Nitrogen content of soil (%)	[0.0020, 0.0025]
$u_2^{\pm}$	Phosphorus content of soil (%)	[0.0009, 0.0011]

(*continued*)

Table 3. (*continued*)

Symbol		
$p_m^{\pm}$	Nitrogen volatilization/denitrification rate of manure (%)	[0.30, 0.35]
$p_f^{\pm}$	Nitrogen volatilization/denitrification rates of fertilizer (%)	[0.10, 0.12]
$c_1^{\pm}$	Maximum allowable dissolved nitrogen loss by runoff (kg/ha)	[2.2, 2.3]
$c_2^{\pm}$	Maximum allowable dissolved phosphorus loss by runoff (kg/ha)	[0.20, 0.22]

References

1. Ren, C., Zhang, X., Reis, S., Gu, B.: Socioeconomic barriers of nitrogen management for agricultural and environmental sustainability. Agricult. Ecosyst. Environ. **333**, 107950 (2022). https://doi.org/10.1016/j.agee.2022.107950
2. Rashmi, I., et al.: Organic and inorganic fertilizer contaminants in agriculture: impact on soil and water resources. In: Naeem, M., Ansari, A.A., Gill, S.S. (eds.) Contaminants in Agriculture, pp. 3–41. Springer, Cham (2020). https://doi.org/10.1007/978-3-030-41552-5_1
3. Phosphorus Eutrophication and Mitigation Strategies | IntechOpen. https://www.intechopen.com/chapters/62471. Accessed on 11 Dec 2023
4. Ren, H., et al.: A water-energy-food-carbon nexus optimization model for sustainable agricultural development in the Yellow River Basin under uncertainty. Appl. Energy **326**, 120008 (2022). https://doi.org/10.1016/j.apenergy.2022.120008
5. Landscape Planning for Agricultural Nonpoint Source Pollution Reduction I: A Geographical Allocation Framework | Environmental Management Available online: https://doi.org/10.1007/s00267-008-9186-3. Accessed on 11 Dec 2023
6. Li, H., Jin, X., Zhao, R., Han, B., Zhou, Y., Tittonell, P.: Assessing uncertainties and discrepancies in agricultural greenhouse gas emissions estimation in China: a comprehensive review. Environ. Impact Assess. Rev. **106**, 107498 (2024). https://doi.org/10.1016/j.eiar.2024.107498.
7. Chen, Y., et al.: Optimization of agricultural soil and water resources under fuzzy and random uncertainties: synergy and trade-off between equity-based economic benefits, nonpoint pollution and water use efficiency. Agricult. Water Manage. **281**, 108264 (2023). https://doi.org/10.1016/j.agwat.2023.108264
8. Xu, X., Chen, Y., Zhou, Y., Liu, W., Zhang, X., Li, M.: Sustainable management of agricultural water rights trading under uncertainty: an optimization-evaluation framework. Agricult. Water Manage. **280**, 108212 (2023). https://doi.org/10.1016/j.agwat.2023.108212
9. Ji, Y., Huang, G.H., Sun, W.: Inexact left-hand-side chance-constrained programming for nonpoint-source water quality management. Water Air Soil Pollut. **2014**, 225 (1895). https://doi.org/10.1007/s11270-014-1895-z
10. Huang, G.H., Baetz, B.W., Patry, G.G.: Grey integer programming: an application to waste management planning under uncertainty. Eur. J. Oper. Res. **83**, 594–620 (1995). https://doi.org/10.1016/0377-2217(94)00093-R
11. Ji, Y., Huang, G.H., Sun, W.: Nonpoint-source water quality management under uncertainty through an inexact double-sided chance-constrained model. Water Resour. Manag. **29**, 3079–3094 (2015). https://doi.org/10.1007/s11269-015-0983-1
12. Li, Y.P., Huang, G.H., Guo, P., Yang, Z.F., Nie, S.L.: A dual-interval vertex analysis method and its application to environmental decision making under uncertainty. Eur. J. Oper. Res. (2010). https://doi.org/10.1016/j.ejor.2009.01.013

13. Ji, Y., et al.: Inexact fuzzy-flexible left-hand-side chance-constrained programming for agricultural nonpoint-source water quality management. Sci. Total. Environ. **854**, 158565 (2023). https://doi.org/10.1016/j.scitotenv.2022.158565
14. Liu, J., Li, Y.P., Huang, G.H., Zeng, X.T.: A dual-interval fixed-mix stochastic programming method for water resources management under uncertainty. Resour. Conserv. Recycl. **88**, 50–66 (2014). https://doi.org/10.1016/j.resconrec.2014.04.010

Combustion Performance of Molded Fuels Prepared from Low-Temperature Pyrolysis of Sludge, Plastics and Pine Wood

Shuowu Chen[1,2], Jingfu Wang[1,2](✉), Ying Wu[1,2], and Zewei Yang[1,2]

[1] MOE Key Laboratory of Enhanced Heat Transfer and Energy Conservation, The Faculty of Environment and Life, Beijing University of Technology, Beijing 100124, China
jfwang@bjut.edu.cn

[2] Beijing Key Laboratory of Heat Transfer and Energy Conversion, The Faculty of Environment and Life, Beijing University of Technology, Beijing 100124, China

Abstract. The treatment process of low-temperature pyrolysis may efficiently utilize solid wastes, including plastics, pine wood, and sludge. This study used a hot-press molding device to treat sludge, plastics, and pine wood to low-temperature pyrolysis. Then, the combustion characteristics of the molded fuels were examined using thermogravimetric, elemental, and industrial analyses at various raw material ratios, pyrolysis temperature, and pyrolysis times. According to the results, the molded fuel with a ratio of 75:15:10 had the best combustion performance, with an overall combustion performance index improvement of 63.5% when compared to sludge. Increasing the ratio of plastic to pine might further enhance the molded fuel's combustion performance; When the fuel's pyrolysis temperature was raised, the fuel's maximum combustion rate, integrated combustion performance index, and calorific value all tended to rise before falling; Increase the pyrolysis time of molded fuels, the combustion performance of the fuel is basically unchanged.

Keywords: Sludge · Plastic waste · Pine Wood · Low-Temperature Pyrolysis · Combustion Characteristics

1 Introduction

The need for energy in society grows yearly as the economy grows, and since conventional energy sources are running out, it's critical to discover new, alternative energy sources. As a carbonaceous renewable energy source, biomass is clean, non-polluting, renewable, and environmentally friendly. It is possible to successfully minimize the usage of traditional energy by using biomass in a sensible manner. Sludge comprises a lot of organic matter, colloids, inorganic particles, and other materials, and when it comes to using it as a resource, it possesses the twin characteristics of waste and energy [1]. When it comes to the waste properties of sludge, untreated sludge contains heavy metals, proteins, and other substances. Prolonged sludge piling will cause untreated heavy metals to seep into the ground along soil gaps, contaminating the soil and groundwater, in addition to causing a large number of proteins to corrode by microorganisms and

D. Han et al. (Eds.): ICEPG 2024, ESE, pp. 999–1012, 2025.
https://doi.org/10.1007/978-3-031-86829-0_109

produce polluting gases [2]. Furthermore, plastic waste has a high calorific value, but because of its mass production and long environmental degradation period, it is one of the solid pollutants in the environment [3]. By effectively using waste plastics, we can cut down on the use of conventional energy sources while also minimizing landfills and burning, which contribute to environmental pollution and other issues.

Since it is still difficult to realize the value features of these three categories of waste, low-temperature pyrolysis technology [4] has garnered a lot of interest as a successful treatment due to its clear stabilization, reduction, and harmlessness effects. By low-temperature pyrolysis, high-calorific plastic, clean, non-polluting biomass, and solid waste sludge with a high water content are extruded into molded solid fuels. It can not only significantly reduce environmental pollution, but it can also benefit from the high calorific value of plastics and the synergistic effect of biomass on the co-pyrolysis of sludge. This process can partially combine the benefits of different wastes and optimize the high moisture and low calorific value of sludge, resulting in the preparation of high calorific value and low moisture molding fuels, which have a comparatively broad range of application scenarios. Moreover, the molding fuel has low emissions of air pollutants and flue gas, which may effectively cut carbon dioxide emissions and provide other benefits in the context of China's "dual-carbon" strategy [5].

Chen [6] et al. demonstrated that when thermal power plant sludge was subjected to low-temperature pyrolysis, over 50% of the sludge's energy was converted to biochar at pyrolysis temperature below 400 °C. Prolonged pyrolysis also increased the biochar's fuel ratio to 0.54 while removing over 50% of the sulfur and nitrogen. Lin [7] et al. created molded fuels using activated sludge and wood chips. They observed that an increase in the quantity of wood chips enhanced the fuels' combined combustion performance, and an increase in activated sludge also improved the fuels' igniting qualities. In his study of blended fuels, including sludge and coal, Li [8] discovered that while the ignition and burnout temperatures of the fuel were raised when sludge was added to coal samples, the overall performance of combustion and its intensity were reduced. Fu [9] et al. found that the temperature during pyrolysis has a bigger impact on the fuel mixture's gas-producing components and that co-pyrolysis of sludge and cotton stalks, with an increase in the proportion of cotton stalks, can accelerate the pyrolysis process of the more difficult-to-break-down organic matter in the sludge. Using fungal chaff of Mucuna pruriens to co-pyrolyze polypropylene, Liu [10] discovered that the addition of polypropylene decreased the ignition temperature, the activation energy of the combustion reaction, the integrated combustion index increased with warming, the solid fuel's calorific value, and the amount of ash content. In the co-pyrolysis of dyeing sludge and polyolefin plastics, Ding [11] found that the two polymers interacted between 200 and 600 °C and that increasing the mixture ratios of PP and PE increased the ignition, burnout, and combined combustion indices. Thus, it is evident that low-temperature pyrolysis is a useful technique for enhancing the efficiency of fuel blends containing binary mixes of biomass, polymers, and sludge.

However, there are currently few studies on the combustion properties of ternary mixes, with the majority of research on the application of pyrolysis of solid waste concentrating on the study of binary combinations, which mostly focuses on sludge together with plastics and biomass. In order to investigate the energy use of molded fuels prepared from sludge, plastic, and pine wood, three raw materials—sludge, plastic, and pine wood chips—were blended. The effects of various raw material ratios, pyrolysis temperature, and pyrolysis times on the combustion performance of molded fuels were studied experimentally in this paper.

2 Experimental Method

2.1 Materials Preparation

The experiment's raw components were pine wood, molded plastic, and sludge. The sludge that was utilized came from Beijing Urban Drainage Group's Gaobeidian wastewater treatment plant [12]. The sludge was allowed to dry naturally before being crushed, sieved, and then carried out to a mesh size of 80 for the test. The final plastic was prepared in accordance with the PE: PP:PVC ratio of 6.5:2.5:1, as the composition of plastic waste is complex. Analysis of the proportion of various components of domestic waste revealed that polyethylene (PE) accounted for 48%, polypropylene (PP) for 18%, polyvinyl chloride (PVC) for 7%, and the remaining 27% for other components [13]. The chosen pine wood was purchased through the Alibaba platform. The pine wood chips were sieved to an 80-mesh screen, dried for 24 h at 105 °C in a drying oven, and then tested. Prior to the experiment, a sample was made by combining pine wood, plastic, and crushed and sieved sludge in a certain ratio. After weighing the combined sample, 2 g was added to the hot press apparatus. The test uses the RYM-600YG high temperature hot pressing mold produced by Shanghai Xinnuo Instrument Group Co. A heated mold, a spacer, a press, a thermostat, a heat shield, and a bespoke or standard press bar make up the hot press system. Following sample entry into the hot press system, the heated mold is preheated to conduct low-temperature pyrolysis, the press pressure is adjusted to 4 MPa, the common press bar is nested into the heated mold, and the sample is subsequently extruded. Table 1 shows the industrial, elemental analysis, and calorific value of the feedstock.

Table 1. Industrial analyses of raw materials

Samples	Proximate analysis (wt%)				Ultimate analysis (wt%)				HHV(MJ/kg)
	M_{ad}	A_{ad}	V_{ad}	F_{cad}	C_{ad}	H_{ad}	O_{ad}	N_{ad}	S_{ad}	
sludge	9.13	44.78	39.18	6.91	36.36	3.86	26.59	5.50	0.77	9.02
plastics	0.20	0.00	99.80	0.00	80.28	12.00	0.26	0.06	0.00	41.69
pine wood	1.74	0.95	79.26	18.05	48.15	3.86	43.59	4.40	0.00	16.91

2.2 Thermogravimetric Experiment

The STA 449 F3 instrument, manufactured by NETZSCH Instruments GmbH, was the simultaneous thermal analysis system used in the experiment. The experiment involved weighing approximately 10 mg of sample into the alumina crucible, setting the temperature range of combustion in the software to be 30–800 °C, and increasing the temperature at a rate of 20 °C per minute. The experiment was finished to obtain the TG curve, and the DTG curve was obtained by one time of differentiation of the curve. The software was used to analyze the obtained TG and DTG curves. The program was used to analyze the TG and DTG curves that were acquired. TG/DTG curves are used to calculate the parameters that describe combustion, such as (1) Ignition temperature T_i: the temperature that corresponds to the point where the TG curve's horizontal tangent to the curve's beginning of weight loss meets the vertical line of the DTG weight loss peak [7]. (2) Burnout temperature T_h: The temperature at the end of the DTG curve is taken to be equal to the temperature at -1%/min, which is the burnout temperature [14, 15]; (3) Maximum mass loss rate W_{max}: the point corresponding to the maximum weight loss peak during combustion is obtained by making a horizontal tangent at the maximum weight loss peak of the DTG curve; (4) Mean mass loss rate W_{mean}: indicates the average rate of combustion of a fuel burning from ignition to exhaustion; (5) In order to evaluate the combustion performance of the mixtures, the combined combustion characteristic index S_N, expressed in terms of the formula:

$$S_N = \frac{W_{\text{max}} \times W_{\text{mean}}}{T_{\text{i}}^2 \times T_{\text{b}}} \tag{1}$$

(6) flammability index C: The flammability index, which is a more direct reflection of the pre-reaction capability of the fuel than the ignition temperature, is expressed by the equation:

$$C = \frac{W_{\text{max}}}{T_{\text{i}}^2} \tag{2}$$

2.3 Industrial and Elemental Analysis

Changsha Youxin automated YX-GYFX7703 industrial analyzer was the instrument used in the experiment. 0.5–0.7 g of sample was put into the crucible to measure the moisture and ash content, 0.9–1.1 g of sample was put into the crucible to measure the volatile fraction, and the fixed carbon content of the sample was calculated using the differential subtraction method. Using the Italian Leeman EA3000 elemental analyzer, the experiment was conducted with the CHNS mode selected. The standard curve was created by selecting the L-NS standard sample. Based on the standard curve, 1–2 mg samples were placed into tin capsules for analysis. Each sample was measured three times, and the average value was determined as the outcome.

2.4 Calorific Value, Mass and Energy Yield

(1) Calorific value (heat generation): Beijing Hongda Boyu BYLRY-3000W type automated calorimeter is the instrument utilized in the experiment. Prior to the test, calibrate the heat capacity using benzoic acid. Once the calibration is finished, take 0.9–1.1 g of the sample and place it in the oxygen bomb's crucible. Once the oxygen bomb is filled with oxygen, measure the sample's calorific value.

(2) Mass yield: The mass yield Y_{mass} is used to assess the effect of fuel pyrolysis on fuel mass loss and is expressed by the formula:

$$Y_{\text{mass}} = \frac{m_{T,t}}{m_i} \times 100\% \tag{3}$$

In the formula, m_i denotes the mass of the feedstock without pyrolysis; $m_{T,t}$ denotes the mass of the fuel at the set pyrolysis temperature T and pyrolysis time t.

(3) Energy yield: expressed by the formula Y_{energy}.

$$Y_{\text{energy}} = \frac{HHV_{T,t}}{HHV_0} \times Y_{\text{mass}} \times 100\% \tag{4}$$

In the formula, HHV_0 denotes the high level heat generation when the feedstock is not pyrolyzed; $HHV_{T,t}$ denotes the high level heat generation of the fuel under the conditions of pyrolysis temperature T and pyrolysis time t of the fuel.

3 Experiment Results and Discussion

3.1 Influence of Feedstock Ratios on Low-Temperature Pyrolysis Solid Products

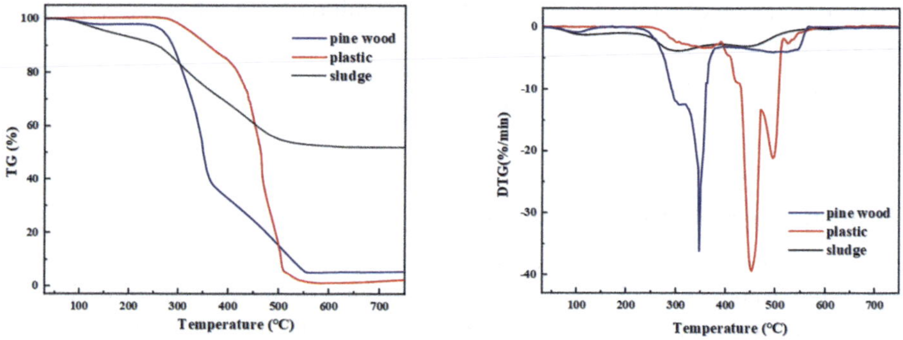

Fig. 1. TG-DTG curves of sludge, plastic and pine wood (20 °C/min)

A significant component influencing how well solid fuels burn is the feedstock ratio. Figure 1 displays the TG-DTG curves for the three feedstocks: sludge, plastic, and pine wood. Since the pyrolysis of sludge, plastic, and pine wood begins at 200 °C, a temperature of 250 °C was chosen for the investigation of mixed fuels. The TG-DTG graphs produced by heating the molding fuels in various ratios at a rate of 20 °C per minute from 30 °C to 750 °C are displayed in Fig. 2. To guarantee that high-calorific value fuels were manufactured simultaneously with a reduced amount of added plastic, plastic ratios of 10% and 15% were established. Molded solid fuels were obtained by pyrolyzing raw materials with varying ratios at a temperature of 250 °C for 20 min. The ratios were generated based on the following: sludge, plastic, and pine wood. The fuel combustion process can be broken down into four main stages, as shown by the analysis

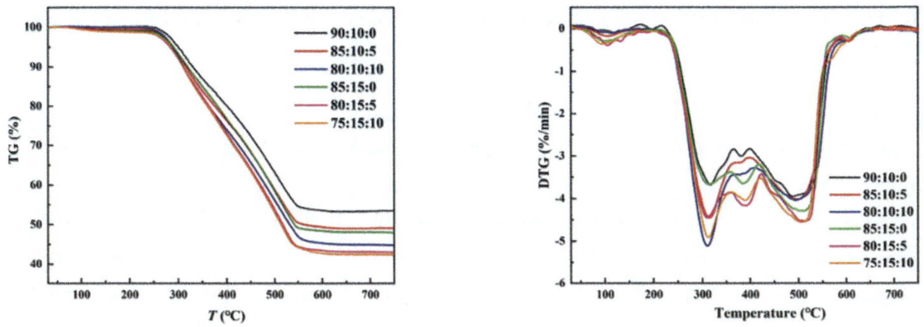

Fig. 2. TG-DTG curves of formed fuel at different feedstock ratios (20 °C/min)

of Fig. 2: the moisture volatilization stage, the exhalation of volatile components that are easily broken down in sludge and plastic (PVC), the combustion of cellulose and lignin in the pine chips, the combustion of the remaining volatile components and fixed carbon in the sludge and pine chips, the combustion stage of the two parts of plastic (PP, PE), and the stage of inorganic salt decomposition at high temperatures. By comparing the feedstock ratios of unadded pine chips (90:10:0, 85:15:0), it was discovered that when the plastic ratios increased, the fuel's maximum weight loss peak moved from the low temperature zone to the high-temperature zone while also decreasing in the low temperature zone. This is due to the fact that when the amount of plastic grows, the proportion of sludge in the fuel decreases, and less organic stuff that decomposes more readily is present in the sludge. Simultaneously, a greater percentage of plastic melted and filled the solid fuel's pore surface, obstructing the pore structure and delaying the precipitation of volatile components. This allowed the fuel to burn more slowly and proceed toward the high temperature zone [3]. In addition, when the ratio of plastics rises, the weight loss peak in the high temperature zone gets bigger. This is because the presence of more volatiles in plastics causes the fuel's weight loss peak in the high temperature region to get bigger. When fuels with identical plastic ratios that contained pine chips (85:10:5, 80:10:10, 80:15:5, 75:15:10) were compared, it was found that adding more pine chips led to larger weight loss peaks in both the high and low temperature zones and a lower temperature at which the fuel reached its maximum weight loss rate. This is specifically due to the fact that pine chips have a lower ignition temperature and a higher combustion rate. Additionally, the energy released during the combustion of pine chips encourages the burning of plastics and sludge, which speeds up combustion and shifts the fuel combustion process overall into the low-temperature zone [16].

The ignition temperature, burnout temperature, maximum burning rate, average burning rate, integrated combustion performance index, and flammability index may all be used to describe the combustion performance of the molded fuels. Table 2 displays the ignition temperatures of sludge, plastic, and pine, which are 263.4 °C, 405.0 °C, and 294.1 °C, respectively. It is evident that plastic had a significantly higher ignition temperature than sludge. When the fuels without pine chips (90:10:0, 85:15:0) were compared, the plastic ratio increased, the plastic's coating impact on the sludge after it softened increased, the fuels' ignition temperatures increased, and the fuels' delayed

Table 2. Combustion characteristic parameters of molded solid fuels with different feedstock ratios

Samples (sludge: plastics: pine wood)	Ti (°C)	Th (°C)	Wmax (%/min)	W_{mean} (%/min)	SN ($K^{-3}\cdot min^{-2}$)	C ($K^{-2}\cdot min^{-1}$)
plastics	405.0	559.3	41.01	10.52	1.13×10^{-6}	2.50×10^{-4}
pine wood	294.1	555.2	37.99	6.30	8.99×10^{-7}	4.39×10^{-4}
sludge	263.4	536.5	3.88	2.80	4.82×10^{-8}	5.59×10^{-5}
90:10:0	271.7	544.8	4.06	3.15	5.23×10^{-8}	5.49×10^{-5}
85:10:5	269.4	542.5	4.48	3.36	6.22×10^{-8}	6.17×10^{-5}
80:10:10	267.6	540.7	5.15	3.52	7.44×10^{-8}	7.19×10^{-5}
85:15:0	302.1	575.2	4.27	3.54	5.56×10^{-8}	4.67×10^{-5}
80:15:5	288.7	561.8	4.56	3.82	6.73×10^{-8}	5.47×10^{-5}
75:15:10	272.4	545.5	5.02	3.83	7.88×10^{-8}	6.76×10^{-5}

ignition occurred. The fuels with the addition of pine chips tended to converge to the sludge's ignition temperature, and their ignition temperatures were much lower when compared to the fuels without pine chips (85:10:5, 80:10:10, 80:15:5, 75:15:10). This is due to the low ignition temperature of the added pine shavings themselves, and the heat produced during their burning creates an environment that facilitates the plastic's ignition and accelerates its combustion [17]. The combined combustion performance index of the fuel improved as the ratio of plastics to pine wood chips grew, and the combined combustion performance index of the fuel was much greater than that of sludge at the same time. It is evident that adding plastic and pine shavings may significantly enhance the fuel's overall combustion performance.

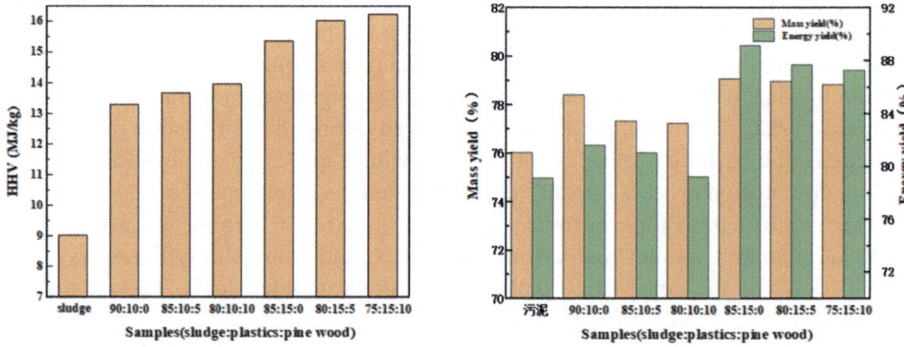

Fig. 3. Calorific value and yield of molded fuels with different feedstock ratios (20 °C/min)

Figure 3 illustrates the calorific value and yield of the molded fuel with varying feedstock percentages. The greatest calorific value of produced fuel (75:15:10) is 16.23 MJ/kg, whereas the calorific value of sludge is only 9.02 MJ/kg. When compared to sludge, the calorific value of the generated fuel (75:15:10) is enhanced by 80%. This suggests that blending three distinct raw materials in a ratio of 75:15:10 can significantly improve the properties of sludge with low calorific value. The addition of plastic can greatly increase the fuel's calorific value because plastic has a much higher calorific value than both pine shavings and sludge. As a result, the calorific value of the fuel increased significantly when the proportion of plastic was added. However, the effect of pine shavings on the fuel's calorific value was clearly not as good as that of plastic.

Table 3. Industrial and elemental analyses of molded solid fuels at different feedstock ratios

Samples	Mad(%)	A_{ad}(%)	V_{ad}(%)	F_{cad}(%)	C_{ad}(%)	H_{ad}(%)	O_{ad}(%)	N_{ad}(%)	S_{ad}(%)
sludge	9.05	28.39	53.98	8.58	36.36	3.86	26.59	5.50	0.77
90:10:0	2.04	31.65	54.37	11.94	43.53	4.17	12.89	4.92	0.27
85:10:5	2.33	29.99	54.79	12.89	44.17	4.86	13.25	4.63	0.25
80:10:10	2.84	28.29	55.18	13.69	45.22	4.89	13.51	4.49	0.22
85:15:0	2.56	29.43	56.90	11.11	45.63	5.37	10.24	4.62	0.27
80:15:5	1.16	28.25	58.31	12.28	48.13	5.678	10.90	4.53	0.26
75:15:10	1.65	26.11	58.43	13.81	48.24	6.23	11.47	4.39	0.25

Fuel evaluations can be made with the help of industrial and elemental analyses. The compositional analysis of fuels that have been molded at various ratios of source materials is displayed in Table 3. Because fixed carbon and ash were absent from plastic, the fuel's ash and fixed carbon content dropped as the plastic ratio rose. However, as the amount of pine wood grows at the same ratio of plastic, the low ash content of pine wood allows for an increase in the fixed carbon content of the fuel and a drop in ash content. The fuel's volatile fraction increased significantly due to the high volatile fraction of plastic and pine wood, rising by 8.2 percent in comparison to the sludge (75:15:10). The elemental analyses show that while the proportion of C and H elements in the fuel rose as plastic and pine grew, the percentage of O elements in the fuel declined as plastic climbed and increased as pine increased. This is due to the fact that although element O is found in pine wood (45.18% element O content), it is lacking from plastic (80.28% element C concentration).

3.2 Influence of Pyrolysis Temperature on Low-Temperature Pyrolysis Solid Products

The TG-DTG curves produced by pyrolyzing 75:15:10 molded fuel at various pyrolysis temperature for 20 min at a ramp rate of 20 °C/min are shown in Fig. 4. The fuel is mostly between 30 and 200 °C in the moisture volatilization stage. The fuel loses weight

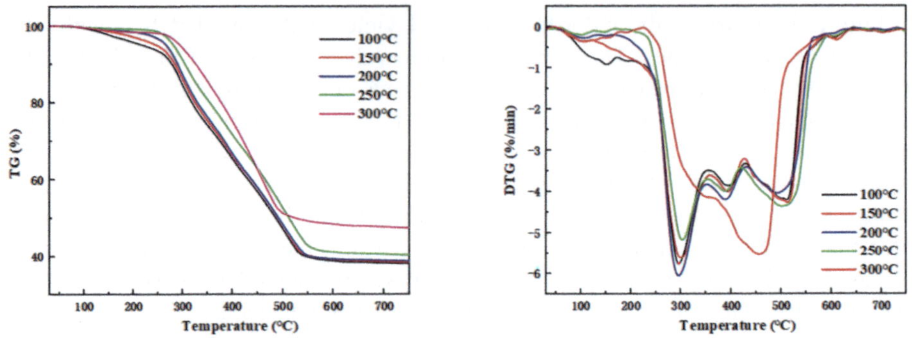

Fig. 4. TG-DTG curves of formed fuel at different pyrolysis temperature (20 °C/min)

Table 4. Combustion characteristic parameters of molded solid fuels at different temperatures

pyrolysis temperature	T_i	T_h	W_{max}	W_{mean}	S_N	C
(°C)	(°C)	(°C)	(%/min)	(%/min)	$(K^{-3}\cdot min^{-2})$	$(K^{-2}\cdot min^{-1})$
100	269.9	542.1	5.53	3.77	8.67×10^{-8}	7.60×10^{-5}
150	268.2	544.9	5.59	3.82	8.91×10^{-8}	7.77×10^{-5}
200	265.3	550.6	6.05	3.87	9.82×10^{-8}	8.59×10^{-5}
250	269.9	559.4	5.17	3.77	7.93×10^{-8}	7.09×10^{-5}
300	302.1	549.4	4.27	3.54	5.55×10^{-8}	4.67×10^{-5}

at a faster pace throughout this period of time and retains more moisture following pyrolysis at lower temperatures. Meanwhile, the DTG curve reveals that there is just one weight loss peak in the fuel with a large range when the pyrolysis temperature is 300 °C. Because most of the fuel's volatile components dissipate due to an excessively high pyrolysis temperature, the largest weight loss peak of fuel combustion occurs in the high temperature stage, over 450 °C. According to Table 4's analysis of the combustion performance characteristics, the fuel's ignition temperature reached 302.1 °C at the 300 °C pyrolysis temperature. This is specifically the high pyrolysis temperature that causes the fuel's volatility to precipitate and makes it difficult to ignite. The fuel's maximum combustion rate and comprehensive combustion performance index simultaneously showed a trend of first increasing and then decreasing with the rise in pyrolysis temperature. The fuel reached its maximum combustion rate and comprehensive combustion performance index at 200 °C.

As the pyrolysis temperature increased, the fuel's calorific value increased first, as seen in Fig. 5, and subsequently began to decline. The fuel achieves its maximum calorific value of 15.87 MJ/kg when the pyrolysis temperature hits 250 °C. A significant percentage of the fuel's volatile content precipitates out as the temperature during pyrolysis rises over 250 °C, causing the fuel's calorific value, mass yield, and energy yield to

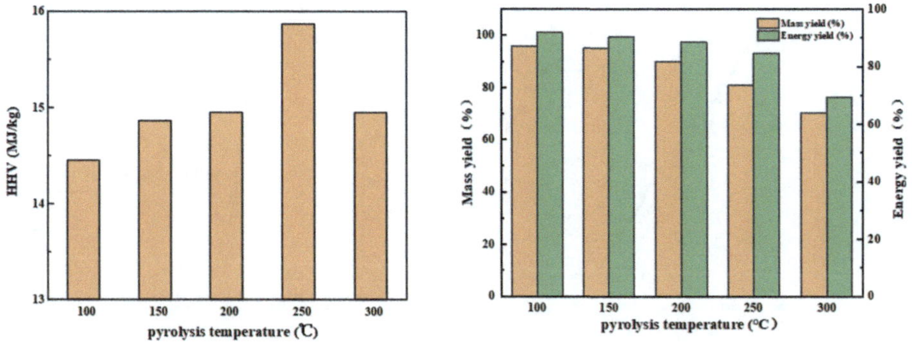

Fig. 5. Calorific value and yield of formed fuel at different pyrolysis temperature (20 °C/min)

Table 5. Industrial and elemental analyses of molded fuels at different pyrolysis temperature

temperature(°C)	Mad(%)	A_{ad}(%)	V_{ad}(%)	F_{cad}(%)	C_{ad}(%)	H_{ad}(%)	O_{ad}(%)	N_{ad}(%)	S_{ad}(%)
100	5.41	20.05	64.46	10.09	43.44	6.70	18.51	2.77	0.63
150	3.82	21.50	63.85	10.84	44.67	6.75	17.92	2.56	0.57
200	2.75	24.97	61.76	10.53	46.23	6.74	16.88	2.77	0.57
250	1.65	25.61	59.97	12.78	48.09	6.24	11.47	2.75	0.39
300	1.60	29.91	51.58	16.91	49.01	5.91	9.62	2.73	0.38

drop sharply. Table 5 illustrates that the content of fixed carbon and volatile matter in the molding fuel does not change significantly when the pyrolysis temperature is below 250 °C. Conversely, when the pyrolysis temperature is above 300 °C, the fuel's volatile matter content abruptly decreases, leading to a rise in fixed carbon content, which rises by 4.13% from 12.78% at 250 °C to 16.91% at 300 °C. This is primarily because, when temperatures rise, a significant percentage of the fuel under analysis volatilizes and the cellulose and hemicellulose in the pine wood chips break down. This leads to a comparatively elevated content of lignin, which is the primary source of coke [18]. The elemental analysis shows that the percentage of C elements in the fuel increases as the pyrolysis temperature rises, while the percentage of H and O elements decreases. This is because the percentage of H and O elements decreases due to water volatilization during the heating process, and when the pyrolysis temperature is between 250 and 300 °C, the fuel's sludge goes through decarboxylation and decarboxylation reactions, which cause the percentage of O elements to decrease. In the meantime, the lignin content in the pyrolysis increases, increasing the percentage of C elements [19].

3.3 Influence of Pyrolysis Times on Low-Temperature Pyrolysis Solid Products

The TG-DTG profiles obtained for varying pyrolysis periods of the 75:15:10 molded fuel, maintained at a temperature of 250 °C at a rate of 20 °C/min, are displayed in Fig. 6. The fuel burned at three different weight loss peaks, and it took ten minutes for the fuel to burn at its highest rate. The easily broken down and volatile components in the

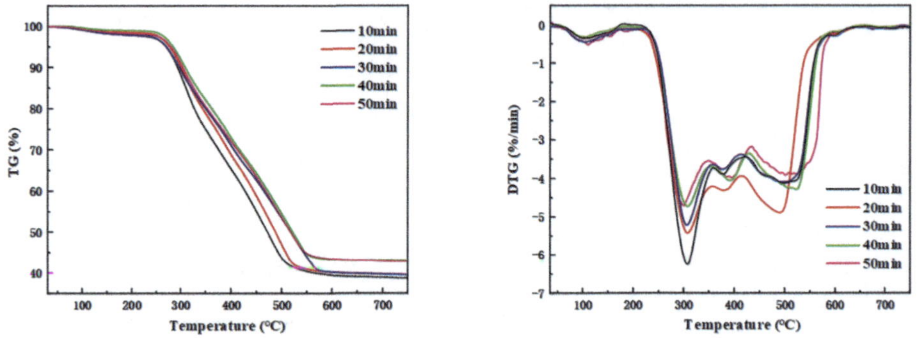

Fig. 6. TG-DTG curves of formed fuel at different pyrolysis times (20 °C/min)

Table 6. Combustion characteristic parameters of formed fuels at different pyrolysis times

pyrolysis times	T_i	T_h	W_{max}	W_{mean}	S_N	C
(min)	(°C)	(°C)	(%/min)	(%/min)	($K^{-3}\cdot min^{-2}$)	($K^{-2}\cdot min^{-1}$)
10	271.6	526.4	6.78	4.20	1.20×10^{-7}	9.19×10^{-5}
20	271.6	531.0	5.63	4.18	9.85×10^{-8}	7.63×10^{-5}
30	261.2	554.8	5.20	3.52	7.76×10^{-8}	7.62×10^{-5}
40	270.4	556.3	4.72	3.16	6.09×10^{-8}	6.46×10^{-5}
50	268.6	572.4	4.71	2.97	5.63×10^{-8}	6.53×10^{-5}

plastic (HCl) and sludge kept fading as the pyrolysis period increased, and the fuel's low temperature peak for weight loss—about 300 °C—was getting smaller. Table 6 shows that the fuel's ignition temperature stayed relatively constant over the course of the various pyrolysis times, but that the fuel's maximum weight loss rate, integrated combustion performance index, and flammability index all peaked after 10 min of pyrolysis. The maximum burning rate, flammability index, and combined combustibility performance index of the fuel exhibit a declining trend as the pyrolysis period exceeds 10 min.

Figure 7 shows that the heat yield of the molded fuel is not significantly affected by extending the pyrolysis duration when it is between 10 and 50 min. The mass yield of the fuel exhibited a declining pattern as the pyrolysis duration increased, but the energy yield first showed a rising trend before declining, peaking at 83.60% after 20 min. Table 7 illustrates that the fuels' volatile percentage and fixed carbon content did not significantly alter as the pyrolysis duration increased. The percentage of C, H, and O components in the fuel did not vary much, according to the elemental analysis. In conclusion, the performance of the molded fuel during combustion is mostly unaffected by the pyrolysis duration.

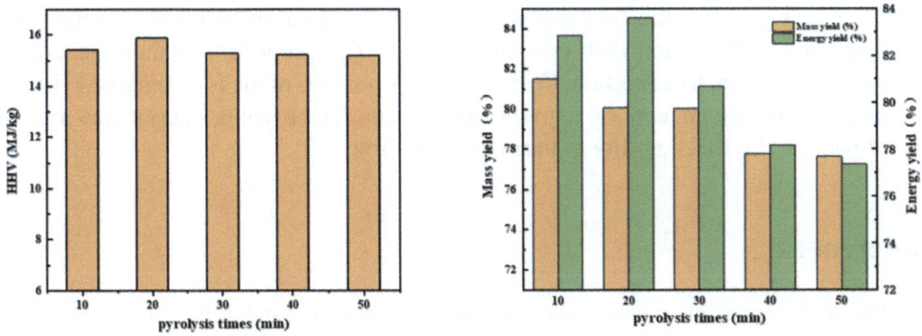

Fig. 7. Calorific value and yield of formed fuel at different pyrolysis times (20 °C/min)

Table 7. Industrial and elemental analyses of formed fuels at different pyrolysis times

pyrolysis times(min)	M_{ad}(%)	A_{ad}(%)	V_{ad}(%)	Fc_{ad}(%)	C_{ad}(%)	H_{ad}(%)	O_{ad}(%)	N_{ad}(%)	S_{ad}(%)
10	1.47	23.04	61.12	14.37	47.81	6.33	13.20	2.84	0.55
20	1.64	22.79	60.88	14.68	49.91	6.50	11.47	2.98	0.58
30	1.65	23.42	60.65	14.28	50.18	6.45	11.35	2.83	0.57
40	1.64	23.75	59.88	14.73	50.23	6.38	10.67	3.05	0.58
50	1.65	23.77	59.85	14.73	50.25	6.20	10.39	2.80	0.55

4 Conclusion

This study employed a hot press molding apparatus to manufacture molded fuel from sludge, plastic, and pine wood chips. The pyrolyzed fuel was then submitted to industrial, elemental, thermogravimetric, and calorific studies, among other analyses. Through studies using various ratios of raw materials, pyrolysis temperature, and pyrolysis periods, the combustion properties of the molded fuel were analyzed. The following findings were reached:

(1) The addition of pine shavings promotes the combustion of sludge and plastics towards the lower temperature zone. The maximum burning rate, overall combustion performance index, calorific value, volatile matter, fixed carbon, C, and H all significantly increased as the ratio of plastic to pine in the molded fuel increased. While the excessive pine chips weakened the increase in mass and energy yields, the addition of plastic significantly increased the fuel's mass and energy yields. In comparison to sludge, the 75:15:10 ratio of the molded fuel increased the total combustion performance index by 63.5% and produced the most heat (16.23 MJ/kg).

(2) The maximum combustion rate, flammability index, comprehensive combustion performance index, and calorific value of the fuel exhibit a tendency to increase and then decrease with increasing pyrolysis temperature; the fuel's mass yield and energy yield decrease with increasing temperature, and they abruptly drop at temperatures above 250 °C; the fuel's volatile fraction and percentage of C elements decrease with increasing temperature, but the percentage of fixed carbon, H, and O increased.

(3) The mass yield of the fuel displayed a declining pattern, while the energy yield displayed a rising and then decreasing trend. The proportion of heat generation, volatile fraction, fixed carbon, C, H, and O components of the fuel remained virtually unchanged with an increase in pyrolysis duration. The pyrolysis period had minimal impact on the fuels' ability to burn after molding.

References

1. Dai, X.H.: Applications and perspectives of sludge treatment and disposal in China. Science **72**(06), 30–34 (2020)
2. Zhao, Y., Zhao, Q.: Research progress of sludge resource utilization technology. Energy Res. Utilization **05**, 40–44 (2023)
3. Li, X., Pan, C., Song, Y., Yuan, X., Cheng, H., Ni, N., et al.: Co-pyrolysis technology and high-value utilization of typical plastic and biomass wastes. Environ. Sci. Res. **36**(09), 1765–1778 (2023)
4. Huang, X., Zhu, C., Guan, Z., Yan, J.: Improvement of low-temperature pyrolysis technology for domestic waste in high-altitude agricultural and pastoral areas. Gansu Sci. Technol. **37**(20), 18–20 (2021)
5. Luo, M., Feng, W., Zi, J.: A comprehensive utilisation technology for sludge treatment. Leather Mak. Environ. Protect. Technol. **2**(16), 24–25, 31 (2021)
6. Chen, R., Sheng, Q., Dai, X., Dong, B.: Upgrading of sewage sludge by low temperature pyrolysis: Biochar fuel properties and combustion behavior. Fuel **300**, 121007 (2021)
7. Lin, J., Wu, X., Wu, L., Li, W., Guan, B.: Study on the preparation of biomass fuel from activated sludge mixed with wood chips and its combustion characteristics. Renew. Energy **37**(02), 159–165 (2019)
8. Li, X.: Experimental Study on Fuel Synthesis from Residual Sludge in Small and Medium-sized Sewage Plants. Chongqing University (2008)
9. Fu, Q., Xu, Y., Yan, Y., Yan, J., Yang, L.: Response surface optimization analysis of co-pyrolysis characteristics of sludge cotton stalks. Combust. Sci. Technol. **29**(06), 699–705 (2023)
10. Liu, L:. Research on the Preparation of High Quality Biofuel by Mixing and Baking of Fungus Chaff of Fungus and Waste Plastic. Northeast Electric Power University (2023)
11. Ding, Z.: Study on the characteristics of synergistic thermal treatment of polyolefin waste plastics and printing and dyeing sludge and the attribution of typical pollutants. Guangdong University of Technology (2022)
12. Beijing Urban Drainage Group Gaobeidian Wastewater Treatment Plant Cleaner Production Project. Energy Conserv. Environ. Protect. (03), 9 (2020)
13. Liu, Y., Ma, X., Chen, D., Zhao, L., Zhou, G.: Co-pyrolysis characteristics and kinetic analysis of typical components of waste plastics. Chin. J. Electr. Eng. **30**(23), 56–61 (2010)
14. Gu, L., Chen, X., Zhao, C., Wu, X.: Thermogravimetric analysis of mixed combustion characteristics of municipal sludge and coal. Thermal Power Engineering, (06), 561–563 (2003)
15. Song, C., Wen, J., Li, Y., Dan, H., Shi, X., Xin, S.: Thermogravimetric assessment of combustion characteristics of blends of lignite coals with coal gangue. In: 3rd Annual International Conference on Mechanics and Mechanical Engineering (MME 2016), vol. 105, pp. 490–495 (2017)
16. Li, W., Liu, H., Jiang, Y.: Construction of solid bridge type biomass pellet fuel and its combustion characteristics. J. Agric. Mach. 1–15 (2024)

17. Xu, X., Pan, R., Chen, R.: Combustion characteristics, kinetics, and thermodynamics of pine wood through thermogravimetric analysis. Appl. Biochem. Biotechnol. **193**(5), 1427–1446 (2021)
18. Zeng, K., He, X., Yang, H., Wang, X., Chen, H.: The effect of combined pretreatments on the pyrolysis of corn stalk. Biores. Technol. **281**, 309–317 (2019)
19. Chanaka Udayanga, W.D., Veksha, A., Giannis, A., Lim, T.-T.: Pyrolysis derived char from municipal and industrial sludge: Impact of organic decomposition and inorganic accumulation on the fuel characteristics of char. Waste Manag. **83**, 131–141 (2019)

Research on Simulation and Optimization of Multi-stage AO Process Based on GPS-X Software

Yuzhou Zhang, Li He(✉), Zongkai Ma, Yawen Luo, Xueping Liu, Zengzhu Wang, Xiaowen Ma, and Yuan An

POWERCHINA Chengdu Engineering Corporation Limited, Chengdu 611130, China
HL_POWERCHINACDY@163.com

Abstract. This study optimized the A^2O process into a multi-stage AO process using GPS-X software and refined operational parameters through Response Surface Methodology (RSM). Initially, the multi-stage AO process was modeled and simulated using the ASM models embedded within the GPS-X software. A sensitivity analysis of various parameters was conducted, and critical parameters were adjusted to enhance model accuracy. RSM was employed to optimize operational key parameters such as dissolved oxygen (DO) in the aerobic tank, influent flow rate ratio (R), and sludge return ratio (r), exploring their impact on biochemical reaction efficiency. The results revealed significant improvements in BOD5 and TN removal rates, reaching 97.1% and 81.7%, respectively, after optimization. Compared to the A^2O process, the multi-stage AO process exhibited notable advantages in reactor capacity, energy consumption, and effluent quality. This study provides a scientific and effective optimization scheme for the upgrade and retrofitting of wastewater treatment plants, demonstrating the broad application potential of the GPS-X combined with RSM in optimizing wastewater treatment processes.

Keywords: Multi-stage AO · A^2O · GPS-X · RSM · Operational parameters

1 Introduction

With the economic development, environmental protection requirements have been steadily increasing. In the context of water environment protection, wastewater treatment plants (WWTPs) play a crucial role [1]. In recent years, more regions have imposed stricter effluent standards on WWTPs. The conventional A^2O process exhibits limitations in nitrogen removal. In contrast, the multi-stage AO process aligns well with the biological nitrogen removal process [2], effectively utilizing carbon sources in the influent water through staged feeding, resulting in higher nitrogen removal efficiency.

The International Water Association (IAWQ) has successively introduced the Activated Sludge Models (ASM) series. With the widespread application of mathematical models and computer simulation technologies in wastewater treatment, the GPS-X software emerged, incorporating the ASM models. This software enables users to

D. Han et al. (Eds.): ICEPG 2024, ESE, pp. 1013–1024, 2025.
https://doi.org/10.1007/978-3-031-86829-0_110

model, simulate, and predict effluent quality for process optimization and fault diagnosis, providing a scientific basis [3].

This study transforms the A^2O process into a multi-stage AO process using GPS-X software. By utilizing the Response Surface Methodology (RSM) and fitting the functional relationship between changing factors and response outcomes through a quadratic regression equation, the study investigates the key impact characteristics of adjustable parameters on process operation. It proposes practical optimization strategies for stable system operation, aiming to enhance the operational efficiency and effluent quality of WWTPs, thereby meeting increasingly stringent environmental protection requirements.

2 Materials and Methods

2.1 WWTP and Water Quality Characteristics

In this study, a WWTP in Sichuan Province, China, was selected. The designe capacity is 50000 m^3/d, A^2O and advanced treatment process was used. The effluent quality meet the Sichuan Provincial Local Standard (DB51/2311-2016). The A^2O biological reactor is designed with a total hydraulic retention time of 14 h, an external recycle ratio of 100%, an internal recycle ratio of 200%, and a sludge age of 15 days. The process flow diagram of the biological treatment system is shown in Fig. 1.

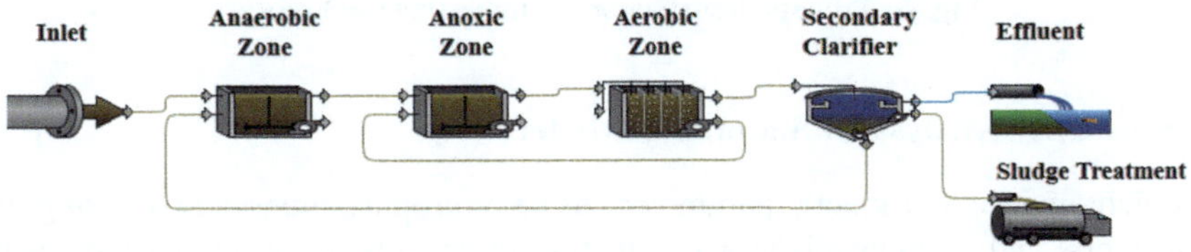

Fig. 1. Process flow diagram of A^2O process

Through continuous monitoring of the influent and effluent of the A^2O biological reactor, the data obtained are presented in Table 1. Regarding TN, the local discharge standard requires a concentration below 10 mg/L [4], which the A^2O biological reactor cannot achieve. Therefore, an advanced treatment process is necessary to degrade TN and meet the discharge requirements. By adjusting the volumes of each section within the biological reactor while maintaining the total reactor volume constant, a three-stage AO process model was established, as shown in Fig. 2. If this process can achieve the standard TN discharge, it will lead to cost savings in investment and energy consumption.

Table 1. Monitoring data of influent and effluent from A^2O biological reactor

Parameter	Unit	Influent	effluent
COD	mg/L	377 ± 67	56 ± 9
BOD_5	mg/L	167 ± 12	14 ± 2
TN	mg/L	40 ± 5.0	12.1 ± 1.5
NH^{4+}-N	mg/L	31.5 ± 3.2	0.4 ± 0.1
TP	mg/L	3.3 ± 0.2	0.6 ± 0.1

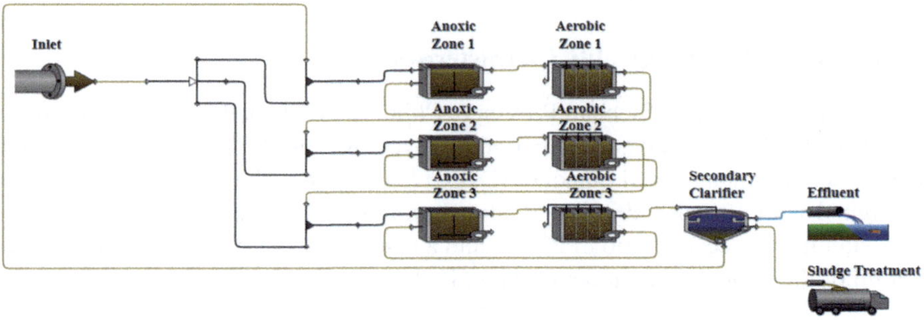

Fig. 2. Process flow diagram of three-stage AO process

2.2 Sensitivity Analysis of Simulation Model

The stoichiometric and kinetic parameters of heterotrophs, ammonia-oxidizing bacteria, nitrifiers, and polyphosphate-accumulating organisms in the model have varying impacts on the effluent water quality. Sensitivity analysis is performed to evaluate these parameters and assess their degrees of influence. The conventional sensitivity calculation method [5] is as follows:

$$S_{i,j} = \left| \frac{\Delta y_j / y_j}{\Delta x_i / x_i} \right| \tag{1}$$

The criteria for judgment are as follows: $S_{i,j} \leq 0.25$ indicates no significant influence of the parameter on the output result; $0.25 \leq S_{i,j} < 1$ indicates some influence; $1 \leq S_{i,j} < 2$ indicates significant influence; and $S_{i,j} \geq 2$ indicates a substantial influence on the output result. In this study, the set parameter values were adjusted by a 10% variation to observe their impact on simulated effluent water quality indicators, with the sensitivity index $S_{i,j}$ used for evaluation.

The Symbols of the GPS-X simulation software is showed in Table 2:

Table 2. The Symbols of the GPS-X

Symbols	
Y_H	Aerobic heterotrophic yield on soluble substrate for heterotrophic biomass (gCOD/gCOD)
Y_A	Anoxic heterotrophic yield on soluble substrate for heterotrophic biomass (gCOD/gCOD)
$Y_{P/O}$	Aerobic yield on PAO growth(gCOD/gCOD)
$Y_{P/A}$	Anoxic yield on PAO growth(gCOD/gCOD)
Y_{PHA}	PHA storage yield(gP/gCOD)
μ_{maxH}	Maximal specific growth rate of heterotrophic biomass on substrate(d^{-1})
K_P	Saturation coefficient for phosphorus(gP/L)
K_{ON}	Oxygen inhibition coefficient for denitrification(mgO_2/L)
b_H	Aerobic heterotrophic decay rate(d^{-1})
μ_{maxA}	Maximum growth rate for ammonia oxidizer(d^{-1})
K_{AA}	Ammonia saturation coefficient forammonia oxidizer(mgN/L)
K_{OA}	Oxygen saturation for ammonia oxidizer(mgO_2/L)
b_A	Ammonia oxidizer aerobic decay rate(d^{-1})
μ_{maxP}	Maximum growth rate of PAO(d^{-1})
K_{PP}	Rate constant for storage of PAO(gP/gPAO/d)
μ_{maxN}	Maximum growth rate of nitrifying bacteria(d^{-1})

2.3 Optimization of Operational Control Parameters Using Software Simulation Combined with RSM

Based on the Box-Behnken design principle of RSM [6], three crucial operational conditions in the multi-stage AO process were selected: dissolved oxygen (DO) in the aerobic tank, influent flow rate ratio (R) in the first stage AO, and sludge return ratio (r) [7]. These factors were represented by X_1, X_2, and X_3, respectively. A 3-factor, 3-level simulation experiment was designed to assess their impact on pollutant removal rates, with the condition settings outlined in Table 3.

Table 3. Response surface factors and level design

factor	level		
	−1	0	1
DO	1 mg/L	2.5 mg/L	4 mg/L
R	10%	35%	60%
r	20%	60%	100%

3 Results and Discussion

3.1 Sensitivity Analysis and Calibration of the Model

The combination of kinetic and stoichiometric parameters in the model constitutes the microbial growth, nitrification, ammonification, phosphorus accumulation, and other microbial processes. Sensitivity analysis can be used to understand the degree of influence of each parameter. In this study, 86 kinetic and stoichiometric parameters were analyzed, and those with sensitivity coefficients ($S_{i,j}$) greater than 0.25 are listed in Table 4. It can be seen that a total of 15 parameters were identified as sensitive. Among them, according to the judgment criteria, 8 parameters have some influence on the output results; 4 parameters have significant influence; and 3 parameters have substantial influence.

As shown in Table 4, the sensitive parameters in the model are related to the growth and decay of heterotrophs, ammonia-oxidizing bacteria, nitrifiers, and polyphosphate-accumulating organisms. Among these parameters, μ_{maxA} and b_A have a very significant impact on the degradation of NH^{4+}-N, and K_{AA} also has an important influence on the concentration of NH^{4+}-N. Additionally, many parameters related to heterotrophs are sensitive, with Y_H having a particularly significant influence on BOD_5 results. Simultaneously, Y_H also affects COD and TN results, making it a crucial research parameter. For TP, μ_{maxP} and K_{PP} in the model parameters also significantly influence its removal. Therefore, it is necessary to reasonably adjust the values of the parameters in the model to make the simulated results closer to the actual results. Focus should be placed on adjusting the parameters with significant influences, while the remaining parameters can retain their default values. Studies have shown that the Y_H value ranges from 0.75 to 0.78 [8]. Therefore, adjusting the Y_H value from the default 0.67 to 0.76 can improve the treatment efficiency of carbon and nitrogen pollutants. For the ammonification process, the μ_{maxA} value was adjusted from the default 0.9 d^{-1} to 0.8 d^{-1}, and the K_{AA} value was adjusted from the default 0.7 to 1.3 mgN/L to make the simulation process more realistic. Simultaneously, the μ_{maxP} value was adjusted from the default 1 d^{-1} to 0.8 d^{-1}, and the K_{PP} value was adjusted from the default 2 to 2.5 gP/(gPAO/d), which is consistent with the typical value range of the ASM2d model [9].

According to the optimization and adjustment of parameters, a comparison of the software simulation results with the actual effluent data is shown in Fig. 3. Through analysis, it can be observed that the simulated values are all within the confidence interval of the actual measured values. There is no significant statistical difference between the simulated and measured values for the variables, indicating that the simulation process is valid.

Table 4. The $S_{i,j}$ values for the most sensitive parameters

parameter	COD	BOD_5	TN	NH^{4+}-N	TP
Y_H	0.35	2.23	0.58		
Y_A		0.79	0.43		
$Y_{P/O}$	0.29				0.86
$Y_{P/A}$					0.32
Y_{PHA}					0.32
μ_{maxH}		1.02			
K_P		0.53			
K_{ON}			0.27		
b_H		0.80			
μ_{maxA}				3.78	
K_{AA}				1.95	
K_{OA}				0.57	
b_A				2.33	
μ_{maxP}					1.11
K_{PP}					1.01

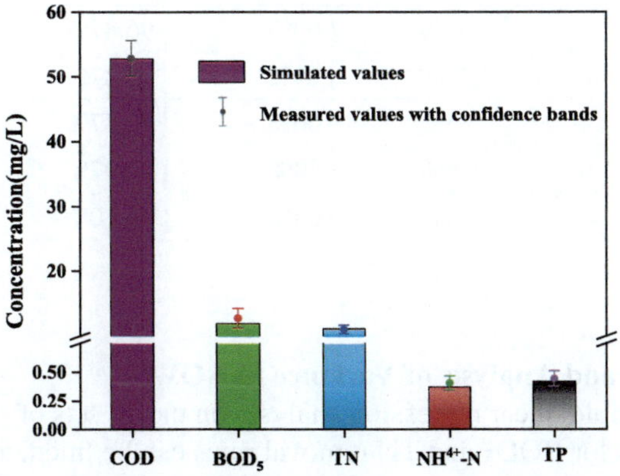

Fig. 3. Comparison chart of simulated and measured effluent Components

3.2 Analysis of Simulation Optimization Results by RSM

Typically, properly functioning wastewater treatment plants exhibit high removal rates for $NH^{4+}N$ and TP. Further removal of TP can be achieved by adding chemical agents. The removal of organic matter and TN mainly depend on biochemical reactions, among which TN is the key controlling factor in the biochemical system. Compared to traditional processes, the multi-stage AO process demonstrates higher TN removal efficiency.

Therefore, by establishing a multi-stage AO model through software and analyzing the impact of variable changes BOD_5 and TN removal rates by using the RSM method, the results are presented in Table 5.

Table 5. Simulation test results

S. No.	factor			Removal rate	
	DO(mg/L)	R	r	De(BOD_5)	De(TN)
1	1	10%	60%	94.60%	68.75%
2	4	35%	100%	97.15%	76.51%
3	4	35%	20%	95.84%	67.81%
4	2.5	10%	20%	92.58%	55.45%
5	4	60%	60%	97.36%	72.51%
6	2.5	35%	60%	96.87%	74.73%
7	1	60%	60%	96.80%	81.73%
8	2.5	60%	20%	95.71%	74.23%
9	2.5	35%	60%	96.87%	74.73%
10	2.5	60%	100%	97.03%	78.39%
11	1	35%	100%	96.57%	81.92%
12	2.5	10%	100%	96.55%	71.54%
13	2.5	35%	60%	96.87%	74.73%
14	2.5	35%	60%	96.87%	74.73%
15	2.5	35%	60%	96.87%	74.73%
16	1	35%	20%	92.97%	74.14%
17	4	10%	60%	96.10%	65.18%

Linear Regression and Analysis of Variance (ANOVA)

By performing multiple linear regression analysis on the 17 sets of simulation test data, the model equations for BOD_5 and TN removal rates can be fitted, as shown in Eqs. (1) and (2).

$$\begin{aligned} De(BOD_5) &= 96.87 + 0.69X_1 + 0.88X_2 + 1.28X_3 - \\ &\quad 0.23X_1X_2 - 0.57X_1X_3 - 0.66X_2X_3 - 0.24X_1^2 - 0.41X_2^2 - 0.99X_3^2 \end{aligned} \tag{2}$$

$$\begin{aligned} De(TN) &= 74.73 - 3.04X_1 + 5.72X_2 + 4.59X_3 - 1.36X_1X_2 + 0.23X_1X_3 \\ &\quad -2.98X_2X_3 + 1.22X_1^2 - 3.97X_2^2 - 0.86X_3^2 \end{aligned} \tag{3}$$

According to the results of ANOVA presented in Table 6, the F-values for the models for BOD_5 and TN are 94.35 and 84.88, respectively, with P-values less than 0.0001 for both. This indicates that the models are highly significant, have low error, and can effectively describe the relationship between the response values and the factors. Additionally, the correlation coefficients (R^2) for BOD_5 and TN are 0.9918 and 0.9909, respectively, both close to 1, indicating a strong correlation. The adjusted R^2 values are 0.9813 and 0.9792, respectively, while the signal-to-noise ratios are 33.04 and 37.04, both greater than 4, suggesting a high degree of model fit. Furthermore, the coefficients of variation (C.V.) for the models are 0.20% and 1.24%, respectively, both less than 4%, indicating that the standard deviation is small relative to the mean response values, resulting in low data dispersion and high model precision [10].

The P-values for the linear terms X_1, X_2, X_3, and t he quadratic terms X_1X_2, X_1X_3, X_2X_3, $X_1{}^2$, $X_2{}^2$, and $X_3{}^2$ in the BOD_5 removal rate model are all less than 0.05, indicating that these parameters have significant effects on the BOD_5 removal rate. Similarly, the P-values for the linear terms X_1, X_2, X_3, and the quadratic terms X_1X_2, X_2X_3, $X_1{}^2$, and $X_2{}^2$ in the TN removal rate model are also all less than 0.05, indicating significant effects of these parameters on the TN removal rate. This further demonstrates the interactive effects of DO, R, and r on both BOD_5 and TN removal rates.

Table 6. Results of ANOVA

Source	Sum of squares		Df	Mean square		F-value		P-value	
	BOD_5	TN		BOD_5	TN	BOD_5	TN	BOD_5	TN
Model	31.83	621.94	9	3.54	69.1	94.35	84.88	<0.0001	
X_1	3.8	73.87	1	3.8	73.87	101.25	90.73	<0.0001	
X_2	6.25	261.29	1	6.25	261.29	166.7	320.93	<0.0001	
X_3	13.01	168.64	1	13.01	168.64	346.97	207.13	<0.0001	
X_1X_2	0.2209	7.37	1	0.2209	7.37	5.89	9.05	0.0456	0.0197
X_1X_3	1.31	0.2116	1	1.31	0.2116	34.98	0.26	0.0006	0.6259
X_2X_3	1.76	35.58	1	1.76	35.58	46.84	43.7	0.0002	0.0003
$X_1{}^2$	0.2527	6.32	1	0.2527	6.32	6.74	7.76	0.0356	0.0271
$X_2{}^2$	0.7078	66.28	1	0.7078	66.28	18.88	81.41	0.0034	<0.0001
$X_3{}^2$	4.15	3.11	1	4.15	3.11	110.66	3.82	<0.0001	0.0914
Residual	0.2624	5.7	7	0.0375	0.8142				
Lack of fit	0.2624	5.7	3	0.0875	1.9				
Pure error	0	0	4	0	0				
Cor total	32.09	621.94	16						

Response Surface Analysis

Using RSM software, 3D surface plots can be created to facilitate a more intuitive observation of the impact of various factors on BOD_5 and TN removal rates, as well as to examine the relationship between the interactions among these factors and the target changes. Response surface plots were generated for any two factors among DO, R, and r, as shown in Fig. 4, which depict the effects of interactions among these three factors on BOD_5 and TN removal rates.

As seen in Fig. 4(a), the BOD_5 removal rate gradually increases with the rise in both DO and R, exhibiting a similar trend. This demonstrates a synergistic effect between DO and R on COD removal, indicating that when oxygen supply is sufficient and a larger fraction of influent is allocated to the first stage of the AO tank, pollutants are more thoroughly degraded through the efficient utilization of multiple anoxic-aerobic tanks. However, based on the analysis of the nitrogen removal process, denitrification requires lower DO concentrations, which is also verified in Fig. 4(d). When the influent flow ratio is fixed, the TN removal rate gradually decreases with increasing DO. Conversely, when DO is constant, the TN removal rate initially increases with R and then gradually decreases after reaching 45%-55%. This may be due to the longer flow path of the initial influent stages allowing for more complete nitrogen degradation, while the insufficient organic matter in later stages hinders the denitrification process. The combined effect of these opposing factors results in the observed changes in TN removal rate. Overall, lower DO and higher R values promote TN removal.

Figure 4(b) and (e) illustrate the impact of DO and r on BOD_5 and TN removal rates. When r is low, the BOD_5 removal rate increases rapidly with increasing DO but plateaus after reaching 3.5 mg/L DO, indicating that excess DO does not significantly enhance organic matter degradation. When DO is constant, the BOD_5 removal rate initially increases rapidly with r but then gradually decreases after r reaches 63%. This is because increased r leads to higher biological loading, prolonging sludge retention time and hindering pollutant decomposition and oxidation, thereby reducing organic matter degradation. Figure 4(b) also shows that TN removal rate increases with decreasing DO and increasing sludge return ratio, indicating a synergistic effect between these two factors.

The trend of BOD_5 removal rate with respect to R and r is similar to that of DO and r, further confirming the significant influence of r in this model. The TN removal rate increases within a certain range with both r and R (Fig. 4(f)) but gradually decreases beyond this range. Notably, changes in flow ratio have a more pronounced effect on TN removal rate when the sludge return ratio is fixed, indicating that R has a more significant impact on TN removal, which validates the advantage of the staged influent characteristic of the multi-stage AO process for TN removal.

By integrating the simulation results of the multi-stage AO model with RSM, optimal operating conditions for BOD_5 and TN removal can be predicted. Under conditions of DO = 1.34 mg/L, RF = 52.9%, and Rw = 85.7%, the model predicts BOD_5 and TN removal rates of 97.1% and 81.7%, respectively. Additionally, COD, NH^{4+}-N, and TP removal rates in the effluent are 87.3%, 99.6%, and 88.2%, respectively, maintaining high removal efficiencies. This demonstrates that combining GPS-X software with

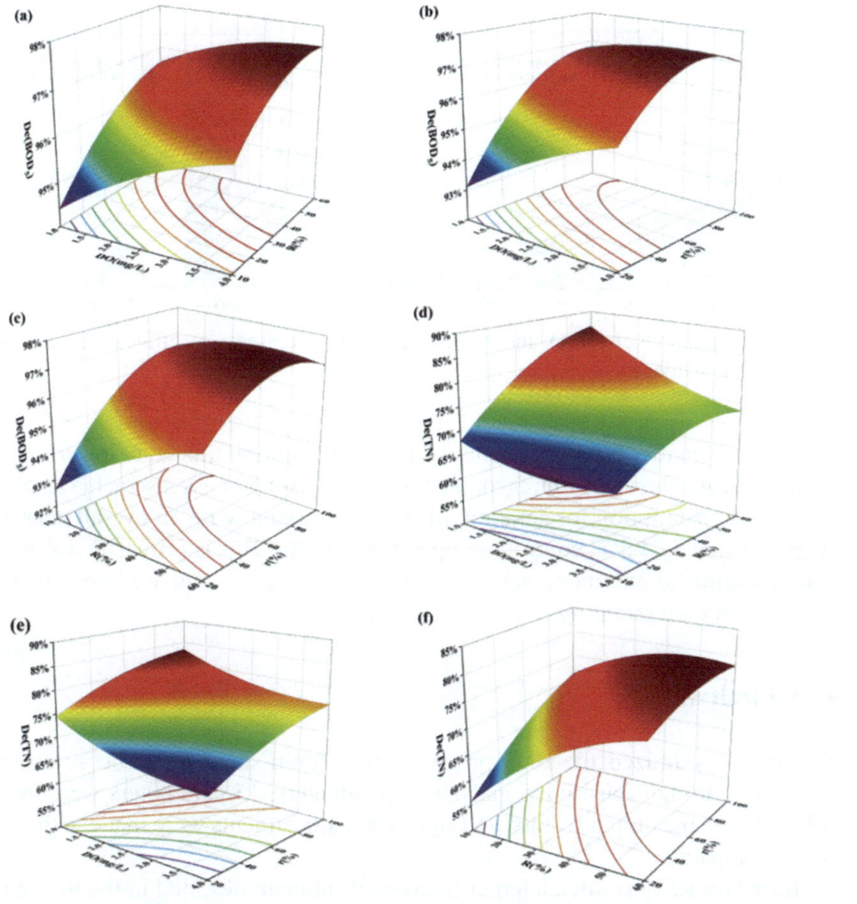

Fig. 4. Response surface analysis of effluent quality with varying factors: (a) DO-R-BOD_5; (b) DO-r-BOD_5; (c) R-r-BOD_5; (d) DO-R-TN; (e) DO-r-BOD_5; (f) DO-R-BOD_5

RSM analysis can provide simple and rapid recommendations for upgrading wastewater treatment plant processes.

3.3 Comparison of Operational Performance Between Multi-Stage AO Process and A^2O Process

The simulated effluent results for treating the same wastewater using both the A^2O and three-stage AO processes are presented in Fig. 5. The three-stage AO process exhibits improved performance in removing COD, BOD_5, TN, and NH^{4+}-N compared to the A^2O process. Notably, the TN concentration in the effluent decreases from 11.1 mg/L to 6.8 mg/L, enabling compliance with discharge standards without relying on advanced treatment processes (such as denitrification filters), highlighting the advantage of the multi-stage AO process in nitrogen removal. Additionally, the multi-stage AO process

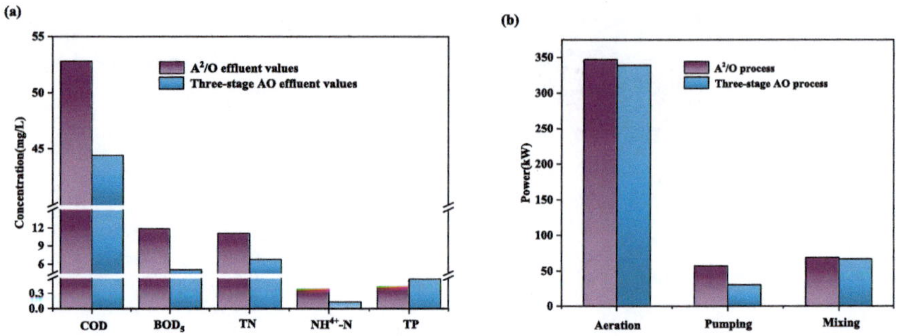

Fig. 5. Comparison of $A^{2/}O$ and three-stage AO processes: (a) effluent results; (b) energy consumption analysis

allows for a smaller bioreactor volume, reduced retention time, and lower initial investment costs. As shown in Fig. 5(b), the three-stage AO process also achieves a reduction in energy consumption for aeration, lifting, and mixing, with an overall energy savings of approximately 4.5%, controlling operational costs. This is primarily due to the better nitrogen removal efficiency achieved without the need for mixed liquor recirculation, thereby reducing pumping energy consumption.

4 Conclusion

This study optimized the A^2O process into a multi-stage AO process using GPS-X software. Through sensitivity analysis of parameters, 15 parameters were identified as influential on the output results, and significant adjustments were made to those with the greatest impact.

Based on RSM results, a larger fraction of influent allocated to the first stage of the AO tank and sufficient oxygen supply enhance pollutants degradation, lower DO and higher sludge return ratios promote TN removal. The results of RSM also demonstrated that the multi-stage AO process increased TN removal efficiency from 70% to 81.7%, enabling TN discharge without additional advanced treatmen.

The multi-stage AO process achieves pollutant degradation in a smaller reactor volume and with lower energy consumption, making it an environmentally friendly technology. This study shows that the combination of GPS-X and RSM methods can provide rapid and scientific optimization schemes for the upgrading of wastewater treatment plants.

References

1. Kehrein, P., van Loosdrecht, M., Osseweijer, P., et al.: A critical review of resource recovery from municipal wastewater treatment plants - market supply potentials, technologies and bottlenecks. Environ. Sci.: Water Res. Technol. **6**(4), 877–891 (2020)
2. Wang, Y., Du, Z., Liu, Y., et al.: The nitrogen removal and sludge reduction performance of a multi-stage anoxic/oxic (A/O) biofilm reactor. Water Environ. Res. **92**(1), 94–105 (2020)

3. Abbasi, N., Ahmadi, M., Naseri, M.: Quality and cost analysis of a wastewater treatment plant using GPS-X and CapdetWorks simulation programs. J. Environ. Manage. **284**, 111993 (2021)
4. Zhou, J., Hu, X., Wang, Q., et al.: Application of multi-stage AO process in dougouhe No.2 underground wastewater treatment plant. China Water Wastewater **38**(20), 48–54 (2022)
5. Flores-Alsina, X., Corominas, L., Neumann, M.B., et al.: Assessing the use of activated sludge process design guidelines in wastewater treatment plant projects: a methodology based on global sensitivity analysis. Environ. Model. Softw. **38**, 50–58 (2012)
6. Liu, C., Liang, L., Han, X., et al.: Optimized preparation of activated carbon from furfural residue using response surface methodology and its application for bisphenol S adsorption. Water Sci. Technol. **85**(3), 811–826 (2022)
7. Hou, Z., Zhou, X., Ronel, K.R.A., et al.: Research progress on the step-feed multi-stage A/O process. Indust. Water Treat. **40**(06), 7–13 (2020)
8. Vital-Jacome, M., Dochain, D., Thalasso, F.: Microrespirometric model calibration applied to wastewater processes. Biochem. Eng. J. **128**, 168–177 (2017)
9. Xiaoqian, D., Jianqiang, Z., Bo, H.: Sensitivity analysis of kinetic parameters of Activated Sludge Model No.2D in two kinds of A2O processes. Environ. Pollut. Control **40**(12), 1347–1358 (2018)
10. Yang, Y., Zhou, Z., Lu, C., et al.: Treatment of chemical cleaning wastewater and cost optimization by response surface methodology coupled nonlinear programming. J. Environ. Manage. **198**(Pt 2), 12–20 (2017)

Common Constraints and Countermeasure for Environmental Impact Assessment of Urban Renewal Planning: A Case Study of Guangzhou City

Yongsheng Liu(✉) and Yaoguang Wu

Guangzhou Research Institute of Environmental Protection Science Co., Ltd., Guangzhou 510620, Guangdong, China
15975612091@163.com

Abstract. Urban renewal planning, as a pivotal component within the national territorial spatial planning system, has not yet fully integrated its environmental impacts into the existing regulatory and technical frameworks for planning environmental impact assessment (EIA). EIA possesses the advantage of mitigating environmental impacts and enhancing environmental quality by addressing them at their source and within the early stages of the prevention system. Drawing upon the implementation status of urban renewal planning in Guangzhou as a case study, this paper emphasizes the imperative need for conducting EIA specifically for urban renewal planning. It undertakes a comprehensive analysis of the common ecological and environmental constraints encountered in urban renewal planning, encompassing regional environmental quality, ecological systems, resource utilization, environmental protection infrastructure, and other aspects of environmental management. Furthermore, the paper proposes a series of targeted countermeasures and recommendations aimed at addressing these constraints and fostering the eco-friendly implementation of urban renewal planning. These suggestions are intended to serve as a valuable reference for policymakers, planners, and stakeholders involved in advancing sustainable urban development through urban renewal initiatives. By doing so, this study contributes to the ongoing discourse on integrating environmental considerations into urban planning processes and promoting environmentally responsible urban renewal practices.

Keywords: Urban Renewal Planning · Environmental Impact Assessment · Constraint factors · Countermeasures and suggestions

1 Introduction

Urban renewal planning, as an important component of the land spatial planning system, has not yet incorporated environmental impacts into the existing regulations and technical system for environmental impact assessments of plans (hereinafter referred to as "planning EIA"). This has led to a lack of scientific and accurate assessment of the environmental feasibility of urban renewal planning. Currently, there is limited research

D. Han et al. (Eds.): ICEPG 2024, ESE, pp. 1025–1032, 2025.
https://doi.org/10.1007/978-3-031-86829-0_111

on the planning EIA for urban renewal, with only a few studies focusing on environmental impact assessments for detailed plans [1]. In practical applications, planning EIAs for urban renewal often fail to fully consider the characteristics of this type of planning, merely referencing general technical specifications for planning EIAs, resulting in a lack of specificity.

As a city that was the first in the country to initiate pilot urban renewal projects, Guangzhou actively coordinates the allocation of new land and existing land resources, promoting the revitalization and reuse of approximately 300 square kilometers of inefficient existing land [2]. Since 2019, Guangzhou has been conducting environmental impact assessments in parallel with its urban renewal planning. This article uses Guangzhou's urban renewal planning as a case study to analyze the characteristics of urban renewal planning and common constraining factors, and to propose targeted countermeasures and recommendations from the perspective of planning EIA.

2 The Necessity of EIA in Urban Renewal Planning

China's territorial spatial planning system consists of "five levels and three categories." The "five levels" refer to the national, provincial, municipal, county, and township levels, while the "three categories" include overall planning, detailed planning, and related special planning [3]. According to the implementation of urban renewal planning in Guangzhou, urban renewal plans fall under the category of special planning. The renewal plans mainly encompass several types, including medium and long-term urban renewal plans, area planning schemes, and urban renewal project implementation plans. The medium and long-term urban renewal plans provide guiding principles, goal setting, strategic deployment, and specific action plans, outlining the scale and key areas of renewal; the annual urban renewal plan is comprehensively developed by relevant municipal departments based on area development planning, covering area planning, project execution plans, and funding allocation schemes; area planning schemes formulate overall plans for urban renewal zones and their projects for regions included in the renewal implementation plan; while the urban renewal project implementation plans are specifically developed by district-level governments based on the annual plan [4]. The relationship of urban renewal planning within the territorial spatial planning system is illustrated in Fig. 1.

Currently, our country is engaged in the construction of a Beautiful China. Areas such as old villages, old towns, and old factories, which suffer from inefficient land use, poor environmental quality, and unattractive urban appearances, urgently need to undergo renewal and transformation. This is essential to address shortcomings, restore ecosystems, and improve both the ecological environment and urban aesthetics. The purpose of planning environmental impact assessments is to evaluate the environmental feasibility of implementing plans, propose targeted optimization and adjustment suggestions, and develop countermeasures to mitigate adverse environmental impacts, thereby achieving an organic unity between urban renewal and ecological environment protection.

Currently, the main implementation areas of planning environmental impact assessments (EIAs) are still concentrated on the development and utilization planning of industrial parks and zones, while there is relatively little practical experience in environmental impact assessments for urban renewal planning [5]. Officially incorporating urban renewal planning into the planning EIA system allows for a comprehensive assessment of its feasibility from a top-level design perspective, thereby providing solid support for promoting the construction of a beautiful China with high quality during the urban renewal process. Guangzhou is one of the earlier cities in the country to carry out urban renewal planning EIAs and has gradually played a positive role in areas such as source prevention and ecological restoration. The practical work in Guangzhou demonstrates that assessing the feasibility of urban renewal planning from a top-level design perspective and proposing optimization and adjustment suggestions from the standpoint of ecological environmental improvement, as well as countermeasures to mitigate adverse environmental impacts, provide important support for the high-quality advancement of creating a green and beautiful Guangzhou during the implementation of planning. The technical flowchart for EIA of Urban Renewal Planning is illustrated in Fig. 2.

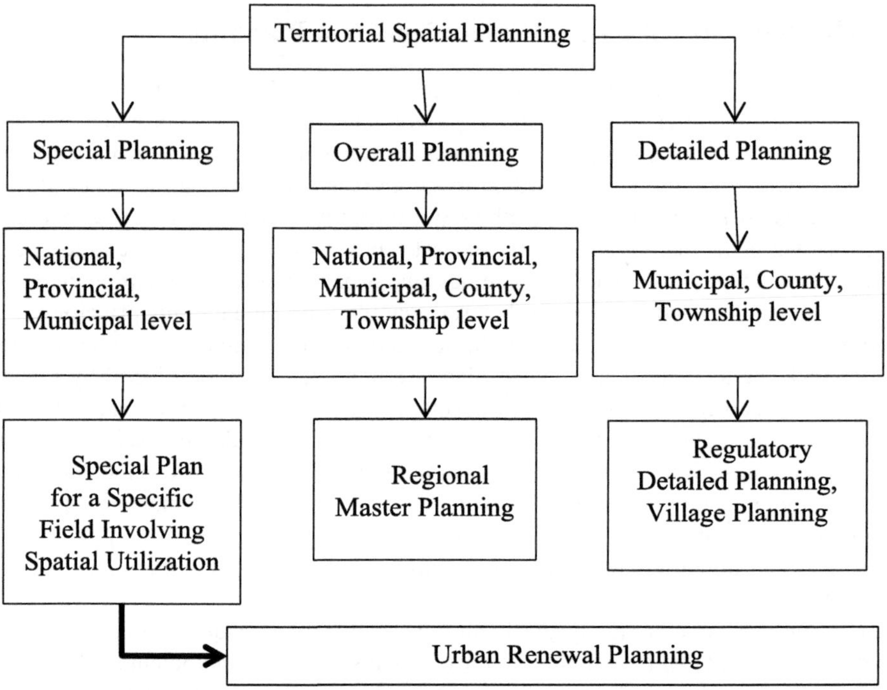

Fig. 1. The relationship diagram of urban renewal planning within the territorial spatial planning system.

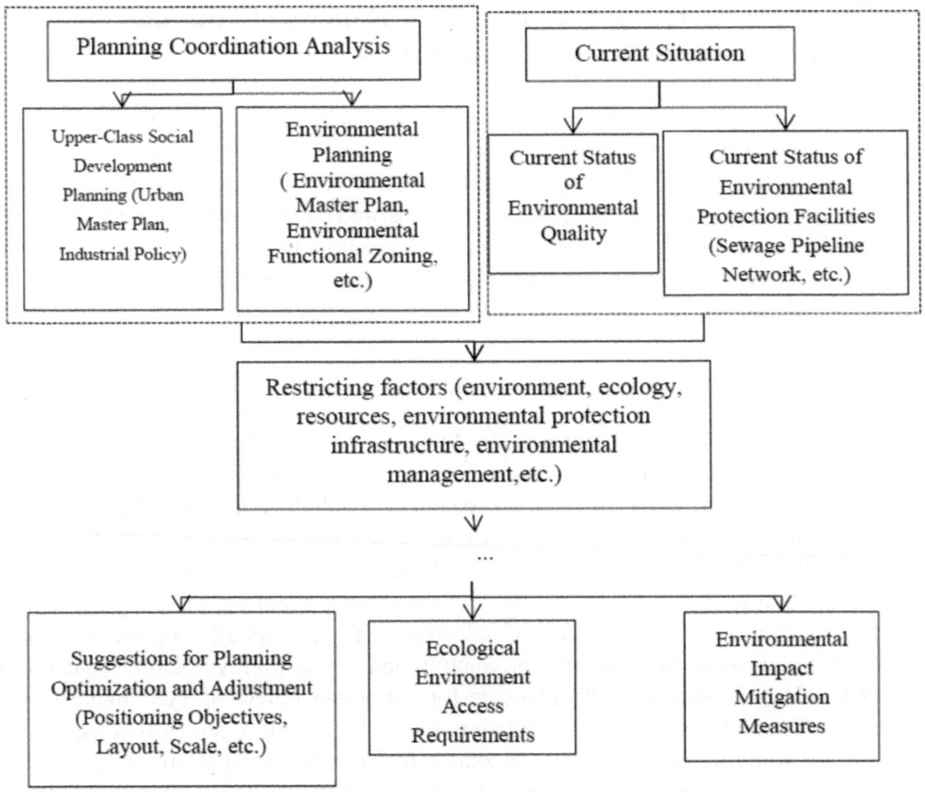

Fig. 2. Technical Flowchart for EIA of Urban Renewal Planning

3 The Importance of Constraints in EIA for Urban Renewal Planning

In conducting environmental impact assessments for urban renewal planning, identifying and properly addressing constraining factors becomes a key link connecting preliminary planning with subsequent implementation. Through in-depth and detailed analysis of these constraining factors, we can more accurately optimize renewal and transformation strategies, thereby ensuring a harmonious coexistence between development activities and environmental protection. Furthermore, we will comprehensively evaluate the coordination between the renewal and transformation plans and environmental protection plans, thoroughly outlining the current status of the renewal area in terms of infrastructure construction, resource utilization, ecological environment quality, and public demands. Next, we need to combine the issues present in the renewal area with the urban renewal planning scheme to deeply identify and analyze those factors that may constrain the development and construction of the urban renewal plan, and based on this analysis, propose targeted and feasible solutions and recommendations.

4 Factors Restricting EIA in Urban Renewal Planning and Recommended Countermeasures

Through in-depth research and analysis of over 50 urban renewal cases in Guangzhou, we have identified common restrictive factors in the environmental impact assessment (EIA) process of urban renewal planning. These factors encompass various aspects, including regional environmental quality, ecology, resources, and environmental protection infrastructure. In response to these issues, we propose corresponding countermeasures from the perspective of ecological and environmental regulations and environmental management requirements.

4.1 Factors Restricting Regional Environmental Quality

(1) Air quality standards are not met. In 2022, ozone levels failed to meet standards in most areas of Guangzhou, with localized exceedances of NO_2 detected [6]. The short-term remediation of ozone exceedances is challenging, and the number of days with acceptable air quality has not yet reached expected targets, leading to constraints on atmospheric environmental quality protection requirements.

Recommended countermeasures: Urban renewal planning should adjust industrial structures, restrict the entry of industries with high volatile organic compound (VOC) emissions, and limit the production and use of highly volatile organic raw materials by businesses. We advocate for the advancement of VOC management measures and the retrofitting of existing boilers for low nitrogen emissions.

(2) Surface water pollutants exceed standards. In 2022, the water quality rate of excellent standards at monitoring points for surface water in Guangzhou was 85.0%, showing continuous improvement in river water quality in recent years. Key rivers, including the upper and middle sections of Liuxi River, the Guangzhou section of the Pearl River, Huangpu Waterway, Dongjiang North River, Jiaomen Waterway, Hongqili Waterway, and Humen Waterway, generally exhibited good water quality, while the west channel of the Pearl River Guangzhou section, Bainihua River, and Shijing River experienced mild pollution. Particularly, areas undergoing renewal of old villages and factories showed significant exceedances of water pollution.

Recommended countermeasures: Urban renewal planning should outline and plan for the construction of a comprehensive regional sewage collection network. Discharges from outdated villages and factories must be strictly prohibited, and industrial and domestic sewage should be collected and treated separately, directing all wastewater to sewage treatment plants for compliant discharge.

(3) Insufficient attention to soil environmental risks in certain areas. Monitoring indicates that the overall soil environmental quality across monitoring points in Guangzhou is generally at a clean (safe) level, with a few points classified as remaining clean or mildly polluted [7, 8].

Recommended countermeasures: Urban renewal planning for old villages should implement soil environmental quality assessments to understand historical land use conditions within the urban renewal scope and evaluate potential environmental risks. The implementation of urban renewal planning for old factories should adhere to the Soil Environmental Protection Law and other regulations to conduct investigations, assessments, risk control, and remediation of closed and relocated industrial enterprises.

4.2 Ecological Restrictive Factors

(1) The scope of urban renewal planning occupies ecological protection red lines.
Recommended countermeasures: The planning area should prioritize avoidance of ecological protection red line zones. If buildings or structures currently exist, consideration should be given to their removal followed by ecological restoration to improve the ecological functions of the red line area.
(2) The planning area for urban renewal encroaches on general ecological spaces.
Recommended countermeasures: The implementation scope of urban renewal planning should prioritize avoidance of general ecological space areas, ensuring that development activities do not impair the ecological functions of these areas.

4.3 Resource Restrictive Factors

Urban renewal planning encroaches on basic farmland.

Recommended countermeasures: In accordance with the Basic Farmland Protection Regulations and other legal requirements, priority should be given to avoiding areas of basic farmland, and development activities must not encroach upon these areas.

4.4 Restrictions on Environmental Protection Infrastructure in Planning Areas

Insufficient and uneven distribution of environmental protection infrastructure. Particularly, in some old villages, the sewage pipeline facilities are inadequate, resulting in situations where sewage cannot be fully collected into the municipal pipeline system, leading to direct discharge into surrounding rivers and channels.

Recommended countermeasures: Based on the scale of urban renewal planning, targeted acceleration of the expansion and construction of regional sewage pipelines and wastewater treatment plants should be undertaken. In the EIA for urban renewal planning, an analysis of the sewage pipeline infrastructure and capacity constraints should be conducted, prioritizing the construction of municipal sewage pipelines.

4.5 Other Environmental Management Constraints

(1) Conflicts between "work and residence" in certain areas are evident.
Due to longstanding issues such as the extensive and disorderly expansion of construction space within old villages, unreasonable regional layout planning, and ineffective execution of planning, many old villages and towns have a mixed production and living space, highlighting the conflict between production and residence. In areas such as Huangpu District, Baiyun District, Huadu District, and Zengcheng District in Guangzhou, industrial clusters are mixed with residential areas, where storage and logistics, industrial material storage, and production enterprises are intermingled with residential neighborhoods or adjacent living areas. This dispersed

layout is challenging to manage, exacerbating the conflicts between production and residence.

Recommendations: Urban renewal planning should further optimize functional layouts, scientifically arrange production, living, and ecological spaces, and refine urban construction land planning classifications while implementing meticulous management. Especially during the renovation of old villages adjacent to industrial parks in Guangzhou, such as Yunpu Industrial Park, a buffer zone between production and living spaces should be established to avoid the problem of mixed production and residence.

(2) Management of construction waste still requires strengthening.

The implementation of renewal and renovation plans will involve the demolition of existing buildings, generating a large amount of construction waste, necessitating enhanced control over its treatment and disposal.

Recommendations: The environmental impact assessment of urban renewal planning should quantitatively account for the amount of construction waste generated, promote the resource utilization of construction waste, and implement detailed classification of construction waste guided by end disposal. This includes classified collection, storage, and disposal, establishing a detailed record system to eliminate vague classifications, ensuring precise and effective management, and creating a foundational institutional guarantee for the utilization and disposal of construction waste.

5 Conclusion

As an indispensable component of land spatial planning, urban renewal planning has yet to fully integrate environmental impact considerations into existing planning and environmental assessment regulations and technical frameworks. To enhance the environmental feasibility of urban renewal planning, we should actively promote and implement environmental assessments for urban renewal projects, starting from the sources of environmental impacts to establish an effective prevention system aimed at mitigating environmental impacts and improving environmental quality.

This paper first defines the positioning and core content of urban renewal planning, emphasizing the necessity of conducting environmental assessments for such planning. It comprehensively reviews the constraints related to regional environmental quality, ecology, resources, environmental protection infrastructure, and other aspects of environmental management, and proposes specific countermeasures in response to these constraints. This study aims to provide strong references and guidance for the widespread implementation of environmental assessments in urban renewal planning.

References

1. Zhao, G., Li, C.: Thoughts on the technological reform of detailed planning under the framework of territorial spatial planning. Urban Plan. Stud. **4**, 37–46 (2019)
2. Guangzhou Municipal People's Government: Special Plan for Urban Renewal in Guangzhou (2021–2035) (2024)

3. The Central Committee of the Communist Party of China and The State Council: Several Opinions on Establishing a National Spatial Planning System and Supervising its Implementation (2019)
4. Yang, H.: The overlapping and integration of urban renewal planning within the national spatial planning system. Planners **8**, 26–31 (2021)
5. Liu, L., Zhang, M., Han, L., et al.: Several issues urgently needed to be solved in the environmental impact assessment of industrial park planning and suggestions for countermeasures. China Environ. Manag. **12**, 47–51 (2020)
6. Guangzhou Municipal Bureau of Ecology and Environment: 2022 Guangzhou Ecological and Environmental Status Report (2023)
7. Chen, Z., Zhou, Z., Hu, D., et al.: Pilot monitoring of soil environmental quality and ecological risk assessment in the built-up area of Guangzhou city. Arid Environ. Monit. **33**, 120–124 (2019)
8. Wu, J., Deng, Y., Lin, L., et al.: Current status, problems, and countermeasures of soil pollution risk control and remediation in construction land in Guangzhou city. Environ. Monit. Manag. Technol. **33**(1–4), 14 (2021)

Quantifying the Uncertainty of Evapotranspiration by Combining Observations and Bayesian Three-Cornered Hat Method in Northwest China

Yong Yang[1], Huaiwei Sun[1](✉), Zhiwen You[1], Xiaoxiao Li[1], Hui Qin[1], and Siyue Li[2]

[1] School of Civil and Hydraulic Engineering, Huazhong University of Science and Technology, Wuhan 430074, China
hsun@hust.edu.cn

[2] School of Environmental Ecology and Biological Engineering, Hubei Key Laboratory of Novel Reactor and Green Chemical Technology, Wuhan Institute of Technology, Wuhan 430205, China

Abstract. Evapotranspiration (ET) plays a crucial role in water resource management in arid and semi-arid regions. However, accurate estimation of ET is subject to significant uncertainties due to discrepancies among various ET methods. This study utilized the generalized Three-Cornered Hat (TCH) method to quantify ET uncertainty from four datasets: ERA5L, GLEAM, GLDAS, and MEP, spanning the period from 1980 to 2018. In-situ observations and the water balance approach were employed for validation at both site and basin scales. Results indicated substantial discrepancies in the spatial patterns of uncertainty among the four ET products, with GLEAM demonstrating the best performance. The uncertainty of the ET methods was ranked as follows: GLEAM < GLDAS < MEP < ERA5L. Based on these results, we applied the Bayesian Three-Cornered Hat (BTCH) method to produce an integrated ET product, achieving the highest ET accuracy. The BTCH method reduced the root mean square error (RMSE) from 80 mm/year to 16 mm/year and increased the Kling-Gupta Efficiency (KGE) value from 0.33 to 0.75 at the basin scale, outperforming all individual ET products. Overall, TCH effectively identifies uncertainty across various ecosystems, and BTCH significantly enhances ET reliability. This study provides a new benchmark for ET estimation, which may serve as a reference for water resource management in arid and semi-arid regions.

Keywords: Evapotranspiration · Three Cornered Hat · Uncertainty · Tarim Basin

1 Introduction

Evapotranspiration (ET) is a critical component of the hydrological cycle, encompassing the processes of evaporation from soil and water bodies and transpiration from vegetation [1]. This dynamic process is pivotal in energy exchange, influencing climate patterns, and regulating atmospheric moisture levels [2]. ET also plays a key role in maintaining

D. Han et al. (Eds.): ICEPG 2024, ESE, pp. 1033–1043, 2025.
https://doi.org/10.1007/978-3-031-86829-0_112

ecosystem health, as it directly impacts soil moisture and vegetation growth [3]. Therefore, accurate estimation of ET is essential for effective water resource management, particularly in arid regions where efficient agricultural irrigation and sustainable water planning are crucial [4]. Various methods have been developed to estimate ET, resulting in numerous large-scale ET products such as the fifth generation of ECMWF reanalysis datasets (ERA5 and ERA5-Land) [5], the Global Land Data Assimilation System (GLDAS) [6], the Global Land Evaporation Amsterdam Model (GLEAM) [7], and the Maximum Entropy Production theory-based ET product (MEP) [8]. These products have been widely applied and extensively validated, offering valuable insights into regional ET estimation and water cycles.

Despite the widespread use of these ET products, quantifying their uncertainties remains challenging. Variations in algorithms, input variables, and spatial-temporal resolutions across different ET products contribute to significant discrepancies in their applicability across diverse vegetation types, ecosystems, and climatic conditions [9]. This variability introduces substantial uncertainty in obtaining accurate and reliable ET estimates from individual products. To address this, multi-model ensemble methods have been applied to improve accuracy. For instance, Bayesian Model Averaging (BMA) can mitigate errors inherent in individual models but requires ground-based observations as prior knowledge [10], which limits its application in data-sparse regions lacking observations. To overcome this limitation, the generalized Three-Cornered Hat (TCH) method has been used to assess uncertainty across different ET methods and does not rely on observational data as prior knowledge [11]. Additionally, the development of the Bayesian Three-Cornered Hat (BTCH) method allows for the integration of multiple ET methods and further enhances the accuracy and reliability of ET estimates [12].

In regions with limited observational data, such as the arid and semi-arid areas of Northwestern China, obtaining accurate ET data is crucial for water resource management and ecosystem stability. This study aims to (1) assess the accuracy of various ET methods and quantify their uncertainties across different ecosystems, and (2) use the BTCH method to generate a new fused ET dataset and validate its applicability at the watershed scale.

2 Data Materials

2.1 Study Area

The northwest region is situated in the hinterland of the Eurasian continent (73°15′E–112°58′E longitude, 34°00′N–50°10′N latitude) and includes five provinces in China: Xinjiang, Ningxia, Gansu, Shaanxi, and Qinghai, covering a total area of approximately 3.2 million km^2 (Fig. 1). This region experiences a typical continental climate with scarce and unevenly distributed precipitation. Most areas receive less than 200 mm of annual precipitation, and strong evaporation exacerbates water scarcity, making it one of the driest regions in the world with very fragile ecosystems. Ecosystem information is obtained from the Geographical Information Monitoring Cloud Platform (http://www.dsac.cn/) based on Landsat images with a 1 km spatial resolution. The region is classified into seven main ecosystems with the following proportions: cropland (7.3%), forest

(5.0%), grassland (36.0%), water/wetland (3.5%), desert (0.4%), settlement (33.8%), and bare land (14.0%).

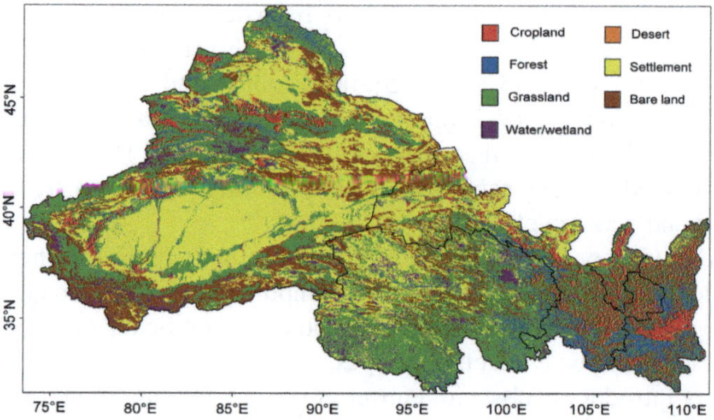

Fig. 1. Study area of the Northwestern region of China, showing the spatial distribution of the seven main ecosystems.

2.2 Actual ET Products

In this study, four ET datasets are used to analyze ET uncertainty in Northwestern China: ERA5-Land, GLDAS, GLEAM, and MEP. Detailed information about the methods, spatial-temporal resolutions, time coverage, and source references is provided in Table 1. All ET products are on a monthly temporal scale with long-term records, and the period from 1978 to 2018 was chosen for analysis. To ensure consistent spatial resolution, the datasets for GLDAS, GLEAM, and MEP were resampled to 0.1 degree using bilinear interpolation.

Table 1. The information of four ET datasets from different methods

Products	Method	Resolution	Time span	Reference
ERA5-Land	ECMWF land surface model	0.1°/monthly	1950-now	[5]
GLDAS	Noah Land Surface Model	0.25°/monthly	1978–2023	[6]
GLEAM (v3.8a)	Modified Priestly-Taylor equation	0.25°/monthly	1980–2022	[7]
MEP	Maximum Entropy Production theory	0.25°/monthly	1978–2018	[8]

2.3 Eddy Covariance Observations

Observations of ET from eddy-covariance (EC) sites were used to validate the accuracy of the four ET products. Four EC sites in Northwestern China were selected from the

Heihe River Basin observatory network [13]. The selected EC sites are: (1) Bajitan station (100.30°E, 38.92°N) with Reaumuria land cover, having records from May 2012 to April 2015; (2) Daman station (100.37°E, 38.86°N) with Maize land cover, having records from May 2012 to December 2018; (3) Huazhaizi Desert Steppe station (100.32°E, 38.77°N) with Kalidium foliatum land cover, having records from June 2012 to December 2018; and (4) Zhangye Wetland station (100.45°E, 38.98°N) with Wetland land cover, having records from June 2012 to December 2018. The observation data were gap-filled using the MDS algorithm implemented in the R package REddy-Proc [14]. Observations were aggregated from half-hourly to daily, and then from daily to monthly to match the monthly ET data from the four ET products. To address the scale mismatch between grid cells and ground observations, ET values extracted from the grid cell containing the EC station were downscaled using the bias-correction approach outlined in [15].

3 Methods

3.1 Bayesian Three-Cornered Hat (BTCH) Method

The generalized Three-Cornered Hat (TCH) method was used to quantify the uncertainty of different ET datasets on a per-pixel basis without requiring prior observational knowledge [11]. Compared to the commonly used Triple Collocation (TC) method, the TCH method has the advantage of accounting for error cross-correlation among the different datasets and can integrate more than three products simultaneously. This capability significantly simplifies the estimation process and enhances the reliability of the derived uncertainty results [16]. The theory behind the TCH method is described in [17], and detailed calculation procedures can be found in [11].

The Bayesian Three-Cornered Hat (BTCH) method is used to generate an integrated ET product by combining the four ET datasets. Maximum likelihood estimation (MLE) is employed to merge these datasets by incorporating the uncertainties obtained from the TCH method. More detailed information about the BTCH method can be found in [12]. Ultimately, the BTCH integrated ET product is calculated as the weighted average of the individual ET products.

3.2 Basin Water Balance Equation

The water balance budget equation at the basin scale is expresses as,

$$ET_{WB} = P - \Delta S - R \tag{1}$$

where ET_{WB} is the ET estimated from the water balance method, P is precipitation, and R is the river runoff. ΔS represents the total storage change, calculated as the difference in terrestrial water storage between two consecutive months. This study applies the water balance method to estimate ET in the Tarim River Basin, as it is the driest basin in China and exhibits significant ET uncertainty among various ET products. The Tarim River Basin is located in southern Xinjiang, with longitudinal coordinates ranging from 71°E to 94°E and latitudinal coordinates from 34°N to 44°N. The ET derived from the water balance approach serves as the benchmark for assessing the accuracy of all ET products.

The ET-WB estimate is obtained from the ET-WB dataset [18], and its accuracy has been validated across 168 globally distributed basins. For this study, the monthly basin ET estimates were aggregated to a yearly scale to mitigate uncertainty caused by negative values.

3.3 Evaluation Metrics

The statistical evaluation metrics used to assess the performance of ET products include: the correlation coefficient (R), the coefficient of determination (R^2), the root mean square error (RMSE), and the Kling-Gupta Efficiency (KGE). Values of R, R^2, and KGE that are closer to 1 indicate better consistency and higher accuracy, and a smaller RMSE signifies lower error and higher accuracy [19].

4 Results and Discussions

4.1 Validations of ET Accuracy at Site Scale

The spatial patterns of annual mean ET exhibit significant discrepancies among the four ET products (Fig. 2). The annual mean ET values are ranked as follows: ERA5L (24 mm/month), MEP (22 mm/month), GLDAS (18 mm/month), and GLEAM (15 mm/month). ERA5L shows the highest ET values, particularly in Shaanxi Province, Ningxia Province, the Hexi/Gansu Corridor, and the Ili and Zhungeer Basins in Xinjiang. The MEP product displays a relatively uniform spatial pattern but notably higher ET values in the Tarim Basin. In contrast, GLEAM shows the lowest ET values among the four products, especially in the Tarim Basin and Shaanxi Province.

To validate the accuracy of the four ET products, comparisons were made against eddy-covariance observations at four stations (Fig. 3). Generally, all ET products showed an overestimation at the Bajitan and Huazhaizi stations, with regression slopes greater than 1. Among the products, ERA5L had the highest agreement coefficient (R^2) at the Bajitan and Huazhaizi sites, the MEP product performed best at the Daman site, and GLEAM had the highest performance at the Zhangye site.

4.2 ET Uncertainty of Four Products Across Different Ecosystems

The validation at only four stations is insufficient to comprehensively evaluate the performance of the four ET methods. Therefore, an evaluation over a larger spatial scale for the entire region was conducted using the TCH method (Fig. 4). Generally, GLEAM exhibited the lowest ET uncertainty across nearly the entire study region, while ERA5L displayed the highest uncertainty in the Ili Basin and Zhungeer Basin. The GLDAS product showed similar uncertainty characteristics to the MEP product, except in the Tarim Basin, which is attributable to the fact that the main input variable for the MEP model was derived from the GLDAS-Noah dataset [8]. According to the density frequency plot statistics, the arithmetic mean values of uncertainty are ranked as follows: ERA5L > MEP > GLDAS > GLEAM.

To further investigate ET uncertainty across various ecosystems, this study compared the average uncertainty values for the four ET products (Fig. 5). Generally, ERA5L

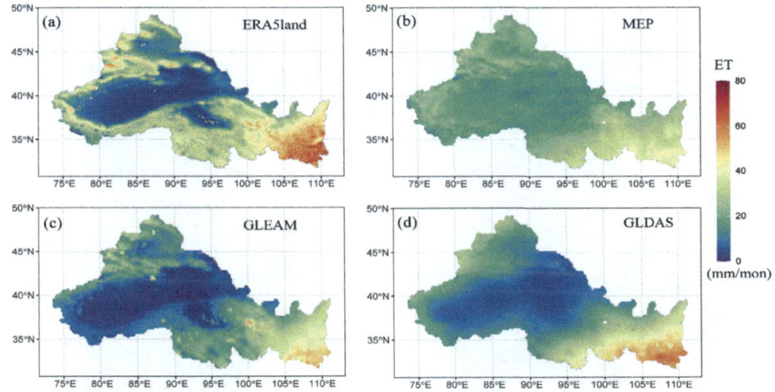

Fig. 2. The spatial distribution patterns of the multi-year monthly average ET in Northwestern China from 1980 to 2018.

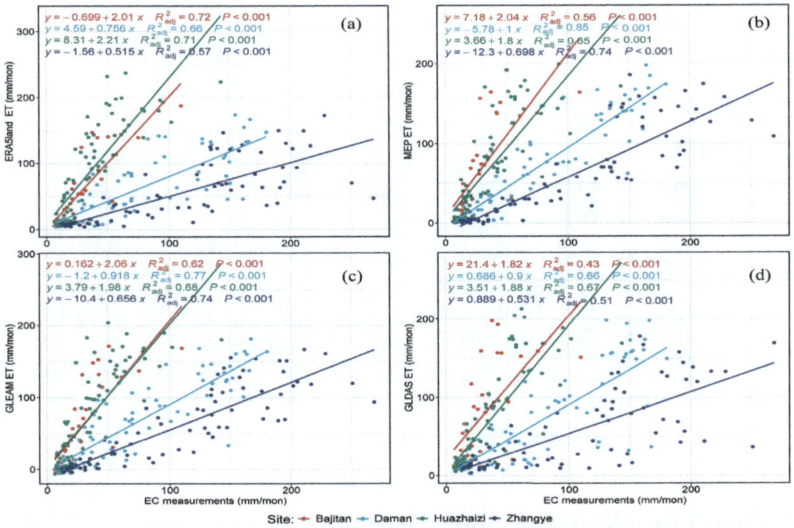

Fig. 3. Comparisons of ET derived from four ET products against observations at four eddy-covariance sites: (a) ERA5L, (b) MEP, (c) GLEAM, and (d) GLDAS. The different colors of the regression lines indicate comparisons at the four observation sites.

exhibited the highest uncertainty, while GLEAM demonstrated the lowest. For each ecosystem, GLEAM consistently performed best with the lowest uncertainty, followed by MEP, with ERA5L showing the highest uncertainty. This aligns with the results from the validation with EC observations (Fig. 3), where the TCH method-derived uncertainties revealed a similar pattern. For example, all ET products showed higher ET errors in wetland and desert ecosystems compared to cropland ecosystems, which is consistent with the underestimation of ET at Zhangye station and overestimation at the Bajitan and Huazhaizi stations in desert ecosystems.

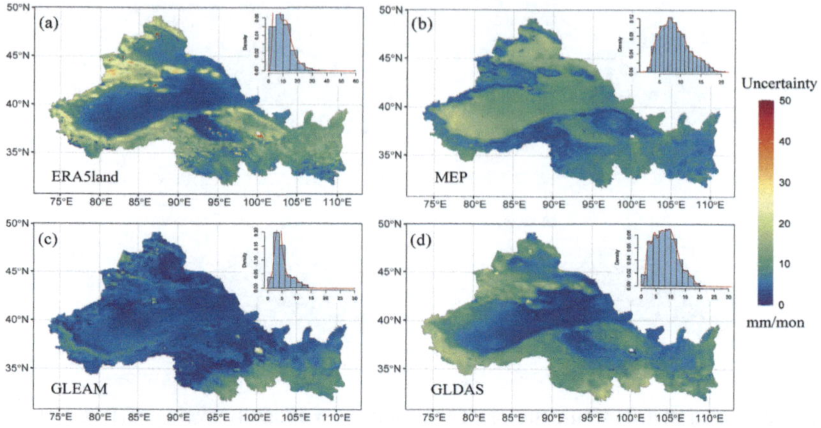

Fig. 4. Spatial patterns of ET uncertainty for the four ET products as obtained from the TCH method. The subplot in the top-right shows the density frequency distribution of all grid cells.

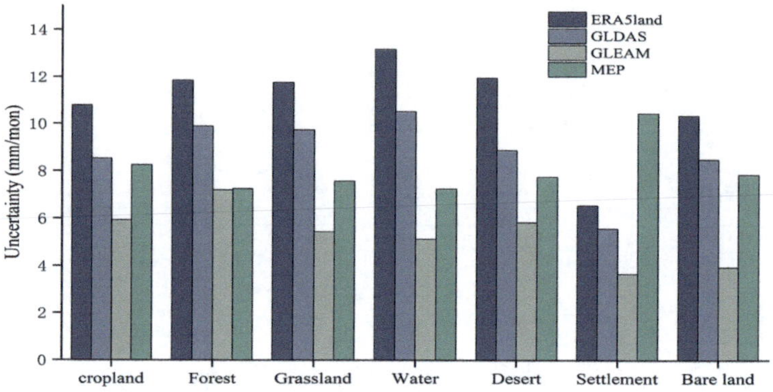

Fig. 5. ET uncertainty of different methods across the seven ecosystems.

4.3 The New Ensemble ET from BTCH Method

Figure 6a shows the best-performing ET product at each grid cell spatially. The proportion of grids with the lowest uncertainty, defined as the best-performing product, is ranked as follows: GLEAM (57%), MEP (19%), GLDAS (13%), and ERA5L (11%). Spatially, GLEAM accounts for the largest number of best-performing areas, particularly at latitudes above 35°. MEP is dominant in the southeastern regions, including Shaanxi Province and the Kunlun Mountains south of the Tarim Basin. In the Tarim Basin, the driest region in China, significant discrepancies in ET estimation are observed among different methods. According to the spatial distribution of ecosystems (Fig. 1), GLDAS and ERA5L perform better in settlement ecosystems, MEP excels in bare land ecosystems, and GLEAM is most effective in grasslands, as well as most areas of bare land and settlement ecosystems. This performance variation is mainly due to overestimated transpiration in water-limited short vegetation caused by soil water stress [20]. In contrast, the GLEAM model accounts for transpiration stress through parameters such as soil moisture content at different depths and vegetation optical depth [7].

As illustrated, accurate ET estimation in the Tarim Basin is challenging due to varying uncertainties among different models under heterogeneous underlying conditions. Therefore, applying the fusion method of BTCH is essential to reduce ET estimation uncertainty by integrating the strengths of various ET methods. The integrated ET dataset produced in this study (Fig. 6b) provides a new pattern and benchmark for the Northwestern China region, but validation at a larger spatial scale is needed to confirm its applicability.

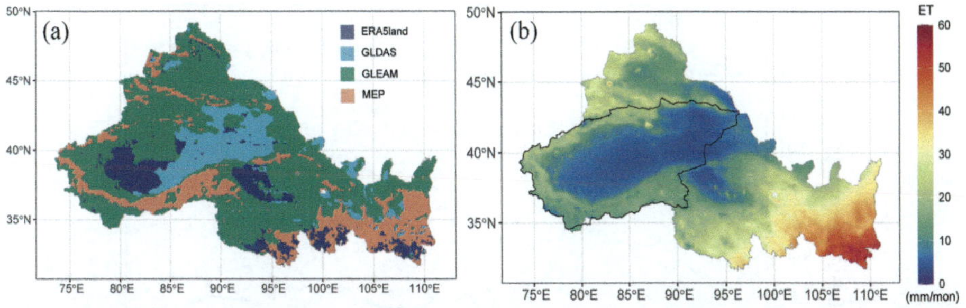

Fig. 6. (a) Spatial patterns of the best-performing ET product based on the lowest uncertainty as determined by the TCH method. (b) Mean annual ET estimates integrated using the BTCH method from 1980 to 2018. The area marked with black lines represents the Tarim River Basin.

4.4 Evaluation of Ensemble ET at Basin Scale

The accuracy of ET estimates from five different methods (MEP, GLDAS, GLEAM, ERA5L, and BTCH) was validated using the water balance approach in the Tarim Basin (Fig. 7). The annual mean basin ET values from 2003 to 2018 are as follows: MEP (189 mm/year), ERA5L (168 mm/year), GLDAS (143 mm/year), BTCH (124 mm/year), WB (110 mm/year), and GLEAM (86 mm/year). MEP shows the highest ET, while GLEAM shows the lowest. Compared to the water balance method's ET, the RMSE values for the individual ET products range from 28 mm/year for GLEAM to 80 mm/year for MEP. However, the BTCH method significantly reduces error in the Tarim Basin, with an RMSE of 16 mm/year, which is notably lower than that of the best-performing individual ET product (GLEAM). This indicates that the BTCH method outperforms all individual methods and enhances ET accuracy at the basin scale.

Compared to the best-performing GLEAM method, the KGE value for the BTCH integrated ET increased from 0.33 to 0.79, indicating a significant improvement in overall performance. While GLEAM underestimated ET in the Tarim Basin, and the other three products overestimated it, the BTCH integrated ET showed a slight overestimation relative to the water balance estimate. Both BTCH and WB exhibited consistent annual variability in ET, with peak and valley values of 130 mm/year (in 2010) and 90 mm/year (in 2009), respectively. Overall, ET in the Tarim Basin declined from 2003 to 2009 but showed an increasing trend thereafter.

The errors produced by different methods can be explained by investigating the sources of ET error due to input variables used in ET calculation algorithms [9]. ERA5 overestimates ET primarily because of an overestimation of net radiation input from the ERA-Interim scheme. GLDAS's overestimation results from parameterization errors in the modified Penman-Monteith equation within the land surface model and underestimation of wind speed in Northwest China. Conversely, GLEAM underestimates ET mainly due to lower net radiation input and smaller values of the empirical parameter (α) in the Penman-Taylor equation [9].

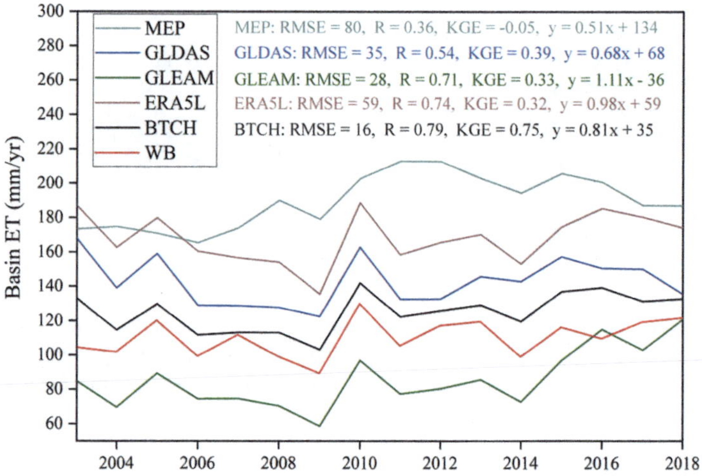

Fig. 7. Annual average ET estimates for the Tarim Basin from 2003 to 2018 using different methods: MEP, GLDAS, GLEAM, ERA5L, BTCH, and the water balance approach (WB). The WB approach is used as the benchmark for the "true" ET value.

5 Conclusions

In this paper, ET uncertainty from four datasets (ERA5L, MEP, GLEAM, and GLDAS) was evaluated at in-situ sites and the basin scale using the generalized TCH method. The BTCH method was applied to create a new ensemble ET estimate, and the water balance method was used to validate ET in the Tarim River Basin. The main conclusions are as follows:

(1) Compared with eddy-covariance observations at four sites, nearly all ET methods overestimated ET at desert stations. However, the MEP and GLEAM products performed better in wetland ecosystems.
(2) The four ET products exhibited significant spatial discrepancies in uncertainty. The GLEAM product demonstrated the lowest ET uncertainty across most of Northwestern China and performed best across all ecosystems.

(3) The BTCH method integrating four ET products has achieved the highest ET estimation accuracy. It reduced RMSE from 80 mm/year to 16 mm/year compared to direct ET estimates from the water balance approach and outperformed the best individual product, GLEAM. The BTCH method effectively enhances ET accuracy and robustness by leveraging the strengths of the different ET methods involved.

Acknowledgements. This work was supported by the Third Xinjiang Scientific Expedition Program (Grant No. 2022xjkk0105) (H.S.). The authors acknowledge funding from the NSFC project (52079055, 52011530128, 52039004), Key Water Conservancy Research Project of Hubei Province, Project Number: HBSLKY202328. H.S. acknowledges funding from the NSFC-STINT projects (No. 202100-3211 and CH2019-8281).

References

1. Wang, K., Dickinson, R.E.: A review of global terrestrial evapotranspiration: observation, modeling, climatology, and climatic variability. Rev. Geophys. **50**(2) (2012)
2. Fisher, J.B., et al.: The future of evapotranspiration: global requirements for ecosystem functioning, carbon and climate feedbacks, agricultural management, and water resources. Water Resour. Res. **53**(4), 2618–2626 (2017)
3. Sun, H., et al.: Assessment of long-term water stress for ecosystems across China using the maximum entropy production theory-based evapotranspiration product. J. Clean. Prod. **349**, 131414 (2022)
4. Sun, H., et al.: Revisiting the role of transpiration in the variation of ecosystem water use efficiency in China. Agric. For. Meteorol. **332**, 109344 (2023)
5. Muñoz-Sabater, J., et al.: ERA5-Land: a state-of-the-art global reanalysis dataset for land applications. Earth Syst. Sci. Data **13**(9), 4349–4383 (2021)
6. Rodell, M., et al.: The global land data assimilation system. Bull. Am. Meteor. Soc. **85**(3), 381–394 (2004)
7. Martens, B., et al.: GLEAM v3: satellite-based land evaporation and root-zone soil moisture. Geosci. Model Dev. **10**(5), 1903–1925 (2017)
8. Yang, Y., Sun, H., Zhu, M., Wang, J., Zhang, W.: An R package of maximum entropy production model to estimate 41 years of global evapotranspiration. J. Hydrol. **614**, 128639 (2022)
9. Xu, C., Wang, W., Hu, Y., Liu, Y.: Evaluation of ERA5, ERA5-Land, GLDAS-2.1, and GLEAM potential evapotranspiration data over mainland China. J. Hydrol.: Reg. Stud. **51**, 101651 (2024)
10. Yao, Y., et al.: Bayesian multimodel estimation of global terrestrial latent heat flux from eddy covariance, meteorological, and satellite observations. J. Geophys. Res.: Atmos. **119**(8), 4521–4545 (2014)
11. Liu, J., et al.: Uncertainty analysis of eleven multisource soil moisture products in the third pole environment based on the three-corned hat method. Remote Sens. Environ. **255**, 112225 (2021)
12. He, X., et al.: A Bayesian three-cornered hat (BTCH) method: improving the terrestrial evapotranspiration estimation. Remote Sens. **12**(5), 878 (2020)
13. Liu, S., et al.: A dataset of energy, water vapor, and carbon exchange observations in oasis–desert areas from 2012 to 2021 in a typical endorheic basin. Earth Syst. Sci. Data **15**, 4959–4981 (2023)

14. Wutzler, T., et al.: Basic and extensible post-processing of eddy covariance flux data with REddyProc. Biogeosciences **15**(16), 5015–5030 (2018)
15. Koppa, A., Rains, D., Hulsman, P., Poyatos, R., Miralles, D.G.: A deep learning-based hybrid model of global terrestrial evaporation. Nat. Commun. **13**(1), 1912 (2022)
16. Gruber, A., Su, C.H., Crow, W.T., Zwieback, S., Dorigo, W.A., Wagner, W.: Estimating error cross-correlations in soil moisture data sets using extended collocation analysis. J. Geophys. Res.: Atmos. **121**(3), 1208–1219 (2016)
17. Tavella, P., Premoli, A.: Estimating the instabilities of N clocks by measuring differences of their readings. Metrologia **30**(5), 479 (1994)
18. Xiong, J., et al.: ET-WB: water-balance-based estimations of terrestrial evaporation over global land and major global basins. Earth Syst. Sci. Data **15**(10), 4571–4597 (2023)
19. Legates, D.R., McCabe, G.J., Jr.: Evaluating the use of "goodness-of-fit" measures in hydrologic and hydroclimatic model validation. Water Resour. Res. **35**(1), 233–241 (1999)
20. Hajji, I., Nadeau, D.F., Music, B., Anctil, F., Wang, J.: Application of the maximum entropy production model of evapotranspiration over partially vegetated water-limited land surfaces. J. Hydrometeorol. **19**(6), 989–1005 (2018)

Urban Light Pollution Control and Night Sky Protection Planning

Weiting Ling[1] and Ming Zhao[2(✉)]

[1] Fuzhou No. 8 Middle School, Fuzhou 350108, Fujian, China
[2] College of Economics and Management, Minjiang University, Fuzhou 350108, Fujian, China
Zhaoming1028@163.com

Abstract. Human society is facing a series of severe challenges, including climate change, resource scarcity, population growth, and environmental degradation. Among these urgent issues, light pollution, an invisible form of pollution in modern society, has increasingly become a focus of public attention and concern for decision-makers. This is due to its wide-ranging effects, complex mechanisms, and profound impacts on both humans and ecosystems.

The concept of night sky protection is an ecological initiative aimed at reducing light pollution and safeguarding the night sky and its landscapes. It advocates for minimizing damage to nighttime environments and ecosystems. Night sky planning serves as an effective method to protect the ecological environment, reduce energy waste, and promote environmental education.

This article found that excessive lighting, inefficient lighting technology, improper lighting design, lack of planning and management in lighting, and lack of public awareness are the main reasons for urban light pollution. Measures such as using shaded lamps, adjusting the brightness of lighting, utilizing warm-colored light sources, developing and implementing light pollution regulations, and designating night protection areas are effective in controlling urban light pollution. At the same time, it is necessary to effectively control and reduce urban light pollution through methods of night sky protection planning. These methods include light environment planning, the issuance of night sky protection regulations, night sky environmental education, and other aspects of night sky protection planning.

Keywords: Urban Light Pollution · Light Pollution Control · Night Sky Protection Planning

1 Introduction

Light pollution, also known as artificial light pollution or night sky light pollution, refers to the excessive, improper, or unnecessary artificial light sources generated by human activities. These activities lead to human-induced changes in outdoor natural light levels, thereby disrupting the natural brightness of the night sky. Urban lighting, streetlights, external building lighting, billboards, and factory lighting are all examples of excessive artificial light sources. These sources, generated by human social life and production, are significant contributors to light pollution.

D. Han et al. (Eds.): ICEPG 2024, ESE, pp. 1044–1050, 2025.
https://doi.org/10.1007/978-3-031-86829-0_113

Light pollution disrupts the biological rhythms of animals and plants, affecting behaviors such as migration, reproduction, and foraging. It may also impact human circadian rhythms, leading to decreased sleep quality, changes in hormone levels, and even increasing the risk of certain types of cancer. Additionally, light pollution results in energy waste, increasing reliance on energy and fuel, which exacerbates climate change issues. The impacts of light pollution span multiple areas, including biological habitats, human health, and environmental ecology.

This article provides an in-depth analysis of the sources and forms of urban light pollution. It suggests targeted strategies for managing this issue and outlines a distinct approach for urban night sky planning. Drawing on international experiences in night sky preservation, the goal is to achieve effective control of urban light pollution through strategic planning of the urban night sky.

2 The Causes and Forms of Urban Light Pollution

The causes of urban light pollution are complex and multifaceted, presenting a multidimensional issue. This article aims to analyze the reasons behind urban light pollution's formation, clarify the core issues that should be the focus of future light pollution control efforts, and identify the main directions for practical action. By doing so, it seeks to provide clear guidance for governments to develop effective strategies and measures.

2.1 Causes of Urban Light Pollution

Excessive Lighting
Excessive lighting is a major source of light pollution, encompassing the following aspects: a. Urban municipal lighting: The brightness of lighting in urban streets, squares, and public areas often exceeds safety and functional needs. b. Industrial lighting: High-intensity lighting in industrial parks, used for safety or operational purposes, may spill over into surrounding areas. c. Outdoor recreational lighting: Lighting at sports venues and amusement parks is sometimes left on during non-operating hours, contributing to unnecessary light pollution. d. Building reflections: Decorative lighting on commercial and residential buildings, particularly glass curtain walls and LED screens, can produce intense light reflections. e. Traffic lighting: Vehicle headlights, taillights, and traffic signals also contribute to urban light pollution.

Inefficient Lighting Technology
On one hand, the use of inefficient or inappropriate lighting technologies, such as high-brightness, high-color-temperature lighting devices or improper light source colors, may generate more blue light and negatively impact the environment. On the other hand, lighting devices without effective shading and light control designs can scatter light into non-target areas, exacerbating light pollution.

Improper Lighting Design
Unreasonable direction and angle settings of lighting equipment can lead to light shining directly into the sky or adjacent areas, a primary cause of light pollution. Additionally,

lighting designs that do not incorporate strategies to reduce light pollution, such as the lack of appropriate shading devices or reflectors like full-cut off luminaires or downward-facing luminaires, can result in ineffective control of light pollution.

Lack of Planning and Management in Lighting

A key issue is the absence of standards for urban lighting layout and intensity, leading to the unreasonable placement of lighting facilities. Furthermore, there is often a lack of supervision over the use of lighting facilities, including control over lighting duration, brightness, and color.

Lack of Public Awareness

Excessive artificial light can disrupt normal sleep patterns and increase the risk of sleep disorders and depression, among other health issues. Currently, there is limited public awareness of light pollution and its environmental and health impacts. Moreover, there is a need for increased government education and guidance for residents and businesses on responsible lighting use.

2.2 Forms of Light Pollution

Outdoor lighting pollution at night primarily takes the following forms: glare, light trespass, skyglow, spill light, reflected light, and light clutter, as illustrated in Fig. 1.

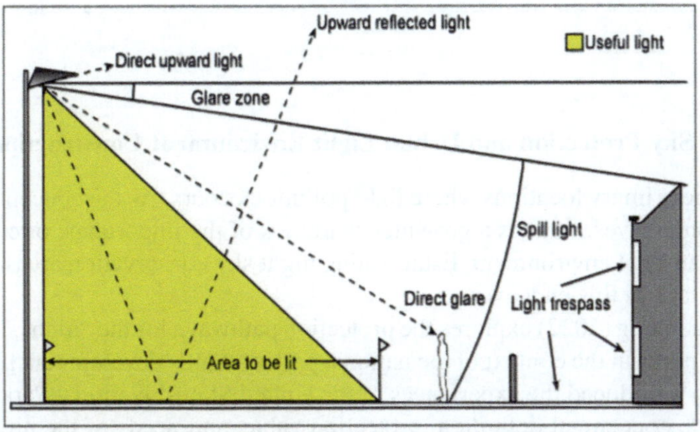

Fig. 1. Forms of light pollution. Photo modified from [1].

3 Night Sky Protection and Urban Light Environment Construction

3.1 The Concept of Night Sky Protection

The concept of night sky protection is an ecological initiative aimed at reducing light pollution and safeguarding the night sky and its landscapes. This concept advocates for minimizing damage to nocturnal environments and ecosystems, encouraging people to actively seek out the starry night sky, fostering a harmonious coexistence between humans and nature during the night, and restoring the tranquil night sky that connects humans with the stars. It is an effective means of protecting natural habitats, reducing energy waste, and promoting environmental education.

Fundamentally, the notion of night sky conservation upholds the quest for a harmonious coexistence between humanity and the natural world, aiming to restore the nocturnal environment to its inherent ecological state. It represents a form of civilization that is based on ecological conservation, sustainable development, humanistic values, and environmental education concepts, and is closely intertwined with the essence of ecological civilization, As shown in Table 1.

Table 1. Theoretical Basis of Night Sky Protection Concept

Theoretical Basis	Principles
Ecological Conservation Concept	Follow the principle of minimal environmental intervention
Sustainable Development Concept	Follow the principles of energy saving and sustainability
Humanism Thought	Follow the human-centered principle
Environmental Education Concept	Follow the principle of night sky environmental education

3.2 Night Sky Protection and Urban Light Environment Construction

Cities are the primary locations where light pollution occurs. As the concept of night sky conservation evolves, there is a growing awareness of the importance of constructing a healthy urban light environment. Establishing night sky conservation areas within cities is a crucial task in this endeavor.

Han Xiaomeng (2022) explores the protection pathways for the "night sky," focusing on national parks in the context of the national park system's development [2]. Zhong Le et al. (2019) introduced the experiences of the United States' National Park Service [3], including the conceptual definition, protection value, and needs of the night sky. They analyzed the advantages of the national park system units for protection and summarized management and safeguard measures from various aspects, such as sky monitoring, planning and design, visitor management, event organization, value display, legal protection, and international cooperation. Zhu Anni et al. (2021) proposed the concept of constructing night sky parks in China, drawing on global research on nights in nature reserves [4].

On August 31, 2021, Shenzhen issued and implemented the "Urban Lighting Special Plan (2021–2035)," marking a departure from traditional urban lighting planning approaches. The plan shifts the focus from key guiding lighting construction to emphasizing lighting control. For the first time, it integrates night sky protection areas into the urban lighting planning framework, with the goal of spearheading sustainable urban lighting development through the establishment of night parks.

4 City Light Pollution Control and Night Sky Protection Planning

The concept of night sky protection, as a significant achievement in the prevention of light pollution, not only offers effective relief strategies and best practices but also guides the developmental direction of light pollution prevention efforts.

4.1 Main Measures of Light Pollution Control

Use Shading Lamps and Adjust the Brightness of Lighting
In 1970, Tucson, USA, passed outdoor lighting regulations to protect the observing conditions of the nearby Kitt Peak National Observatory. These regulations specified the installation of street lights with shielded fixtures to reduce light spill into the sky, thereby decreasing light pollution. In 2006, the city updated these ordinances, and from 2014 to 2017, it replaced 18,000 street lights—initially high-pressure sodium and low-pressure sodium lamps—with 3000K white LED lights, resulting in a 7% reduction in sky brightness. Paris also implemented a policy to reduce the intensity of public lighting by turning off some street and building lights every night [5]. This measure not only significantly reduces light pollution but also saves a considerable amount on electricity costs.

Use Warm-Colored Light Sources
The blue light emitted by lighting equipment is a primary contributor to skyglow. Warm light sources produce less blue light compared to cold white or daylight sources (those with a color temperature above 5000K), which results in less light pollution. Additionally, the spectral distribution of warm light sources causes less interference with astronomical observations because the light they emit is within the visible light range, minimizing impact on the infrared waveband of astronomical telescopes [6]. In summary, utilizing warm light sources helps maintain the natural state of the night sky and preserves the natural beauty of the starry sky.

Develop and Implement Light Pollution Regulations
The formulation and enforcement of light pollution regulations support the achievement of environmental sustainability goals and play a positive role in increasing public awareness of light pollution issues. They also promote societal attention to the protection of night skies. In 1958, Flagstaff, Arizona, became the world's first city to pass the 'Flagstaff Lighting Ordinance,' which requires the use of fully shielded light fixtures and limits unnecessary lighting. This was done to protect the nearby Lowell Observatory from the effects of light pollution.

Authenticate Night Protection Areas
To ensure that every citizen has the right to enjoy the starry sky, some countries and local governments have established 'night sky oases' to protect the ecological environment

in those areas from excessive artificial light impact. They also actively certify night sky reserves. In densely populated areas near these night sky oases, such as cities or towns, it is required to comply with regulations that reduce lighting levels to minimize light diffusion or light intrusion. This includes lowering lighting levels to a minimum and directing light sources only where needed. These measures not only save a considerable amount of energy but also help promote the conservation of natural resources.

4.2 Night Sky Planning

From an implementation perspective, night sky planning primarily aims to minimize interference with the night sky environment and restore the night sky ecology of Earth's surface area. This is achieved through measures such as avoiding excessive lighting, designing light environments reasonably, and using energy-efficient and environmentally friendly lighting equipment. Specific implementation paths include, but are not limited to, establishing night sky reserves, planning light landscape environments, promulgating relevant regulations for night sky protection—such as light pollution management methods and lighting management regulations—and conducting environmental education, As shown in Table 2.

Table 2. Main Implementation Paths of Night Protection Planning

Implementation path		Content
Light Environment Planning	Light Landscape Management	Determine the principles and guidelines for light landscape management to provide leverage and guidance for shaping the light landscape of night protection areas and constructing the light environment
	Lighting Zoning	Divide the area into core zones and buffer zones based on actual conditions, and conduct light invasion monitoring in different zones
	Intelligent Lighting Management System	Further improve lighting efficiency, save energy, and reduce the maintenance and operation costs of lighting systems
	Lighting Restoration Project	Conduct comprehensive audits of lighting facilities, and carry out lighting restoration to ensure that the lighting meets the guiding principles of night protection
	Lighting Management Plan	The Lighting Management Plan (LMP) is a written policy approved by the management of the night protection area, setting out the minimum requirements for lighting within these areas

(continued)

Table 2. (*continued*)

Implementation path		Content
Issuance of night Protection Related Regulations	Light Pollution Management Measures	By formulating specific policies and regulations to limit excessive lighting and the inappropriate use of lighting sources, we can provide guidance for night sky protection work in China and clarify goals
	Lighting Management Regulations	
Night Sky Environmental Education		Encourage public participation and education to increase public awareness and support for night protection. Mobilize various sectors of society to participate in the call for night protection

5 Conclusions

This article discusses light pollution as a hidden form of pollution in modern society, gradually attracting the attention of the public and decision-makers. The focus of the article is to elaborate on the extensive impact and complex mechanisms of light pollution on humans and ecosystems, and to propose night sky protection as an ecological initiative to address light pollution. By exploring specific measures for night sky protection planning, the article points out the main causes of current urban light pollution and provides effective methods for controlling light pollution.

References

1. Li, M.: Research on Measurement, Experiment and Evaluation of Main Light Pollution in Urban Lighting. Tianjin University (2009)
2. Han, X.: Exploration of the protection path of "dark night sky" from the perspective of national parks. J. Hebei Environ. Eng. Coll. **32**(06), 47–50+61 (2022)
3. Zhong, L., Yang, R., Zhao, Z.: Half of the national park is the dark night: American experience and Chinese path of dark night sky research. Landsc. Archit. **26**(06), 85–90 (2019)
4. Zhu, A., Zhou, T., Zhong, Y.: Construction of dark night parks in China: reflections based on global research of dark nights in nature reserves. Nat. Reserves **1**(01), 42–51 (2021)
5. Kyba, C.C.M., et al.: Artificially lit surface of Earth at night increasing in radiance and extent. Sci. Adv. **3**(11) (2017)
6. Rodrigo-Comino, J., Seeling, S., Seeger, M.K., et al.: Light pollution: a review of the scientific literature. Anthropocene Rev. **10**(2), 367–392 (2023)

The Study on Evaluating the Participation of Residents in Ecological and Environmental Protection in the Henan Section of the Yellow River Basin

Chuwen Huang[1,2] and Yuan Liu[1,2](✉)

[1] School of Water Conservancy and Transportation, Zhengzhou University, Zhengzhou 450001, Henan, People's Republic of China
ly18638238878@163.com

[2] Yellow River Laboratory, Zhengzhou University, Zhengzhou 450001, Henan, People's Republic of China

Abstract. The ecological protection and high-quality development of the Yellow River Basin have been designated as a national strategy in China, with ecological issues in the basin assuming a focal point of national concern. Due to the comprehensive influence of climate change, landform shifts, and human activities, serious ecological environmental challenges are being confronted in the Yellow River Basin. Residents play a direct role as participants and beneficiaries in the ecological environmental protection of the Yellow River. Therefore, the discussion of people's participation in ecological environmental protection (PEEP) in the Yellow River Basin is deemed necessary. Therefore, the concept of PEEP was presented in this paper, then the evaluation Indicator system of PEEP was established in the Yellow River Basin. An improved grey target evaluation model, utilizing relative entropy, was constructed to determine the weights of the indices. Taking 8 cities in the Yellow River Basin (Henan Section) as examples, the people PEEP was evaluated, and the results indicated that ZhengZhou (2.136) scored the highest, while PuYang (1.201) scored the lowest, highlighting significant spatial differences. Finally, the paper proposed several countermeasures to enhance PEEP effects among residents along the Yellow River, aiming to offer scientific insights for the government in formulating ecological and environmental protection policies for the Yellow River Basin residents.

Keywords: The Yellow River basin · Ecological protection · Relative entropy · Improved grey target model · Participation evaluation

1 Introduction

As the second largest river in China, the Yellow River has increasingly faced challenges such as flooding, soil erosion, ecological degradation, channel breakage, and water resource shortages in recent years, attributed to abrupt climate change and expanding human activities [1, 2]. While China has achieved significant progress in implementing

D. Han et al. (Eds.): ICEPG 2024, ESE, pp. 1051–1066, 2025.
https://doi.org/10.1007/978-3-031-86829-0_114

Table 1. (*continued*)

Form	Classification	Numbers	Ratios
	master	140	5.61%
Address	SMX	144	5.75%
	LY	452	18.10%
	JY	45	1.80%
	JZ	239	9.56%
	ZZ	643	25.70%
	XX	391	15.64%
	KF	333	13.33%
	PY	253	10.11%

3 Method

3.1 Conceptual Definition of PEEP

(1) Ecological awareness

Two main concepts of ecological awareness exist:

Firstly, there is an awareness of people's thoughts, attitudes, and behavioral tendencies towards ecological environmental protection, which primarily derives from their thoughts and inclinations without specific behaviors in environmental protection being exhibited.

The second aspect pertains to the concept of ecological environmental protection awareness, which encompasses behaviors directly related to ecological environmental protection. Public awareness of environmental protection reflects individuals' understanding, attitudes, behaviors, and habits concerning ecological environmental protection. In summary, the article defines ecological environmental protection awareness as the combined impact of people's attitudes and behaviors toward environmental protection in their daily production and life activities.

(2) Participation of the population in ecological environmental protection

The current definition of the degree of PEEP is broadly conceived, typically encompassing the proportion of individuals engaging in ecological protection behaviors relative to the total population of a specific area. However, participation in ecological protection constitutes more than just individual actions. It entails residents' engagement with ecological and environmental issues, encompassing attitudes, behaviors, awareness of problems, and levels of knowledge pertaining to ecological and environmental protection. This systemic approach encompasses both breadth and depth.

3.2 Indicator System of PEEP

To comprehend the status of residents' participation in ecological environmental protection within the Yellow River Basin, an indicator system has been developed comprising four levels: awareness of ecological environmental issues (AOEEP) among residents, knowledge levels regarding ecological environmental protection (KLEEP), attitudes toward ecological environmental preservation (ATEEP), and behavioral practices in ecological environmental protection (BPEEP).

The AOEEP reflects their concern and sensitivity to environmental issues in the study area. It primarily includes awareness of the destructiveness of the ecological environment and its causes, awareness of the sustainability of the ecological environment, and awareness of the responsibility for ecological environmental protection. Two indicators are comprised: the level of knowledge regarding environmental issues (LKEI) and the importance ascribed to environmental protection (IAEP).

KLEEP is an important indicator for reflecting residents' participation in ecological and environmental protection. The depth of knowledge in environmental protection theory forms the basis for residents' practice of ecological and environmental protection behaviors. The residents' basic cognitive level in addressing environmental protection issues and their general knowledge of ecological and environmental protection to some extent reflect residents' attitudes towards environmental protection, which indirectly influences their ecological protection behaviors. Three indicators are encompassed: the knowledge of policies (TKOP), understanding of watershed issues (UOWI), and awareness of ecological indicators (AOEI).

ATEEP is a psychological tendency that influences the personal behavior of the population regarding issues related to environmental protection. It primarily encompasses the individual's willingness to ecologically protect the environment, concerns about ecological pollution, the balance of ecological benefits, and attitudes towards personal efforts in ecological environmental protection. It includes indicators of the extent to which environmental problems impact economic development (EPID) and the willingness to engage in environmental protection (WEEP).

BPEEP is a crucial gauge of individuals' awareness levels regarding ecological and environmental protection [11]. This report primarily surveys whether residents have familiarized themselves with the "Speech of Chinese Leaders at the Symposium on Ecological Protection and High-Quality Development of the Yellow River Basin." It also investigates whether residents would proactively discourage enterprises from polluting the ecological environment of the Yellow River Basin. Furthermore, it assesses residents' participation in activities aimed at protecting the ecological environment of the Yellow River, thereby measuring the extent of their engagement in ecological and environmental protection behaviors. It includes two indicators, namely, pollution dissuasion efforts (NPDE) and environmental protection participation behaviors (EPPB). The indicator system for PEEP is shown in Fig. 2.

comprehensive management projects for the Yellow River, the current challenge lies in promoting the harmonious coexistence of its protection and development [3, 4].

The ecosystem of the Yellow River Basin encompasses organisms and environments across its upper, middle, and lower reaches, including main tributaries and coastal areas. The natural resources and ecosystems along the Yellow River exhibit significant differentiation [5–7]. Moreover, while the design and engineering of ecological environmental protection in the Yellow River Basin are crucial, it is even more essential to increase public awareness of ecological environmental protection and translate it into tangible actions. This represents a pivotal factor in advancing ecological environmental protection along the Yellow River [8–10]. Currently, there is insufficient investigation and research on residents' understanding of and participation in ecological protection within the Yellow River Basin.

In conclusion, in addition to conducting rigorous scientific research on the ecological environment protection of the Yellow River Basin, there exists an obligation and responsibility to promote the concept of ecological environment protection, investigate the ecological protection awareness among residents in the Yellow River Basin, analyze the factors influencing residents' ecological protection behaviors, and propose corresponding solutions. Consequently, this study introduces the concept of residents' PEEP, defines its scope, and establishes an Indicator system to measure residents' involvement in the Yellow River Basin. Utilizing relative entropy and improved grey-target evaluation models, the study evaluates residents' participation across eight cities along the Yellow River (Henan section). The outcomes aim to inform and enhance environmental protection initiatives in the region.

2 Study Area and Data

2.1 Study Area

Eight cities along the Yellow River in Henan Province ZhengZhou (ZZ), LuoYang (LY), KaiFeng (KF), JiYuan (JY), JiaoZuo (JZ), XinXiang (XX), SanMenXia (SMX) and PuYang (PY) were taken as the study area (Fig. 1). A questionnaire was designed to investigate the current situation within this area. Through this survey, a comprehensive assessment was conducted regarding residents' PEEP in the Henan section of the Yellow River Basin. Based on the findings, measures were proposed to enhance residents' awareness of PEEP.

In 2022, the total GDP of the eight cities amounted to 3,130.3 billion yuan, constituting 54% of the province's GDP. The combined population of these cities reached 40,799,000, representing 41% of the province's total population. Among these cities are ZZ, recognized as the economic hub of Henan Province, alongside the historic cities of LY and KF, as well as prominent "Coal and Energy Cities" and the traditional industrial center of JZ. This study investigates residents' PEEP across these eight cities along the Yellow River in Henan Province. It examines residents' awareness of ecological protection along the river, analyzes factors influencing their engagement in ecological protection efforts, and proposes actionable recommendations to address current challenges.

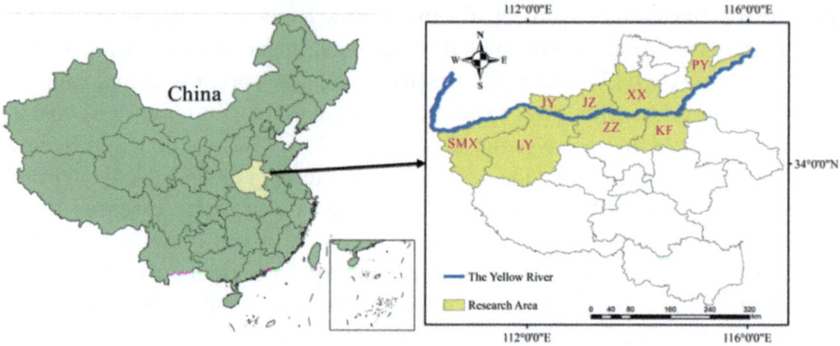

Fig. 1. Geographical location of the study area

2.2 Questionnaire Design and Survey Sample Presentation

In the study area, the latest information was collected through online questionnaires, offline interviews, and field visits. The study and research encompassed individuals from various educational backgrounds, genders, and age groups. A total of 2,500 offline questionnaires were distributed. Ultimately, 2,395 questionnaires were retrieved, yielding a recovery rate of 95.8%. Details of the sample distribution are presented in the Table 1 below:

The questionnaire was structured around three primary aspects: public environmental awareness, the current state of environmental science, and the reception among diverse demographics toward environmental science. It consists of 20 questions, including one open-ended question, seven multiple-choice questions, and twelve single-choice questions. Additionally, respondents were asked to provide detailed personal information and other relevant details.

Table 1. Sample distribution of the survey on residents' PEEP in the Yellow River basin

Form	Classification	Numbers	Ratios
Genders	Male	1307	52.28%
	Female	1193	47.72%
Age	Under 18	148	5.91%
	18–29	798	31.93%
	29–50	1195	47.81%
	Over 50	359	14.35%
Educational background	Senior High School and below	1829	73.16%
	bachelor's degree	531	21.23%

(continued)

weight coefficient of the empowerment method. The objective function min $Q(W^C)$ is solved though using the Lagrange multiplier method [15], and the optimal solution of the model can be obtained from the necessary conditions for the existence of extreme values:

$$W_i^C = \sum_{k=1}^{4} \alpha_k u_{ki} \tag{6}$$

Calculate the confidence level α_k for the k-th assignment result:

According to the concept and properties of relative entropy, the problem of assembling weights $z = [z_1, z_2, \cdots z_m]$ obtained as a result of various assignments can be transformed into a mathematical planning problem:

$$\min H(z) = \sum_{k=1}^{4} \sum_{i=1}^{m} z_i \ln \frac{z_i}{u_{ki}} \tag{7}$$

① The Lagrange function is introduced to determine the global optimal solution of the above model $z^* = [z_1^*, z_2^* \cdots z_m^*]$:

$$z_i^* = \frac{\prod_{k=1}^{4} [u_{ki}]^{\frac{1}{4}}}{\sum_{i=1}^{m} \prod_{k=1}^{4} [u_{ki}]^{\frac{1}{4}}} \tag{8}$$

② Calculate the progress $h(u_k, z^*)$ of the weighted result and the aggregate weight $z^* = [z_1^*, z_2^* \cdots z_m^*]$, The reliability of each weighting result is obtained α_k:

$$\begin{cases} \alpha_k = \frac{h(u_k, z^*)}{\sum_{k=1}^{4} h(u_k, z^*)} \\ h(u_k, z^*) = \sum_{i=1}^{m} u_{ki} \ln \frac{u_{ki}}{z_i^*} \end{cases} \tag{9}$$

By substituting the result of credibility weight in the above formula into Eq. (6), the combined weight of each Indicator can be obtained:

$$W_i^C = \sum_{k=1}^{4} \frac{\sum_{i=1}^{m} u_{ki} (\ln u_{ki} - \ln \prod_{k=1}^{4} [u_{ki}]^{\frac{1}{4}} - \ln \sum_{i=1}^{m} \prod_{k=1}^{4} [u_{ki}]^{\frac{1}{4}})}{\sum_{k=1}^{4} \sum_{i=1}^{m} u_{ki} (\ln u_{ki} - \ln \prod_{k=1}^{4} [u_{ki}]^{\frac{1}{4}} - \ln \sum_{i=1}^{m} \prod_{k=1}^{4} [u_{ki}]^{\frac{1}{4}})} u_{ki} \tag{10}$$

An Improved Grey Target Model

The grey-target model defines the ideal optimum as the bull's-eye, typically using Euclidean distance to compute the distance from each result to this bull's-eye (bull's-eye distance). This approach ranks the merits of evaluation schemes based on their proximity to the ideal. However, the traditional grey-target model struggles to ensure independence among indicators [12]. To address this, the model is enhanced by incorporating the Mahalanobis distance [13]. Simultaneously, the TOPSIS model incorporates positive and negative ideal solutions [14] to calculate positive and negative bull's-eye distances using the Mahalanobis distance for each evaluation scheme, respectively. The specific steps are outlined as follows:

Step 1: Calculate positive and negative bull's-eye

$$\begin{aligned} R_n^+ &= (\max_{1\le i\le m}(x_{i1}), \max_{1\le i\le m}(x_{i2}), \cdots, \max_{1\le i\le m}(x_{i=n})) \\ &= (r_1^+, r_2^+, \cdots, r_n^+) \\ R_n^- &= (\min_{1\le i\le m}(x_{i1}), \min_{1\le i\le m}(x_{i2}), \cdots, \min_{1\le i\le m}(x_{i=n})) \\ &= (r_1^-, r_2^-, \cdots, r_n^-) \end{aligned} \tag{11}$$

Step 2: Calculate positive and negative Mahalanobis distance

$$\begin{aligned} &\begin{cases} e_{ij} = W^C r_{ij} - W^C r_j^+ \\ E_i = (e_{i1}, e_{i2} \cdots, e_{im})^T \end{cases} \\ &\begin{cases} f_{ij} = W^C r_{ij} - W^C r_j^- \\ F_i = (f_{i1}, f_{i2} \cdots, f_{im})^T \end{cases} \end{aligned} \tag{12}$$

where E_i and F_i represent the weighted difference matrices from scheme i to the positive and negative fiducials, respectively.

Step 3: Calculate the positive and negative bull's-eye distances for scheme i as:

$$\mu_i^+ = \sqrt{E_i^T G^{-1} E_i} \tag{13}$$

$$\mu_i^- = \sqrt{F_i^T G^{-1} F_i} \tag{14}$$

where G is the covariance matrix of matrix R.

Step 4: Calculate relative distance from the bull's-eye μ_i^*:

$$\mu_i^* = \frac{\mu_i^-}{\mu_i^+ + 1} \tag{15}$$

If the evaluation Indicator is larger, it indicates that the i-th program is closer to the positive bull's-eye and simultaneously farther from the negative bullseye, suggesting a superior program effect.

The Calculation of the Composite Evaluation Indicator (CEI)

According to formula (1–2), the value r_{ij} after standardization of the Indicator value of each evaluation object and the combined weight W^C are calculated according to formula (10), and the comprehensive evaluation Indicator Z_i of each evaluation object can be obtained:

$$Z_i = \sum_{j=1}^{n} r_{ij} \cdot W_j^C (i = 1, 2, \cdots, m) \tag{16}$$

The calculation flowchart is shown in Fig. 3.

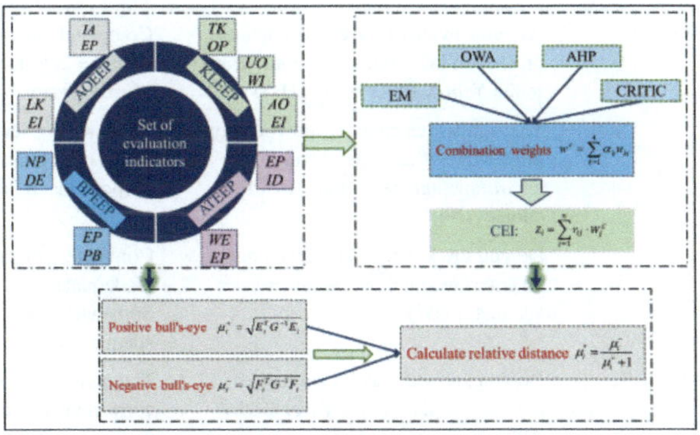

Fig. 3. The calculation flowchart

4 Results and Discussion

4.1 Analysis of the Status of Evaluation at the Indicator Level

Specific values were quantified through Table 2 and mean values were calculated by combining the results of the questionnaire. Combined with 3.3.1 method to normalise the results of the questionnaire, assigning values to each indicator (B1-B15) through the mean value of the results of each question, and combining with 3.3.2 method to calculate the weights of the indicators B1-B15, the results of the questionnaire are shown in Fig. 4, and the results of the indicator quantification are shown in Fig. 5.

Table 2. Survey questionnaire quantitative table

normative layer	Level 1 indicators	Level2 indicators	Criteria for quantification
AOEEP	LKEI (A1)	What do you think about the current water quality of the Yellow River? (B1)	Continually worsening = 1; Gradually worsening = 2; Remains unchanged = 3; Gradually improving = 4; Continually improving = 5
		Have you paid attention to environmental news or reports? (B2)	Never = 1; Rarely = 2; Sometimes = 3; Often = 4; Always = 5
	IAEP (A2)	What do you think is the impact of protecting the Yellow River on the ecological environment? (B3)	No impact = 1; Little impact = 2; Large impact = 3
		What do you think is the level of ecological awareness of residents in the Yellow River region? (B4)	Very low = 1; Low = 2; Average = 3; High = 4; Very high = 5

(continued)

Table 2. (*continued*)

normative layer	Level 1 indicators	Level2 indicators	Criteria for quantification
KLEEP	UOWI (A3)	Are you aware of the environmental protection laws and regulations related to the Yellow River Basin? (B5)	Completely unaware = 1; Slightly aware = 2; Somewhat aware = 3; Mostly aware = 4; Fully aware = 5
		Through which channels do you acquire knowledge of ecological and environmental protection? (multiple choice) (B6)	TV & radio; Newspapers & magazines; Online media; Community propaganda; Score 1–5
		Do you think the government's policies on environmental protection are adequate? (B7)	Completely inadequate = 1; Inadequate = 2; Average = 3; Somewhat adequate = 4; Very adequate = 5
	AOEI (A4)	Do you know the major sources of pollution in the Yellow River Basin? (B8)	The number of identified sources corresponds to the score, up to a maximum of 5 points
	TKOP (A5)	Which ecological environment indicators of the Yellow River are you familiar with? (multiple choice) (B9)	Yellow River water quality category, water resources, air quality, vegetation coverage, soil and water conservation rate, etc.; the number of identified indicators corresponds to the score, up to a maximum of 5 points
ATEEP	EPID (A6)	Do you think the environmental pollution of the Yellow River Basin affects residents' production and life? (B10)	No impact = 1; Little impact = 2; Significant impact = 3
		Compared to other areas, how do you rate the economic development speed of the Yellow River region? (B11)	Lagging = 1; Slow = 2; Normal = 3; Fast = 4; Very fast = 5
	WEEP (A7)	Are you willing to participate in community-organized environmental protection activities? (B12)	Completely unwilling = 1; Unwilling = 2; Neutral = 3; Willing = 4; Very willing = 5
BPEEP	NPDE (A8)	If a company pollutes the ecological environment of the Yellow River Basin, would you intervene? (B13)	Will not intervene = 1; Verbal intervention = 2; Active intervention until the company stops = 3
	EPPB (A9)	Do you promote the importance of ecological and environmental protection to those around you? (B14)	Never = 1; Rarely = 2; Sometimes = 3; Often = 4; Always = 5
		Have you participated in activities to protect the ecological environment of the Yellow River? (B15)	The number of activities you participated in corresponds to the score, up to a maximum of 5 points

4.2 Analysis of the Status of Evaluation at the Normative Level

(1) AOEEP

With the improvement of residents' quality of life and the government's enhancement of the Yellow River's water environment, significant environmental issues have been addressed. Consequently, residents' concern regarding water pollution extends beyond the immediate aquatic environment. During the survey, it was found that a majority of residents believe that the condition of the Yellow River's water body has gradually improved. Specifically, 55% of residents perceive continuous

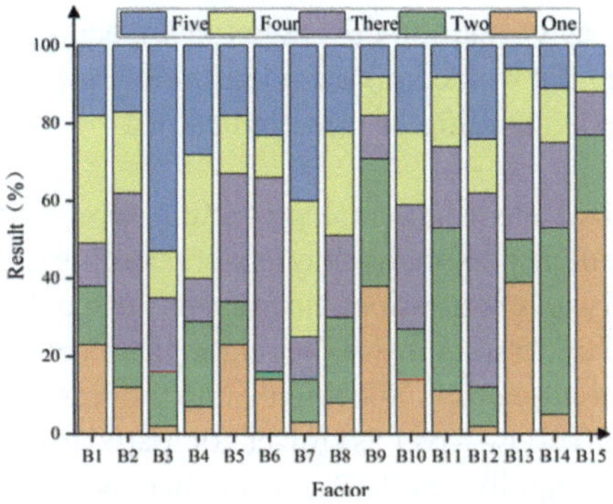

Fig. 4. Results of the questionnaire survey

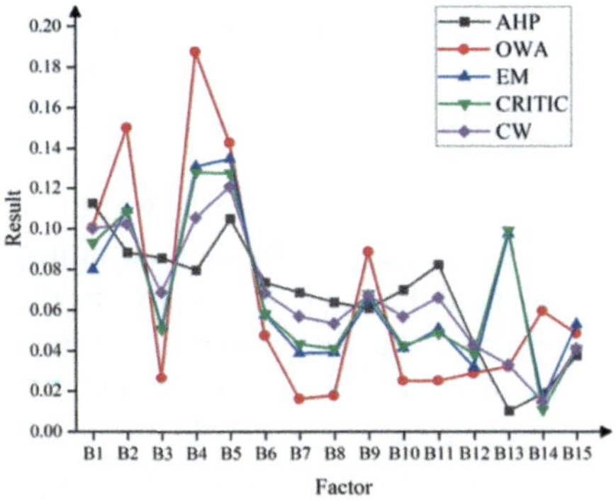

Fig. 5. Indicator weighting charts

improvement in the water quality of the Yellow River, while 23% believe it has remained stable. The remaining 22% perceive a deterioration in water quality. A significant majority (49%) of residents believe that the protection of the Yellow River significantly impacts ecological conservation efforts, encompassing issues such as soil erosion, flood risk, and biodiversity. Most residents hold a critical view of land ecological protection along the Yellow River, with 34% perceiving average awareness and effectiveness of local ecological protection efforts. Moreover, 27% express satisfaction with the current environmental conditions and the effectiveness of protection measures. Conversely, 22% indicate a severe deficiency in environmental protection awareness and substantial ecological damage in the area. Finally, 17% of residents believe that environmental awareness is strong and has resulted in satisfactory outcomes.

(2) KLEEP

Here is the revised paragraph, adhering to academic journal standards with a focus on passive voice where appropriate and ensuring grammatical correctness:

According to the statistical survey, residents have taken the initiative to acquire knowledge related to ecological and environmental protection. However, the depth and breadth of this knowledge are not sufficiently scientific and comprehensive. The questionnaire primarily addresses residents' basic understanding of environmental protection and systematically analyzes the sources and channels through which residents along the Yellow River acquire ecological and environmental protection knowledge. Based on the analysis of questionnaire results, it was found that 60% of residents acquire information on Yellow River-related policies through TV news, bus advertisement boards, and other mediums. Regarding laws such as the Water Law of the People's Republic of China, the Ecological Environment Protection Law of the People's Republic of China, the Yellow River Water Dispatching Regulations, and other relevant water conservation or environmental protection laws and regulations, residents generally exhibit a limited understanding: 68% do not understand these laws, 27% have heard of them but lack clarity on specifics, and only 8% are well-informed. Concerning the primary indicators of the Yellow River's ecological environment, "Yellow River water quality and quantity" are considered the most crucial (97%), followed by air quality (95%). Due to insufficient promotion of ecological knowledge, indicators such as vegetation coverage (45%), ecological diversity (32%), and soil erosion (39%) are not well comprehended among residents. Overall, there is a low general level of understanding of these indicators.

(3) ATEEP

Residents' attitudes towards ecological environmental protection represent a psychological inclination that influences their personal behaviors concerning environmental issues. According to the analysis of questionnaire results, it has been observed that there is a need for improved understanding of the relationship between economic development and environmental protection. Specifically, 56% of residents perceive economic development along the Yellow River as slower compared to other regions, while 32% believe it is progressing rapidly, and 12% consider it to be lagging behind. Regarding the impact of environmental pollution in the Yellow River Basin on residents' production and daily life, 42% of respondents generally acknowledge its influence as moderate, 30% believe it has a significant impact, and 28% perceive no discernible impact.

(4) BPEEP

Based on the statistical findings, the overall environmental protection behavior among residents shows limited engagement. For instance, when asked whether they would proactively discourage individuals or businesses from polluting the ecological environment of the Yellow River Basin, a significant majority of residents (69%) tend to overlook this issue. Conversely, 21% of residents indicate they would initiate dissuasion efforts, but may cease if unsuccessful, while 10% persist in dissuading until the behavior ceases.

Regarding participation in activities aimed at protecting the ecological environment of the Yellow River, responses vary: 22% of residents opt for individual organizations, 36% have never participated, 20% prefer collective organizations, and another 22% engage in both collective and individual organizations.

4.3 Results of PEEP in the Yellow River Basin

According to the methodology outlined in Sect. 3, the comprehensive evaluation results of residents' participation in ecological environmental protection in the Yellow River Basin were calculated. The results of positive and negative bull's-eye distances for each city (Fig. 6) indicate a significant disparity in ZZ City, whereas PY exhibits the lowest disparity. Furthermore, the change lines of positive and negative bull's-eye distances are nearly symmetrical, with a correlation coefficient of −0.9916, suggesting a high degree of similarity in evaluating the strengths and weaknesses of the program based on positive and negative perspectives.

Based on the comprehensive evaluation results of residents' participation in ecological environmental protection in each city (Fig. 7), it is evident that the top three cities in the Henan section of the Yellow River Basin are Zhengzhou (2.136), Luoyang (1.903), and Xinxiang (1.849). Conversely, the bottom three cities are Sanmenxia (1.445), Kaifeng (1.360), and Puyang (1.201). These findings highlight significant disparities in residents' engagement in ecological environmental protection across cities in the Yellow River Basin. There appears to be a correlation with each city government's initiatives and planning for ecological environmental protection.

In recent years, Zhengzhou City has accelerated its urbanization and modernization efforts along the Yellow River. Residents of Zhengzhou City are actively encouraged to take a leading role among all cities. Consequently, residents' participation in ecological environmental protection within the Yellow River Basin has notably increased.

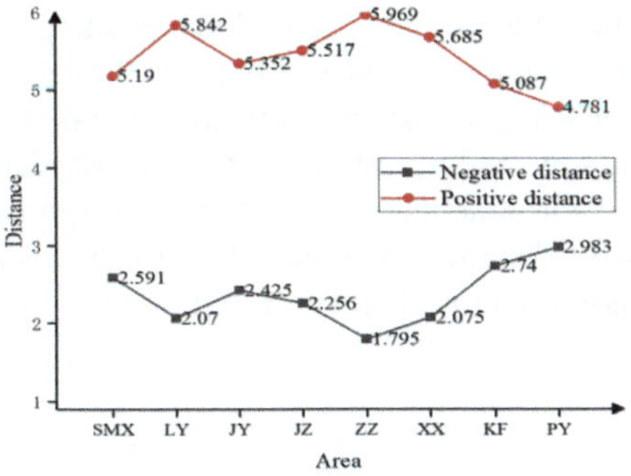

Fig. 6. Bull's-eye results

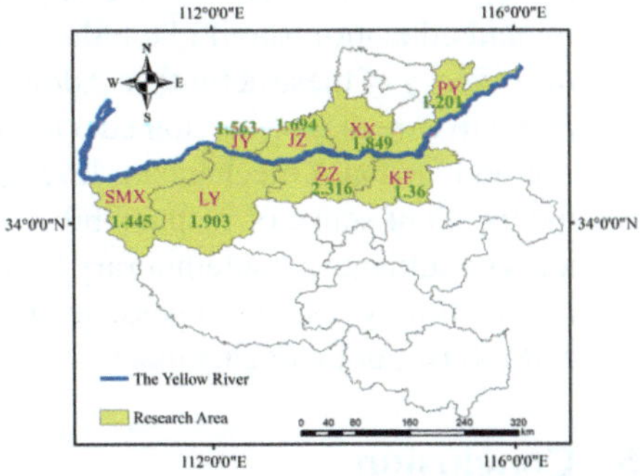

Fig. 7. Comprehensive evaluation results for PEEP

4.4 Countermeasures to Enhance the Ecological and Environmental Protection Awareness of Residents in the Yellow River Basin

(1) Enhancing green production capacity and balancing economic development and ecological protection

The key to enhancing awareness among watershed residents of ecological and environmental protection lies in addressing the issue of income growth, as economic empowerment instills confidence in residents to engage in ecological protection. It is essential to assess the interplay between environmental conservation, resource management, and the adoption of a green industrial structure to foster sustainable economic and ecological development. Implementing the principles of green development involves strategically planning industrial growth within the watershed, promoting both green concepts and practices to establish an economic foundation for bolstering residents' ecological awareness. Integration of both internal and external resources across the Yellow River Basin is crucial to harmonizing ecological conservation with livelihood improvement. Additionally, effective government guidance includes measures such as formulating tax incentives for the environmental protection sector and other related inducements to support green industry development. Establishing a green development financing platform aims to create more employment opportunities for residents. Simultaneously, leveraging the natural beauty along the Yellow River banks through sustainable eco-tourism development and moderating agricultural product chains will further encourage residents to conscientiously embrace ecological and environmental protection.

(2) Strengthening publicity for environmental protection by various means

Publicity and education regarding ecological environment protection represent a crucial approach to enhancing residents' awareness of environmental conservation.

Such efforts should integrate new media and innovative approaches, employing various measures to consolidate resources for environmental education under governmental leadership, with active cooperation from subordinate units and departments in joint implementation. Survey results reveal significant disparities in environmental protection knowledge and practical skills among residents across different cultural and age groups. Therefore, the implementation of environmental protection publicity and education should be tailored to individuals and specific locations to ensure the efficacy of these activities. Additionally, there is a need to strengthen the capacity of environmental protection education departments, establish effective communication and feedback mechanisms leveraging new media platforms, and foster residents' independent sense of responsibility for environmental stewardship. This approach aims to cultivate a contemporary framework for environmental education characterized by timely responsiveness, diverse dissemination channels, professional content, and robust public engagement.

5 Conclusion

To comprehensively assess residents' engagement in ecological environmental protection within the Yellow River Basin, this paper first defines the concept of residents' participation in ecological environmental protection. Subsequently, it establishes a model for evaluating such participation specifically within the Yellow River Basin. The study evaluates residents' engagement across eight cities in the Henan section of the Yellow River Basin and derives the following conclusions:

(1) There are pronounced disparities in residents' involvement in ecological environmental protection across cities within the Yellow River Basin, correlating closely with each city government's initiatives in ecological environmental protection. Zhengzhou, for instance, is actively implementing strategies and initiatives for ecological protection within the Yellow River Basin, resulting in residents' heightened participation compared to other cities.

(2) Based on the survey results, residents' environmental protection behaviors demonstrate a need for improvement, with participation levels not reaching optimal levels. To address this, the government must proactively foster the Yellow River green industry, strengthen the foundation of residents' environmental awareness, enhance environmental education efforts, and cultivate a favorable environmental protection ethos. Additionally, residents should integrate the concept of green development into their daily lives and actively engage in green practices.

Acknowledgement. The authors would like to thank the anonymous reviewers for their valuable comments. This work was funded by the Project of National Natural Science Foundation of China (51879241).

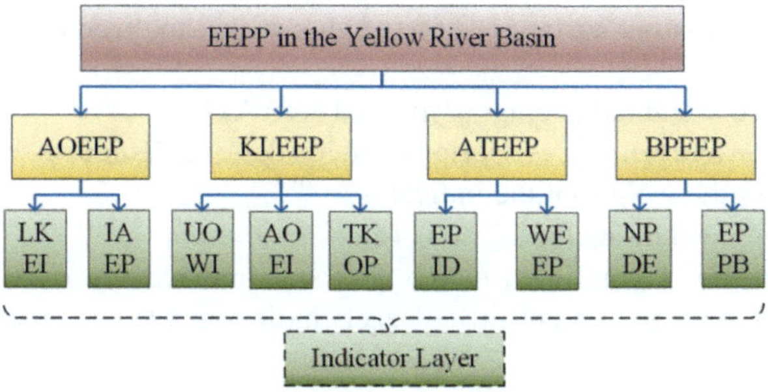

Fig. 2. Indicator system for PEEP

3.3 Enhanced Grey Target Evaluation Model Based on Relative Entropy Correlation

Standardisation of Data

The indicators, which varied in unit scales, were standardized to derive a relative affiliation matrix R. $R = \left(r_{ij}\right)_{m \times n}(i = 1, 2, \cdots, m; j = 1, 2, \cdots, n)$.

Indicators of the larger-is-better type:

$$r_{ij} = \left(x_{ij} - x_{\min}\right)/(x_{\max} - x_{\min}) \tag{1}$$

Indicators of the larger-is-better type:

$$r_{ij} = \left(x_{\max} - x_{ij}\right)/(x_{\max} - x_{\min}) \tag{2}$$

where $\left(x_{ij}\right)_{\max}$ and $\left(x_{ij}\right)_{\min}$ denote the maximum and minimum values of the indicator x_{ij} across different objects, respectively.

Calculation of Combined Weights

Two common approaches for determining indicator weights are subjective and objective. The subjective approach tends to be more arbitrary and reliant on individual judgment, while the objective approach pays less attention to regional variability. Combined weights (CW) integrate the decision maker's preference for attributes by harmonizing these different methods, thereby mitigating the subjectivity in weight assignment (Zou et al., 2020). In information theory or probability theory, relative entropy quantifies the difference between two probability distributions (Lin et al., 2021; Liu et al., 2022). Here, combining the weight vector with those obtained from each method of assignment constitutes independent discrete distributions. Relative entropy serves to measure the "distance" between these discrete distributions, with values approaching zero indicating greater similarity between the combined weights and each assignment method. CW specific calculation steps are as follows:

Step 1: The calculation of the subjective weight (SW) and objective weight (OW).

① Analytic hierarchy process (AHP) [2].

(1) Compare the relative importance of the evaluation Indicators in pairs; (2) Construct judgment matrix and carry out consistency test; (3) Calculate the ranking weight

vector of the comparison matrix; (4) The weights are normalized to obtain the weights w^{AHP}.

② The ordered weighted averaging method (OWA) [5].

(1) Realization and ranking of scoring decision data; (2) Establish the ranking weight vector; (3) Obtain the weight of the indicator w^{OWA}。

③ The entropy weight method (EM) [16].

(1) Establish the original data matrix and standardized processing; (2) Calculate the characteristic gravity and entropy of each Indicator; (3) Calculate the weights of indicators w^{EN}。

④ The criteria importance though intercriteria correlation method (CRITIC) [13].

(1) Calculate the volatility of data after standardization; (2) Calculate data conflict; (3) Calculate Indicator weights w^{CRI}.

Step 2: Calculation of combined weights

To ascertain the combined weights of evaluation criteria, this study developed an optimal mathematical model designed to minimize the sum of squared deviations between the weight vectors derived from the combined method and multiple other methods. The objective of this model is to establish the optimal CW for the evaluation criteria. The principle of relative entropy is introduced to compute the aggregated weights based on various weighting methods. Subsequently, the proximity between each weight result and the aggregated weights is calculated to assess the reliability of each weighting outcome.

To obtain the optimal weight vector $W^C = \left[W_1^C, W_2^C, \cdots, W_m^C\right]$, it is necessary to minimize the deviation between it and the four weight vectors. A combination weight optimization objective function is introduced based on the idea of deviation function.

$$\min Q(W^C) = \sum_{k=1}^{4}\sum_{i=1}^{m}\alpha_k(W_i^C - u_{ki})^2 \tag{3}$$

$$s.t.\begin{cases}\sum\limits_{i=1}^{m} W_i^C = 1, W_i^C > 0 \\ \sum\limits_{i=1}^{m} u_{ki} = 1 \\ \sum\limits_{k=1}^{4} \alpha_k = 1\end{cases} \tag{4}$$

where, W_i^C is the CW of Indicator, u_{ki} is the weight calculated by the k-th weighting method of the i-th Indicator, $i = 1\sim m$, $k = 1\sim K$, form the weight set u:

$$u = \begin{bmatrix} u_{11} & \cdots u_{1i} & \cdots u_{1m} \\ \vdots & \cdots \vdots & \cdots \vdots \\ u_{k1} & \cdots u_{ki} & \cdots u_{km} \\ \vdots & \cdots \vdots & \cdots \vdots \\ u_{K1} & \cdots u_{Ki} & \cdots u_{Km} \end{bmatrix} \tag{5}$$

In this paper, four assignment methods are used, $k = 1–4$,u_{1i}, u_{2i}, u_{3i}, u_{4i} corresponds to the weights w_i^{AHP}, w_i^{OWA}, w_i^{EM}, w_i^{CRI} of the four methods described above. α_k is the

References

1. Li, M.H., et al.: Provenance analysis of yellow river terraces highlights causes of siltation and natural hazards in the North China plain. J. Geophys. Res.-Earth Surf. **128**(11), 14 (2023b). https://doi.org/10.1029/2023jf007246
2. Liu, L.Y.L., Xiao, C., Liu, Y.H.: Projected water scarcity and hydrological extremes in the yellow river basin in the 21st century under SSP-RCP scenarios. Water **15**(3), 14 (2023). https://doi.org/10.3390/w15030446
3. Li, J., Zhang, Y., Li, H.: The impact of the Three-North Shelter Forest Program on the ecological benefits of the Yellow River Basin: an empirical verification based on DID method. Rev. Adhes. Adhes. **11**(3), 198–213 (2023a). https://doi.org/10.47750/raa/11.3.11
4. Ren, L., et al.: Initiatives to clarify mechanisms of hydrological evolution in human influenced Yellow River Basin. Water Sci. Eng. **16**(2), 117–121 (2023)
5. Ding, Y., Li, X., Wang, D., Xu, J.M., Yu, Y.: Study on spatial and temporal differences of water resource sustainable development and its influencing factors in the yellow river basin, China. Sustainability **15**(19), 20 (2023). https://doi.org/10.3390/su151914316
6. Li, P.Y.: Meeting the environmental challenges. Hum. Ecol. Risk Assess. **26**(9), 2303–2315 (2020). https://doi.org/10.1080/10807039.2020.1797472
7. Ma, L., Zhao, X.C., Yan, B.J., Fan, J.R., Wang, M.M., Liu, M.Q.: How does the coupling coordination between high-quality development and eco-environmental carrying capacity in the Yellow river basin over time? Front. Environ. Sci. **12**, 12 (2024). https://doi.org/10.3389/fenvs.2024.1403265
8. Deng, X.J., Hu, S., Zhan, C.S.: Attribution of vegetation coverage change to climate change and human activities based on the geographic detectors in the Yellow River Basin, China. Environ. Sci. Pollut. Res. **29**(29), 44693–44708 (2022). https://doi.org/10.1007/s11356-022-18744-8
9. Li, P., Wei, C.P.: A novel grey target decision method based on a cobweb area model for standard interval grey numbers. J. Grey Syst. **31**(3), 29–44 (2019)
10. Ma, L.J., et al.: Variations of runoff and sediment and their response to human activities in the source region of the Yellow River, China. Environ. Earth Sci. **80**(17), 12 (2021). https://doi.org/10.1007/s12665-021-09850-w
11. Liu, S.H., Liu, Y.J., Yang, C.S., Deng, L.: Relative entropy of distance distribution based similarity measure of nodes in weighted graph data. Entropy **24**(8), 23 (2022). https://doi.org/10.3390/e24081154
12. Lin, H., Pan, T.H., Chen, S.: Comprehensive evaluation of urban air quality using the relative entropy theory and improved TOPSIS method. Air Qual. Atmos. Health **14**(2), 251–258 (2021). https://doi.org/10.1007/s11869-020-00930-7
13. Gao, X., Ma, Y., Zhou, W.: Analysis of software trustworthiness based on FAHP-CRITIC method. J. Shanghai Jiaotong Univ. Sci. (Germany) **29**(3), 588–600 (2024). https://doi.org/10.1007/s12204-022-2496-4
14. Akram, M., Sultan, M., Al-Kenani, A.N.: Group decision analysis based on complex <i>m</i>-polar fuzzy <i>N</i>-soft environment. Math. Probl. Eng. **2022**, 24 (2022). https://doi.org/10.1155/2022/4917408
15. Altman, T., Boulos, P.F.: A hypoquadratic convergence method for Lagrange multipliers. Comput. Math. Appl. **27**(1), 25–31 (1994). https://doi.org/10.1016/0898-1221(94)90003-5
16. Lee, S.: Determination of priority weights under multiattribute decision-making situations: AHP versus Fuzzy AHP. J. Constr. Eng. Manag. **141**(2), 9 (2015). https://doi.org/10.1061/(asce)co.1943-7862.0000897

Bridging Gaps in Disaster Preparedness: Integrating Environmental Governance and Public Awareness in Cambodia's Disaster-Prone Regions

Panhapin Theang[1] and Nimol Thuon[1,2](✉)

[1] University of Science and Technology of China, Anhui, China
nimol.thuon@gmail.com
[2] One to Many Research Center, Phnom Penh, Cambodia

Abstract. Cambodia faces increasing challenges from climate-induced natural disasters, including floods, droughts, and typhoons, which are exacerbated by weak environmental governance and limited public awareness. This study investigates the integration of environmental governance and public awareness to improve disaster preparedness and response strategies. Using survey data from 270 respondents across 16 disaster-prone provinces, the research identifies key factors influencing disaster preparedness, such as past disaster experiences, risk perception, and access to information. Findings show that effective governance frameworks and enhanced public awareness through education and community-based initiatives significantly improve disaster management outcomes. The study emphasizes the need for stronger collaboration between local authorities and communities to implement region-specific disaster strategies, offering valuable insights for policymakers seeking to strengthen disaster resilience in Cambodia.

Keywords: Cambodia · Disaster preparedness · Climate · Natural disasters

1 Introduction

Cambodia's vulnerability to natural disasters, particularly floods, droughts, and typhoons, has intensified due to climate change, leading to more frequent and severe events. The nation geographical features around the Mekong River and climatic conditions exacerbate these risks, affecting millions and causing substantial socio-economic disruptions. Despite efforts led by the National Committee for Disaster Management (NCDM), gaps in policy implementation and limited community outreach persist, as reflected in Cambodia's high ranking on the WorldRiskIndex 2021 [1].

This study examines how integrating environmental governance with public awareness can improve disaster preparedness by using quantitative survey data from 270 respondents in disaster-prone areas that experienced natural disaster in 2020. According to the Theory of Reasoned Action [2], personal experiences, knowledge and severity perception alter subjective norms and attitudes that leads to stronger intention to engage in

D. Han et al. (Eds.): ICEPG 2024, ESE, pp. 1067–1072, 2025.
https://doi.org/10.1007/978-3-031-86829-0_115

preparedness behaviors. Therefore, the study utilizes questions based on United Nations Office for Disaster Risk Reduction (UNISDR) Strategic Framework [3] that includes these variables to determine effective environmental governance, public engagement and preparedness behaviors.

The significant impact of these variables shows how authorities and stakeholders can develop more effective strategies that not only raise awareness but also facilitate actionable preparedness behaviors. Empowering communities through experience-sharing, risk communication, and active participation can lead to more sustainable environmental practices and improved governance outcomes on vulnerable populations [4].

This study offers insights for policymakers on the role of decentralization governance in strengthen community resilience and implement targeted public awareness initiatives for effective disaster management and risk reduction.

2 Methodology

2.1 Research Design

This study employs a quantitative survey approach to analyze the relationship between public awareness and disaster preparedness in Cambodia. The data was collected from 270 respondents across 16 disaster-prone provinces through Likert scale survey questionnaire. The survey includes 40 questions that were developed based on UNISDR Strategic Framework [3] that includes disaster experiences, public awareness, authority engagement, risk perception, and preparedness behaviors. The data collection was supported by local NGOs and community organizations to ensure the inclusion of rural respondents. The study then conducts multiple linear regression using STATA to examine the correlations between the critical factors that can guide policy improvements for enhancing disaster resilience in the region.

2.2 Sample and Procedure

The cross-sectional data for this investigation were gathered from 270 respondents that have experienced natural disaster in 2020 across 16 provinces in the central plain, northwestern, coastal and eastern region. Most of the affected respondents are located in the central plain region (46.3%). The gender is equally distributed with most of the participations are older than 35 years (93.3%). Around 50% of the respondents have a small household size, and majority had attained high school education (59%). In terms of employment, 81.11% were employed full-time and most respondents (57.04%) reported a monthly income of less than $250 per month. The Kaiser-Meyer-Olkin Measure sampling adequacy test was also used in the study to determine the sample size and partial correlation strength and reports the critical value of 0.85 ($p < 0.001$).

3 Results and Analysis

3.1 Reliability and Validity Analysis

The study conducted Exploratory Factor Analysis to determine the concept items reliability, discriminant validity, and convergent validity. As shown in Fig. 1 the validated items were included after achieving a loading of higher than 0.70. Subsequently, the study uses Cronbach's alpha (CA) and composite reliability (CR) to assess each constructs construct-reliability. The figure reports all the items achieve substantial reliability as they all exceed the cutoff points of 0.70, which indicates that the conceptual model is reliable. Lastly, the figure also reports that all the variables exceed the cutoff points of 0.50 for the average variance extracted (AVE), a measure of the concept items acceptable convergent validity.

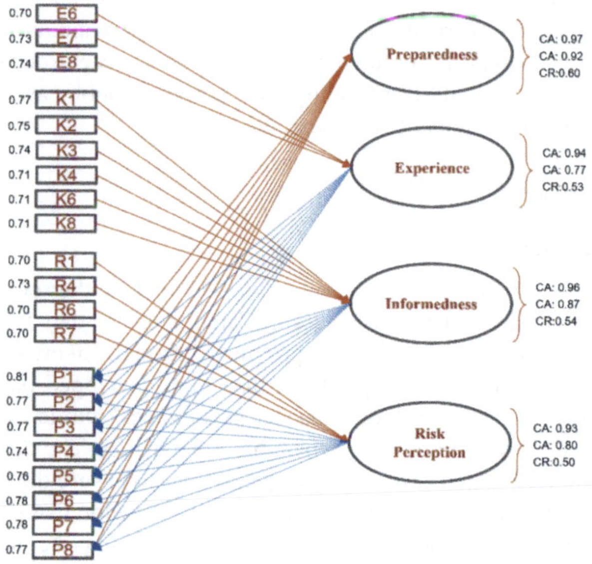

Fig. 1. Exploratory factor analysis and reliability and validity analysis

3.2 Correlation Matrix and Multicollinearity

The study conducted Pearson's bivariate correlation to estimate the degree to which all the variables have a linear connection. The key variables experience, informedness, risk perception and preparedness indicate positive and significant relationship. The control variables indicate demonstrate minimal positive association, whereas income has notable negative correlations with experience and informedness. The study obtains the value of less than the threshold of 10 for the variance inflation factors (VIF), which indicates no sign of multicollinearity.

3.3 Multiple Linear Regression

The regression analysis reveals that experience, informedness, and risk play significant roles in predicting preparedness. As shown in Table 1, in model (2), experience significantly increases one's level of readiness. ($\beta = 1.559$, $p < 0.001$), indicating that greater experience strongly enhances preparedness. Similarly, in model (3), informedness shows a significant positive association with preparedness ($\beta = 0.778$, $p < 0.001$), suggesting that being more informed increases one's preparedness. Risk also emerges as a significant factor, with a strong positive effect in both models ($\beta = 1.331$, $p < 0.001$ in model (4) and $\beta = 0.795$, $p < 0.001$ in model (5)), implying that individuals perceiving higher risks are more likely to exhibit greater preparedness. As these three variables are incorporated together in model (5), their effects remain statistically significant, underscoring the importance of experience, informedness, and risk in shaping preparedness. The model R-squared also improves across models, reaching 0.619 in the final model, indicating that the inclusion of these key variables enhances the explanatory power of the model.

4 Findings and Discussion

The findings are crucial not only for disaster management but also for enhancing environmental governance effectiveness in Cambodia. Firstly, communities frequently affected by floods and droughts tend to develop adaptive strategies and resistant infrastructure that shows community-led environmental decision-making. Secondly, the strong correlation between public awareness and disaster preparedness behaviors indicates that targeted communication strategies can significantly enhance community resilience. So, investing in locally tailored campaigns and participatory risk communication will improve engagement and foster a culture of preparedness. Lastly, heightened risk perception motivates individuals and communities to participate actively in disaster preparedness activities, such as developing emergency plans and participating in community drills [5]. Yet the findings reveal that despite recognizing environmental risks, individuals may feel powerless due to limited resources or distrust in governance structures. The findings highlight significant gaps in governance and public awareness that undermine Cambodia's disaster preparedness as effective governance is hampered by a lack of coordination between national and local authorities. By addressing the barriers that prevent individuals from translating risk perception into action, authorities can foster greater public engagement and collaboration in environmental management [6].

5 Practical Contributions

This research provides practical guidance for enhancing Cambodia's disaster management capacity:

Experience Exercises. Establish forums for sharing experience and preparedness strategies and regularly conduct disaster drills to reinforce experimental learning.

Public Awareness. Implement multi-channel campaigns using local languages and trusted sources to reach rural populations.

Training and Education. Launch community-based training programs targeting marginalized groups to increase participation in disaster preparedness activities.

Risk Communication. Develop a risk communication strategy that convey the severity and immediacy of environmental risks (visual tools and scenario simulations).

Table 1. Multiple linear regression analysis

Variables	(1)	(2)	(3)	(4)	(5)
	Preparedness				
Gender	−1.46	−0.20	0.18	0.65	0.73
	(−1.23)	(−0.24)	−0.21	−0.80	−0.91
Age	−1.56***	−1.13**	−0.86**	−0.79**	−0.75**
	(−3.73)	(−3.30)	(−2.33)	(−2.27)	(−2.27)
Edu.	−1.01	0.06	−0.20	−0.41	−0.02
	(−0.98)	−0.08	(−0.26)	(−0.63)	(−0.02)
H. Size	0.91	0.18	−0.20	−0.44	−0.47
	−1.17	−0.31	(−0.33)	(−0.81)	(−0.95)
Employ.	−2.72*	−1.75	−2.09**	−2.56**	−2.07**
	(−1.81)	(−1.64)	(−2.06)	(−2.32)	(−2.18)
Income	−4.65***	−0.67	−1.19	−0.47	0.33
	(−4.93)	(−0.75)	(−1.36)	(−0.68)	−0.45
Region	−0.87*	−1.14**	−0.59	−0.77*	−0.84**
	(−1.65)	(−2.55)	(−1.30)	(−1.79)	(−2.00)
Exp.		1.56***			0.62**
		−11.81			−3.15
Informed			0.78***		0.20*
			−10.57		−1.85
Risk				1.33***	0.80***
				−12.92	−4.49
Constant	47.09***	21.39***	21.26**	18.16***	12.87**
	−11.15	−5.10	−5.01	−4.36	−3.08
Obs.	270	270	270	270	270
R^2	0.16	0.50	0.48	0.56	0.62

Robust t-statistics in parentheses
*** $p < 0.001$, ** $p < 0.05$, * $p < 0.1$

Policy Reform. Empower local authorities and communities to participate in decision-making, enhancing the relevance and efficacy of environmental policies.

Monitoring and Evaluation. Set up mechanisms to monitor and continuously improve environmental governance initiatives based on community feedbacks.

6 Limitations and Future Work

There are some limitations to the study that pave ways for future research. First, the sampling distribution is not uniform across all regions, which may affect the generalizability of the findings. For instance, the Central Plain Region comprises 46.3% of the sample, while the Eastern Region accounts for only 14%. Given that different regions may experience varying impacts from natural disasters and have distinct governance strategies, it is challenging to generalize the results. Future research should consider a more detailed breakdown by region to enhance understanding of specific governance and preparedness behaviors.

Secondly, there are challenges related to data collection and availability. This study was limited to data from a single timeframe (2020). To develop a more comprehensive understanding of changes and improvements in strategies and governance effectiveness over time, future studies are encouraged to collect and compare data across multiple timeframes. This approach would provide richer insights into the dynamic nature of disaster management and environmental governance.

7 Conclusion

The study concludes that integrating environmental governance and public awareness is crucial for enhancing disaster preparedness in Cambodia. While past disaster experiences and public awareness campaigns significantly contribute to higher preparedness levels, gaps in governance and information dissemination, especially in rural areas, limit overall effectiveness. Strengthening local governance, improving coordination between national and local agencies, and expanding public awareness initiatives are key to addressing these issues. Additionally, promoting risk perception and engaging communities through targeted training and real-time warning systems will further boost preparedness. Implementing these strategies will not only improve disaster management but also support long-term resilience against climate change impacts.

References

1. Radtke, K., Weller, D.: The WorldRiskIndex 2021. WorldRiskReport (2021). https://weltrisikobericht.de/
2. Fishbein, M.: A theory of reasoned action: some applications and implications (1979)
3. United Nations Office for Disaster Risk Reduction (UNISDR): UNISDR Strategic Framework 2016–2021 (2015). https://www.unisdr.org/
4. Geng, Q., Lo, K.: Geopolitical strategies and transnational environmental governance: a comparative study of international NGOs in Cambodia. Geoforum **154**, 104063 (2024). https://doi.org/10.1016/j.geoforum.2024.104063
5. Bubeck, P., Botzen, W.J.W., Aerts, J.C.: A review of risk perceptions and other factors that influence flood mitigation behavior. Risk Anal.: Int. J. **32**(9), 1481–1495 (2012). https://doi.org/10.1111/j.1539-6924.2011.01783.x
6. Wachinger, G., Renn, O., Begg, C., Kuhlicke, C.: The risk perception paradox—Implications for governance and communication of natural hazards. Risk Anal. **33**(6), 1049–1065 (2013). https://doi.org/10.1111/j.1539-6924.2012.01942.x

Innovative Research on E-Waste Recycling Models Guided by Public Values Based on Big Data

Yu Xiang[1], Jinyi Qi[1], and Jianmin Liu[2,3](✉)

[1] Guangxi University of Finance and Economics, Nanning 530007, China
[2] Guangxi Key Laboratory of Big Data in Finance and Economics, Guangxi University of Finance and Economics, Nanning 530007, China
jianminliu2007@163.com
[3] School of Big Data and Artificial Intelligence, Guangxi University of Finance and Economics, Nanning 530007, China

Abstract. This study investigates innovative e-waste recycling models that incorporate public values and leverage big data analytics to optimize processes. It explores how aligning recycling practices with environmental ethics and technological innovation can enhance public participation and improve recycling efficiency. By emphasizing the role of public values in shaping e-waste recycling policies and utilizing big data for process improvements, the research advocates for a sustainable, community-centered approach. This approach not only fosters environmental stewardship but also ensures ethical technological progress, aiming for a more effective and socially responsible e-waste management system.

Keywords: E-waste Recycling · Public Values · Big Data Analytics · Environmental Ethics

1 Introduction

The unprecedented proliferation of electronic devices in the contemporary digital age has generated a significant rise in e-waste production, thereby presenting formidable challenges to environmental sustainability and public health. E-waste refers to discarded electronic devices, including computers, mobile phones, and household appliances, many of which contain hazardous substances capable of severely polluting the environment if improperly managed. As society increasingly depends on technological innovations, addressing the management and recycling of e-waste has become a critical global issue.

The rapid pace of technological progress has resulted in a markedly reduced lifespan for electronic devices, driving a substantial increase in e-waste generation. This trend is compounded by improper disposal practices, particularly in developing nations where informal recycling methods prevail, often leading to severe environmental and health consequences. Toxic elements commonly found in e-waste, such as lead, mercury, and cadmium, contaminate soil and water sources, leading to long-term ecological

D. Han et al. (Eds.): ICEPG 2024, ESE, pp. 1073–1080, 2025.
https://doi.org/10.1007/978-3-031-86829-0_116

degradation. Moreover, exposure to these hazardous materials during informal recycling processes poses significant health risks to workers, particularly in low-income regions.

Given these pressing concerns, the question arises: how can innovative solutions, guided by public values and enhanced through big data analytics, improve the efficiency and effectiveness of e-waste recycling processes? Furthermore, what role do public values play in shaping the policies and initiatives associated with e-waste recycling, and how can these values be harnessed to promote sustainable practices? This study seeks to address these urgent issues by investigating new approaches to e-waste recycling that integrate public values with advanced big data methodologies. Through a comprehensive analysis of environmental ethics, technological innovation, and public engagement, this research aims to uncover the transformative potential of combining public values with big data in e-waste recycling, thus promoting a more sustainable and ethically responsible approach to managing electronic waste.

2 Literature Review

2.1 The Role of Big Data in E-Waste Management

The management of electronic waste has faced unprecedented challenges in recent years due to rapid technological advancements and the growing consumption of electronic devices worldwide. As the volume of e-waste continues to escalate, innovative strategies are essential to enhance recycling efficiency. One of the most promising tools for addressing this challenge is big data analytics, which offers a range of solutions for optimizing waste management processes.

Predictive Analytics in E-Waste Generation. Big data analytics, particularly predictive algorithms, plays a crucial role in forecasting e-waste generation trends. By analyzing historical data, including sales patterns, consumer behaviors, and product lifecycles, predictive analytics allows for accurate estimations of future e-waste volumes. This predictive capability is invaluable for policymakers and waste management organizations, enabling them to allocate resources more effectively, develop recycling infrastructure, and implement targeted awareness campaigns. Khoshand et al. (2023) employed a modified adaptive neuro-fuzzy inference system (MANFIS) to estimate e-waste generation accurately. Their study demonstrated that selecting key input parameters and optimizing the structure of MANFIS led to reliable and precise forecasts, which in turn informed circular economy planning and resource recovery efforts [1]. Likewise, Mohsin et al. (2022) developed a forecasting framework (CH-X12/STL-X) that effectively handled time-series data with seasonal variations, further enhancing predictive accuracy [2].

Supply Chain Optimization. Big data analytics also optimizes the e-waste supply chain by tracking electronic products from production to disposal. Utilizing Internet of Things (IoT) sensors and Radio Frequency Identification (RFID) tags, real-time data on product movement and condition is collected and analyzed, leading to improved logistics, reduced transportation costs, and minimized environmental impact. This technology ensures the secure transfer of e-waste to recycling facilities, streamlining the entire recycling process. Khan and Ahmad (2022) proposed a blockchain-based IoT system for monitoring post-production processes of electronic devices, emphasizing transparency and security. Their system ensures traceability through smart contracts and distributed ledgers, contributing to more efficient e-waste management practices [3].

Intelligent Recycling Processes. The integration of big data into recycling processes also enables automation through intelligent systems. Machine learning algorithms, trained on extensive datasets of electronic components, can accurately identify and categorize recyclable materials. Robotic systems equipped with sensors enhance the dismantling of electronic products, ensuring precise separation and recovery of valuable materials. Verma (2023) highlighted the application of deep learning techniques within big data analytics for waste management, illustrating how intelligent recycling can revolutionize the industry [4]. Additionally, Shevchenko et al. (2021) developed a smart reverse logistics system that connects end-of-life electronics holders with local recycling infrastructures, utilizing IoT tools to facilitate efficient waste collection and transportation [5].

2.2 Public Values and Environmental Protection

Public values, which reflect societal expectations and priorities, play a crucial role in shaping environmental protection policies. The theory of public value emphasizes the importance of aligning public services and initiatives with societal needs and values, ensuring that they contribute meaningfully to the public good.

Public Values and Environmental Awareness. He et al. (2011) investigated how geographical location influences environmental awareness, comparing student groups from Shanghai and Gansu in China [6]. While regional differences in attitudes toward local environmental issues were observed, both groups demonstrated strong pro-environmental attitudes regarding global concerns. This suggests that environmental awareness can transcend regional disparities, with public values driving a collective desire to improve environmental stewardship.

Public Values and Policy Formulation. Demski et al. (2015) identified six public value clusters that influence energy policy acceptance, including environmental preservation, social equity, and security [7]. By examining these value frameworks, policymakers can better understand the underlying factors that shape public support or opposition to environmental initiatives, thus facilitating more effective and equitable policy development.

Public Values and Community Engagement. Community-based natural resource management (CBNRM) has increasingly become integrated into state-regulated frameworks, requiring local communities to engage with public policies. Ojha et al. (2016) highlighted the importance of intermediaries and networking in helping communities navigate these regulatory landscapes, demonstrating that public engagement is essential for the success of environmental initiatives [8].

The integration of big data analytics and public values into e-waste management presents a promising path forward for addressing the environmental and health challenges associated with electronic waste. By harnessing the power of predictive analytics, supply chain optimization, and intelligent recycling systems, waste management can become more efficient and sustainable. Simultaneously, the alignment of public values with environmental policies fosters community engagement and ensures that recycling initiatives are both ethical and effective.

3 Research Methods

3.1 Data Collection

This study utilized a range of big data sources to compile a comprehensive dataset. These sources included social media platforms, online forums, government databases, and sensor networks. The diverse datasets were aggregated to provide a rich foundation for analysis.

3.2 Research Design

A mixed-methods approach was employed, integrating qualitative and quantitative methodologies. Six indicators related to public value in e-waste recycling were selected. Statistical methods refined these indicators, identifying highly representative factors and assigning corresponding weights for analysis.

The study applied factor analysis to filter the six public value indicators. Based on disciplinary knowledge, objective data distribution patterns were analyzed to select relevant indicators. Redundant or low-contributing indicators were eliminated, ensuring efficiency and reliability in data analysis.

Environmental Awareness Index. This index measures the public's awareness and concern about e-waste management and its environmental impact.

Sustainability Index. This index evaluates the contribution of recycling models to social, economic, and environmental sustainability.

Community Engagement Index. This index quantifies public participation in e-waste recycling initiatives.

Public Satisfaction Index. This index measures public satisfaction with e-waste recycling services, focusing on service quality and convenience.

Accessibility and Equity Index. This index assesses the accessibility and fairness of e-waste recycling services across different communities.

Transparency Index. This index evaluates the openness and transparency of the recycling system, particularly regarding policies and outcomes.

4 Results

4.1 Reliability Analysis

A confirmatory factor analysis was performed on six factors and 34 items, with a sample size of 379. The reliability coefficient (Cronbach's Alpha = 0.953) indicates high internal consistency (see Table 1).

Table 1. Cronbach's Alpha

items	Number of items	Cronbach's Alpha
34	379	0.953

Table 2. Convergent validity

Factor	AVE	CR
Environmental Awareness	0.677	0.829
Sustainability	0.634	0.889
Community Engagement	0.696	0.855
Public Satisfaction	0.757	0.932
Accessibility and Equity	0.738	0.919
Transparency	0.733	0.872

From the results of the above measurement model analysis (see Table 2), it is evident that all the constructs in this study model have good reliability, convergent validity and discriminant validity, so the structural model can be further analyzed to understand the validation of the hypothesis are supported or not.

4.2 Weight Calculation

The weight calculation was conducted using the Analytic Hierarchy Process (AHP). The judgment matrix passed the consistency test with a CR of 0.000, confirming the consistency of calculated weights (see Table 3).

Table 3. Weight

Index	Weights	CI
Environmental Awareness	15.76	0.000
Sustainability	17.07	
Community Engagement	18.74	
Public Satisfaction	18.22	
Accessibility and Equity	13.51	
Transparency	16.7	

4.3 Theoretical Analysis

The analysis highlighted the critical role of public environmental awareness, sustainability, and community engagement. The high weights of these indices underscore their influence on optimizing recycling models through big data insights. Transparency and accessibility are pivotal for fostering trust and ensuring equitable access to services, further contributing to the model's success.

Environmental Awareness Index (Weight: 15.76). The Environmental Awareness Index plays a crucial role in e-waste recycling models driven by public values. Its significant weight indicates the strong influence of public awareness on decision-making. Big data analytics enables the tracking of public knowledge, attitudes, and behaviors concerning environmental conservation, informing targeted awareness campaigns and promoting the sustainable development of recycling models.

Sustainability Index (Weight: 17.07). The Sustainability Index emphasizes the importance of balancing social, economic, and environmental sustainability in e-waste recycling. Data on resource efficiency, carbon footprints, and economic returns guide continuous improvements, promoting a synergy between social responsibility and economic benefits.

Community Engagement Index (Weight: 18.74). The Community Engagement Index highlights the necessity of active public involvement in recycling activities. Big data helps uncover patterns of participation, enabling the design of initiatives that boost recycling rates and strengthen community cohesion.

Public Satisfaction Index (Weight: 18.22). The Public Satisfaction Index underscores the importance of meeting public expectations regarding recycling services. By analyzing public satisfaction in areas such as service quality and convenience, the model fosters trust, increases engagement, and supports sustainable development.

Accessibility and Equity Index (Weight: 13.51). The Accessibility and Equity Index ensures fairness in service provision. Big data identifies underserved regions, optimizing the location of recycling points to provide equitable access to all residents, thereby promoting social justice.

Transparency Index (Weight: 16.7). The Transparency Index emphasizes the significance of openness in recycling processes. Big data ensures real-time disclosure of policies, fostering trust between stakeholders and driving participation in sustainable recycling initiatives.

5 Innovation Path of E-Waste Recycling Model Based on Public Values and Big Data

Guided by public values, the innovation of the e-waste recycling model based on big data follows a clear, strategic path.

Big data analytics offers a pathway to enhance e-waste recycling by analyzing public awareness, identifying waste generation patterns, and facilitating community engagement. By understanding public concerns, targeted environmental campaigns can be developed to enhance awareness. Moreover, the analysis of waste generation trends allows for efficient network planning and resource utilization. Public participation, crucial to recycling success, can be supported through mobile applications and data-driven insights into user needs, increasing engagement and satisfaction.

Firstly, big data can be used to assess public environmental awareness, providing insights into attitudes and demands regarding e-waste. This understanding allows for the creation of targeted campaigns to raise awareness and promote improvements in recycling models.

Secondly, big data analysis reveals e-waste generation patterns across regions, facilitating efficient planning of recycling networks to optimize resource use. Additionally, predictive analytics can forecast future e-waste trends, enabling real-time adjustments to strategies for more informed decision-making.

Public participation is another crucial factor. By analyzing engagement trends, we can develop mobile applications that offer convenient recycling information and appointment services, increasing public involvement. Furthermore, big data helps measure public satisfaction with recycling services, driving continuous improvements in service quality.

Equity and accessibility are key considerations. Data analysis can identify underserved areas and help optimize the location of service points, ensuring fair access to recycling services for all communities, thereby promoting social justice.

Lastly, transparency plays a vital role in advancing the model. Big data ensures the real-time disclosure of policies, processes, and outcomes, fostering trust between governments, businesses, and the public. This openness enhances public participation, furthering the model's sustainability.

By integrating public values with big data analytics, the e-waste recycling model becomes more efficient and sustainable, maximizing resource reuse and significantly contributing to societal and environmental well-being.

In conclusion, the combination of public values and big data analytics marks a pivotal step toward sustainable e-waste management. Aligning the recycling process with values such as environmental awareness, community engagement, and transparency, this model addresses critical environmental issues while fostering social responsibility and inclusivity. By incorporating diverse perspectives and leveraging data-driven insights, the framework optimizes recycling strategies, pointing toward a future where environmental protection, community participation, and technological innovation converge to create a more sustainable planet.

Acknowledgments. This research was supported by Innovation Project of Guangxi Graduate Education (JGY2024351), Guangxi First-class Discipline statistics Construction Project Fund, Guangxi Key Laboratory of Big Data in Finance and Economics Project Fund.

References

1. Khoshand, A., Karami, A., Rostami, G., Emaminejad, N.: Prediction of e-waste generation: application of modified adaptive neuro-fuzzy inference system (MANFIS). Waste Manage. Res. **41**(2), 389–400 (2023)

2. Mohsin, A.K.M., Hongzhen, L., Masum Iqbal, M., Salim, Z.R., Hossain, A., Al Kafy, A.: Forecasting e-waste recovery scale driven by seasonal data characteristics: a decomposition-ensemble approach. Waste Manage. Res. **40**(7), 870–881 (2022)
3. Khan, A.U.R., Ahmad, R.W.: A blockchain-based IoT-enabled E-Waste tracking and tracing system for smart cities. IEEE Access **10**, 86256–86269 (2022)
4. Verma, J.: Deep technologies using big data in: energy and waste management. In: Deep Learning Technologies for the Sustainable Development Goals: Issues and Solutions in the Post-COVID Era, pp. 21–39. Singapore: Springer Nature Singapore (2023)
5. Shevchenko, T., Saidani, M., Danko, Y., Golysheva, I., Chovancová, J., Vavrek, R.: Towards a smart E-waste system utilizing supply chain participants and interactive online maps. Recycling **6**(1), 8 (2021)
6. He, X., Hong, T., Liu, L., Tiefenbacher, J.: A comparative study of environmental knowledge, attitudes and behaviors among university students in China. Int. Res. Geogr. Environ. Educ. **20**(2), 91–104 (2011)
7. Demski, C., Butler, C., Parkhill, K.A., Spence, A., Pidgeon, N.F.: Public values for energy system change. Glob. Environ. Chang. **34**, 59–69 (2015)
8. Ojha, H.R., et al.: Delocalizing communities: changing forms of community engagement in natural resources governance. World Dev. **87**, 274–290 (2016)

Research on Power Battery Recycling Models: Analysis and Comparison Based on ASME Process Diagnostic Method

Xiaoda Liu[1(✉)] and Gao Yuan[2]

[1] Beijing Jiaotong University, Beijing, China
735564490@qq.com
[2] Shandong Energy Group Materials Co, Ltd., Luzhong Branch, Tai'an, Shandong, China

Abstract. With the rapid development of the new energy industry, the issues of power battery recycling and reuse have become increasingly prominent. However, a large number of power batteries have unclear destinations and are not properly treated, causing environmental pollution. This paper analyzes various power battery recycling models, including the traditional model, the automaker-led alliance model, the battery manufacturer-led model, and the NIO battery swap station model, using the ASME process diagnostic method for comparison. It ultimately recommends promoting the battery manufacturer-led recycling model to enhance consumer satisfaction, recycling efficiency, and reduce the number of batteries entering unregulated enterprises, thus minimizing environmental pollution.

Keyword: Recycling Model · ASME Process Diagnostic Method · New Energy Vehicles

1 Introduction

With the rapid development of the new energy industry, the annual sales of new energy vehicles (NEVs) in China have been steadily increasing [1]. However, the batteries used in NEVs typically have a lifespan of 5 to 8 years. Once they reach the end of their lifecycle, replacing and handling these batteries becomes a significant challenge for both companies and society [2]. If retired power batteries are not handled properly, they may be abandoned, wasting their remaining value and causing environmental pollution. Additionally, they could be collected by unregulated enterprises and reprocessed improperly, which not only disrupts the market but also poses safety risks. The high residual value of retired power batteries and the severe consequences of improper handling make researching effective battery recycling models crucial. In 2023, the Ministry of Industry and Information Technology released the "Administrative Measures for the Comprehensive Utilization of NEV Power Batteries (Draft for Public Comment)," which clearly established the extended producer responsibility (EPR) system as the core of the recycling framework, requiring automakers to take primary responsibility for battery recycling.

D. Han et al. (Eds.): ICEPG 2024, ESE, pp. 1081–1087, 2025.
https://doi.org/10.1007/978-3-031-86829-0_117

Nevertheless, various issues remain in the process of recycling power batteries, especially with a large number of used batteries being directed to "black workshops" rather than legitimate recycling companies. In practice, consumers tend to prefer non-regulated channels for battery recycling, while the recycling capacity of these unregulated entities is insufficient. This indicates that the main challenge lies in consumers' low acceptance of formal recycling channels, rather than a lack of logistics capacity from regulated recyclers.

There are multiple power battery recycling models in the Chinese market. In recent years, scholars have widely explored these models from the perspectives of the government, automakers, and battery manufacturers, aiming to identify which model best suits battery recycling by achieving low costs and high recovery rates. Analysis of recycling models from the consumer perspective remains relatively scarce. This study identifies recycling service duration as the key factor influencing consumer satisfaction. Using the ASME process analysis method, the study quantifies and evaluates consumer satisfaction across different recycling models, allowing for a comparison and exploration of power battery recycling models for new energy vehicles. This method provides insights into value-added and non-value-added activities within the recycling process, offering a deeper understanding of consumer experiences and supporting improvements in recycling models.

2 Status and Analysis of Recycling Models

The recycling models in the Chinese market and their characteristics are as follows:

Traditional Recycling Model: This model relies on 4S stores as temporary transfer stations. Automakers collect power batteries from these stores on a fixed schedule and transport them to recycling centers for further processing.

Automaker-Led Alliance Recycling Model: This model utilizes automakers' nationwide sales and service networks to organize and coordinate the collection of used power batteries, while battery recycling companies handle the subsequent processing. Unlike the traditional model, the automaker network replaces 4S stores, actively driving power battery recycling instead of passively waiting for the next scheduled collection [3].

Battery Manufacturer-Led Recycling Model: Battery manufacturers employ various methods to collect used power batteries, such as establishing their own collection points, utilizing automaker sales networks, or partnering with small auto repair shops. Compared to the previous two models, this approach offers a greater number of initial collection points and a wider coverage. Additionally, battery manufacturers possess stronger logistics networks to efficiently transport the collected batteries [4].

NIO Battery Swap Station Recycling Model: Unlike other recycling models, this model uses battery swap stations as the first node in the battery recycling process. After a battery swap, the automaker conducts a basic inspection and categorization of the replaced battery. Batteries that meet health standards are recharged and reused, while those that do not are sent to recycling companies for further processing. The advantage of this model is that it ensures a high recycling rate and quick service. However, the high construction and maintenance costs of swap stations lead to higher vehicle prices compared to other brands.

3 Comparison of Recycling Models

This paper uses the ASME process diagnostic method to represent consumer satisfaction with the service duration across different recycling models. The ASME method classifies the various stages of a business process into value-added activities and non-value-added activities. Value-added activities are those for which consumers are willing to pay, while non-value-added activities are the opposite. The higher the proportion of value-added activities in the entire process, the higher the consumer satisfaction.

The traditional recycling model serves as the foundation for the remaining models, but this paper focuses on analyzing and comparing the other three recycling models.

The automaker-led alliance recycling model follows the recycling process flow, as depicted in Fig. 1.

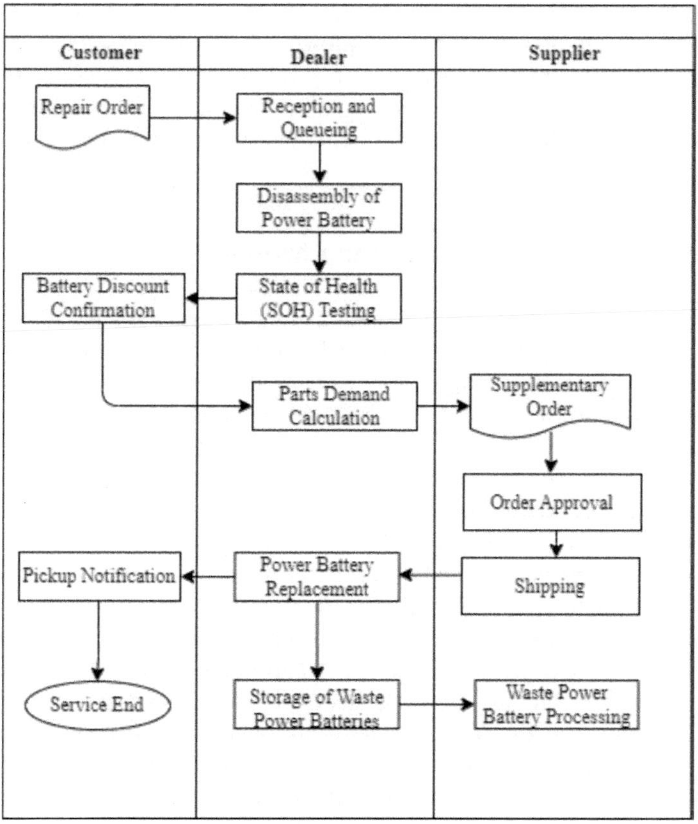

Fig. 1. Automaker-Led Recycling Model Flowchart

When estimating the duration of activities in the process (Measured in minutes; The durations of different brands, different sites, and different processes vary. The values below are estimated based on the results of the investigation.), the waiting time caused by the connection between activities is not included. Each activity should be labeled based on its type: value-added, non-value-added, or other. "A" represents a value-added activity, "N" represents a non-value-added activity, and "O" represents other. As illustrated in Table 1 below.

Table 1. Automaker-Led Recycling Process Diagnostic Table

Task	Time	Type
Place Order	1	A
Reception and Queueing	120	N
Disassembly of Power Battery	20	A
State of Health(SOH) Testing	15	A
Battery Discount Confirmation	10	N
Parts Demand Calculation	20	N
Supplementary Order	5	N
Order Approval	5	N
Shipping	180	O
Power Battery Replacement	30	A
Pickup Notification	1	N
Storage of Waste Power Batteries	0	O
Waste Power Battery Processing	0	A
Total	407	

According to the Table 1, the total duration of value-added activities (TA1) is 66 min, and the total duration of non-value-added activities (TN1). The proportion of value-added activities accounts for 16.22% of the total process duration. Specifically, TA1 / (TN1 + TA1) * 100% = 26.72%. The main reason for the low proportion of value-added activities is that when the vehicle needs to replace the power battery, it is often in a non-operational state, requiring the customer to spend a significant amount of time transporting the vehicle to a 4S shop. Moreover, this model does not support the 4S shop holding battery inventory, meaning that when a customer requests a battery replacement, the shop must wait for restocking from the automaker.

The recycling process flow for the Battery Manufacturer-Led Power Battery Recycling Model is depicted in Fig. 2.

The duration of activities in the process should be estimated using the same standards mentioned above, as shown in Table 2.

According to the Table 2, the total duration of value-added activities (TA2) is 66 min, and the total duration of non-value-added activities (TN2). The proportion of value-added activities accounts for 19.88% of the total process duration. Specifically, TA2 / (TN2 + TA2) * 100% = 43.42%. The main reason for the increase in the proportion of value-added activities is that the time customers spend transporting their vehicles to

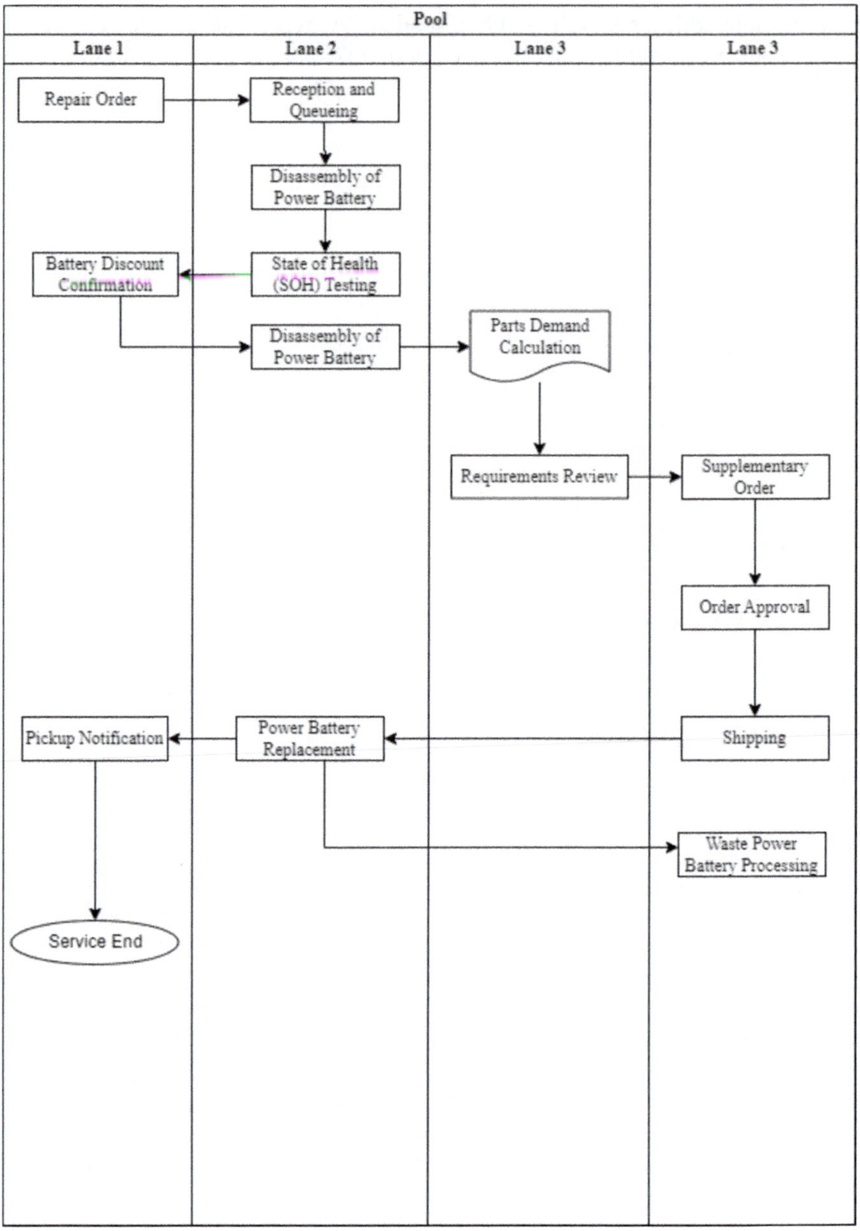

Fig. 2. Manufacturer-Led Recycling Model Flowchart

the service point is significantly reduced. However, this model still requires restocking from the automaker when customers request a power battery replacement.

The NIO battery swap station recycling model no longer requires ASME process diagnostic analysis. This model integrates power battery recycling into the user's daily

Table 2. Manufacturer-Led Recycling Process Diagnostic

Task	Time	Type
Place Order	1	A
Reception and Queueing	40	N
Disassembly of Power Battery	20	A
State of Health(SOH) Testing	15	A
Battery Discount Confirmation	10	N
Calculate Required Parts	20	N
Requirements Review	5	N
Supplementary Order	5	N
Order Approval	5	N
Shipping	180	O
Power Battery Replacement	30	A
Pickup Notification	1	N
Waste Power Battery Processing	0	A
Total	332	

routine, and the entire process of battery replacement is classified as a value-added activity.

Overall, the automaker-centric recycling model clearly defines the boundaries of collaboration between upstream and downstream partners in the supply chain, with responsibilities well-defined. However, consumer satisfaction with the service duration under this model is low, and in practice, the trade-in value of old power batteries for customers tends to be lower, which makes this model less appealing to consumers.

On the other hand, the battery manufacturer-led power battery recycling model is more flexible and conveniently meets consumers' needs for power battery replacement.

The NIO battery swap station recycling model fully satisfies consumer expectations for service duration. However, in practice, consumers are dissatisfied with the increase in vehicle prices caused by this model, and there is resistance to the fact that the replaced battery is still a used battery.

4 Conclusion

The NIO battery swap station recycling model integrates battery replacement and recycling into consumers' daily vehicle usage, significantly improving service efficiency and customer satisfaction. However, this model also has its limitations, such as high construction and maintenance costs leading to increased vehicle prices, as well as consumers' resistance to the fact that the replaced power batteries are still used batteries. To promote this recycling model in the future, it is essential for both companies and the government to guide public awareness to enhance consumer acceptance.

From the perspective of consumer service market satisfaction, the battery manufacturer-led recycling model is more suitable for promotion. This model is more flexible and convenient, and it can fully leverage logistics and transportation capabilities.

Of course, no matter which recycling model is promoted, government support and advancement are indispensable [2].

References

1. Kang, H., Li, J.: Current status and future prospects of battery recycling in China. J. Clean. Prod. **278**, 123467 (2021). https://doi.org/10.1016/j.jclepro.2020.123467
2. Wang, X., Zhang, Z.: A comparative study of battery recycling models: from consumer perspective. Resour. Conserv. Recycl. **168**, 105298 (2021). https://doi.org/10.1016/j.resconrec.2020.105298
3. Chen, X.: Exploration of new energy vehicle power battery recycling models. Era Automobile **15**, 74–76 (2023)
4. Guan, Y., He, W.: Exploration of new energy vehicle battery recycling models from an enterprise perspective. Automobile Knowl. **24**(06), 8–10 (2024)

An Integrated Production-Waste-Harm Model Based on Combination Prediction and Dynamic Programming

Yuqin Feng(✉)

Shanghai Jiao Tong University, No. 800 Dongchuan Road, Minhang District, Shanghai, China
1216210268@sjtu.edu.cn

Abstract. In recent years, the problem of plastic waste has attracted widespread attention, and its continued growth poses a serious threat to the environment and human health. To address this plastic crisis, we propose the Production-Waste-Hazard (PWH) model for simulating the life cycle of plastics and its associated impacts. It was found that by 2020, when the share of single-use plastics is 27.8%, the annual waste generated is expected to be 10.97 million tons, with a maximum mitigation of 3.43 million tons. In addition, we extend the PWH model to an integrated PWH (IPWH) model by adding multiple influences such as plastic substitutes, policy interventions, increased environmental awareness among citizens, and removal of accumulated plastic waste in the ocean and on land. We set minimum achievable goals for plastic waste reduction through dynamic programming: goal one is to reduce annual waste from single-use plastics by 2125, and goal two is to reduce plastic hazards by 2131.

Keywords: Plastic waste · Single-use or disposable plastic · Dynamic Programming

1 Introduction

The large-scale production of plastics has experienced significant growth since the 1950s, primarily driven by the rising demand for plastic products across various sectors, including domestic, industrial, and medical applications following World War II. The advantages of plastics, such as affordability, versatility, lightweight nature, and chemical resistance, have contributed to their widespread adoption.

However, the persistence of plastics in the environment, attributed to their corrosion-resistant characteristics, poses substantial challenges, as they can remain intact for up to a century. This situation has raised considerable concerns regarding plastic waste. For instance, in 2015, global primary plastic production reached 407 million tonnes, with approximately three-quarters—around 302 million tonnes—culminating in waste, of which merely 9% underwent proper recycling. A significant portion of this mismanaged plastic waste comprises single-use or disposable items. The consequences of such mismanagement are profound; plastics improperly discarded into marine environments regularly injure marine life, entangling species like seabirds and sea turtles.

D. Han et al. (Eds.): ICEPG 2024, ESE, pp. 1088–1095, 2025.
https://doi.org/10.1007/978-3-031-86829-0_118

2 PWH Plastics' Life-Circle Simulation Model

In this section, we devise a model PWH (Production-Waste-Harm) to simulate the process of annual Production, usage and consequent Waste, and the Harm, or pollution done to the environment by plastics. In this model, we distinguish 2 kinds of plastics: single-use or disposable plastics and other plastics.

Input data collected and used in the PWH model include annual plastic production, recycle rate, incinerate rate, discarded rate and different plastic's life-time information from [1] and [5].

2.1 Production Model

Based on the assumption above, we admit that the rate of single-use or disposable plastics production is equal to β1, therefore we have,

$$AP_s(t) = \beta_1 \times AP(t) \tag{1}$$

$$AP_o(t) = (1 - \beta_1) \times AP(t) \tag{2}$$

2.2 Waste Model

In our model we define the "AnnualWaste" of a year as the plastics (produced year or years before) whose usage terminate on that year [2].

Thus previous annual total production in Eq. 1, 2 enable us to calculate the annual waste by summing up all types of previously-produced plastics that reach their end of life-time this year [1].

$$AW_s(t) = \sum_{j=1}^{t-1950} AP_s(t-1)D_1(j) \tag{3}$$

$$AW_o(t) = \sum_{j=1}^{t-1950} \sum_{i=2}^{8} AP_{o,i}(t-j)D_i(j) \tag{4}$$

where the distribution of different plastics' life-time will be extrapolated in Sect. 2.4.

2.3 Harm Model

Mismanaged waste is material which is at high risk of entering the ocean via wind or tidal transport, or carried to coastlines from inland waterways. Mismanaged waste is the sum of material which is either littered or inadequately disposed [3].

In our model, it is equal to wastes that are neither recycled nor incinerated. Thus we define the annual "Harm" done to the environment has the total amount of mismanaged plastic wastes,

$$H(t) = AW_s(t) \times (1 - i_s) + AW_o(t) \times (1 - i_o - r_o) \tag{5}$$

2.4 Plastics Classification and Life-Time Distribution

In our model, we classify all plastics produced into several categories. And we collect data of their production rates, as well as each category's average life-time and standard deviation [4].

We adopt the Log Normal Distribution as the model of plastic life-times. This is based on the fact that Log Normal Distribution can be used to describe the lives of industrial units, of which we can make an analogy with lives of plastics. Thus, with the average plastic life-times and standard deviations provided, we are able to calculate the probability distribution of their lifetimes as follows Table 1.

Since we know,

$$E(X) = e^{\mu + \frac{\sigma^2}{2}} \tag{6}$$

$$var(X) = \left(e^{\sigma^2} - 1\right) \cdot e^{2\mu + \sigma^2} \tag{7}$$

Table 1. Plastics' categories and their life-time

Category	Production	Average Life Time	Standard Deviation
Packing	35.9%	0.5	0.1
Transportation	6.6%	13	3
Building and Construction	16.0%	35	7
Electronic	4.4%	8	2
Consumer and Institutional Products	10.3%	3	1
Industrial Machinery	0.7%	20	3
Textile	11.5%	5	1.5
Other	14.5%	5	1.5

we are able to calculate μ and σ for programming with

$$\mu = \ln(E(X)) - \frac{1}{2}\ln\left(1 + \frac{var(X)}{E(X)^2}\right) \tag{8}$$

$$\sigma^2 = \ln\left(1 + \frac{var(X)}{E(X)^2}\right) \tag{9}$$

Finally, we have Eq. 10 and draw the probability distribution shown in Fig. 1

$$D(x; \mu; \sigma) = \frac{1}{\sqrt{2\pi}} \cdot e^{-\frac{(\ln x - \mu)^2}{2\sigma^2}} \tag{10}$$

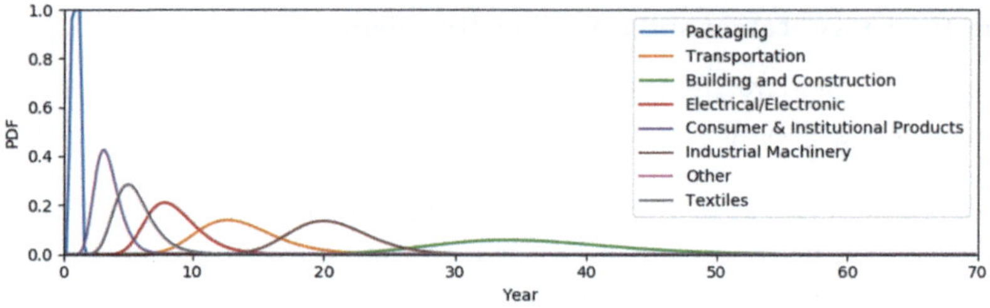

Fig. 1. Probability distribution of different plastics' lifetimes

2.5 Implementation of PWH Model

We observe that although the rate of single-use or disposable plastics among all plastic production(35.9%) is considerably lower than the rate of other plastics(64.1%), the annual wastes resulting from the two(AWs(t), AWo(t)) are close. Moreover, the former cannot be recycled, contributing to a worse case in the annual harm. We also observe from the Fig. 1 that the growth rates continue to accelerate and thus show no signs of slowing down.

Maximum Levels In this task, we aim to maximize the mitigation of single-use plastic wastes as follows: change β_1 of this year; calculate the new $AW_s(t+1)_{\beta_1}$ associated; calculate its maximum difference with $AW_s(t+1)_{original}$ before changing β_1.
To conclude,

$$\begin{aligned} max\ Mitigation = AW_s(t+1)_{\beta_1=35.9\%} - AW_s(t+1)_{\beta_1} \\ s.t.\ 0 \leq \beta_1 \leq 1, \\ H(t+1)_{\beta_1} \leq H(t+1)_{\beta_1=35.9\%} \times (1-10\%) \end{aligned} \tag{11}$$

Where,

$$\begin{aligned} H(t+1)_{\beta_1} = AW_s(t+1)_{\beta_1} \times (1 - i_s(t+1)) \\ +AW_o(t+1)_{\beta_1} \times (1 - i_o(t+1) - r_o(t+1)) \end{aligned} \tag{12}$$

$$AW_s(t+1)_{\beta_1} = AW_s(t+1)_{\beta_1=35.9\%} + (\beta_1 - 35.9\%) \times AP(t) \tag{13}$$

Method of Optimization in Programming According to the regressive prediction of the model, we adopt the idea of Dynamic Programming to calculate a maximum H(t) within any time span. We take each year's predicted H as a new state, and make new decisions according to the variation of parameters to predict the value of the objective function, in order to achieve an optimal choice.

Equation 12, 13 are based on the following modeling assumptions: annual waste of single-use plastic result from the previous year's production; other plastic wastes have a life-time longer than a year and therefore we can ignore their relationship with last year's production.

We solve the optimization problem of Eq. 11 with the PWH model. Solutions are provided in Table 2.

Table 2. Solution of PWH model

Year	Single-Use Ratio	Annual Single-Use Waste (million tonnes)	Maximum Mitigation (million tonnes)
2016	27.6%	10.28	3.09
2017	27.6%	10.62	3.19
2018	27.6%	10.97	3.30
2019	27.8%	11.41	3.32
2020	27.8%	11.77	3.43
2021	27.8%	12.14	3.54
2022	28.0%	12.61	3.56
2023	28.0%	13.00	3.67
2024	28.2%	13.48	3.68

3 IPWH(Integrated Production-Waste-Harm) Model

3.1 Alternatives of Plastic Material

Currently, plastics' substitutes are partly recyclable materials such as paper or glass and partly biodegradable materials such as PLA polyesters, starch-based Polymers, etc. [6] Meanwhile, some bent on making conventional thermoplastics biodegradable by adding additives called prodegradant concentrates (PDCs) [6].

In our integrated model, we discuss the influence of the aforementioned types of alternatives.

We also consider the different percentage of single-use plastics between other plastics that are replaced by alternatives. We set the initial alternative ratios at 2015 ($\alpha_{s,initial}$, $\alpha_{o,initial}$). Specifically, we acknowledge that single-use plastics are easier to be replaced, because they are more widely used; they are more easily manufactured.

$$0 \le \alpha_s \le \alpha_o \le 1 \tag{14}$$

Thus we are able to calculate the annual production and waste of plastic alternatives.

$$AP_a(t) = AP(t) \times \beta_1 \times \alpha_s + AP(t) \times \left(1 - \beta_1\right) \times \alpha_o \tag{15}$$

$$AW_a(t) = \sum_{j=1}^{t-2015} AP(t-1)\beta_1\alpha_s D_1(j) + \sum_{i=2}^{8}\sum_{j=1}^{t-2015} AP(t-j)\left(1-\beta_1\right)\alpha_o D_i(j) \tag{16}$$

3.2 Policy Intervention

Policies have been implemented across nations to alleviate the plastic waste crisis. Notably, the number of policies regulating single-use or disposable plastics at the national level increased steeply after 2015 [7].

We assume that annual demand of products (that were previously made from plastic) remains its current growth trend. Since policy intervention decreases plastic products, it results directly to the increase of amount of plastics replaced by alternatives. In other words, α_s and α_o are positively related to p(t). We devise this relationship as follows.

$$p(t) = e^{\alpha_p \times (t-2016)} \tag{17}$$

a_p is a constant that controls the growth rate of policies' strength.

$$\begin{aligned} \alpha_s &= min(\alpha_{s,initial} \times p(t), 0.8) \\ \alpha_o &= min(\alpha_{o,initial} \times p(t), 0.5) \end{aligned} \tag{18}$$

3.3 Citizens' Eco-Awareness

We imitate the gradual process of citizens gaining environmental awareness with c(t). And in this model, we focus on the impact of citizens' awareness on the recycling process. It functions similarly with the policies' impact on the alternatives' ratio. Specifically, r_a and r_o are positively related to c(t). We devise c(t) and its relationship with r as follows.

$$c(t) = e^{\alpha_c \times (t-2016)} \tag{19}$$

a_c is a constant that controls the growth rate of citizens' eco-awareness.

$$\begin{aligned} r_\alpha(t) &= min(r_{\alpha,initial}(t) \times c(t), 1) \\ r_o(t) &= min(r_{o,initial} \times c(t), 0.5) \end{aligned} \tag{20}$$

3.4 Harm Recovery

We define this process as "Harm Recovery", and we use $\eta(t)$ to denote the annual removal of accumulated plastic wastes.

$$H(t) = H(t)_{original} - \eta(t) \tag{21}$$

$$\eta(t) = \eta_{ocean}(t) + \eta_{land}(t) \tag{22}$$

We separate the annual removal into removal from ocean and from land. For the ocean removal, in the past, manta trawlers, a hybrid between a fish trawler and a plankton tow, were used for plastic collection from oceans [8]. Currently, an estimaed.

7.25 million t of plastic debris could be collected and removed from the ocean [9]. We assume the growth rate of removal to be linear and their equations. B_{ocean}, b_{land} in Eq. 23 and 24 are constants that describe the exact growth rate of annual removal [10].

$$\eta_{ocean}(t) = b_{ocean} \times t + 7.25 \times 10^6 \tag{23}$$

$$\eta_{land}(t) = b_{land} \times t \tag{24}$$

3.5 Implementation of IPWH Model

$AW_S(t)$ and H(t) will continue to grow exceedingly fast and will not converge if no measure is adopted.

Thus we set 2 different targets for our IPWH model.

- Target 1: $AW_S(t)$ starts to decrease.
- Target 2: H(t) starts to decrease.

Our calculation with IPWH shows that Target 1 can be achieved by year 2125, and Target 2 by 2131 with the following parameter values. Specifically, as shown in Figs. 2 and 3. The settings of the relevant parameters are shown in Table 3.

Table 3. Some parameter settings

Parameter	$\alpha_{s,initial}$	$\alpha_{o,initial}$	α_p	α_c	b_{ocean}	b_{land}
Value	0.1	0.05	0.014	0.007	$2.5*10^5$	2.5*105

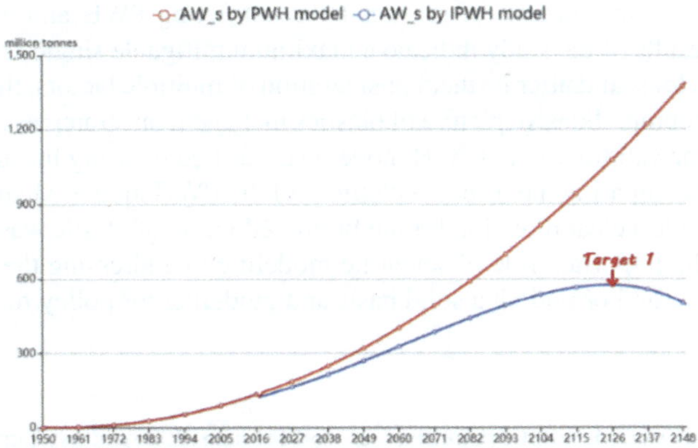

Fig. 2. Comparison of $AW_S(t)$ by PWH and IPWH model

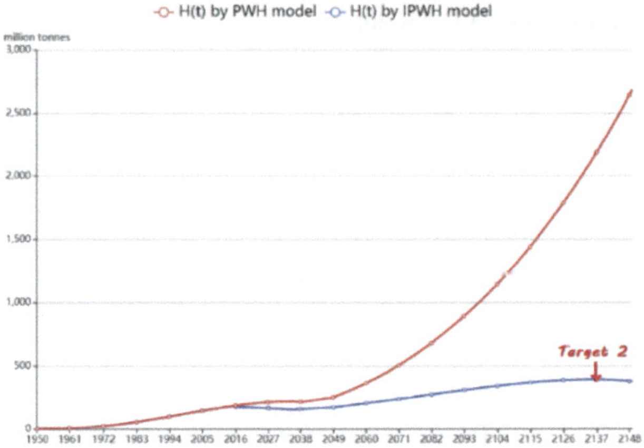

Fig. 3. Comparison of H(t) by PWH and IPWH model

4 Conclusion

This study provides an effective method to analyze the impact relationship between production, waste and hazards of plastics by constructing PWH and extended IPWH models. The results of the study indicate a maximum mitigable single-use plastic waste of 3.43 million tons, and after further consideration of multiple factors, the IPWH model reveals a more comprehensive picture of plastics management strategies and their potential impacts. The validity of the IPWH model was verified by using linear regression for prediction and comparing the results with those of the PWH model. Meanwhile, the two targets set provide a clear direction for the future reduction of plastic waste and hazards, emphasizing the important role of scientific modeling in addressing the environmental crisis of plastics, and providing a solid basis and guideline for policy formulation.

References

1. Ritchie, H., Roser, M.: Plastic pollution. our world in data (2020). https://ourworldindata.org/plastic-pollution
2. Bai, M., Zhu, L., An, L., Peng, G., Li, D.: Estimation and prediction of plastic waste annual input into the sea from china (2018)
3. for Disease Control and Prevention. National report on human exposure to environmental chemicals (2018). https://www.cdc.gov/exposurereport/index.html
4. Wikipedia. plastic pollution. https://en.wikipedia.org/wiki/Plastic_pollution
5. Geyer, R., Jambeck, J.R., Law, K.L.: Production, use, and fate of all plastics ever made (2017)
6. Trimarchi, M., Kiger, P.J., Giuggio, V.M.: Top 10 eco-friendly substitutes for plastic (2019). https://science.howstuffworks.com/environmental/green-tech/sustainable/5-plastic-substitutes3.htm
7. UNEP, SINGLE-USE PLASTICS: A Roadmap for Sustainability (2018)
8. Li, W.C., Tse, H.F., Fok, L.: Plastic waste in the marine environment: a review of sources, occurrence and effects (2016)
9. Song, Y.K., et al.: Large accumulation of micro-sized synthetic polymer particles in the sea surface microlayer (2014)
10. D. E. N. A. Service. Plastic waste: ecological and human health impacts (2011)

Advantages and disadvantages: Rapid reaction, can quickly generate a large number of $CH_3{}^{131}I$ gas into the ventilation system, but dimethyl sulfate is a highly toxic chemical, with strong corrosion, $Na^{131}I$ as a radioactive chemical, radiation risk to people, high management costs.

3.2 Freon/Sample Cup Analysis

Principle: Using freon as the test medium, mechanical leakage detection of iodine adsorbers is carried out, and activated carbon samples are sent to the laboratory for analysis after the test.

Advantages and disadvantages: Freon as a medium avoids the use of radioactive ^{131}I nuclides and highly toxic chemicals in field work, reducing the risk of toxic radiation to personnel, but the use of freon is limited by environmental protection, and the laboratory analysis of activated carbon samples is prone to radiation risks caused by radioactive residue of samples.

3.3 Chlorotrimethylsilane Test Method

Principle: Using chlorotrimethylsilane/sodium iodide as de-alkylated tenth stage and trimethyl phosphonoacetate reaction in acetonitrile solution to prepare airborne radioactive methyl iodine.

Advantages and disadvantages: the use of chlorotrimethylsilane instead of dimethyl sulfate as methylation reagent reduces the toxicity to personnel, but its reaction byproducts are easy to clog equipment and difficult to promote.

3.4 Isotope Replacement Method

Principle: The non-radioactive methyl iodine alcohol solution is mixed with the radioactive sodium iodide solution, and configured into the generating liquid, and the radioactive $CH_3{}^{131}I$ tracer gas is obtained by iodine isotope replacement.

Advantages and disadvantages: The same advantages with the chlorotrimethylsilane test method, and there is no risk of clogging the equipment, but there is still a radiation risk.

3.5 Cyclohexane Test Method

Principle: Non-toxic and non-radioactive cyclohexane was used as the tracer gas to carry out mechanical leakage test on the iodine adsorbent, and the iodine adsorbent was replaced with the activated carbon sample cup sampling analysis results.

Advantages and disadvantages: Avoid the use of highly toxic chemicals and radioactive substances on site, but because it can only carry out mechanical leakage work on the iodine adsorber, it cannot directly judge the purification efficiency of the iodine adsorber, so the iodine adsorber must be replaced every year, resulting in a lot of solid waste and replacement costs, only for the main control ventilation system.

4 Introduction to the Principle of Non-methylated Iodine Test

The current iodine adsorber purification efficiency test mainly uses radioactive methyl iodine efficiency test method (isotope replacement method), that is, non-radioactive $CH_3{}^{127}I$ is used to configure isotope replacement liquid with ethanol and water in a certain ratio, and reacts with $Na^{131}I$ to replace radioactive $CH_3{}^{131}I$. The reaction formula is as follows:

$$CH_3^{127}I + Na^{131}I = CH_3^{131}I \uparrow + Na^{127}I$$

During the test, the iodine test generator was used to establish a negative pressure environment, so that the configured radioactive source containing radioactive $CH_3{}^{131}I$ was injected into the system, and the upstream and downstream samples were respectively collected in the system, and the purification efficiency of the system for radioactive iodine was finally calculated.

The principle of the non-methylated iodine test method is to replace the traditional radioactive $CH_3{}^{131}I$ tracer with a non-radioactive $CH_3{}^{127}I$ gas tracer for the efficiency test of the iodine adsorber. $CH_3{}^{127}I$ and $CH_3{}^{131}I$ are both methyl iodine molecules, which have the same physical and chemical properties when ^{131}I has not decayed, and have similar adsorption mechanism when absorbed by activated carbon in theory, mainly including physical adsorption, chemical adsorption, etc. However, since $CH_3{}^{127}I$ and $CH_3{}^{131}I$ are still molecules composed of different I nuclides, $CH_3{}^{127}I$ and $CH_3{}^{131}I$ have similar adsorption mechanism. The nuclear grade iodine adsorber is also impregnated with potassium iodide ($K^{127}I$) to further improve the adsorption and purification effect of airborne radioactive iodine. When $CH_3{}^{131}I$ is used as the test tracer, the following substitution reactions still exist:

$$CH_3^{131}I(g) + K^{127}I(surface) \rightarrow CH_3^{127}I(g) + K^{131}(surface)$$

However, since the process required by displacement adsorption is more complex than that of chemical complex adsorption, it is generally believed that the proportion of this adsorption effect is small. Because non-radioactive ^{127}I nuclide is used in the non-radioactive test, there is no such nuclide replacement reaction, which is ultimately reflected in the test results that the adsorption performance of the iodine adsorber for $CH_3{}^{127}I$ is slightly weaker than that of $CH_3{}^{131}I$, and the theoretical test efficiency coefficient is also lower than that of the radioactive method.

5 Introduction of Non-methylated Iodine Tester by Photoelectric Method

5.1 Principle of Photoelectric Signal Conversion

The concentration variation of non-methylated iodine tracer gas after passing through the iodine adsorber is very low, generally only nmol/mol level content, only accurate trace methyliodine measurement technology can achieve the measurement of the tracer concentration downstream of the iodine adsorber, and then evaluate the purification capacity of the iodine adsorber.

Application of Photoelectric Method in Purifying Efficiency Test of Iodine Adsorber in Nuclear Power Plant

Yao Tian[1], Weiye Yin[2(✉)], Shuai Zhou[1], Zhaojie Li[2], Sijun Fan[2], Shuai Yang[2], Chen Wang[1], and Tingyuan He[1]

[1] Yangjiang Nuclear Power Co., Ltd., Yangjiang, Guangdong, China
[2] Suzhou Nuclear Power Research Institute Co., Ltd., Suzhou, Jiangsu, China
ywyhit2017@163.com

Abstract. Based on the theory that non-methylated iodine replaces traditional radioactive methyliodine in the efficiency test of iodine adsorber in nuclear power plant, a non-methylated iodine tester is developed with high-precision photoelectric sensor as the main component. By comparing with the traditional radioactive test method, the advantages of the non-radioactive test method were verified, and combined with the comparison data between the field work and the traditional radioactive methyl iodine test method, the improvement direction was provided for the new efficiency test of iodine adsorbers in nuclear power plants.

Keywords: Methyl iodine · Non-discharge test method · Photoelectric method · on-line

1 Introduction

When a nuclear hazard event occurs in a nuclear power plant, the high efficiency particle filter and iodine adsorbers [1– 3] in the ventilation system can respectively quickly purify various radioactive aerosols and radioactive iodine-containing molecules released by radionuclides in the nuclear power plant. Since radioactivity is extremely harmful to human body [4–6], the iodine adsorbers play an irreplaceable role in the ventilation system. For this reason, nuclear power plants need to conduct annual efficiency tests on iodine adsorbers. The current method is the radioactive methyl iodine efficiency test method. This method can directly evaluate the efficiency of iodine adsorbers from the perspective of iodine adsorber design requirements, but it also has many shortcomings, such as radioactive hazards, inability to on-line, complicated processes, and large demand for test personnel. Therefore, the non-methylated iodine test method using the photoelectric method as a test means has provided an improved direction for the efficiency test of iodine adsorbers in the future after the theoretical basis research [7].

D. Han et al. (Eds.): ICEPG 2024, ESE, pp. 1096–1105, 2025.
https://doi.org/10.1007/978-3-031-86829-0_119

2 Photoelectric Sensor Measurement Principle and Other Field Applications

2.1 A Subsection Sample

Photoelectric technology in gas measurement technology has more than 40 years of development history, and the performance of continuous improvement and perfection, has been widely used in chemistry, biology, medicine, environmental protection and other technical science and technology fields. The photoelectric method consists of a high frequency, high energy deep ultraviolet lamp, an ionized gas chamber, an I/V conversion circuit and a back-end filter circuit. The photoelectric method adopts the principle of photoionization of organic gas to detect the concentration of the measured gas [8]. The ionization potential (IP) required for the ionization of different substances is different. If the ionization potential of the substance to be measured is lower than the energy provided by the external environment, it is considered that the substance can be ionized under this energy. When there are components of the measured gas in the ionization chamber, the high-energy ultraviolet lamp will ionize the measured gas into charged particles, charged particles with charge density is proportional to the measured gas concentration, you can measure the corresponding gas concentration, the sensor is non-destructive to the detected substance, so it can be used with other detectors to expand the analysis. Similarly, the range of energy also makes the photoelectric method non-selective, and the detected charge flow is the sum of the charge flow of all ionizable gas molecules. The photoelectric method has the advantages of high measurement accuracy (μmol/mol level), miniaturization of equipment and instruments, convenient and fast operation, real-time online detection, etc., which is very in line with the principles of safety and environmental protection, economic and efficient, convenient and fast experimental scheme.

3 Introduction to the Purification Efficiency Test of Existing Iodine Adsorbers in Nuclear Power Plants

In the process of nuclear power generation, fission will produce a radionuclide ^{131}I of iodine, which will accumulate in the thyroid gland after ingestion, causing harm to the human body. The current iodine adsorbers used in the ventilation system of nuclear power plants are mainly XZ-1200 folding iodine adsorbers [9] (type I iodine adsorbers), whose main filling material is granular activated carbon, which has good adsorption performance for methyl iodine gas.

There are five kinds of iodine adsorber test methods in China, which are dimethyl sulfate method, Freon + sample cup analysis, trimethylchlorosilane method, isotope replacement method and cyclohexane method.

3.1 Dimethyl Sulfate Test Method Subsection Sample

Principle: Using dimethyl sulfate as methylation reagent to react radioactive $Na^{131}I$ solution to produce radioactive $CH_3{}^{131}I$, the efficiency test of iodine adsorber was carried out.

In the non-methyl-iodine tester, the photoelectric sensor is the main component for specific analysis of the methyl iodine concentration in the sampled gas. The working principle is to use the photoelectric sensor for concentration-voltage signal conversion. The measurement principle of the photoelectric sensor is as follows: The sensor uses ultraviolet light to ionize the tracer gas, collect positive and negative charges on the collector plate, and the amount of charge generated is proportional to the concentration of the gas. The calibration curve is obtained through the standard concentration gas, and the tracer concentration can be calculated according to the measured charge size.

According to the relevant standards [10–12], the National Metrology Institute calibrates the photoelectric sensor of the non-methylated iodine tester every year.

5.2 Non-methyl-Iodine Tester Sampling Loop

The sampling loop protects sampling metal hoses, sensor mounting boxes, sensor flowmeters and regulating valves, main flow regulating valves, main sampling flowmeters, sampling thermometers, sampling pressure gauges, vacuum injection pumps, compressed air pressure regulating valves, etc.

The sensor mounting box is the core component of measurement, in which the photoelectric sensor and circuit board are arranged (see Fig. 1). The large space sensor mounting box makes the sensor measurement end face and the surrounding environment are in the sample gas, avoiding the measurement inaccuracy caused by the leakage of the sensor measurement section. The sensor mounting box is equipped with pressure and temperature measuring probes to monitor changes in sampling pressure and temperature in real time.

Fig. 1. Sensor installation box diagram

The main flow control valve and the sensor flow control valve cooperate with each other to make the sensor flow reach the required 200–300 ml/min, and the main sampling flow is greater than 3 m^3/h. The sampling power comes from the negative pressure generated by the vacuum jet pump. The vacuum injection pump uses compressed air to work, and the compressed air regulating valve is used to change the intake pressure and flow of the injection pump.

5.3 Power Supply Circuit

The circuit includes lithium battery, adjustable resistor, voltage regulator capacitor, selective switch, etc. The lithium battery is the voltage source that powers the sensor. When the third gear selection switch is reversed to side A (the two sides are distinguished by A and B), the sensor is powered, and the sensor power supply indicator lights up. After the power supply voltage is divided by the adjustable resistance, it can meet the corresponding voltage requirements of the sensor, and the voltage regulator can avoid voltage fluctuations caused by startup and compliance fluctuations. When the three-speed selection switch stops in the middle position, the lithium battery and the voltage regulator capacitor stop supplying power to the sensor, and the sensor power supply indicator is turned off. When the third gear selection switch is reversed to the B side, connect the external charging power supply, the charging indicator lights up, and charge the lithium battery.

5.4 Data Acquisition and Analysis System

The voltage signal of the test sensor is transmitted to the computer analysis software through the 24-bit isolated USB high-speed data acquisition card. The acquisition card collects the voltage data of the upstream and downstream channels of the iodine adsorber, and the data is processed and calculated by the special software. The special software can realize the measurement channel voltage correction, data filtering, data storage, upstream voltage measurement U, downstream voltage measurement u, upstream methyl iodine concentration measurement C, downstream methyl iodine concentration measurement c. According to the above data, the purification efficiency and purification coefficient are calculated and presented in the data recording interface in real time.

5.5 Integration

The integrated non-methylated iodine test system includes two upstream and downstream testers, a data collector and a computer, as well as supporting analysis software. The tester panel mainly contains four data display modules: digital thermometer, digital barometer, sensor digital flowmeter and total flowmeter.

6 Performance Test Results and Field Test Plan

6.1 Performance Test Results

Through laboratory testing and analysis and National Metrology Institute's calibration verification of non-methyl-iodine test equipment, calibration data are obtained as shown in Table 1 below.

Table 1. Table of calibration data

Item	Technical index	Low concentration nmol/mol level	High concentration μmol/mol level
Quantitative repeatability	≤2.0%	0.2%	0.5%
Stability	≤3.0%	1.6%	2.4%
linearity	R ≥ 0.99	0.9995	0.9999
Relative deviation of integral concentration	/	−1.5%	0.7%
Maximum indication deviation	/	0.6 nmol/mol	0.07 μmol/mol
Instrument detection limit	/	0.14 nmol/mol	/

This can be seen from the data in Table 1 above. The upstream and downstream photoelectric sensors have high confidence and excellent linear data presentation at both low and high concentrations. The downstream detection limit is only 0.14 nmol/mol. Considering the influence of relative pressure P and absolute temperature T on the measurement results, the requirements of downstream methyl iodine concentration detection in the purification efficiency test of iodine adsorber in the field ventilation system are met.

6.2 Field Test Plan

The non-discharge test method is mainly composed of three parts: gas cylinder, upstream sampling tester and downstream sampling tester.

After the fan in the on-site ventilation system is started, power on the upper and downstream sampling tester and connect it with the sampling port for the purification efficiency test of the upstream and downstream iodine adsorbent of the ventilation system. Adjust the main flow rate of the tester and the flow rate of the sensor to the rated value by using the onsite compressed air.

Connect the upper and lower test instrument with the USB acquisition card through the data cable, and use the computer side for data acquisition.

After the upstream and downstream voltage stabilize, the theoretical injection flow calculated according to the system air volume and methyl iodine gas concentration of gas cylinder before the test, control the opening of the pressure reducing valve, and inject gaseous methyl iodine into the system (after the injection personnel wait for the instruction to stop injection, slowly close the pressure reducing valve and restore the original working condition on site).

Open the non-methylated iodine test software, conduct online data collection, and record the upstream and downstream relative pressure and standard temperature data.

After the test is over, recover the upstream and downstream tester, pipe fittings, data lines, etc., restore the pre-test conditions, and finish the work.

After returning to the laboratory, the test data are integrated and processed. The calculation method of time integral accumulation in non-radioactive method is theoretically the same as the method of cumulative sampling with activated carbon box in traditional radioactive method technical scheme, that is, the total amount of time accumulation. During the sampling time, the upstream and downstream tracer concentration is integrated and calculated according to the system air volume, and the total amount of tracer delivery and the total downstream leakage after passing through the iodine adsorbent can be shown, so as to calculate the purification coefficient E or purification efficiency η of the iodine adsorbent. E and η are calculated as follows:

$$E = \frac{\int CQdH}{\int cQdh} \tag{1}$$

$$\eta = \eta(1 - 1/E) * 100\% \tag{2}$$

Note: C and c are the methyl iodine concentration after upstream and downstream photoelectric data conversion, Q is the flow rate, H and h are the upstream and downstream sampling time.

6.3 Field Test Comparison Results

In order to verify the consistency of evaluation between non-methylated iodine method and traditional radioactive test method, nine groups of comparative tests were conducted, and the test data were shown in Table 2 below.

Table 2. Table of the comparative test data

Test 1			
Quantitative repeatability	Purification coefficient	Purification efficiency	Standard: 99%
R	760	99.868%	qualified
NR	602	99.833%	qualified
Test 2			
Quantitative repeatability	Purification coefficient	Purification efficiency	Standard: 99.7%
R	816	99.877%	qualified
NR	524	99.809%	qualified
Test 3			
Quantitative repeatability	Purification coefficient	Purification efficiency	Standard: 99.7%
R	648	99.846%	qualified
NR	627	99.841%	qualified
Test 4			
Quantitative repeatability	Purification coefficient	Purification efficiency	Standard: 99.7%
R	731	99.863%	qualified
NR	594	99.832%	qualified
Test 5			
Quantitative repeatability	Purification coefficient	Purification efficiency	Standard: 99.7%
R	703	99.858%	qualified
NR	438	99.772%	qualified
Test 6			
Quantitative repeatability	Purification coefficient	Purification efficiency	Standard: 99%
R	352	99.716%	qualified
NR	347	99.712%	qualified
Test 7			
Quantitative repeatability	Purification coefficient	Purification efficiency	Standard: 99%
R	571	99.825%	qualified
NR	322	99.689%	qualified
Test 8			
Quantitative repeatability	Purification coefficient	Purification efficiency	Standard: 99%
R	1889	99.947%	qualified
NR	1781	99.944%	qualified
Test 9			
Quantitative repeatability	Purification coefficient	Purification efficiency	Standard: 99%
R	380	99.737%	qualified
NR	355	99.718%	qualified

As can be seen from the data in Table 2 above, the results of the non-radioactive method are smaller than those of the radioactive method, which accords with the theoretical analysis of conservative decision making. Moreover, the results of judging whether the iodine adsorber is qualified by the two methods are consistent, and there is no case of premature replacement due to conservative test data.

7 Conclusion

The non-radiomethyl-iodine tester uses photoelectric sensor as the core component, and integrates the theory of on-line and non-radiomethyl-iodine test. The non-methylated iodine tester not only has the advantages of online testing, but also makes the test equipment convenient, intuitive and digital, essentially eliminating the potential radiation risk of the iodine adsorber efficiency test, and can greatly improve the efficiency of the iodine adsorber efficiency test in nuclear power plants.

As the photoelectric sensor in the non-methylated iodine tester is an important data acquisition device for the final test results, the measurement accuracy of the tester will also improve the accuracy of the test results with the progress and upgrading of the photoelectric sensor technology, and its application in the purification efficiency test of iodine adsorbers in nuclear power plants has a long-term development prospect.

References

1. Shilian, et al.: Analysis of the dispersion timeline and isotope activity ratio characterization of airborne radionuclides released from the Fukushima Daiichi nuclear power plant accident. J. ACS Earth Space Chem. **5**(10), 2938–2946 (2021)
2. Yumiko, Y., et al.: Risk perception regarding implementation of iodine thyroid blocking during a nuclear disaster of mothers living near a nuclear power station in Japan. Endocr. J. **68**(5), 553–560 (2021)
3. Eriko, S., Christopher, E., Polivka, B.J.: Lessons from Fukushima: potassium iodide after a nuclear disaster. A. J. Nurs. **121**(2), 63–67 (2021)
4. NB/T 20039.11–2011, Specification for nuclear air and gas handling - Ventilation, air conditioning and air purification - Part 11: Iodine sorbers (Type I). Chinese standard
5. NB/T 20039.12–2011, Specification for nuclear air and gas handling - Ventilation, air conditioning and air purification - Part 12: Iodine sorbers (Type II). Chinese standard
6. NB/T 20039.13–2012, Specification for nuclear air and gas handling - Ventilation, air conditioning and air purification - Part 13: Iodine sorbers (Type III). Chinese standard
7. Gaoxin, Z., Sijun, F., Jiyi, S., Wei, C., Weiye, Y.: Effect of CH3127I tracer Method on purification efficiency of iodine adsorber in nuclear power plant. Nucl. Electr. Detect. Technol. **44**(04), 738–743 (2024)
8. Jinren, W., Zhaoli, D.: Application of photoelectric sensor in gas concentration measurement. J. Integr. Circuit Appl. **39**(03), 130–131 (2022)
9. Liu, Q., et al.: Development of XZ-1200 folding iodine adsorber. Radiat. Prot. (03), 39–45 (1998)
10. JJG 1055–2009, on-line gas chromatograph. Chinese standard
11. JJG 700–2016, Gas chromatograph. Chinese standard
12. DIN EN ISO 6145–7–2011, Gas analysis. The calibration gas mixture was prepared by dynamic capacity method. Part 7: Heat and mass flow controllers (ISO 6145–7–2009). German standard

Characteristics of Spatial and Temporal Changes in Phytoplankton Community Structure and Environmental Drivers in the Yalong River Basin

Long Shi[1], Dili Li[2], Long Yan[2(✉)], and Nan Wang[1]

[1] College of Water Resources and Environmental Engineering, Changchun Institute of Technology, Changchun 130000, China

[2] State Key Laboratory of Water Cycle Simulation and Regulation, China Institute of Water Resources and Hydropower Research, Beijing 100038, China

{lidl,yanlong}@iwhr.com

Abstract. In order to understand the current status of the phytoplankton in the Yalong River after years of operation of the terraced hydropower stations, surveys of phytoplankton and environmental factors were carried out at 29 sections of the Yalong River Basin in May (spring) and November (autumn) in 2023, to analyse the characteristics of the changes in the structure of the phytoplankton community and the corresponding relationship with the environmental factors. A total of 132 species of phytoplankton were collected, belonging to 6 phylums, 7 classes, 17 orders, 28 families and 42 genera, mainly composed of diatoms with relative abundance of 81%, and the dominant species group included 8 species of phytoplankton such as Diatoma. Tenue, Achnanthes. sp, etc., of which the number of species in the spring was obviously higher than that in the autumn; 34 species of zooplankton were identified, belonging to 3 phylums, 4 classes, 10 orders, 20 families and 28 genera, mainly composed of Diatoma. Tenue, Achnanthes. sp and other species. 28 genera, mainly composed of Protozoa phylum with a relative abundance of 62%, and the dominant species group included 4 species of zooplankton such as Difflugia avellana, Difflugia globulosa, etc.; there were seasonal and spatial differences in the dominant species of phytoplankton and their biomass and diversity indices. Phytoplankton biomass was much higher in spring than in autumn, while zooplankton was much higher in autumn than in spring; the overall upstream phytoplankton biomass was higher than that in the middle and lower reaches. From the results of the diversity index evaluation, it was concluded that the water quality of Yalong River in spring was better than that in autumn, and the water quality of the upper reaches was better than that of the middle and lower reaches; the results of the redundancy analysis (RDA) showed that the dominant species of phytoplankton were mainly closely related to NH_3-N, NO_3^--N, WT, and Chla; and that the dominant species of zooplankton were mainly closely related to TN and DTN.

Keywords: Zooplankton · Spatial and temporal variability · Environmental factors · Community structure

D. Han et al. (Eds.): ICEPG 2024, ESE, pp. 1106–1117, 2025.
https://doi.org/10.1007/978-3-031-86829-0_120

1 Introduction

Phytoplankton is one of the main biological groups in the aquatic environment, mainly including phytoplankton and zooplankton, which plays a decisive role in the material cycle and energy flow in aquatic ecosystems [1, 2]. Phytoplankton is the primary producer in the aquatic environment, and its community structure is directly or indirectly affected by environmental factors such as physical, chemical and biological factors [3], and at the same time, its species composition and distribution can objectively reflect the change law of the aquatic environment, which is an important indicator, therefore, phytoplankton is often used as an important indicator to evaluate the quality of water bodies in many studies [4]. Zooplankton, as primary consumers of aquatic ecosystems, play an important role in the food chain and are useful indicators of predation pressure by filter-feeding fish [5]. The adaptive capacity and sensitivity of zooplankton to environmental changes are related to their species [6], e.g., the eco-dynamic processes of zooplankton at low trophic levels are very sensitive to environmental changes [7]. Currently, scholars have been investigating the relationship between phytoplankton community structure and environmental factors in different waters through canonical correspondence analysis (CCA) [8–10], redundancy analysis (RDA) [11], cluster analysis [12, 13], Pearson's correlation analysis [11–13], stepwise regression analysis [14], analysis of similarity (ANOSIM) [15], and nonmetric multidimensional scaling analysis (NMDS) [16], and so on. The relationship between the community structure of plants and animals and environmental factors in different waters was investigated.

The Yalong River is the largest first-class tributary of the Jinsha River and one of the eight major tributaries of the Yangtze River. At present, the number of hydropower stations planned and constructed in Yalong River reaches 21, of which, 10 terrace power stations in the upper reaches, 6 terrace power stations in the middle reaches and 5 terrace power stations in the lower reaches. Habitat perturbation and alteration caused by construction and operation of hydropower stations, as well as unfavourable factors such as obstruction and fragmentation of terrace dams, have obvious impacts on the population composition and number of zooplankton plants and animals [17–19]. In addition, reservoir operation brings about changes in the physicochemical properties of water bodies such as altered water transparency, gas supersaturation, reduced dissolved oxygen and water temperature stratification, which also have an impact on the reproduction of zooplankton and flora [19]. Most of the studies on the flora and fauna of the Yalong River have been carried out upstream and downstream of a single hydropower station or on a typical section of the middle and lower reaches [10, 12–15], and there has not yet been a survey and analysis of flora and fauna in the whole basin. Moreover, the built and under-construction step power stations are distributed in the middle and lower reaches, and the upper Yalong River is in a state less disturbed by human beings, but in the future, with the continuous development and construction of hydropower stations, the aquatic ecological environment, the structure of the phytoplankton community, and the spatial and temporal distribution of the upper Yalong River will undergo a drastic change [16]. Therefore, it is necessary to carry out a basin-wide sampling survey of zooplankton to analyse the physicochemical characteristics of the water body, species composition, diversity index, and the relationship between community structure and environmental factors [17–19]. In order to provide basic information for improving the effectiveness and precision of aquatic ecological protection measures.

2 Study Area

Yalong River originates from the Bayan Har Mountain in the eastern part of the Tibetan Plateau, geographic location between 96°52′~102°48′ longitude, latitude 26°32′~33°58′ between the trunk of the total length of 1535 km, the natural drop of 3830m, the annual runoff of 60.9 billion m^3. As shown in Fig. 1. The river section upstream of the Lianghekou Hydropower Station is considered to be the upper reaches Yalong River across the dimension range is very large, and the upstream and downstream elevation has a huge difference, so that the Yalong River upstream and downstream of the climatic characteristics of the difference is significant, the northern part of the basin is dry and cold continental climate, the central and southern part of the subtropical climate of dry and wet distinct.

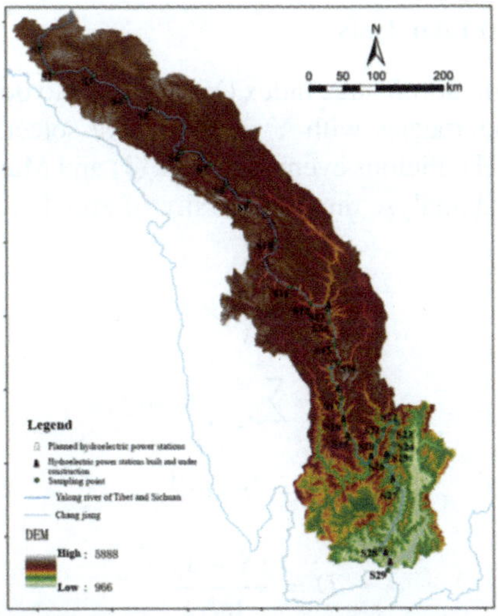

Fig. 1. Schematic diagram of Yalong River region

3 Research Methods

3.1 Sample Collection and Processing

In this study, two sampling surveys were conducted in the Yalong River in May 2023 (spring) and November 2023 (autumn), with 29 sampling points evenly distributed in the upper, middle and lower reaches. The 25# (64 μm) phytoplankton nets and 13# (112 μm) zooplankton nets were moistened and washed with river water, and the sampling was chosen to be conducted at the river flow location near the centre of the river. A total of 50 L of water samples were collected for zooplankton and 10 L for phytoplankton, which were filtered into 100 ml sampling bottles with 25 ml of formaldehyde and 25 ml of Rugo's reagent respectively and transported to the laboratory as soon as possible, siphoned off for 48 h, and then concentrated to 30 mL for a further 24 h. Before microscopic counting, the water samples were shaken thoroughly with a rapid mixer, and a rapid aspirate of 0.1 was placed in the 0.1 mL. Plankton counting frame, in the microscope (Leica DM 5000B) 10 × 40 times under the identification and counting, each sample counting 2 times, take its average, each counting results and the average value of the difference should be within 15%, otherwise increase the number of counting.

To determine the physicochemical properties of the water body at each sampling point, water temperature (WT), pH, dissolved oxygen (DO) and conductivity (PC) were measured on site by YSI water quality analyser; another 1L water sample was collected and sent to the laboratory to test the water samples for total nitrogen (TN), chlorophyll a (Chl.a), total carbon (TC), total dissolved nitrogen (DTN), ammonia nitrogen (NH_3-N), nitrate nitrogen (NO_3-N) and dissolved total carbon (DTC).

3.2 Data Processing and Analysis

In this paper, McNaughton dominance index (Y) was used to determine the composition of dominant species, and species with Y > 0.02 were selected as dominant species. Shannon-Wiener index (H), Pielous evenness index (J) and Margalef richness index (D) were used to evaluate and analyse the biodiversity of zooplankton. The formulae were calculated as follows:

$$\mathrm{Y} = \frac{n_i}{N} \times f_i \tag{1}$$

$$H' = -\sum\nolimits_{i=1}^{S} p_i \ln p_i \tag{2}$$

$$J' = H'/lnS \tag{3}$$

$$D = \frac{(S-1)}{\ln N} \tag{4}$$

Where: n_i is the total biomass of species i; N is the total biomass of all species; f_i is the frequency of occurrence of species i; p_i is the relative density of species i at the sampling site; S is the number of total taxonomic units of the species at the sampling site; and N is the number of individuals at the sampling site

4 Results

4.1 Physical and Chemical Characteristics of Water Bodies

The physicochemical characteristics of water bodies in different seasons are shown in Table 1, except for NH_3-N and pH, other environmental factors showed significant seasonal differences, from the perspective of time, the concentrations of DO, PC, TN, and

NO_3^--N were significantly higher than those in spring in autumn, and the concentrations of WT, DTC, and TC were significantly higher than those in spring. From a spatial point of view, the sampling points upstream of the Lianghekou Hydropower Station show obvious differences from those in the middle and lower reaches, which may be related to the regulation of the reservoir and the dam's barrier effect.

Table 1. Spatial and temporal changes of environmental factors in Yalong River

	May 2023 (spring)		November 2023 (autumn)	
	upper reaches	middle and lower reaches	upper reaches	middle and lower reaches
WT/°C	7.66 ± 2.7	14.81 ± 2.92	0.79 ± 0.54	11.78 ± 3.49
pH	8.81 ± 0.11	8.74 ± 0.09	8.41 ± 0.1	8.31 ± 0.08
DO/mg·L^{-1}	6.21 ± 0.39	6.92 ± 0.27	9.02 ± 0.63	8.84 ± 1.06
NH_3-N/mg·L^{-1}	0.05 ± 0.01	0.07 ± 0.04	0.14 ± 0.03	0.09 ± 0.01
PC/us·cm^{-1}	475.35 ± 62.45	310.47 ± 37.43	576.01 ± 94.07	301.26 ± 65.27
TN/mg·L^{-1}	0.74 ± 0.13	0.64 ± 0.1	1.33 ± 0.29	0.72 ± 0.14
DTN/mg·L^{-1}	0.7 ± 0.13	0.49 ± 0.09	0.94 ± 0.47	0.38 ± 0.25
NO_3^--N/mg·L^{-1}	0.54 ± 0.14	0.31 ± 0.07	0.93 ± 0.13	0.48 ± 0.11
DTC/mg·L^{-1}	1.44 ± 0.4	4.79 ± 0.81	1.43 ± 0.31	2.105 ± 0.625
TC/mg·L^{-1}	49.5 ± 6.76	36.11 ± 3.22	37.43 ± 2.44	29.52 ± 5.08
Chla/ug·L^{-1}	0.958 ± 0.6	0.758 ± 0.4	0.83 ± 0.23	0.614 ± 0.43

4.2 Species Composition and Community Structure

The species composition of phytoplankton in Yalong River is shown in Fig. 2. 126 species of phytoplankton, belonging to 6 phylums, 7 classes, 17 orders, 28 families and 41 genera were collected in May. Among them, diatoms were absolutely dominant, with 103 species, accounting for 81.7%. 29 species of phytoplankton, belonging to 6 phylums, 7 phylums, 10 orders, 20 families, and 26 genera, were collected in November. Among them, diatoms were absolutely dominant, with 22 species, accounting for 76%.A total of 14 species of zooplankton, belonging to 3 phylums, 5 phylums, 8 orders, 10 families, and 13 genera, were collected in May. Protozoa was absolutely dominant with a total of 8 species, accounting for 57.1%. 32 species of zooplankton were collected in November, belonging to 3 phylums, 4 phyla, 10 orders, 20 families and 26 genera. Protozoa was absolutely dominant with a total of 28 species, accounting for 88%.

A total of 25766×10^4 phytoplankton/L were collected in May from 29 sampling sections of Yalong River, of which a total of 8 species with dominant degree of $Y > 0.02$ in the upstream section belonged to the diatom phylum; a total of 5 species with dominant degree of $Y > 0.02$ in the middle and downstream sections also belonged to the diatom phylum; a total of 5503 × 10^4 phytoplankton/L were collected in November, of which a total of 4 species were dominant in the upstream section and all of them belonged to the phylum Diatom; the total number of dominant species in the middle and lower reaches was 4, all belonged to the phylum Diatom; a total of 673.3 ind./L of zooplankton was collected in May, of which the dominant species in the upstream section and in the middle and lower reaches were 3, all belonged to the phylum Protozoa; in November, a total of 2247.7 ind./L was collected, of which the total number of dominant species in the upstream section and in the middle and lower reaches were 3, all belonged to the phylum Protozoa, and the results are shown in Table 2.

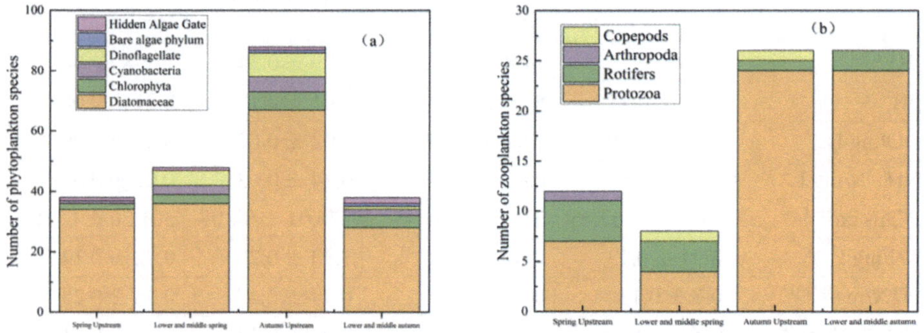

Fig. 2. Species composition of phytoplankton and zooplankton in Yalong River

The trends of the density and biomass of planktonic organisms in the sampling sections were basically consistent, showing an upward trend. The density of planktonic animals as a whole showed a downward trend. The biomass of planktonic plants in spring was much higher than in autumn, while that of planktonic animals in autumn was much higher than in spring, which may be related to the food chain among planktonic organisms [19]. Planktonic animals feed on planktonic plants, and the decrease in the number of planktonic animals in spring increases the chance of survival for planktonic plants. Conversely, as the number of planktonic animals increases in autumn, the number of planktonic plants gradually decreases.

4.3 Analysis of Diversity Indices

By calculating the biodiversity index of phytoplankton at each sampling section, from the temporal analysis, the Margalef richness index and Shannon-Wiener diversity index were significantly higher than that of November in May, and the Pielou evenness index was relatively stable in different periods; from the spatial analysis, the Margalef richness index and Shannon-Wiener diversity index showed an overall decreasing trend from upstream to downstream. Wiener diversity indices showed an overall decreasing trend from upstream to downstream, and all the upstream unbuilt hydropower plant sites were higher than the downstream built sites, which may be due to the influence of downstream hydropower plants, while the upstream is less affected by human activities; Pielou evenness index did not differ significantly in space.

Table 2. Dominant species of Yalong River zooplankton

	Dominant species	May-23		Nov-23	
		upper reaches	middle and lower reaches	upper reaches	middle and lower reaches
Phytoplankton	*Diatoma. Tenue*	0.104	0.022	0.186	0.094
	Achnanthes exigua	0.049	0	0	0.064
	Achnanthes.sp	0.087	0.02	0.032	0
	Navicula.cincta	0.063	0.021	0.178	0
	Navicula.cuspidata	0.047	0	0.102	0.1
	Asterionella formosa Hass	0.038	0.15	0	0
	Fragilaria.crotonensis	0.03	0.056	0	0.128
	Cyclotella.Meneghiniana	0.056	0	0	0
Zooplankton	*Difflugia avellana*	0.535	0.638	0.229	0.252
	Difflugia globulosa	0.139	0.153	0.272	0.242
	Rhagadostoma completum	0.02	0	0	0.196
	Centropyxis aculeata	0	0.02	0.148	0

By calculating the biodiversity index of zooplankton at each sampling section, Margalef richness index, Shannon-Wiener diversity index fluctuation trend is basically the same and the fluctuation is larger; Pielou evenness index in May in the middle of the river S14 section before no obvious difference, since S15 section after the fluctuation is relatively larger, November fluctuation The fluctuation in November was relatively stable. The diversity indices showed a decreasing trend from upstream to downstream.

Table 3. Biodiversity indices of zooplankton in different seasons in Yalong River

Form	Dates	*H*	*J*	*D*	*S*
Phytoplankton	May	2.27 ± 0.65a	0.69 ± 0.18a	3.97 ± 0.73a	24.37 ± 7.14a
	November	1.46 ± 0.47a	0.79 ± 0.25a	1.19 ± 0.46b	7 ± 2.24b
Zooplankton	May	0.95 ± 0.41a	0.83 ± 0.24a	0.67 ± 0.37a	3.13 ± 1.33a
	November	1.81 ± 0.2b	0.86 ± 0.04a	1.73 ± 0.38b	8.48 ± 2.1b

Biodiversity index is one of the indicators for evaluating the quality condition of water bodies, and scholars have already carried out studies on water quality evaluation based on it. In this study, the biodiversity index of phytoplankton community was used to evaluate the water quality of Yalong River, and the larger values of the three items indicated that the higher the diversity and the better the water quality. From the Shannon-Wiener diversity index, the phytoplankton was β-medium polluted water quality in both spring and autumn; the Pielou evenness index showed that the Yalong River was clean-oligopolluted in both, and in autumn was in the excess of the clean type; and the Margalef abundance index found that the Yalong River belonged to the clean type in the spring, but was medium-polluted in the autumn. Overall, the diversity indices of Yalong River in spring and autumn were not significantly different, and both of them were in clean-oligopollution type and approaching to clean type, as shown in Table 3. Among them, the evaluation results of diversity index in spring were better than that in autumn. The large amount of exogenous pollutants such as nutrients and organic matter brought by the rapid development of industry and agriculture in the surrounding areas may be the reason for the deterioration of the water quality of the Yalong River [13, 19].

4.4 Environmental Impact Factor Analysis

The results of Pearson correlation analyses in this study showed that the spring ZA of zooplankton biomass, PA and Chla were all significantly positively correlated; ZH, ZD, and with DTC and were all significantly negatively correlated; PJ, and PD were all positively correlated with WT and DO; PH and ZH were significantly positively correlated with PC and TC; PH was significantly positively correlated with NO_3^--N and TC, Chla; PJ was significantly positively correlated with PC, DTN and NO_3^--N; ZD was significantly positively correlated with TC and Chla; PD was significantly positively correlated with TC. There was no significant correlation between phytoplankton biomass and environmental factors in autumn; PH had a significant positive correlation with PC, NO_3^--N, DTC, and TC, and a significant negative correlation with WT; ZH had a significant positive correlation with PC, TN, NO_3^--N, and TC; PJ had a significant positive correlation with DO, TC, and Chla; ZD had a significant positive correlation with PC and TC, and a significant negative correlation.As shown in Fig. 3a.

In this study, there was no significant correlation between zooplankton and phytoplankton bioindicators, and there was a significant positive correlation between zooplankton biomass and phytoplankton biomass in autumn, indicating that there are seasonal differences in the downward control effect of zooplankton on phytoplankton in the Yalong River, and that the biodiversity of zooplankton in autumn is closely related to phytoplankton, which is probably due to the fact that autumn favours zooplankton and plants, and that the zooplankton's feeding pressure on the phytoplankton. Large number of studies have shown that the concentration of DO and the growth of phytoplankton are significantly and positively correlated. In this study, the environmental factors NO_3^--N, DO, and TN were found to be unfavourable to the growth of zooplankton but favourable to the growth of phytoplankton. The biomass of phytoplankton in spring and autumn is positively correlated with Chla, and it has been shown that the change of the existing phytoplankton stock in the water body can be reflected by the concentration of Chla, and the zooplankton filter feeds on phytoplankton, so it can be deduced that the DTN, and 5 m/s. Pool Type 2 has overflow with inlet velocities of 2 m/s, 4 m/s, and 5 m/s, Pool Type 3 has overflow with inlet velocities of 4 m/s and 5 m/s. Under the same inlet velocity, the overflow rate of pool type 1 is the highest (excluding the case of no overflow), indicating that the lower the height of the overflow pipe, the greater overflow volume under the same inlet velocity. Under the same pool type, the overflow rate increases with the increase of water inlet velocity. As the water inlet velocity increases, the jet pressure of the water will gradually increase, and the jet intensity of the upward swirling water will also increase constantly, causing an increase in overflow pipe pressure and flow, and an increase in overflow rate.

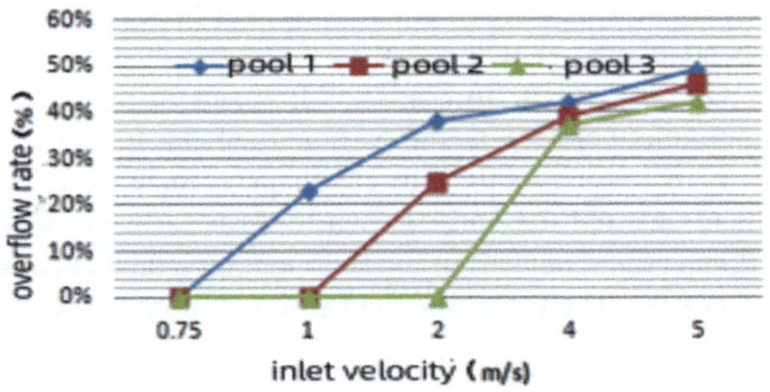

Fig. 4. The overflow rate of different pool type under different inlet velocity

4.3 Comparative Analysis of Particle Removal Rates

Analyze the particle removal rates of different switching well types under different inlet velocities for four different particle diameters. The results are shown in Fig. 5. It can be seen from the figure that the overall particle removal rates of the three pool types are not significantly different. Because the influence range of the axial circulation intensity of the inlet water flow under hydraulic natural conditions is limited.

The removal rate of particles with a diameter of 1 um is very low in all three types of pools, ranging from 8.9% to 10.9%, and the removal rate does not change much with the increase of inlet flow rate. Because the particle diameter and mass of 1 um is too small, it is subjected to very little centrifugal force, and is greatly affected by water flow. Under the action of water flow in all directions, it is easy to follow the flow of water flow and cannot achieve effective separation in various situations.

Overall, the removal rate of basic pool type 3 is relatively high for other diameters, indicating that the height of the overflow pipe has an impact on the internal flow movement of the switching well, which affects the particle removal rate. The lower the height of the overflow pipe, the closer it is to the bottom of sedimentation tank. Under the guidance of the overflow pipe, more particle mixture will flow away with the overflow pipe, causing poor separation of particles and water and a decrease in removal rate. However, the influence of the height of the overflow pipe on the removal rate is also related to the inlet velocity and overflow situation. When the inlet velocity is small (0.75 m/s) with no overflow situation, the removal rates of the three types of pools are the same. At a velocity of 2 m/s, the removal rates of the three types of pools differ the most. However, when the inlet velocity is large (5 m/s), the removal rates of the three types of tanks are not significantly different. Because the inlet velocity is too high, the jet intensity and axial circulation are large, the particles didn't settle in time and flowed out from the outlet, resulting in a relatively small removal rate.

Except for particles with a diameter of 1 um, the removal rate reaches maximum at an inlet velocity of 2 m/s. When the inflow velocity is low, the horizontal and axial circulation intensity formed by the water flow is small, and the centrifugal force on the particles is small, which is not conducive for particles being thrown towards the outer wall and settling, resulting in a lower particle removal rate. When the inlet velocity is too high (4 m/s or 5 m/s), the particle movement path will be significantly shortened, and it will be in a suspended state, which is not easy to settle into the sand hopper. At this time, the particles will not settle in time and will flow out through the outlet or overflow port. Under the same inlet velocity, as the particle diameter increases, the removal rate of particles gradually increases. As the particle diameter increases, the centrifugal force on the particles increases and they are thrown towards the wall of the sedimentation tank. They spiral down along the wall of the sedimentation tank and are separated at the bottom of the sand hopper. The larger the diameter, the more particles sink.

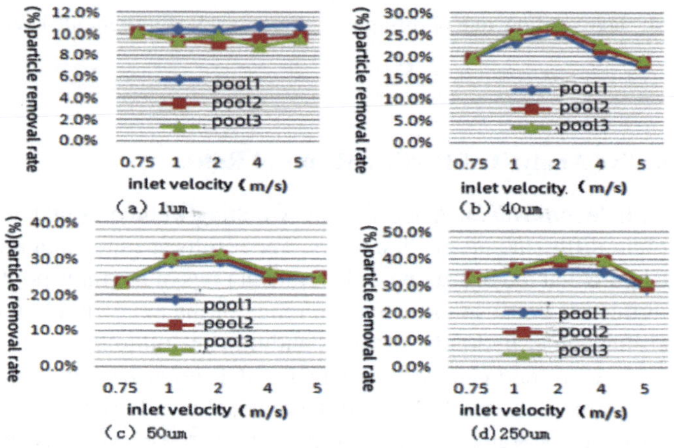

Fig. 5. Particle removal rate of different pool types and diameters under different inlet velocity

5 Conclusion

(1) The inlet velocity of overflow for each type of pool is different. Under the same pool type, the overflow rate increases with the increase of inlet velocity. Under the same inlet velocity, the overflow rate of pool type 1 is the highest (excluding the case of no overflow).

DOC, NO_3^--N, DO, and TN affect the community structure of phytoplankton, which indirectly affects the community structure of zooplankton.As shown in Fig. 3b.

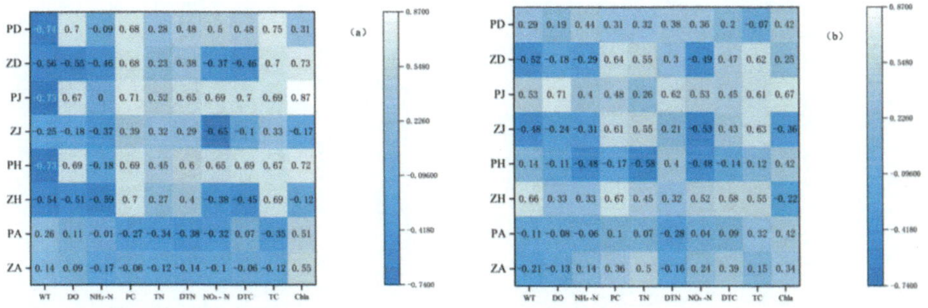

Fig. 3. Thermograms of correlation between biological indicators of Yalong River Yalong River zooplankton and environmental factors in spring (a) autumn (b)

4.5 Redundancy Analyses

Meneghiniana were negatively correlated with ammonia nitrogen and positively correlated with Chla, and Asterionella formosa Hass was positively correlated with DTC; the variability in the correlation between animal dominant species and environmental factors was more pronounced in the spring of May, in which Rhagadostoma completum was positively correlated with TC, PC, and NH_3-N. Difflugia globulosa was positively correlated with TN, DTN and NO_3^--N, and Difflugia avellana was positively correlated with WT; in November the phytoplankton dominant species Fragilaria. Crotonensis was positively correlated with NH_3-N and negatively correlated with Chla, with results opposite to those of spring, and Navicula.cincta was positively correlated with DO, WT, and Chla, and Diatoma. Tenue was positively correlated with TN, DTN, and NO_3^--N; in November the dominant zooplankton species Centropyxis aculeata was positively correlated with DTC, and both Difflugia avellana and Difflugia globulosa were positively correlated with DTC, DTN, TC, and TN were positively correlated.As shown in Fig. 4.

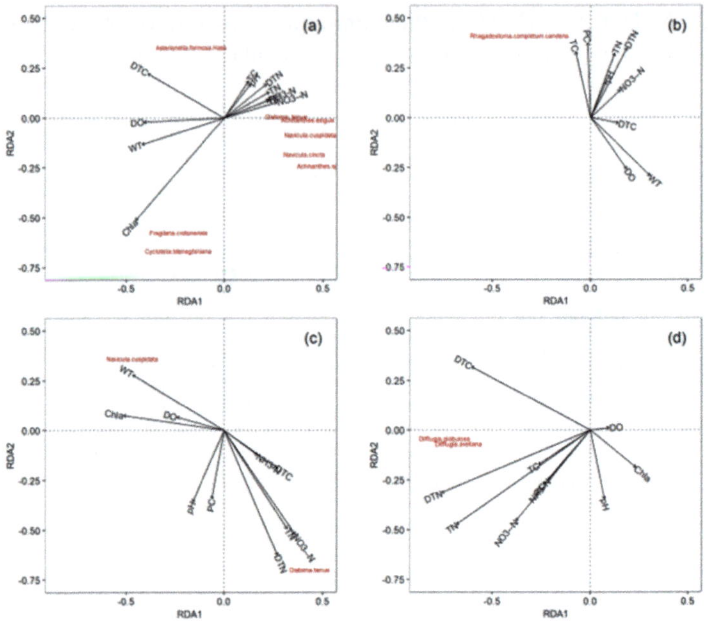

Fig. 4. RDA ranking of dominant species of phytoplankton versus environmental factors in Yalong River for May plants (a) May animals (b) November plants (c) November animals (d)

5 Conclusions

(1) During the two surveys on the zooplankton and flora of the Yalong River in May and November 2023, a total of 132 species of phytoplankton were collected, with diatoms and cyanobacteria dominating, the number of species in the spring was obviously higher than that in the autumn; a total of 34 species of zooplankton were collected, with Protozoa dominating. The dominant species were all Protozoa, of which the number of species in autumn was significantly higher than that in spring; the upstream biomass was higher than that in the middle and lower reaches.

(2) Through the indicator role of zooplankton and water quality evaluation using biodiversity index, it was concluded that the diversity index of Yalong River in spring and autumn seasons was not significantly different, and both of them were in the clean-oligopollution type and approaching to the clean type. The diversity index in spring was better than that in autumn, and the water quality in the upper reaches was better than that in the middle and lower reaches.

(3) The environmental factors affecting the community structure of the zooplankton are WT, DTN, DOC, NO_3^--N, DO, TN and Chla. The environmental factors affecting the community structure of the zooplankton in different seasons are different, so it is necessary to take into account the relationship between the interactions between the zooplankton and the environmental factors in different periods, and to take effective measures to prevent and control the environmental pollution of the Yalong River.

(4) Redundancy analysis (RDA) yielded that the correlation between dominant species of phytoplankton and environmental factors in the Yalong River varied greatly in different seasons, and the dominant species of phytoplankton were mainly closely related to NH_3-N, NO_3^--N, WT and Chla; and the dominant species of zooplankton were mainly closely related to TN and DTN.

Acknowledgment. This work was supported by the National Key Research Program of China (No. 2022YFC3202003), the National Natural Science Foundation of China (No. U2240202, 52122902), and China Yangtze Power Company Limited (Z542402001).

References

1. Dou, Y., Huo, D., Jiang, Z., et al.: Study of the community characteristics of plankton and their correlation with the environmental factors in surface water in Haihe Estuary. Ecol. Environ. Sci. **25**(4), 647–655 (2016)
2. Bo, F., Yang, H., Zuo, Z., et al.: Relationship between phytoplankton community in summer and environmental factors in Shanghai park lakes. Chin. J. Ecol. **28**(07), 1259 (2009)
3. Yun, L., Feizhou, C.: Are zooplankton useful indicators of water quality in subtropical lakes with high human impacts? Ecol. Indic. **113**, 106167 (2020). ISSN 1470-160X, https://doi.org/10.1016/j.ecolind.2020.106167
4. Zhibing, W., Yongfeng, H., Jinling, G., et al.: Temporal and spatial variation of phytoplankton community structure in the main stream of the Jinsha river. Resour. Environ. Yangtze Basin **29**, 1356–1365 (2020)
5. Zhang, Y., Gao, W., Li, Y., et al.: Characteristics of the phytoplankton community structure and water quality evaluation in autumn in the Huaihe River (China). Int. J. Environ. Res. Public Health **18**(22), 12092 (2021)
6. Wei, H., Zhao, W., Luo, X., et al.: Simulation of spatial distribution and seasonal variation of plankton in the Arctic Ocean. Haiyang Xuebao **41**(9), 65–79 (2019)
7. Yang, Z., Bu, J., Li, Z., et al.: A novel index-based method associated with aquatic ecosystem for evaluating river longitudinal connectivity: a case study for cascade dams in the Yalong River, China. Ecol. Ind. **154**, 110903 (2023)
8. Wang, L., Tan, L., Cai, Q.: Distinct differences of vertical phytoplankton community structure in mainstream and a tributary bay of the Three Gorges Reservoir, China. Front. Plant Sci. **15**, 1381798 (2024)
9. Li, F., Li, T., Gan, W., et al.: Seasonal and spatial variations of environmental factors and phytoplankton communities, and their correlations in the Guandi Reservoir, 2021
10. Li, F., Zeng, R., Liu, X., et al.: The seasonal change of phytoplankton community structure and water environmental factors in ertan reservoir, 2021
11. Gao, X., Song, J.: Phytoplankton distributions and their relationship with the environment in the Changjiang Estuary, China. Mar. Pollut. Bull. **50**(3), 327–335 (2005)
12. Hu, Y., Liu, X., Xing, W., et al.: Marker gene analysis reveals the spatial and seasonal variations in the eukaryotic phytoplankton community composition in the Yangtze River, Three Gorges Reservoir, China. J. Plankton Res. **41**(6), 835–848 (2019)
13. Jiang, Y., Wang, B., Yang, H., et al.: Community structure of phytoplankon and its relation with water quality in Dongjiang River. Ecol. Environ. **20**(11), 1700 (2011)
14. Zhao, H.J., Wang, Y., Yang, L.L., Yuan, L.W., Peng, D.C.: Relationship between phytoplankton and environmental factors in landscape water supplemented with reclaimed water, Ecol. Indic. **58**, 113–121 (2015). ISSN 1470-160X, https://doi.org/10.1016/j.ecolind.2015.03.033

15. Han, X., Pan, B., Zhao, G., et al.: Phytoplankton community structure and distribution characteristics in the source region of the Yangtze River. Resour. Environ. Yangtze Basin **28**(11), 2621–2631 (2019)
16. Ruzhong, L., Kefeng, L., Jing, Q.: Nitrogen and phosphate pollution characteristics and eutrophication evaluation for typical urban landscape waters in Hefei City. Environ. Sci. **35**(5), 1718–1726 (2014)
17. Du, G.S., Wang, J.T., Wu, D.W., et al.: Structure and density of the phytoplankton community of Miyun reservoir. Chin. J. Plant Ecol. **25**(4), 501 (2001)
18. Cao, J., Hou, Z., Li, Z., et al.: Succession of phytoplankton functional groups and their driving factors in a subtropical plateau lake. Sci. Total. Environ. **631**, 1127–1137 (2018)
19. Nwankwegu, A.S., et al.: Nutrient addition bioassay and phytoplankton community structure monitored during autumn in Xiangxi Bay of Three Gorges Reservoir, China. Chemosphere **247**, 125960 (2020). ISSN 0045–6535, https://doi.org/10.1016/j.chemosphere.2020.125960

Design and Simulate the Swirling Hydraulic Switching Well of the Intercepting and Abandon Flow Based on Fluent

Jing Zhang(✉)

Key Laboratory of Estuarine Ecological Security and Environmental Health, Xiamen University Tan Kah Kee College, Zhangzhou 363105, Fujian, China
Zhangjin@xujc.com

Abstract. The country has invested a large amount of manpower and material resources to control water pollution in Dianchi Lake in recent years. Although the water environment of Dianchi Lake has improved, there are still many problems. So far, the sewage interception and control systems has been built. In the entire system framework, in order to prevent backflow and large-scale overflow of the main canal, and to reduce the deposit of organic particles and silt in the intercepted rainwater on the main canal, it is crucial to carry out sand settling treatment and choose appropriate control methods to discard the rainwater in the rainwater main pipe before connecting to the main canal. Develop and design a cyclone hydraulic control switching well system of interception and abandon flow on this study, which controls the interception and abandon flow according to the water volume of the main canal. Using the three-dimensional design software SOLIDWORKS and simulation software FLUENT to construct a switching well model and simulate solid-liquid two-phase flow and studying the relationship between particle diameter, inlet flow rate, overflow pipe height, and particle removal rate. Optimal pool structure and corresponding operating conditions parameters of the switching well were finally obtained. At the same time, a practical survey was conducted on the site, and suitable sampling points for rainwater main channels were selected for rainwater particle size detection, some of the which were selected for parameter settings in the switching well simulation.

Keywords: Initial rainwater pollution control · Swirling hydraulic switching well of interception and abandon flow · Fluent · Particle Removal Rate · Overflow Rate

1 Introduction to Switching Well

1.1 Research Background

The research area is located on the east bank of Dianchi Lake, in Chenggong New District. The sources of initial rainwater surface runoff erosion are extensive and complex [1]. The pollution sources in which are relatively diverse and the composition are complex. At present, in order to effectively control pollution sources, building a sewage interception main channel surrounded the lake, which is divided into sewage and rainwater. However, its pipeline network and related system are not perfect.

D. Han et al. (Eds.): ICEPG 2024, ESE, pp. 1118–1124, 2025.
https://doi.org/10.1007/978-3-031-86829-0_121

The surface runoff water quality complies with the following rules: the concentration of pollutants is relatively high in the early stage of rainfall runoff, but rapidly decreases in the middle and later stages. Therefore, the initial rainwater has a high pollution load [2]. At the same time, there is a good correlation between COD, TN, TP and SS in surface runoff [3], indicating that most pollutants in the initial rainwater exist in the form of particle adsorption [4], and inevitably some sewage flows into the rainwater, which contains suspended solids and sediment. If these particulate matter flow directly into the main canal through rainwater collection pipes without effective sand settling treatment, it will cause sedimentation in the main canal. In addition, Backflow and large-area overflow may occur in the main canal and it also causes significant impact loads on the rainwater treatment station, if there is no capacity to receive water collected from the rainwater inlet pipe. Therefore, It is crucial to design and study of sediment treatment and abandon flow equipment for the main pipe rainwater leading to the interception main canal, which is also the foundation for the normal operation of the main canal.

1.2 The Control and Working Principle of Switching Well

The switching well adopts a fully hydraulic sand removal method, which is installed at a certain position where the rainwater main pipe leads to the main canal. The Fig. 1 and Fig. 2 show the specific composition. The opening of the electric gate valve of switching well is controlled through the water level of the main canal. When the water level in the main canal is low, the electric gate valve remains normally open. After the sand settling treatment of the switch well, most of the rainwater enters the main canal to realize the closure. When there is no capacity for the main canal to receive water from the initial rainwater collection main pipe, and the water level is high, the electric gate valve is closed until the outlet of the switching well outlet pipe is completely blocked by the valve plate. The water level inside the switching well gradually rises. After the sand settling treatment of the switch well, the rainwater is discharged through the overflow pipe to achieve abandoned flow, which plays a role in peak shaving and staggering and reduces the load of subsequent water treatment in the main canal.

The basic working principle of the switching well is based on centrifugal settling [5]. After entering the switching well at a certain speed and pressure, the solid-liquid mixture will undergo a spiral swirling motion. Due to the centrifugal effect generated by the spiral rotation, the centrifugal effect will be much greater than the weight effect of the mixture itself [6], so centrifugal settling plays a dominant role. Because the density of solid and liquid phases is different, the centrifugal force is also different. As a result, solid particles with higher density will be thrown towards the pool wall and sink into the sand hopper, while light particles (water phase) will flow out through the outlet pipe or overflow pipe [7].

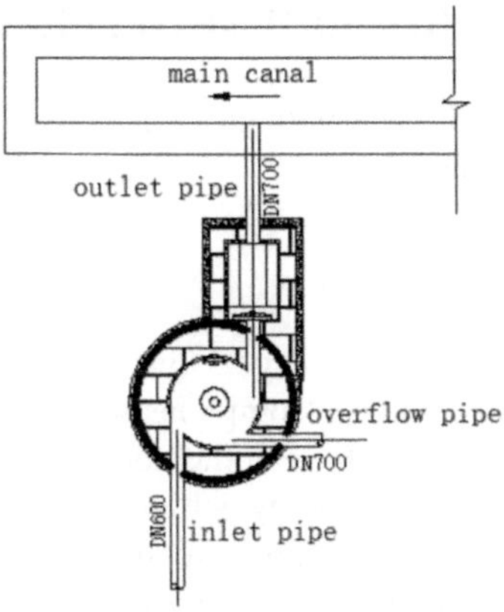

Fig. 1. Switching well plan

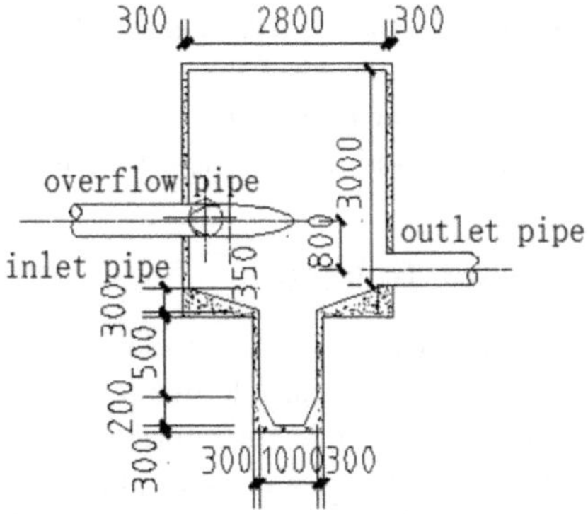

Fig. 2. Switching well sectional view

2 Switching Well Model Construction

The overflow height is usually an important parameter of switching wells, which directly affects indicators such as abandoned flow rate and particle removal rate. Therefore, according to the different net height between the bottom of the overflow pipe and the inlet pipe, the three-dimensional model of the switching well is divided into three basic pool types.

The net height difference between the bottom of overflow pipe and inlet pipe of basic tank type I is 0 m. The bottom of Overflow pipe is half of the inlet pipe for the basic pool type two, that is, the net height difference at the bottom is 0.3 m, and the bottom of overflow pipe for the basic pool type three is at the top of the inlet pipe, that is, the net height difference at the bottom is 0.6 m. The other dimensions of the model are the same, with an angle of 90 degrees between the overflow pipe and the inlet pipe, and a pool diameter to depth ratio of 0.9.

3 Determination of Simulation Parameters for Switching Wells

3.1 Inlet Flow Rate

Due to the minimum design flow velocity of 0.75 m/s for rainwater pipes and the maximum design flow velocity of 5 m/s for non-metallic pipes. The range of inlet flow velocity is between 0.75 m/s and 5 m/s (including these two values). Other inlet flow rates are set at 1 m/s, 2 m/s and 4 m/s.

3.2 Particle Diameter

The particle diameter (particle size) parameter is based on the particle size analysis data of the four-control sampling of the main rainwater canal using a laser particle size analyzer. Select typical indicators for particle size analysis, including D10, D40, D50, and D90. According to the particle size characteristic index, simulate separately four particle sizes of 1 um, 40 um, 50 um, and 250 um.

4 Analysis of Switching Well Simulation Results Based on Fluent

Simulate the basic pool structure of different overflow heights with the outlet valve open, and obtain the optimal overflow height basic pool structure.

4.1 Particle Trajectory

The simulated motion trajectories of particles are shown in Fig. 3 (taking the inlet flow rate 5 m/s and particle diameter 50 um as an example). Comprehensive analysis shows that the particle trajectories of the three pool types have some commonalities. The particle motion trajectories are all in a spiral vortex shape, which are regular and have good swirling properties. As the inflow velocity increases, the pressure increases, the particles tend to spray upwards in a swirling flow, and their trajectories gradually fill the entire pool. Part of the particles swirl horizontally or downwards, and finally most of the particles flow out with the outlet and overflow, while a small part sinks into the sand hopper. As the particle diameter increases, more particles are thrown towards the wall of the pool, and the residence time of particles on the outer wall of the pool is short, while the residence time of swirling particles inside the outer wall is long. The number of particles in the sand hopper also increases with the increase of particle diameter. When the inlet water velocity is high, the trend of particles sinking downwards becomes less and less obvious, and they are mostly in a suspended state in the sand hopper.

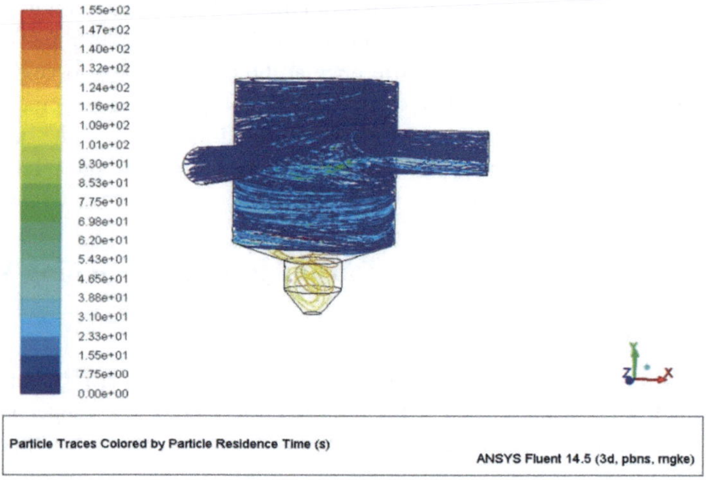

Fig. 3. Particles trajectory under the inlet velocity of 5 m/s and particle diameter 50 um

4.2 Comparative Analysis of Overflow Rate

The overflow rates of the three pool types are shown in Fig. 4. Among the simulated inflow velocities, Pool Type 1 has overflow with inlet velocities of 1 m/s, 2 m/s, 4 m/s,

(2) The particle motion trajectory of the basic pool type is spiral swirl, which is relatively regular and has good particle swirl properties. As the particle diameter increases, more particles are thrown towards the wall of the pool, and the residence time of the wall particles is shorter, while the residence time of the swirling particles inside the pool is longer.
(3) The particle removal rate ranges from 8% to over 40%, the height of the overflow pipe has a small impact on the particle removal rate, while the inlet flow rate and particle diameter size are important factors affecting the particle removal rate. For particles with a diameter other than 1 um, the removal rate of tank type three is relatively high, but the height of its overflow pipe is very high, which has a significant impact on downstream flow when the overflow pipe is full. Moreover, its overflow rate is low and the abandon flow rate is small. However, the removal rate of pool type one is relatively low. Based on the above analysis, select basic pool type two as the optimal basic pool type.
(4) The three types of pools mainly remove particles with diameters of 50 um and 250 um. The optimal particle removal rate is achieved at an inlet velocity of 2 m/s, with an optimal particle removal rate of 40.9% and an optimal overflow rate of 49%.

References

1. Liu, X.D., Zhou, Y.L., Ji, D.P., et al.: Discussion on the scale of initial rainfall interception in cities. Water Wastewater Eng. (49), 512–517 (2023)
2. Zhou, C.T., Wang, M.Y., Xing, Y.X., et al.: Research progress of urban initial stormwater pollution and treatment measures. Water Purif. Technol. **41**(7), 17–26 (2022)
3. Cristina, C., Tramonte, J., Sansalone, J.J.: A granulometry-based selection methodology for separation of traffic-generated particles in urban highway snowmelt runoff. Water Air Soil Pollute **136**, 33–53(2002)
4. Gu, Y.: Analysis of urban surface runoff pollution characteristics in the main urban area of kunming. Environ. Eng. **7**(7), 2588–2595 (2013)
5. Shao, C., Yan, T.N.: Improvement and experimental effect of cyclone desander. Coal Geol. Explor. **34**(5), 71–74 (2006)
6. Zhang, X.Q., Yan, T.N.: Research on the mechanism of cyclone desanding and development of automatic cyclone desander. Explor. Eng. (Geotech. Drill. Excav. Eng.) (11), 34–36(2008)
7. Yuan, H.X., Wang, Y.J.: Application of cyclone separation technology in petroleum and petrochemical industry. Filter. Sep. **12**(3), 33–36 (2002)

Research on Surface Litter Target Detection Algorithm Based on SSM-YOLOv9

Lizhen Fan, Jiangang Dong(✉), and Hailin Ma

School of Software, Xinjiang University, Urumqi 830091, Xinjiang, China
137384753@qq.com

Abstract. Surface litter target detection plays an important role in timely detection and early warning of river pollution, providing strong support for the realisation of round-the-clock uninterrupted monitoring and providing data support for river litter clean-up. River rubbish has the characteristics of small target and easy to cover, which brings difficulties for target detection. In order to solve the above problems and improve the detection accuracy, the river trash detection algorithm based on improved YOLOv9 is proposed. Aiming at the problem of small targets of water surface rubbish, a multi-scale spatial pyramid module is introduced to extract finer granularity rubbish features; aiming at the problem that river rubbish will have mutual occlusion in the actual river environment, the SEAM module is added to highlight the rubbish region information in the image and weaken the background region information; the shapeIoU replaces the CIoU in the original model to reduce the rubbish bounding box regression of the small targets The IoU value of the sample is affected by the shape of the manually labelled box to further improve the detection accuracy. Experimental validation is conducted on the self-constructed visible water surface litter dataset DSGD and infrared water surface litter dataset ISGD as well as the dataset FLOW-IMG, which is publicly available for inland river floating object detection, respectively, and the results show that the detection accuracies reach 85.8%, 71.1%, and 86.8%, which are 4.4%, 3.5%, and 0.8% higher than those of the original YOLOv9 model, respectively. Compared with the existing mainstream target detection models, SSM-YOLOv9 all show better water surface trash recognition and meet the real-time requirements.

Keywords: River Ecological Protection · Target Detection · YOLOv9 · Surface Litter · Small Target Detection

1 Introduction

River ecological protection is an important part of ecological civilisation construction. River rubbish seriously endangers the ecosystem of the river and the quality of the water environment, not only destroying the natural beauty of the river, but also directly affecting the ecological balance of the water body. Plastic, metal and other rubbish will slowly decompose in the river, releasing harmful chemicals that pollute the water quality and endanger the health of aquatic plants and animals, and the various kinds of rubbish in the river can be used as a kind of visual indicator to measure the water quality. At the

D. Han et al. (Eds.): ICEPG 2024, ESE, pp. 1125–1137, 2025.
https://doi.org/10.1007/978-3-031-86829-0_122

same time, rubbish clogs the river channel, increasing the risk of flooding and posing a threat to the life and property safety of people living in areas along the river [1].

The Opinions on Comprehensively Implementing the River Length System issued by the State Council in 2016 states that a daily supervision and inspection system should be established and dynamic supervision should be implemented. Manual inspection usually requires a lot of manpower and time, especially when conducting a comprehensive inspection in a large river network, which is less efficient and difficult to cover all areas. Due to the influence of human factors, inspectors may miss some hidden rubbish accumulation points, resulting in rubbish not being cleared for a long time. Drone inspections are costly and cannot be monitored in real time due to range time issues. With the continuous development of the Internet of Things technology, 5G communication technology and machine vision, using data sharing and exchange, target intelligent detection and other technologies to carry out garbage inspection in waters and rivers will become a more efficient solution.

Current target detection algorithms are mainly classified into two categories: one is the two-stage model (Two-Stage) represented by the R-CNN [2] series: Jiang et al. [3] proposed an improved Faster-RCNN water-bleaching rubbish detection algorithm, MP-Faster-RCNN, which combines null convolution with ResNet-50 as a backbone feature extraction network, expands the receptive field to extract more small target features, exploiting multi-scale features in the region generating network (RPN), and adding polarised self-attention before the RPN to achieve more accurate target frame localisation. Bai et al. [4] proposed a neural network-based approach to improve the accuracy of trash identification and classification by combining an improved Faster RCNN target detection model with Resnet50 Yi et al. [5] proposed a CA-Faster R-CNN based detection and localisation algorithm for images containing objects. For images containing objects, the proposed algorithm uses Faster R-CNN to detect and classify the objects and CA network to locate the objects. Since the two-stage model requires two processing steps (generation of candidate regions and classification regression), it is difficult to achieve real-time detection, and it is not effective for river litter detection. Another category is the one-stage model represented by the YOLO [6] series: Chen L et al. [7] proposed a lightweight water surface litter detection model based on YOLOv5, which replaces the original network structure of YOLOv5 with the Shufflenetv2 + SE network structure, and uses the GSConv module instead of the Conv module segment in Neck and the VoV- GSCSP module instead of C3 module to further improve the detection accuracy. Li L et al. [8] applied the newly proposed DS method to the original YOLOv7 network for multi-scale feature fusion, and at the same time applied the CBAM attention mechanism module to the network backbone to improve the detection accuracy.

With the update of YOLO series, the accuracy and efficiency of target detection is gradually improved, for the characteristics of river floating rubbish target with small size, dispersed, and easy to be affected by the complex situation of the river such as water surface reflections, waves, and vegetation occlusion, etc., this paper proposes the Garbage Target Detection Algorithm based on YOLOv9, which extracts finer-grained rubbish features of the water surface of the small target by introducing the Multi-scale Spatial Pyramid Module, and highlighting the rubbish region information in the image by adding the SEAM module, which is used to solve the problem that the river rubbish will exist in mutual occlusion, and finally replacing the CIoU in the original model with the shapeIoU to reduce the IoU value of the rubbish bounding box regression samples of the small targets by the influence of the shape of the manually marked boxes, which further improves the detection accuracy, respectively, in the self-constructed visible water surface rubbish dataset DSGD and the infrared water surface litter dataset ISGD and the dataset FLOW-IMG for public inland river floating object detection for experimental validation.

2 Related Work

2.1 YOLOv9 Target Detection Model

YOLOv9 [9] In order to improve the efficiency of information integration and propagation in model training, YOLOv9 introduces a novel lightweight network architecture called the Generalised Efficient Layer Aggregation Network (GELAN).GELAN combines CSPNet and ELAN [10] in order to efficiently aggregate information from the network, to reduce the loss of information in the propagation process and to enhance the information interaction between layers. Due to its lower number of parameters and computational complexity, this architecture is particularly suitable for river litter detection in environments with limited computational resources.

The improved SSM-YOLOv9 network structure is shown in Fig. 1, where the MSAM block at the dashed box as well as the SEAM block are the improved modules in this paper.

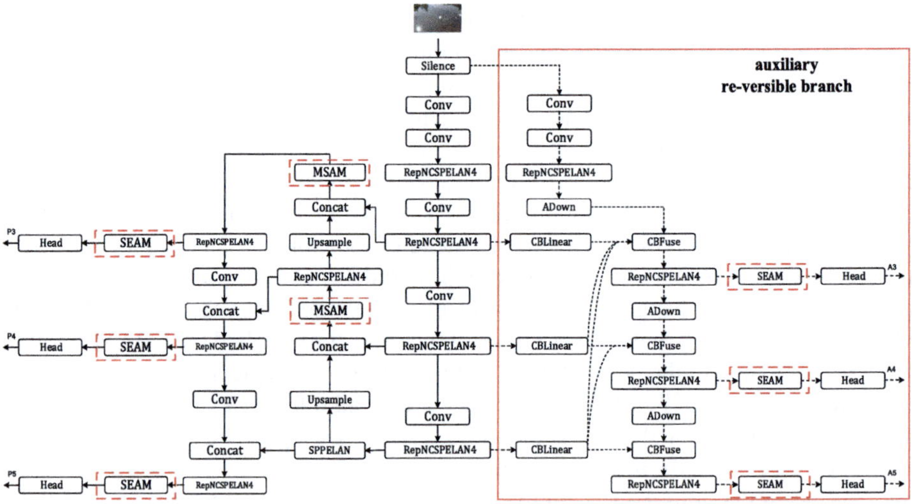

Fig. 1. YOLOv9 network architecture diagram

2.2 Multi-scale Spatial Pyramid Module

River litter targets are small, occupy fewer pixels and require fine-grained feature extraction to identify their edges and shapes. In the YOLOv9 algorithm, the traditional feature pyramid network structure (FPN) is used to achieve multi-scale target detection, but due to the single way of multi-layer feature fusion, feature resolution limitation and the lack of dynamic weight adjustment, it leads to certain shortcomings when dealing with fine-grained features. To solve these problems, this paper introduces the Multi-scale Spatial Pyramid Attention (MSPA) mechanism [11], which effectively overcomes these defects through finer-grained feature extraction, dynamic weight adjustment, and structural regularisation, and significantly improves the recognition of fine features and small targets. It contains three core components: hierarchical feature convolution (HPC) module, spatial pyramid recalibration (SPR) module, and Softmax operation. The HPC module is used to extract multi-scale spatial information. The SPR module is responsible for learning the channel attention weights in order to establish cross-dimensional interactions. The Softmax operation is used to recalibrate the attention weights of the channels in order to establish long-range channel dependencies. The MSPA structure is shown in Fig. 2.

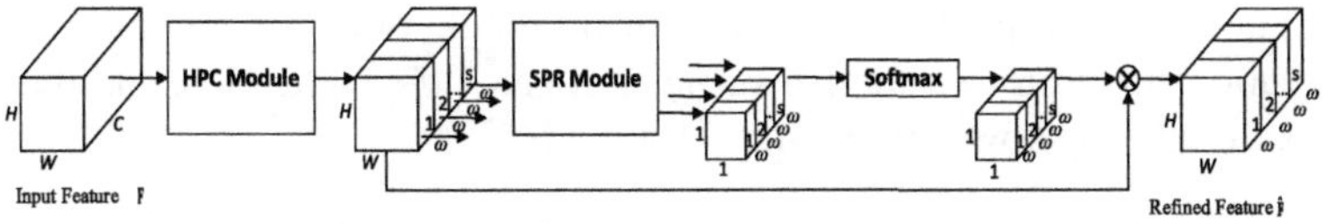

Fig. 2. MSPA structure diagram

The HPC module first splits the input feature map in the channel dimension by the Split operator $\mathbf{F} \in \mathbb{R}^{C \times H \times W}$ uniformly into s feature map subsets, a step that ensures that each subset contains a different portion of channel features. Afterwards, for each feature map subset, it is summed element by element with the previous augmented feature map subset to perform feature fusion, which enhances the feature representation and enables the model to learn richer features at multiple levels. Afterwards, a 3 × 3 standard convolution operation is performed on each subset of the fused feature maps to extract spatial features, and Batch Normalization is applied to normalise the feature maps to improve training speed and model stability. Finally, all the enhanced feature map subsets are stitched together to obtain the complete enhanced feature map, and the structure of the HPC module is shown in Fig. 3.

$$\hat{\mathbf{F}}_i = \begin{cases} \mathcal{T}_i(\mathbf{F}_i), & i = 1 \\ \mathcal{T}_i\left(\mathbf{F}_i \oplus \hat{\mathbf{F}}_{i-1}\right), & 1 < i \leq s \end{cases} \quad (1)$$

Eq: $\mathcal{T}_i(\cdot)$ denotes the 3 × 3 standard convolution and batch normalisation operation; $\oplus$ denotes the element-by-element summation operation; $\hat{\mathbf{F}}_i$ denotes the first i subset of the enhanced feature map.

$$\hat{\mathbf{F}} = \mathrm{Concat}\left(\left[\hat{\mathbf{F}}_1, \hat{\mathbf{F}}_2, \ldots, \hat{\mathbf{F}}_s\right]\right) \quad (2)$$

Eq: Concat([·, ·, . . . , ·]) denotes the splicing of enhanced feature subsets operation by operation; $\hat{\mathbf{F}}$ denotes the complete augmented feature map obtained after splicing.

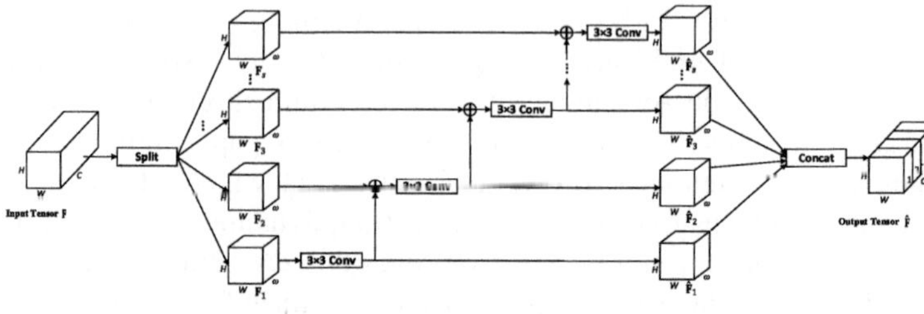

Fig. 3. Structure of HPC module

The Spatial Pyramid Recalibration (SPR) module consists of two main components: the Spatial Pyramid Aggregation (SPA) block and the Channel Interaction (CI) block. The enhanced feature map generates channel descriptors through 1 × 1 global average pooling and 2 × 2 local average pooling to capture global and local feature information, which are merged by upsampling and weighted fusion. Afterwards the CI block extracts inter-channel relationships through two point-by-point convolutional layers and a Sigmoid activation function to establish stronger channel dependencies, and the structure of the SPR module is shown in Fig. 4.

$$\mathbf{Z}_i = SPA\left(\hat{\mathbf{F}}_i\right) = \psi_{re}\left(\alpha \otimes T_{up}\left(P\left(\hat{\mathbf{F}}_i, 1\right)\right) \oplus \beta \otimes T_{up}\left(P\left(\hat{\mathbf{F}}_i, 2\right)\right)\right) \tag{3}$$

Eq: SPA(·) denotes the SPA block; $\psi_{\mathrm{re}}(\cdot)$ denotes the adjustment of the tensor to a vector; P(·, ·) denotes the average pooling layer; $\mathrm{T_{up}}(\cdot)$ denotes up-sampling; α and β are two learnable floating-point parameters optimised by stochastic gradient descent.

$$\mathbf{V}_i = CI(\mathbf{Z}_i) = \sigma\left(f_2^{1\times1}\left(\mathrm{Re}LU\left(f_1^{1\times1}(\mathbf{Z}_i)\right)\right)\right) \tag{4}$$

Eq: $CI(\cdot)$ denotes the CI block; $\mathrm{Re}LU(\cdot)$ denotes the ReLu function; $\sigma(\cdot)$ denotes the Sigmoid activation function, which is responsible for mapping the output to the range (0, 1); $f_1^{1\times1}(\cdot)$ and $f_2^{1\times1}(\cdot)$ denote two point-by-point convolutional layers.denote two point-by-point convolutional layers.

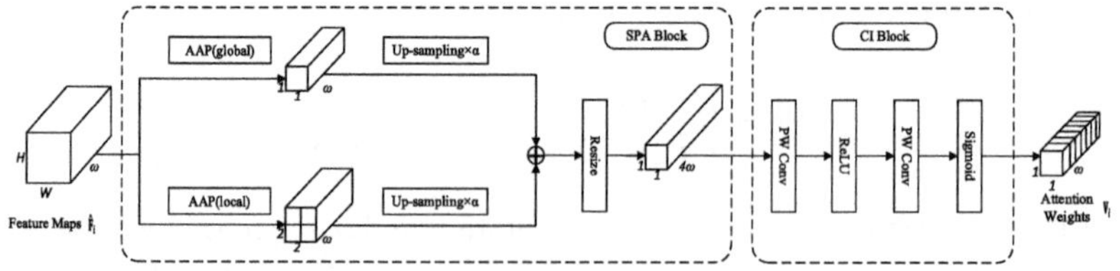

Fig. 4. Structure of SPR module

2.3 SEAM Module

In the actual river environment, the river trash will have mutual occlusion, which will lead to the reduction of the accuracy of the river trash detection model. In order to solve the above problems, this paper introduces the SEAM module proposed by Yu et al. [12] to improve the rubbish feature extraction ability, connecting the SEAM module with the two sets of three different scales of the main branch and the auxiliary reversible branch of the main branch and the three detector heads of the output part in YOLOv9, highlighting the rubbish region information in the image and weakening the background region information. The structure of the SEAM module is shown in Fig. 5. Shown.

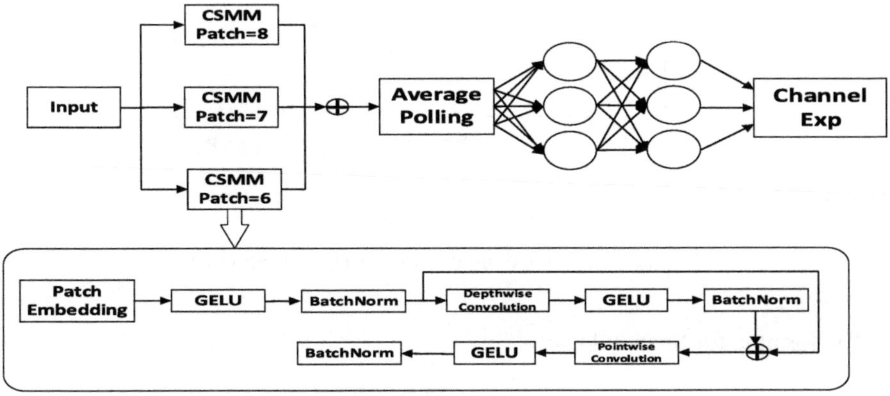

Fig. 5. SEAM module structure

The first part of the SEAM module is a depth-separable convolution with residual connectivity, in order to compensate for the loss of information relations between channels by depth-separable convolution, the outputs of different depth convolutions are combined by point-by-point convolution, and finally the information of each channel is fused by a two-layered fully-connected network, so that the network can enhance the connectivity between all channels. The output of the fully connected layer is processed by an exponential function to extend the value domain from [0, 1] to [1,e], and the above exponential normalisation improves the tolerance of the results to positional errors through monotonic mapping. The output of the SEAM module is used as a feature to multiply with the original feature, which allows the model to effectively deal with the problem of occlusion of the target by river litter when.

2.4 ShapeIoU

The traditional IoU method considers the influence of the distance, shape and angle of the manually labelled box (Ground Truth Box) and the anchor box (Anchor Box) on the bounding box regression, but ignores the fact that the shape and scale of the bounding box itself also have an influence on the bounding box regression, and because of the small dimensions of the river litter floats, the IoU values of the bounding box regression samples are more significantly affected by the manually labelled box's shape, resulting in the detection accuracy cannot be further improved.

To solve the above problems, the ShapeIoU method proposed by Zhang et al. [13] is introduced in this paper to replace the CIoU method in the original algorithm. The schematic diagram of the ShapeIoU method is shown in Fig. 6. Shown.

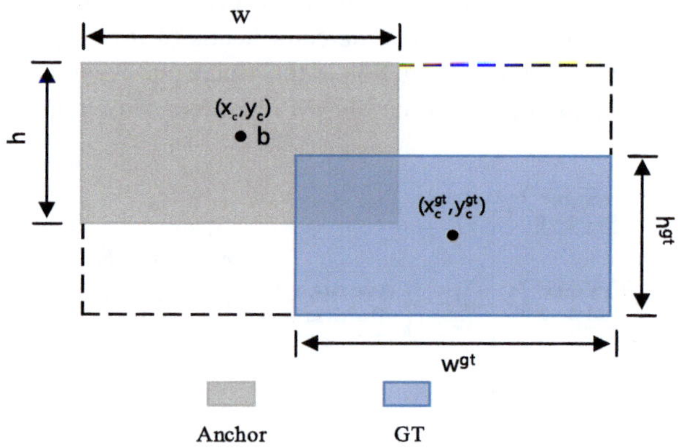

Fig. 6. Schematic diagram of the ShapeIoU approach

The formula for ShapeIoU is:

$$IoU = \frac{|B \cap B^{gt}|}{|B \cup B^{gt}|} \tag{5}$$

Eq: B and B^{gt} denote the prediction and manual labelling frames respectively.

$$ww = \frac{2 \times (w^{gt})^{\text{scale}}}{(w^{gt})^{\text{scale}} + (h^{gt})^{\text{scale}}} \tag{6}$$

$$hh = \frac{2 \times (h^{gt})^{\text{scale}}}{(w^{gt})^{\text{scale}} + (h^{gt})^{\text{scale}}} \tag{7}$$

Eq: w^{gt} and h^{gt} denote the width and height of the manually labelled box, respectively. Scale denotes the hyperparameter related to the scale of the target in the dataset, for smaller targets, scale takes a larger value, this model uses scale = 1.5, the ww and hh represent the horizontal and vertical weight coefficients, respectively, and their values are related to the shape of the manual labelling box.

$$\text{distance}^{\text{shape}} = hh \times \left(x_c - x_c^{gt}\right)^2 / c^2 + ww \times \left(y_c - y_c^{gt}\right)^2 / c^2 \tag{8}$$

Eq: $\text{distance}^{\text{shape}}$ denotes the distance loss of the shape, and (x_c, y_c) denotes the centroid coordinates of the prediction box; $\left(x_c^{gt}, y_c^{gt}\right)$ denotes the coordinates of the centre point of the manual labelling frame; c denotes the length of the line connecting the centre point of the prediction frame and the centre point of the manual labelling frame.

$$\Omega^{\text{shape}} = \sum_{t=w,h} (1 - e^{-\omega_i})^{\theta}, \; \theta = 4 \tag{9}$$

$$\begin{cases} \omega_w = hh \times \frac{|w-w^{gt}|}{\max(w,w^{gt})} \\ \omega_h = ww \times \frac{|h-h^{gt}|}{\max(h,h^{gt})} \end{cases} \tag{10}$$

$$L_{\text{Shape-IoU}} = 1 - \text{IoU} + \text{distance}^{\text{shape}} + 0.5 \times \Omega^{\text{shape}} \tag{11}$$

Eq: Ω^{shape} denotes the loss of value of the shape; $L_{\text{Shape-IoU}}$ denotes the overall result of the loss function calculation.

3 Data Sets and Evaluation Indicators

3.1 Floater Data Sets

The dataset is the basis for evaluating the performance of target detection algorithms, and its quality and size have an important impact on the fitting and generalisation ability of the model [*1]. In this experiment, surveillance cameras were set up at the Huangtugang hydrological monitoring station in Xinxiang City, Henan Province, and samples of floating trash on the surface of the Weihe River and the Communist Drainage Canal under visible light conditions and infrared conditions were collected in order to construct a reliable floating trash dataset. The data were cleaned and, to improve the generalisation ability of the trash target detection algorithm, the collected sample data were subjected to data enhancement operations of horizontal flipping, random contrast and random luminance to obtain 6560 floating trash RGB images as well as 6972 floating trash infrared images, and the original samples were randomly partitioned into a training set and a test set according to the ratio of 8:2. The open-source image annotation tool LableImg was used to manually annotate the sample graphics, which were classified into "plastic-garbage", "melonpeel", "bottle", "trash", "garbage", "trash", "trash", "trash", "trash", "trash", "trash", "trash", "trash", "trash", "trash", and "trash". "bottle", "box", the dataset format is Pascal VOC, and the annotation file is saved in XML format, which is recorded as visible surface rubbish dataset as DSGD and infrared surface rubbish dataset as ISGD.

In addition, the experiments in this paper also use the FLOW-IMG-Img sub-dataset released by Ouka Smart Hublot to validate the reliability of the model. FLOW-IMG is the first dataset for inland river floating object detection, which contains 2,000 images with a total of 5,271 "bottles" labelled as targets. Some sample images are shown in Fig. 7(a, b, c).

3.2 Experimental Environment and Important Parameters

The experimental device operating system for this study is Window10, GPU is a single NVIDIA RTX 4060, and the model code uses Pytorch version 2.1.0 + cu121 in python 3.9.18 environment. The learning rate was set to 0.01, the optimiser was Adam and the batch size was set to 4.

(a) (b) (c)

Fig. 7. (a) Example of DSGD visible surface litter dataset; (b) Example of ISGD infrared surface litter dataset; (c) Example of FLOW-IMG surface litter dataset

3.3 Performance Evaluation Indicators

In this paper, the floating rubbish average precision mAP (mean Average Pricision) and FPS (Frames Per Second) are used as the evaluation indexes of the model performance. mAP is calculated by the accuracy rate P (precision) and the recall rate R (recall), the higher the value of mAP is, the better the effect of target detection and recognition is. The higher the value of mAP, the better the target detection and recognition effect. The accuracy P, recall R and mAP are calculated as follows:

$$P = \frac{TP}{TP + FP} \tag{12}$$

$$R = \frac{TP}{TP + FN} \tag{13}$$

$$AP = \int_0^1 P(R)\mathrm{d}R \tag{14}$$

$$mAP = \frac{\sum_{i=1}^{N} AP_i}{N} \tag{15}$$

$$FPS = \frac{1000}{pre + ms + NMS} \tag{16}$$

where N denotes the number of floating target species; TP denotes the number of floaters correctly detected; FP denotes the number of non-floaters incorrectly detected as floaters;

FN denotes the number of floaters that are not detected; pre denotes the image pre-processing time, including image scaling and filling, channel transformation, dimension enhancement processing, etc.; ms denotes the inference speed, which refers to the time from preprocessed image input to the model output result; NMS denotes the post-processing time used to transform the model output result, etc.

4 Data Sets and Evaluation Indicators

4.1 Experimental Comparison of YOLOv9 Network Models Before and After Improvement

Comparison of the P-R diagrams obtained by applying the YOLOv9 network model to the detection of river trash targets before and after the improvement is shown in Fig. 8.The size of the area enclosed with the coordinates of the P-R diagram represents the accuracy value of identifying river trash. It can be clearly seen that the improved YOLOv9 model achieves a better detection effect in detecting the river trash, and the accuracy value is significantly higher than that of the pre-improved YOLOv9 model. YOLOv9 model, the experimental effect before model improvement is shown in Fig. 9, and the experimental effect before model improvement is shown in Fig. 10.

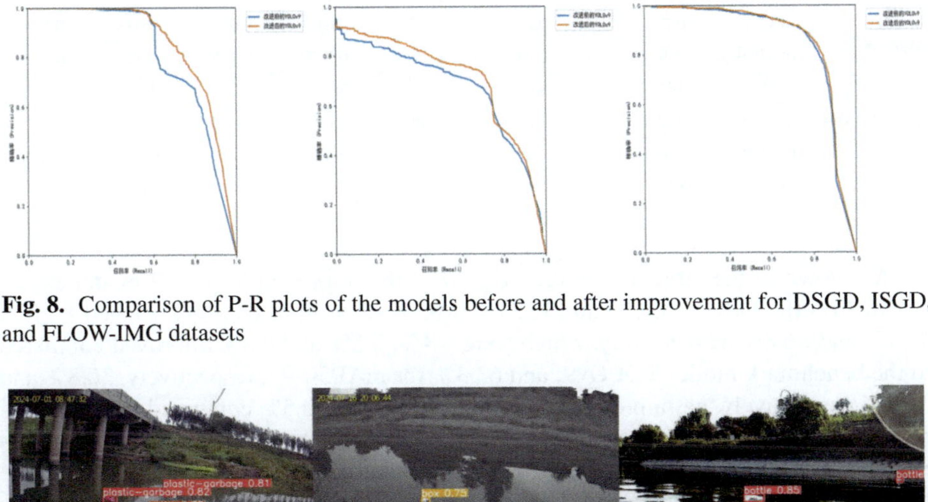

Fig. 8. Comparison of P-R plots of the models before and after improvement for DSGD, ISGD, and FLOW-IMG datasets

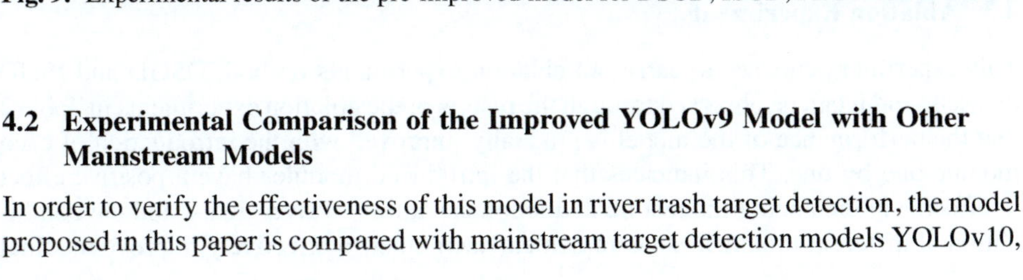

Fig. 9. Experimental results of the pre-improved model for DSGD, ISGD, FLOW-IMG datasets

4.2 Experimental Comparison of the Improved YOLOv9 Model with Other Mainstream Models

In order to verify the effectiveness of this model in river trash target detection, the model proposed in this paper is compared with mainstream target detection models YOLOv10,

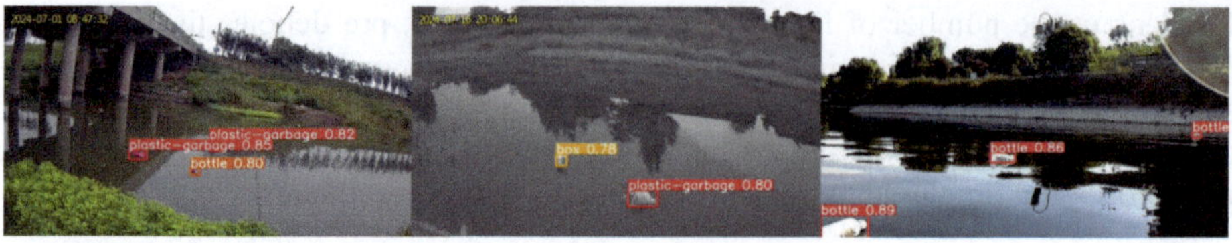

Fig. 10. Experimental results of the improved model for DSGD, ISGD, FLOW-IMG datasets

YOLOv8, YOLOv7, etc., and the environment configurations and training parameters are kept consistent, and the results are shown in Table 1.

Table 1. Performance evaluation results of 7 models on 3 datasets

Data set	Norm	Modelling						
		SSM-YOLOv9	YOLOv10	YOLOv9	YOLOv8	YOLOv7	YOLOv5	Faster RCNN
DSGD Dataset	**mAP50**	**0.858**	0.827	0.814	0.823	0.771	0.779	0.740
	mAP50-95	**0.633**	0.590	0.581	0.670	0.491	0.514	0.536
	FPS	**49**	41	52	45	43	35	13
ISGD Dataset	**mAP50**	**0.711**	0.661	0.676	0.613	0.654	0.628	0.576
	mAP50-95	**0.368**	0.343	0.359	0.317	0.348	0.326	0.286
	FPS	**48**	41	63	50	43	38	17
FLOW-IMG Dataset	**mAP50**	**0.868**	0.814	0.860	0.857	0.688	0.832	0.646
	mAP50-95	**0.434**	0.390	0.429	0.432	0.292	0.414	0.336
	FPS	**93**	87	96	78	68	61	31

As shown in the table above, the mAP50 of the improved YOLOv9 model experimented on the DSGD, ISGD and FLOW-IMG surface litter datasets reached 85.8%, 71.1% and 86.8%, respectively, which were 4.4%, 3.5% and 0.8% improved compared to the benchmark model YOLOv9, and 63.3% for mAP50-95, respectively, 36.8% and 43.4%, respectively, an improvement of 5.2%, 0.9% and 0.5% compared to the benchmark model YOLOv9. Although the FPS is slightly reduced compared with the benchmark model, it still achieves a better real-time monitoring speed, and the above table shows that the performance of the river trash target detection proposed in this paper is better than other models.

4.3 Ablation Experiments

This experiment chooses to carry out ablation experiments on both DSGD and ISGD datasets, and it can be observed through the results of the ablation experiments in Table 2 that the performance of the model is gradually improved with the introduction of each module one by one. This indicates that the introduced modules have a positive effect in effectively enhancing the feature extraction capability of river floats, and the combination of different modules can complement each other to effectively improve the final performance of the model. This experimental result also proves the superiority of the proposed method in solving the task of river litter target detection.

Table 2. Statistics of ablation experiment results

Data set	YOLOv9	MSPA	SEAM	ShapeIoU	map50
DSGD data set	√				0.814
	√	√			0.830
	√	√	√		0.847
	√	√	√	√	0.858
ISGD data set	√				0.676
	√	√			0.684
	√	√	√		0.699
	√	√	√	√	0.711

5 Conclusion

Aiming at the problem of difficult detection of river trash targets, this paper proposes a water surface trash target detection method based on improved YOLOv9, the SSM-YOLOv9 model. By introducing the multi-scale spatial pyramid module, the ability of the model to handle fine-grained features is improved, so that the model can better identify small target trash on the water surface. By introducing the SEAM attention mechanism module, the ability of the model to highlight the rubbish region information in the image and weaken the background region information is improved, so that the model can better identify the water surface rubbish that occludes each other. Finally, the CIoU method in the original algorithm is replaced with the ShapeIoU method to further improve the detection accuracy of the model. The experimental results demonstrate that the detection accuracy of the model proposed in this paper is significantly improved and meets the speed requirement for real-time detection of water surface litter. The follow-up work will consider the application of multi-spectral cameras to surface garbage shooting, so as to achieve all-weather real-time monitoring, which can better adapt to the intelligent monitoring and early warning of important water sources in the future.

References

1. Huang, F.Y., An, X.L., Li, L., et al.: Effects of domestic waste landfill on antibiotic resistance genes in rivers. China Environ. Environ. Sci. **37**(01), 203–209 (2017)
2. Girshick, R., Donahue, J., Darrell, T., et al.: Rich feature hierarchies for accurate object detectionand semantic segmentation. In: 2016 IEEE Conference on Computer Vision and Pattern Recognition, pp. 580–587 (2014)

3. Jiang, Z., Wu, B., Ma, L., et al.: Faster-RCNN water-floating garbage recognition based on multi-scale feature and polarised self-attention. J. Comput. Appl. **44**(3), 938 (2024)
4. Bai, J., Yang, S.: Detection method of marine floating garbage based on improved faster R-CNN. Highlights Sci. Eng. Technol. **15**, 120–127 (2022)
5. Yi, Z., Yao, D., Li, G., et al.: Detection and localisation for lake floating objects based on CA-faster R-CNN. Multimed. Tools Appl. **81**(12), 17263–17281 (2022)
6. Redmon, J., Divvala, S., Girshick, R., et al.: You only look once: unified, real-time objectdetection. In: Proceedings of CVPR 2015, pp. 779–778, 2015
7. Chen, L., Zhu, J.: Water surface garbage detection based on lightweight YOLOv5. Sci. Rep. **14**(1), 6133 (2024)
8. Li, L., Li, Y., Jiang, Z., et al.: Real-Time Detection of Surface Floating Garbage Based on Improved YOLOv7. In: International Conference on Intelligent. Springer Nature, Singapore, pp. 573–582 (2023)
9. Wang, C.Y., Yeh, I.H., Liao, H.Y.M.: Yolov9: learning what you want to learn using programmable gradient information. arXiv preprint arXiv:2402.13616, 2024
10. Wang, C.Y., Liao, H.Y.M., Wu, Y.H., et al.: CSPNet: a new backbone that can enhance learning capability of CNN. In: Proceedings of the IEEE/CVF Conference on Computer Vision and Pattern Recognition Workshops, pp. 390–391 (2020)
11. Yu, Y., Zhang, Y., Cheng, Z., et al.: Multi-scale spatial pyramid attention mechanism for image recognition: an effective approach. Eng. Appl. Artif. Intell. **133**, 108261 (2024)
12. Yu, Z., Huang, H., Chen, W., et al.: Yolo-facev2: a scale and occlusion aware face detector. arxiv 2022. arxiv preprint arxiv:2208.02019, 2019
13. Zhang, H., Zhang, S.: Shape-iou: more accurate metric considering bounding box shape and scale. arxiv preprint arxiv:2312.17663, 2023

Study on Urban Landscape Green Space - -Species Diversity of Woody Plants in Ma'anshan Forest Park

Huan Dai(✉), Rangjie Wu, Shijing He, and Li Liu

Wuhan Institute of Design and Sciences, Wuhan 430205, China
191735729@qq.com

Abstract. According to the current situation of woody plant diversity in Ma'anshan Forest Park, Wuhan, the species diversity of woody plants in Ma'anshan Forest Park was investigated by means of field investigation and typical sample plot method. Fifteen sample plots in the selected area were investigated, and the plants in the sample plots were recorded. And calculate the species richness index, diversity index, evenness index, frequency and other values. A total of 122 plant species belonging to 87 genera and 49 families were recorded. Among them, there are 75 species of trees belonging to 55 genera and 31 families, and 42 species of shrubs belonging to 29 genera and 17 families. The mean values of Margalef richness index in tree layer and shrub layer were 1.7134 and 1.5559, respectively. The mean values of Shannon-Winner diversity index in tree layer, shrub layer and herb layer were 0.8011, 0.6726 and 0.5989, respectively.

Keywords: Ma'anshan Forest Park · Wuhan · Plant Community · Species Diversity of Woody Plants

1 Preface

Urban biodiversity is the basis for improving the function of urban green space ecosystem and the health of green space ecosystem, and it is also an important symbol of urban greening level. Urban plant diversity is a concrete manifestation of urban biodiversity and an important basis for enriching urban landscape, improving urban ecological environment and promoting the sustainable development of urban green space system [1]. As an important natural part of the urban ecosystem, the urban park is a key ecological strategic point to break the homogeneous artificial environment of the city and create a heterogeneous natural landscape. It has ecological, economic and social benefits. Woody plants are the main components of forest parks, occupy a very important position in the forest landscape, and are the backbone of building forest landscapes. It is also the basis of urban biodiversity and an important part of urban ecosystem (Cui, 2018). The woody plants in the forest park not only have important ecological functions in purifying the air, regulating the climate and beautifying the city, but also play an important role in creating forest landscape and maintaining the stability of the ecosystem. However, urban parks are greatly affected by human disturbance, and the plant communities in parks tend to be

D. Han et al. (Eds.): ICEPG 2024, ESE, pp. 1138–1148, 2025.
https://doi.org/10.1007/978-3-031-86829-0_123

more fragile. Therefore, it is of great significance to study the plant community structure and woody plant diversity of urban park green space for urban ecological protection construction and sustainable development [2].

The species composition and distribution characteristics of urban plants in Beijing were investigated and studied. It was found that the species richness of park plants was the highest in each functional area of the city and played an important role in the species bank. The diversity and number of plant species in parks in different periods have been increased. The gap in the number of existing species between parks is small. Both native species and exotic species have increased, but the growth rate of exotic species is greater. In the park, woody plant diversity and species richness directly determine the microhabitat and its quality in the park, and have an important impact on community stability and other non-plant organisms. Most of the domestic research on plant diversity in urban parks mainly covers the composition and endemism of plant species diversity in parks, plant community types and diversity in parks, plant diversity and landscape research in parks, and biodiversity assessment indicators in parks. Urban parks play an important role in maintaining plant diversity [3].

2 Research Area and Method

2.1 Overview of the Research Area

Luomaanshan Forest Park is located in the eastern suburbs of Wuhan City, Hubei Province. It is bounded by Luoyu East Road in the south, East Lake in the north, Jiufeng Port in the east, and Yujia Lake in the west. It belongs to the national key scenic spot-East Lake Eco-tourism Scenic Area of the flute scenic spot. Ma'anshan Forest Park was built in 1993 and officially opened to the outside world in 1995, with a total area of 713 hectares. There are 17 peaks in the park, with a forest coverage rate of 80%. The main peak of Ma'anshan is 136.02 m above sea level, with a middle depression, two ends protruding, and the shape is like a saddle. The park is named after it.

Wuhan belongs to the north subtropical monsoon climate, which has the characteristics of abundant rainfall, sufficient heat, rain and heat in the same season, light and heat in the same season, cold in winter and hot in summer, and distinct four seasons. The annual average temperature in Wuhan is 15.8 °C–17.5 °C, the extreme maximum temperature is 41.3 °C, and the extreme minimum temperature is −18.1 °C. The annual frost-free period is generally 211 days–272 days, the total annual sunshine hours are 1810 h–2100 h, the total annual radiation is 104 kcal/square centimeter −113 kcal/square centimeter, the annual precipitation is 1150 mm–1450 mm, and the rainfall is concentrated from June to August every year.

2.2 Investigation Method

Based on the investigation of the whole park of Ma'anshan Forest Park, 15 samples were selected from six areas of Lingxiao Pavilion, Meishan, Taiyu Mountain, Zhaolong Mountain, South Gate of the park and East Lake for investigation. Using the typical sample method, we selected 15 representative $30 \times 30\ m^2$ arbor quadrats in Ma'anshan

2.2 Experimental Design

Study Area

One pumping well and three observation wells were arranged in the study area, among which the pumping well was numbered ZW1, the three observation wells were numbered GW1, GW2 and GW3 in turn. GW1 and GW2 were 3 m far from the pumping well and GW3 was 6 m far from the pumping well. The pumping well used Φ200 mm steel pipe and the well depth is 7.10 m. The observation well was made of Φ110 mm PVC pipe with a depth of 7.30 m. All wells were filled with gravel around the perimeter, filled with bentonite at the top and sealed with cement. The well placement in the study area is shown in Fig. 1.

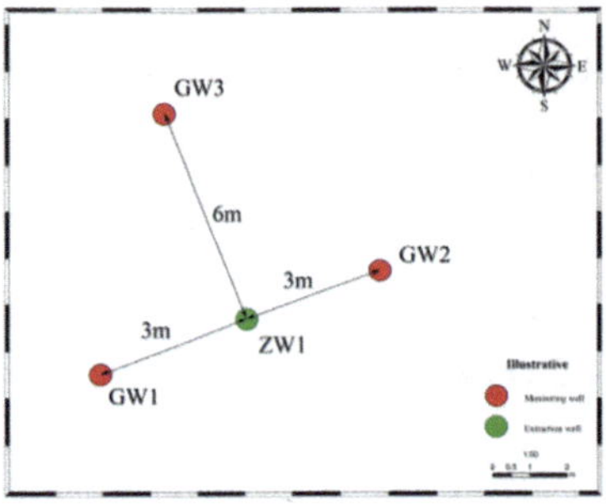

Fig. 1. Layout of wells in the study area.

Experimental Method

(1) The pumping well were pumped at a design flow rate of 29 L/h and real-time observation of the amount of water pumped as well as the water level in the pumping well and observation wells was carried out. Before the pumping started, the initial water level and well depth of the pumping well and the observation wells were measured. After the start of pumping water level observation at certain intervals, the initial observation frequency of 10 min, 30 min once, gradually transitioned to 1 h once, 2–7 h once, the middle of the 6–10 h once, the middle and late gradually transitioned to 12–24 h observation once.

(2) Observe groundwater levels before stopping pumping and conducted water quality sampling of pumping well and observation wells. After the concentration of pollutants in the pumping well was trailed and lasts for a certain time, the pumping was stopped.

Testing Indicators

On-site water quality analysis indicators were temperature (air temperature and groundwater temperature), pH, ORP and specific conductance; Laboratory sampling and analysis concentrations for benzene, toluene and 1,2-DCA. The detection method refers to "Water quality-Determination of volatile organic compounds-Purge and trap/gas chromatography-mass spectrometer (HJ 639–2012)" [22].

3 Results and Discussion

3.1 Changes of Pollutants in Pumping Well Over Time

When the groundwater level stabilized during the experiment, the radius of influence of the P&T was 3 m. Referring to "Standard for groundwater quality (GB/T 14848–2017)", Class IV water [21] was selected as the remediation target of polluted groundwater, as shown in Table 1. Groundwater in the study area was sampled and analyzed twice at 3-day interval prior to the extraction-treatment experiment. It can be seen from Table 2 that benzene, toluene and 1, 2-DCA in groundwater generally exceed the remediation target and the maximum exceedance times were 19, 32 and 5 times. Groundwater has a temperature of about 19 °C, pH neutral, in a reducing environment.

Table 1. Remediation target values of the study area [21].

No.	Pollutants	c (mg/L)
1	Benzene	≤ 0.12
2	Toluene	≤ 1.40
3	1,2-DCA	≤ 0.04

Table 2. Initial water quality analysis results of pumping well and observation wells.

No.	Well number	Water temperature (°C)	pH	Specific conductance (mS/cm)	ORP (mV)	Benzene (mg/L)	Toluene (mg/L)	1,2-DCA (mg/L)
1	ZW1	19.3	7.65	3.45	−219	1.85	45.06	0.17
		19.5	7.35	4.05	−223	1.94	43.33	0.19
2	GW1	19.1	7.30	3.61	−129	2.36	43.54	0.17
		19.3	7.32	5.31	−132	2.21	45.54	0.16
3	GW2	18.9	7.81	4.02	−135	1.03	19.85	0.07
		19.0	7.47	5.12	−139	1.42	19.71	0.08
4	GW3	18.8	7.56	8.98	−98	1.05	27.44	0.06
		18.9	7.45	7.98	−101	1.11	28.36	0.06
5	remediation target	/	/	/	/	0.12	1.4	0.04

After 552 h of continuous pumping, the pumping was stopped, and the groundwater sampling analysis was carried out immediately in the pumping well and the observation wells. The results showed that benzene and 1, 2-DCA were not detected in the pumping well and the observation wells. Toluene was detected but the concentration was lower than the remediation target value.

Forest Park, Wuhan. Each arbor quadrat contained two $5 \times 5\ m^2$ shrub quadrats and five $1 \times 1\ m^2$ herb quadrats. The species and trees of woody plants were investigated, and the species diversity of woody plants in Ma'anshan Forest Park was analyzed by frequency, species richness index, Shannon-Wiener diversity index and Pielou evenness index.

2.3 Data Processing

Calculation of Species Importance

- Important value of trees and shrubs = (relative density + relative coverage + relative frequency)/3
- Herb important value = (relative coverage + relative frequency)/2
- In the formula, relative density = (number of individuals of a species/number of individuals of all species) × 100%
- Relative coverage = (coverage of a species/total coverage of all species) × 100%
- Relative frequency = (the frequency of a species/the frequency of all species and) × 100%

Species Diversity Calculation

- Margalef richness index formula: R = (S-1)/lnN;
- shannon-Wiener diversity index: H = -$\sum$ pilnpi, pi = ni/N;
- simpson diversity index: D = 1-$\sum$ pi 2;
- pielou evenness index: E = H/lnS.
- The number of S-species;
- n-the sum of the number of individuals of all species;
- pi - the relative abundance of the i-th species;
- ni - the number of individuals of the i-th species.

3 Results and Analysis

4 Species Composition Analysis

According to the statistics of the survey data, there are 122 species of woody plants belonging to 87 genera and 49 families in the quadrat. As shown in Table 1, there are 75 species of trees belonging to 55 genera and 31 families, 44 species of evergreen trees and 30 species of deciduous trees, accounting for 61.48% of plant species. There are 42 species of shrubs belonging to 29 genera and 17 families, 27 species of evergreen shrubs, 15 species of deciduous shrubs, accounting for 34.43% of plant species, and 4 species of bamboo belonging to 3 genera and 1 family. From the analysis of plant species composition in the park, the dominant families are Fagaceae (8 genera and 11 species), Rosaceae (9 genera and 13 species), Pinaceae (3 genera and 4 species), Lauraceae (4 genera and 6 species), Leguminosae (5 genera and 7 species), Taxodiaceae (3 genera and 4 species), Oleaceae (4 genera and 6 species), as shown in Table 2.

Table 1. Species composition table of woody plants in Ma'anshan Forest Park

No.	Form	Family	Genus	Species
1	Trees	31	55	75
2	Shrubs	17	29	42
3	Bamboos	1	3	4

Table 2. Dominant families ordination of woody plants in Ma'anshan Forest Park

No.	Family name	Number of genera	Number of species
1	Rosaceae	9	13
2	Fagaceae	8	11
3	Leguminosae	5	7
4	Lauraceae	4	6
5	Oleaceae	4	6
6	Pinaceae	3	4
7	Taxodiaceae	3	4

4.1 Analysis on the Distribution Frequency of Woody Plants in Ma'anshan Forest Park

In the tree layer, the frequency of masson pine was the highest (66.67%), followed by camphor (60%), and the frequency of holly and quercus was higher (53.33%). There were 8 species of tree plants that appeared only once in the plot, accounting for 6.78% of the tree plant species. The frequency of small fruit roses in shrubs was the highest (46.67%), and the frequency of Rosa laevigata, Vitex negundo, and White horse bone was higher (40%). There were 6 shrub species that appeared only once in the plot, accounting for 14.28% of the shrub species. On the whole, the species of shrubs in the quadrat are less.

4.2 Analysis of Woody Plant Species Diversity Index in Ma'anshan Forest Park

The measurement of biodiversity has different methods and scales, which can be generally divided into three scales: α diversity, β diversity and γ diversity [4]. We generally measure species diversity at the α diversity scale. Alpha diversity is used to measure the number of species in the community and the relative abundance between species, reflecting the co-existence results of species in the community by competing for resources or using the same habitat. Generally speaking, community diversity refers to α diversity, that is, species diversity within the community.

Analysis of Species Importance Value
A total of 122 plant species belonging to 87 genera and 49 families were recorded in 15

quadrats of 900 m^2. Among them, there are 75 species of trees belonging to 55 genera and 31 families, and 42 species of shrubs belonging to 29 genera and 17 families. In the tree layer, the top five important values are Pinus massoniana, Cinnamomum camphora, Ilex, Quercus, Lindera bungeana, as shown in Table 3. Among them, the relative density, relative frequency and relative coverage of Pinus massoniana are the largest, which is the dominant tree species in the whole park. The relative density and relative frequency of camphor are the second, but the relative coverage is only 0.4259, indicating that although it appears more frequency and trees, but most of the growth conditions are not good, the growth space is limited. Although the relative density and relative frequency of holly are not too high, the relative coverage is very high, indicating that it adapts to the local climate environment and grows well. In the shrub layer, the top five important values are Rosa cymosa, Vitex negundo, Rosa laevigata, Symplocos paniculata, Serissa serissoides, as shown in Table 4. In the herb layer, the top five important values are Quercus aliena, Rhus chinensis, Rubus idaeus, Mallotus apelta, Smilax china, as shown in Table 5.

Table 3. Ma'anshan Forest Park tree layer plant important value sorting (take the first five)

No.	Species	Relative density	Relative frequency	Relative cover	Importance value
1	*Pinus massoniana*	0.8951	0.6288	0.8143	0.7794
2	*Ilex chinensi*	0.6562	0.5714	0.7761	0.6679
3	*Cinnamomu-m camphora*	0.8493	0.6032	0.4259	0.6261
4	*Lindera glauca*	0.6874	0.5000	0.6452	0.6108
5	*Quercus aliena*	0.4434	0.5714	0.5927	0.5358

Table 4. Ma'anshan Forest Park shrub layer plant important value sorting (take the first five)

No.	Species	Relative density	Relative frequency	Relative cover	Importance value
1	*Rosa cymosa*	0.8750	0.5333	0.5672	0.6585
2	*Rosa laevigata*	0.8713	0.5000	0.4563	0.6091
3	*Vitex negundo*	0.7143	0.5000	0.6029	0.6057
4	*Symplocos paniculata*	0.6375	0.4682	0.5978	0.5678
5	*Serissa foetida*	0.5950	0.5000	0.3764	0.4904

Table 5. Ma'anshan Forest Park herb layer plant importance value sorting (take the first five)

No.	Species	Relative frequency	Relative cover	Importance value
1	*Rubus corchorifolius*	0.6667	0.6818	0.6742
2	*Mallotus apelta*	0.5328	0.5471	0.5399
3	*Smilax china*	0.5083	0.2083	0.3583
4	*Rhus chinensis*	0.5153	0.1632	0.3393
5	*Quercus aliena*	0.4832	0.1271	0.3052

Species Richness Analysis

The average Margalef species richness index of tree layer, shrub layer and herb layer were 1.7134, 1.5559 and 0.9506, respectively. The Margalef species richness index of tree layer was the highest. As shown in Table 6, the top five tree layer richness were plot 15 (2.2425), plot 4 (2.2042), plot 7 (2.1982), plot 1 (2.0851) and plot 12 (2.0512), respectively. Plots 1, 4 and 7 are located near Lingxiao Pavilion; sample 12 is located near Meishan Park; therefore, there are many tree species in Lingxiao Pavilion of the park. The top five shrub richness were sample plot 12 (2.3297), sample plot 3 (1.9850), sample plot 8 (1.9244), sample plot 11 (1.8728) and sample plot 14 (1.8205).

Table 6. Richness index of tree and shrub layer in Ma'anshan Forest Park

Plot	Tree layer	Shrub layer	Herbaceous layer
1	2.0851	1.4701	1.4427
2	2.5010	1.6876	1.0986
3	1.8519	1.9850	0.6068
4	2.2042	1.6743	------
5	1.6739	1.6690	1.1333
6	1.2577	0.9102	1.4427
7	2.1983	1.2581	1.1765
8	1.4807	1.9244	1.3654
9	1.6541	1.7015	------
10	1.2307	1.2277	0.9102
11	1.6653	1.8728	1.3654
12	2.0512	2.3297	0.9861
13	0.9209	0.3899	------
14	1.6829	1.8205	0.9102
15	2.2425	1.5005	1.8204

Analysis of Species Diversity Index
Species diversity is an important index to measure the complexity of community structure and function. The species diversity index of plants is large, indicating that the plant community type is composed of multiple tree species in this area, with rich species and complex composition [5]. The Shannon-Winner index can more stably reflect the diversity of the community because it is not affected by the size of the plot. According to Table 7, the average Shannon-Winner index of the tree layer is 0.8011, 0.6726, and 0.5989, respectively, and the tree layer is the highest. The top five plots of Shannon-Winner index of tree layer were plot 4 (0.9476), plot 15 (0.9369), plot 12 (0.8593), plot 1 (0.8471) and plot 14 (0.8408), respectively. The plots 1, 4 and 5 were located near Lingxiao Pavilion, the plot 12 was near Meishan, and the plot 15 was near Taiyu Mountain, indicating that the tree layer near Lingxiao Pavilion was rich. According to the Shannon-Winner index of the shrub layer in Table 6, the top five plots are plot 12 (0.8638), plot 3 (0.8144), plot 11 (0.8142), plot 2 (0.7932) and plot 8 (0.0.7721), respectively. The plots 11 and 12 are near Meishan, the plots 2 and 3 are near Lingxiaoge, and the plot 8 is near Zhaolongshan, so the species richness in the shrub layer near Meishan is higher.

Table 7. Shannon-Winner index of arbor and shrub layer in Ma'anshan Forest Park

Plot	Tree layer	Shrub layer	Herbaceous layer
1	0.8471	0.6689	0.6341
2	0.8358	0.7932	0.6636
3	0.8167	0.8144	0.4620
4	0.9476	0.7159	------
5	0.8408	0.6683	0.8661
6	0.6913	0.5159	0.6931
7	0.8029	0.5970	0.8669
8	0.7635	0.7721	0.8956
9	0.7751	0.7306	------
10	0.6686	0.6135	0.6641
11	0.7985	0.8142	0.7796
12	0.8593	0.8638	0.6924
13	0.6047	0.2997	------
14	0.8283	0.7663	0.6931
15	0.9369	0.7322	1.0725

The Simpson diversity index is used to reflect the probability that the two plant individuals in the community belong to the same species. According to Table 8, the average Simpson index of the tree, shrub and grass layers is 0.8011, 0.6911 and 0.5989, respectively, and the Simpson index of the tree layer is the highest. The top five sample plots of Simpson diversity index of arbor layer were sample plot 15 (0.8593), sample plot

11 (0.8584), sample plot 4 (0.8542), sample plot 3 (0.8442) and sample plot 5 (0.8324). Sample plot 3, 4 and 5 were located near Lingxiaoge, sample plot 11 was located near Meishan, and sample plot 15 was located near Taiyushan, indicating that the probability of individual tree layer plants belonging to the same species near Lingxiaoge was higher. The top five plots of Simpson diversity index of shrub layer were plot 12 (0.8408), plot 11 (0.8310), plot 2 (0.8245), plot 3 (0.8218) and plot 8 (0.795), respectively. Among them, plot 11 and 12 were located near Meishan, plot 12 and 3 were located near Lingxiaoge, and plot 8 was located near Zhaolongshan, indicating that the shrubs near Meishan had a higher probability of belonging to the same species.

Table 8. Simpson index of arbor and shrub layer in Ma'anshan Forest Park

Plot	Tree layer	Shrub layer	Herbaceous layer
1	0.8103	0.4889	0.5000
2	0.8234	0.8245	0.4711
3	0.8487	0.8218	0.3333
4	0.8542	0.7701	------
5	0.8324	0.5925	0.4856
6	0.7635	0.6529	0.5000
7	0.8196	0.6956	0.4784
8	0.7957	0.7950	0.5862
9	0.7827	0.7820	------
10	0.7238	0.7219	0.4422
11	0.8583	0.8311	0.4835
12	0.8240	0.8408	0.4453
13	0.7684	0.4970	------
14	0.8158	0.7257	0.4422
15	0.8592	0.6479	0.6733

Species Evenness Analysis

In the tree layer, shrub layer and herb layer, the mean values of Pielou evenness index were 0.3645, 0.3791 and 0.7439, respectively. According to Table 9, the top five plots of tree layer evenness were plot 2 (0.4019), plot 5 (0.3827), plot 3 (0.3813), plot 14 (0.3770) and plot 13 (0.3769), respectively. Among them, plot 2, 3 and 5 were near Lingxiaoge, plot 13 and 14 were near Taiyu Mountain. It shows that the distribution of species in the tree layer near Lingxiao Pavilion is relatively uniform; the top five plots of shrub layer evenness were plot 13 (0.4324), plot 2 (0.4076), plot 12 (0.3931), plot 3 (0.3916) and plot 11 (0.3915), respectively. Among them, plot 11 and 12 were located near Meishan, plot 2 and 3 were located near Lingxiaoge, and plot 13 was located near Taiyushan, indicating that the distribution of shrub layer species near Meishan was relatively uniform. The top

five of herb layer uniformity were plot 8 (1.2921), plot 7 (1.2507), plot 14 (0.9999), plot 6 (0.9999), plot 15 (0.9762).It can be seen that the herb uniformity near Taiyu Mountain is high.

Table 9. Species evenness index of arbor and shrub layer in Ma'anshan Forest Park

Plot	Tree layer	Shrub layer	Herbaceous layer
1	0.3532	0.3733	0.9148
2	0.4019	0.4076	0.6040
3	0.3547	0.3916	0.6665
4	0.3813	0.3678	------
5	0.3827	0.3601	0.7884
6	0.3553	0.3722	0.9999
7	0.3231	0.3709	1.2507
8	0.3671	0.3713	1.2921
9	0.3527	0.3459	------
10	0.3436	0.3812	0.9581
11	0.3634	0.3915	0.7096
12	0.3583	0.3931	0.9989
13	0.3757	0.4324	------
14	0.3769	0.3652	0.9999
15	0.3770	0.3631	0.9762

5 Conclusions and Discussion

Luomaanshan Forest Park is located in the eastern suburbs of Wuhan City, Hubei Province, south of Luoyu East Road, north of East Lake, east of Jiufeng Port, west of Yujia Lake. It belongs to the national key scenic spot-East Lake Ecotourism Scenic Area of the flute scenic spot. It has good geographical conditions, convenient transportation, beautiful environment, warm and humid climate, and large forest coverage. It is an important ecological play park in Wuhan.

The park is divided into four parts: south, middle, northwest and northeast. The sample 1, 2, 3, 4, 5 and 7 of the survey are located in the south of the park, the sample 8 and 9 are located in the middle of the park, the sample 10, 11 and 12 are located in the northwest, and the sample 14 and 15 are located in the northeast of the park. The sample 6 and the sample 13 are used as the control group, which are the samples with large artificial interference. The sample 6 is located near the south gate of the park, while the sample 13 is located in the northeast near the East Lake area [6]. A total of 49 families, 87 genera and 122 species of plants were recorded in the park. In general, because the

forest in the park is a secondary succession forest, Pinus massoniana, as a pioneer tree species planted artificially, has the highest planting density and planting frequency.

The proportion characteristics of arbor, shrub and grass in the whole park are also consistent with the characteristics of typical Pinus massoniana forest, which is a warm evergreen deciduous forest. Most of the plant species in the park are native and deciduous tree species, which can adapt to the existing geographical environment. However, the single genus and species of plants are concentrated in a small range, and the common plants such as Rosaceae are concentrated in a large range, and the stability of the biological community is poor. However, there are also cases where tree planting is too dense, resulting in poor relative coverage of tree species such as camphor trees and unsatisfactory growth conditions [7]. Compared with the tree layer and the herb layer, there are fewer herbs in the area with more trees, and there are more herbs in the area with fewer trees, that is, the herb layer is inversely proportional to the tree layer, indicating that most of the plants in the herb layer are the result of natural selection, without too much artificial interference. Specifically, in the southern part of the park sample 2,3 and 4, due to less human interference, the existence of biological isolation belts, the species richness of trees is very high, and the growth of trees is good, so it can not be avoided. The uneven development of the vertical structure, the shrubs under the shade due to the lack of access to sunlight, most of the growth conditions are poor, and the plant coverage is low. In the 11 and 12 plots located in the northwest, due to the more human construction in the northwest, there are more areas for visitors to play, so the density and coverage of trees are smaller than those in the southern plots, but the shrub species richness is very high. Among them, there are not only shrubs that originally exist in the forest, but also shrubs planted by people. Because the planting density of trees is smaller, the opportunities for plants below to contact the sun are more, which also promotes the good development of shrubs. As the control group, the sample 6, due to the artificial selection and planting, is mostly alien species, and basically does not grow in other plots. The sample 13 is a willow community near the East Lake, and there is basically no shrub growth in the sample plot, making it a plant configuration of trees and herbs, which is convenient for people to rest and play.

Therefore, from the perspective of the tree layer, the park should pay attention to the planting density of the trees, leaving a certain interval for the trees to have sufficient growth space. From the perspective of shrubs, it is suggested that the park should plant shade-tolerant and cold-tolerant plants under the shade, appropriately increase some shrubs, form upper, middle and lower structural levels with trees and herbs, and improve the diversity and richness of shrubs [8]. From the perspective of community, taking the difference of species richness index, overall diversity index and evenness index of different scenic spots in forest park as reference, different greening renewal schemes are adopted for different scenic spots to improve the plant diversity of forest park. In order to create a unique plant landscape of Ma'anshan Forest Park, different scenic spots should have different dominant tree species as far as possible [9].

Acknowledgments. This study was funded and supported by the Hubei Provincial Scientific Research Project (B2021374).

References

1. Yao, S., Wang, H.: Investigation and research on woody plant community of retaining wall facade in mountainous city - taking the main urban area of Chongqing as an example. J. Southwest Univ. (Nat. Sci. Edit.) **44**(03), 93–101 (2022)
2. Wu, Z.: Comparative analysis of community characteristics between artificial Acacia forest and monsoon evergreen broad-leaved forest in southern subtropical China. Southern Agriculture **15**(23), 130–131 (2021)
3. Shen, D., Lv, L., Xi, l.: Analysis of diversity and floristic characteristics of woody plants in Qingyang. Forestry Sci. Technol. Newsl. (03), 43–46 (2021)
4. Wang, Y., Qin, J., Ma, J., et al.: Species diversity of woody plants under Pinus massoniana plantation in southern subtropical China. J. Guangxi Norm. Univ. (Nat. Sci. Edit.) **38**(06) : 131–139(2020)
5. Li, L., Tai, P.: Study on the diversity of garden plants in Nanning Flower Park. Guangxi Forestry Sci. **49**(02), 212–217 (2020)
6. Dong, D., Qin, Z., Li, C., et al.: Comparative analysis of woody plant landscape and species diversity between new and old parks in Huainan City. J. Northwest Forestry Coll. **34**(05), 247–254 (2019)
7. Chen, S., Ren, H., Tong, G., et al.: Spatial distribution pattern of woody plant species diversity in Qianjiangyuan National Park. Biodiversity **31**(07), 5–17 (2023)
8. Zhao, J., Song, C., Liu, S.: Characteristics of urban plant species composition-Taking Xiamen City as an example. J. Southwest Univ. (Nat. Sci. Edit.) **40**(07), 1–8 (2018)
9. Zhao, K., Pan, J.: Analysis on the structure and configuration of woody plants in campus green space of Tsinghua University. J. Southwest Forestry Univ. (Soc. Sci.) **2**(01), 87–91 (2018)

Correction to: Environmental Crisis: Pollution and Governance

Dongfei Han, Mukesh Khare, and Siyue Li

Correction to: D. Han et al. (Eds.): *Environmental Crisis: Pollution and Governance*, ESE, https://doi.org/10.1007/978-3-031-86829-0

In the original version of the book, the belated correction in the corresponding author name indication has been updated for the chapters 79, 80, 92 and 96. The correction chapters and the book have been updated with these changes.

The updated version of these chapters can be found at
https://doi.org/10.1007/978-3-031-86829-0_79
https://doi.org/10.1007/978-3-031-86829-0_80
https://doi.org/10.1007/978-3-031-86829-0_92
https://doi.org/10.1007/978-3-031-86829-0_96

D. Han et al. (Eds.): ICEPG 2024, ESE, p. C1, 2025.
https://doi.org/10.1007/978-3-031-86829-0_124

Author Index

D. Han et al. (Eds.): ICEPG 2024, ESE, pp. 1149–1154, 2025.
https://doi.org/10.1007/978-3-031-86829-0

H

J

K

L